MICROBIOLOGY
Laboratory Theory & Application
THIRD EDITION

Michael J. Leboffe
San Diego City College

Burton E. Pierce

MORTON
P U B L I S H I N G

925 W. Kenyon Ave., Unit 12
Englewood, Colorado 80110
www.morton-pub.com

Book Team

Publisher:	Douglas N. Morton
Biology Editor:	David Ferguson
Production Manager:	Joanne Saliger
Production Assistant:	Desiree Coscia
	Patricia Billiot
Typography:	Ash Street Typecrafters, Inc.
Copyediting:	Carolyn Acheson
Illustrations:	Imagineering Media Services, Inc.
Cover Design:	Bookends Design

ISBN 10: 0-89582-830-8
ISBN 13: 978-0-89582-830-9

Library of Congress Control Number 2009944089

10   9   8   7   6   5   4   3

Printed in the United States of America

Preface

This third edition in many ways is like another first edition. We have added 20 new exercises, incorporated four more exercises from *MLTA Brief Edition*, and have substantially rewritten several others. Every exercise has been screened and updated for better clarity, comprehensiveness, and appropriate placement in the manual. We have added three more of the boxed "A Word About . . ." features to offer expanded introductory material in Sections 4, 5, and 8. Finally, we replaced many older photographs, and perhaps most important, employed a new artist to update the illustrations and enhance the overall beauty of the book.

Following are major features of each section.

✦ **Introduction** As in previous editions, the Introduction emphasizes safety. Refinements include chemical safety awareness and examples of organisms to be handled at each BSL level.

✦ **Section 1 Fundamental Skills** Exercise 1-1 (The Glo-Germ™ Hand Wash Education System) was brought over from the *Brief* Edition. This is a fun, eye-opening lab exercise, raising consciousness about how easily the "unseen" can be overlooked. In Exercise 1-4 (Streak Plate Method of Isolation) we introduce alternative methods for streaking.

✦ **Section 2 Microbial Growth** Exercises 2-5 (Evaluation of Media) and 2-12 (Steam Sterilization) have been added from the Brief Edition. Exercise 2-11 (The Effect of Osmotic Pressure on Microbial Growth) has been rewritten to include *Halobacterium*. Exercise 2-13 (The Lethal Effect of Ultraviolet Radiation on Microbial Growth) has been renamed, and the procedure simplified.

✦ **Section 3 Microscopy and Staining** Exercise 3-1 (Introduction to the Light Microscope) has been rewritten to include an activity using the "Letter e" slide and the "Colored Thread" slide, both of which provide opportunities for novice microscopists to learn more about how to operate the light microscope. Two new labs, Exercise 3-4 (Microscopic Examination of Pond Water) and Exercise 3-11 (Parasporal Crystal Stain), round out the changes for this section.

✦ **Section 4 Selective Media** This section has been updated for greater clarity and to address reviewers' concerns. The boxed item "A Word About Selective Media" has been added as an adjunct to the Section introduction, and Bile Esculin Test (Exercise 4-3) has been moved from Section 5 to Selective Media for the Isolation of Gram-Positive Cocci.

✦ **Section 5 Differential Tests** This section has seen some reorganization and the addition of new material. The boxed item "A Word About Biochemical Tests and Acid-Base Reactions," has been added to supplement the introduction and help students better understand the recurring theme of fermentation seen in differential media. Following the introductory material, the section begins with an exercise demonstrating Reduction Potential. It is designed to be an easy and fun introduction to the concept of energy transformations in redox reactions, the understanding of which can be applied to many exercises that follow. Novobiocin and Optochin susceptibility tests have been moved from Gram-positive coccus identification and combined with Bacitracin (Exercise 5-24) in the Antibacterial Susceptibility Testing subsection. Because of popular demand to include both Kligler Iron Agar and Triple Sugar Iron Agar in the Combination

Differential Media subsection, we have written an exercise (5-21) that can be used with either or both media.

◆ **Section 6 Quantitative Techniques** In an ongoing effort to help students understand the dilutions and calculations necessary in quantitative techniques (and to simplify the terminology), we have rewritten the introduction to this section and the theory portion of Exercise 6-1 (Standard Plate Count). The terms "dilution factor" and "final dilution factor" associated with dilutions and platings have been replaced with "dilution" and "sample volume," respectively. We hope this helps.

◆ **Section 7 Medical Microbiology** In this section we have added one new lab demonstrating clinical biofilms (Exercise 7-4). Exercise 7-3 (Antimicrobial Susceptibility Testing) has been rewritten with new antibiotics and an optional exercise to demonstrate the difference between bacteriostatic and bacteriocidal agents. Last, Exercise 7-9 (Identification of Gram-Positive Rods) rounds out the series of three unknown identifications.

◆ **Section 8 Environmental Microbiology** This is the section that has seen the most growth. Of the 13 exercises, 9 are new. Exercise 8-1 (Winogradsky Column) should be done near the beginning of the semester so it can be used as a source of microorganisms in other labs (Exercises 8-6 through 8-8). The next seven exercises provide activities related to the Nitrogen Cycle (Exercises 8-2 through 8-5) and the Sulfur Cycle (Exercises 8-6 through 8-8). The other new lab is Exercise 8-10 (Soil Slide Culture), in which soil microorganisms are grown and then can be viewed in their proper spatial orientation. Also included in this section is a brief explanation of trophic group terminology ("A Word About Trophic Groups").

◆ **Section 9 Food Microbiology** One exercise (that was problematic) has been removed from this section. Otherwise, the content here is unchanged. You can expect a lot of growth in the food microbiology section in subsequent editions of this manual.

◆ **Section 10 Microbial Genetics** This section has been reorganized and three new exercises added to reflect the change in microbiology from traditional methods of identification to more sensitive techniques involving molecular biology. Exercises 10-2 and 10-4 address two important methods of molecular biology: performing a restriction digest and polymerase chain reaction, respectively. Exercise 10-7 (Phage Typing of *E. coli* Strains) introduces the student to using viral susceptibility as a tool in microbial identification.

◆ **Section 11 Hematology and Serology** Two of the seven exercises in this section are new. Exercise 11-6 is a hemagglutination test used in the diagnosis of infectious mononucleosis. Exercise 11-7 is a quantitative ELISA that models the identification and quantification of HIV antibodies.

◆ **Section 12 Eukaryotic Microbes** This section continues to be a survey of the microscopic eukaryotes encountered in a medical microbiology laboratory. One new exercise, Exercise 12-2 (Fungal Slide Culture) provides an opportunity to cultivate fungi in a way that their true structure can be observed microscopically.

◆ **Appendices** Appendix G (Agarose Gel Electrophoresis) contains instructions for preparing and (multiple techniques for) staining agarose gels used in the electrophoresis portion of Exercises 10-2 and 10-5. Many additions have been made to Appendix H (Medium, Reagent, and Stain Recipes) to accompany all the new exercises. The Glossary also has been updated and expanded.

Our book is maturing, becoming more complete and polished. And as we use it, new ideas for presentation or content will occur to us for future changes. That's why new editions are numbered!

Last, though we do get compensated for our work, we are primarily educators. We take great satisfaction that our efforts may in some small way contribute to your successful academic and professional careers.

All our best,
Mike and Burt

Acknowledgments

Thanks so much to all of you who had a part in making this project a success. We have thanked all of you personally; this is our opportunity to let our readers know who you are and why you are so appreciated. Listed below, in no particular order, are the people who went out of their way to give time, advice, space, and patience to support us during this project. We sincerely hope that, with the pages that follow, we have earned that support.

First of all, thanks to Debra Reed, Biology Laboratory Technician at San Diego City College, for her longstanding support of our projects. Deb's assistance spans more than a decade and includes help with test media and cultures, hand modeling for photos, and gentle directions when we occasionally forget where we are. We thank Muu Vu for her assistance modeling for photos, and helping to make valuable contacts. Thanks very much to Brett Ruston for the free lunch and the "loan" of several chemicals used in the Nitrogen and Sulfur cycle exercises. Thanks to lab technician Laura Steininger for running interference by listening to Burt's sustained belly-aching about living in an RV for a month without his dogs, Yancy, Megan, and Beau to keep him company. Thanks also to Ed Sebring of the Chemistry Department who provided a pinch of this or that chemical when we were desperate.

Additionally, thank you to Dr. Carla Sweet for helping with some new photographs as a hand model, Gary Wisehart for assisting with the Winogradsky column, Alicia Leboffe for editorial work, Nathan Leboffe for help in photography, and Dr. Steven J. Byers for piloting Mike around San Diego County so he could get aerial photographs.

Thanks to Dr. Donna DiPaolo, Dr. Anita Hettena, Dr. Roya Lahijani, Erin Rempala, Dr. David Singer, and Gary Wisehart of the San Diego City College Biology Department for patience and understanding about losing territory in the Biology Resource Center due to our set-up and for putting up with some of the foul odors produced by the cultures. Thanks to Sonia Bertschi co-owner and manager of the Jacumba Hotsprings Spa and Lodge in Jacumba, California for a tour of her lovely resort and donation of natural mineral hot springs water.

Thanks to Jerry Davis, San Diego City College Vice President of Administrative Services and Joyce Thurman, San Diego City College Business Services Administrative Secretary, for expediting use of college facilities. Thanks also to Dr. Steve Barlow, Associate Director of the San Diego State University Electron Microscope Facility for his patience and humor as one of us learned how to use the transmission electron microscope.

A very special thank you goes to Dr. Radu Popa, Portland State University Microbiology Professor for taking time on numerous occasions to give expert advice on many issues, but most specifically, sulfur biogeochemical transformations. Also thank you to Jane Boone, Portland State University Biology Laboratory Coordinator, for giving Burt a workspace of his own and occasional needed advice. Both Dr. Popa and Ms. Boone, who were working on their own advanced microbiology laboratory manual during this time, generously provided a collegial and sharing environment.

Thanks to Imagineering Media Services Inc. from Toronto, Ontario, Canada for the new artwork included in this edition. Thanks also to Bob Schram of Bookends Design for the cover design. And, as always, thanks so much to Joanne Saliger at Ash Street

Typecrafters, Inc., who capably (and cool-handedly) managed to produce a beautiful book out of moderately well organized scraps. Thanks to Carolyn Acheson, who copyedited (and removed most of the split infinitives from) the manuscript. We also are indebted to reviewers and students who provided valuable suggestions for improving our book. Special recognition and thanks go to Dr. Amy Warenda Czura of Suffolk County Community College and Dr. Lisa Lyford of University of the Cumberlands for their thorough input.

A special thanks to Dr. Jack G. Chirikjian, Chairman of Edvotek, Inc. for agreeing to the use of his company's kits in Exercises 10-2, 10-4, 11-7, and Appendix G.

We, of course, remain grateful to the Morton Publishing team for their support and patience. Specifically, thanks to Doug Morton, President, Chrissy Morton DeMier, Business Manager, David Ferguson, Acquisitions Editor, Carter Fenton, Sales and Marketing Manager, and Desiree Coscia, Publisher's Assistant.

And as always, thanks to our wives Karen Leboffe and Michele Pierce for their continued support and understanding. If writing a book is difficult, being married to an author is worse.

Contents

Safety and Laboratory Guidelines

Microbiology lab can be an interesting and exciting experience, but at the outset you should be aware of some potential hazards. Improper handling of chemicals, equipment and/or microbial cultures is dangerous and can result in injury or infection. Safety with lab equipment will be addressed when you first use that specific piece of equipment, as will specific examples of chemical safety. Our main concern here is to introduce you to safe handling and disposal of microbes.[1]

Because microorganisms present varying degrees of risk to laboratory personnel (students, technicians, and faculty), people outside the laboratory, and the environment, microbial cultures must be handled safely. The classification of microbes into four biosafety levels (BSLs) provides a set of minimum standards for laboratory practices, facilities, and equipment to be used when handling organisms at each level. These biosafety levels, defined in the U. S. Government publication, *Biosafety in Microbiological and Biomedical Laboratories*, are summarized below and in Table I-1. For complete information, readers are referred to the original document.

BSL-1: Organisms do not typically cause disease in healthy individuals and present a minimal threat to the environment and lab personnel. Standard microbiological practices are adequate. These microbes may be handled in the open, and no special containment equipment is required. Examples include *Bacillus subtilis*, *Escherichia coli*, *Rhodospirillum rubrum*, and *Lactobcillus acidophilus*.

BSL-2: Organisms are commonly encountered in the community and present a moderate environmental and/or health hazard. These organisms are associated with human diseases of varying severity. Individuals may do laboratory work that is not especially prone to splashes or aerosol generation, using standard microbiological practices. Examples include *Salmonella*, *Staphylococcus aureus*, *Clostridium dificile*, and *Borrelia burgdorferi*.

BSL-3: Organisms are of local or exotic origin and are associated with respiratory transmission and serious or lethal diseases. Special ventilation systems are used to prevent aerosol transmission out of the laboratory, and access to the lab is restricted. Specially trained personnel handle microbes in a Class I or II biological safety cabinet (BSC), not on the open bench (see Figure I-1). Examples include *Bacillus anthracis*, *Mycobacterium tuberculolosis*, and West Nile virus.

---

[1] Your instructor may augment or revise these guidelines to fit the conventions of your laboratory.

TABLE **1-1**   Summary of Recommended Biosafety Levels for Infectious Agents

BSL	Agents	Practices	Safety Equipment (Primary Barriers)	Facilities (Secondary Barriers)
1	Not known to consistently cause disease in healthy individuals	Standard microbiological practices	None required	Laboratory bench sink required
2	✦ Agents associated with human disease ✦ Routes of transmission through percutaneous injury, ingestion, mucous membrane exposure	BSL-1 practice plus: ✦ Limited access ✦ Biohazard warning signs ✦ "Sharps" precautions ✦ Biosafety manual defining any needed waste decontamination or medical surveillance policies	Primary barriers: ✦ Class I or II BSCs or other physical containment devices used for all manipulations of agents that cause splashes or aerosols of infectious materials Personal Protective Equipment (PPEs): ✦ Laboratory coats, gloves, face protection, as needed	BSL-1 plus: ✦ Autoclave available
3	✦ Indigenous or exotic agents with potential for aerosol transmission ✦ Disease may have serious or lethal consequences	BSL-2 practice plus: ✦ Controlled lab access ✦ Decontamination of all waste ✦ Decontamination of all lab clothing before laundering ✦ Baseline serum	Primary barriers: ✦ Class I or II BSCs or other physical containment devices used for all open manipulations of agents PPEs: ✦ protective laboratory clothing, gloves, respiratory protection as needed	BSL-2 plus: ✦ Physical separation from access corridors ✦ Self-closing, double-door access ✦ Exhausted air not recirculated ✦ Negative airflow into laboratory
4	✦ Dangerous/exotic agents that pose high risk of life-threatening disease ✦ Aerosol-transmitted lab infections have occurred; or related agents with unknown risk of transmission	BSL-3 practices plus: ✦ Clothing change before entering ✦ Shower on exit ✦ All material decontaminated on exit from facility	Primary barriers: ✦ All procedures conducted in Class III BSCs or Class I or II BSCs in combination with full-body, air-supplied, positive pressure personnel suit	BSL-3 plus: ✦ Separate building or isolated zone ✦ Dedicated supply and exhaust, vacuum, and decontamination systems ✦ Other requirements outlined in the text

Source: Reprinted from *Biosafety in Microbiological and Biomedical Laboratories*, 5th edition (Washington: U.S. Government Printing Office, 2007).

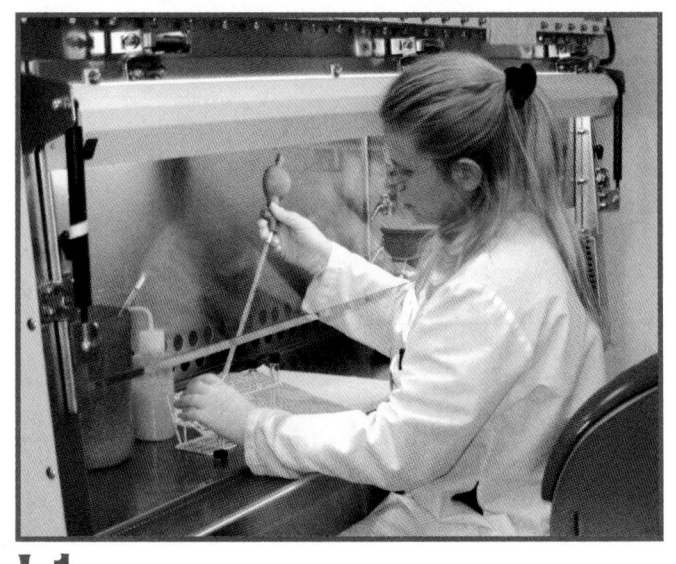

I-1 BIOLOGICAL SAFETY CABINET IN A BSL-2 LABORATORY ✦ In this Class II BSC, air is drawn in from the room and is passed through a HEPA filter prior to release into the environment. This airflow pattern is designed to keep aerosolized microbes from escaping from the cabinet. The microbiologist is pipetting a culture. When the BSC is not in use at the end of the day, an ultraviolet light is turned on to sterilize the air and the work surface. (San Diego County Public Health Laboratory)

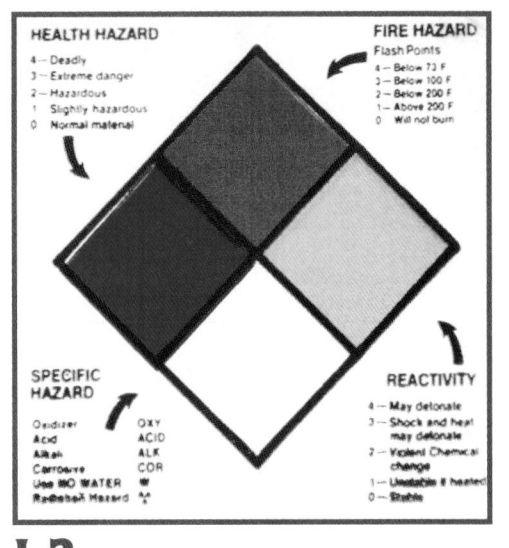

I-2 CHEMICAL HAZARDS ✦ Be aware of these (or similar) labels on the chemicals you handle. Your instructor will advise you on appropriate safety measures to be taken with each.

BSL-4: Organisms have a great potential for lethal infection. Inhalation of infectious aerosols, exposure to infectious droplets, and autoinoculation are of primary concern. The lab is isolated from other facilities, and access is strictly controlled. Ventilation and waste management are under rigid control to prevent release of the microbial agents to the environment. Specially trained personnel perform transfers in Class III BSCs. Class II BSCs may be used as long as personnel wear positive pressure, one-piece body suits with a life-support system. Examples include agents causing hemorrhagic diseases, such as Ebola virus, Marburg virus, and Lassa fever.

The microorganisms used in introductory microbiology courses depend on the institution, objectives of the course, and student preparation. Most introductory courses use organisms that may be handled at BSL-1 and BSL-2 levels so we have followed that practice in designing this set of exercises. Following are general safety rules to reduce the chance of injury or infection to you and to others, both inside and outside the laboratory. Although they represent a mixture of BSL-1 and BSL-2 guidelines, we believe it is best to err on the side of caution and that students should learn and practice the safest level of standards (relative to the organisms they are likely to encounter) at all times. Please follow these and any other safety guidelines required by your college.

Chemical safety is also important in a microbiology laboratory. Be aware of the hazards presented by the chemicals you are handling. Most will be labeled with a sticker as shown in Figure I-2. Numbers are assigned to the degree of health, fire, and reactivity hazard posed by the chemical. There also is a space to enter specific hazards, such as acid, corrosive, and radioactivity.

Student Conduct

✦ To reduce the risk of infection, do not smoke, eat, drink, or bring food or drinks into the laboratory room—even if lab work is not being done at the time.

✦ Do not apply cosmetics or handle contact lenses in the laboratory.

✦ Wash your hands *thoroughly* with soap and water after handling living microbes and before leaving the laboratory each day. Also, wash your hands after removing gloves.

✦ Do not remove any organisms or chemicals from the laboratory.

✦ Lab time is precious, so come to lab prepared for that day's work. Figuring out what to do as you go is likely to produce confusion and accidents.

✦ Work carefully and methodically. Do not hurry through any laboratory procedure.

Basic Laboratory Safety

✦ Wear protective clothing (*i.e.*, a lab coat) in the laboratory when handling microbes. Remove the coat prior to leaving the lab and autoclave it regularly (Figure I-3).

✦ Do not wear sandals or open-toed shoes in the laboratory.

✦ Wear eye protection whenever you are heating chemicals, even if you wear glasses or contacts (Figure I-3).

✦ Turn off your Bunsen burner when it is not in use. In addition to being a fire and safety hazard, it is an unnecessary source of heat in the room.

✦ Tie back long hair, as it is a potential source of contamination as well as a likely target for fire.

✦ If you are feeling ill, go home. A microbiology laboratory is not a safe place if you are ill.

✦ If you are pregnant, immune compromised, or are taking immunosuppressant drugs, please see the instructor. It may be in your best *long-term* interests to postpone taking this class. Discuss your options with your instructor.

✦ If it is your lab's practice to wear disposable gloves while handling microorganisms, be sure to remove them each time you leave the laboratory. The proper method for removal is with the thumb under the cuff of the other hand's glove and turning it inside out without snapping it. Gloves should then be disposed of in the container for contaminated materials. Then, wash your hands.

✦ Wear disposable gloves while staining microbes and handling blood products—plasma, serum, antiserum, or whole blood (Figure I-3). Handling blood can be hazardous, even if you are wearing gloves. Consult your instructor before attempting to work with any blood products.

✦ Use an antiseptic (*e.g.,* Betadine®) on your skin if it is exposed to a spill containing microorganisms. Your instructor will tell you which antiseptic you will be using.

✦ Never pipette by mouth. Always use mechanical pipettors (see Figure C-1, Appendix C).

✦ Dispose of broken glass or any other item that could puncture an autoclave bag in an appropriate sharps or broken glass container (Figure I-4).

✦ Use a fume hood to perform any work involving highly volatile chemicals or stains that need to be heated.

✦ Find the first-aid kit, and make a mental note of its location.

I-3 SAFETY FIRST ✦ This student is prepared to work safely with microorganisms. The lab area is uncluttered, tubes are upright in a test tube rack, and the flame is accessible but not in the way. The student is wearing a protective lab coat, gloves, and goggles, all of which are to be removed prior to leaving the laboratory. Not all procedures require gloves and eye protection. Your instructor will advise you as to the standards in your laboratory.

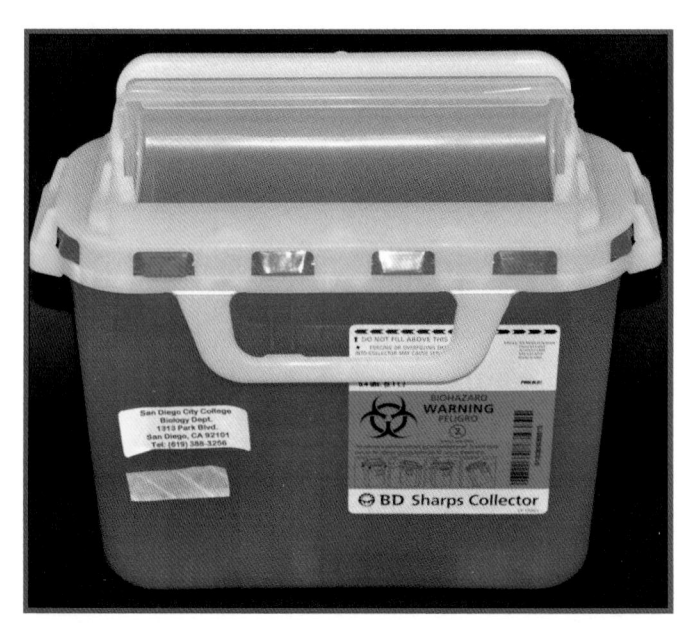

I-4 SHARPS CONTAINER ✦ Needles, glass, and other contaminated items that can penetrate the skin or an autoclave bag should be disposed of in a sharps container. Do not fill above the dashed black line. Notice the autoclave tape in the lower left. The white stripes will turn black after proper autoclaving. Above the autoclave tape is the address of the institution that produced the biohazardous waste.

✦ Find the fire blanket, shower, and fire extinguisher, note their locations, and develop a plan for how to access them in an emergency.

✦ Find the eye wash basin, learn how to operate it, and remember its location.

Reducing Contamination of Self, Others, Cultures, and the Environment

✦ Wipe the desktop with a disinfectant (*e.g.,* Amphyl® or 10% chlorine bleach) before *and* after each lab period. Never assume that the class before you disinfected the work area. An appropriate disinfectant will be supplied. Allow the disinfectant to evaporate; do not wipe it dry.

✦ Never lay down culture tubes on the table; they always should remain upright in a tube holder (Figure I-3). Even solid media tubes contain moisture or condensation that may leak out and contaminate everything it contacts.

✦ Cover any culture spills with paper towels. Soak the towels immediately with disinfectant, and allow them to stand for 20 minutes. Report the spill to your instructor. When you are finished, place the towels in the container designated for autoclaving.

✦ Place all nonessential books and papers under the desk. A cluttered lab table is an invitation for an accident that may contaminate your expensive school supplies.

✦ When pipetting microbial cultures, place a disinfectant-soaked towel on the work area. This reduces contamination and possible aerosols if a drop escapes from the pipette and hits the tabletop.

Disposing of Contaminated Materials

In most instances, the preferred method of decontaminating microbiological waste and reusable equipment is the autoclave (Figure I-5).

✦ Remove all labels from tube cultures and other contaminated *reusable* items and place them in the autoclave container so designated. This will likely be an open autoclave pan to enable cleaning the tubes, and other items following sterilization.

✦ Dispose of plate cultures (if plastic Petri dishes are used) and other contaminated nonsharp *disposable* items in the autoclave container so designated (Figure I-6). Petri dishes should be taped closed. (**Note:** To avoid recontamination of sterilized culture media and other items, autoclave containers are designed

I-5 AN AUTOCLAVE ✦ Media, cultures, and equipment to be sterilized are placed in the basket of the autoclave. Steam heat at a temperature of 121°C (produced at atmospheric pressure plus 15 psi) for 15 minutes is effective at killing even bacterial spores. Some items that cannot withstand the heat, or have irregular surfaces that prevent uniform contact with the steam, are sterilized by other means.

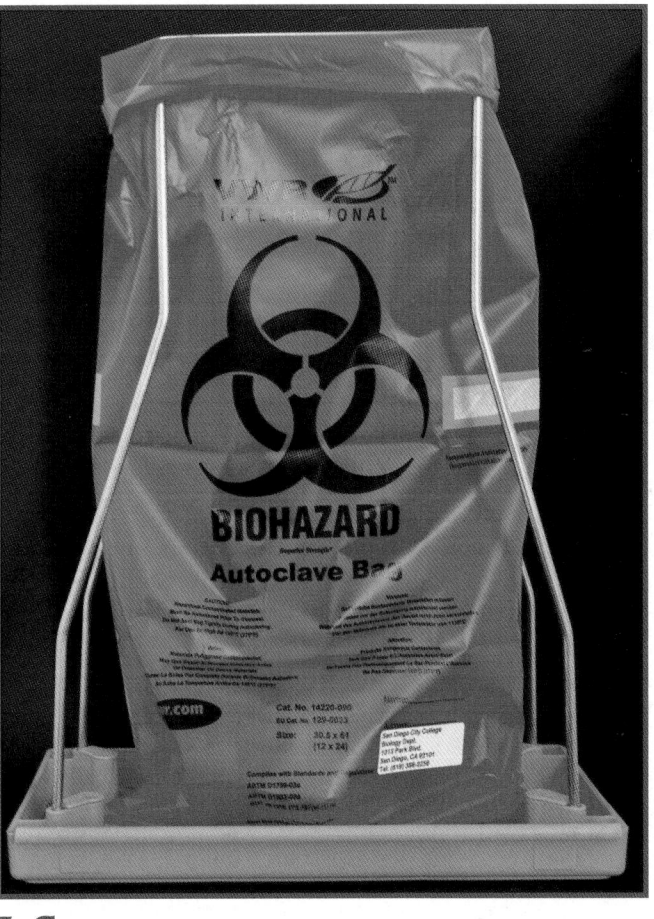

I-6 AN AUTOCLAVE BAG ✦ Nonreusable items (such as plastic Petri dishes) are placed in an autoclave bag for decontamination. Petri dishes should be taped closed. Do not overfill or place sharp objects in the bag. Notice the autoclave tape at the middle right. The white stripes will turn black after proper autoclaving. At the lower right is the address of the institution that produced the biohazardous waste.

to be permanently closed, autoclaved, and discarded. Therefore, do not place reusable and nonreusable items in the same container.

✦ Dispose of all blood product samples and disposable gloves in the container designated for autoclaving.

✦ Place used microscope slides of bacteria in a sharps container designated for autoclaving, or soak them in disinfectant solution for at least 30 minutes before cleaning or discarding them. Follow your laboratory guidelines for disposing of glass.

✦ Place contaminated broken glass and other sharp objects (anything likely to puncture an autoclave bag) in a sharps container designated for autoclaving (Figure I-4). Uncontaminated broken glass does not need to be autoclaved, but should be disposed of in a specialized broken glass container.

References

Barkley, W. Emmett, and John H. Richardson. 1994. Chapter 29 in *Methods for General and Molecular Bacteriology*, edited by Philipp Gerhardt, R. G. E. Murray, Willis A. Wood, and Noel R. Krieg. American Society for Microbiology, Washington, DC.

Collins, C. H., Patricia M. Lyne, and J. M. Grange. 1995. Chapters 1 and 4 in *Collins and Lyne's Microbiological Methods*, 7th ed. Butterworth-Heineman, Oxford.

Darlow, H. M. 1969. Chapter VI in *Methods in Microbiology*, Volume 1, edited by J. R. Norris and D. W. Ribbins. Academic Press, Ltd., London.

Fleming, Diane O., and Debra L. Hunt (Editors). 2000. *Laboratory Safety—Principles and Practices*, 3rd ed. American Society for Microbiology, Washington, DC.

Koneman, Elmer W., Stephen D. Allen, William M. Janda, Paul C. Schreckenberger, and Washington C. Winn, Jr. 1997. *Color Atlas and Textbook of Diagnostic Microbiology*, 5th ed. Lippincott-Raven Publishers, Philadelphia and New York.

Power, David A., and Peggy J. McCuen. 1988. Pages 2 and 3 in *Manual of BBL™ Products and Laboratory Procedures*, 6th ed. Becton Dickinson Microbiology Systems, Cockeysville, MD.

Wilson, Deborah E., and L. Casey Chosewood. 2007. U. S. Department of Health and Human Services, *Biosafety in Microbiological and Biomedical Laboratories*, 5th ed. U. S. Government Printing Office, Washington, DC.

A Word About Experimental Design

Like most sciences, microbiology has descriptive and experimental components. Here we are concerned with the latter. Science is a philosophical approach to finding answers to questions. In spite of what you may have been taught in grade school about THE "Scientific Method," science can approach problems in many ways, rather than in any *single* way. The nature of the problem, personality of the scientist, intellectual environment at the time, and good, old-fashioned luck all play a role in determining which approach is taken. Nevertheless, in experimental science, one component that is always present is a **control** (or controls).

A controlled experiment is one in which all **variables** except one—the **experimental variable**—are maintained without change. This is the only way the results can be considered reliable. By maintaining all variables except one, other potential sources of an observed event can be eliminated. Then (presumably), a **cause and effect relationship** between the event and the experimental variable can be established. If the event changes when the experimental variable changes, we provisionally link that variable and the event. Alternatively, if there is no observed change, we can eliminate the experimental variable from involvement with the event.

Throughout science experimentation—and this book—you will see the word *control*. Controls are an essential and integral part of all experiments. As you work your way through the exercises in this book, pay attention to all the ways controls are used to improve the reliability of the procedure and your confidence in the results.

Microbiological experimentation often involves tests that determine the ability of an organism to use or produce some chemical, or to determine the presence or absence of a specific organism in a sample. Ideally, a positive result in the test indicates that the microbe has the ability or is present in the sample, and a negative result indicates a lack of that ability or absence in the sample (Figure I-7). The tests we run, however, have limitations and occasionally may give **false positive** or **false negative** results. An inability to detect small amounts of the chemical or organism in question would yield a false negative result and would be the result of inadequate **sensitivity** of the test (Figure I-7). An inability to discriminate between the chemical or organism in question and similar chemicals or organisms would yield a false positive result and would be the result of inadequate **specificity** of the test (Figure I-7). Sensitivity and specificity can be quantified using the following equations:

$$\text{Sensitivity} = \frac{\text{True Positives}}{\text{True Positives} + \text{False Negatives}}$$

$$\text{Specificity} = \frac{\text{True Negatives}}{\text{True Negatives} + \text{False Positives}}$$

The closer sensitivity and specificity are to a value of one, the more useful the test is. As you perform the tests in this book, be mindful of each test's limitations, and be open to the possibility of false positive and false negative results.

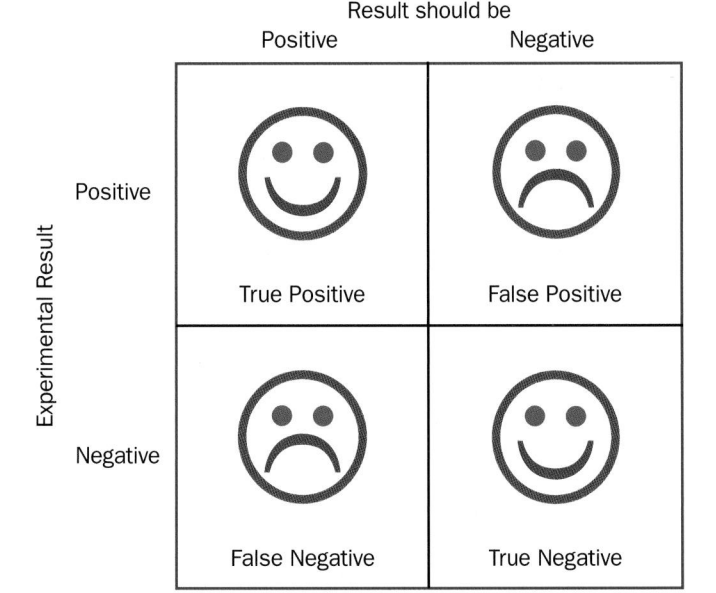

I-7 LIMITATIONS OF EXPERIMENTAL TESTS ✦ Ideally, tests should give a positive result for specimens that are positive, and a negative result for specimens that are negative. False positive and false negative results do occur, however, and these are attributed to inadequate specificity and inadequate sensitivity, respectively, of the test system.

References

Forbes, Betty A., Daniel F. Sahm, and Alice. S. Weissfeld. 1998. Chapter 5 in *Bailey & Scott's Diagnostic Microbiology*, 10th ed. Mosby-Year Book, St. Louis.

Lilienfeld, David E. and Paul D. Stolley. 1994. Page 118 in *Foundations of Epidemiology*, 3rd ed. Oxford University Press, New York.

Mausner, Judith S., and Shira Kramer. 1985. Pages 217–220 in *Epidemiology: An Introductory Text*, 2nd ed. W.B. Saunders Company, Philadelphia.

Data Presentation: Tables and Graphs

In microbiology, we perform experiments and collect data, but it is often difficult to know what the data mean without some method of organization. Tables and graphs allow us to summarize data in a way that makes interpretation easier.

Tables

A table is often used as a preliminary means of organizing data. As an example, Table I-2 shows the winning times for each male and female age division in a half-marathon race. Again, the aim of a table is to provide information to the reader. Notice the meaningful title, the column labels, and the appropriate measurement units. Without these, the reader cannot completely understand the table and your work will go unappreciated! Data tables are provided for you on the Data Sheets for each exercise in this book, but you may be required to fill-in certain components (units, labels, *etc.*) in addition to the data.

TABLE **I-2** Winning Half-Marathon Times By Sex and Age Division

Male Runners		Female Runners	
Winner's Age (Years)	Winning Time (Minutes)	Winner's Age (Years)	Winning Time (Minutes)
15	73	15	88
25	67	24	82
31	67	30	82
35	71	39	84
40	71	42	85
52	78	50	109
62	95	62	108
70	123	70	126

Graphs

Table I-2 does give the information, but what it is telling us may not be entirely clear. It appears that the times increase as runners get older, but we have difficulty determining if this is truly a pattern. That is why data also are presented in graphic form at times; a graph usually shows the relationship between variables better than a table of numbers.

X–Y Scatter Plot The type of graph you will be using in this manual is an "*X–Y* Scatter Plot," in which two variables are graphed against each other. Figure I-8 shows the same data as Table I-2, but in an *X–Y* Scatter Plot form.

Notice the following important features of the graph in Figure I-8:

✦ *Title:* The graph has a meaningful title—which should tell the reader what the graph is about. A title of "Age vs. Winning Time" is vague and inadequate.

✦ *Dependent and independent variables:* The graph is read from left to right. In our example, we might say for the male runners, "As runners get older, winning times get longer." *Winning time* depends on *age*, so winning time is the *dependent* variable and age is the *independent* variable. By convention, the independent variable is plotted on the *x*-axis and the dependent variable is plotted on the *y*-axis. (Age does *not* depend on the winning time.) By way of comparison, notice the consequence of plotting age on the *y*-axis and winning time on the *x*-axis: "As runners get slower, they get older"—which doesn't reflect the actual relationship between the variables.

✦ *Axis labels:* Each axis is labeled, including the appropriate units of measure. "Age" without units is meaningless. Does the scale represent months? years? centuries?

✦ *Axis scale:* The scale on each axis is uniform. The distance between marks on the axis is always the same and represents the same amount of that variable. (But increments on the *x*-axis don't have to equal those on the *y*-axis, as shown.) The size of each increment is up to the person making the graph and is dictated by the magnitude and range of the data. Most of the time, we choose a length for the axis that fills the available graphing space.

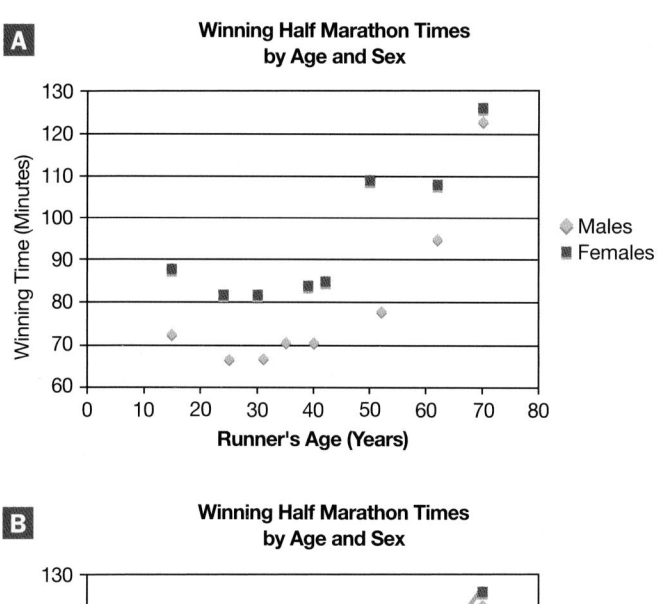

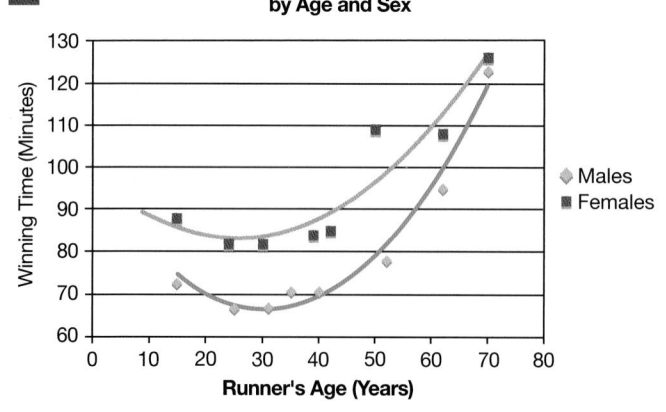

I-8 SAMPLE X–Y SCATTER PLOT ✦ A graph often shows the relationship between variables better than a table of numbers. Examine this sample and identify the essential components of a quality graph (see text). **A** Presentation of data without a best-fit line is acceptable if there are not enough data points to justify illustrating a trend. **B** Shown here are the same data but with a trend line. Notice that the points do not fall directly on the line but, rather, that the line gives the general trend of the data. "Connecting the dots" is not appropriate.

✦ *Axis range:* The scale for an axis does not have to begin with "0." Use a scale that best presents the data. In this case, the smallest *y*-value was 67 minutes, so the scale begins at 60 minutes.

✦ *Multiple Data Sets and the Legend:* The two data series (male and female times) are plotted on the same set of axes, but with different symbols that are defined in the legend at the right. The symbols shown differ in color *and* shape, but one of these is adequate.

✦ *Best-fit Line:* If a line is to be drawn at all, it should be an average line for the data points, not one that "connects the dots" (Figure I-8B). Notice that the points are not necessarily *on* the line. The purpose of a best-fit line is to illustrate the general trend of the data, not the specifics of the individual data points. (Be assured that most graphs in your textbooks where a smooth line is shown were experimentally determined and the lines are derived from points scattered around the line.) There is a mathematical formula that allows one to compute the slope and *y*-intercept of the **trend line** if the relationship is linear or a **best-fit line** if the relationship is nonlinear (as in the half marathon times example), but this is beyond our needs. For our purposes, a hand-drawn trend line that looks good is good enough. (If you use a computer graphing program, then it will produce the trend line without you doing any of the math—the best of all situations!)

Bar Graphs Bar graphs are used to illustrate one variable. Using a bar graph to show the relationship between winning times and ages is inappropriate. Examine Figure 1-9A. Notice that the space each bar fills is meaningless; that is, the only important part of the bar is the top—which is the value used in the *X–Y* scatter plot. Appropriate use of a bar graph would be the distribution of student performance on an exam (Figure 1-9B). Notice that the space each bar fills has meaning. Each student in a particular group adds height to the bar.

Data Presentation:
Be Creative, But Complete

There is no single correct way to produce a graph for a particular data set. Actually, most people working independently would graph the same data set in different ways (*e.g.*, different scales, colors, wording of the title and axis labels), but the essential components listed above would have to be there. You will be asked to graph some of the data you collect. Be sure your graphs tell a complete and clear story of what you've done.

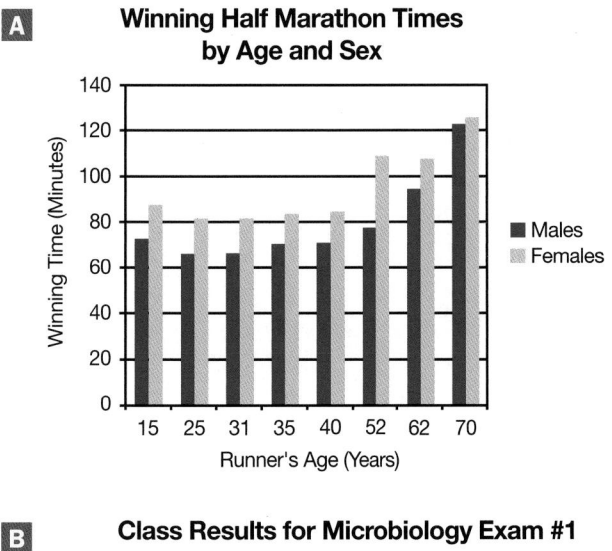

A Winning Half Marathon Times by Age and Sex

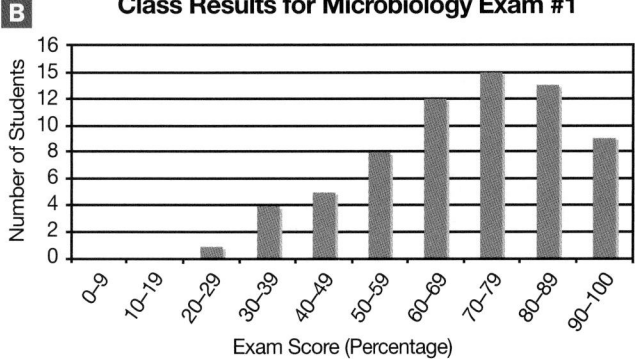

B Class Results for Microbiology Exam #1

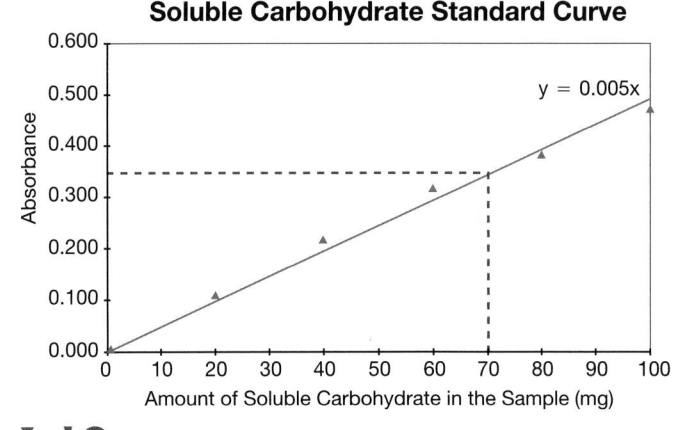

Soluble Carbohydrate Standard Curve

I-10 A STANDARD CURVE ✦ Standard curves are used to determine the value of some unknown in a sample. In this example, the absorbances of six solutions with known soluble carbohydrate concentrations were determined experimentally, plotted (blue triangles), and used to make a trend line (blue line). Absorbance of a sample with an unknown soluble carbohydrate concentration then was determined to be 0.35 by the same experimental method. The point where the y value (0.35) intersects the trend line gives the x value that corresponds to it (red dashed lines): 70 µg of soluble carbohydrate. The equation for the trend line ($y = 0.005x$) is also given and can be used to determine the x value by substituting 0.35 for y.

1-9 BAR GRAPHS ✦ A bar graph is appropriate to present data involving a single variable. **A** Plotting the winning half marathon times from Table I-2 using a bar graph is inappropriate because the only meaningful point is at the top. **B** A bar graph is useful in presenting data of a single variable, such as the number of students earning a specific score on their microbiology exam.

Standard Curve Besides making interpretation easier, graphs sometimes are used to establish an experimental value. Each point on the trend line represents a correlation between the x and y values (Figure I-10). If we know one value, we can read the other off the graph. We use this process with something called a "standard curve" or "calibration curve." To produce a standard curve, samples with a known amount of the independent variable (*e.g.*, soluble carbohydrate concentration) are subjected to the experiment. The resulting data (the y values—*e.g.*, absorbances) are plotted and a trend line is drawn.

Now a sample with an *unknown* amount of the independent variable can be subjected to the same experimental procedure to determine its y value. Once the y value is known, the corresponding x value is read directly off the graph to determine the unknown amount. (If the relationship is linear, the "trend" line is described by the equation $y = mx + b$. Once the y value is determined experimentally, the x value can be calculated by substitution.)

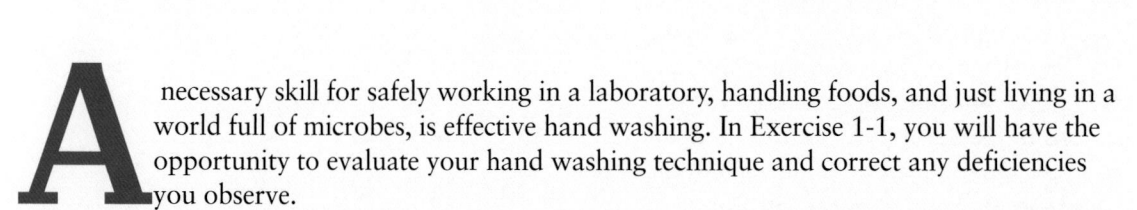

Fundamental Skills for the Microbiology Laboratory

A necessary skill for safely working in a laboratory, handling foods, and just living in a world full of microbes, is effective hand washing. In Exercise 1-1, you will have the opportunity to evaluate your hand washing technique and correct any deficiencies you observe.

Bacterial and fungal **cultures** are grown and maintained on or in solid and liquid substances called **media**. Preparation of these media involves weighing ingredients, measuring liquid volumes, calculating proportions, handling basic laboratory glassware, and operating a pH meter and an autoclave. In Exercise 1-2 you will learn and practice these fundamental skills by preparing a couple of simple growth media. When you have completed the exercise, you will have the skills necessary to prepare almost any medium if given the recipe.

A third fundamental skill necessary for any microbiologist is the ability to transfer microbes from one place to another without contaminating the original culture, the new medium, or the environment (including the microbiologist). This **aseptic** (sterile) transfer technique is required for virtually all procedures in which living microbes are handled, including isolations, staining, and differential testing. Exercises 1-3 through 1-5 present descriptions of common transfer and inoculation methods. Less frequently used methods are covered in Appendices B through D. ✦

EXERCISE 1-1

Glo Germ™ Hand Wash Education System

✦ Theory

The concept of good hand hygiene has evolved from a controversial beginning (in the early 1800s) to an accepted practice that is still problematic. Studies designed to test the efficacy of various agents often have subjects wash for an unrealistic amount of time (longer than workers routinely wash on the job), test artificially contaminated hands (or not), and use different standards of evaluation, making comparisons difficult. We still are left with the question: "What works best?"

Although hand washing has been identified as an important, easily performed act that minimizes transfer of pathogens to others, uniform compliance with hand-washing standards has been difficult to attain. Factors that contribute to noncompliance include heavy workloads, skin reactions to the agent (*e.g.,* plain or antimicrobial soap, iodine compounds, alcohol), skin dryness from frequent washing, and many others (see Boyce and Pittet, 2002). Alcohol-based hand rubs, in many instances, have replaced conventional hand-washing agents because they are more effective than soap and water, require less time, produce fewer skin reactions, and have been shown to result in a higher level of compliance by health care workers.

The Glo Germ™ Hand Wash Education System was developed to train people to wash their hands more effectively. The lotion (a powder also is available) contains minute plastic particles (artificial germs) that fluoresce when illuminated with ultraviolet (UV) radiation but are invisible with normal lighting. Initially the hands are covered with the lotion, but the location and density of the germs is unknown because of the normal room lighting. After washing, a UV lamp is shined on the hands. Wherever the "germs" remain, hand washing was not effective. This provides immediate feedback to the washers as to the effectiveness of their hand washing and provides information about where they have to concentrate their efforts in the future.

✦ Application

Effective hand washing to minimize direct person-to-person transmission of pathogens by health-care professionals and food handlers is essential. It also is critical to laboratorians handling pathogens to minimize transmission to others, inoculation of self, and contamination of cultures.

✦ In This Exercise

You will cover your hands with nontoxic, synthetic fluorescent "germs" and compare the degree of contamination before and after hand washing to evaluate your hand-washing technique and demonstrate the difficulty in removing hand contaminants.

✦ Materials[1]

Per Student Group
✦ one bottle of Glo Germ™ lotion-based simulated germs
✦ one ultraviolet penlight

🖐 Procedure

1 Shake the lotion bottle well.

2 Have your lab partner apply 2–3 drops of gel on the palms of both of your hands.

3 Rub your hands together, thoroughly covering your hand surfaces, including the backs and between the fingers. Spread the lotion up to the wrists on both sides. Also, scratch the palms with all your fingernails.

---

[1] Available from Glo Germ™, PO Box 537, Moab, UT 84532. 1-800-842-6622 (USA). Online: http://www.glogerm.com/

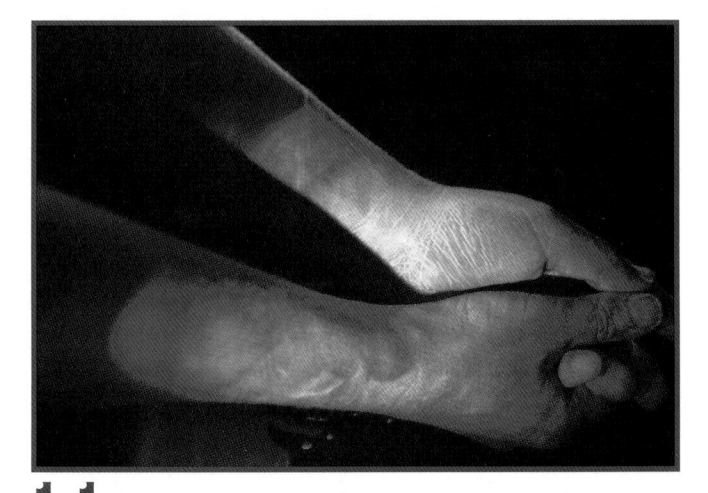

1-1 HANDS COVERED WITH GLO GERM™ PRIOR TO WASHING ✦
Shown are properly prepared hands covered with the fluorescent Glo Germ™ lotion prior to washing. Note the thorough coverage, including the back of the hands and under the fingernails.

4 Have your lab partner shine the UV light on your hands to see the extent of coverage with the lotion. *Do not look directly at the lamp*. This works best in an area with limited ambient light. Do *not* handle the light yourself because you will contaminate it with the artificial germs.

5 Have your lab partner turn on warm water at a sink for you. Then wash your hands with soap and warm water as thoroughly as you can for at least 20 seconds. Use a fingernail brush if you have one. When you are finished, have your lab partner turn off the water and hand you a fresh paper towel. Dry your hands.

6 Have your lab partner shine the UV light on your hands once more. Examine the hand surfaces contaminated by the artificial germs.

7 Now that you know where the artificial germs remain, wash your hands once more to remove as many as possible. As before, have your lab partner turn the water on and off for you.

8 Repeat the experiment with your lab partner, but with the roles reversed.

9 Record your results on the Data Sheet, and answer the questions.

10 After recording your results, shine the UV lamp once more on your Data Sheet and pen/pencil to see how much of the lotion was transferred to these. *Do not look directly at the lamp*.

References

Boyce, John M., and Didier Pittet (2002). Centers for Disease Control and Prevention. Guideline for Hand Hygiene in Health-Care Settings: Recommendations of the Healthcare Infection Control Practices Advisory Committee and the HICPAC/SHEA/APIC/IDSA Hand Hygiene Task Force. MMWR 2002;51 (No. RR-16), pages 1–45.

Glo Germ™. Package insert for the Glo Germ™ Hand Wash Education System. Glo Germ™, PO Box 537, Moab, UT 84532.

Basic Growth Media

To cultivate microbes, microbiologists use a variety of growth media. Although these media may be formulated from scratch, they more typically are produced by rehydrating commercially available powdered media. Media that are routinely encountered in the microbiology laboratory range from the widely used, general-purpose growth media, to the more specific selective and differential media used in identification of microbes. In Exercise 1-2 you will learn how to prepare simple general growth media. ✦

EXERCISE 1-2

Nutrient Broth and Nutrient Agar Preparation

✦ Theory

Nutrient broth and nutrient agar are common media used for maintaining bacterial cultures. To be of practical use, they have to meet the diverse nutrient requirements of routinely cultivated bacteria. As such, they are formulated from sources that supply carbon and nitrogen in a variety of forms—amino acids, purines, pyrimidines, monosaccharides to polysaccharides, and various lipids. Generally, these are provided in digests of plant material (phytone) or animal material (peptone and others). Because the exact composition and amounts of carbon and nitrogen in these ingredients are unknown, general growth media are considered to be **undefined**. They are also known as **complex media**.

In most classes (because of limited time), media are prepared by a laboratory technician. Still, it is instructive for novice microbiologists to at least gain exposure to what is involved in media preparation. Your instructor will provide specific instructions on how to execute this exercise using the equipment in your laboratory.

✦ Application

Microbiological growth media are prepared to cultivate microbes. These general growth media are used to maintain bacterial stock cultures.

✦ In This Exercise

You will prepare 1-liter batches of two general growth media: nutrient broth and nutrient agar. Over the course of the semester, a laboratory technician will probably do this for you, but it is good to gain firsthand appreciation for the work done behind the scenes!

✦ Materials

Per Student Group

✦ one 2-liter Erlenmeyer flask for each medium made

✦ three or four 500 mL Erlenmeyer flasks and covers (can be aluminum foil)

✦ stirring hotplate

✦ magnetic stir bars

✦ all ingredients listed below in the recipes (or commercially prepared dehydrated media)

✦ sterile Petri dishes

✦ test tubes (16 mm × 150 mm) and caps

✦ balance

✦ weighing paper or boats

✦ spatulas

✦ Medium Recipes

Nutrient Broth

✦ beef extract	3.0 g
✦ peptone	5.0 g
✦ distilled or deionized water	1.0 L

pH 6.6–7.0 at 25°C

Nutrient Agar

✦ beef extract	3.0 g
✦ peptone	5.0 g
✦ agar	15.0 g
✦ distilled or deionized water	1.0 L

pH 6.6–7.0 at 25°C

✦ Preparation of Medium

Day One

To minimize contamination while preparing media, clean the work surface, turn off all fans, and close any doors that might allow excessive air movements.

Nutrient Agar Tubes

1 Weigh the ingredients on a balance (Figure 1-2).

2 Suspend the ingredients in one liter of distilled or deionized water in the two-liter flask, mix well, and boil until fully dissolved (Figure 1-3).

3 Dispense 7 mL portions into test tubes and cap loosely (Figure 1-4). If your tubes are smaller than those listed in Materials, adjust the volume to fill 20% to 25% of the tube. Fill to approximately 50% for agar deeps.

4 Sterilize the medium by autoclaving for 15 minutes at 121°C (Figure 1-5).

5 After autoclaving, cool to room temperature with the tubes in an upright position for agar deep tubes. Cool with the tubes on an angle for agar slants (Figure 1-6).

6 Incubate the slants and/or deep tubes at 35 ± 2°C for 24 to 48 hours.

Nutrient Agar Plates

1 Weigh the ingredients on a balance (Figure 1-2).

2 Suspend the ingredients in one liter of distilled or deionized water in the two-liter flask, mix well, and boil until fully dissolved (Figure 1-3).

3 Divide into three or four 500 mL flasks for pouring. Smaller flasks are easier to handle when pouring plates. Don't forget to add a magnetic stir bar and to cover each flask before autoclaving.

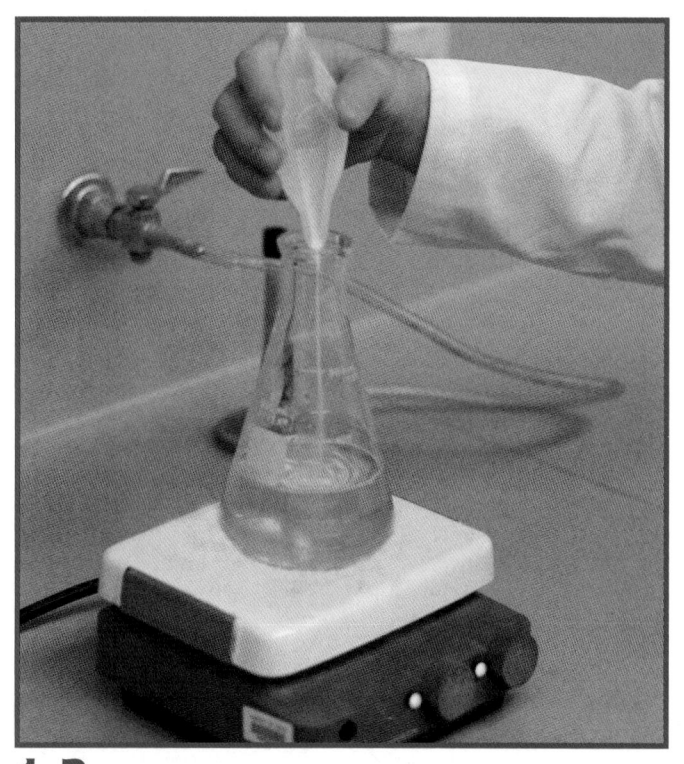

1-3 MIXING THE MEDIUM ✦ The powder is added to a flask of distilled or deionized water on a hotplate. A magnetic stir bar mixes the medium as it is heated to dissolve the powder.

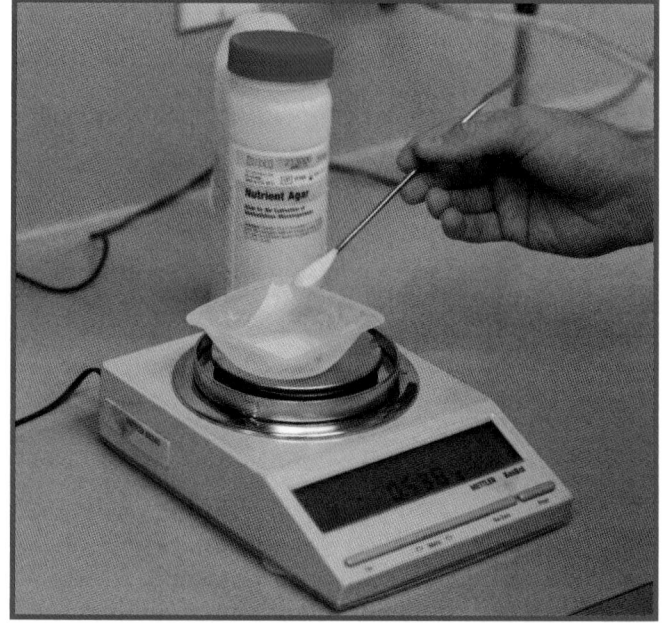

1-2 WEIGHING MEDIUM INGREDIENTS ✦ Solid ingredients are weighed with an analytical balance. A spatula is used to transfer the powder to a tared weighing boat. Shown here is dehydrated nutrient agar, but the weighing process is the same for any powdered ingredient.

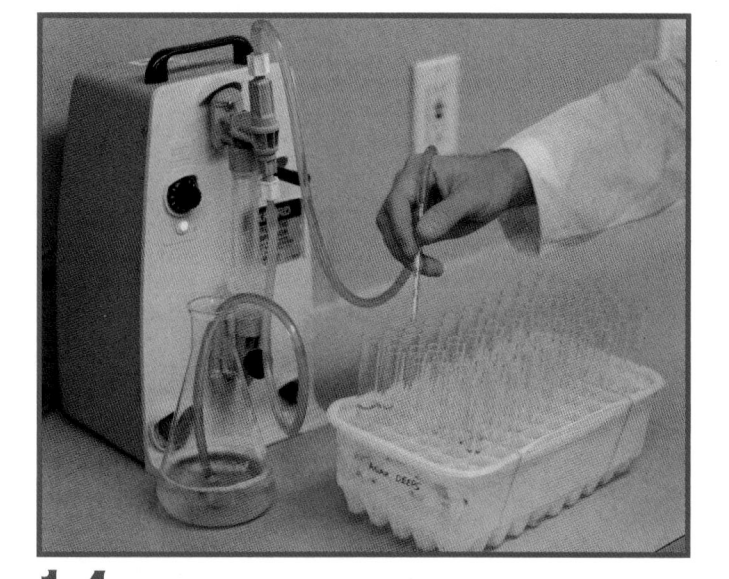

1-4 DISPENSING THE MEDIUM INTO TUBES ✦ An adjustable pump can be used to dispense the appropriate volume (usually 7–10 mL) into tubes. Then, loosely cap the tubes.

1-5 AUTOCLAVING THE TUBED MEDIUM ✦ The basket of tubes is sterilized for 15 minutes at 121°C in an autoclave. When finished, the tubes are cooled.

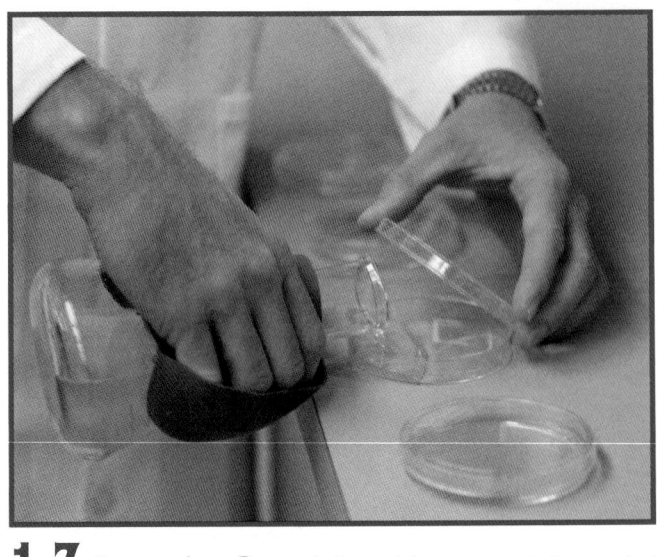

1-7 POURING AGAR PLATES ✦ Agar plates are made by pouring sterilized medium into sterile Petri dishes. The lid is used as a shield to prevent airborne contamination. Once poured, the dish is gently swirled so the medium covers the base. Plates are then cooled and dried to eliminate condensation.

1-6 TUBED MEDIA ✦ From left to right: a broth, an agar slant, and an agar deep tube. The solid media are liquid when they are removed from the autoclave. Agar deeps are allowed to cool and solidify in an upright position, whereas agar slants are cooled and solidified on an angle.

4 Autoclave for 15 minutes at 121°C to sterilize the medium.

5 Remove the sterile agar from the autoclave, and allow it to cool to 50°C while you are stirring it on a hotplate.

6 Dispense approximately 20 mL into sterile Petri plates (Figure 1-7). *Be careful! The flask will still be hot, so wear an oven mitt.* While you pour the agar, shield the Petri dish with its lid to reduce the chance of introducing airborne contaminants. If necessary, *gently* swirl each plate so the agar completely covers

the bottom; do not swirl the agar up into the lid. Allow the agar to cool and solidify before moving the plates.

7 Store these plates on a counter top for 24 hours to allow them to dry prior to use.

Nutrient Broth

1 Weigh the ingredients on a balance (Figure 1-2).

2 Suspend the ingredients in one liter of distilled or deionized water in the two-liter flask. Agitate and heat slightly (if necessary) to dissolve them completely (Figure 1-3).

3 Dispense 7 mL portions into test tubes and cap loosely (Figure 1-4). As with agar slants, if your tubes are smaller than those recommended in Materials, add enough broth to fill them approximately 20% to 25%.

4 Sterilize the medium by autoclaving for 15 minutes at 121°C (Figure 1-5).

Day Two

1 Examine the tubes and plates for evidence of growth.

2 Record your observations on the Data Sheet.

Reference

Zimbro, Mary Jo and David A. Power. 2003. Pages 404–405 and 408 in *DIFCO™ & BBL™ Manual—Manual of Microbiological Culture Media.* Becton, Dickinson and Company, Sparks, MD.

Aseptic Transfers and Inoculation Methods

As a microbiology student, you will be required to transfer living microbes from one place to another aseptically (*i.e., without contamination of the culture, the sterile medium, or the surroundings*). While you won't be expected to master all transfer methods right now, you will be expected to perform most of them over the course of the semester. Refer back to this section as needed.

To prevent contamination of the sample, inoculating instruments (Figure 1-8) must be sterilized prior to use. Inoculating loops and needles are sterilized immediately before use in an incinerator or Bunsen burner flame. The mouths of tubes or flasks containing cultures or media are also incinerated at the time of transfer by passing their openings through a flame. Instruments that are not conveniently or safely incinerated, such as Pasteur pipettes, cotton applicators, glass pipettes, and digital pipettor tips are sterilized inside wrappers or containers by autoclaving prior to use.

Aseptic transfers are not difficult; however, a little preparation will help assure a safe and successful procedure. Before you begin, you will need to know where the sample is coming from, its destination, and the type of transfer instrument to be used. These exercises provide step-by-step descriptions of different transfer methods. In an effort to avoid too much repetition, skills that are basic to most transfers are described in detail once under "The Basics" and mentioned only briefly as they apply to transfers in the discussion of "Specific Transfer Methods." These are printed in regular type. New material in each specific transfer will be introduced in blue type. Certain less routine transfer methods are discussed in Appendices B through D. ✦

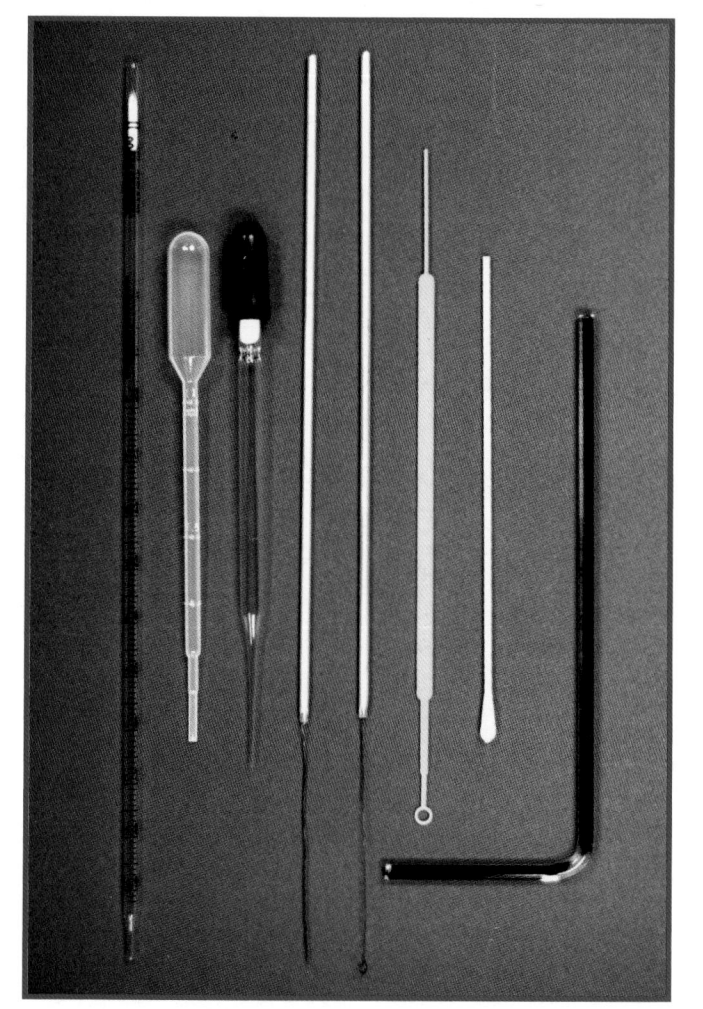

1-8 INOCULATING INSTRUMENTS ✦ Any of several different instruments may be used to transfer a microbial sample, the choice of which depends on the sample source, its destination, and any special requirements imposed by the specific protocol. Shown here are several examples of transfer instruments. From left to right: serological pipette (see Appendix C), disposable transfer pipette, Pasteur pipette, inoculating needle, inoculating loop, disposable inoculating needle/loop, cotton swab (see Appendix B and Exercise 1-4), and glass spreading rod (see Exercise 1-5).

EXERCISE 1-3
Common Aseptic Transfers and Inoculation Methods

✦ Application—The Basics

The following is a listing of general techniques and practices and is not presented as sequential.

✦ *Minimize potential of contamination.* Do not perform transfers over your books and papers because you may inadvertently and unknowingly contaminate them. Put them safely away.

✦ *Be organized.* Arrange all media in advance and clearly label them with your name, the date, the medium and the inoculum. Tubes are typically labeled with tape or paper held on with rubber bands; you may write directly on the base of Petri plates. Be sure not to place labels in such a way as to obscure your view of the inside of the tube or plate.

✦ *Take your time.* Work efficiently, but *do not hurry.* You are handling potentially dangerous microbes.

✦ *Place all media tubes in a test tube rack when not in use* whether they are or are not sterile. Tubes should never be laid on the table surface (Figure 1-9).

✦ *Hold the handle of an inoculating needle or loop like a pencil* in your dominant hand and relax (Figure 1-9)!

✦ *Adjust your Bunsen burner* so its flame has an inner and an outer cone (Figure 1-10).

✦ *Sterilize a loop/needle by incinerating it in the Bunsen burner flame* (Figure 1-11). Pass it through the tip of

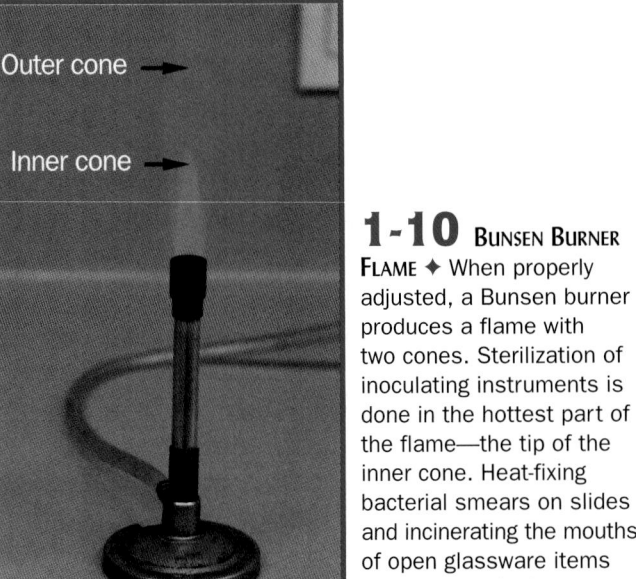

1-10 BUNSEN BURNER FLAME ✦ When properly adjusted, a Bunsen burner produces a flame with two cones. Sterilization of inoculating instruments is done in the hottest part of the flame—the tip of the inner cone. Heat-fixing bacterial smears on slides and incinerating the mouths of open glassware items may be done in the outer cone.

1-11 FLAMING LOOP ✦ Incineration of an inoculating loop's wire is done by passing it through the tip of the flame's inner cone. Begin at the wire's base and continue to the end, making sure that all parts are heated to a uniform orange color. Allow the wire to cool before touching it or placing it on/in a culture. The former will burn you; the latter will cause aerosols of microorganisms.

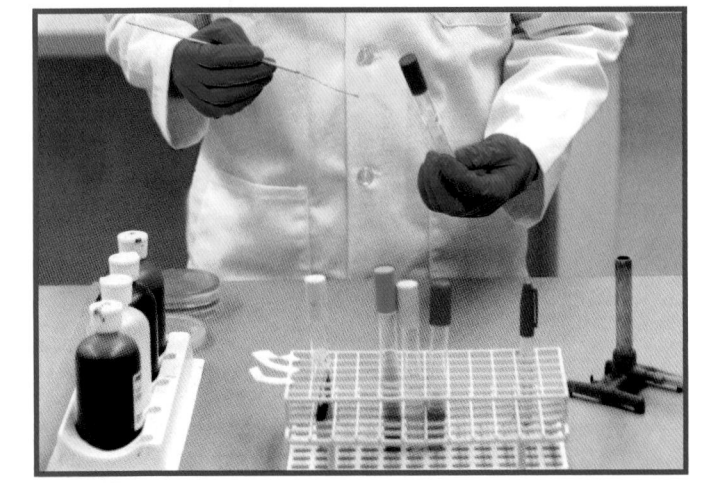

1-9 MICROBIOLOGIST AT WORK ✦ Materials are neatly positioned and not in the way. To prevent spills, culture tubes are stored upright in a test tube rack. They are never laid on the table. The microbiologist is relaxed and ready for work. Notice he is holding the loop like a pencil, not gripping it like a dagger.

the flame's inner cone, holding it at an angle with the loop end pointing downward. Begin flaming about 2 cm up the handle, then proceed down the wire by pulling the loop backward through the flame until the entire wire has become uniformly orange-hot. Flaming in this direction limits aerosol production by allowing the tip to heat up more slowly than if it were thrust into the flame immediately.

✦ *Hold a culture tube in your nondominant hand* and move it, not the loop, as you transfer. This will minimize aerosol production from loop movement.

✦ *Grasp the tube's cap with your little finger* and remove it by pulling the *tube* away from the cap. Hold the cap in your little finger during the transfer (Figure 1-12). (The cap should be loosened prior to transfer, especially if it's a screw-top cap.) When replacing the cap, move the tube back to the cap to keep your loop hand still. The replaced cap doesn't have to be on firmly at this time—just enough to cover the tube.

✦ *Flame tubes* by passing the open end through the Bunsen burner flame two or three times (Figure 1-13).

✦ *Hold open tubes at an angle* to minimize the chance of airborne contamination (Figure 1-14).

✦ *Suspend bacteria in a broth* with a vortex mixer prior to transfer (Figure 1-15). Be sure not to mix so vigorously that broth gets into the cap or that you lose control of the tube. Start slowly, then gently increase

the speed until the tip of the vortex reaches the bottom of the tube. Alternatively, broth may be agitated by drumming your fingers along the length of the tube several times (Figure 1-16). Be careful not to splash the broth into the cap or lose control of the tube.

✦ *When opening a plate, use the lid as a shield* to minimize the chances of airborne contamination (Figure 1-17).

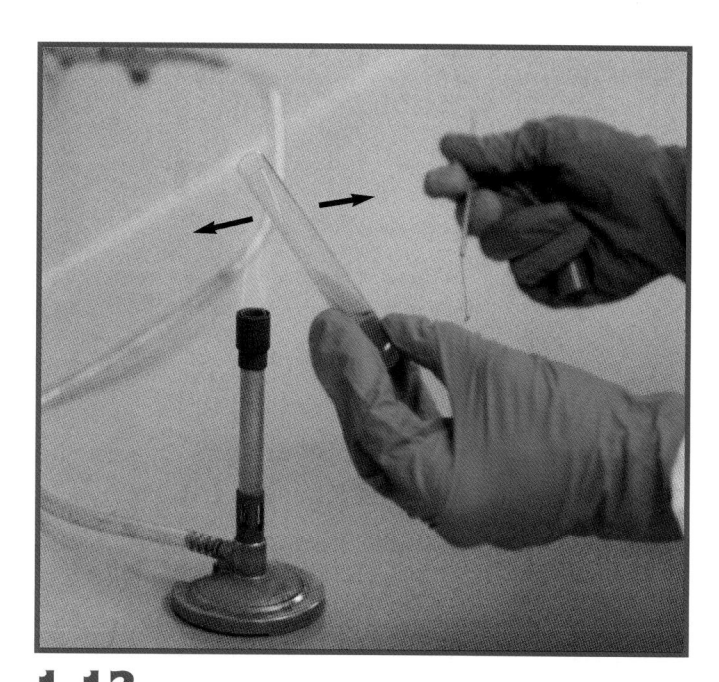

1-13 FLAMING THE TUBE ✦ The tube's mouth is passed quickly through the flame a couple of times to sterilize the tube's lip and the surrounding air. Notice that the tube's cap is held in the loop hand.

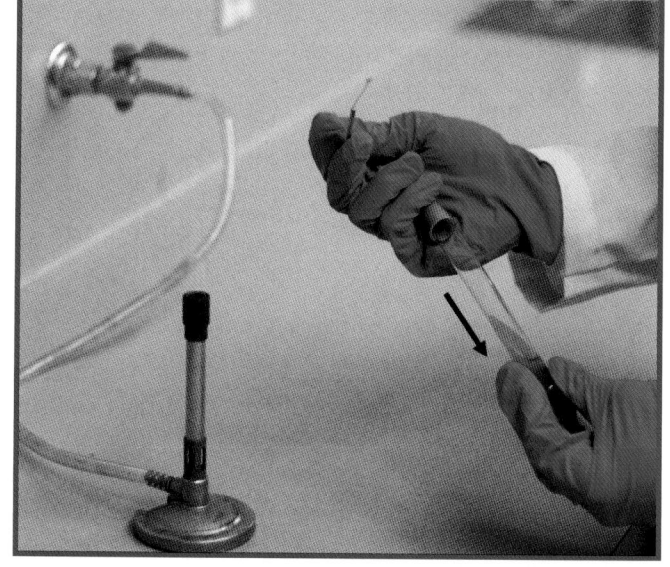

1-12 REMOVING THE TUBE CAP ✦ The loop is held in the dominant hand and the tube in the other hand. Remove the tube's cap with your little finger of your loop hand by pulling the tube away with the other hand; keep your loop hand still. Hold the cap in your little finger during the transfer. When replacing the cap, move the tube back to the cap to keep your loop hand still. The replaced cap doesn't have to be on firmly at this time—just enough to cover the tube.

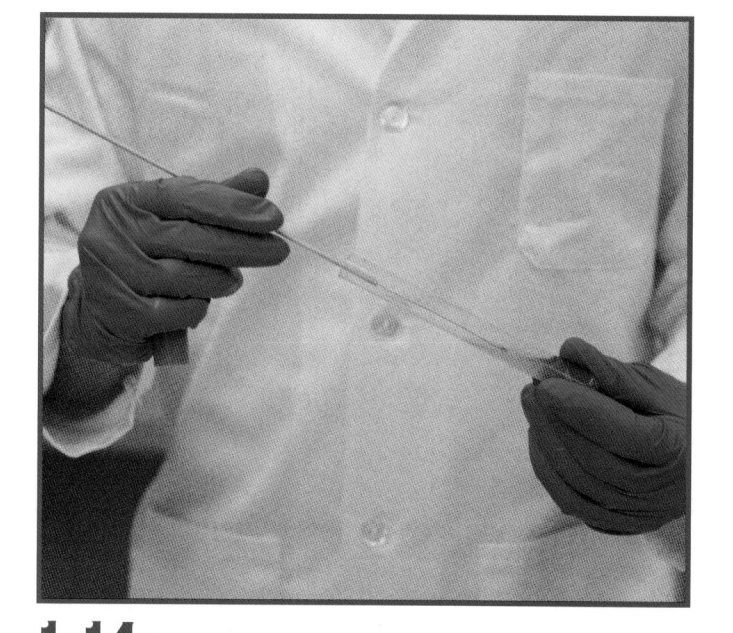

1-14 HOLDING THE TUBE AT AN ANGLE ✦ The tube is held at an angle to minimize the chance that airborne microbes will drop into it. Notice that the tube's cap is held in the loop hand.

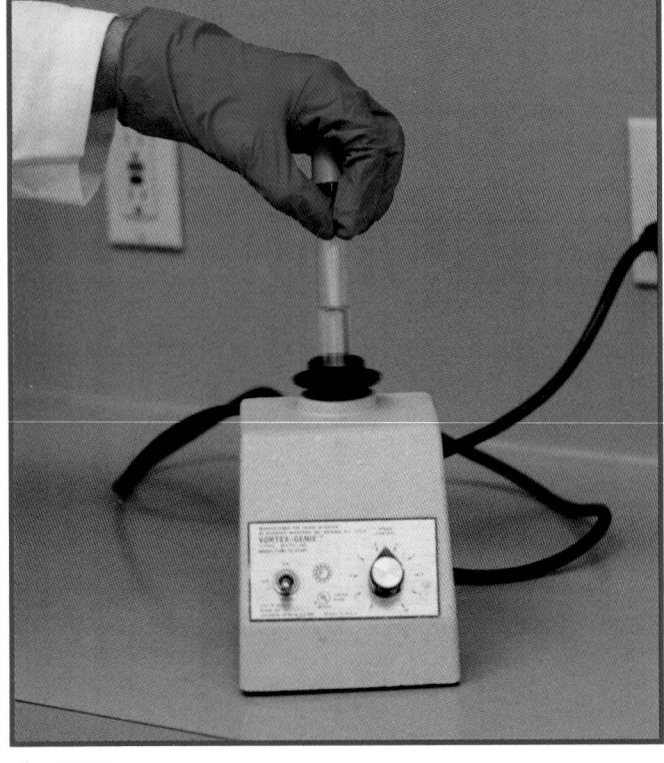

1-15 VORTEX MIXER ✦ Bacteria are suspended in a broth with a vortex mixer. The switch on the left has three positions: on (up), off (middle), and touch (down). The rubber boot is activated when touched only if the "touch" position is used; "on" means the boot is constantly vibrating. On the right is a variable speed knob. Caution must be used to prevent broth from getting into the cap or losing control of the tube and causing a spill.

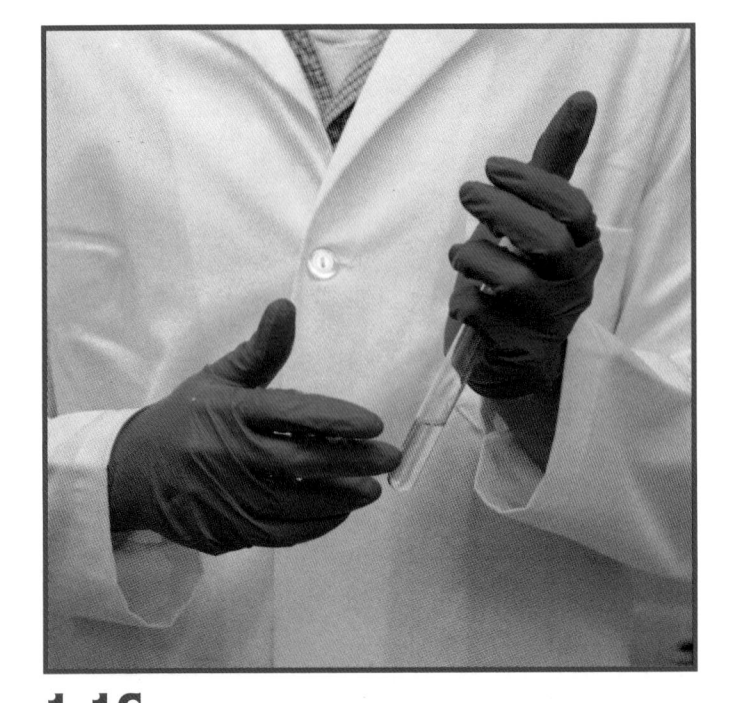

1-16 MIXING BROTH BY HAND ✦ A broth culture always should be mixed prior to transfer. Tapping the tube with your fingers gets the job done safely and without special equipment.

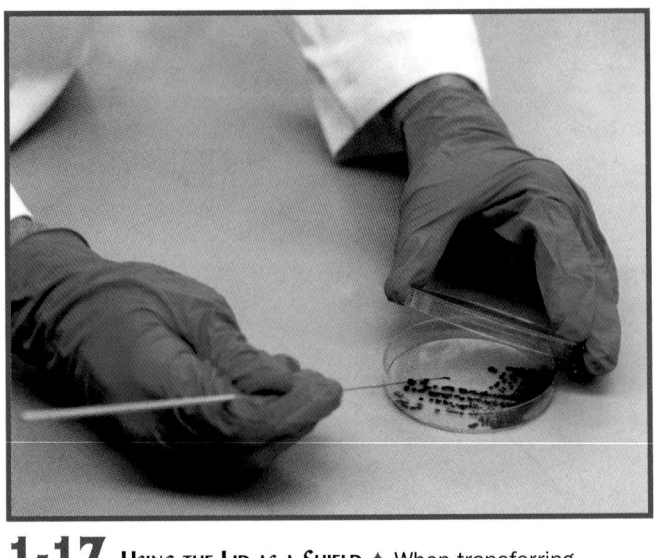

1-17 USING THE LID AS A SHIELD ✦ When transferring bacteria to or from a Petri dish, the lid is used to cover the surface of the agar to minimize airborne contamination.

✦ Application—Specific Transfer Methods

Transfers occur in two basic stages:

1. obtaining the sample to be transferred, and
2. transferring the sample (inoculum) to the sterile culture medium.

These two stages may be combined in various ways. The following descriptions are organized to reflect that flexibility. (Recall that steps in the transfer *not* covered in the Basics Section are printed in blue type.)

Transfers Using an Inoculating Loop or Needle

Inoculating loops and needles are the most commonly used instruments for transferring microbes between all media types—broths, slants, or plates can be the source, and any can be the destination. For ease of reading and because loops and needles are handled in the same way, we refer only to loops in the following instructions.

Obtaining a Sample with an Inoculating Loop or Needle

✦ From a Broth

1 Suspend bacteria in the broth with a vortex mixer (Figure 1-15) or by agitating the tube with your fingers (Figure 1-16).

2 Flame the loop (Figure 1-11).

3 Remove and hold the tube's cap with the little finger of your loop hand (Figure 1-12).

4 Flame the open end of the tube by passing it quickly through a flame two or three times (Figure 1-13).

5 Hold the open tube at an angle to prevent airborne contamination (Figure 1-14).

6 Holding the loop hand still, move the tube up the wire until the tip is in the broth. Continuing to hold the loop hand still, *remove the tube* from the wire (Figure 1-18). There should be a film of broth in the loop (Figure 1-19). Be especially careful not to catch the loop tip on the tube lip. This springing action of the loop creates bacterial aerosols.

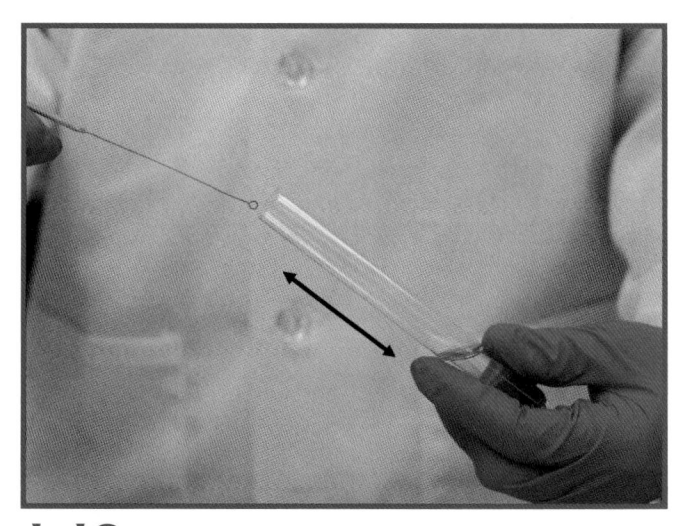

1-18 Loop In/Out of Broth ✦ The open tube is held at an angle to minimize airborne contamination. When placing a loop into a broth tube or removing it, keep the loop hand still and move the tube. Be careful not to catch the loop on the tube's lip when removing it. This produces aerosols that can be dangerous or produce contamination.

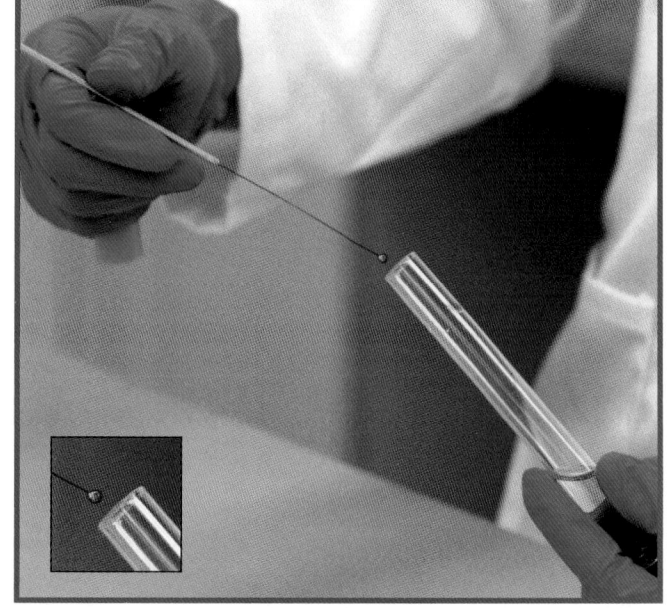

1-19 Removing the Loop from Broth ✦ Notice the film of broth in the loop (see inset). Be careful not to catch the loop on the lip of the tube when removing it. This would produce aerosols that can be dangerous or produce contamination.

7 Flame the tube lip as before. Keep your loop hand still.

8 Keeping the loop hand still (remember, it has growth on it), move the tube to replace its cap.

What you do next depends on the medium to which you are transferring the growth. Please continue with the appropriate inoculation section.

✦ From a Slant

1 Flame the loop (Figure 1-11).

2 Remove and hold the culture tube's cap with the little finger of your loop hand (Figure 1-12).

3 Flame the open end of the tube by passing it quickly through a flame two or three times (Figure 1-13).

4 With the agar surface facing upward, hold the open tube at an angle to prevent airborne contamination (Figure 1-14).

5 Holding the loop hand still, move the tube up the wire until the wire tip is over the desired growth (Figure 1-20). Touch the loop to the growth and obtain the smallest visible mass of bacteria. Then, holding the loop hand still, *remove the tube* from the wire. Be especially careful not to catch the loop tip on the tube lip. This springing action of the loop creates bacterial aerosols.

6 Flame the tube lip as before. Keep your loop hand still.

7 Keeping the loop hand still (remember—it has growth on it), move the tube to replace its cap.

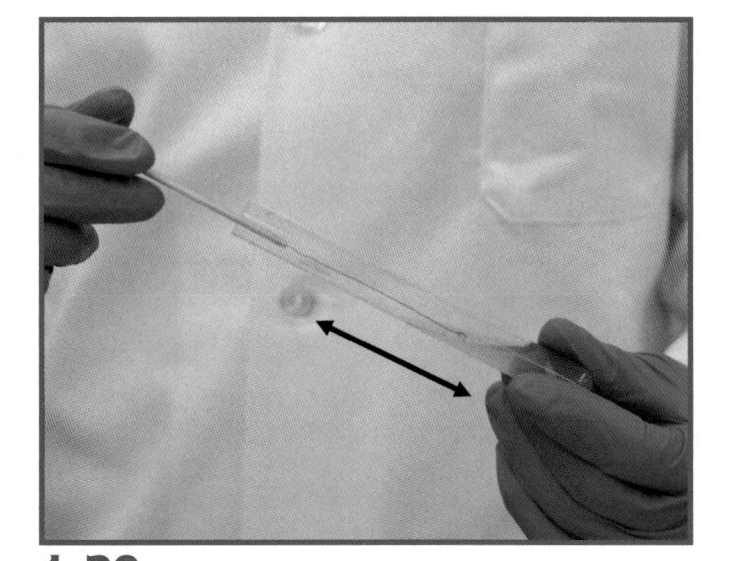

1-20 A Loop and an Agar Slant ✦ When placing a loop into a slant tube or removing it, the loop hand is kept still while the tube is moved. Hold the tube so the agar is facing upward.

What you do next depends on the medium to which you are transferring the growth. Please continue with the appropriate inoculation section.

✦ From an Agar Plate

1 Flame the loop (Figure 1-11).

2 Lift the lid of the agar plate, but continue to use it as a cover to prevent contamination from above (Figure 1-17).

3 Touch the loop to an uninoculated portion of the plate to cool it. (Placing a hot wire on growth may cause the growth to spatter and create aerosols.) Obtain a small amount of bacterial growth by gently touching a colony with the wire tip (Figure 1-17).

4 Carefully remove the loop from the plate and hold it still as you replace the lid.

What you do next depends on the medium to which you are transferring the growth. Please continue with the appropriate inoculation section.

Inoculating Media with an Inoculating Loop or Needle

✦ Fishtail Inoculation of Agar Slants

Agar slants generally are used for growing stock cultures that can be refrigerated after incubation and maintained for several weeks. Many differential media used in identification of microbes are also slants.

1 Remove the cap of the sterile medium with the little finger of your loop hand and hold it there (Figure 1-12).

2 Flame the tube by passing it quickly through the flame a couple of times. Keep your loop hand still (Figure 1-13).

3 Hold the open tube on an angle to minimize airborne contamination. Keep your loop hand still (Figure 1-14).

4 With the agar surface facing upward, carefully move the tube over the wire. Gently touch the loop to the agar surface near the base.

5 Beginning at the bottom of the exposed agar surface, drag the loop in a zigzag pattern as you withdraw the tube (Figure 1-21). Be careful not to cut the agar surface, and be especially careful not to catch the loop tip on the tube lip as you remove it. This springing action of the loop creates bacterial aerosols.

6 Flame the tube mouth as before. Keep your loop hand still.

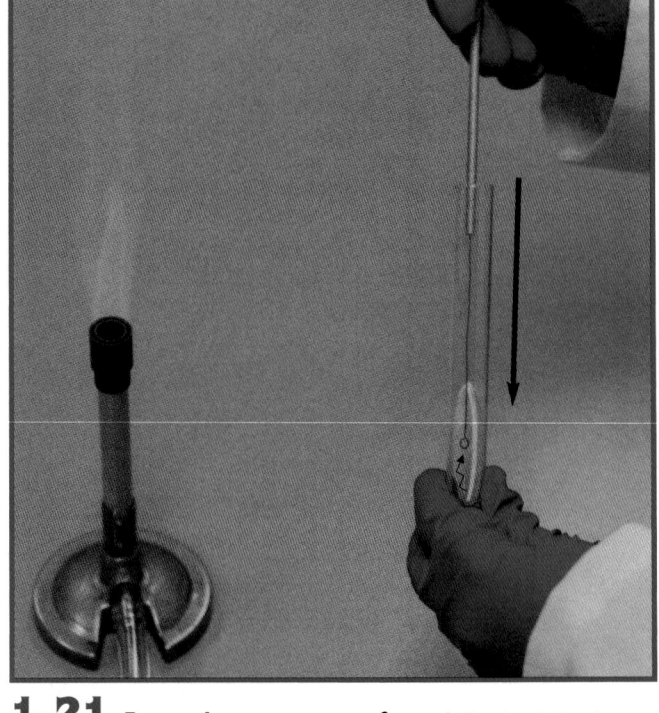

1-21 FISHTAIL INOCULATION OF A SLANT ✦ Begin at the base of the slant surface and gently move the loop back and forth as you withdraw the tube. Be careful not to cut the agar. After completing the transfer, sterilize the loop.

7 Keeping the loop hand still (remember—it has growth on it), move the tube to replace its cap.

8 Sterilize the loop as before by incinerating it in the Bunsen burner flame. It is especially important to flame it from base to tip now because the loop has lots of bacteria on it.

9 Label the tube with your name, date, medium, and organism. Incubate at the appropriate temperature for the assigned time.

Inoculating Broth Tubes

Broth cultures are often used to grow cultures for use when fresh cultures or large numbers of cells are desired. Many differential media are also broths.

1 Remove the cap of the sterile medium with the little finger of your loop hand and hold it there (Figure 1-12).

2 Sterilize the tube by quickly passing it through the flame a couple of times. Keep your loop hand still (Figure 1-13).

3 Hold the open tube on an angle to minimize airborne contamination. Keep your loop hand still (Figure 1-14).

4 Carefully move the broth tube over the wire (Figure 1-22). Gently swirl the loop in the broth to dislodge microbes.

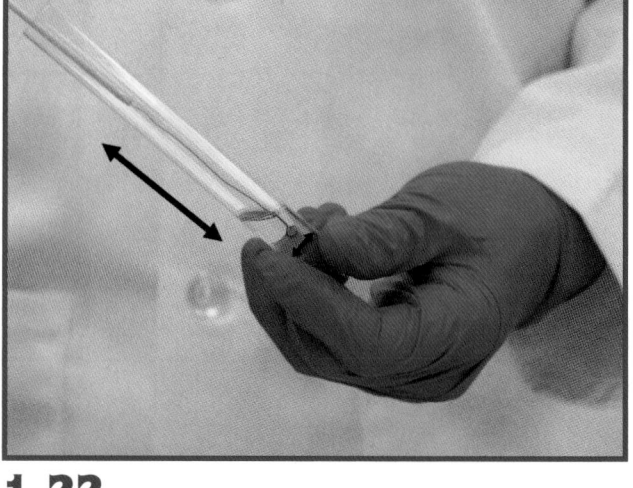

1-22 INOCULATION OF A BROTH ✦ When entering or leaving the tube, move the tube and keep the loop hand still. Gently swirl the loop in the broth to transfer the organisms.

5 Withdraw the tube from over the loop. Before completely removing it, touch the loop tip to the glass to remove any excess broth (Figure 1-23). Then be especially careful not to catch the loop tip on the tube lip when withdrawing it. This springing action of the wire creates bacterial aerosols.

6 Flame the tube lip as before. Keep your loop hand still.

7 Keeping the loop hand still (remember—it has growth on it), move the tube to replace its cap.

8 Sterilize the loop as before by incinerating it in the Bunsen burner flame. It is especially important to flame it from base to tip now because the loop and wire have lots of bacteria on them.

9 Label the tube with your name, date, medium, and organism. Incubate at the appropriate temperature for the assigned time.

✦ In This Exercise

You will perform some simple aseptic transfers: slant to slant and broth, broth to slant and broth, and plate to slant and broth. Master these, and you are well on your way to becoming a microbiologist!

✦ Materials

Per Student

✦ inoculating loop (one per student)
✦ four Nutrient Broth tubes
✦ four Nutrient Agar slants
✦ marking pen and labels
✦ vortex mixer (optional)

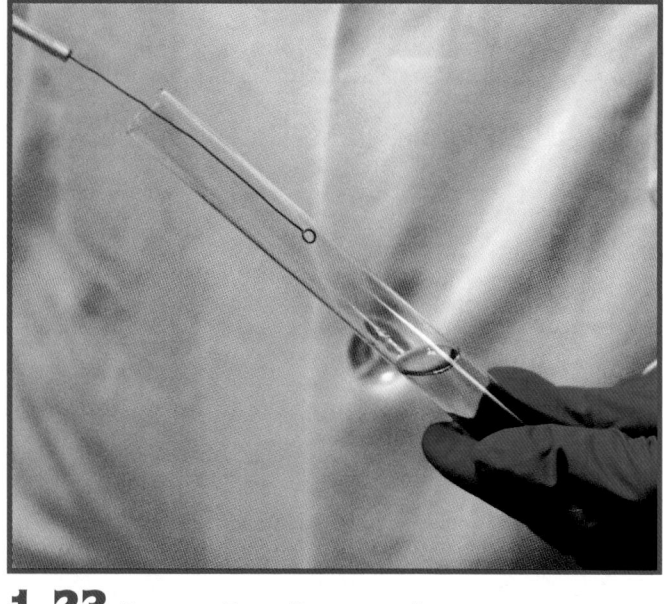

1-23 REMOVING EXCESS BROTH FROM LOOP ✦ Before removing it from the tube, touch the loop to the glass to remove excess broth. Failing to do so will result in splattering and aerosols when sterilizing the loop in a flame.

✦ Nutrient Agar slant cultures of:
  ✦ *Escherichia coli*
  ✦ *Micrococcus luteus*
✦ Nutrient Broth culture of:
  ✦ *Micrococcus luteus*
✦ Nutrient Agar plate culture of:
  ✦ *Micrococcus luteus*

 Procedure

Lab One

1 To make the transfers listed, refer to the appropriate section in "Specific Transfer Methods," beginning on page 20.
   a. From the *E. coli* slant to a nutrient agar slant and nutrient broth using an inoculating loop.
   b. From the *M. luteus* slant to a nutrient agar slant and nutrient broth using an inoculating loop.
   c. From the *M. luteus* broth to a nutrient agar slant and nutrient broth using an inoculating loop.
   d. From the *M. luteus* plate to a nutrient agar slant and nutrient broth. (For this transfer choose a well-isolated colony and touch the center with the loop as in Figure 1-17).

2 Label all tubes clearly with your name, the organisms' names, their source medium (slant, broth, or plate), and the date.

3 Incubate *M. luteus* at 25°C and *E. coli* at 35 ± 2°C until next class.

Lab Two

1 Remove your cultures from the incubators and examine the growth. Record your observations and answer the questions on the Data Sheet.

2 Your instructor may ask you to save your cultures for later use. Otherwise, dispose of all materials in the appropriate autoclave containers.

References

Barkley, W. Emmett and John H. Richardson. 1994. Chapter 29 in *Methods for General and Molecular Bacteriology*. American Society for Microbiology, Washington, DC.

Claus, G. William. 1989. Chapter 2 in *Understanding Microbes—A Laboratory Textbook for Microbiology*. W. H. Freeman and Company, New York, NY.

Darlow, H. M. 1969. Chapter VI in *Methods in Microbiology*, Volume 1. Edited by J. R. Norris and D. W. Ribbins. Academic Press, Ltd., London.

Fleming, Diane O. 1995. Chapter 13 in *Laboratory Safety—Principles and Practices*, 2nd ed. Edited by Diane O. Fleming, John H. Richardson, Jerry J. Tulis, and Donald Vesley. American Society for Microbiology, Washington, DC.

Koneman, Elmer W., Stephen D. Allen, William M. Janda, Paul C. Schreckenberger and Washington C. Winn, Jr. 1997. Chapter 2 in *Color Atlas and Textbook of Diagnostic Microbiology*, 5th ed. Lippincott-Raven Publishers, Philadelphia.

Murray, Patrick R., Ellen Jo Baron, Michael A. Pfaller, Fred C. Tenover, and Robert H. Yolken. 1995. *Manual of Clinical Microbiology*, 6th ed. American Society for Microbiology, Washington, DC.

Power, David A. and Peggy J. McCuen. 1988. *Manual of BBL™ Products and Laboratory Procedures*, 6th Ed. Becton Dickinson Microbiology Systems, Cockeysville, MD.

EXERCISE 1-4

Streak Plate Methods of Isolation

✦ Theory

A microbial culture consisting of two or more species is said to be a **mixed culture,** whereas a **pure culture** contains only a single species. Obtaining isolation of individual species from a mixed sample is generally the first step in identifying an organism. A commonly used **isolation technique** is the **streak plate** (Figure 1-24).

In the streak plate method of isolation, a bacterial sample (always assumed to be a mixed culture) is streaked over the surface of a plated agar medium. During streaking, the cell density decreases, eventually leading to individual cells being deposited separately on the agar surface. Cells that have been sufficiently isolated will grow into **colonies** consisting only of the original cell type. Because some colonies form from individual cells and others from pairs, chains, or clusters of cells, the term **colony-forming unit (CFU)** is a more correct description of the colony origin.

Several patterns are used in streaking an agar plate, the choice of which depends on the source of inoculum and microbiologist's preference. Although streak patterns range from simple to more complex, all are designed to separate deposited cells (CFUs) on the agar surface so individual cells (CFUs) grow into isolated colonies. A quadrant streak is generally used with samples suspected of high cell density, whereas a simple zigzag pattern may be used for samples containing lower cell densities.

✦ Application

The identification process of an unknown microbe relies on obtaining a pure culture of that organism. The streak plate method produces individual colonies on an agar plate. A portion of an isolated colony then may be transferred to a sterile medium to start a pure culture.

Following are descriptions of streak techniques. As in Exercise 1-3, basic skills are printed in the regular black type and new skills are printed in blue.

Inoculation of Agar Plates
Using the Quadrant Streak Method

This inoculation pattern is usually performed as the initial streak for isolation of two or more bacterial species in a mixed culture with suspected high cell density.

1 Obtain the sample of mixed culture with a sterile loop.

2 You have two options at this point. Use whichever is more comfortable for you or is required by your instructor.

  a. Leave the sterile agar plate on the table and lift the lid slightly, using it as a shield from airborne contamination (as in Figure 1-17).

  or,

  b. Place the plate lid down on the table (Figure 1-25A). Then remove the base and hold it in the air on an angle (Figure 1-25B).

3 Starting at the edge of the plate lightly drag the loop back and forth across the agar surface as shown in Figure 1-26A. Be careful not to cut the agar surface.

4 Remove the loop and replace the lid.

5 Sterilize your loop as before. It is especially important to flame it from base to tip now because the loop has lots of bacteria on it.

6 Rotate the plate a little less than 90°.

7 Let the loop cool for a few moments (or you can touch an open part of the agar), then perform

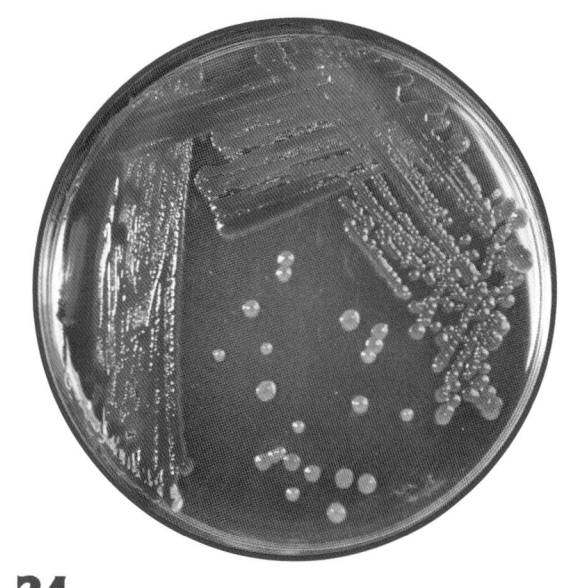

1-24 STREAK PLATE OF *SERRATIA MARCESCENS* ✦ Note the decreasing density of growth in the four streak patterns. On this plate, isolation is first obtained in the fourth streak. Cells from an individual colony may be transferred to a sterile medium to start a pure culture.

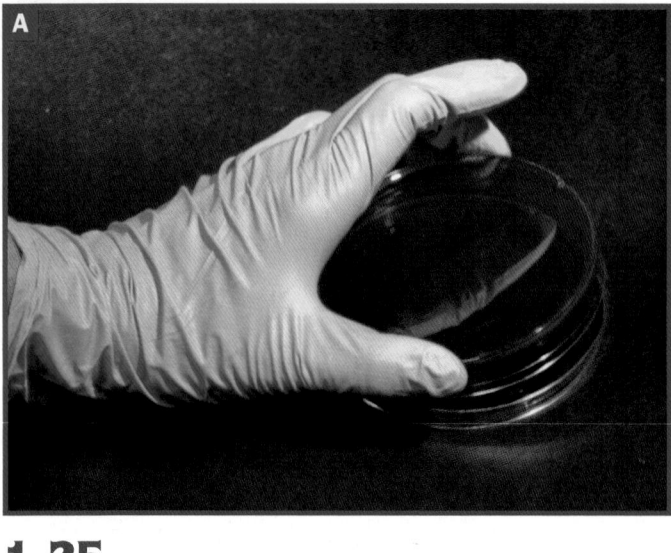

1-25 STREAK PLATE INOCULATION ✦ **A** Some microbiologists prefer to hold the Petri dish in the air when performing a streak plate. To do this, place the plate lid down on the table and lift the base from it, holding it on an angle. **B** Perform the streak as described in the text and as shown in Figure 1-26.

another streak with the sterile loop beginning at one end of the first streak pattern (Figure 1-26B). Intersect the first streak only two or three times.

8 Sterilize the loop, then repeat with a third streak beginning in the second streak (Figure 1-26C).

9 Sterilize the loop, then perform a fourth streak beginning in the third streak and extending into

the middle of the plate. Be careful not to enter any streaks but the third (Figure 1-26D).

10 Sterilize the loop.

11 Label the plate's base with your name, date and sample inoculated.

12 Incubate the plate in an inverted position for the assigned time at the appropriate temperature.

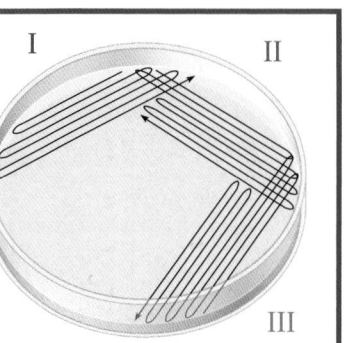

1-26A BEGINNING THE STREAK PATTERN ✦ Streak the mixed culture back and forth in one quadrant of the agar plate. Do not cut the agar with the loop. Flame the loop, then proceed.

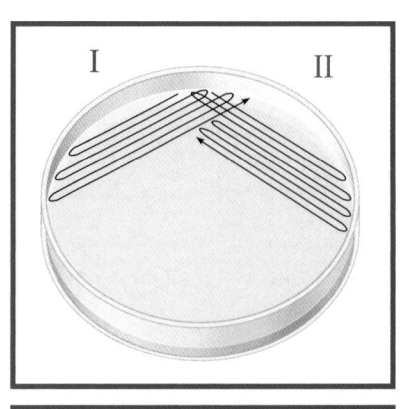

1-26B STREAKING AGAIN ✦ Rotate the plate nearly 90° and touch the agar in an uninoculated region to cool the loop. Streak again using the same wrist motion. Flame the loop.

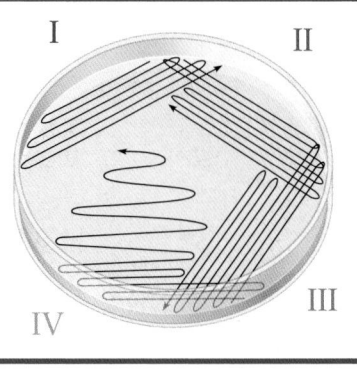

1-26C STREAKING YET AGAIN ✦ Rotate the plate nearly 90° and streak again using the same wrist motion. Be sure to cool the loop prior to streaking. Flame again.

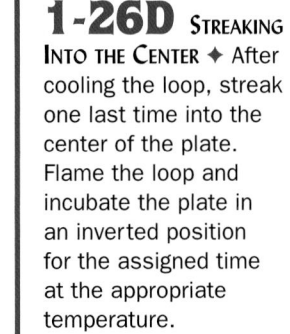

1-26D STREAKING INTO THE CENTER ✦ After cooling the loop, streak one last time into the center of the plate. Flame the loop and incubate the plate in an inverted position for the assigned time at the appropriate temperature.

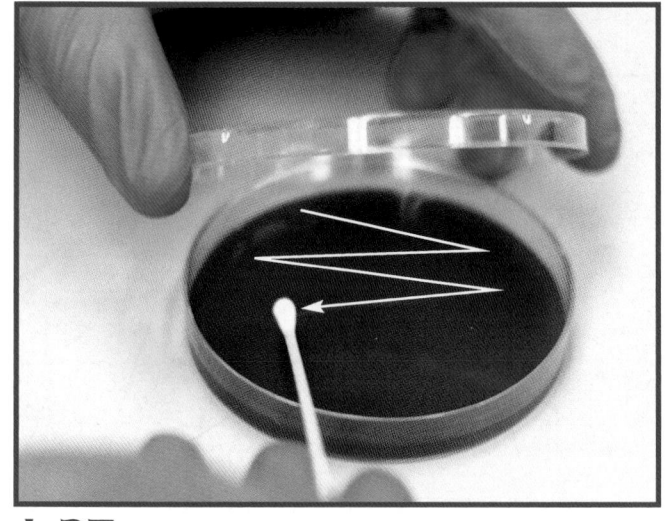

1-27 ZIGZAG INOCULATION ✦ Use the cotton swab to streak the agar surface to get isolated colonies after incubation. Be careful not to cut the agar. Properly dispose of the swab in a biohazard container.

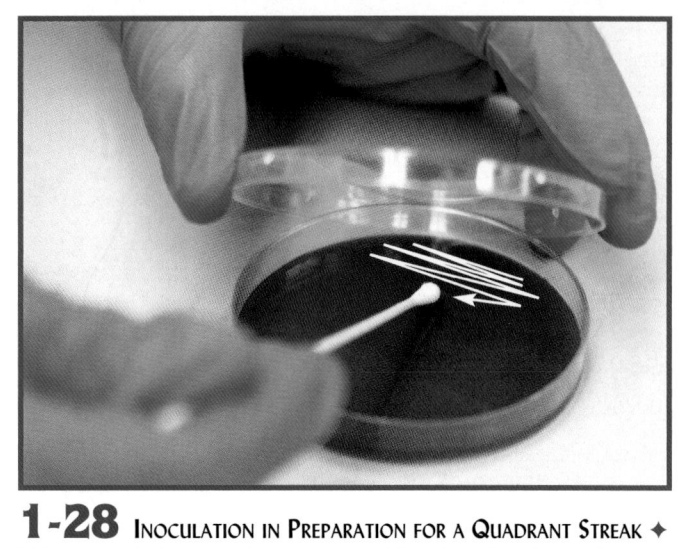

1-28 INOCULATION IN PREPARATION FOR A QUADRANT STREAK ✦ If the sample is expected to have a high density of organisms, streak one edge of the plate with the swab. Then continue with the quadrant streak, using a loop. Be careful not to cut the agar. Properly dispose of the swab in a biohazard container.

Zigzag Inoculation of Agar Plates Using a Cotton Swab

This inoculation pattern is usually performed when the sample does not have a high cell density and with pure cultures when isolation is not necessary.

1 Hold the swab comfortably in your dominant hand and lift the lid of the Petri dish with the other. Use the lid as a shield to protect the agar from airborne contamination. Alternatively, the plate can be held in the air as shown in Figure 1-25B.

2 Lightly drag the cotton swab across the agar surface in a zigzag pattern. Be careful not to cut the agar surface (Figure 1-27).

3 Replace the lid.

4 Dispose of the swab according to your lab's practices (generally in a sharps container).

5 Label the base of the plate with your name, date, and sample.

6 Incubate the plate in an inverted position for the assigned time at the appropriate temperature.

Inoculation of Agar Plates with a Cotton Swab in Preparation for a Quadrant Streak Plate

This inoculation pattern is usually performed as the initial streak for isolation of two or more bacterial species in a mixed culture with suspected high cell density.

1 Hold the swab comfortably in your dominant hand and lift the lid of the Petri dish with the other. Use the lid as a shield to protect the agar from airborne contamination (Figure 1-28).

2 Lightly drag the cotton swab back and forth across the agar surface in one quadrant of the plate (Figure 1-28). This replaces the first streak as shown in Figure 1-26A.

3 Dispose of the swab according to your lab's practices (generally in a sharps container).

4 Further streaking is performed with a loop as shown in Figures 1-26B through 1-26D.

5 Label the plate's base with your name, date and sample.

6 Incubate the plate in an inverted position for the assigned time at the appropriate temperature.

✦ In This Exercise

You will learn how to isolate individual organisms from a mixed culture, the first step in producing a pure culture. Three related streaking techniques will be used, the choice of which is determined by the anticipated cell density of the sample.

✦ Materials

Per Student Pair

✦ inoculating loop (one per student)

✦ four Trypticase Soy Agar (TSA) or Nutrient Agar (NA) plates

✦ two sterile cotton swabs in sterile distilled water

✦ broth cultures of:
  ◆ *Micrococcus luteus*
  ◆ *Staphylococcus epidermidis*

 **Procedure**

Lab One

1 Using a pencil, practice quadrant-streaking the plate on the Data Sheet before trying it with living bacteria. *Hint:* Keep your wrist relaxed.

2 Each student should transfer a loopful of *M. luteus* or *S. epidermidis* to a sterile Trypticase Soy Agar or Nutrient Agar plate and follow the diagrams in Figure 1-26 to streak for isolation. Avoid digging into or cutting the agar, which ruins the plate and may create dangerous aerosols. Label the plate with your name, the date, and the organism.

3 Each student should use the cotton swab to sample an environmental source (see Appendix B), then do a simple zigzag streak on a Trypticase Soy Agar or Nutrient Agar plate (Figure 1-27). Dispose of the swabs in an autoclave container. Label the plate with your name, the date, and the sample source.

4 Tape the four plates together, invert them, and incubate them at 25°C for 24 to 48 hours.

Lab Two

1 After incubation, examine the plates for isolation.

2 Compare your streak plates with your lab partner's plates and critique each other's technique. Remember, a successful streak plate is one that has isolated colonies; the pattern doesn't have to be textbook quality—it's just that textbook quality provides you with a greater chance of getting isolation.

References

Collins, C. H. and Patricia M. Lyne. 1995. Chapter 6 in *Collins and Lyne's Microbiological Methods,* 7th ed. Butterworth-Heineman.

Delost, Maria Dannessa. 1997. Chapter 1 in *Introduction to Diagnostic Microbiology.* Mosby, Inc., St. Louis, MO.

Forbes, Betty A., Daniel F. Sahm, and Alice S. Weissfeld. 2002. Chapter 1 in *Bailey & Scott's Diagnostic Microbiology,* 11th ed. Mosby-Yearbook, St. Louis, MO.

Koneman, Elmer W., Stephen D. Allen, William M. Janda, Paul C. Schreckenberger and Washington C. Winn, Jr. 1997. Chapter 2 in *Color Atlas and Textbook of Diagnostic Microbiology,* 5th ed. J. B. Lippincott Company, Philadelphia, PA.

Power, David A. and Peggy J. McCuen. 1988. Pages 2 and 3 in *Manual of BBL™ Products and Laboratory Procedures,* 6th ed. Becton Dickinson Microbiology Systems, Cockeysville, MD.

Spread Plate Method of Isolation

✦ Theory

The spread plate technique is a method of isolation in which a diluted microbial sample is deposited on an agar plate and spread uniformly across the surface with a glass rod. With a properly diluted sample, cells (CFUs) will be deposited far enough apart on the agar surface to grow into individual colonies.

✦ Application

After incubation, a portion of an isolated colony can be transferred to a sterile medium to begin a pure culture. The spread plate technique also has applications in quantitative microbiology (see Section 6).

Following is a description of the spread plate technique. As in the previous exercises, basic skills are printed in the regular black type and new skills are printed in blue.

Spread Plate Technique

1 Arrange the alcohol beaker, Bunsen burner, and agar plate as shown in Figure 1-29. This arrangement minimizes the chances of catching the alcohol on fire.

2 Lift the plate's lid and use it as a shield to protect from airborne contamination.

3 Using an appropriate pipette, deposit the designated inoculum volume on the agar surface. (Please see Appendices C and D for use of pipettes.) From this point, the remainder of steps should be completed within about 15 seconds to prevent the inoculum from soaking into the agar.

4 Properly dispose of the pipetting instrument used to inoculate the medium, because it is contaminated. Each lab has its own specific procedures and your instructor will advise you what to do.

5 Remove the glass spreading rod from the alcohol and pass it through the flame to ignite the alcohol (Figure 1-30). Remove the rod from the flame and allow the alcohol to burn off completely. Do not leave the rod in the flame; the combination of the alcohol and brief flaming are sufficient to sterilize it. *Be careful not to drop any flaming alcohol on the work surface. Be especially careful not to drop flaming alcohol back into the alcohol beaker.*

6 After the flame has gone out on the glass rod, lift the lid of the plate and use it as a shield from airborne contamination. Then touch the rod to the agar surface away from the inoculum to cool it.

1-30 FLAMING THE GLASS ROD ✦ Remove the glass spreading rod from the alcohol and pass it through the flame to ignite the alcohol. Remove the rod from the flame and allow the alcohol to burn off completely. Do not leave the rod in the flame; the combination of the alcohol and brief flaming are sufficient to sterilize it. *Be careful not to drop any flaming alcohol on the work surface or back into the alcohol beaker.*

1-29 SPREAD PLATE SET-UP ✦ The spread plate technique requires a Bunsen burner, a beaker with alcohol, a glass spreading rod, and the plate. Position these components in your work area as shown: isopropyl alcohol, flame, and plate. This arrangement reduces the chance of accidentally catching the alcohol on fire.

7 To spread the inoculum, hold the plate lid with the base of your thumb and index finger and use the tip of your thumb and middle finger to rotate the base (Figure 1-31). At the same time, move the rod in a back-and-forth motion across the agar surface. After a couple of turns, do one last turn with the rod next to the plate's edge. Alternatively, place the plate on a rotating platform and spread the inoculum (Figure 1-32).

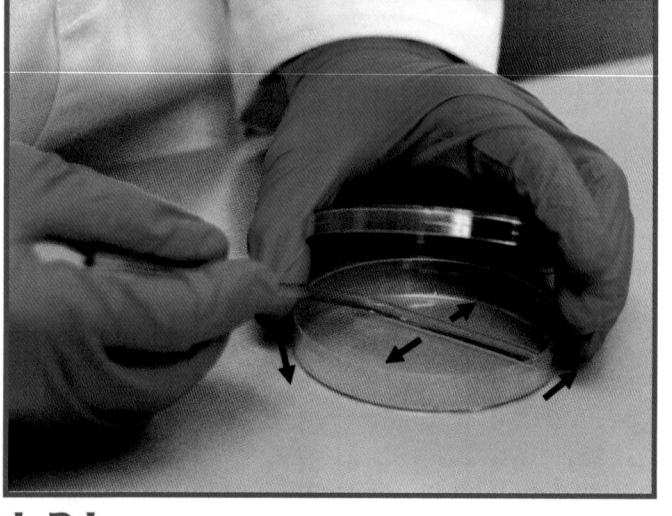

1-31 SPREADING PLATE INOCULUM ✦ After the flame has gone out on the rod, lift the lid of the plate and use it as a shield from airborne contamination. Then, touch the rod to the agar surface away from the inoculum in order to cool it. To spread the inoculum, hold the plate lid with the base of your thumb and index finger, and use the tip of your thumb and middle finger to rotate the base. At the same time, move the rod in a back-and-forth motion across the agar surface. After a couple of turns, do one last turn with the rod next to the plate's edge.

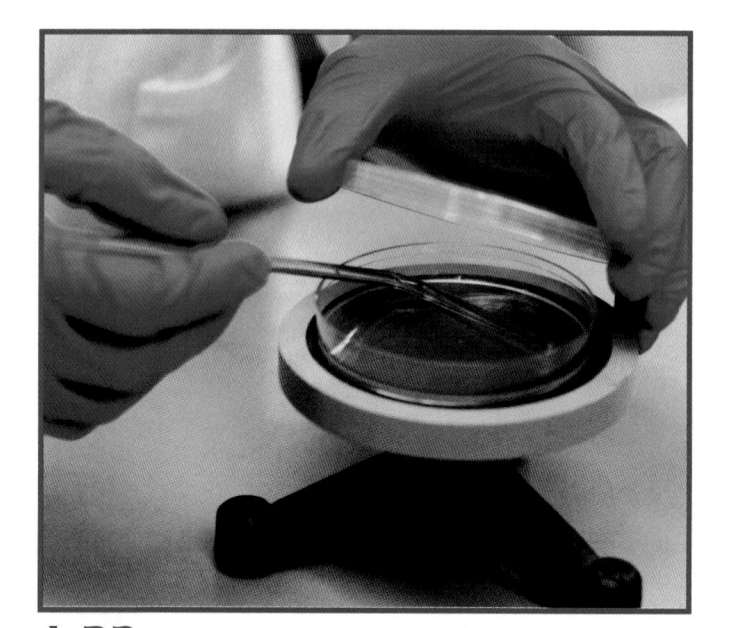

1-32 INOCULATING WITH A TURNTABLE ✦ A turntable makes it easier to rotate the plate during the spread plate technique.

8 Remove the rod from the plate and replace the lid.

9 Return the rod to the alcohol in preparation for the next inoculation. There is no need to flame it again.

10 Label the plate base with your name, date, organism, and any other relevant information.

11 Incubate the plate in an inverted position at the appropriate temperature for the assigned time. (If you plated a volume of inoculum greater than 0.5 mL, wait a few minutes and allow it to soak in before inverting the plate.)

✦ In This Exercise

You will perform a spread plate inoculation. In the context of this exercise, it is used as an isolation procedure, but it also can be used in quantifying cell densities in broth samples.

✦ Materials

Per Student Pair

- ✦ inoculating loop (each student)
- ✦ six sterile plastic transfer pipettes
- ✦ 500 mL beaker with 50 mL of isopropyl alcohol
- ✦ glass spreading rod
- ✦ Bunsen burner and striker
- ✦ four Nutrient Agar plates
- ✦ one sterile microtube
- ✦ four capped microtubes with about 1 mL sterile distilled or deionized water (dH$_2$O)
- ✦ vortex mixer (optional)
- ✦ broth cultures of:
  - ◆ *Escherichia coli*
  - ◆ *Serratia marcescens*

 Procedure

Lab One

1 Using a different pipette for each, transfer a few drops of *E. coli* and *S. marcescens* to a microtube. Cap the tube and mix well with a vortex mixer. *Or* use the second pipette to mix well by gently drawing and dispensing the mixture in and out of the tube a couple of times. Do not spray the mixture!

2 Label the four microtubes containing sterile dH$_2$O "A," "B," "C," and "D."

3 Label the four Nutrient Agar plates "A," "B," "C," and "D."

4 Transfer a loopful of the mixture from the microtube to Tube A and mix well with the loop or a vortex mixer. If using a vortex mixer, be sure to cap the tube.[1]

5 Transfer a loopful of the mixture in Tube A to Tube B and mix well with the loop or a vortex mixer.

6 Transfer a loopful of the mixture in Tube B to Tube C and mix well with the loop or a vortex mixer.

7 Transfer a loopful of the mixture in Tube C to Tube D and mix well with the loop or a vortex mixer.

8 Using a sterile transfer pipette, place a couple of drops of sample from Tube A on Plate A. Spread the inoculum with a glass rod as described in "Spread Plate Technique" and as shown in Figures 1-30 through 1-32. Let the plate sit for a few minutes.

9 Repeat Step 8 for Tubes B, C, and D and Plates B, C, and D, respectively.

10 Tape the four plates together (be sure they are facing the same direction), invert them and incubate them at 25°C for 24 to 48 hours.

---

[1] Because this is not a quantitative procedure, it is not necessary to flame the loop between transfers.

Lab Two

1 After incubation, examine the plates for isolation. *S. marcescens* produces reddish-orange colonies, and *E. coli* produces buff-colored colonies.

2 Fill in the Data Sheet.

References

Clesceri, WEF, Chair; Arnold E. Greenberg, APHA; Andrew D. Eaton, AWWA ; and Mary Ann H. Franson. 1998. Pages 9–38 in *Standard Methods for the Examination of Water and Wastewater*, 20th edition. Joint publication of American Public Health Association, American Water Works Association and Water Environment Federation. APHA Publication Office, Washington, DC.

Downes, Frances Pouch, and Keith Ito. 2001. Page 57 in *Compendium of Methods for the Microbiological Examination of Foods*. American Public Health Association. Washington DC.

Gerhard, Philipp, R. G. E. Murray, Willis A. Wood, and Noel R. Kreig. 1994. Pages 255–257 in *Methods for General and Molecular Bacteriology*. American Society for Microbiology, Washington, DC.

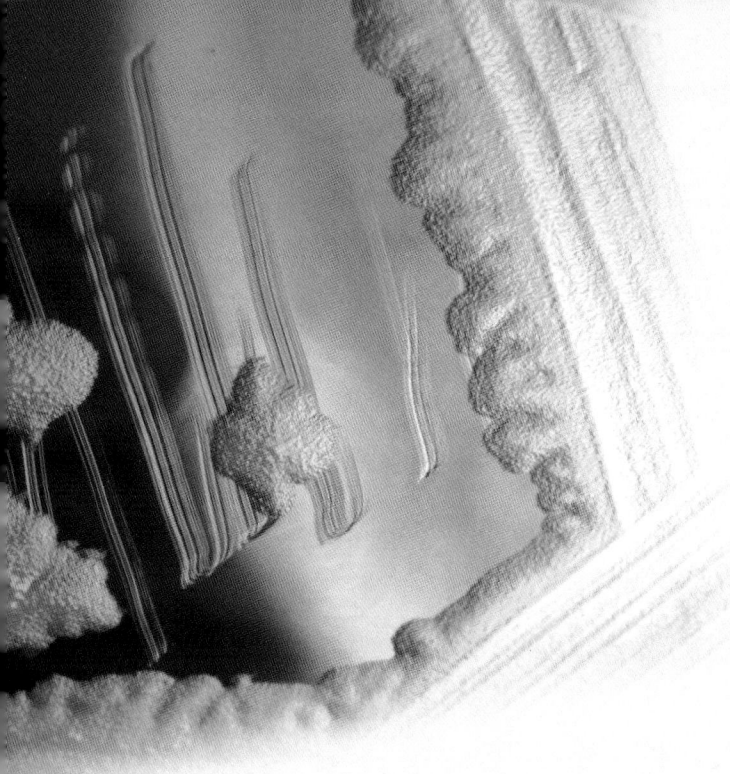

Microbial Growth

Microorganisms are extraordinarily diverse. Every species demonstrates a unique combination of characteristics, some of which can be easily observed. In this section we illustrate some of those characteristics and factors that affect them.

Allowing for variables, such as the growth medium and incubation conditions, much can be determined about an organism by simply looking at the colonies it produces. Distinguishing different growth patterns is an important skill—one that you can use as you progress through the semester. Note the growth characteristics of all the organisms provided for your laboratory exercises, and jot them down or even sketch them. When the time comes to identify your unknown species, you will find your records very useful.

You will begin this section with an exercise intended to sensitize you to the vast microbial population living all around us. Then you will examine some microbial growth characteristics and cultivation methods. Next you will look at some environmental factors affecting microbial growth, including pH, oxygen, temperature, and osmotic pressure. Finally, you will examine some physical and chemical microbial control agents and systems. ✦

Diversity and Ubiquity of Microorganisms

Microorganisms are found everywhere that other forms of life exist. They can be isolated from soil, bodies of water, even from the air. As unwanted parasites or colonizers, some microorganisms cause diseases or infections. Most, however, are harmless saprophytes; they simply live in, on, or around plants and animals and decompose dead organic matter. In so doing, they perform the essential function of nutrient recycling in ecosystems.

In this section, you will grow microorganisms from seemingly uninhabited sources. Then you will learn to identify the various growth characteristics that these "invisible" cohabitants produce when they are cultivated in broth and on solid media. ✦

EXERCISE 2-1   Ubiquity of Microorganisms

✦ Theory

In the literature on microorganisms, a frequently encountered phrase is "ubiquitous in nature." This means that the organism being considered can be found just about everywhere. More specifically, the organism likely could be isolated from soil, water, plants, and animals (including humans). Although the word *ubiquitous* doesn't apply to every species, it does apply to many, and certainly to microorganisms as a group.

Many microorganisms are **free-living**—they do not reside on or in a specific plant or animal **host** and are not known to cause disease. They are **nonpathogenic**. Other microorganisms are **pathogens** and generally are associated with their host (or hosts). Even many of the **commensal** or **mutualistic** strains inhabiting our bodies are **opportunistic pathogens**. That is, they are capable of producing a disease state if introduced into a suitable part of the body. Any area, including sites outside the host organism, where a microbe resides and serves as a potential source of infection is called a **reservoir**.

✦ Application

This exercise is designed to demonstrate the ubiquitous nature of microorganisms and the ease with which they can be cultivated. (It should be noted that although we can find living microorganisms virtually everywhere and confirm their presence by cultivation, molecular techniques developed over the last two or three decades

demonstrate that the organisms successfully grown in the lab represent a minute fraction of those still uncultivable.

✦ In This Exercise

Today, you will work in small groups to sample and culture several locations in your laboratory. Your instructor may have other locations outside of the lab to sample as well. Remember that even relatively "harmless" bacteria, when cultivated on a growth medium, are in sufficient numbers to constitute a health hazard. Treat them with care.

✦ Materials

Per Student Group

- ✦ eight Nutrient Agar plates
- ✦ one sterile cotton swab

 Procedure

Lab One

1 Number the plates 1 through 8.

2 Open plate number 1 and expose it to the air for 30 minutes or longer. Set it aside and out of the way of the other plates.

2-1 SIMPLE STREAK PATTERN ON NUTRIENT AGAR ✦ Roll the swab as you inoculate the plate. Do not press so hard that you cut the agar.

3 Use the cotton swab to sample your desk area and then streak plates 2 and 3 in the pattern shown in Figure 2-1. (Pressing very lightly, roll the swab on the agar as you streak it.)

4 Cough several times on the agar surface of plate 4.

5 Rub your hands together, and then touch the agar surface of plate 5 lightly with your fingertips. (A light touch is sufficient; touching too firmly will crack the agar.)

6 Remove the lid of plate 6 and vigorously scratch your head above it. (Keep this plate away from plate 1 to avoid cross-contamination.)

7 Leave plates 7 and 8 covered; do not open them.

8 Label the base of each plate with the date, type of exposure it has received, and the name of your group.

9 Invert all plates and incubate them for 24 to 48 hours at the following temperatures:

Plates 1, 2, and 8:	25°C
Plates 3, 4, 5, 6, and 7:	37°C

Lab Two

1 Using the plate diagrams on the Data Sheet, draw the growth patterns on each of your agar plates. Be sure to label them according to incubation time, temperature, and source of inoculum.

2 Save these plates in a refrigerator for use in Exercise 2-2.

References

Forbes, Betty A., Daniel F. Sahm, and Alice S. Weissfeld. 2002. Chapter 10 in *Bailey & Scott's Diagnostic Microbiology*, 11th ed. Mosby, St. Louis.

Holt, John G. (Editor). 1994. *Bergey's Manual of Determinative Bacteriology*, 9th ed. Williams and Wilkins, Baltimore.

Winn, Washington C., *et al.* 2006. *Koneman's Color Atlas and Textbook of Diagnostic Microbiology*, 6th ed. Lippincott Williams & Wilkins, Baltimore.

Varnam, Alan H., and Malcolm G. Evans. 2000. *Environmental Microbiology*. ASM Press, Washington, DC.

2-2 Colony Morphology

✦ Theory

When a single bacterial cell is deposited on a solid nutrient medium, it begins to divide. One cell makes two, two make four, four make eight . . . one million make two million, and so on. Eventually a visible mass of cells—a **colony**—appears where the original cell was deposited. Color, size, shape, and texture of microbial growth are determined by the genetic makeup of the organism, but also greatly influenced by environmental factors including nutrient availability, temperature, and incubation time.

The basic categories of colony morphology are colony shape, margin (edge), elevation, texture, and pigment production (color).

1. *Shape* may be described as **round**, **irregular**, or **punctiform** (tiny, pinpoint).

2. The *margin* may be **entire** (**smooth**, with no irregularities), **undulate** (wavy), **lobate** (lobed), **filamentous**, or **rhizoid** (branched like roots).

3. *Elevations* include **flat, raised, convex, pulvinate** (very convex), and **umbonate** (raised in the center).

4. *Texture* may be **moist, mucoid,** or **dry.**

5. *Pigment production* (color), is another useful characteristic and may be combined with optical properties such as **opaque, translucent, shiny,** or **dull.**

✦ Application

Recognizing different bacterial growth morphologies on agar plates is a useful step in the identification process. Once purity of a colony has been confirmed by an appropriate staining procedure, cells can be cultivated and maintained on sterile media for a variety of purposes.

✦ In This Exercise

Today you will be viewing colony characteristics on the plates saved from Exercise 2-1 and (if available) prepared streak plates provided by your instructor. Colony characteristics may be viewed with the naked eye or with the assistance of a colony counter (Figure 2-2). Figures 2-3 through 2-28 show a variety of bacterial colony forms and characteristics. Where applicable, contrasting environmental factors are indicated.

2-2 **COLONY COUNTER** ✦ Subtle differences in colony shape and size can best be viewed on the colony counter. The **transmitted light** and magnifying glass allow observation of greater detail; however, colony color is best determined with **reflected light**. The grid in the background is a counting aid.

✦ Materials

Per Student Group

✦ colony counter (optional)

✦ dissecting microscope (optional)

✦ metric ruler

✦ plates from Exercise 2-1

✦ (Optional) streak plate cultures of any of the following:
  ◆ *Micrococcus luteus*
  ◆ *Corynebacterium xerosis*
  ◆ *Lactobacillus plantarum*
  ◆ *Mycobacterium smegmatis* (BSL-2)
  ◆ *Bacillus subtilis*
  ◆ *Proteus mirabilis* (BSL-2)

✎ Procedure

Using the terms in Figure 2-3, describe the colonies on your plates. Measure colony diameters (in mm) with a ruler and include them with your descriptions in the table on the Data Sheet. It may be helpful to use a colony counter (Figure 2-2). (**Note:** Remember that many microorganisms are opportunistic pathogens, so be sure to handle the plates carefully. Do not open plates containing fuzzy growth, as a fuzzy

Elevation

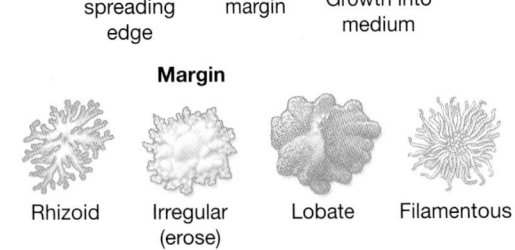

Convex   Umbonate   Plateau   Flat

Raised   Raised,   Flat, raised   Growth into
          spreading   margin   medium
          edge

Margin

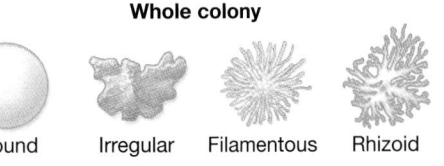

Smooth,   Rhizoid   Irregular   Lobate   Filamentous
entire   (erose)

Whole colony

Round   Irregular   Filamentous   Rhizoid

2-3 A SAMPLING OF BACTERIAL COLONY FEATURES ✦ These terms are used to describe colonial morphology. Descriptions also should include color, surface characteristics (dull or shiny), consistency (dry, butyrous-buttery, or moist) and optical properties (opaque or translucent).

appearance suggests fungal growth containing spores that can spread easily and contaminate the laboratory and other cultures. If you are in doubt, check with your instructor.)

2 Unless you have been instructed to save today's cultures for future exercises, discard all plates in an appropriate autoclave container.

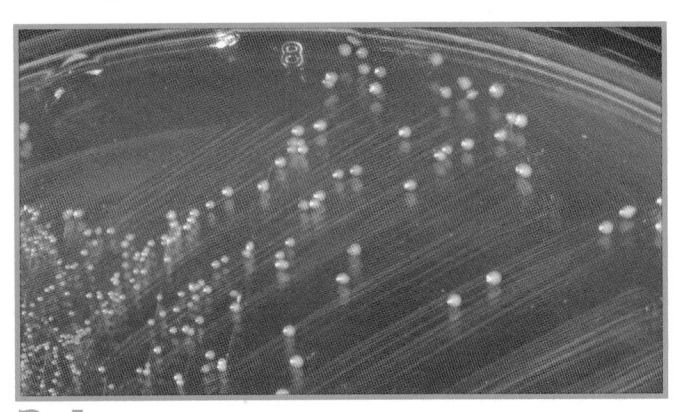

2-4 ENTEROCOCCUS FAECIUM GROWN ON NUTRIENT AGAR ✦ The colonies are white, circular, convex, and have an entire margin. *E. faecium* (formerly known as *Streptococcus faecium*) is found in human and animal feces.

2-5 STAPHYLOCOCCUS EPIDERMIDIS GROWN ON SHEEP BLOOD AGAR ✦ The colonies are white, raised, circular, and entire. *S. epidermidis* is an opportunistic pathogen.

2-6 CHROMOBACTERIUM VIOLACEUM GROWN ON SHEEP BLOOD AGAR ✦ *C. violaceum* produces shiny, purple, convex colonies. It is found in soil and water, and rarely produces infections in humans.

2-7 PROVIDENCIA STUARTII GROWN ON NUTRIENT AGAR ✦ The colonies are shiny, buff, and convex. *P. stuartii* is a frequent isolate in urine samples obtained from hospitalized and catheterized patients.

2-8 KLEBSIELLA PNEUMONIAE GROWN ON NUTRIENT AGAR ✦ The colonies are mucoid, raised, and shiny.

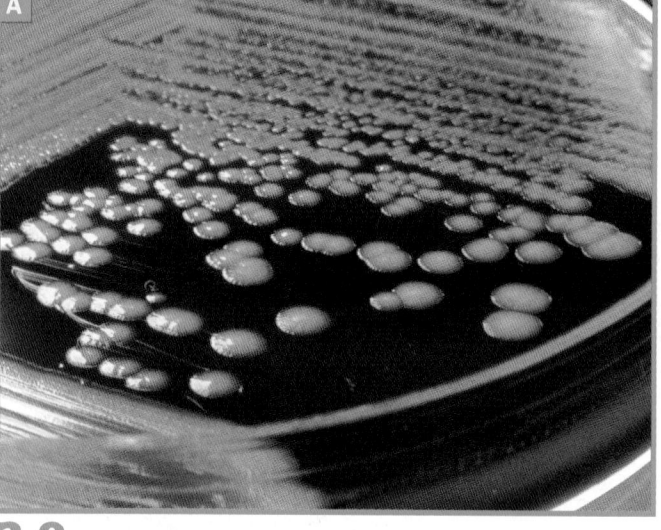

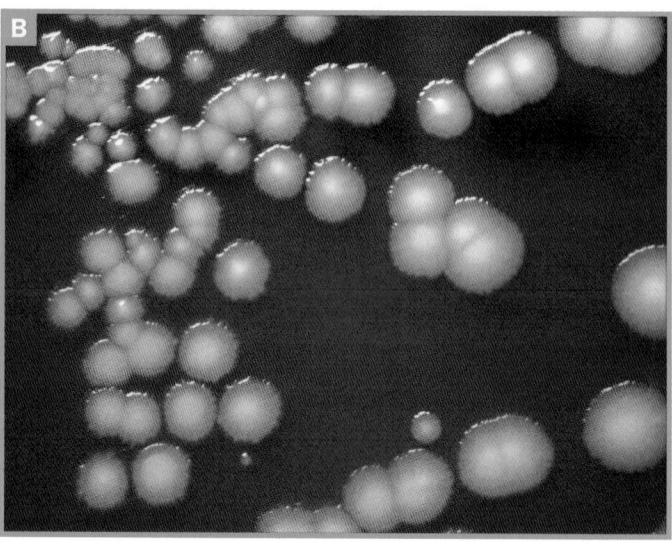

2-9 ALCALIGENES FAECALIS COLONIES ON SHEEP BLOOD AGAR ✦ The colonies of this opportunistic pathogen are umbonate with an opaque center and a spreading edge. **A** Side view: Note the raised center. **B** Close-up of the *A. faecalis* colonies showing spreading edge.

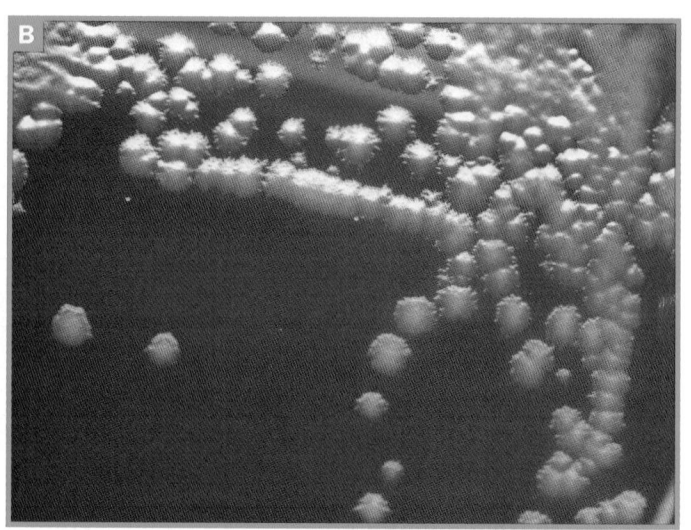

2-10 COMPARISON OF TWO BACILLUS SPECIES ✦ The colonies are dry, dull, raised, rough-textured, and gray. **A** *Bacillus cereus* on Sheep Blood Agar. **B** *Bacillus anthracis* on Sheep Blood Agar.

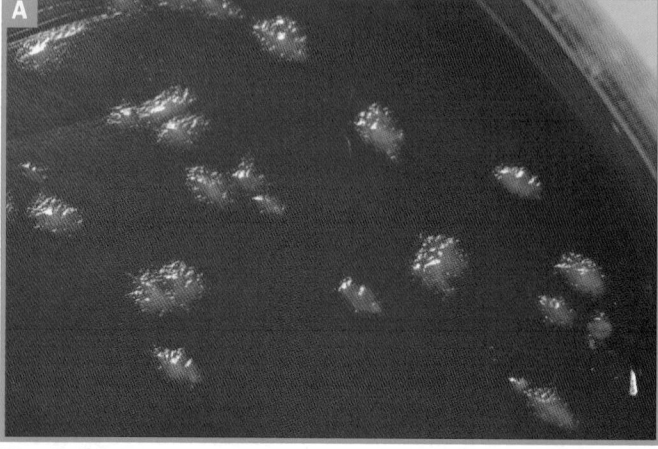

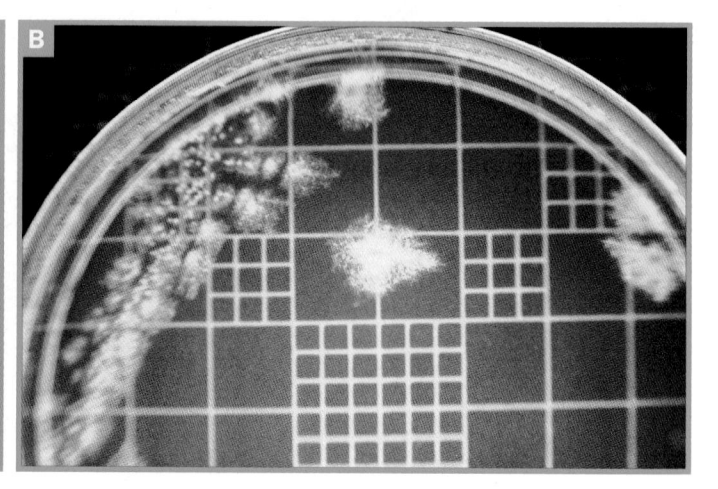

2-11 COMPARISON OF CLOSTRIDIUM SPOROGENES COLONIES GROWN ON DIFFERENT MEDIA ✦ The colonies are irregular and rhizoid. *C. sporogenes* is found in soils worldwide. **A** *C. sporogenes* grown anaerobically on Sheep Blood Agar and viewed with reflected light. **B** *C. sporogenes* grown anaerobically on Nutrient Agar and viewed with transmitted light.

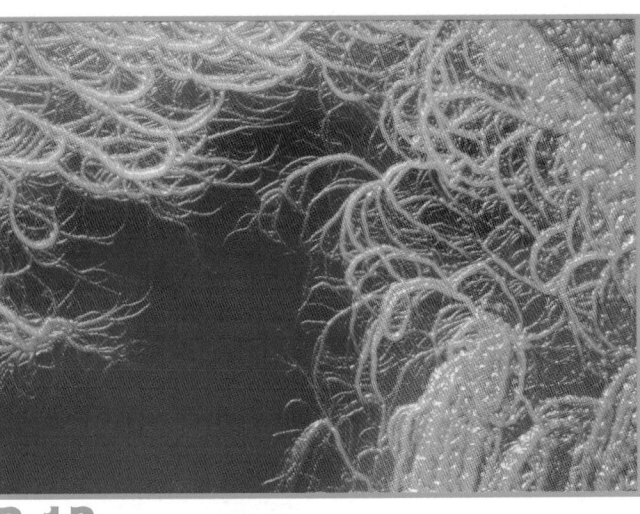

2-12 FILAMENTOUS GROWTH ✦ This is an unknown soil organism grown on Sheep Blood Agar.

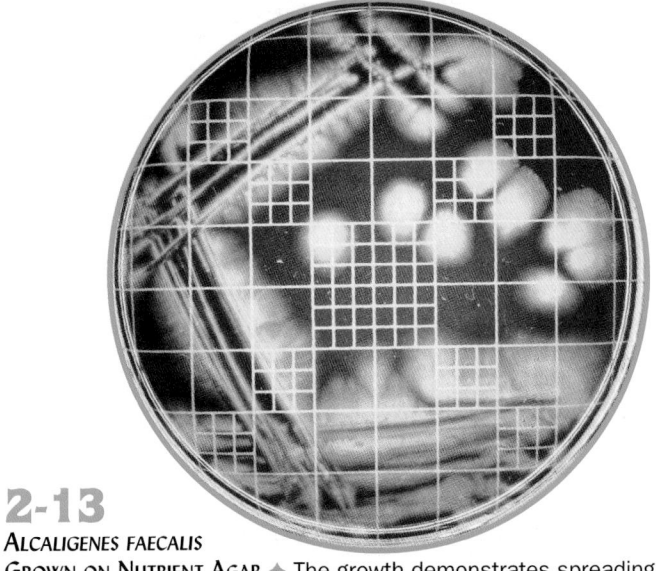

2-13

ALCALIGENES FAECALIS
GROWN ON NUTRIENT AGAR ✦ The growth demonstrates spreading attributable to motility and is translucent. Compare with the growth in Figure 2-9.

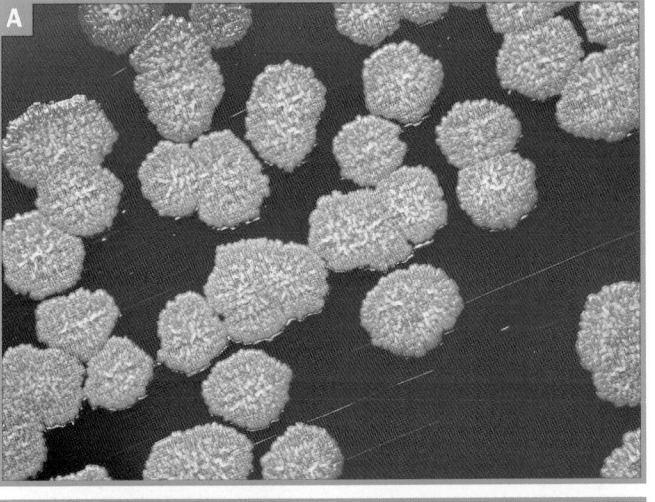

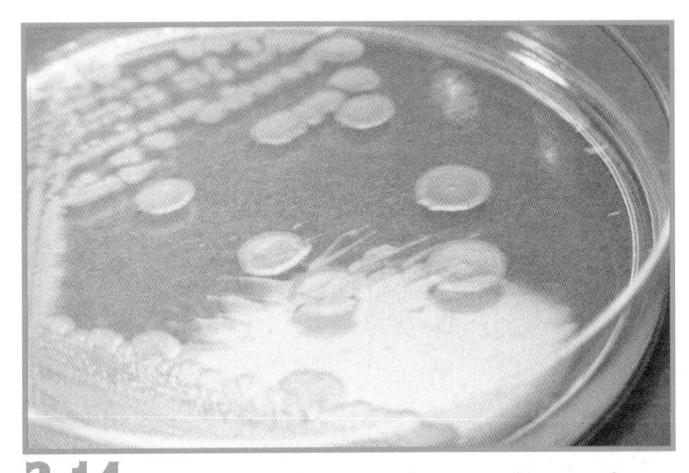

2-14 *BACILLUS SUBTILIS* COLONIES GROWN ON NUTRIENT AGAR VIEWED FROM THE SIDE ✦ *B. subtilis* produces colonies with a raised margin and a dull surface. Compare with *B. cereus* and *B. anthracis* in Figure 2-10 and B. subtilis in Figure 2-15.

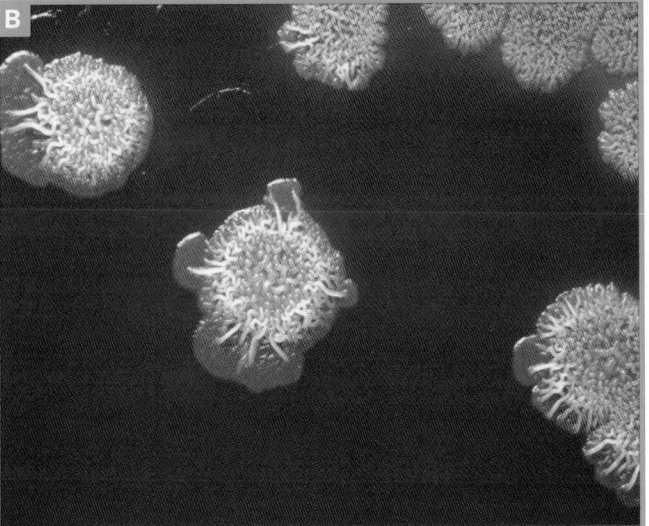

2-15 EFFECT OF AGE ON COLONY MORPHOLOGY ✦ A Close-up of *Bacillus subtilis* on Sheep Blood Agar after 24 hours' growth. B Close-up of *Bacillus subtilis* on Sheep Blood Agar after 48 hours' growth. Note the wormlike appearance.

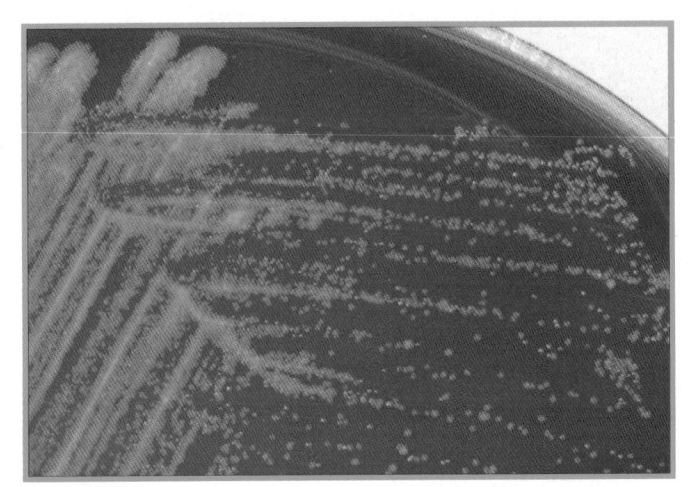

2-16 *MYCOBACTERIUM SMEGMATIS* GROWN ON SHEEP BLOOD AGAR ✦ The colonies of this slow-growing relative of *M. tuberculosus* are punctiform.

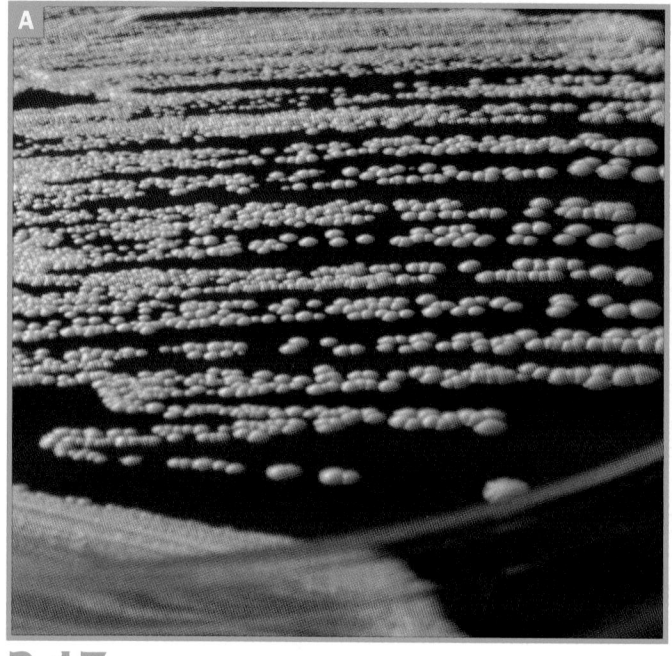

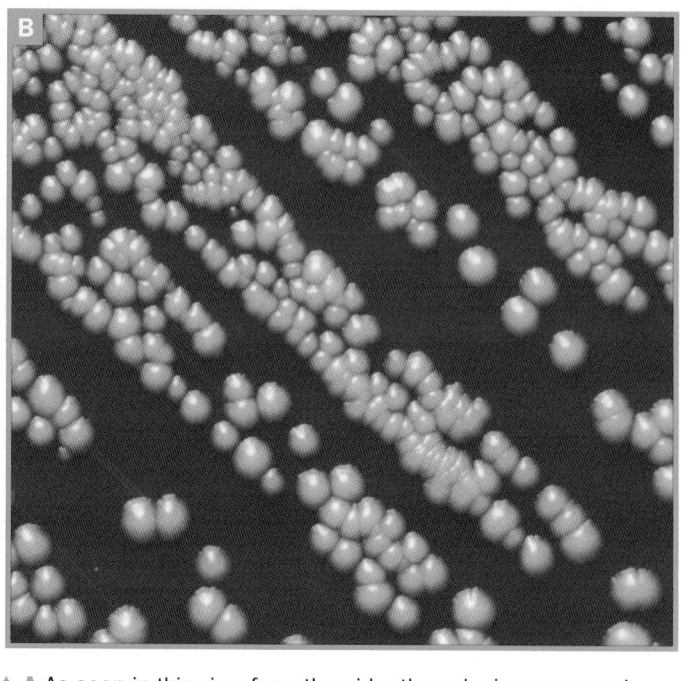

2-17 CORYNEBACTERIUM XEROSIS GROWN ON SHEEP BLOOD AGAR ✦ A As seen in this view from the side, the colonies are round, dull, buff, and convex. B Close-up of circular *C. xerosis* colonies. *C. xerosis* is rarely an opportunistic pathogen.

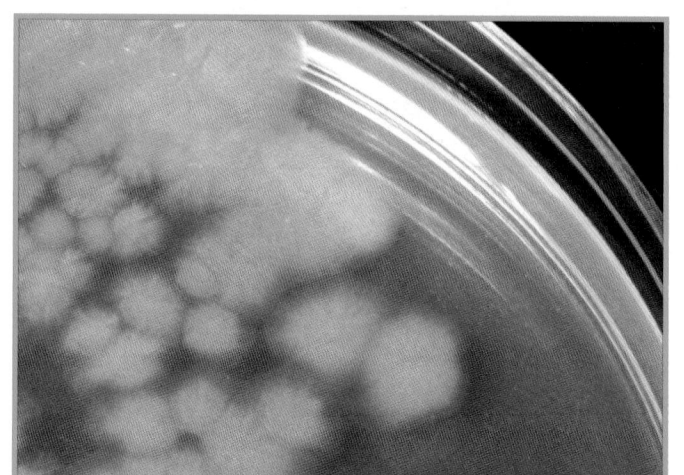

2-18 ERWINIA AMYLOVORA COLONIES GROWN ON NUTRIENT AGAR ✦ Note the irregular shape and spreading edges. *E. amylovora* is a plant pathogen.

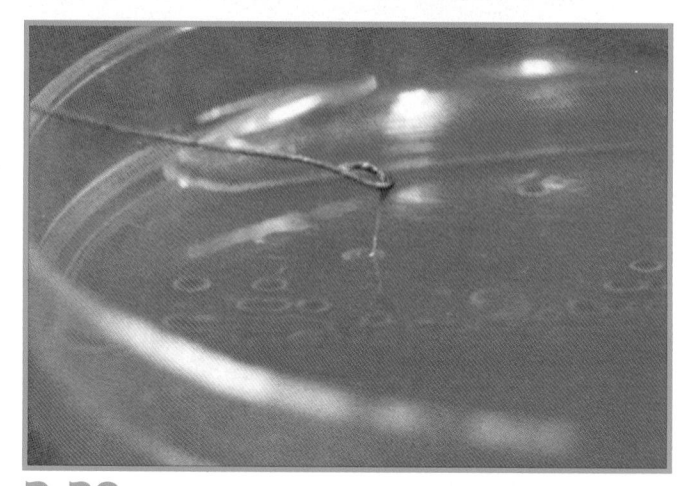

2-19 SWARMING GROWTH PATTERN ✦ Members of the genus *Proteus* will swarm at certain intervals and produce a pattern of concentric rings because of their motility. This photograph demonstrates the swarming behavior of *P. vulgaris* on DNase agar.

2-20 MUCOID COLONIES OF PSEUDOMONAS AERUGINOSA ✦ *Pseudomonas aeruginosa* grown on Endo agar illustrates a mucoid texture. *P. aeruginosa* is found in soil and water and can cause infections of burn patients.

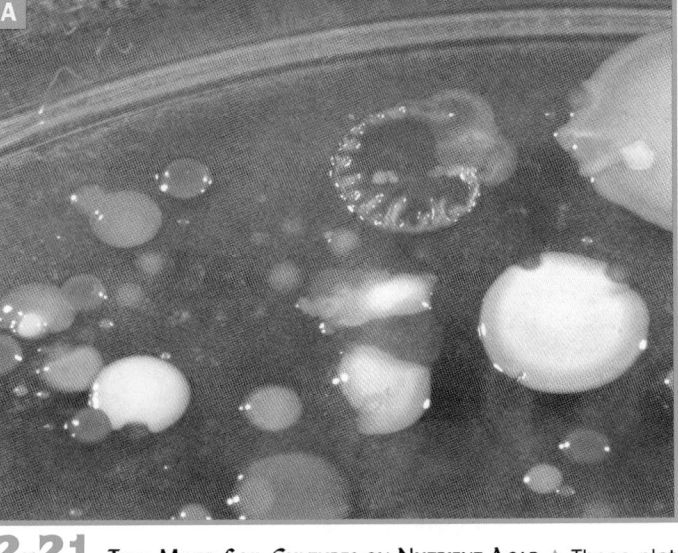

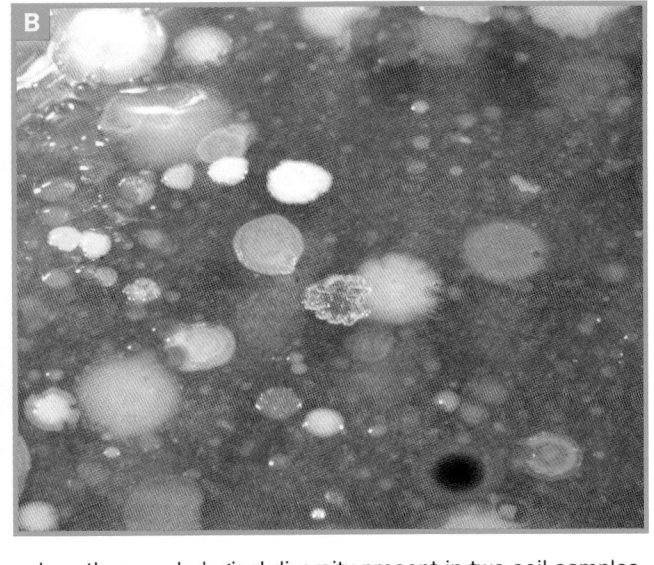

2-21 TWO MIXED SOIL CULTURES ON NUTRIENT AGAR ✦ These plates show the morphological diversity present in two soil samples.

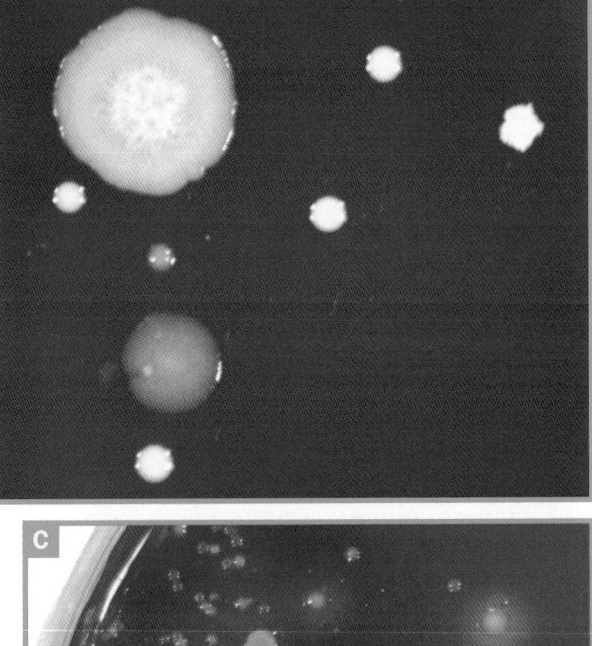

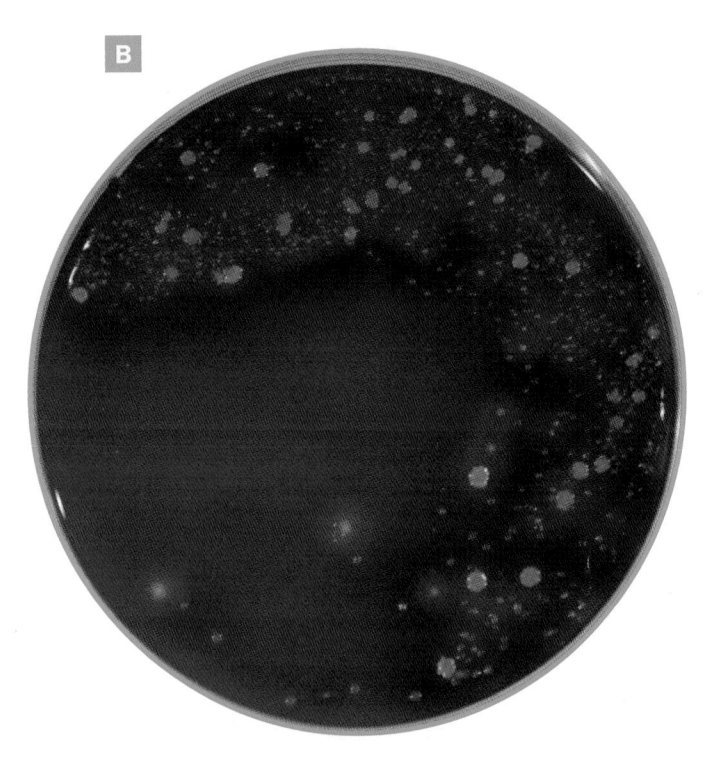

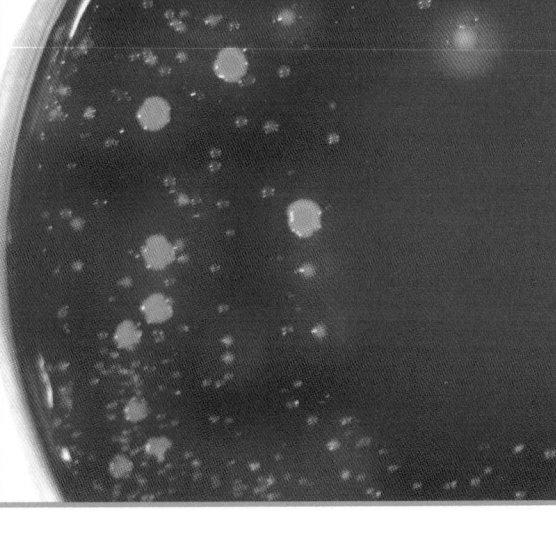

2-22 TWO THROAT CULTURES ON SHEEP BLOOD AGAR ✦
A Several different species are growing on this plate.
B Note the darkening of the agar from partial hemolysis
(α-hemolysis) of the sheep red blood cells (RBCs). C This
is a close-up of the same plate as in B. Note the weak
β-hemolysis of the white colony. β-hemolysis is complete
lysis of RBCs. White growth on a blood agar plate demonstrating β-hemolysis is characteristic of *Staphylococcus
aureus*. See Exercise 5-25 for more information on hemolytic reactions.

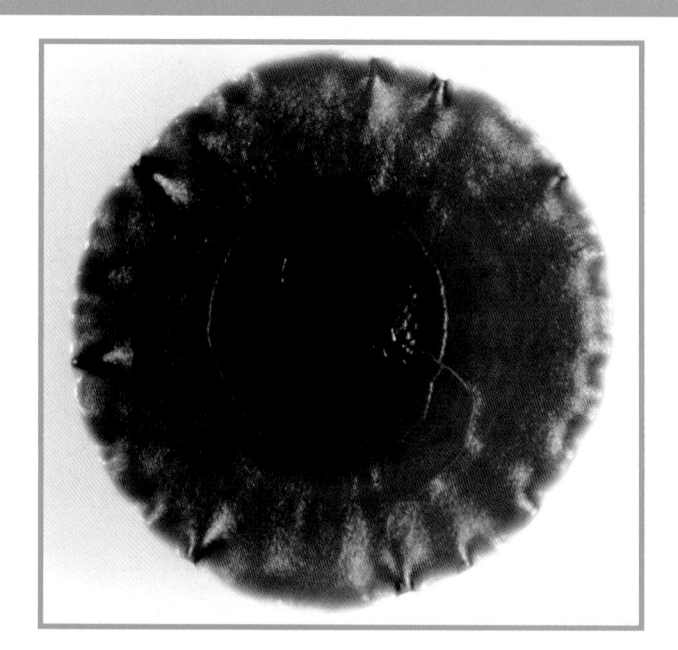

2-23 CHROMOBACTERIUM VIOLACEUM COLONY ✦
This is a magnified *C. violaceum* colony (approximately X10) after one week of incubation on Trypticase Soy Agar. Compare this colony with those in Figure 2-6.

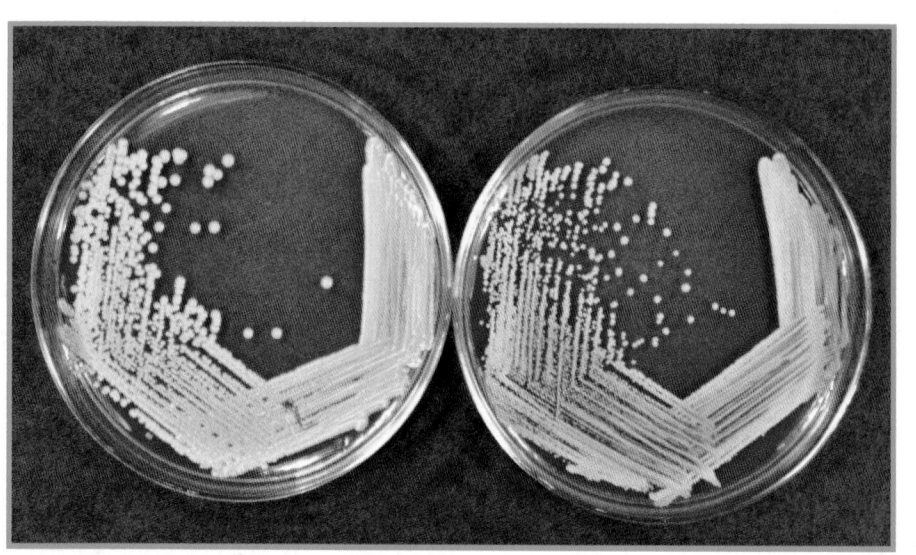

2-24 PIGMENT PRODUCTION ✦
Closely related species may look very different, as seen in these plates of *Micrococcus luteus* (left) and *Kocuria rosea*, formerly *Micrococcus roseus* (right).

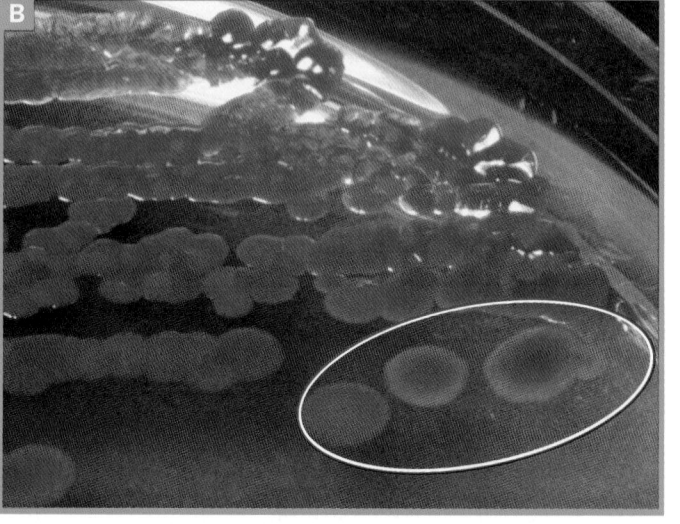

2-25 INFLUENCE OF AGE ON PIGMENT PRODUCTION ✦ A *Serratia marcescens* grown on Nutrient Agar after 24 hours' growth.
B The same plate of *S. marcescens* after 48 hours' growth. Note in particular the change in the three colonies in the lower right (encircled).

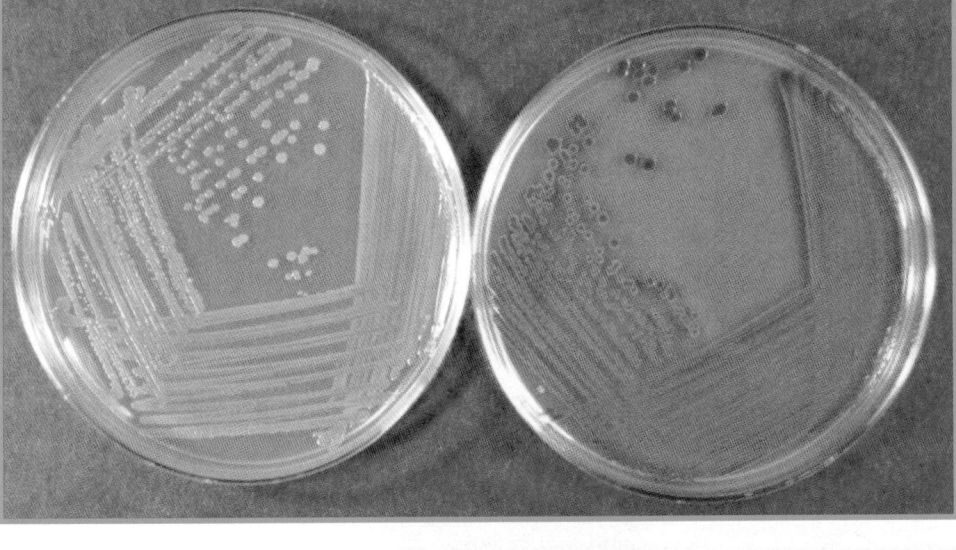

2-26 INFLUENCE OF TEMPERATURE ON PIGMENT PRODUCTION ✦ Pigment production also may be influenced by temperature. *Serratia marcescens* produces less orange pigment when grown at 37°C (left) than when grown at 25°C (right).

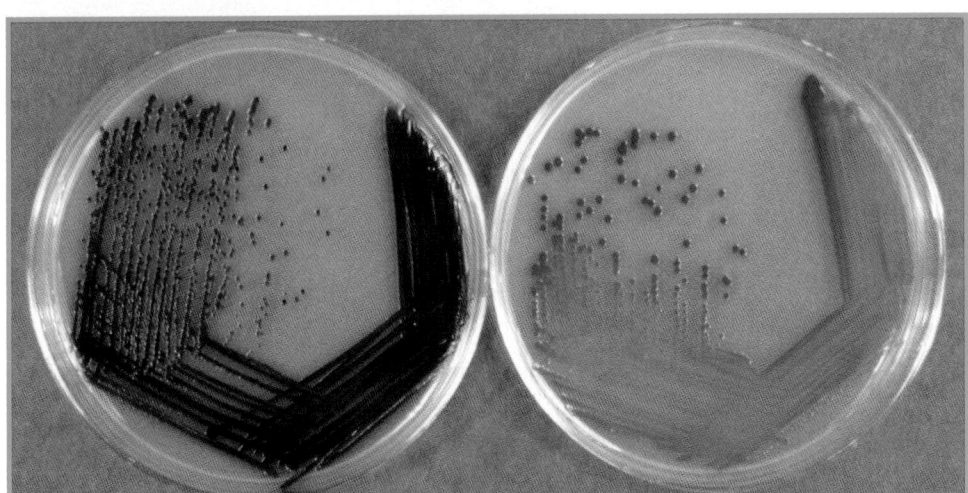

2-27 INFLUENCE OF NUTRIENT AVAILABILITY ON PIGMENT PRODUCTION ✦ Pigment production may be influenced by environmental factors such as nutrient availability. *Chromobacterium violaceum* produces a much more intense purple pigment when grown on Trypticase Soy Agar (left) than when grown on Nutrient Agar (right), a less nutritious medium.

2-28

DIFFUSIBLE PIGMENT ✦ Here, *Pseudomonas* is growing on Trypticase Soy Agar. *P. aeruginosa* often produces a characteristic diffusible, blue-green pigment.

References

Claus, G. William. 1989. Chapter 14 in *Understanding Microbes —A Laboratory Textbook for Microbiology*. W. H. Freeman and Co., New York.

Collins, C. H., Patricia M. Lyne, and J. M. Grange. 1995. Chapter 6 in *Collins and Lyne's Microbiological Methods*, 7th ed. Butterworth-Heineman, Oxford, England.

Forbes, Betty A., Daniel F. Sahm, and Alice S. Weissfeld. 2002. *Bailey & Scott's Diagnostic Microbiology*, 11th ed. Mosby-Yearbook, St. Louis.

Winn, Washington C., *et al.* 2006. *Koneman's Color Atlas and Textbook of Diagnostic Microbiology*, 6th ed. Lippincott Williams & Wilkins, Baltimore.

2-3 Growth Patterns on Slants

✦ Theory

Agar slants are useful primarily as media for cultivation and maintenance of stock cultures. Organisms cultivated on slants, however, do display a variety of growth characteristics. We offer these more as an item of interest than one of diagnostic value.

Most of the organisms you will see in this class produce **filiform** growth (dense and opaque with a smooth edge). All of the organisms in Figure 2-29 are filiform and **pigmented**. Most species in the genus *Mycobacterium* produce **friable** (crusty) growth (Figure 2-30—*Mycobacterium phlei*). Some motile organisms produce growth with a **spreading edge** (Figure 2-30—*Alcaligenes faecalis*). Still others produce **translucent** or **transparent** growth (Figure 2-30—*Lactobacillus plantarum*).

✦ Application

Although not definitive by themselves, growth characteristics on slants can provide useful information when attempting to identify an organism.

✦ In This Exercise

In Exercise 2-2, you examined growth on agar plates. Today you will be looking at the growth patterns of six organisms on prepared agar slants. If your instructor had you save your plates from Exercise 2-2, you will compare the growth patterns on the two different types of media.

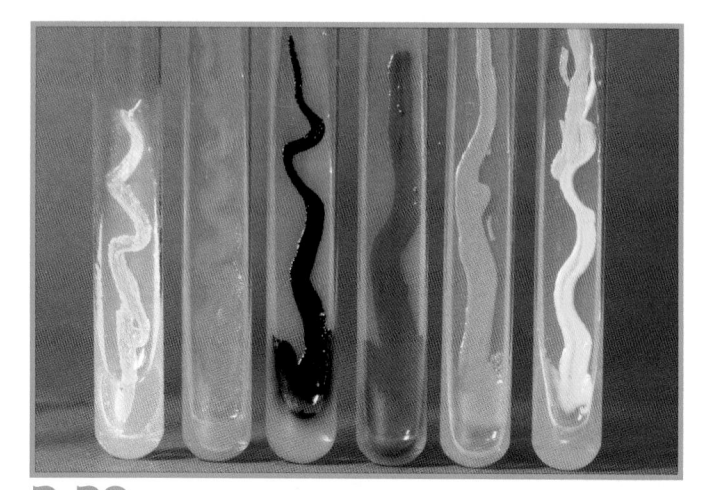

2-29 PIGMENT PRODUCTION ON SLANTS ✦ From left to right: *Staphylococcus epidermidis* (white), *Pseudomonas aeruginosa* (green), *Chromobacterium violaceum* (violet), *Serratia marcescens* (red/orange), *Kocuria rosea* (rose), *Micrococcus luteus* (yellow).

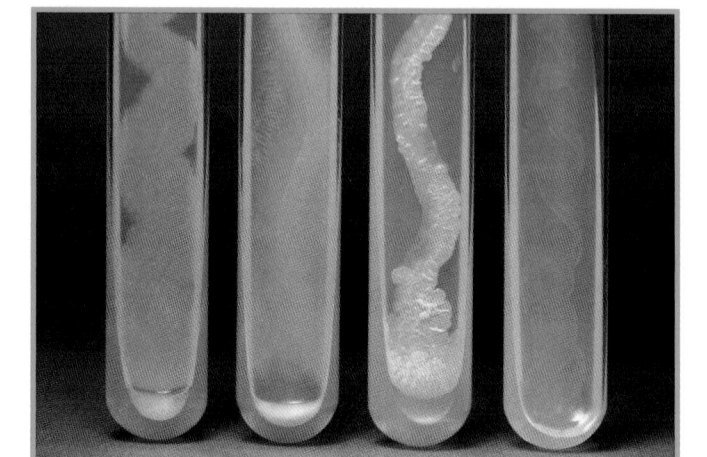

2-30 GROWTH TEXTURE ON SLANTS ✦ From left to right, *Bacillus* spp. (flat, dry), *Alcaligenes faecalis* (spreading edge), *Mycobacterium phlei* (crusty/friable), *Lactobacillus plantarum* (transparent, barely visible).

✦ Materials

Per Student Group

✦ Tryptic Soy Agar or Brain Heart Infusion Agar Slant cultures of:
  ◆ *Micrococcus luteus*
  ◆ *Corynebacterium xerosis*
  ◆ *Lactobacillus plantarum*
  ◆ *Mycobacterium smegmatis* (BSL-2)
  ◆ *Bacillus subtilis*
  ◆ *Proteus mirabilis* (BSL-2)

✦ Uninoculated tube of the medium used for cultures.

 Procedure

Lab One

1 Examine the slants and describe the different growth patterns on the Data Sheet. Include a sketch of a representative portion of each.

2 (Optional) Compare the slant growth to that on the agar plates from Exercise 2-2.

3 (Optional) Save the tubes for comparison with broths in Exercise 2-4.

Reference

Claus, G. William. 1989. Chapter 17 in *Understanding Microbes—A Laboratory Textbook for Microbiology*. W. H. Freeman and Co., New York.

Growth Patterns in Broth

✦ Theory

Microorganisms cultivated in broth display a variety of growth characteristics. Some organisms float on top of the medium and produce a type of surface membrane called a **pellicle**. Others sink to the bottom as **sediment**. Some bacteria produce **uniform fine turbidity**, and others appear to clump in what is called **flocculent** growth. Refer to Figures 2-31 and 2-32.

2-31 GROWTH PATTERNS IN BROTH ✦ From left to right in pairs (of similar species): *Enterobacter aerogenes* and *Citrobacter diversus*—motile members of *Enterobacteriaceae* (uniform fine turbidity—UFT), *Enterococcus faecalis* and *Staphylococcus aureus*—nonmotile Gram-positive cocci (sediment), *Mycobacterium phlei* and *Mycobacterium smegmatis* (relatives of *Mycobacterium tuberculosis*)—nonmotile with a waxy cell wall (pellicle).

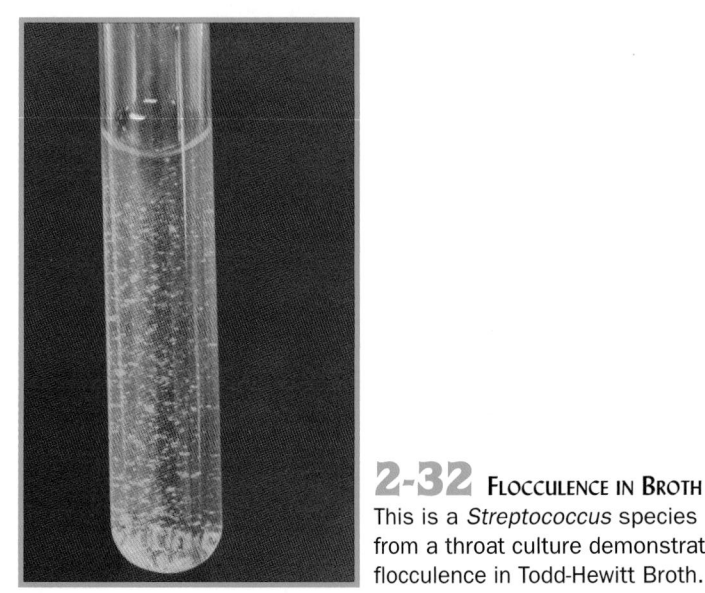

2-32 FLOCCULENCE IN BROTH ✦ This is a *Streptococcus* species from a throat culture demonstrating flocculence in Todd-Hewitt Broth.

✦ Application

Bacterial genera, and frequently individual species within a genus, demonstrate characteristic growth patterns in broth that provide useful information when attempting to identify an organism.

✦ In This Exercise

Today, you will be examining the growth characteristics of six different bacteria in broth. If you were instructed to save plates or slants from previous exercises, you will compare the growth patterns on the different types of media.

✦ Materials

Per Student Group

✦ Brain–Heart Infusion (BHI) Broth cultures of:
  ✦ *Micrococcus luteus*
  ✦ *Corynebacterium xerosis*
  ✦ *Lactobacillus plantarum*
  ✦ *Mycobacterium smegmatis* (BSL-2)
  ✦ *Bacillus subtilis*
  ✦ *Proteus mirabilis* (BSL-2)

✦ (Optional) your slants from Exercise 2-3

✦ one sterile uninoculated BHI Broth control

 ## Procedure

Examine the tubes and compare them with the control. Describe the different growth patterns in the table on the Data Sheet.

References

Claus, G. William. 1989. Chapter 17 in *Understanding Microbes—A Laboratory Textbook for Microbiology*. W. H. Freeman and Co., New York.

Forbes, Betty A., Daniel F. Sahm, and Alice S. Weissfeld. 2002. *Bailey & Scott's Diagnostic Microbiology*, 11th ed. Mosby-Yearbook, St. Louis.

Environmental Factors Affecting Microbial Growth

Bacteria have limited control over their internal environments. Whereas many eukaryotes have evolved sophisticated internal control mechanisms, bacteria are almost completely dependent on external factors to provide conditions suitable for their existence. Minor environmental changes can dramatically change a microorganism's ability to transport materials across the membrane, perform complex enzymatic reactions, and maintain critical cytoplasmic pressure.

One way to observe bacterial responses to environmental changes is to artificially manipulate external factors and measure growth rate. Understand that growth rate, as it is used in this manual, is synonymous with reproductive rate. Optimal growth conditions, as might be expected, result in faster growth and greater cell density (as evidenced by greater turbidity) than do less than optimal conditions.

In this series of laboratory exercises, you will examine the effects of oxygen, temperature, pH, osmotic pressure, and the availability of nutritional resources on bacterial growth rate. You also will learn some methods for cultivating anaerobic bacteria. When possible, you will attempt to classify organisms based on your results. ✦

EXERCISE

2-5 Evaluation of Media

✦ Theory

Living things are composed of compounds from four biochemical families:

1. proteins,
2. carbohydrates,
3. lipids, and
4. nucleic acids.

Even though all organisms share this fundamental chemical composition, they differ greatly in their ability to make these molecules. Some are capable of making them out of the simple carbon compound carbon dioxide (CO_2). These organisms, called **autotrophs,** require the least "assistance" from the environment to grow. The remaining organisms, called **heterotrophs,** require preformed organic compounds from the environment.

Some heterotrophs are metabolically flexible and require only a few simple organic compounds from which to make all their biochemicals. Others require a greater portion of their organic compounds from the environment. An organism that relies heavily on the environment to supply ready-made organic compounds is referred to as **fastidious.** Heterotrophs vary greatly in their dependence on the environment to supply organic

compounds and, as such, range from highly fastidious to **nonfastidious.** Autotrophs are less fastidious than the most nonfastidious heterotrophs.

Successful cultivation of a microbe in the laboratory requires an ability to satisfy its nutritional needs. The absence of a single required chemical resource prevents its growth. In general, the more fastidious the organism, the more ingredients a medium must have. **Undefined** media (also known as complex media) are composed of extracts from plant or animal sources and are rich in nutrients. Even though the exact composition of the medium and the amount of each ingredient are unknown, undefined media are useful in growing the greatest variety of culturable microbes. A **defined** medium (or chemically defined medium) is one in which the amount and identity of every ingredient is known. Defined media typically support a narrower range of organisms.

✦ Application

The ability of a microbiologist to cultivate a microorganism requires some knowledge of its metabolic needs. One quick way to make this determination is to transfer it to a variety of media containing different nutritional components and observe how well it grows.

✦ In This Exercise

Today you will evaluate the ability of three media—
Brain–Heart Infusion Broth, Nutrient Broth, and Glucose
Salts Medium—to support bacterial growth. You will
do this by visually comparing the density of growth
(turbidity) between organisms in the various broths. It
is important to make your inoculations as uniform as
possible. Make sure that your loop is fully closed so
it will hold enough broth to produce a film across its
opening (similar to a toy loop for blowing bubbles).
Inoculate each medium with a single loopful of broth.

✦ Materials

Per Student Group

✦ five tubes each of:

- ◆ Brain–Heart Infusion Broth
- ◆ Nutrient Broth
- ◆ Glucose Salts Medium

✦ Broth cultures of:

- ◆ *Escherichia coli*
- ◆ *Lactococcus lactis*
- ◆ *Moraxella catarrhalis* (BSL-2)
- ◆ *Staphylococcus epidermidis*

✦ Medium Recipes

Nutrient Broth

◆ Beef extract	3.0 g
◆ Peptone	5.0 g
◆ Distilled or deionized water	1.0 L

Brain–Heart Infusion Broth

◆ Calf Brains, Infusion from 200g	7.7 g
◆ Beef Heart, Infusion from 250g	9.8 g
◆ Proteose peptone	10.0 g
◆ Dextrose	2.0 g
◆ Sodium chloride	5.0 g
◆ Disodium phosphate	2.5 g
◆ Distilled or deionized water	1.0 L

Glucose Salts Medium

◆ Glucose	5.0 g
◆ Sodium chloride	5.0 g
◆ Magnesium sulfate	0.2 g
◆ Ammonium dihydrogen phosphate	1.0 g
◆ Dipotassium phosphate	1.0 g
◆ Distilled or deionized water	1.0 L

 Procedure

Lab One

1 Inoculate each medium with a single loopful of each
organism. Leave one of each tube uninoculated.

2 Incubate all tubes (including the uninoculated ones)
at $35 \pm 2°C$ for 24–48 hours.

Lab Two

1 Mix all tubes well and examine them for turbidity.
Score relative amounts of growth using "0" for no
growth, and "1," "2," and "3" for successively
greater degrees of growth.

2 Record your results on the Data Sheet. (**Note:** Results
may vary. Record what you see, not what you expect.)

References

Delost, Maria Dannessa. 1997. Page 144 in *Introduction to Diagnostic
Microbiology: A Text and Workbook.* Mosby, St. Louis.

Forbes, Betty A., Daniel F. Sahm, and Alice S. Weissfeld. 2002. *Bailey &
Scott's Diagnostic Microbiology,* 11th ed. Mosby-Yearbook, St. Louis.

Zimbro, Mary Jo, and David A. Power, Eds. 2003. *Difco™ and BBL™
Manual—Manual of Microbiological Culture Media.* Becton Dickinson
and Co., Sparks, MD.

Aerotolerance

Most microorganisms can survive within a range of environmental conditions, but not surprisingly, tend to produce growth with the greatest density in the areas where conditions are most favorable. One important resource influencing microbial growth is oxygen. Some organisms require oxygen for their metabolic needs. Some other organisms are not affected by it at all. Still other organisms cannot even survive in its presence. This ability or inability to live in the presence of oxygen is called **aerotolerance**.

Most growth media are sterilized in an autoclave during preparation. This process not only kills unwanted microbes, but removes most of the free oxygen from the medium as well. After the medium is removed from the autoclave and allowed to cool, the oxygen begins to diffuse back in. In tubed media (both liquid and solid) this process creates a gradient of oxygen concentrations, ranging from **aerobic** at the top, nearest the source of oxygen, to **anaerobic** at the bottom. Because of microorganisms' natural tendency to proliferate where the oxygen concentration best suits their metabolic needs, differing degrees of population density will develop in the medium over time that can be used to visually examine their aerotolerance.

Obligate (strict) aerobes, organisms that require oxygen for respiration, grow at the top where oxygen is most plentiful. **Facultative anaerobes** grow in the presence *or* absence of oxygen. When oxygen is available, they respire aerobically. When oxygen is not available, they either respire anaerobically (reducing sulfur or nitrate instead of oxygen) or ferment an available substrate. Refer to Appendix A and Section Five for more information on anaerobic respiration and fermentation. Where an oxygen gradient exists, facultative anaerobes grow throughout the medium but appear denser near the top. **Aerotolerant anaerobes**, organisms that don't require oxygen and are not adversely affected by it, live uniformly throughout the medium. Aerotolerant anaerobes are fermentative even in the presence of free oxygen. **Microaerophiles**, as the name suggests, survive only in environments containing lower than atmospheric levels of oxygen. Some microaerophiles called **capnophiles** can survive only if carbon dioxide levels are elevated. Microaerophiles will be seen somewhere near the middle or upper middle region of the medium. Finally, **obligate (strict) anaerobes** are organisms for which even small amounts of oxygen are lethal and, therefore, will be seen only in the lower regions of the medium, depending on how far into the medium the oxygen has diffused. ✦

EXERCISE

2-6   Agar Deep Stabs

✦ Theory

Agar deep stabs are prepared with Tryptic Soy Agar (TSA) enriched with yeast extract to promote growth of a broad range of organisms. As stated in the introduction to this section, oxygen, which is removed from the medium during preparation and autoclaving, immediately begins to diffuse back into the medium as the agar cools and solidifies. This process creates an oxygen gradient in the medium, ranging from aerobic at the top to anaerobic at the bottom.

Agar deeps are prepared with 10 mL instead of the customary 7 mL used in slants. This extra depth ensures (if inoculated soon after preparation) that the bottom portion of the medium is anaerobic. The agar is stab-inoculated with an inoculating needle to introduce as little air as possible. The location of growth that develops indicates the organism's aerotolerance (Figure 2-33).

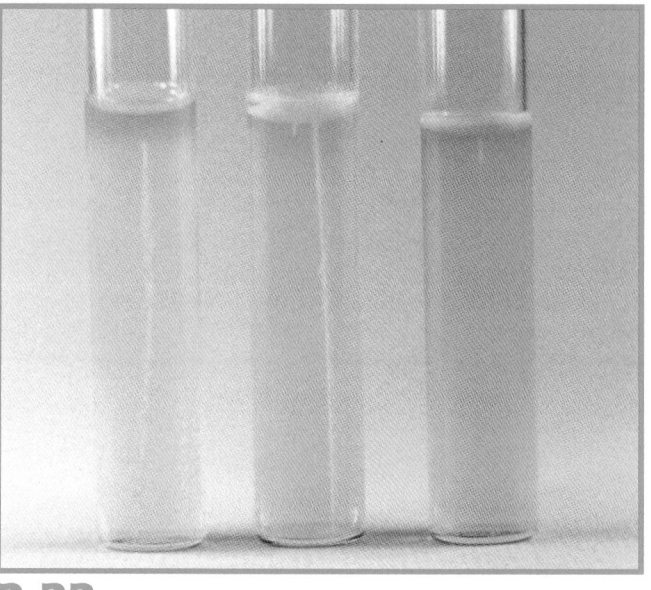

2-33 AGAR DEEP STAB TUBES ✦ From left to right: strict anaerobe (possible microaerophile), facultative anaerobe, and strict aerobe.

✦ Application

This test is a good visual indicator of oxygen tolerance in microorganisms.

✦ In This Exercise

Today you will stab-inoculate three agar tubes to determine the aerotolerance categories of three bacteria.

✦ Materials

Per Student Group

✦ four agar deep stab tubes

✦ inoculating needle

✦ fresh cultures of:

   ✦ *Clostridium sporogenes*
   ✦ *Alcaligenes faecalis* (BSL-2)
   ✦ *Staphylococcus aureus* (BSL-2)

✦ Medium Recipe

Enriched TSA

✦ Tryptone	15.0 g
✦ Soytone	5.0 g
✦ Sodium Chloride	5.0 g
✦ Yeast Extract	5.0 g
✦ Agar	15.0 g
✦ Distilled or deionized water	1.0 L

 Procedure

Lab One

1 Obtain 4 agar deep stab tubes.

2 With a heavy inoculum on your inoculating needle, carefully stab the agar tubes with the test organisms. Refer to Appendix B for the proper stab technique. (***Note:*** To minimize the introduction of unwanted oxygen to the medium, try to avoid lateral motion of the needle during inoculation.)

3 Stab the fourth tube with your sterile needle.

4 Label each tube with your name, the date, the medium, and the name of the organism.

5 Incubate the tubes at 35 ± 2°C for 24 to 48 hours.

Lab Two

1 Examine the tubes and determine the aerotolerance category of each organism.

2 Record your results on the Data Sheet.

References

Forbes, Betty A., Daniel F. Sahm, and Alice S. Weissfeld. 2002. Chapter 10 in *Bailey & Scott's Diagnostic Microbiology*, 11th ed. Mosby, St. Louis.

Koneman, Elmer W., Stephen D. Allen, William M. Janda, Paul C. Schreckenberger, and Washington C. Winn, Jr. 1997. Chapter 14 in *Color Atlas and Textbook of Diagnostic Microbiology*, 5th ed. J. B. Lippincott Co. Philadelphia.

EXERCISE 2-7

Fluid Thioglycollate Medium

✦ Theory

Fluid Thioglycollate Medium is prepared as a basic medium (as used in this exercise) or with a variety of supplements, depending on the specific needs of the organisms being cultivated. As such, this medium is appropriate for a broad variety of aerobic and anaerobic, fastidious and nonfastidious organisms. It is particularly well adapted for cultivation of strict anaerobes and microaerophiles.

Key components of the medium are yeast extract, pancreatic digest of casein, dextrose, sodium thiogly-collate, L-cystine, and resazurin. Yeast extract and pancreatic digest of casein provide nutrients; sodium thioglycollate and L-cystine reduce oxygen to water; and resazurin (pink when oxidized, colorless when reduced) acts as an indicator. A small amount of agar is included to slow oxygen diffusion.

As mentioned in the introduction to aerotolerance, oxygen removed during autoclaving will diffuse back into the medium as the tubes cool to room temperature. This produces a gradient of concentrations from fully aerobic at the top to anaerobic at the bottom. Thus, fresh media will appear clear to straw-colored with a pink region at the top where the dye has become oxidized (Figure 2-34). Figure 2-35 demonstrates some basic bacterial growth patterns in the medium as influenced by the oxygen gradient.

✦ Application

Fluid Thioglycollate Medium is a liquid medium designed to promote growth of a wide variety of fastidious micro-organisms. It can be used to grow microbes representing all levels of oxygen tolerance; however, it generally is associated with the cultivation of anaerobic and micro-aerophilic bacteria.

✦ In This Exercise

Today you will be inoculating Fluid Thioglycollate Medium to determine the aerotolerance categories of three bacteria. Your instructor may have additional organisms or may have you collect environmental samples to test.

✦ Materials

Per Student Group

✦ four Fluid Thioglycollate Medium tubes
✦ fresh cultures of:
  ◆ *Alcaligenes faecalis* (BSL-2)
  ◆ *Clostridium sporogenes*
  ◆ *Staphylococcus aureus* (BSL-2)

✦ Medium Recipe

Fluid Thioglycollate Medium

◆ Yeast extract	5.0 g
◆ Pancreatic digest of casein	15.0 g
◆ Dextrose	5.5 g
◆ Sodium chloride	2.5 g
◆ Sodium thioglycollate	0.5 g
◆ L-cystine	0.5 g
◆ Agar	0.75 g
◆ Resazurin	0.001 g
◆ Distilled or deionized water	1.0 L

 Procedure

Lab One

1 Obtain four Fluid Thioglycollate tubes and label them with your name, the date, medium, and organism.

2 Using your loop, inoculate three broths with the organisms provided. (**Note:** When inoculating Thioglycollate Broth, it helps to dip the loop all the way to the bottom of the tube and gently mix the broth with the loop as you remove it. Finish mixing by gently rolling the tube between your hands.) Do not inoculate the fourth tube.

3 Incubate the tubes at $35 \pm 2°C$ for 24 to 48 hours.

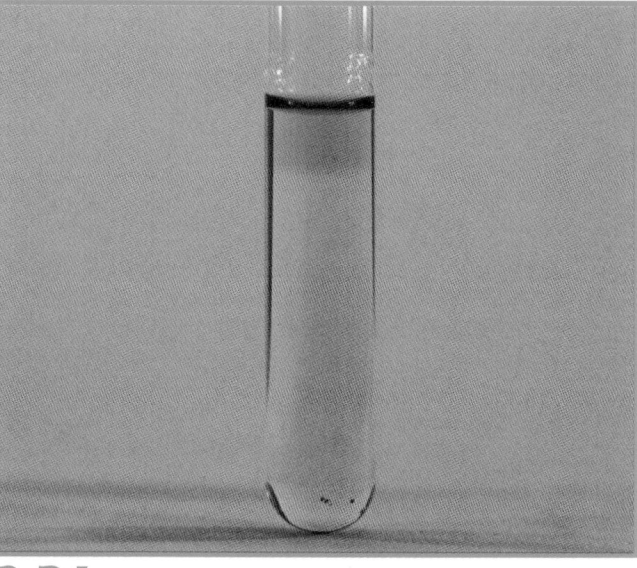

2-34 AEROBIC ZONE IN THIOGLYCOLLATE BROTH ✦ Note the pink region in the top (oxidized) portion of the broth resulting from the indicator resazurin. In the bottom (reduced) portion, the dye is colorless and the medium is its typical straw color.

2-35 GROWTH PATTERNS IN THIOGLYCOLLATE MEDIUM ✦ Growth patterns of a variety of organisms are shown in these Fluid Thioglycollate Broths. Pictured from left to right are: aerotolerant anaerobe, facultative anaerobe, strict anaerobe, strict aerobe, and microaerophile. Compare these tubes with the uninoculated broth in Figure 2-34.

Lab Two

1 Check the control tube for growth to assure sterility of the medium. Note any changes that may have occurred as a result of incubation, especially in the colored region at the surface.

2 Using the control as a comparison, examine and note the location of the growth in all tubes.

3 Enter your observations and interpretations in the chart provided on the Data Sheet.

References

Allen, Stephen D., Christopher L. Emery, and David M. Lyerly. 2003. Chapter 54 in *Manual of Clinical Microbiology*, 8th ed. Edited by Patrick R. Murray, Ellen Jo Baron, James H. Jorgensen, Michael A. Pfaller, and Robert H. Yolken. ASM Press, American Society for Microbiology, Washington, DC.

Forbes, Betty A., Daniel F. Sahm, and Alice S. Weissfeld. 2002. Chapter 10 in *Bailey & Scott's Diagnostic Microbiology*, 11th ed. Mosby, St. Louis.

Winn, Washington C., *et. al.* 2006. *Koneman's Color Atlas and Textbook of Diagnostic Microbiology*, 6th ed. Lippincott Williams & Wilkins, Baltimore.

Zimbro, Mary Jo and David A. Power, Eds. 2003. *Difco™ and BBL™ Manual—Manual of Microbiological Culture Media*. Becton Dickinson and Company, Sparks, MD.

EXERCISE 2-8

Anaerobic Jar

✦ Theory

The GasPak® Anaerobic System by BBL™ is a plastic jar in which to create anaerobic, microaerophilic, or CO_2-enriched conditions depending on the specific needs of the bacteria being cultivated. The components required for anaerobic growth include a chemical gas generator packet (envelope) containing sodium borohydride and sodium bicarbonate, and a paper indicator strip saturated with methylene blue to confirm the absence of oxygen (Figure 2-36). Methylene blue is colorless when reduced and blue when oxidized. Also included in the packet is a small amount of palladium to act as a catalyst for the reaction that will produce the necessary conditions inside the jar.

After the inoculated media (typically agar plates) are placed inside the jar, the opened gas generator envelope is placed inside along with the anaerobic indicator strip. Water is added to the envelope and the jar lid immediately is fastened down. The sodium borohydride and sodium bicarbonate in the envelope react with the water to produce hydrogen and carbon dioxide gases. The palladium catalyzes a reaction between the hydrogen and free oxygen in the jar to produce water, as shown in Figure 2-37. Removal of free oxygen produces anaerobic conditions

$$2H_2 + O_2 \xrightarrow{\text{Pd}} 2H_2O$$

2-37 CONVERSION OF H_2 AND O_2 TO WATER USING A PALLADIUM CATALYST ✦

in the jar within approximately an hour, as evidenced by a white indicator strip and moisture on the inside of the jar.

Figure 2-38 illustrates some typical growth patterns under aerobic and anaerobic conditions.

✦ Application

This procedure provides a means of cultivating anaerobic and microaerophilic bacteria.

✦ In This Exercise

Today each group will inoculate two Nutrient Agar plates with three organisms. (Your instructor may have additional organisms for you to test.) One of the plates will go into the jar, and the second plate will not. Both plates will be incubated at $35 \pm 2°C$ for 24 to 48 hours. Following incubation, the growth on the two plates will be examined and compared.

✦ Materials

Per Class

✦ one anaerobic jar with gas generator packet[1]

Per Group

✦ two Nutrient Agar plates

✦ fresh broth cultures of:
  ◆ *Alcaligenes faecalis* (BSL-2)
  ◆ *Clostridium sporogenes*
  ◆ *Staphylococcus aureus* (BSL-2)

2-36 THE ANAEROBIC JAR ✦ Note the white methylene blue strip and the open packet, which has discharged H_2 and CO_2 gases. The palladium, contained in the packet, catalyzes the conversion of H_2 and O_2 as shown in Figure 2-36.

---

[1] Available from Becton Dickinson Microbiology Systems, Sparks, MD
http://www.bd.com

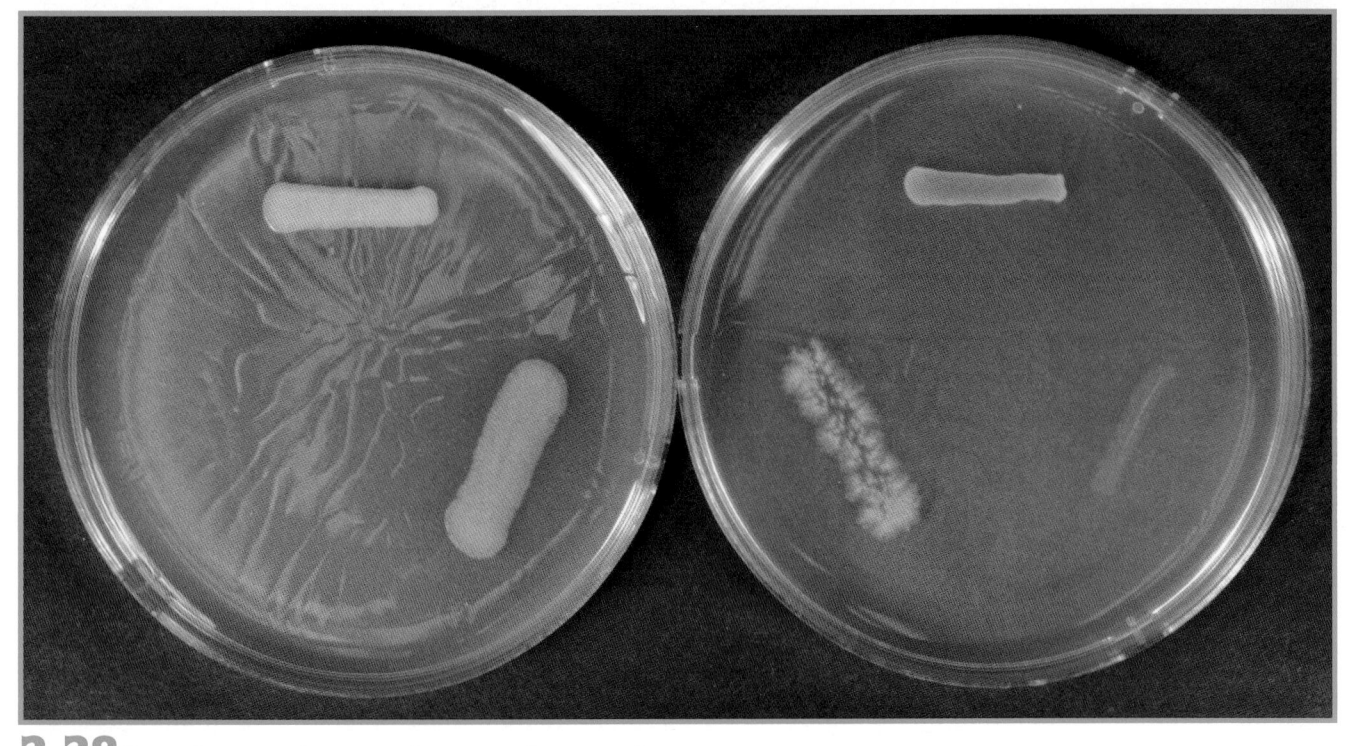

2-38 PLATES INCUBATED INSIDE AND OUTSIDE THE ANAEROBIC JAR ✦ Both Nutrient Agar plates were spot-inoculated with the same facultative anaerobe (top), strict aerobe (right), and strict anaerobe (left). The plate on the left was incubated outside the jar, and the plate on the right was incubated inside the jar. Note the relative amounts of growth of the three organisms.

Procedure

Lab One

1 Obtain two Nutrient Agar plates. Using a marking pen, divide the bottom of each plate into three sectors.

2 Label each plate with your name, the date, and organism by sector.

3 Using your loop, inoculate the sectors of both plates with the organisms provided. (**Note:** Inoculate with single streaks about one centimeter long.) Tape the lids in place.

4 Place one plate in the anaerobic jar in an inverted position.

5 When all groups have placed their plates in the jar, discharge the packet as follows (or follow the instructions for your system):
   a. Stick the methylene blue strip on the wall of the jar.
   b. Open the packet and add 10 mL of distilled water.
   c. Place the open packet in the jar with the label facing inward.
   d. Immediately close the jar.

6 Place the second plate and the anaerobic jar in the $35 \pm 2°C$ incubator for 24 to 48 hours. Be sure the plate is inverted.

Lab Two

1 Examine and compare the growth on the plates. (**Note:** Density of growth will be the most useful basis for comparison.)

2 Record your results and interpretations in the chart provided on the Data Sheet.

References

Allen, Stephen D., Christopher L. Emery, and David M. Lyerly. 2003. Chapter 54 in *Manual of Clinical Microbiology*, 8th ed. Edited by Patrick R. Murray, Ellen Jo Baron, James H. Jorgensen, Michael A. Pfaller, and Robert H. Yolken. ASM Press, American Society for Microbiology, Washington, DC.

Forbes, Betty A., Daniel F. Sahm, and Alice S. Weissfeld. 2002. Chapter 10 in *Bailey & Scott's Diagnostic Microbiology*, 11th ed. Mosby, St. Louis.

Winn, Washington C., *et. al.* 2006. *Koneman's Color Atlas and Textbook of Diagnostic Microbiology*, 6th ed. Lippincott Williams & Wilkins, Baltimore.

Zimbro, Mary Jo, and David A. Power, Eds. 2003. *Difco™ and BBL™ Manual—Manual of Microbiological Culture Media*. Becton Dickinson and Company, Sparks, MD.

The Effect of Temperature on Microbial Growth

✦ Theory

Bacteria have been discovered living in habitats ranging from –10 degrees Celsius to more than 110 degrees Celsius. The temperature range of any single species, however, is a small portion of this overall range. As such, each species is characterized by a minimum, maximum, and optimum temperature—collectively known as its **cardinal temperatures** (Figure 2-39). Minimum and maximum temperatures are, simply, the temperatures below and above which the organism will not survive. Optimum temperature is the temperature at which an organism shows the greatest growth over time—its highest growth rate.

Organisms that grow only below 20 degrees Celsius are called **psychrophiles**. These are common in ocean, Arctic, and Antarctic habitats where the temperature remains permanently cold with little or no fluctuation. Organisms adapted to cold habitats that fluctuate from about 0 degrees to above 30 degrees Celsius are called **psychrotrophs**. Bacteria adapted to temperatures between 15 degrees and 45 degrees Celsius are known as **mesophiles**.

Most bacterial residents in the human body, as well as numerous human pathogens, are mesophiles. **Thermophiles** are organisms adapted to temperatures above 40 degrees Celsius. Typically, they are found in composting organic material and in hot springs. Thermophiles that will not grow at temperatures below 40

degrees are called **obligate thermophiles**; those that will grow below 40 degrees are known as **facultative thermophiles**. Bacteria isolated from hot ocean floor ridges living between 65 and 110 degrees Celsius are called **extreme thermophiles**. Extreme thermophiles grow best above 80 degrees Celsius. Figure 2-40 illustrates bacterial temperature ranges and classifications.

✦ Application

This is a qualitative procedure designed for observing the effect of temperature on bacterial growth. It allows an estimation of the cardinal temperatures for a single species.

✦ In This Exercise

Today you will examine the growth characteristics of four organisms at five different temperatures. In addition, you will observe the influence of temperature on pigment production.

✦ Materials

Per Class

✦ Five incubating devices set at 10°C, 20°C, 30°C, 40°C, and 50°C. These devices may be any combination of the following:
   ◆ refrigerator,
   ◆ incubator,

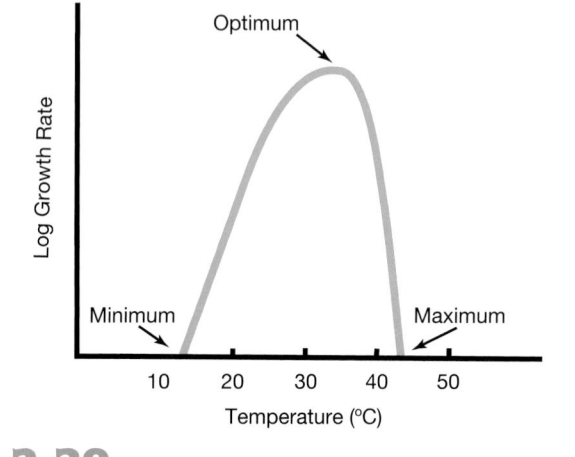

2-39 TYPICAL GROWTH RANGE OF A MESOPHILE ✦ The "minimum" and "maximum" are temperatures beyond which no growth takes place. The "optimum" is the temperature at which growth rate is highest.

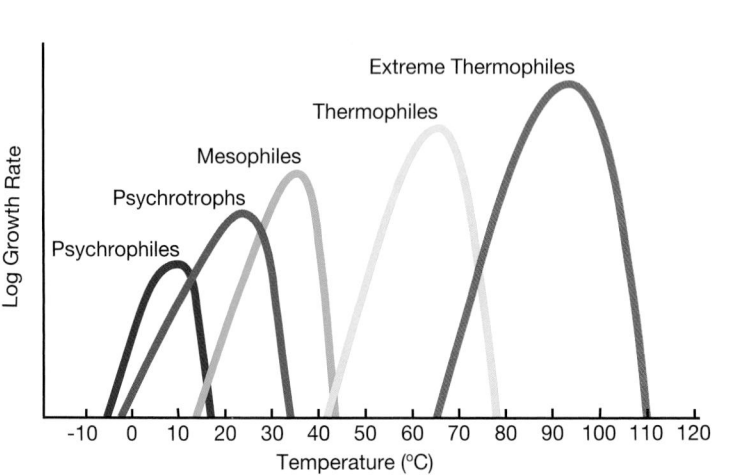

2-40 THERMAL CLASSIFICATIONS OF BACTERIA ✦ Refer to the text for a description of each category.

◆ hot water bath, or

◆ cold water bath

Per Student Group

✦ twenty sterile Nutrient Broths

✦ two Trypticase Soy Agar (TSA) plates

✦ four sterile transfer pipettes

✦ fresh Nutrient Broth cultures of:

  ◆ *Escherichia coli*

  ◆ *Serratia marcescens*

  ◆ *Bacillus stearothermophilus*

  ◆ *Pseudomonas fluorescens*

Procedure

Lab One

1 Obtain 20 Nutrient Broths—one broth for each organism at each temperature. Label them accordingly. Also obtain two TSA plates and label them 20°C and 35°C, respectively.

2 Mix each culture thoroughly before making the following transfers. Using a sterile pipette, transfer a *single drop* of each broth culture to its appropriate Nutrient Broth tube. (**Note:** Because you will be comparing the amount of growth in the Nutrient Broth tubes, you must be sure to begin by transferring the same volume of culture to each one. Use the same pipette for all transfers with a single organism.)

3 Using a simple zigzag pattern (as in Exercise 1-4), inoculate each plate with *Serratia marcescens*.

4 Incubate all tubes in their appropriate temperatures for 24 to 48 hours. Incubate the plates in the 20°C and 35°C incubators in an inverted position.

Lab Two

1 Clean the outside of all tubes with a tissue, and place them in a test tube rack organized into groups by organism.

2 Shake each broth gently until uniform turbidity is achieved.

3 Compare all tubes in a group to each other. Rate each as 0, 1, 2, or 3, according to its turbidity (0 is clear and 3 is highly turbid). Record these in the Broth Data chart on the Data Sheet.

4 Examine the plates incubated at different temperatures, compare the growth characteristics, and enter your results in the Plate Data chart on the Data Sheet.

5 Using the data from the Broth data chart, determine the cardinal temperatures and classification of each of the four organisms.

6 On the graph paper provided on the Data Sheet, plot the data (numeric values versus temperature) for the four organisms.

References

Forbes, Betty A., Daniel F. Sahm, and Alice S. Weissfeld. 2002. Chapters 2 and 10 in *Bailey & Scott's Diagnostic Microbiology*, 11th ed. Mosby-Yearbook, St. Louis.

Holt, John G., Ed. 1994. *Bergey's Manual of Determinative Bacteriology*, 9th ed. Williams and Wilkins, Baltimore.

Moat, Albert G., John W. Foster, and Michael P. Spector. 2002. Pages 597–601 in *Microbial Physiology*, 4th ed. Wiley-Liss, New York.

Prescott, Lansing M., John P. Harley, and Donald A. Klein. 2005. Chapter 6 in *Microbiology*, 6th ed., WCB McGraw-Hill, Boston.

Varnam, Alan H., and Malcolm G. Evans. 2000. *Environmental Microbiology*. ASM Press, Washington, DC.

White, David. 2000. Pages 384–387 in *The Physiology and Biochemistry of Prokaryotes*, 2nd ed. Oxford University Press, New York.

Winn, Washington C., *et al.* 2006. *Koneman's Color Atlas and Textbook of Diagnostic Microbiology*, 6th ed. Lippincott Williams & Wilkins, Baltimore.

The Effect of pH on Microbial Growth

✦ Theory

The conventional means of expressing the concentration (or activity) of hydrogen ions in a solution is "pH". The term pH, which stands for "pondus hydrogenii" (variably defined as hydrogen power or hydrogen potential), was invented in 1909 by the Danish biochemist, Søren Peter Lauritz Sørensen. The 0–14 pH range he developed is a logarithmic scale designed to simplify acid and base calculations that otherwise would be expressed as molar values. Sørensen's formula for the calculation of pH is expressed as follows:

$$pH = -\log [H^+]$$

For example, an aqueous solution containing 10^{-6} moles of disassociated hydrogen ions per liter would be converted using Sørensen's formula as follows:

$$pH = -\log [H^+]$$
$$pH = -\log 10^{-6}M\ H^+$$
$$pH = 6$$

Pure water contains 10^{-7} moles of hydrogen ions per liter and has a pH of 7. As hydrogen ions increase, the solution becomes more acidic and the pH decreases (see Table 2-1).

Bacteria live in habitats throughout the pH spectrum; however, the range of most individual species is small. Like temperature and salinity, pH tolerance is used as a means of classification. The three major classifications are

1. **acidophiles:** organisms adapted to grow well in environments below about pH 5.5,
2. **neutrophiles:** organisms that prefer pH levels between 5.5 and 8.5, and
3. **alkaliphiles:** organisms that live above pH 8.5.

Under normal circumstances, bacteria maintain a near-neutral internal environment regardless of their habitat; pH changes outside an organism's range may destroy necessary membrane potential (in the production of ATP) and damage vital enzymes beyond repair. This **denaturing** of cellular enzymes may be as minor as

TABLE **2-1** pH SCALE

Solution classification Acidity/ Alkalinity	pH	H+ Concentration In Moles/Liter	Common Examples	Organismal classification
	0	10^0	Nitric acid	
	1	10^{-1}	Stomach acid	
	2	10^{-2}	Lemon juice	
	3	10^{-3}	Vinegar, cola	
	4	10^{-4}	Tomatoes, orange juice	
	5	10^{-5}	Black coffee	Acidophiles
Acidic	6	10^{-6}	Urine	
Neutral	7	10^{-7}	Pure water	Neutrophiles
Alkaline	8	10^{-8}	Seawater	
	9	10^{-9}	Baking soda	Alkaliphiles
	10	10^{-10}	Soap, milk of magnesia	
	11	10^{-11}	Ammonia	
	12	10^{-12}	Lime water [Ca(OH)$_2$]	
	13	10^{-13}	Household bleach	
	14	10^{-14}	Drain cleaner	

conformational changes in the proteins' tertiary structure, but usually is lethal to the cell.

Acids from carbohydrate fermentation and alkaline products from protein metabolism are sufficient to disrupt microbial enzyme integrity when grown *in vitro*. This is why buffers made from weak acids such as hydrogen phosphate are added to bacteriological growth media. In solution, buffers are able to alternate between weak acid ($H_2PO_4^-$) and conjugate base (HPO_4^{2-}) to maintain H^+/OH^- equilibrium.

$$H^+ + HPO_4^{2-} \longrightarrow H_2PO_4^-$$

$$OH^- + H_2PO_4^- \longrightarrow HPO_4^{2-} + H_2O$$

✦ Application

This is a qualitative procedure used to estimate the minimum, maximum, and optimum pH for growth of a bacterial species.

✦ This Exercise

Today you will cultivate and observe the effects of pH on four organisms. Then you will classify them based on your results.

✦ Materials

Per Student Group

✦ five of each pH adjusted Nutrient Broth as follows: pH 2, pH 4, pH 6, pH 8, and pH 10

✦ four sterile transfer pipettes

✦ fresh Nutrient Broth cultures of:
  ♦ *Lactobacillus plantarum*
  ♦ *Lactococcus lactis*
  ♦ *Enterococcus faecalis* (BSL-2)
  ♦ *Alcaligenes faecalis* (BSL-2)

✦ Medium Recipe

pH-Adjusted Nutrient Broth

♦ Beef extract	3.0 g
♦ Peptone	5.0 g
♦ Distilled or deionized water	1.0 L
♦ NaOH or HCl as needed to adjust pH	

Procedure

Lab One

1. Obtain five tubes of each pH broth—one of each pH per organism (20 tubes total). Label them accordingly.

2. Mix each culture thoroughly before making the following transfers. Using a sterile pipette, transfer a *single drop* of each broth culture to its appropriate Nutrient Broth tube. (**Note:** Because you will be comparing the amount of growth in the Nutrient Broth tubes, you must be sure to begin by transferring the same volume of culture to each tube. Use the same pipette for all transfers with a single organism.)

3. Incubate all tubes at $35 \pm 2°C$ for 48 hours.

Lab Two

1. Clean the outside of all tubes with a tissue and place them in a test tube rack organized into groups by organism.

2. Shake each broth gently until uniform turbidity is achieved.

3. Compare all tubes in a group to each other. Rate each one as 0, 1, 2, or 3 according to its turbidity (0 is clear and 3 is highly turbid). Enter your observations in the chart on the Data Sheet. (**Note:** Some color variability may exist between the different pH broths; therefore, base your conclusions solely on turbidity, not color.)

4. Determine the range and classification of each test organism. Record the information on the Data Sheet.

5. On the graph paper provided on the Data Sheet, plot the data (numeric values versus pH) of the four organisms.

References

Forbes, Betty A., Daniel F. Sahm, and Alice S. Weissfeld. 2002. Chapter 10 in *Bailey & Scott's Diagnostic Microbiology*, 11th ed. Mosby, St. Louis.

Holt, John G., Ed. 1994. *Bergey's Manual of Determinative Bacteriology*, 9th ed. Williams and Wilkins, Baltimore.

Varnam, Alan H., and Malcolm G. Evans. 2000. *Environmental Microbiology*. ASM Press, Washington, DC.

Winn, Washington C., *et al.* 2006. *Koneman's, Color Atlas and Textbook of Diagnostic Microbiology*, 6th ed. Lippincott Williams & Wilkins, Baltimore.

EXERCISE
2-11
The Effect of Osmotic Pressure on Microbial Growth

✦ Theory

Water is essential to all forms of life. It is not only the principal component of cellular cytoplasm, but also an essential source of electrons and hydrogen ions. Prokaryotes, like plants, require water to maintain cellular **turgor pressure**. Whereas eukaryotic cells burst with a constant influx of water, prokaryotes require water to prevent shrinking of the cell membrane, resulting in separation from the cell wall—an occurrence known as **plasmolysis**.

Many bacteria regulate turgor pressure by transporting in and maintaining a relatively high cytoplasmic potassium or sodium ion concentration, thereby creating a concentration gradient that promotes inward **diffusion** of water. For bacteria living in saline habitats, the job of maintaining turgor pressure is continuous because of the constant efflux of water.

Irrespective of a cell's efforts to control its internal environment, natural forces will cause water to move through its semipermeable membrane from an area of low **solute** concentration to an area of high solute concentration. In a solution in which solute concentration is low, water concentration is high, and *vice versa*. Therefore, water moves through a cell membrane from where its concentration is high to where its concentration is low. This process is called **osmosis**, and the force that controls it is called **osmotic pressure**.

Osmotic pressure is a quantifiable term and refers, specifically, to the ability of a solution to *pull water*

toward itself through a semipermeable membrane. If a bacterial cell is placed into a solution that is **hyposmotic** (a solution having low osmotic pressure), there will be a *net* movement of water into the cell. If an organism is placed into a **hyperosmotic** solution (a solution having high osmotic pressure), there will be a net movement of water out of the cell. For a bacterial cell in an **isosmotic** solution (a solution having osmotic pressure equal to that of the cell), water will tend to move in both directions equally; that is, there is no net movement (Figure 2-41).

Bacteria constitute a diverse group of organisms and, as such, have evolved many adaptations for survival. Microorganisms tend to have a distinct range of salinities that are optimal for growth, with little or no survival outside that range. For example, some bacteria called **halophiles** grow optimally in NaCl concentrations of 3% or higher. **Extreme halophiles** are organisms with specialized cell membranes and enzymes that require salt concentrations from 15% up to about 25% and will not survive where salinity is lower (Figure 2-42). Except for a few **osmotolerant** bacteria, which will grow over a wide range of salinities, most bacteria live where NaCl concentrations are less than 3%.

✦ Application

This is a qualitative procedure used to demonstrate bacterial tolerances to NaCl.

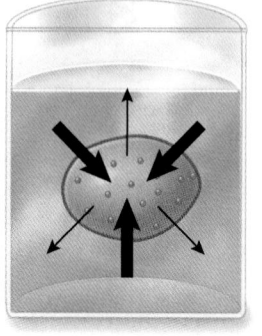

Hypotonic

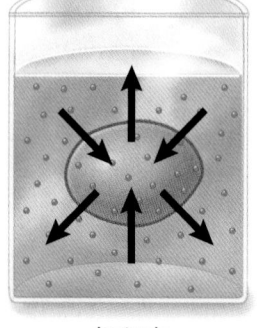

Isotonic

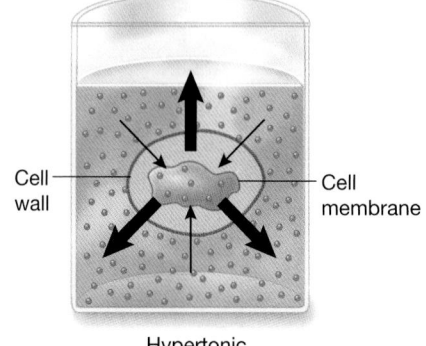

Hypertonic

2-41 **THE EFFECT OF OSMOTIC PRESSURE ON BACTERIAL CELLS** ✦ This osmosis diagram illustrates the movement of water into and out of cells. The labels refer to the osmotic pressure of the solution outside the cell. In a hyposmotic environment, the cell has greater osmotic pressure, so the net movement of water (arrows) will be into the cell. In an isosmotic environment, there is no net movement because the osmotic pressure of the cell and that of the environment are equal. (Actually, the water is moving equally in both directions.) In a hyperosmotic environment, the osmotic pressure of the environment is greater, so the net movement is outward and results in plasmolysis. Note the shrinking membrane (CM) and the rigid cell wall (CW) in the hyperosmotic solution.

2-42 A SALTERN IN SAN DIEGO BAY ✦ Salterns are low pools of saltwater used in the harvesting of salt. As water evaporates, the saltwater becomes saltier and saltier, until only salt remains. This can then be purified and sold. The colors in the pools result from different communities of halophilic microorganisms that are associated with different salinities as the pools dry out.

✦ In This Exercise

You will be growing three microorganisms at a variety of NaCl concentrations to determine the tolerance range of salinities and optimum salinity for each organism.

✦ Materials

Per Student Group

✦ four tubes each of saline medium prepared with 0%, 5%, 10%, 15%, 20%, and 25% NaCl

✦ three sterile transfer pipettes

✦ fresh cultures of:
  ✦ *Escherichia coli*
  ✦ *Halobacterium*
  ✦ *Staphylococcus aureus* (BSL-2)

✦ Medium Recipe

Modified Halobacterium Broth

✦ Sodium chloride	0 g, 50 g, 100 g, 150 g, 200 g, or 250 g
✦ Magnesium sulfate, heptahydrate	20.0 g
✦ Trisodium citrate, dihydrate	3.0 g
✦ Potassium chloride	2.0 g
✦ Casamino acids	5.0 g
✦ Yeast extract	5.0 g
✦ Deionized water	1.0 L

Adjust pH to 7.2 using 5 M or concentrated HCl.

Procedure

Lab One

1 Obtain four tubes of each *Halobacterium* Broth—one of each concentration per organism plus a control. Label them accordingly.

2 Mix each culture thoroughly before making the following transfers. Using a sterile pipette, transfer a *single drop* of each broth culture to its appropriate *Halobacterium* Broth tube. (Because you will be comparing the amount of growth in the *Halobacterium* Broth tubes, take care to begin by transferring the same volume of culture to each tube. Use the same pipette for all transfers with a single organism.)

3 Incubate all tubes at $35 \pm 2°C$ for 48 hours.

Lab Two

1 Clean the outside of all tubes with a tissue, and place them in a test tube rack organized into groups by organism.

2 Shake each broth gently until uniform turbidity is achieved.

3 Compare all tubes in a group to each other and to the control. Rate each one as 0, 1, 2, or 3 according to its turbidity (0 is clear and 3 is very turbid.) (**Note:** The different broths may show some color variability. Check the uninoculated controls and, thus, base your conclusions solely on turbidity, not color.

4 Enter your results in the chart provided on the Data Sheet.

5 On the graph paper provided with the Data Sheet, plot the numeric value of turbidity versus the salt concentration for all three organisms.

References

Forbes, Betty A., Daniel F. Sahm, and Alice S. Weissfeld. 2002. Chapter 10 in *Bailey & Scott's Diagnostic Microbiology*, 11th ed. Mosby, Inc., St. Louis.

Hauser, Juliana T. 2006. *Techniques for Studying Bacteria and Fungi.* Carolina Biological Supply Company, 2700 York Road, Burlington, NC.

Holt, John G., Ed. 1994. *Bergey's Manual of Determinative Bacteriology*, 9th ed. Williams and Wilkins, Baltimore.

Koneman, Elmer W., Stephen D. Allen, William M. Janda, Paul C. Schreckenberger, and Washington C. Winn, Jr. 1997. *Color Atlas and Textbook of Diagnostic Microbiology*, 5th ed. J. B. Lippincott Co., Philadelphia.

Moat, Albert G., John W. Foster, and Michael P. Spector. 2002. Pages 582–587 in *Microbial Physiology*, 4th ed. Wiley-Liss, New York.

Varnam, Alan H., and Malcolm G. Evans. 2000. *Environmental Microbiology.* ASM Press, Washington, DC.

White, David. 2000. Pages 388–394 in *The Physiology and Biochemistry of Prokaryotes*, 2nd ed. Oxford University Press, New York.

Control of Pathogens: Physical and Chemical Methods

Every patient in a hospital or other clinical setting has the right to expect that he or she will not contract a disease or infection while in that institution's care. Every person donating blood at a blood bank or mobile center has the right to expect that all materials and surfaces they come in contact with will be free of pathogens. Workers in health clinics, hospitals, medical laboratories, and public health laboratories have the right to assume that reasonable precautions have been and are being taken to protect their safety while in the workplace.

These are only a few of the many reasons why the importance of understanding and use of pathogen control systems cannot be overemphasized. Fortunately, with relatively few exceptions, the above-described conditions exist in this and other developed countries largely because of the dedication of thousands of employees and the oversight of dozens of international, governmental, and private organizations such as the World Health Organization (WHO), Centers For Disease Control and Prevention (CDC), Food and Drug Administration (FDA), Environmental Protection Agency (EPA), American Public Health Association (APHA), and Association of Official Analytical Chemists (AOAC). These and many other federal and private organizations are responsible for the proper testing, registration, and classification of the substances or systems used to prevent the spread of pathogens.

These substances or systems, both chemical and physical, are referred to broadly as germicides. Some germicides are specific in nature and typically include the name of the target pathogen, such as "tuberculocide," "virucide," or "sporocide." Most germicides are broad-spectrum and, thus, target a wide variety of pathogens. Although some overlap occurs, germicidal systems fall into three categories: decontamination, disinfection, or sterilization.

1. **Decontamination** is the lowest level of control and is defined as "reduction of pathogenic microorganisms to a level at which items are safe to handle without protective attire." Decontamination usually includes physical cleaning with soaps or detergents, and removal of all (ideally) or most organic and inorganic material. Proper cleaning of all instruments and surfaces is considered the critical first step toward disinfection or sterilization because, to be fully effective, a disinfectant or sterilant must come in direct contact with all pathogens present. Materials left to dry on a surface or apparatus can actually shield pathogens from a disinfecting or sterilizing agent or otherwise neutralize it.

2. **Disinfection** is the next level of control and is divided into three sublevels—low, medium, and high—based on effectiveness against specific control pathogens or their surrogates. All sublevels kill large numbers, if not all, of the targeted pathogens but typically do not kill large numbers of spores. Some high-level disinfectants are called chemical sterilants because they have the ability to kill all vegetative cells and spores.

    Disinfectants typically are liquid chemical agents but can also be solid or gaseous. Other disinfection methods include dry heat, moist heat, and ultraviolet light. Disinfectants that are designed to reduce or eliminate pathogens on or in living tissue are called antiseptics. For obvious safety reasons, antiseptics are subject to additional testing to minimize the risks of side-effects. Some antiseptics are considered drugs and, therefore, are regulated by the FDA.

3. **Sterilization** is the complete elimination of viable organisms including spores and, as such, is the highest level of pathogen control. Sterilization can be achieved by some chemicals, some gases, incineration, dry heat, moist heat, ethylene oxide gas, ionizing radiation (Gamma, X-ray, and electron-beam), low-temperature plasma (utilizing a combination of chemical sterilants and ultraviolet radiation in a vacuum chamber), or low-temperature ozone (utilizing bottled oxygen, water, and electricity in a chamber to produce a lethal level of ozone).

In this unit you will examine both physical and chemical means of pathogen control. The following exercises illustrate the germicidal effects of UV radiation, disinfection, antisepsis, and steam (moist heat) sterilization. (For more information on microbial control, refer to Exercise 1-1 GloGerm, and to Exercise 7-3 Antimicrobial Susceptibility Test.) ✦

EXERCISE 2-12

Steam Sterilization[1]

✦ Theory

Of the many methods or agents that have been developed for sterilizing surgical and dental instruments, microbiological media, infectious waste, and other materials not harmed by moisture or heat, steam is still the most effective and most common. The device used most commonly for this purpose is called a steam sterilizer, or **autoclave**. Autoclaves are relatively safe, easy to operate, and, if used properly, effective at killing all microbial vegetative cells and spores.

Under atmospheric pressure, water boils at 100°C (212°F). At pressures above atmospheric pressure, water must be heated above 100°C before it will boil. Much like home pressure cookers, which create pressure and high temperatures to shorten cooking times, autoclaves use super-heated steam under pressure to kill heat-resistant organisms. Examples of heat-resistant organisms include members of the spore-producing genera—*Bacillus*, *Geobacillus*, and *Clostridium*.

In the microbiology laboratory, sterilizing temperature usually is set at between 121° and 127°C (250° and 260°F); however, sterilizing time can vary according to the size and consistency of the material being sterilized.[2] At a minimum, to be sure that all vegetative cells and spores have been killed, items being processed must reach optimum temperature for at least 15 minutes. This includes items deep inside the autoclave container that may be partially insulated from the steam by surrounding items. Understandably, larger loads take longer to process than smaller loads do. (Certain sensitive applications, such as microbiological media preparation, in which formula integrity must be maintained and when specific growth inhibiting ingredients are included, lower times and temperatures are acceptable.)

To maintain laboratory safety and comply with laws regarding infectious waste disposal, sterilizers must be checked regularly for operating effectiveness. Special thermometers placed in an autoclave can record the maximum temperature reached inside the chamber but do not measure how low the temperature dips during the normal cycling of the heating elements. Specialized color-coded autoclave tape can be a fairly good indicator that sterilization is complete, but the only way, with certainty, to determine that sterilization has been achieved is by using a device called a **biological indicator** (Figure 2-43).

Biological indicators, as the name suggests, are test systems that contain something living. A typical biological indicator that is particularly useful for testing autoclaves is one that contains **bacterial spores**. Bacterial spores, the dormant form of an organism, are highly resistant to

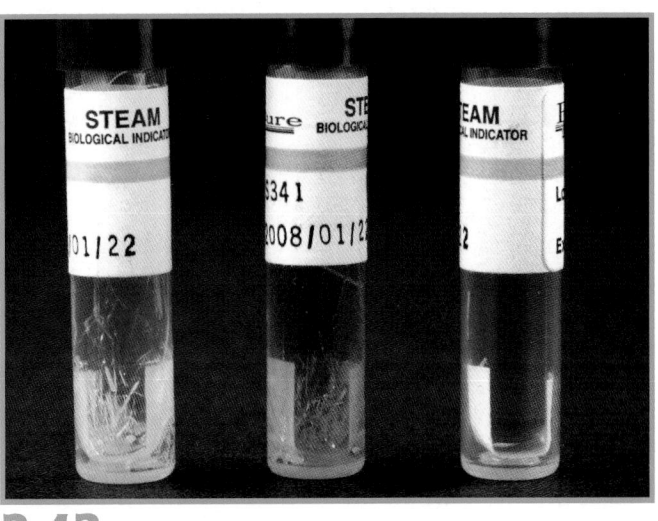

2-43 AUTOCLAVE BIOLOGICAL INDICATORS ✦ These indicator vials contain an ampule of fermentation broth and a filter paper strip containing spores of *Geobacillus stearothermophilus*—a spore-forming organism capable of withstanding high temperatures. The vial in the center was autoclaved for 15 minutes at 121°C, cooled, pinched to crush the inner glass ampule, and incubated. The purple color (compared to the uncrushed negative control on the right) indicates that an acidic condition from fermentation does not exist. This suggests that the organism has been killed by the autoclaving. Note the gray-colored band on the label. This chemical indictor changes from blue to gray upon autoclaving. The ampule in the negative control was not crushed, so the spores in the filter paper never made contact with the broth and, thus, provides a color example of unfermented broth. The ampule on the left is a positive control to verify the viability of the organism used in the system. It was not autoclaved (as evidenced by the blue band on the label) but was crushed and incubated. The development of yellow color indicates that the organism in the system is viable and that the lack of yellow color in the center vial is a result of autoclaving.

---

[1] Steam sterilization is not yet considered reliable at inactivating prion proteins, such as those that cause Bovine Spongiform Encephalopathy, the so-called Mad Cow Disease, and its variant, Creutzfeldt-Jakob disease.

[2] In clinics, hospitals, or other locations where surgical instruments are being processed, the World Health Organization (WHO) recommends a minimum processing time and temperature of 134°C for 18 minutes.

both chemical and physical means of control. Therefore, if an autoclave kills the spores in the test system, it is safe to assume that it will destroy other microbes as well. This is important not only for safety reasons but is of legal importance as well. Public health and safety agencies maintain compliance with hazardous waste disposal regulations by requiring regular testing of autoclaves used to process biohazardous material.

A typical system, and the one selected for today's lab, includes a small heat-resistant plastic vial containing a glass ampule of sterile fermentation broth. Also inside the vial, but outside of the ampule is a strip of filter paper containing bacterial spores. The vial is placed inside the autoclave and heated at 121°C for 15 minutes. After autoclaving, the vial is cooled and crushed with a special device that breaks the inner ampule without damaging the plastic vial. Breaking the ampule allows the fermentation broth to come in contact with the bacterial spores in the filter paper. The vials are then incubated at 55°C for 48 hours. If the spores have been killed in the autoclave, incubation will produce no growth. If they have not been killed, they will germinate and ferment the substrate in the broth. A pH-indicating dye, included in the broth, will reveal any acid produced (during fermentation) with a distinctive color change. No color change during incubation is, thus, an indication that sterilization is complete, the spores have been killed, and the autoclave is operating properly. (For more information on fermentation, refer to Section 5 and Appendix A.)

✦ Application

Biological indicators are available in many forms and commonly used to test the efficiency of steam sterilizers.

✦ In This Exercise

You are going to use bacterial spores and their resistance to steam sterilization to test the effectiveness of your lab's autoclave. This procedure is written for a product called BTSure Biological Indicator[3], but can be applied to other brands as well. BTSure Biological Indicators include a pH indicator that turns the broth from purple to yellow under the acidic conditions produced by fermentation.

✦ Materials

Per Class

✦ one (or more) steam autoclave(s)

✦ incubator set at 55°C

Per Student Group

✦ four BTSure Biological Indicators

✦ one autoclave pan

 Procedure

1 Obtain four BTSure Biological Indicator vials and label them #1, #2, #3, and #4 with a Sharpie® or other permanent marker. Do not label the vial with tape or paper, as this will insulate it from the steam.

2 Place vial #1 on its side uncovered in an autoclave pan.

3 Place vial #2 inside a container (or multiple containers), as well insulated as can be achieved with materials provided by your instructor. We recommend placing the vial inside a screw-capped test tube inside other tubes—two, three, or four layers deep. Place this vial in the autoclave pan with vial #1.

4 Do nothing with vials #3 and #4 as yet.

5 Place the autoclave pan in the autoclave.

6 Follow your instructor's guidelines to add water to the chamber, set the temperature at 121°C (250°F), and set the timer at 15 minutes. When instructed to do so, close and start the autoclave.

7 When autoclaving is complete, all of the steam has been vented, and the machine has been allowed to cool slightly, remove your pan (while protecting your hands with appropriate gloves) and allow the contents to cool to room temperature.

8 Keeping the vials in an upright position, use the crushing device to squeeze vials #1, #2, and #3 (one at a time) until you hear the glass ampule inside the vial break.

9 Place these vials along with vial #4, again in an upright position, into the 55°C incubator for 48 hours.

10 After incubation, examine the vials for color changes.

11 Using Table 2-2 as a guide, record your results in the table provided on the Data Sheet.

References

McDonnell, Gerald E. 2007. *Antisepsis, Disinfection, and Sterilization: Types, Action, And Resistance.* ASM Press, American Society for Microbiology, Washington, DC.

Widmer, Andreas F., and Reno Frei. 2007. Chapter 7 in *Manual of Clinical Microbiology*, 9th ed. Edited by Patrick R. Murray, Ellen Jo Baron, James H. Jorgensen, Marie Louise Landry, and Michael A. Pfaller. ASM Press, American Society for Microbiology, Washington, DC.

---

[3] BTSure Biological Indicators are available from Barnstead/Thermolyne, Inc., 2555 Kerper Blvd., Dubuque, IA 52004-0797, 800/553-0039.

TABLE **2-2** AUTOCLAVE BIOLOGICAL INDICATOR TEST RESULTS AND INTERPRETATIONS ✦ (Assume the ampules inside the vials have been crushed.)

Broth Color	Interpretation
Purple	No fermentation/acid production in the medium. The organism is dead.
Yellow	Fermentation/acid production in the medium. The organism is alive.

EXERCISE 2-13

The Lethal Effect of Ultraviolet Radiation on Microbial Growth

✦ Theory

Ultraviolet radiation (UV light) is a type of **electromagnetic energy**. Like all electromagnetic energy, UV travels in waves and is distinguishable from all others by its **wavelength**. Wavelength is the distance between adjacent wave crests and is typically measured in nanometers (nm) (Figure 2-44).

Ultraviolet light is divided into three groups categorized by wavelength:

UV-A, the longest wavelengths, ranging from 315 to 400 nm

UV-B, wavelengths between 280 and 315 nm

UV-C, wavelengths ranging from 100 to 280 nm (These wavelengths are most detrimental to bacteria. Bacterial exposure to UV-C for more than a few minutes usually results in irreparable DNA damage and death of the organism. For a discussion on the mutagenic effects of UV and DNA repair, refer to Exercise 10-5.)

✦ Application

Ultraviolet light is commonly used to disinfect laboratory work surfaces.

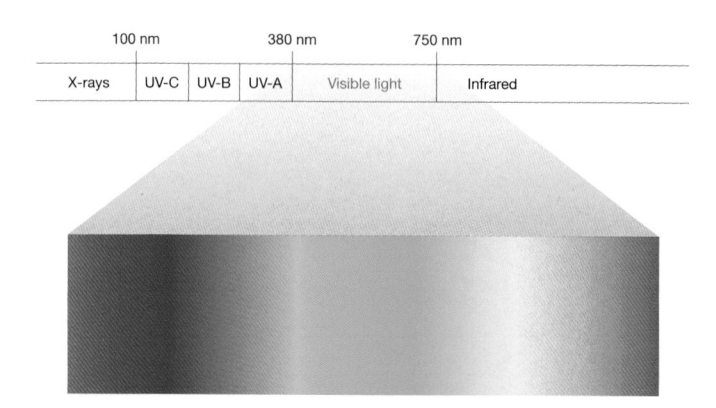

2-44 ELECTROMAGNETIC SPECTRUM ✦ The shortest and highest energy wavelengths are those of gamma rays, starting at about at 10^{-5} nm. Radio waves, at the other end of the spectrum, can be one kilometer or longer. Between about 100 nm and 380 nm (just above visible light) is the sliver known as ultraviolet light.

✦ In This Exercise

Today you will compare the effect of UV exposure on two cultures—*Bacillus subtilis* and *Escherichia coli*. Because of the large number of plates to be treated, the work will be divided among six groups of students. Refer to Table 2-3 for assignments.

✦ Materials

Per Student Group

✦ ultraviolet lamp with appropriate shielding

✦ cardboard to cover plates

✦ two Tryptic Soy Agar (TSA) plates (four for group one)

✦ stopwatch or electronic timer

✦ sterile cotton swabs (two per group; four for group 1)

✦ 48-hour Nutrient Broth cultures of:
  ◆ *Bacillus subtilis*
  ◆ *Escherichia coli*

✎ Procedure

Lab One

1. Enter your group number and exposure time (from Table 2-3) on the Data Sheet.

2. Obtain two TSA plates and label the bottom of each with the name of the organism to be inoculated and your group number. Draw a line to divide the plates in half, and label the sides "A" and "B."

3. Dip a sterile cotton swab into the broth of one culture and wipe the excess on the inside of the tube. Inoculate the appropriate plate by spreading the organism over the entire surface of the agar. Do this by streaking the plate surface completely three times, rotating it one-third turn between streaks. When incubated, this will form a bacterial lawn.

4. Repeat step 3 with the other organism and plate.

5. Place a paper towel on the table next to the UV lamp and soak it with disinfectant.

6. Place your plates under the UV lamp and set the covers, open side down, on the disinfectant-soaked

TABLE **2-3** Group Assignments by Number

Organism	No UV	5 minutes	10 minutes	15 minutes	20 minutes	25 minutes	30 minutes
B. subtilis	1	1	2	3	4	5	6
E. coli	1	1	2	3	4	5	6

towel. Cover the B half of the plates with the cardboard as shown in Figure 2-45.

7 Turn on the lamp for the prescribed time. *Caution: Be sure the protective shield is in place and do not look at the light while it is on!*

8 Immediately replace the covers, invert the plates and incubate them at $35 \pm 2°C$ for 24 to 48 hours.

Lab Two

1 Remove your plates from the incubator and observe for growth. Side B should be covered with a bacterial lawn. If this is not the case, see your instructor.

2 Record the growth on side A of each plate in the table on the Data Sheet. Enter "0" if you observe no growth, "1" for poor growth, "2" for moderate growth, and "3" for abundant growth.

3 Using the results from other groups, complete the class data chart on the Data Sheet.

4 On the graph paper provided with the Data Sheet, construct a graph representing growth versus UV exposure time for all three organisms.

2-45 **PLATES SHIELDED FOR UV EXPOSURE** ✦ Place the two plates under the UV lamp with the covers removed and the cardboard shield covering half of each plate as shown. Make sure the Petri dish covers are placed open side down on a disinfectant-soaked towel.

References

Lewin, Benjamin. 1990. *Genes IV*. Oxford University Press, Cambridge, MA.

Varnam, Alan H., and Malcolm G. Evans. 2000. *Environmental Microbiology*. ASM Press, Washington, DC.

EXERCISE
2-14

Chemical Germicides: Disinfectants and Antiseptics

✦ Theory

Chemical germicides are substances designed to reduce the number of pathogens on a surface, in a liquid, or on or in living tissue. Germicides designed for use on surfaces (floors, tables, sinks, countertops, surgical instruments, *etc.*) or liquids are called **disinfectants**. Germicides designed for use on or in living tissue are called **antiseptics**.

Before a new substance can be registered by either the FDA or EPA and allowed on the market, it must be tested and classified according to its effectiveness against pathogens. The Use-Dilution Test, published by the Association of Official Analytical Chemists (AOAC), is one of many commonly used tests for this purpose.

The Use-Dilution Test is a standard procedure used to measure the effectiveness of disinfectants specifically against *Staphylococcus aureus*, *Salmonella enterica* serovar Cholerasuis, and *Pseudomonas aeruginosa*. In the standard procedure, glass beads or stainless steel cylinders coated with living bacteria are exposed to varying concentrations (dilutions) of test germicides then transferred to a growth medium. After a period of incubation, the medium is examined for growth. If a solution is sufficient to prevent microbial growth at least 95% of the time, it meets the required standards and is considered a usable dilution of that germicide for a specific application. Today's exercise is an adaptation of this method.

✦ Application

This procedure is used to test the effectiveness of germicides against *Staphylococcus aureus* and *Pseudomonas aeruginosa*.

✦ In This Exercise

Today you will examine the effectiveness of four germicides—two common household disinfectants and two over the counter antiseptics. The disinfectants selected for the exercise are household bleach and Lysol® Brand II Disinfectant. The antiseptics are hydrogen peroxide and isopropyl alcohol. The organisms used for the test are *Staphylococcus aureus* and *Escherichia coli*.

You will first coat the beads with bacteria, expose them to three concentrations of your germicide, and then use them to inoculate sterile Nutrient Broth. If, during exposure to the germicide, all of the bacteria on the bead are killed the broth inoculated with that bead will remain clear. If any of the bacteria survive the germicide exposure, they will reproduce during incubation and make the broth turbid. With these results, you will determine the effective concentration (dilution) of your germicide.

The tasks for the exercise are divided among eight groups of students. Each group will be responsible for one organism and three dilutions of one germicide. Refer to Table 2-4 for your assignments.

Finally, this is an interesting exercise with moderate amount of work involved. If you do not hurry and are careful to use aseptic technique, you will be rewarded with reliable data at the end.

✦ Materials

Per Class (See Table 2-4)

Disinfectants

✦ 0.01%, 0.1%, and 1.0% household bleach

✦ 25%, 50%, and 100% Lysol® Brand II Disinfectant

Antiseptics

✦ 0.03%, 0.3%, and 3% hydrogen peroxide (3% is full strength as purchased at the pharmacy)

✦ 10%, 30%, and 50% isopropyl alcohol (70% is full strength as purchased at the pharmacy)

Per Student Group

✦ 100 mL flask of sterile deionized water

✦ three concentrations of one germicide (listed above)

✦ five sterile 60 mm Petri dishes

TABLE **2-4** Group Assignments

Germicide	*Staphylococcus aureus*	*Escherichia coli*
Bleach	Group 1	Group 2
Lysol	Group 3	Group 4
Hydrogen Peroxide	Group 5	Group 6
Isopropyl Alcohol	Group 7	Group 8

◆ one sterile glass 100 mm Petri dish containing filter or bibulous paper

◆ one container of sterile ceramic or glass beads[1]

◆ sterile transfer pipette

◆ seven sterile Nutrient Broth tubes

◆ needle-nose forceps (or appropriate device for aseptically picking up beads)

◆ small beaker with alcohol (for flaming forceps)

◆ fresh broth cultures of (one per group):
  ◆ *Staphylococcus aureus* (BSL-2)
  ◆ *Escherichia coli*

Procedure[2]

Timing is important in this procedure. Read through it and make a plan before you begin so your transfers and soaking times are done uniformly and are consistent with those of other groups.

Lab One

1 Enter the name of your organism here:

_____.

2 Enter the name of your germicide here:

_____.

3 Obtain all of the necessary items for your group as listed in Materials.

4 Place the materials properly on your workspace as shown in Figure 2-46. Label all seven broths with your group name or number. Also number three of the broths 1–3 also as shown in the diagram. Label the last four broths controls #1, #2, #3, and #4 as shown in the diagram.

5 Label three of the 60 mm plates with the name and concentration of your germicide. Label the other two plates "#4 sterile water" and "#5 sterile water."

6 Add approximately 15 mL of each germicide concentration to its respective plate. Add approximately 15 mL sterile water into each of the other two plates.

7 Mix your bacterial culture until uniform turbidity is achieved.

8 Aseptically transfer one loopful of culture broth to control #1.

9 Alcohol-flame your forceps and aseptically drop four beads into the broth culture. (*Note:* Store your forceps in the beaker of alcohol between transfers. When it is time to make a transfer, remove the forceps and burn off the alcohol. When finished, return the forceps to the beaker.)

10 After 1 minute, decant the broth into a beaker of disinfectant. Remove as much of the broth as possible without losing the beads in the disinfectant.

11 Dispense the beads onto the sterile filter paper. This can be done by tapping the mouth of the tube on the paper. If this doesn't work, remove the beads with a sterile inoculating loop.

12 Using sterilized forceps spread the beads apart on the paper and allow them to dry. Do not roll them around as this may remove bacteria.

13 After 10 minutes, place one bead in each of the three germicide plates. Mark the time here:

_____.

14 Place the fourth bead in plate #4 (sterile water).

15 Place a sterile bead in plate #5 (sterile water).

16 After 10 minutes from the time marked above, remove the beads from the solutions, in the same order as they were added, and place them in their respective Nutrient Broths. Mix the broths immediately to disperse any residual disinfectant on the beads.

17 Incubate all seven broths at $35 \pm 2°C$ for 48 hours.

Lab Two

1 Remove all broth tubes from the incubator. Gently mix the controls and examine them for evidence of growth. Enter your results on the Data Sheet. Control #1 should have produced growth (turbidity) and Control #2 should not have produced growth. If both of these conditions have been met, you may proceed. If not, see your instructor.

2 Using controls #1 and #2 as comparisons, examine broths containing beads exposed to germicide (1, 2, and 3). Using "G" to indicate growth and "NG" to indicate no growth, enter your results in both the individual data chart and the class data chart on the Data Sheet.

3 Again using controls #1 and #2 as comparisons, examine controls #3 and #4. Using "G" to indicate growth and "NG" to indicate no growth, enter your results on the Data Sheet.

4 Answer the questions on the Data Sheet.

---

[1] Sterilized #8 seed beads from a craft store will work for this purpose.

[2] This procdure has been modified from its original form and is to be used for instructional purposes only.

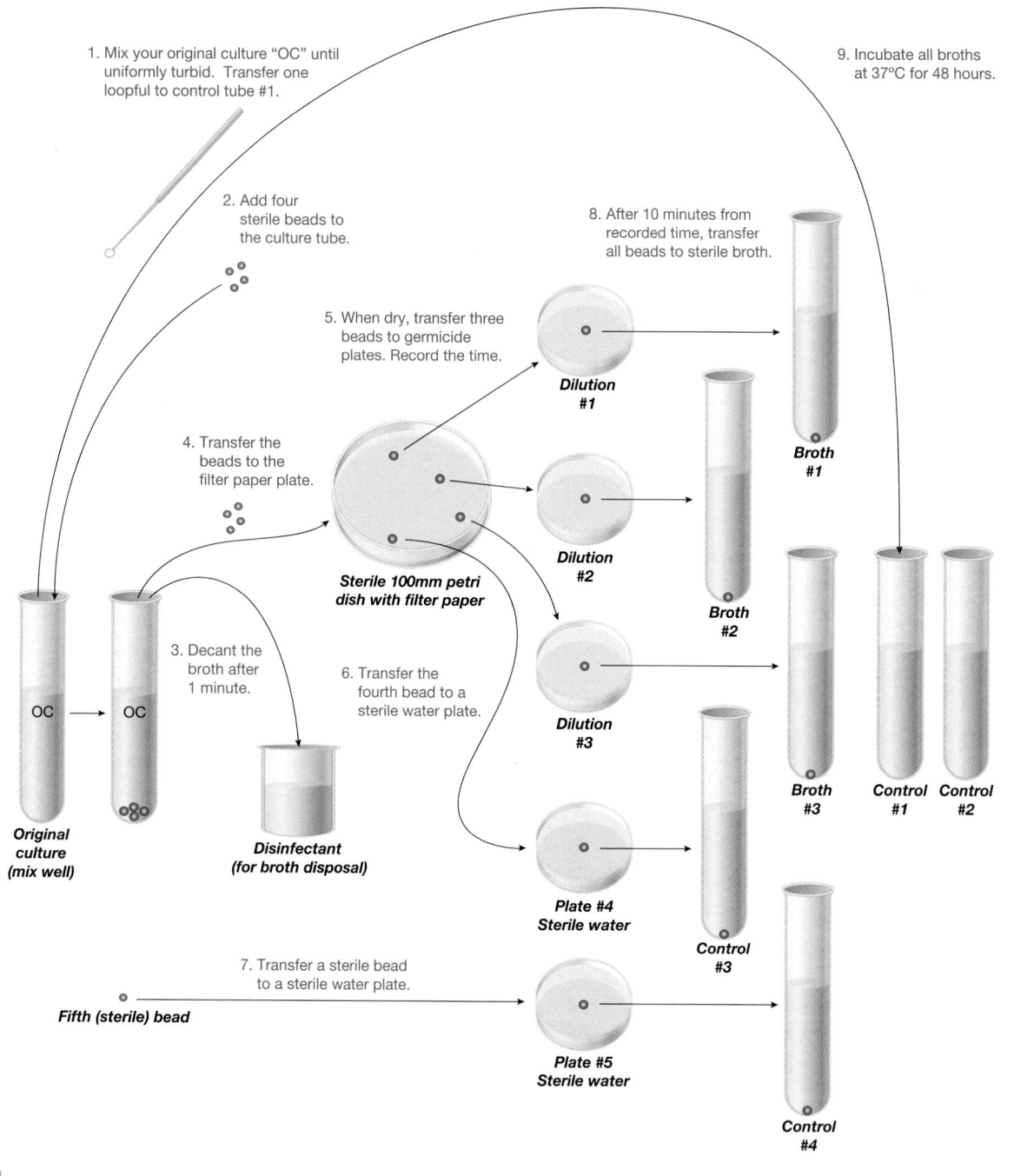

1. Mix your original culture "OC" until uniformly turbid. Transfer one loopful to control tube #1.

9. Incubate all broths at 37°C for 48 hours.

2. Add four sterile beads to the culture tube.

8. After 10 minutes from recorded time, transfer all beads to sterile broth.

5. When dry, transfer three beads to germicide plates. Record the time.

Dilution #1

4. Transfer the beads to the filter paper plate.

Dilution #2

Broth #1

Sterile 100mm petri dish with filter paper

Dilution #3

Broth #2

3. Decant the broth after 1 minute.

6. Transfer the fourth bead to a sterile water plate.

Broth #3   Control #1   Control #2

OC   OC

Original culture (mix well)

Disinfectant (for broth disposal)

Plate #4 Sterile water

Control #3

7. Transfer a sterile bead to a sterile water plate.

Fifth (sterile) bead

Plate #5 Sterile water

Control #4

2-46 PROCEDURAL DIAGRAM FOR CHEMICAL GERMICIDES ✦

References

McDonnell, Gerald E. 2007. *Antisepsis, Disinfection, and Sterilization: Types, Action, And Resistance.* ASM Press, American Society for Microbiology, Washington, DC.

Widmer, Andreas F. and Reno Frei. 2007. Chapter 7 in *Manual of Clinical Microbiology,* 9th ed. Edited by Patrick R. Murray, Ellen Jo Baron, James H. Jorgensen, Marie Louise Landry and Michael A. Pfaller. ASM Press, American Society for Microbiology, Washington, DC.

Microscopy and Staining

Microbiology as a biological discipline would not be what it is today without microscopes and cytological stains. Our ability to visualize, sometimes in great detail, the form and structure of microbes too small or transparent to be seen otherwise is attributable to developments in microscopy and staining techniques. In this section you will learn (or refine) your microscope skills. Then you will learn simple and more sophisticated bacterial staining techniques. ✦

Microscopy

The earliest microscopes used visible light to create images and were little more than magnifying glasses. Today, more sophisticated compound light microscopes (Figure 3-1) are used routinely in microbiology laboratories. The various types of light microscopy include bright-field, dark-field, fluorescence, and phase contrast microscopy (Figure 3-2). Although each method has specific applications and advantages, the one used most commonly in introductory classes and clinical laboratories is bright-field microscopy. Many research applications use electron microscopy because of its ability to produce higher quality images of greater magnification. ✦

EXERCISE

3-1 Introduction to the Light Microscope

✦ Theory

Bright-field microscopy produces an image made from light that is transmitted through a specimen (Figure 3-2A). The specimen restricts light transmission and appears "shadowy" against a bright background (where light enters the microscope unimpeded). Because most biological specimens are transparent, the contrast between the specimen and the background can be improved with the application of stains to the specimen (Exercises 3-5 through 3-11 and 3-13). The "price" of the improved contrast is that the staining process usually kills cells. This is especially true of bacterial-staining protocols.

Image formation begins with light coming from an internal or an external light source (Figure 3-3). It passes through the **condenser** lens, which concentrates the light and makes illumination of the specimen more uniform. **Refraction** (bending) of light as it passes through the **objective lens** from the specimen produces a magnified **real image**. This image is magnified again as it passes through the **ocular lens** to produce a **virtual image** that appears below or within the microscope. The amount of magnification that each lens produces is marked on the lens (Figure 3-4A and Figure 3-4B). Total magnification of the specimen can be calculated by using the following formula:

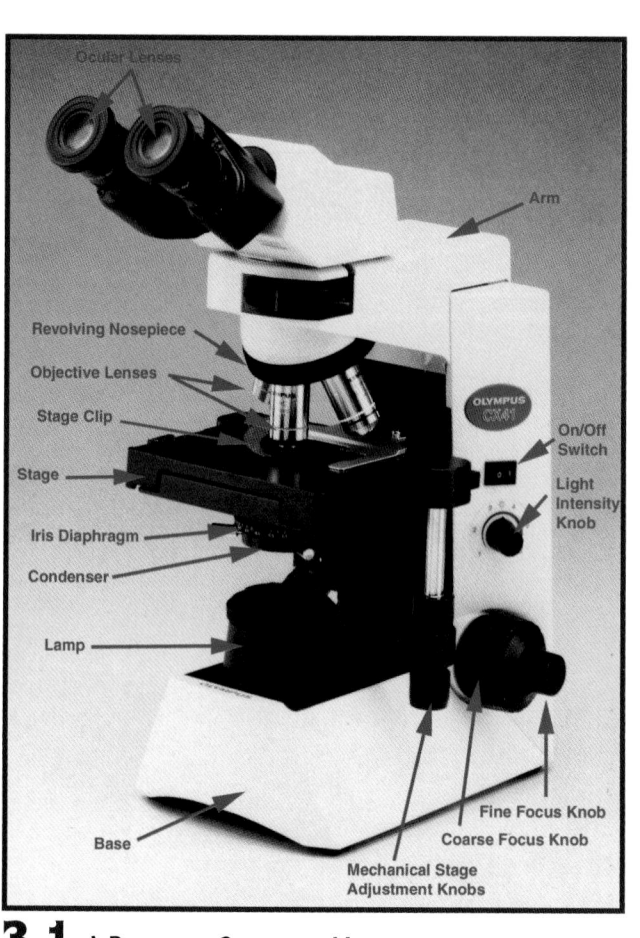

3-1 A BINOCULAR COMPOUND MICROSCOPE ✦ A quality microscope is an essential tool for microbiologists. Most are assembled with exchangeable component parts and can be customized to suit the user's specific needs.

Photograph courtesy of Olympus America Inc.

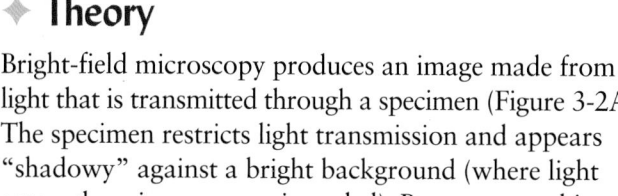

$$\begin{array}{c} \text{Total} \\ \text{Magnification} \end{array} = \begin{array}{c} \text{Magnification by the} \\ \text{Objective Lens} \end{array} \times \begin{array}{c} \text{Magnification by the} \\ \text{Ocular Lens} \end{array}$$

The practical limit to magnification with a light microscope is around 1300X. Although higher magnifications

3-2 TYPES OF LIGHT MICROSCOPY ✦

A This is a bright-field micrograph of an entire diatom (called a "whole mount"). Because of its thickness, the entire organism will not be in focus at once. Continually adjusting the fine focus to clearly observe different levels of the organism will give a sense of its three-dimensional structure. The bright rods around the diatom are bacteria. **B** This is a dark-field micrograph of the same diatom. Notice that dark field is especially good at providing contrast between the organism's edge and its interior and the background. Notice also that the bacteria are not visible, though this would not always be the case. **C** This is a phase contrast image of the same diatom. Different parts of the interior and its detail are visible than what is seen in the other two micrographs. Also, notice the bacteria are dark. **D** This is a fluorescence micrograph of *Mycobacterium kansasii*. The apple green is one of the characteristic colors of fluorescence microscopy.

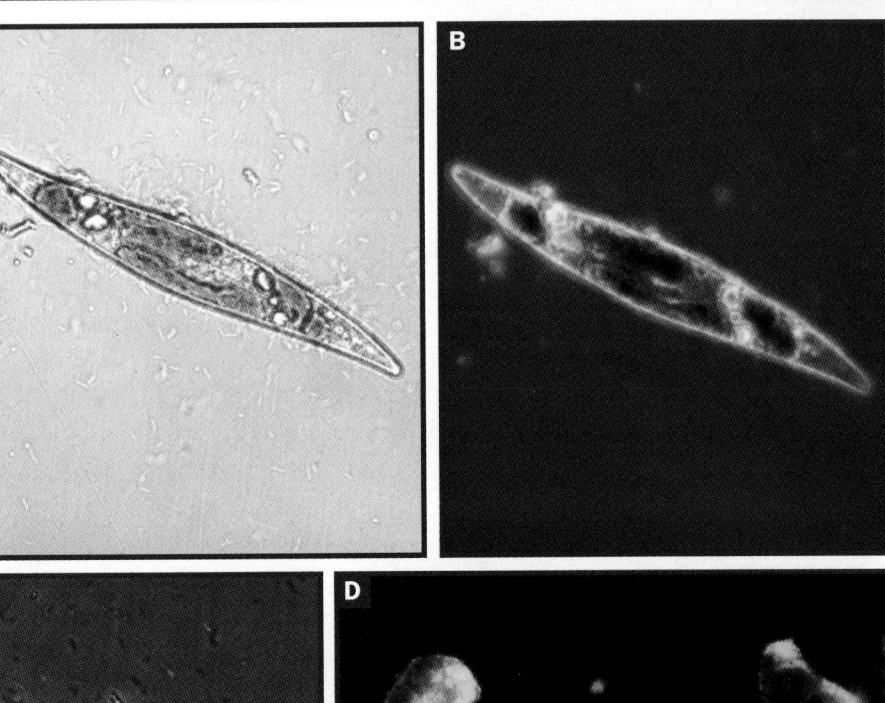

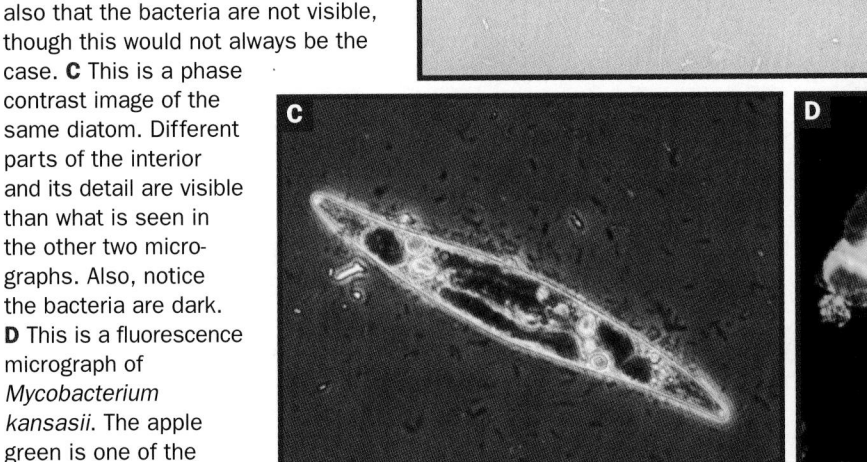

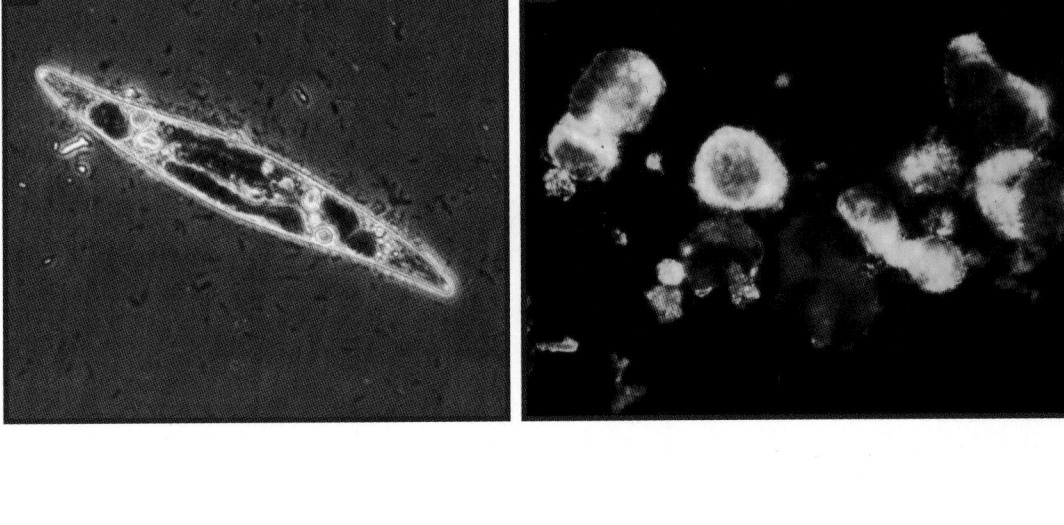

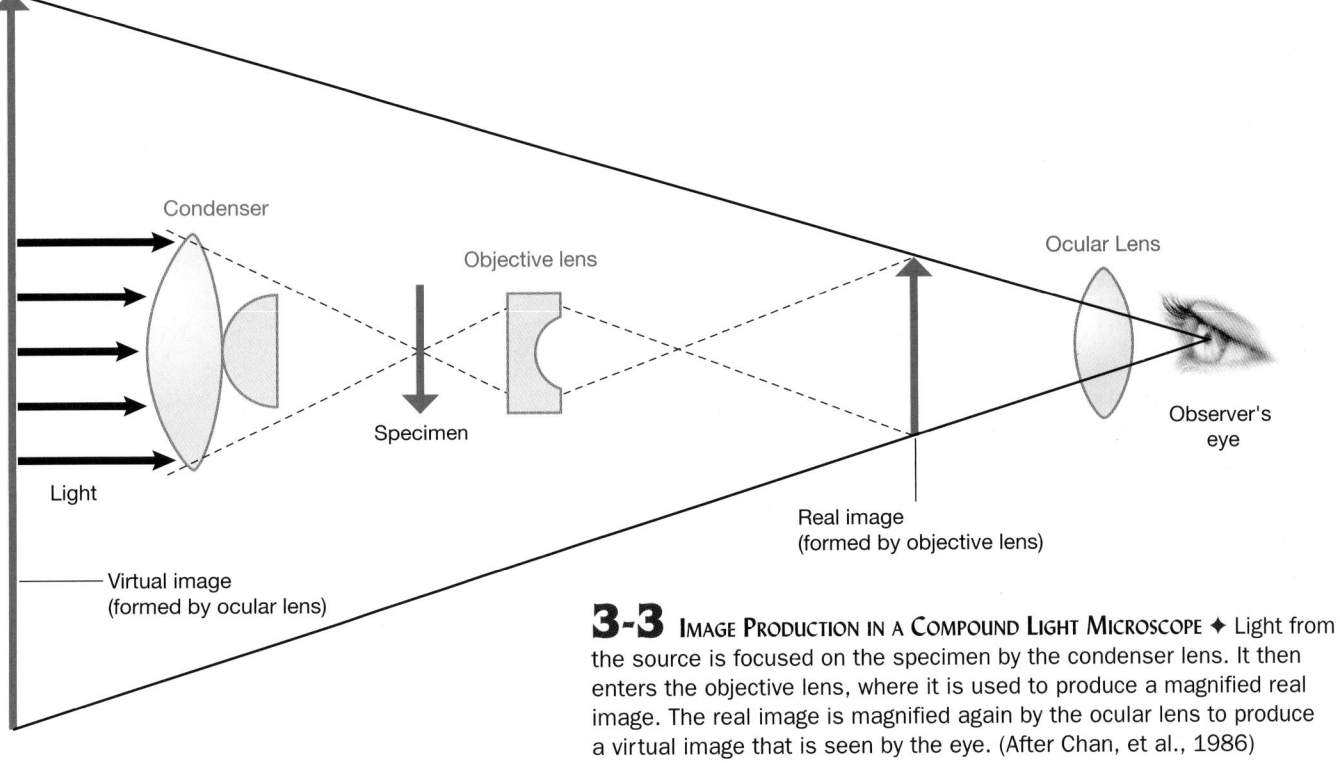

3-3 IMAGE PRODUCTION IN A COMPOUND LIGHT MICROSCOPE ✦

Light from the source is focused on the specimen by the condenser lens. It then enters the objective lens, where it is used to produce a magnified real image. The real image is magnified again by the ocular lens to produce a virtual image that is seen by the eye. (After Chan, et al., 1986)

Condenser

Objective lens

Ocular Lens

Light

Specimen

Real image
(formed by objective lens)

Observer's eye

Virtual image
(formed by ocular lens)

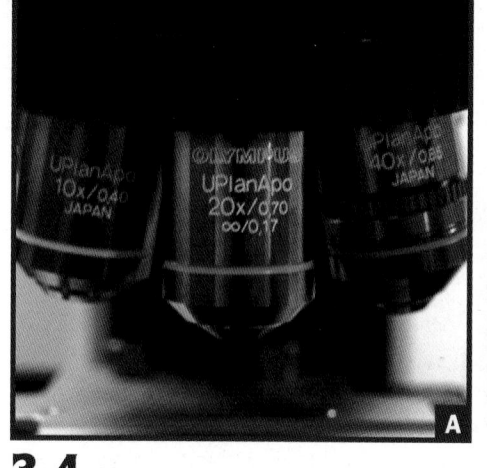  

3-4 MARKINGS OF MAGNIFICATION AND NUMERICAL APERTURE ON MICROSCOPE COMPONENTS ✦ **A** Three plan apochromatic objective lenses on the nosepiece of a light microscope. Plan means the lens produces a flat field of view. *Apochromatic* lenses are made in such a way that chromatic aberration is reduced to a minimum. From left to right, the lenses magnify 10X, 20X, and 40X, and have numerical apertures of 0.40, 0.70, and 0.85. The 20X lens has other markings on it. The mechanical tube length is the distance from the nosepiece to the ocular and is usually between 160 to 210 mm. However, this 20X lens has been corrected so the light rays are made parallel, effectively creating an infinitely long mechanical tube length (∞). This allows insertion of accessories into the light path without decreasing image quality. The thickness of cover glass to be used is also given (0.17 ± 0.01 mm). Also notice the standard colored rings for each objective: yellow for 10X, green for 20X (or 16X), and light blue for 40X (or 50X). **B** A 10X ocular lens. **C** A condenser (removed from the microscope) with a numerical aperture of 1.25. The lever at the right is used to open and close the iris diaphragm and adjust the amount of light entering the specimen.

are possible, image clarity is more difficult to maintain as the magnification increases. Clarity of an image is called **resolution** (Figure 3-5). The **limit of resolution** (or **resolving power**) is an actual measurement of how far apart two points must be for the microscope to view them as being separate. Notice that resolution improves as the limit of resolution is made smaller.

The best limit of resolution achieved by a light microscope is about 0.2 μm. (That is, at its absolute best, a light microscope cannot distinguish between two points closer together than 0.2 μm.) For a specific microscope, the actual limit of resolution can be calculated using the following formula:

$$D = \frac{\lambda}{NA_{condenser} + NA_{objective}}$$

where D is the minimum distance at which two points can be resolved, λ is the wavelength of light used, and $NA_{condenser}$ and $NA_{objective}$ are the numerical apertures of the condenser lens and objective lenses, respectively. Because numerical aperture has no units, the units for D are the same as the units for wavelength, which typically are in nanometers (nm).

Numerical aperture is a measure of a lens's ability to "capture" light coming from the specimen and use it to make the image. As with magnification, it is marked on the lens (Figure 3-4A and Figure 3-4C). Using immersion oil between the specimen and the oil objective lens

increases its numerical aperture and, in turn, makes its limit of resolution smaller. (If necessary, oil also may be placed between the condenser lens and the slide.) The result is better resolution.

The light microscope may be modified to improve its ability to produce images with contrast without staining, which often distorts or kills the specimen. In **dark-field microscopy** (Figure 3-2B), a special condenser is used so only the light reflected off the specimen enters the objective. The appearance is of a brightly lit specimen against a dark background, and often with better resolution than that of the bright-field microscope.

Phase contrast microscopy (Figure 3-2C) uses special optical components to exploit subtle differences in the refractive indices of water and cytoplasmic components to produce contrast. Light waves that are in phase (that is, their peaks and valleys exactly coincide) reinforce one another, and their total intensity (because of the summed amplitudes) increases. Light waves that are out of phase by exactly one-half wavelength cancel each other and result in no intensity—that is, darkness. Wavelengths that are out of phase by any amount will produce some degree of cancellation and result in brightness that is less than maximum but more than darkness. Thus, contrast is provided by differences in light intensity that result from differences in refractive indices in parts of the specimen that put light waves more or less out of phase. As a result, the specimen appears as various levels of "darks" against a bright background.

3-5 RESOLUTION AND LIMIT OF RESOLUTION ✦ The headlights of most automobiles are around 1.5 m apart. As you look at the cars in the foreground of the photo, it is easy to see both head-lights as separate objects. The automobiles in the distance appear smaller (but really aren't) as does the apparent distance between the headlights. When the apparent distance between automobile headlights reaches about 0.1 mm, they blur into one because that is the limit of resolution of the human eye.

Fluorescence microscopy (Figure 3-2D) uses a fluorescent dye that emits fluorescence when illuminated with ultraviolet radiation. In some cases, specimens possess naturally fluorescing chemicals and no dye is needed.

✦ Application

Light microscopy (used in conjunction with cytological stains) is used to identify microbes from patient specimens or the environment. It also may be used to visually examine a specimen for the presence of more than one type of bacteria, or for the presence of other cell types that indicate tissue inflammation or contamination by a patient's cells.

✦ In this Exercise

Today you will become familiar with the operation and limitations of your light microscope. You also will examine two practice slides to learn about microscope functioning.

✦ Materials

- ✦ compound light microscope
- ✦ lens paper
- ✦ nonsterile cotton swabs
- ✦ lens-cleaning solution or 95% ethanol
- ✦ letter "e" slide
- ✦ colored threads slide

✦ Instructions for Using the Microscope

Proper use of the microscope is essential for your success in microbiology. Fortunately, with practice and by following a few simple guidelines, you can achieve satisfactory results quickly. Because student labs may be supplied with a variety of microscopes, your instructor may supplement the following procedures and guidelines with instructions specific to your equipment. Refer to Figure 3-1 as you read the following (if working independently), or follow along on your microscope as your instructor guides you. (**Note:** This is a thorough treatment of microscope use, and not all parts may be immediately relevant to your laboratory. Refer back to this exercise as necessary.)

Transport

1 Carry your microscope to your workstation using both hands—one hand grasping the microscope's arm and the other supporting the microscope beneath its base.

2 Gently place the microscope on the table.

Cleaning

1 Lens paper is used for gently cleaning the condenser and objective lenses. Light wiping is usually enough. If that still doesn't clean the lens, call your instructor.

2 To clean an ocular, moisten a cotton swab with cleaning solution and gently wipe in a spiral motion starting at the center of the lens and working outward. Follow with a dry swab in the same pattern.

Operation

1 Raise the substage condenser to a couple of milli-meters below its maximum position nearly even with the stage and open the iris diaphragm.

2 Plug in the microscope and turn on the lamp. Adjust the light intensity slowly to its maximum.

3 Using the nosepiece ring, move the scanning objective (usually 4×) or low power objective (10×) into position. Do not rotate the nosepiece by the objectives as this can damage the objective lenses and cause them to unscrew from the nosepiece.

4 Place a slide on the stage in the mechanical slide holder and center the specimen over the opening in the stage.

5 If using a binocular microscope, adjust the distance between the two oculars to match your own inter-pupillary distance.

6 Adjust the iris diaphragm and condenser position to produce optimum illumination, contrast, and image. (As a rule, use the maximum light intensity combined with the smallest aperture in the iris diaphragm that produces optimum illumination. Remember: This is bright-field microscopy, so don't close down the iris diaphragm too much.)

7 Use the coarse-focus adjustment knob to bring the image into focus. (*Note:* For most microscopes, the distance from the nosepiece opening to the focal plane of each lens has been standardized at 45 mm. This makes the lenses **parfocal** and gives the user an idea of where to begin focusing.) Bring the image into sharpest focus using the fine-focus adjustment knob. Then observe the specimen with your eyes relaxed and slightly above the oculars to allow the images to fuse into one. If you are using a monocular microscope, keep both eyes open anyway to reduce eye fatigue.

8 If you are using a binocular microscope, adjust the oculars' focus to compensate for differences in visual acuity of your two eyes. Close the eye with the adjustable ocular and bring the image into focus with the coarse- and fine-focus knobs. Then, using only the eye with the adjustable ocular, focus the image using the ocular's focus ring.

9 Scan the specimen to locate a promising region to examine in more detail.

10 If you are observing a nonbacterial specimen, progress through the objectives until you see the degree of structural detail necessary for your purposes. You will have to adjust the fine focus and illumination for each objective. Before advancing to the next objective, be sure to position a desirable portion of the specimen in the center of the field or you will risk "losing" it at the higher magnification.

11 If you are working with a bacterial smear, you will have to use the oil immersion lens.

12 To use the oil immersion lens, work through the low (10×), then high dry (40×) objectives, adjusting the fine focus and illumination for each. Before advancing to the next objective, be sure to position a desirable portion of the specimen in the center of the field or you risk "losing" it at the higher magnification.

    When the specimen is in focus under high dry, rotate the nosepiece to a position midway between the high dry and oil immersion lenses. Then place a drop of immersion oil on the specimen. *Be careful not to get any oil on the microscope or its lenses, and be sure to clean it up if you do.* Rotate the oil

lens so its tip is submerged in the oil drop. Be careful not to trap any air between the slide and the oil objective. If you do, rotate the oil lens into and out of position a couple of times to pop the bubble. (*Note:* Do not move the stage down to add oil to the slide or the specimen will no longer be in focus. On a properly adjusted microscope, the oil and the high dry lenses have the same focal plane. Therefore, when a specimen is in focus on high dry, the oil lens, although longer, will also be in focus and won't touch the slide when rotated into position.) Focus and adjust the illumination to maximize the image quality.

13 When you are finished, lower the stage (or raise the objective) and remove the slide. Dispose of the freshly prepared slides in a jar of disinfectant or a sharps container; return permanent slides to storage.

Storage

When you are finished for the day:

1 Move the scanning objective into position.

2 Center and lower the mechanical stage.

3 Lower the light intensity to its minimum, then turn off the light.

4 Wrap the electrical cord according to your particular lab rules.

5 Clean any oil off the lenses, stage, *etc.* Be sure to use only cotton swabs or lens paper for cleaning any of the optical surfaces of the microscope (see "Cleaning," above).

6 Return the microscope to its appropriate storage place.

 **Procedure**

1 Get out your microscope and record the magnifications and numerical aperture values in the chart on the Data Sheet.

2 Clean your microscope lenses as outlined in "Instructions for Using the Microscope."

3 Plug in the microscope and position the scanning objective over the stage. Make condenser and lamp adjustments appropriate for scanning power.

4 Pick up the letter "e" slide and examine it without the microscope. Record the orientation of the letter when the slide label is on the left. Record the appearance of the letter on your Data Sheet.

5 Place the slide on the stage in the same position as you examined it with your naked eyes. Now, center the "e" in the field and examine it with the scanning objective. Record its orientation as viewed with the microscope on your Data Sheet.

6 Now, move the *stage* to the right and record the direction the *image* moves.[1]

7 Position the "e" in the center of the field again. Move the *stage* toward you and record on your Data Sheet the direction the *image* moves. Then remove the slide from the microscope.

8 Examine the colored thread slide without the microscope (Figure 3-6). See if you can tell where in the stack of three threads each color resides. That is, is the red thread on the top, bottom, or middle? Do the same for the yellow and blue threads. Record your observations on the Data Sheet.

9 Now, place the slide on the microscope and determine the order of the threads using the low and high power objectives. Record your observations on the Data Sheet.

3-6 CHALLENGE OF THE THREADS ✦ Even with the microscope, determining the order of threads from top to bottom is a challenge. This will require patience and use of the fine focus! Making it worse, not all the slides will be the same. Good luck!

References

Abramowitz, Mortimer. 2003. *Microscope Basics and Beyond.* Olympus America Inc., Scientific Equipment Group, Melville, NY.

Ash, Lawrence R., and Thomas C. Orihel. 1991. Pages 187–190 in *Parasites: A Guide to Laboratory Procedures and Identification.* American Society for Clinical Pathology (ASCP) Press, Chicago.

Bradbury, Savile, and Brian Bracegirdle. 1998. Chapter 1 in *Introduction to Light Microscopy.* BIOS Scientific Publishers Limited, Oxford, United Kingdom.

Forbes, Betty A., Daniel F. Sahm, and Alice S. Weissfield. 2002. Pages 119–121 in *Bailey & Scott's Diagnostic Microbiology,* 11th ed. Mosby, St. Louis.

---

[1] If your microscope doesn't have a mechanical stage, move the slide with your hands in the appropriate direction.

EXERCISE 3-2

Calibration of the Ocular Micrometer

✦ Theory

An **ocular micrometer** is a type of ruler installed in the microscope eyepiece, composed of uniform but unspecified graduations (Figure 3-7). As such, it must be calibrated before any viewed specimens can be measured. The device used to calibrate ocular micrometers is called a **stage micrometer**. As illustrated in Figure 3-8, a stage micrometer is a type of microscope slide containing a ruler with 10 µm and 100 µm graduations. (Other measuring instruments may be used in place of a stage micrometer, as shown in Figure 3-9.)

When the stage micrometer is placed on the stage, it is magnified by the objective being used; therefore, the size of the graduations (relative to the ocular micrometer divisions) increases as magnification increases. Consequently, the *value* of ocular micrometer divisions decreases as magnification increases. For this reason, calibration must be done for each magnification.

As shown in Figure 3-10, the stage micrometer is placed on the stage and brought into focus such that it

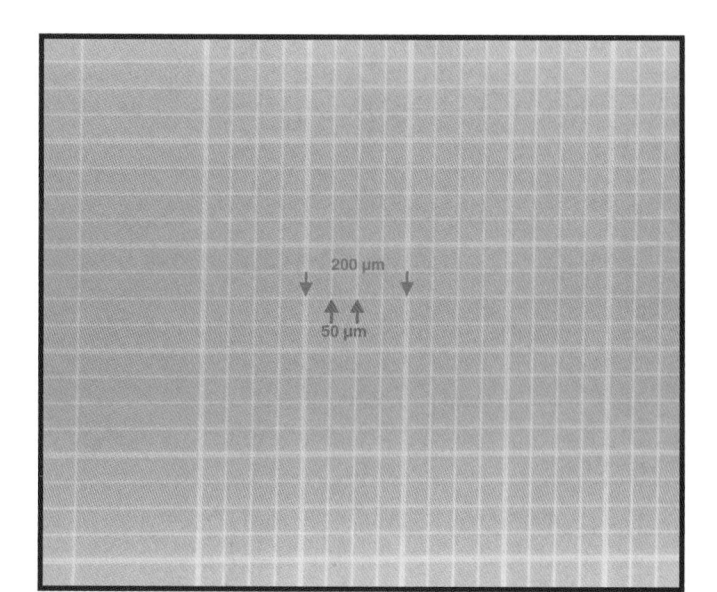

3-9 A HEMACYTOMETER ✦ Any instrument with markings of known distance apart may be used as a stage micrometer. The hemacytometer is a grid with lines 50 µm apart (red arrows). A larger grid is formed by lines 200 µm apart (blue arrows). Use any horizontal line as the micrometer, with the smallest divisions 50 µm apart.

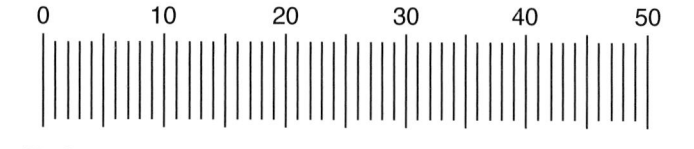

3-7 AN OCULAR MICROMETER ✦ The ocular micrometer is a scale with uniform increments of unknown size. It has to be calibrated for each objective lens on the microscope.

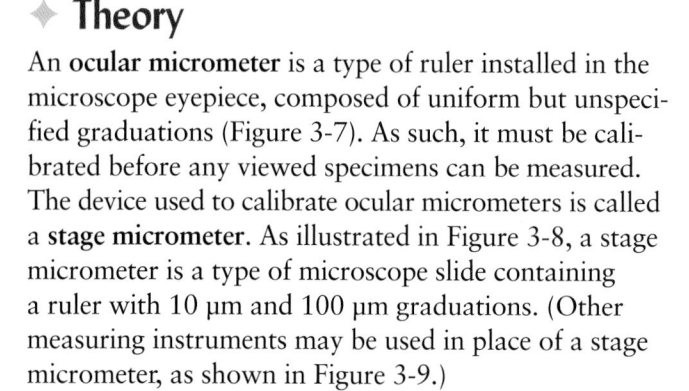

A

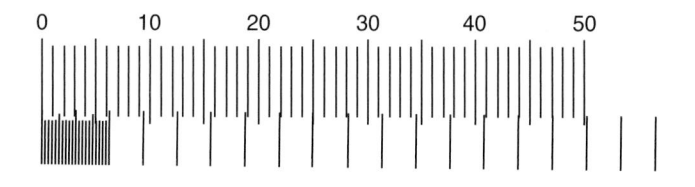

3-10 WHAT THEY LOOK LIKE IN USE ✦ When properly aligned, the ocular micrometer scale is superimposed over the stage micrometer scale. Notice that they line up at their left ends.

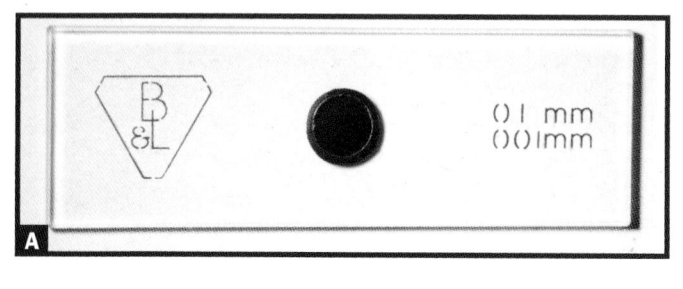

B

3-8 A STAGE MICROMETER ✦ **A** The stage micrometer is a microscope slide with a microscopic ruler engraved into it (not visible in the dark center of the slide). The markings on this micrometer indicate that the major increments are 0.1 mm (100 µm) apart. There is also a section of the scale that is marked off in 0.01 mm (10 µm) increments. **B** This drawing represents what the stage micrometer on the slide in **A** looks like. The micrometer is 2200 µm long. The major divisions are 100 µm apart. The 200 µm at the left are divided into 10 µm increments.

is superimposed by the ocular micrometer. Then the first (left) line of the ocular micrometer is aligned with one of the marks on the stage micrometer. (The line chosen on the stage micrometer depends on the power of the lens being calibrated. Lower powers use the large graduations; higher powers use the smaller graduations on the left. Figure 3-10 illustrates proper alignment with the scanning objective.)

Notice in Figure 3-10 that line 25 of the ocular micrometer and the eighth major line of the stage micrometer are perfectly aligned. This indicates that 25 ocular micrometer divisions (also called ocular units, or OU) span a distance of 800 µm because the stage micrometer lines are 100 µm apart. Notice also that line 47 of the ocular micrometer is aligned with the fifteenth major stage micrometer line. This means that 47 ocular units span 1500 µm. These values have been entered for you in Table 3-1.

To determine the value of an ocular unit on a given magnification, divide the distance (from the stage micrometer) by the corresponding number of ocular units.

$$\frac{800 \ \mu m}{25 \ ocular \ units} = 32 \ \frac{\mu m}{OU}$$

$$\frac{1500 \ \mu m}{47 \ ocular \ units} = 32 \ \frac{\mu m}{OU}$$

As shown in Table 3-1, it is customary to record more than one measurement. Each measurement is calculated separately. If the calculated ocular unit values differ, use their arithmetic mean as the calibration for that objective lens.

As mentioned previously, each magnification must be calibrated. Because of its short working distance, calibrating the oil immersion lens may be difficult to accomplish using the stage micrometer. It also may be difficult because the distance between stage micrometer lines is too large. If this is the case, its value can be calculated using the calibration value of one of the other lenses. Refer to Table 3-2 for the total magnifications for each objective lens on a typical microscope. Notice that the magnification of the oil immersion lens is 10 times greater than the low-power lens. This means that objects viewed on the stage (stage micrometer *or* specimens) appear 10 times larger when changing from low power

TABLE **3-1** Sample Data from Figure 3-10

Stage Micrometer	Ocular Micrometer
800 µm	25 OU
1500 µm	47 OU

TABLE **3-2** Total Magnifications for Different Objective Lenses of a Typical Microscope and the Calibrations of the Ocular Micrometer for the Scanning Objective. Calculate the remaining calibratons as described in the text.

Power	Total Magnification	Calibration (µm/OU)
Scanning	40X	32
Low Power	100X	
High Dry Power	400X	
Oil Immersion	1000X	

to oil immersion. But, because the magnification of the ocular micrometer does not change, an ocular division now covers only one-tenth the distance. Thus, the size of an ocular unit using the oil immersion lens can be calculated by dividing the calibration for low power by 10.

Ocular micrometer values can be calculated for any lens using values from any other lens, and they provide a good check of measured values. As calculated from Figure 3-10, 32 µm/OU was the calibration for the scanning objective. For practice, calculate the low, high dry and oil immersion calibrations. Write the values in the Table 3-2.

Once you have determined the ocular unit values for each objective lens, use the ocular micrometer as a ruler to measure specimens. For instance, if you determine that under the scanning objective a cell is 5 ocular units long, the cell's actual length would be determined as follows (using the sample values from Table 3-2):

Cell Dimension = Ocular Units × Calibration

Cell Dimension = 5 Ocular Units × 32 µm/OU

Cell Dimension = 160 µm

Be sure to include the proper units in your answer!

✦ Application

The ability to measure microbes is useful in their identification and characterization.

✦ In This Exercise

This lab exercise involves calibrating the ocular micrometer on your microscope. Actual measurement of specimens will be done in subsequent lab exercises as assigned by your instructor.

✦ Materials

Per Student

✦ compound microscope equipped with an ocular micrometer

✦ stage micrometer

Procedure

Following is the general procedure for calibrating the ocular micrometer on your microscope. Your instructor will notify you of any specific details unique to your laboratory.

1 Check your microscope and determine which ocular has the micrometer in it.

2 Move the scanning objective into position.

3 Place the stage micrometer on the stage and position it so its image is superimposed by the ocular micrometer and the left-hand marks line up.

4 Examine the two micrometers and, as described above, record two or three points where they line up exactly. Record these values on the Data Sheet and calculate the value of each ocular unit.

5 Change to low power and repeat the process.

6 Change to high dry power and repeat the process.

7 Change to the oil immersion lens and repeat the process. If this cannot be done (either because the stage micrometer lines are too far apart or the slide is too thick for the oil lens to be rotated into position), complete the calibration from the value of another lens.

8 Compute average calibrations for each objective lens and record these on the Data Sheet.

9 As long as you keep this microscope throughout the term, you may use the calibrations you recorded without recalibrating the microscope.

References

Abramoff, Peter, and Robert G. Thompson. 1982. Pages 5 and 6 in *Laboratory Outlines in Biology—III*. W. H. Freeman and Co., San Francisco.

Ash, Lawrence R., and Thomas C. Orihel. 1991. Pages 187–190 in *Parasites: A Guide to Laboratory Procedures and Identification*. American Society for Clinical Pathology (ASCP) Press, Chicago.

EXERCISE 3-3

Microscopic Examination of Eukaryotic Microbes

✦ Theory

Cells are divided into two major groups—the **prokaryotes**[1] and the **eukaryotes**—based on size and complexity. These and other differences are summarized in Table 3-3 and shown in Figure 3-11. The prokaryotes are further divided into two domains—the Archaea and the Bacteria. The eukaryotes belong to a single domain and are divided into as few as four and as many as eight kingdoms. The four kingdoms are: Protista, Fungi, Animalia, and Plantae. (The eight eukaryotic kingdom system breaks up the protists into five kingdoms.) Figure 3-12 provides a phylogenetic tree of these groups based on RNA comparisons. In this lab, you will observe simple eukaryotic microorganisms of various types: protists (protozoans and algae) and yeasts.

Protist Survey: Protozoans and Algae

Protozoans are unicellular eukaryotic heterotrophic microorganisms. A typical life cycle includes a vegetative **trophozoite** and a resting **cyst** stage. Some have additional stages, making their life cycles more complex.

One protozoan classification scheme recognizes the following groups: Phylum Sarcomastigophora (including

---

[1] The validity of the term "prokaryote" has recently been called into question. We continue to use it because of the uncertainty of the change.

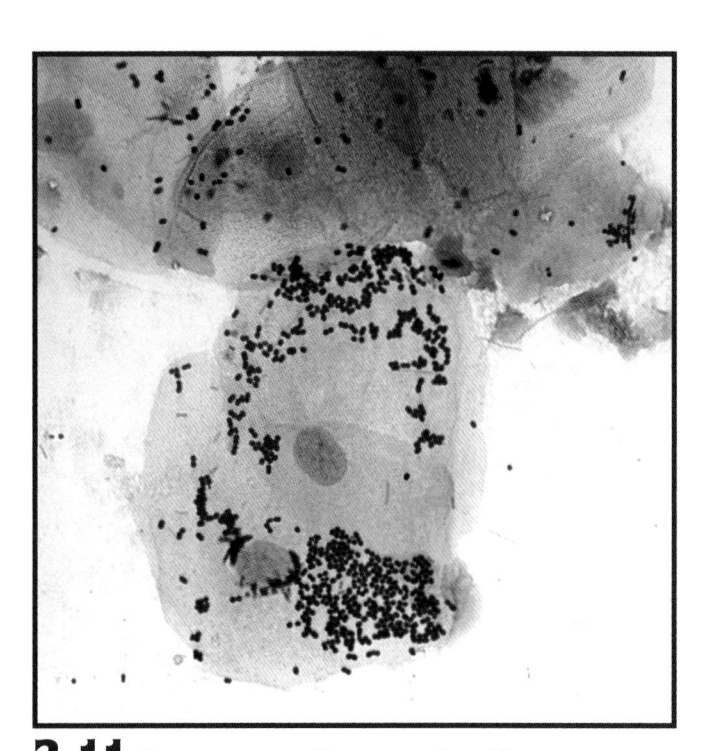

3-11 PROKARYOTIC AND EUKARYOTIC CELLS (GRAM STAIN, X1000) ✦ This is a direct smear specimen taken from around the base of the teeth below the gum line. The large, pink cells are human epithelial cells and are eukaryotic (notice the prominent nuclei). The small, purple cells are prokaryotic bacteria. Typically, prokaryotic cells range in size from 1 to 5 µm, whereas eukaryotic cells are in the 10 to 100 µm range.

TABLE **3-3** Summary of Major Prokaryotic and Eurayotic Features

Character	Prokaryotes	Eukaryotes
Organismal groups	Bacteria and Archaeans	Plants, animals, fungi and protists
Typical size	1–5 µm	10–100 µm
Membrane-bound Organelles (including a nucleus)	Absent	Present
Ribosomes	70S (30S and 50S subunits)	80S (40S and 60S subunits)
Microtubules	Absent	Present
Flagellar movement	Rotary	Whip-like
DNA	Single, circular molecule called a chromosome	Two to many linear molecules; each is a chromosome
Introns	Absent (sometimes in Archaens)	Common
Mitotic division	Absent	Present

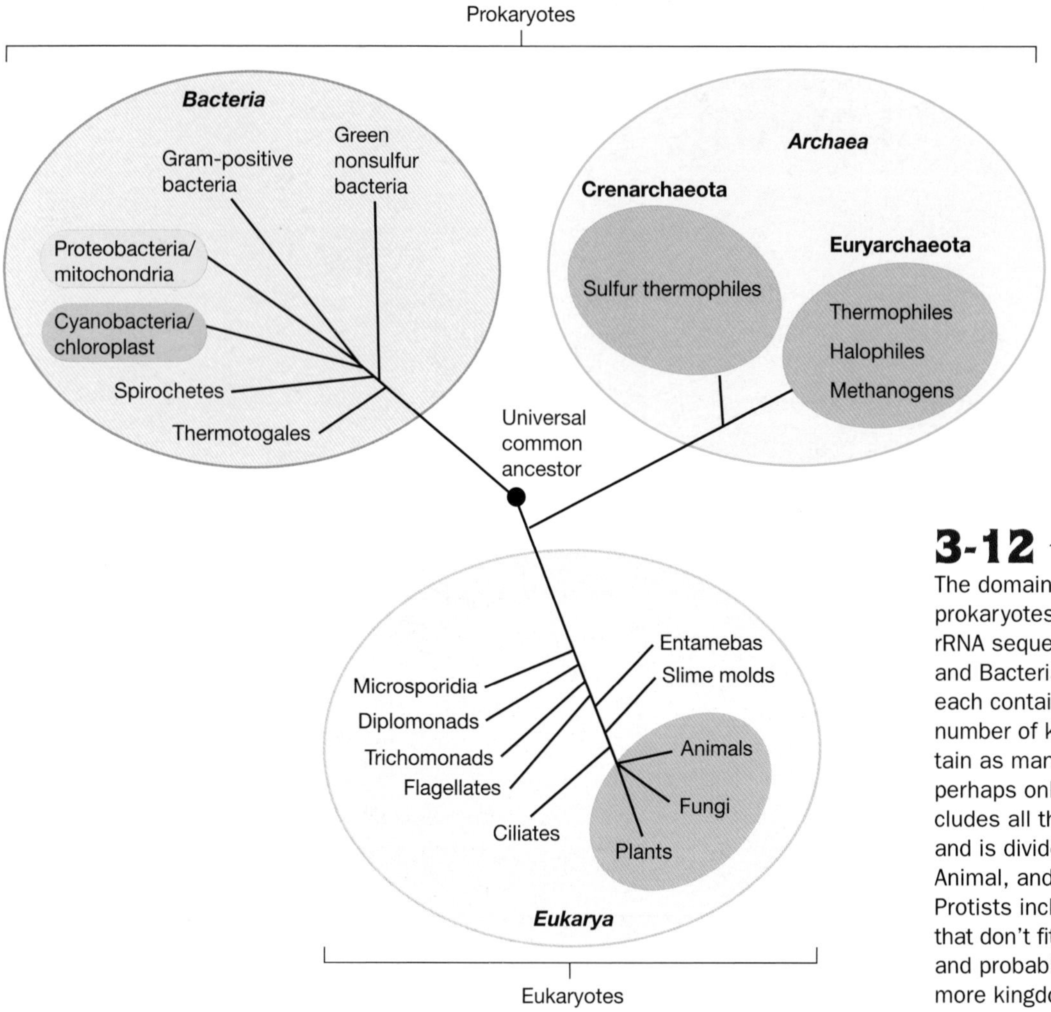

Prokaryotes

Bacteria

Gram-positive bacteria

Green nonsulfur bacteria

Proteobacteria/mitochondria

Cyanobacteria/chloroplast

Spirochetes

Thermotogales

Archaea

Crenarchaeota

Sulfur thermophiles

Euryarchaeota

Thermophiles

Halophiles

Methanogens

Universal common ancestor

Microsporidia

Diplomonads

Trichomonads

Flagellates

Ciliates

Entamebas

Slime molds

Animals

Fungi

Plants

Eukarya

Eukaryotes

3-12 THE THREE DOMAINS OF LIFE ✦ The domains are based on 16S (in prokaryotes) and 18S (in eukaryotes) rRNA sequencing results. The Archaea and Bacteria are prokaryotic domains, each containing an as yet undetermined number of kingdoms. (Bacteria may contain as many as 50 kingdoms, Archaea perhaps only 3.) Kingdom Eukarya includes all the eukaryotic organisms and is divided into the familiar Plant, Animal, and Fungus Kingdoms. The Protists include all the other eukaryotes that don't fit into the first three kingdoms and probably will be broken up into 5 or more kingdoms as we learn more.

Subphylum Mastigophora [the flagellates] and Subphylum Sarcodina [the amoebas]), Phylum Ciliophora (the ciliates), and Phylum Apicomplexa (sporozoans and others). Sarcodines move by forming cytoplasmic extensions called **pseudopods**. Division is by binary fission. Ciliates owe their motility to the numerous **cilia** covering the cell. Reproduction is by transverse fission. Members of Mastigophora are characterized by one or more flagella and division by longitudinal fission. Sporozoans are typically nonmotile and usually have complex life cycles involving asexual reproduction in one host and sexual reproduction in another. Figures 3-13 through 3-15 show protozoan representatives of Mastigophora, Sarcodina, and Ciliophora.

Algae comprise a diverse group of simple, photosynthetic eukaryotic organisms with uncertain relatedness. **Green algae** (Division Chlorophyta) are common in freshwater and are usually unicellular or colonial. *Spirogyra* (Figure 3-16), characterized by its spiral chloroplast, and the large, spherical colonies of *Volvox* (Figure 3-17) are examples.

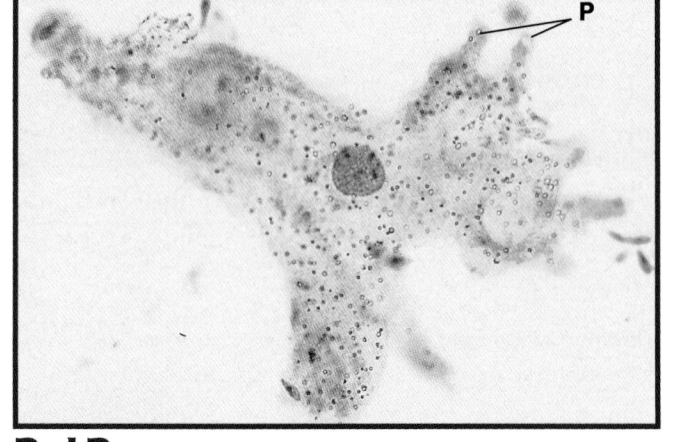

3-13 AMOEBA, A SARCODINE (X55) ✦ Note the numerous pseudopods (P) used for movement and capturing food.

Other algal groups are the red algae, brown algae, golden-brown algae, yellow-green algae, and diatoms. These groups are differentiated by their photosynthetic pigments, motility, cell wall material and storage carbohydrate.

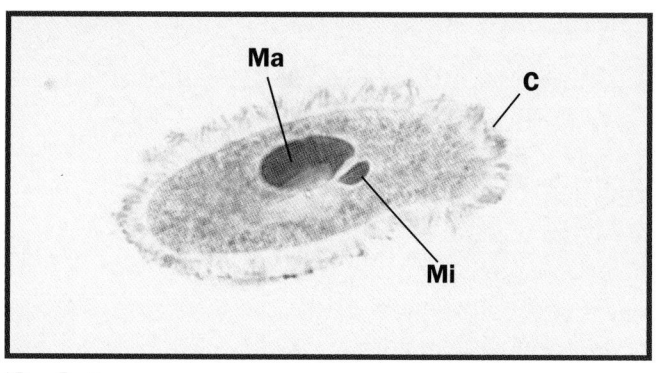

3-14 PARAMECIUM BURSARIA, A CILIATE (X300) ✦ Note the cilia (C) around the edge of the cell. The macronucleus (Ma) and micronucleus (Mi) are also visible. Cilia serve to move the cell as well as to sweep food into its "mouth."

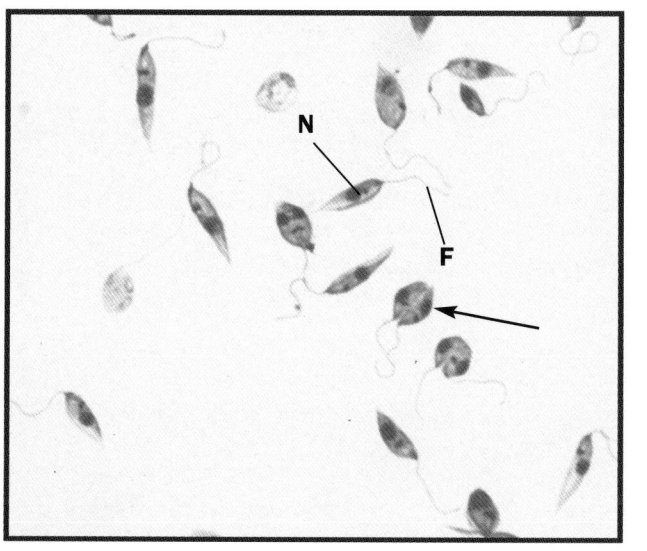

3-15 LEISHMANIA DONOVANI (X650) ✦ Notice the anterior flagellum (F) and the nucleus (N). Two of the cells (arrow) are dividing.

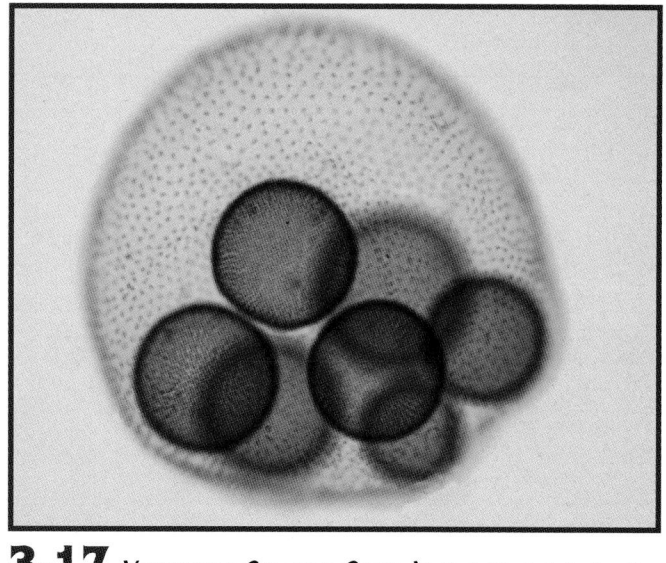

3-17 VOLVOX IS A COLONIAL GREEN ALGA ✦ Flagellated cells are joined together to form the hollow, spherical colony. Daughter colonies are visible within the larger ones.

Fungal Survey: Yeasts and Molds

Members of the Kingdom Fungi are nonmotile eukaryotes. Their cell wall is usually made of the polysaccharide chitin, not cellulose as in plants. Unlike animals (that ingest then digest their food), fungi are **absorptive heterotrophs**. That is, they secrete **exoenzymes** into the environment, then absorb the digested nutrients. Most are **saprophytes** that decompose dead organic matter, but some are **parasites** of plants, animals, or humans. Fungi are informally divided into unicellular **yeasts** and filamentous **molds** based on their overall appearance. Brewer's yeast *(Saccharomyces cerevisiae)* is shown in Figure 3-18.

In this lab you will observe representative protozoans, algae and fungi. Some specimens are on commercially

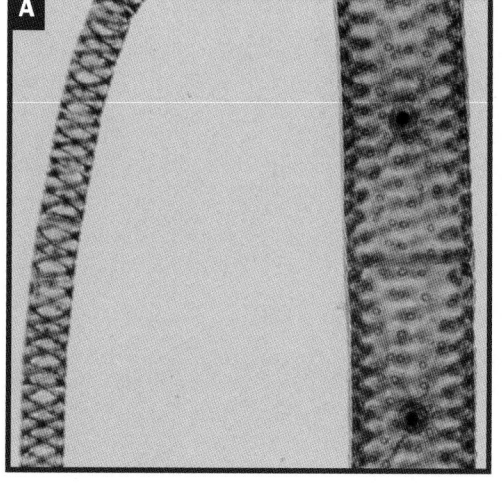
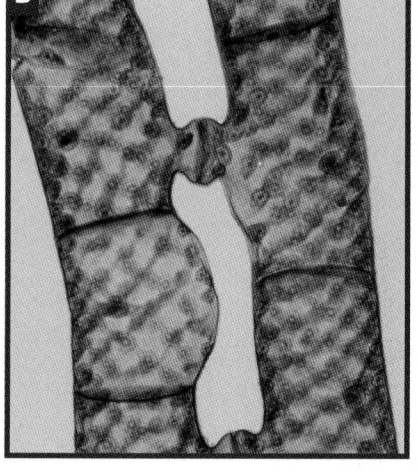
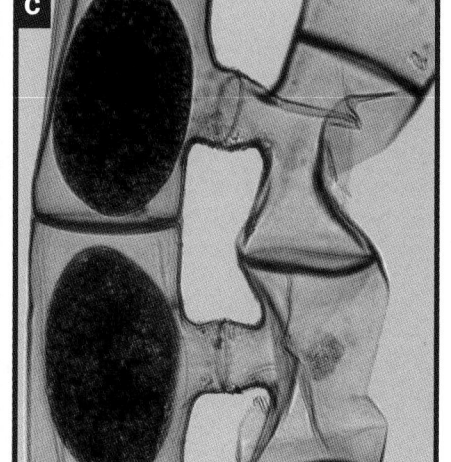

3-16 SPIROGYRA SPP. ✦ Notice the spiral chloroplast and the nucleus in each cell. **A** A vegetative filament of cells. **B** Early conjugation between filaments. **C** Conjugation is completed with the formation of zygote.

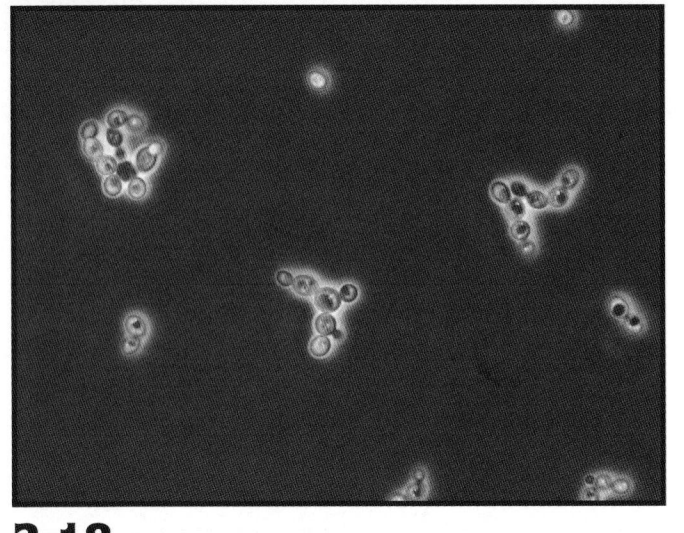

3-18 SACCHAROMYCES CEREVISIAE ✦ The "brewer's yeast" is shown in a wet mount preparation using phase contrast microscopy. The cells are oval with dimensions of 3–8 × 5-10 µm. Short chains of cells (pseudohyphae) are visible in this field.

prepared slides, whereas others need to be prepared as **wet mounts** (Figure 3-18) of living cultures. In a wet mount, a drop of water is placed on the slide and the organisms are introduced into it. Or, if the organism is already in a liquid medium, then a drop of medium is placed on the slide. A cover glass is placed over the preparation to flatten the drop and keep the objective lens from getting wet. A stain may or may not be applied to add contrast.

✦ Application

Familiarity with eukaryotic cells not only rounds out your microbiological experience, it is important to be able to differentiate eukaryotic from prokaryotic cells when examining environmental or clinical specimens.

✦ In This Exercise

Most of this manual is devoted to prokaryotes, but in this exercise you will be given the opportunity to examine various eukaryotic microorganisms. This will not only serve to familiarize you with simple eukaryotes, but also give you practice using the microscope, measuring specimens, and making wet-mount preparations.

✦ Materials

Per Class

✦ Prepared slides or living cultures of a variety of protists and fungi (such as, *Amoeba, Paramecium, Leishmania*—prepared only, *Spirogyra, Volvox,* diatoms, and *Saccharomyces.*)

Per Student

✦ clean glass slides and cover glasses
✦ compound microscope with ocular micrometer
✦ cytological stains (*e.g.*, methylene blue, I_2KI)
✦ methyl cellulose
✦ immersion oil
✦ cotton swabs
✦ lens paper
✦ lens cleaning solution

Procedure

1 Obtain a microscope and place it on the table or workspace. Check to be sure the stage is all the way down and the scanning objective is in place.

2 Begin with a prepared slide. Clean it with a tissue if it is dirty, then place it on the microscope stage. Center the specimen under the scanning objective.

3 Follow the instructions given in Exercise 3-1, to bring the specimen into focus at the highest magnification that allows you to see the entire structure you want to view.

4 Practice scanning with the mechanical stage until you are satisfied that you have seen everything interesting to see. Sketch what you see in the table provided on the Data Sheet.

5 Measure cellular dimensions and record these in the table provided on the Data Sheet.

6 Repeat with as many slides as you have time for.

7 Prepare wet mounts of available specimens by following the Procedural Diagram in Figure 3-19. Methyl cellulose may be added to the wet mount if you have fast swimmers. Sketch what you see and record cellular dimensions in the chart provided on the Data Sheet.

8 When you are finished observing specimens, blot the oil from the oil immersion lens (if used) with a lens paper and do a final cleaning with a cotton swab and alcohol or lens cleaning solution. Dry the lens with a clean swab.

9 Return all lenses and adjustments to their storage positions before putting the microscope away.

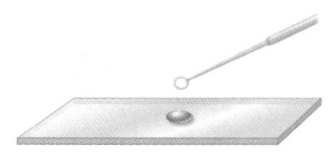

1. Place a drop of water on a clean slide using an inoculating loop. This is unnecessary if observing a liquid culture.

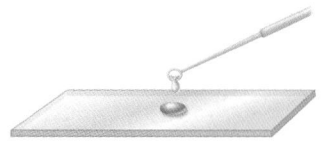

2. Add a drop of specimen to the water.

3. Gently lower the cover glass onto the drop with your index finger and thumb, or use a loop. Be careful not to trap bubbles.

If not staining...　　　　　　　　　If staining...

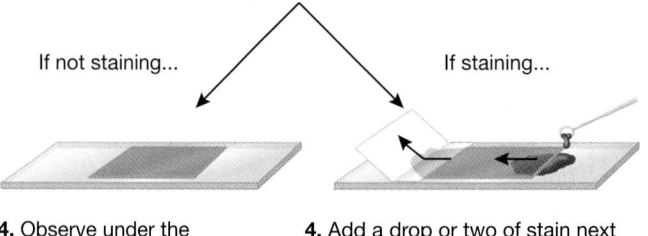

4. Observe under the microscope.

4. Add a drop or two of stain next to the cover glass. Draw the stain under the cover glass with a piece of paper in contact with the cover glass edge on opposite side.

5. Observe under the microscope.

3-19 PROCEDURAL DIAGRAM: WET MOUNT ✦ Use these instructions to make wet mounts of living specimens. The preparation may be stained or not, but staining will eventually kill the organisms. Of course, so will drying. You can expect a wet mount to last only 15–20 minutes.

References

Campbell, Neil A., and Jane B. Reece. 2005. Chapters 25 and 28 in *Biology*, 7th ed. Pearson Education/Benjamin Cummings Publishing Co., San Francisco.

Freeman, Scott. 2005. Chapter 27 in *Biological Science*, 2nd ed. Pearson Education/Prentice Hall. Upper Saddle River, NJ.

Madigan, Michael T. and John M. Martinko. 2006. Chapter 11 in *Brock's Biology of Microorganisms*, 11th ed. Pearson Education/Prentice Hall, Upper Saddle River, NJ.

EXERCISE 3-4

Microscopic Examination of Pond Water

✦ Theory

At some time in your life, you undoubtedly have passed a lake or river and casually noticed that a green goo was growing on the water's surface or that the mud was dark in color. Perhaps you were put off by a foul smell. Microorganisms cause all of these—and more. Today you will see the previously unseen microbes that caught your attention.

Bacteria are the smallest organisms that you will see in your preparation (Figure 3-20). Without staining, they will appear as small (about 10 μm or less), transparent cells in the shapes of rods (bacilli), spheres (cocci), and spirals (spirilla). Sometimes they will be joined together in clusters or chains. Watch for swimming bacteria. Follow their path and see if they swim in a straight line or move along an irregular course.

Cyanobacteria are in the Domain Bacteria. They are easily seen without staining because of their combination of photosynthetic pigments, which confer on them a bluish-green color. In fact, they were formerly known as "blue-green algae." When they are single-celled, they are about the size of bacteria. But when they are found in chains, called **trichomes**, they are easily visible at high dry magnification. Some trichomes have an extracellular sheath. Also, watch for slow, gliding motility of the trichomes. Trichomes often have specialized cells, including

heterocysts, which are nitrogen-fixing cells, and **akinetes**, which are resistant spores (Figures 3-21 to 3-29).

Protists are a heterogeneous group of simple eukaryotic microorganisms. Formerly they were placed into a single kingdom called Protista. Modern molecular evidence has revealed that protists are more diverse than previously thought, and their taxonomy and phylogeny

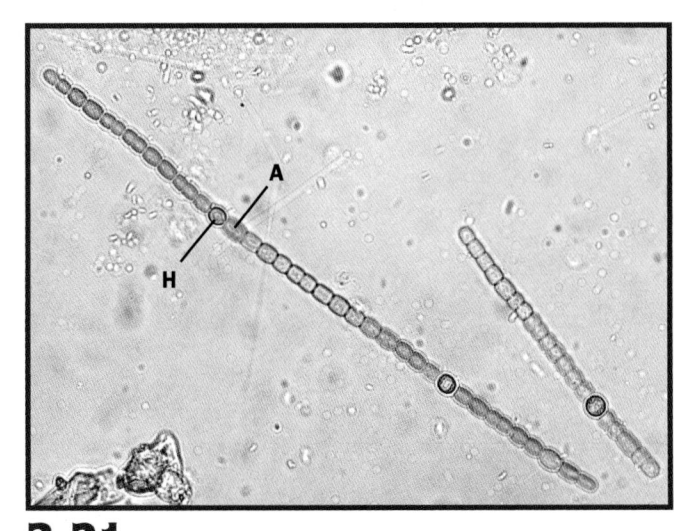

3-21 ANABAENA ✦ The trichomes of this gliding cyanobacterium may possess thick-walled spores called **akinetes–A** and specialized, nitrogen-fixing cells called **heterocysts–H**. Trichomes are of variable lengths but are approximately 20 μm in width.

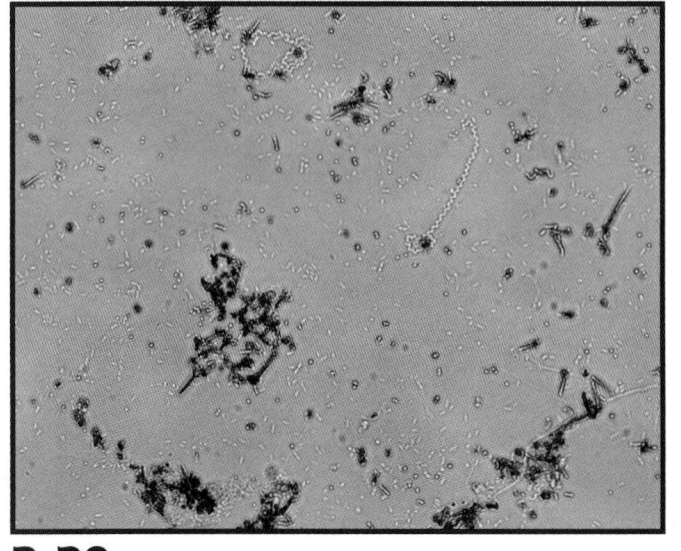

3-20 BACTERIA ✦ Bacterial cells (the small, light objects) are shaped like spheres, rods, or spirals, and all are represented in this micrograph. These are the smallest cells you will see. Most are in the range between 1 and 10 μm.

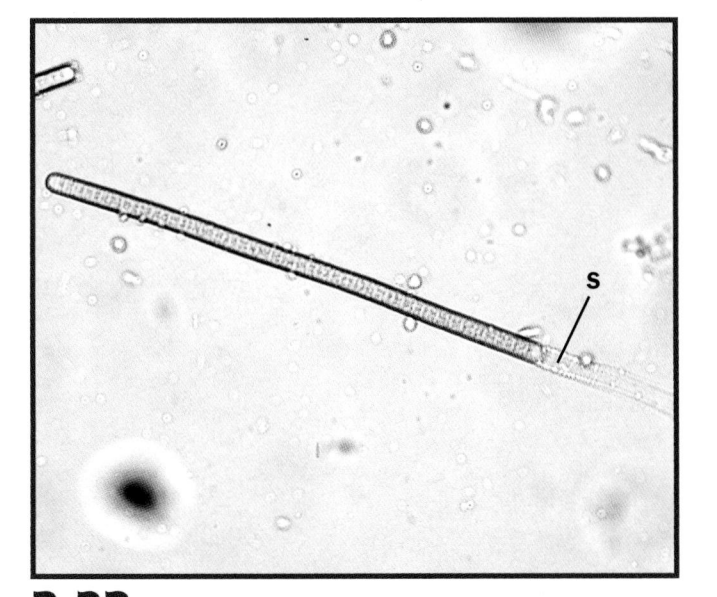

3-22 LYNGBYA ✦ *Lyngbya* trichomes are distinguished from *Oscillatoria* (Figure 3-26) by the distinctive sheath (**S**). The trichomes are approximately 20 μm in width.

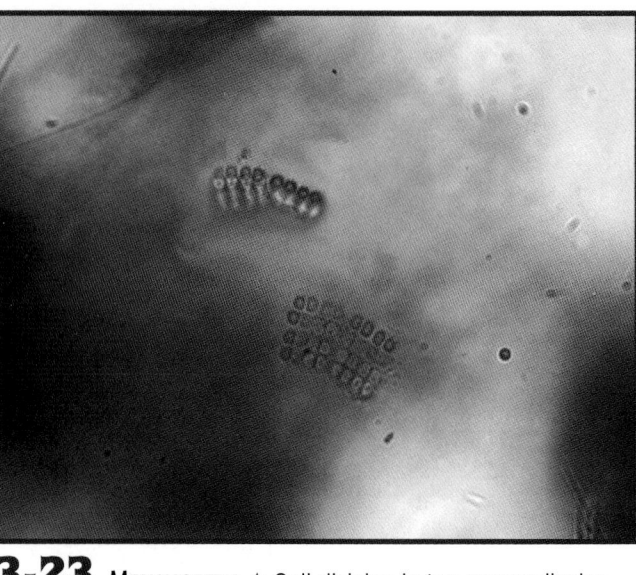

3-23 MERISMOPEDIA ✦ Cell division in two perpendicular directions produces the planar arrangement characteristic of this genus. The cells are enclosed in a mucilaginous sheath and are approximately 1–2 μm in diameter.

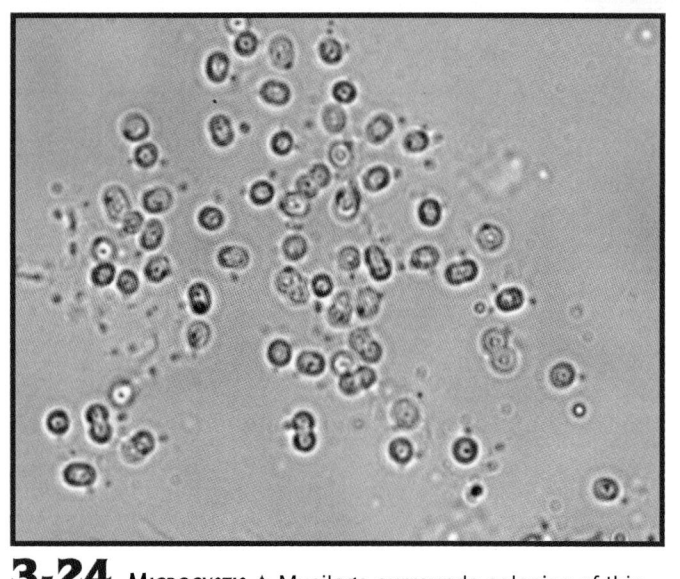

3-24 MICROCYSTIS ✦ Mucilage surrounds colonies of this planktonic genus. The dark spots are gas "pseudovacuoles" that allow it to float in the water. Some species produce an animal toxin. Cells are approximately 3 μm in diameter.

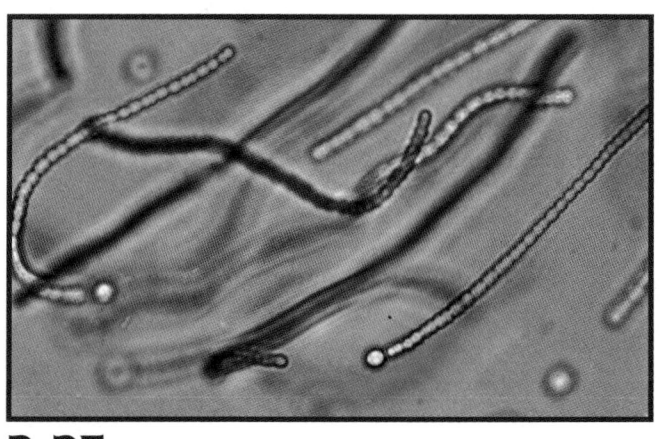

3-25 NOSTOC ✦ This genus is easily identified macroscopically because of its globular colonies and thick, rubbery mucilage. Trichomes are composed of spherical cells and form a tangled mass within the colony. Note the terminal spherical heterocysts. Trichomes are about 5 μm in width.

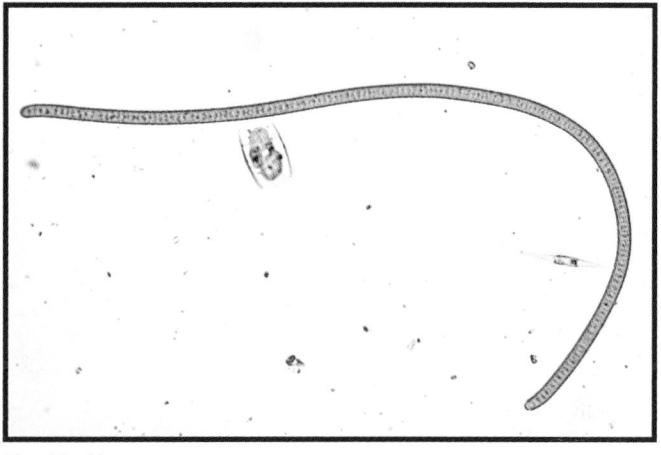

3-26 OSCILLATORIA ✦ Trichomes are formed from disc-shaped cells that are approximately 10 μm in width. Watch for gliding motility.

3-27 SPIRULINA ✦ *Spirulina* is perhaps the most distinctive cyanobacterium because of its helical shape. This genus is sold in health food stores as a dietary supplement.

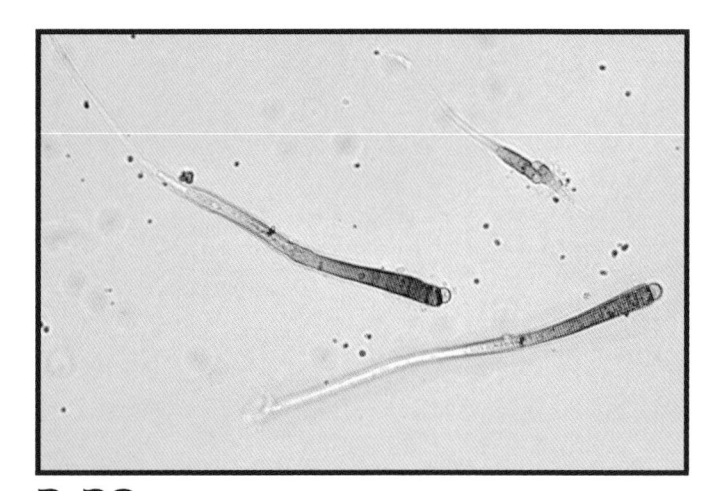

3-28 RIVULARIA ✦ Macroscopically, trichomes form globular, gelatinous masses. Microscopically, they are tapered to a fine point. Note the round, basal heterocysts. Akinetes are absent.

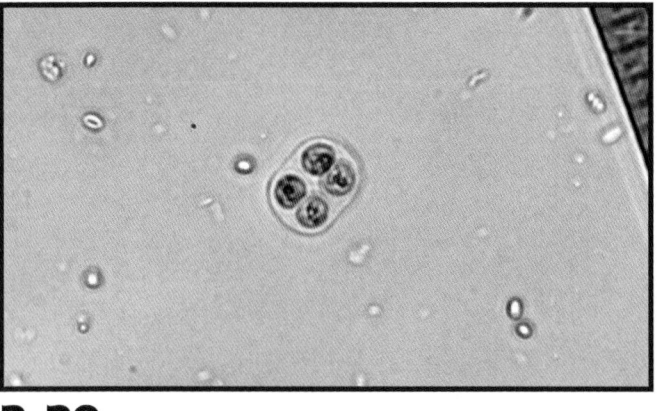

3-29 CHROOCOCCUS ✦ The sheathed cells of this genus occur as singles, pairs, or tetrads, with the latter two being more hemispherical than spherical.

are in a state of flux. Although *all* taxonomic decisions are provisional, protists are now placed into major groups (clades) that are more provisional than usual. Represented are heterotrophs, autotrophs, and mixotrophs (those that are capable of both autotrophic and heterotrophic growth), motile and nonmotile forms, those with cell walls and those without, and those with mitochondria and those with mitochondrial remnants.

The purpose of this lab is to practice using the microscope and view a variety of microbes. Its purpose is not to teach taxonomy. For that reason, coupled with the uncertain status of protist taxonomy, common names will be used for the groups you might observe.

✦ Green algae, **Chlorophyta**, will be seen as single cells, colonies, and branched or unbranched filaments (Figures 3-30 to 3-37). You may also see sheets of cells. Look for green chloroplasts, thick cell walls, vacuoles, and nuclei. Many also exhibit **pyrenoids** in the chloroplasts where photosynthetic products are stored.

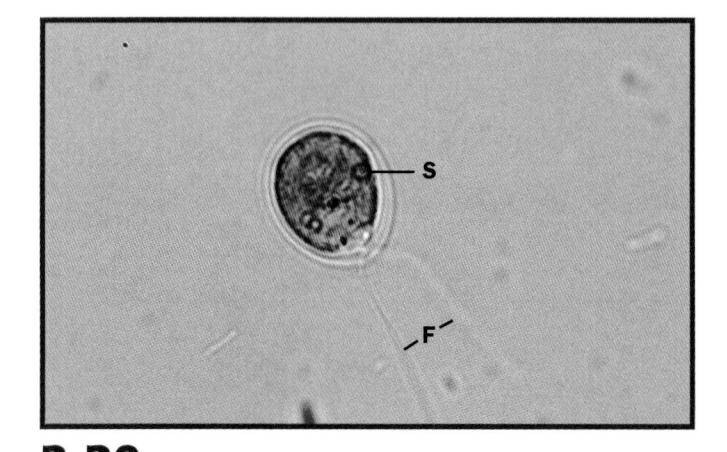

3-30 CHLAMYDOMONAS ✦ This unicellular green alga has two flagella (**F**) and a single, cup-shaped chloroplast. A red-orange stigma (**S**) is found within the chloroplast and is involved in phototaxis. Cells are approximately 30 μm long and 20 μm wide.

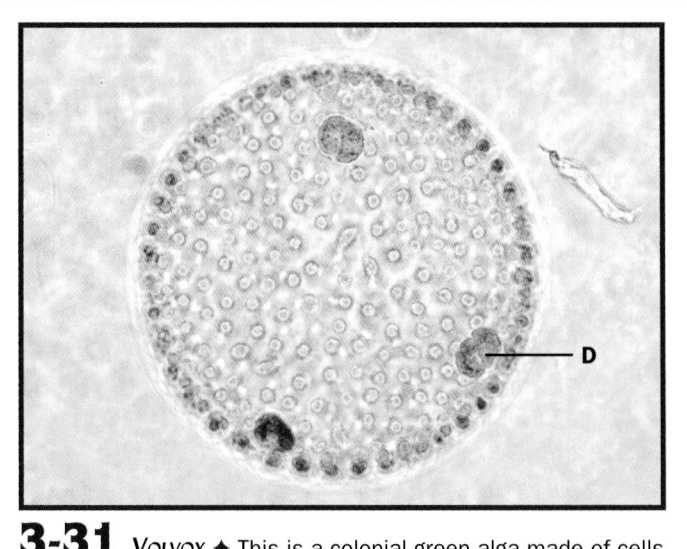

3-31 VOLVOX ✦ This is a colonial green alga made of cells similar in shape to *Chlamydomonas*. Daughter colonies (**D**) form asexually from special cells in the parent colony. Release of the daughter colony involves its eversion as it exits through an enzymatically produced pore. Mature colonies can reach a size visible to the naked eye.

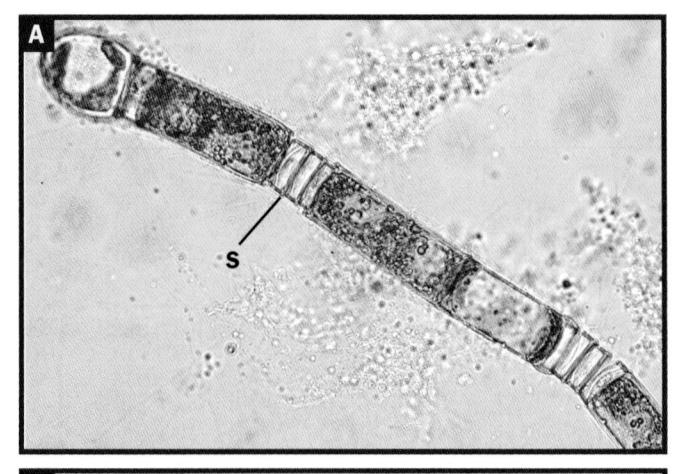

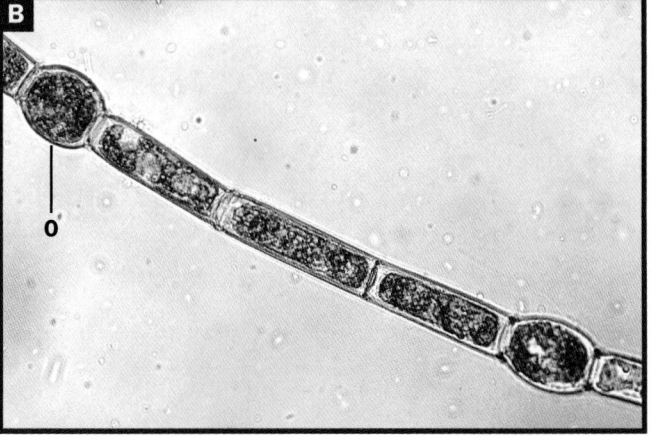

3-32 OEDOGONIUM ✦ **A** *Oedogonium* is a large chlorophyte genus composed of unbranched filaments. Division of cells within the filament results in its elongation and produces distinctive division scars (**S**). **B** The cylindrical cells of this filament are interrupted by two eggs, also called "oogonia" (**O**).

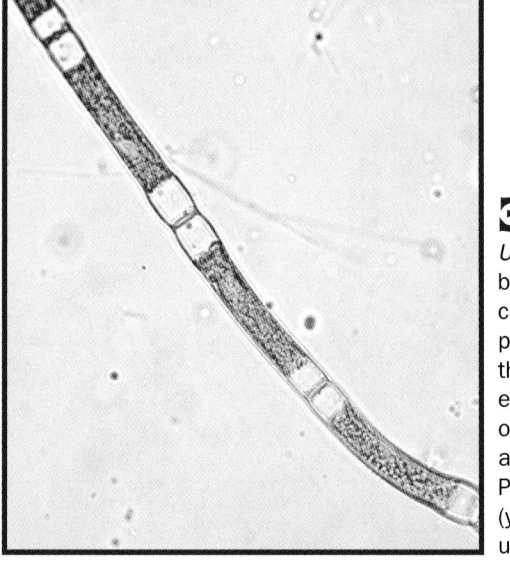

3-33 ULOTHRIX ✦ *Ulothrix* is another un-branched, filamentous chlorophyte. Its chloroplasts are found near the cell wall and form either a complete ring or an incomplete ring around the cytoplasm. Prominent pyrenoids (yellowish dots) are usually present.

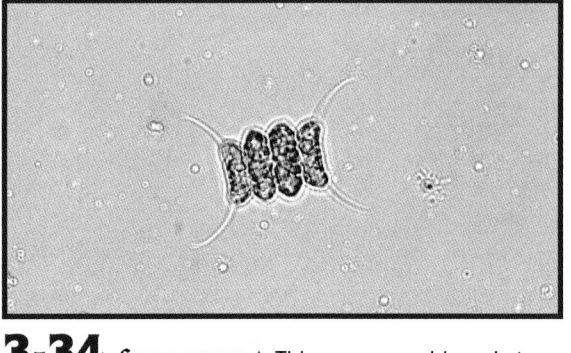

3-34 SCENEDESMUS ✦ This common chlorophyte consists of 4 or 8 cells joined along their edges. The cells on each end have distinctive spines. Asexual reproduction occurs as each cell of the colony divides to produce a new colony within the confines of its cell wall. These are subsequently released as the parental cell wall breaks down. Cells range from 30 to 40 μm across.

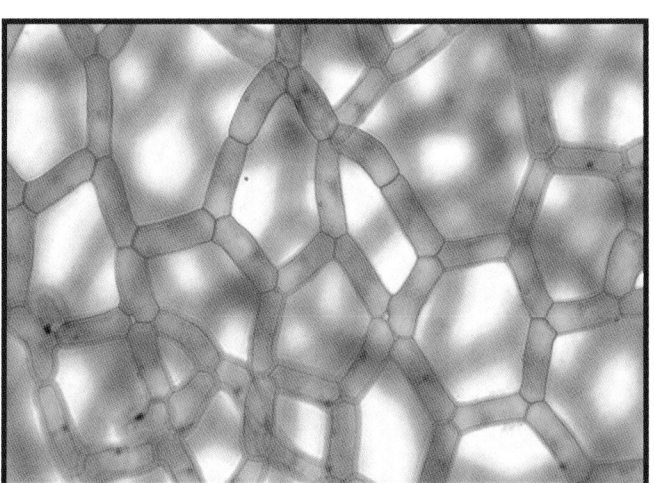

3-35 HYDRODICTYON ✦ The cells of this green alga form open connections with 4 (sometimes more) of its neighbors to produce a complex network of multinucleate cells. This explains its common name: the "water net." Cells may reach 1 cm in length, and each contains a single chloroplast with numerous pyrenoids. This specimen is from a culture.

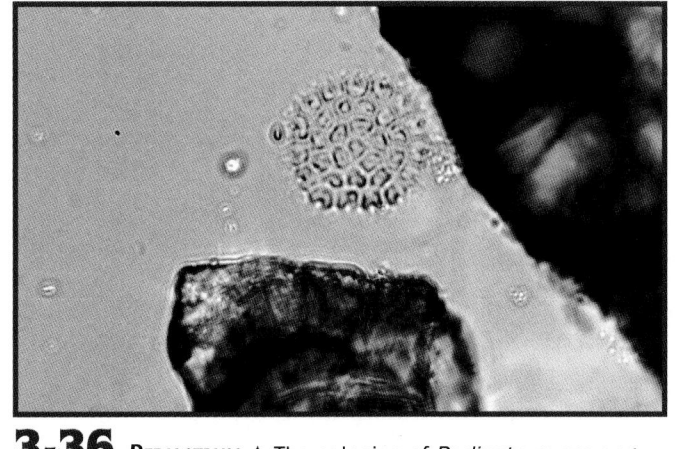

3-36 PEDIASTRUM ✦ The colonies of *Pediastrum* are angular and star-like, and the surface cells often contain bristles that are thought to be used for buoyancy. The cell number and arrangement are constant for each species. Each cell has the potential to produce a colony of the same number and arrangement of cells as the parent colony.

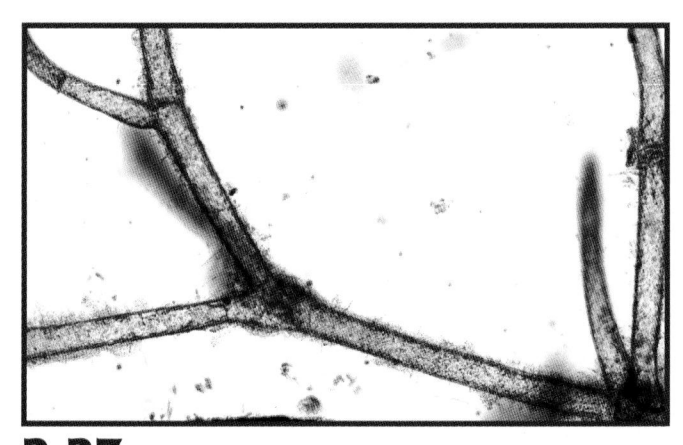

3-37 CLADOPHORA ✦ *Cladophora* is a branched chlorophyte with elongated cells. The single chloroplast has numerous pyrenoids and may look like a network of unconnected pieces.

✦ **Charophyceans** (Figures 3-38 to 3-40) comprise a second, diverse group of chlorophytes. Their distinctions are complex and often not visible to the light microscopist. Two groups are pretty easy to identify, however. These are the desmids, composed of paired "semicells" and species of the genus *Spirogyra* with their spiral chloroplasts.

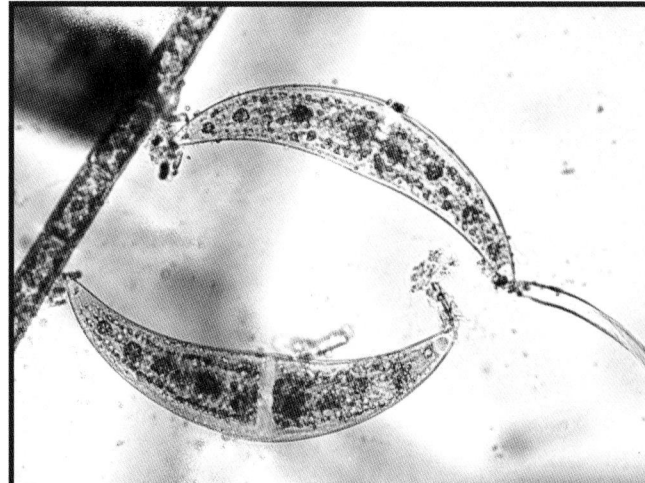

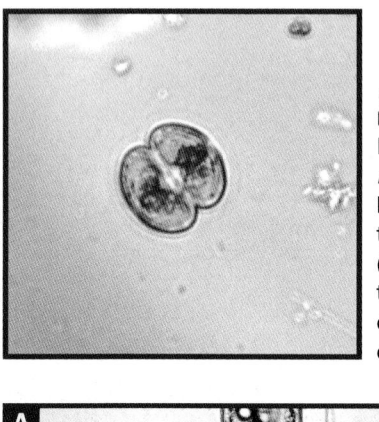

3-38 COSMARIUM ✦ Desmids, such as *Cosmarium*, are characterized by paired "semicells." In this species of *Cosmarium* (one of more than 1000!) two chloroplasts—one at each end—are present in each cell.

3-39 CLOSTERIUM ✦ *Closterium* is another desmid composed of two cells, but lacking the distinct constriction typical of most desmids. Vacuoles are present at the outer ends of the cells.

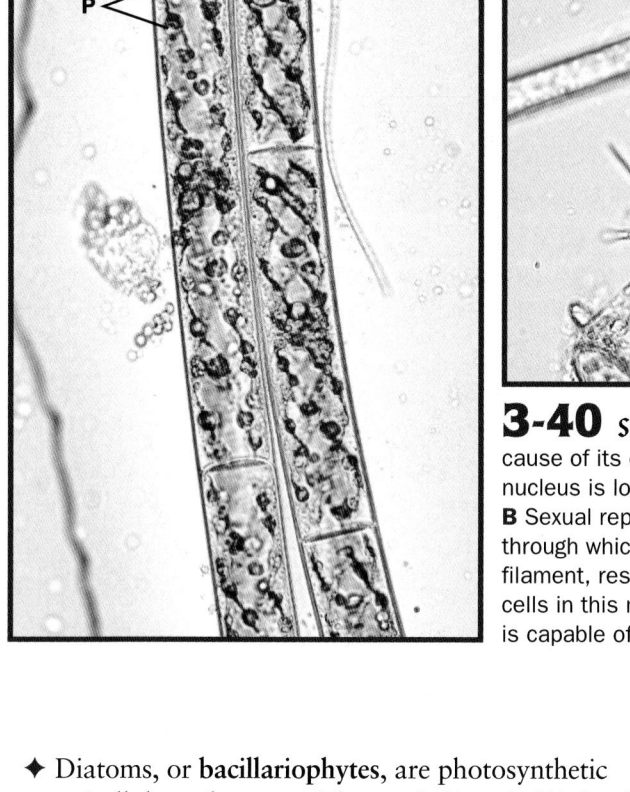

3-40 SPIROGYRA ✦ **A** *Spirogyra* is a well-known filamentous charophyte because of its distinctive spiral chloroplast. Note the numerous pyrenoids (**P**). The nucleus is located in the center of the cell and is obscured by the chloroplast. **B** Sexual reproduction occurs as two parallel filaments form conjugation tubes, through which one cell (a gamete) moves to join the cell (gamete) of the other filament, resulting in a zygote (**Z**). This type of conjugation has happened in four cells in this micrograph. Each zygote undergoes meiosis, and each resulting cell is capable of developing into a new filament.

✦ Diatoms, or **bacillariophytes,** are photosynthetic unicellular eukaryotes (Figures 3-41 to 3-50). Look for the distinctive golden brown color from the pigment **fucoxanthin** located in the **chromoplasts,** which may be of variable shape. Oil droplets are also frequently visible. Cell shapes are either round (**centric**) or bilaterally symmetrical and elongated (**pennate**). The cell wall is made of silica embedded in an organic matrix and consists of two halves, with one half overlapping the other in the same way the lid of a Petri dish overlaps its base. The combination of both halves is called a **frustule.** (Often, diatoms can be identified from the frustules of deceased cells.) Some pennate diatoms are motile and have a central, longitudinal line called a **raphe.** Look for these creeping diatoms!

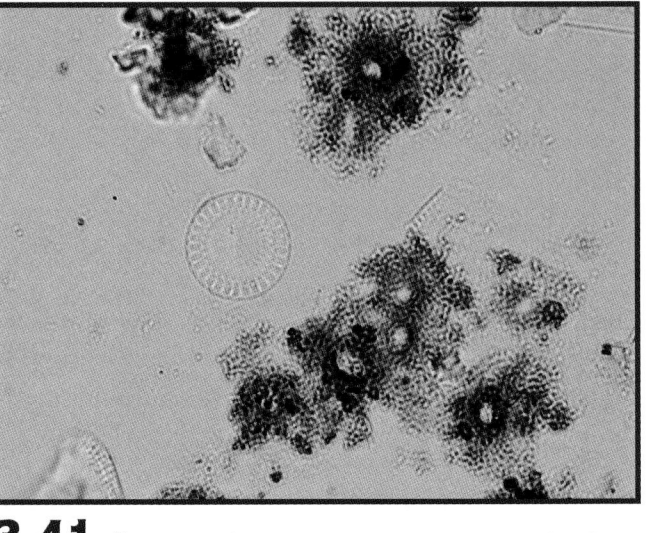

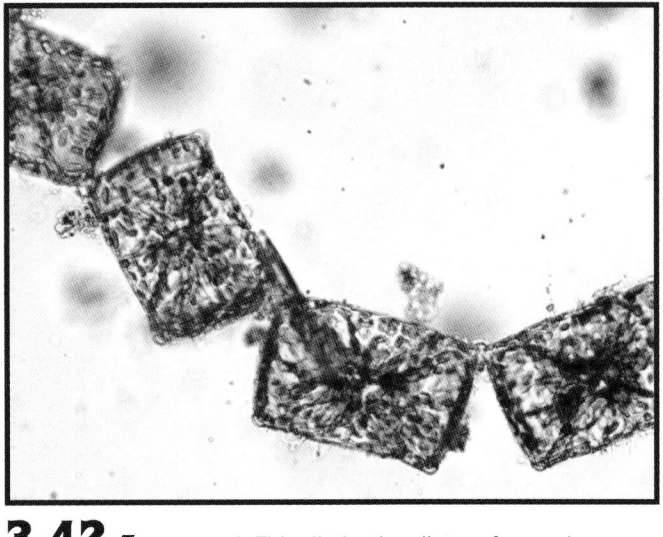

3-41 CYCLOTELLA ✦ In this micrograph we are seeing the empty frustule of *Cyclotella*, a centric diatom. The valve has a peripheral zone of radiate ("like spokes of a wheel") ridges.

3-42 TABELLARIA ✦ This distinctive diatom forms zigzag colonies. Note the golden brown chloroplasts.

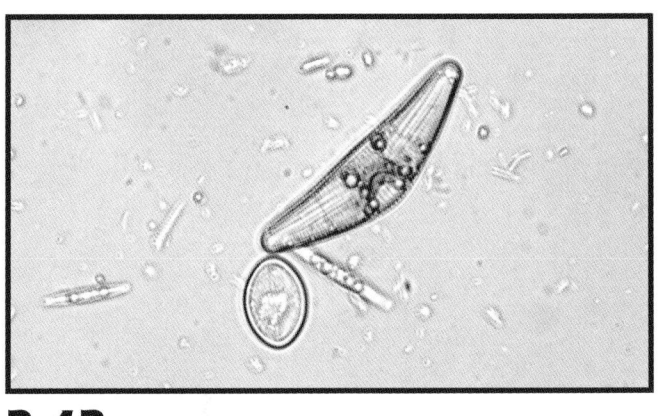

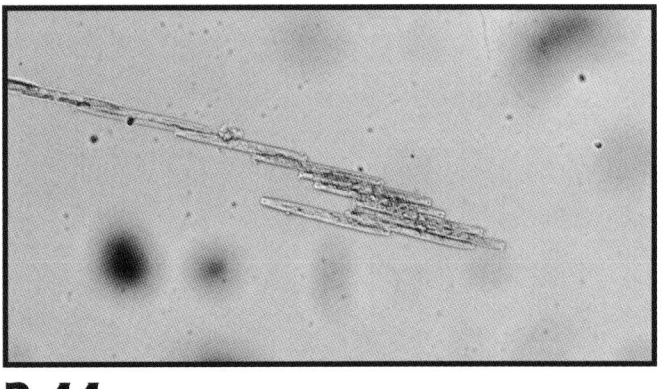

3-43 CYMBELLA ✦ This *Cymbella* species has bilaterally asymmetric valves and an H-shaped chloroplast. Note the prominent oil droplets.

3-44 BACILLARIA, ALSO KNOWN AS NITZSCHIA ✦ Members of this genus glide with individual cells sliding over one another. In this colony, the top cells are in the process of gliding to the right over the lower cells.

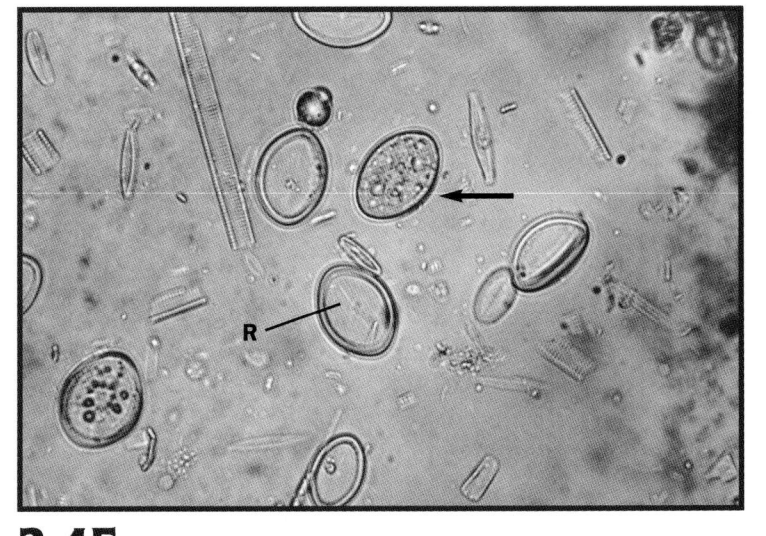

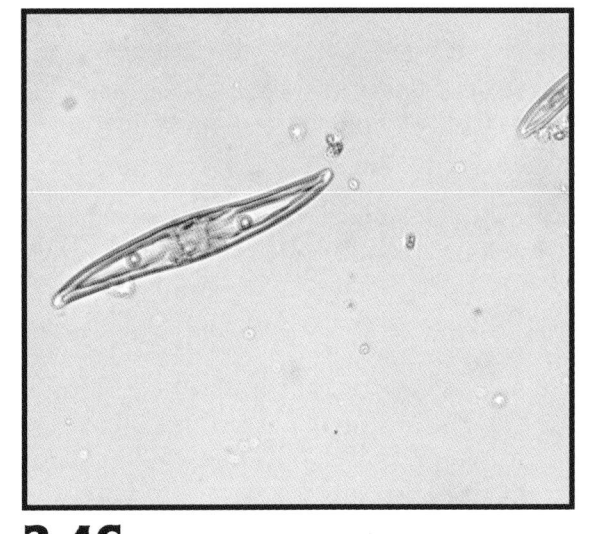

3-45 COCCONEIS ✦ These elliptical diatoms have transverse lines and a thick marginal band. There is also a faint longitudinal line called a **raphe** (**R**). Several empty frustules are visible, as is one living specimen (arrow).

3-46 GYROSIGMA ✦ All species of this genus have sigmoid-shaped cells. Note the golden brown chloroplasts at the edge and two prominent oil droplets.

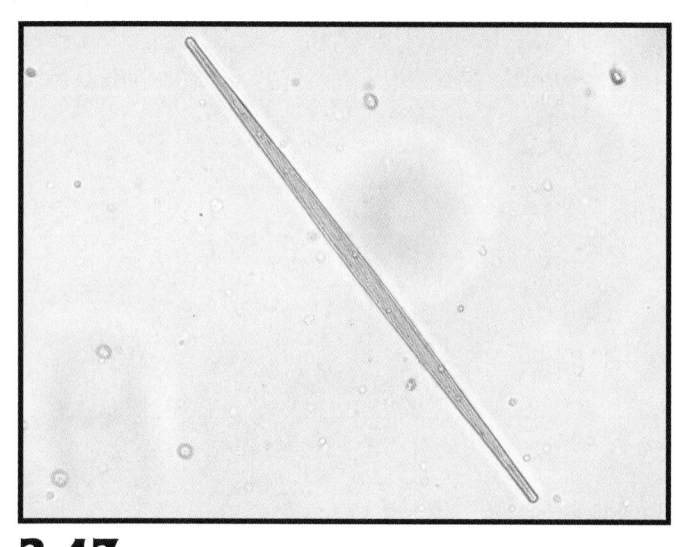

3-47 SYNEDRA ✦ These diatoms are distinctive because of their long, needle-shaped cells. They occur singly or sometimes in groups of cells attached at their ends, radiating outward like spokes of a wheel.

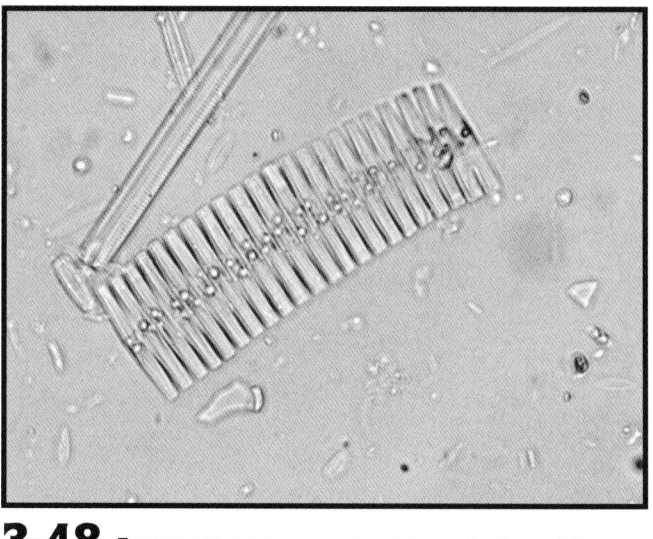

3-48 FRAGILLARIA ✦ These rectangular cells form ribbons with cells attached side-by-side. Note the golden brown chloroplast and oil droplets.

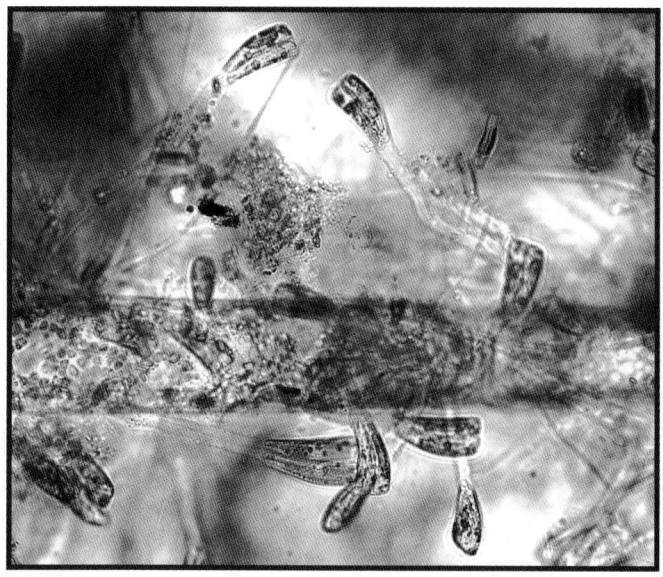

3-49 GOMPHONEMA ✦ Wedge-shaped cells on branched stalks characterize the species of *Gomphonema*.

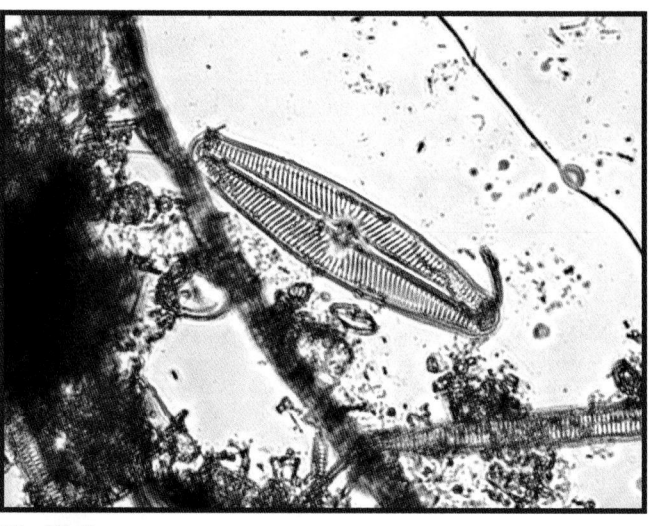

3-50 NAVICULA ✦ All of the many species of *Navicula* are shaped like a cigar or a boat. Cells are identifiable by prominent transverse lines converging on the central open space. A prominent longitudinal line is also present.

✦ **Ciliates** (Figures 3-51 to 3-54) are a diverse unicellular group, but all have cilia for locomotion and feeding. Some are free swimming while others are attached to a surface. Look for an oral groove, contractile vacuoles that pump water out of the cell, and food vacuoles.

3-51 VORTICELLA ✦ **A** *Vorticella* is a genus of stalked, inverted bell-shaped ciliates. A crown of cilia surround the oral groove and are involved in feeding. **B** *Vorticella* is capable of retraction when the stalk coils. If you find a *Vorticella* spend some time watching it until it recoils. It's worth the wait (which shouldn't be more than a minute).

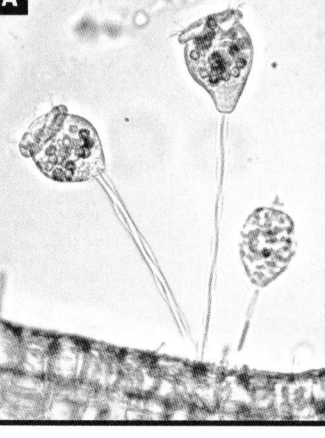
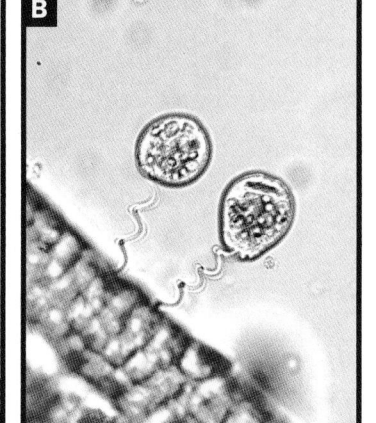

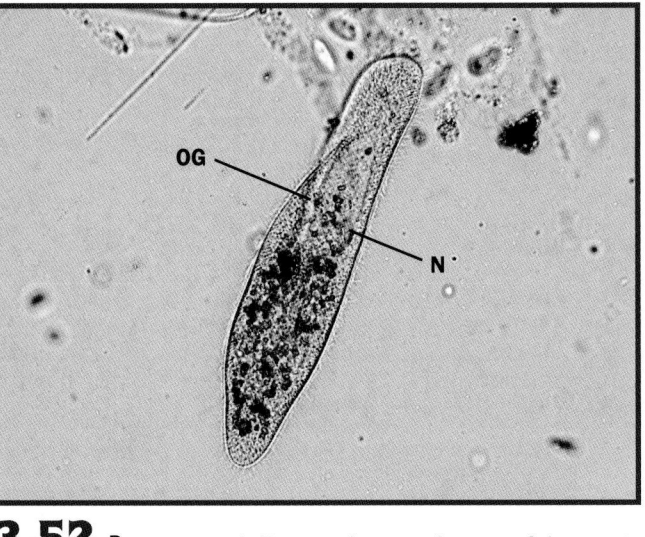

3-52 PARAMECIUM ✦ *Paramecium* may be one of the most famous protists. Notice the cilia and the oral groove (**OG**) present in the overlapping part of the cell. The black structures are food vacuoles at the front of the cell. A nucleus (**N**) is visible just behind the food vacuoles.

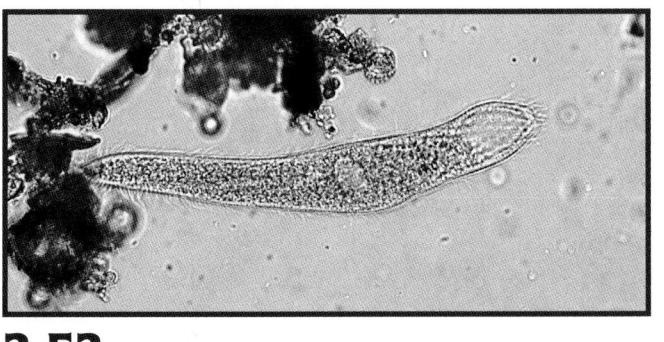

3-53 TRACHELOCERCA ✦ This ciliate has the general appearance of *Paramecium* but is usually more elongated and has a prominent contractile vacuole (used for pumping water out of the cell) in the posterior.

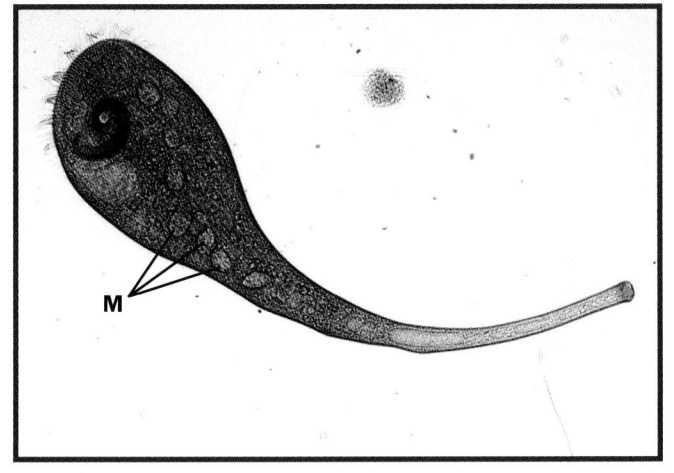

3-54 STENTOR ✦ Its trumpet shape, size (up to 2 mm), and beaded macronucleus (**M**) make *Stentor* an easy ciliate to identify. It is covered with cilia, some of which surround its "mouth" (the white spot inside the dark coiled region). This species is naturally green; this is not a stained specimen.

✦ **Dinoflagellates** (Figure 3-55) are typically unicellular and autotrophic. Most have two flagella: one protruding from the cell and the other positioned in a groove (**G**) encircling the cell.

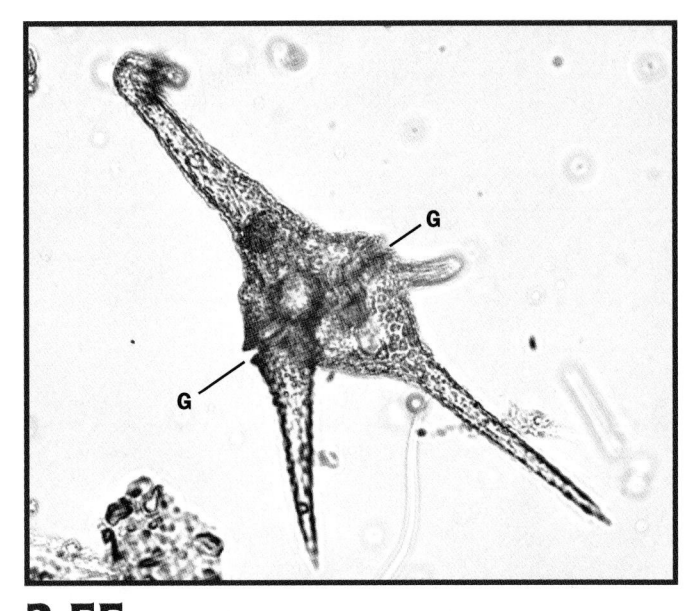

3-55 CERATIUM ✦ Dinoflagellates have two flagella. One flagellum is found in a groove (girdle) (**G**) encircling the cell, whereas the other is elongated and is oriented posteriorly. This species has one anterior horn and three posterior horns and is covered with cellulose plates.

✦ **Euglenids** (Figure 3-56) are green, photosynthetic protists when light is available, but they are capable of heterotrophy when light is not. Look for chloroplasts. One or two flagella are present and emerge from an invagination of the anterior (forward) cytoplasmic membrane. A red photoreceptor called an **eyespot** at the cell's anterior is another distinctive feature.

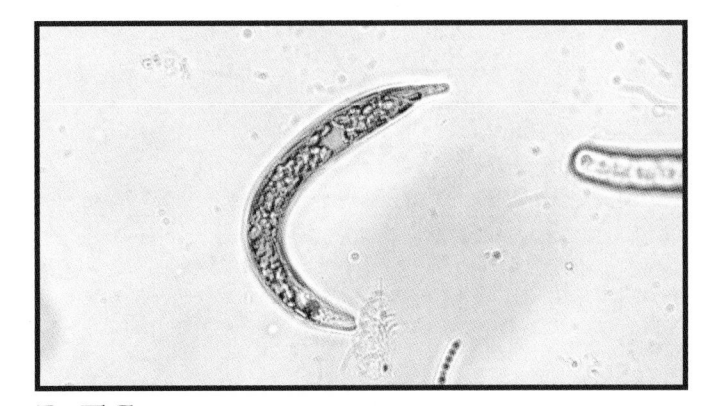

3-56 EUGLENA ✦ *Euglena* is a large genus of mixotrophic flagellates. Most species have chloroplasts, which are discoid in this specimen. A red "eyespot" is located in the colorless anterior of the cell. The single flagellum also emerges from the anterior.

✦ **Amoebas** (Figures 3-57 and 3-58) move by producing temporary, lobe-shaped **pseudopodia** that also serve as feeding structures. In this latter role, they form and subsequently engulf their microorganismal food. Look for pseudopods, food vacuoles, and the nucleus.

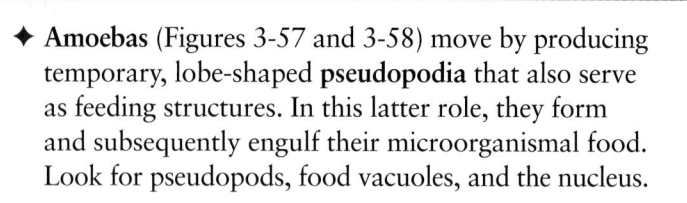
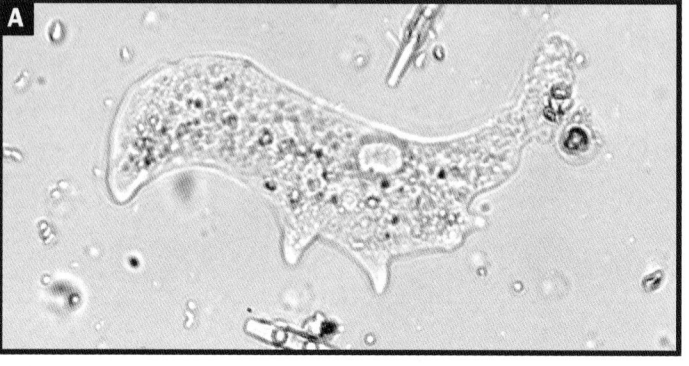

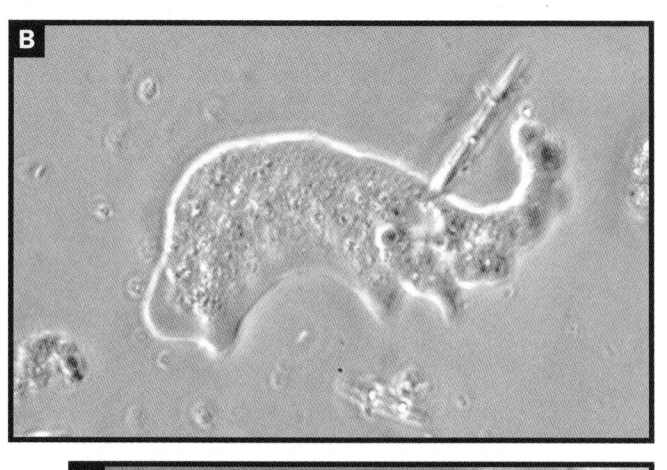

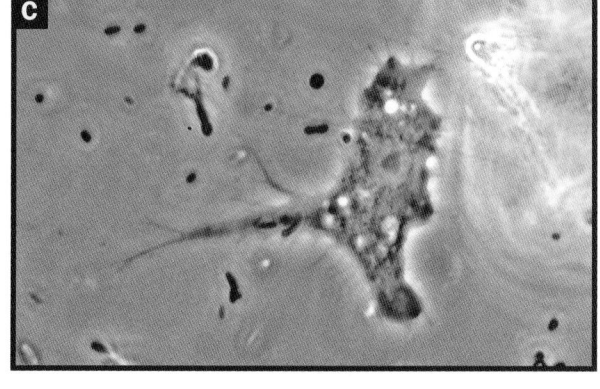

3-57 AMOEBA ✦ Cells with flowing cytoplasm and a constantly changing shape are likely to be amoebas. They move and capture prey by extending pseudopods ("false feet") outward. **A** This bright-field micrograph clearly illustrates the difference between the cytoplasm at the periphery of the cell (ectoplasm) with the cytoplasm in the interior (endoplasm). The nucleus is visible, as are numerous food vacuoles seen as golden spots—probably an indication of this individual's food preferences! **B** This is the same amoeba viewed with phase contrast only 1 or 2 seconds after the previous micrograph (notice that the shape is basically the same and the diatom at the top has moved only a short distance). Phase contrast produces a different texture to the image. Compare the ectoplasm, endoplasm, and nucleus with the bright-field image. **C** This is a totally different amoeba, interesting for its ornate pseudopods.

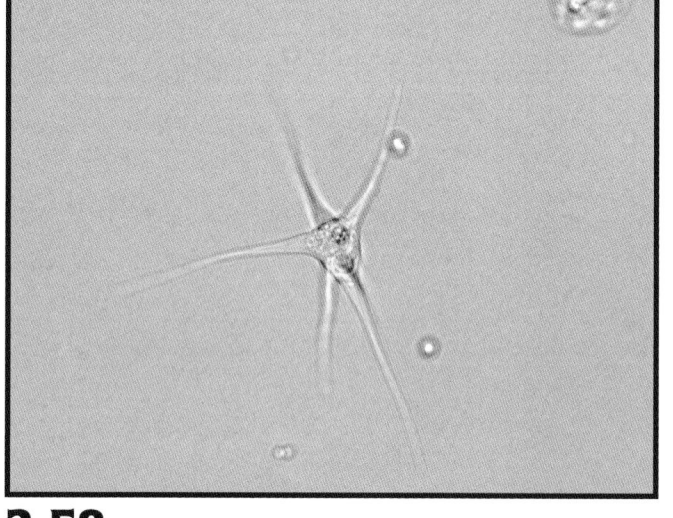

3-58 ASTRAMOEBA ✦ This amoeba possesses long, thin pseudopods that may coil when disturbed. Movement is slow, but the irregular shape tends to remain unchanged for long periods of time. These cells are less than 2 mm across (including pseudopods).

✦ **Heliozoans** (Figure 3-59) are spherical, unicellular, phagocytic protists characterized by projections called **axopodia**, used in feeding. Their sunburst appearance accounts for their name.

3-59 HELIOZOAN ✦ These spherical, sun-shaped (hence, "*helio*") organisms are among the most beautiful protists. They are spherical and have thin extensions called **axopodia**, which are used in feeding. They are not rigid but, rather, are extensions of ectoplasm and can be bent or retracted.

The **invertebrates** (Figures 3-60 to 3-63) include all animal phyla that are not in the phylum Chordata—in other words, the majority of animals. These are discussed in the photograph captions.

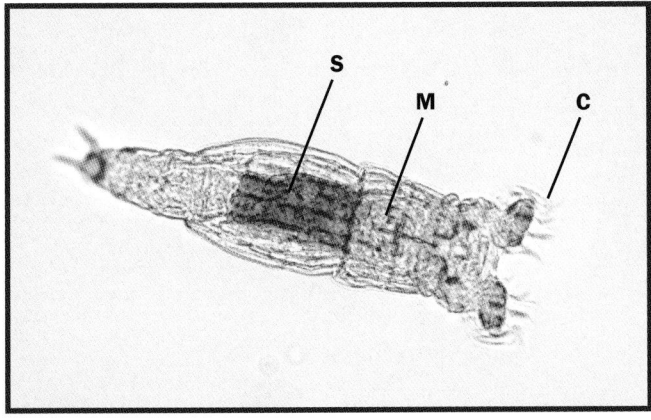

3-60 ROTIFER ✦ These are commonly called "wheel animals," and it takes only a quick glance to see why. At their "head" is a crown, or **corona**, of cilia (**C**), which sweep food particles into their mouth and also provide a means of locomotion. Perhaps the most prominent part of the cylindrical body is the **mastax** (**M**), which is a grinding organ just posterior to the mouth. You will see it move. In some rotifers it can be extended to capture prey. In this unstained specimen, the stomach (**S**) is also easily seen. Also look for the spurs extending from the posterior, which are used for attachment.

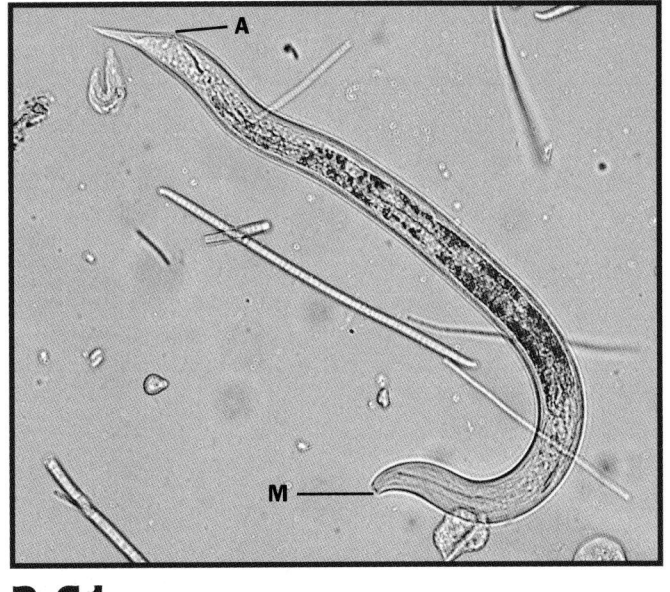

3-61 NEMATODE WORM ✦ Nematodes are also known as roundworms, and they are abundant in all habitats. They are unsegmented and have a complete digestive tract; that is, they have a mouth (**M**) and an anus (**A**). The most notable feature of these worms is how they thrash about. You may see them directly, or your eyes may be drawn to plant debris and soil particles being stirred up. Look more closely and you will probably find a nematode!

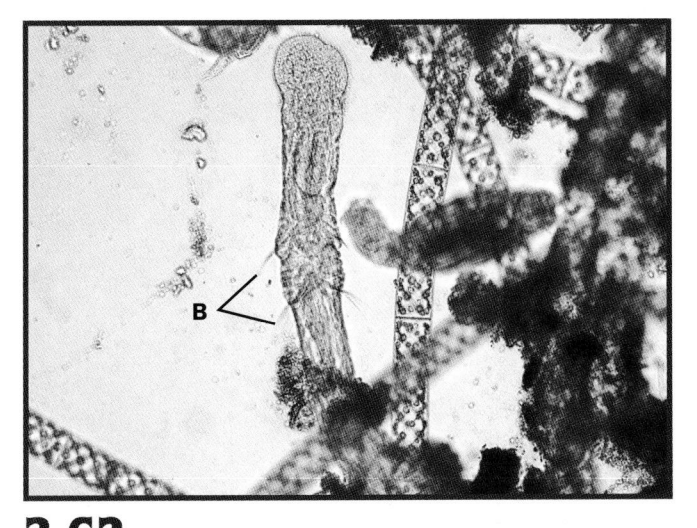

3-62 ANNELID WORM ✦ The bodies of annelid worms are divided into repeating segments, and they also have a complete digestive tract. This is a micrograph of a polychaete worm and is easily identified by the presence of bristles (**B**) on each segment.

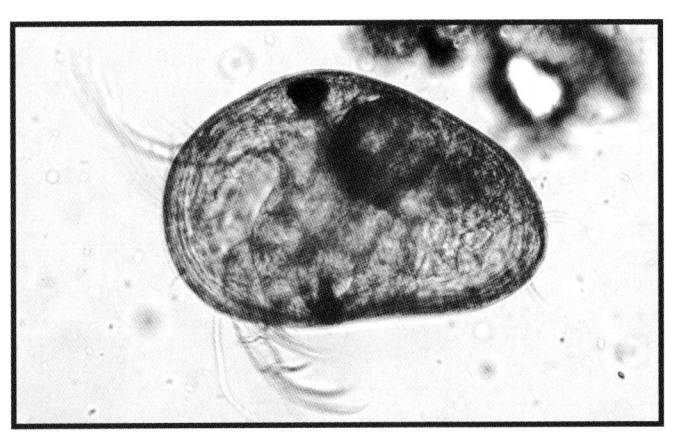

3-63 ARTHROPOD ✦ Arthropods, the largest animal phylum, are characterized by their jointed appendages. They comprise such diverse animals as barnacles, spiders, crabs, insects, and many others. Shown here is a crustacean, probably of the ostracod group, which is closely related to the "water fleas," such as *Daphnia*.

✦ Application

This lab is designed to have you practice using the microscope and experience the awe-inspiring microbial diversity and beauty of microbial life found in natural water sources.

✦ In This Exercise

You will observe pond, river, or other source of freshwater life and identify representative microorganisms from diverse biological taxa such as Bacteria, cyanobacteria, green algae, protists, and a variety of invertebrate animals. Although invertebrates are not considered to be microorganisms, you will see them, so they are covered here.

✦ Materials

Per Student Group

✦ pond water—the goopier the better! Mud and green stuff is good!

✦ microscope slides

✦ cover slips

✦ compound microscope with ocular micrometer

✦ plastic transfer pipettes

✦ one dropper bottle of 1.5% w/v methylcellulose

 Procedure

1 Obtain a clean glass microscope slide and cover slip.

2 Prepare a wet mount slide by following the Procedural Diagram (Figure 3-19). With a transfer pipette, place a drop of pond water on the slide. Take your sample from a region with some goo or mud. You may see some organisms if you take water from the surface, but you'll see more where there is organic material.

3 Carefully place a cover slip over the drop. Try not to trap any air underneath it, but this may be impossible if you have sand grains or other large objects in the drop. If you do, take off the cover slip and try to remove the offending objects. It may take a new drop of water before you replace the cover slip. In some cases, it is just easier to wipe off the slide, start over, and avoid the sand.

4 Examine the water drop using the scanning, low, and high dry objectives.

5 If the swimmers are moving too quickly to observe, remove the cover slip and add a drop of methyl cellulose to slow them. Replace the cover slip (avoiding air bubbles) and continue your observations. Be careful not to get the methyl cellulose on any objective lens or the stage. If you do, clean it up immediately with lens paper for the lenses and tissue for the stage.

6 Use Figures 3-20 to 3-63 to place the organisms you observe into the taxa described in the discussion of Theory.

7 Record your observations and answer the questions on the Data Sheet.

References

Bold, Harold C., and Michael J. Wynne. 1985. *Introduction to the Algae*, 2nd ed. Prentice-Hall, Englewood Cliffs, NJ.

Graham, Linda E., James M. Graham, and Lee W. Wilcox. 2009. *Algae*, 2nd ed. Pearson Benjamin Cummings, San Francisco, CA.

Jahn, Theodore Louis, Eugene C. Bovee, Francis Floed Jahn, John Bamrick, Edward T. Cauley, and Wm G. Jaques. 1978. *How to Know the Protozoa*, 2nd ed. The McGraw-Hill Companies, P.O. Box 182604, Columbus, OH.

Lee, Robert Edward. 2008. *Phycology*, 4th ed. Cambridge University Press, New York, NY.

Madigan, Michael T., John M. Martinko, Paul V. Dunlap, and David P. Clark. 2009. *Brock Biology of Microorganisms*, 12th ed. Pearson Benjamin Cummings, San Francisco, CA.

Needham, James G., and Paul R. Needham. 1962. *A Guide to the Study of Fresh-water Biology*, 5th ed. Holden-Day, San Francisco, CA.

Prescott, G., John Bamrick, Edward T. Cauley, and Wm G. Jaques. 1978. *How to Know the Protozoa*, 3rd ed. The McGraw-Hill Companies, Columbus, OH.

Vinyard, William C. 1979. *Diatoms of North America*. Mad River Press, Eureka, CA.

Bacterial Structure and Simple Stains

In Exercise 3-1 you were introduced to two of the three important features of a microscope and microscopy: magnification and resolution. A third feature is **contrast**. To be visible, the specimen must contrast with the background of the microscope field. Because cytoplasm is essentially transparent, viewing cells with the standard light microscope is difficult without stains to provide that contrast. In this set of exercises, you will learn how to correctly prepare a bacterial smear for staining and how to perform simple and negative stains. Cell morphology, size, and arrangement then may be determined. In a medical laboratory these usually are determined with a Gram stain (Exercise 3-7), but you will be using simple stains as an introduction to the staining process, as well as an introduction to these cellular characteristics.

Bacterial cells are much smaller than eukaryotic cells (Figure 3-11) and come in a variety of morphologies (shapes) and **arrangements**. Determining cell morphology is an important first step in identifying a bacterial species. Cells may be spheres (**cocci**, singular **coccus**), rods (**bacilli**, singular **bacillus**) or spirals (**spirilla**, singular **spirillum**). Variations of these shapes include slightly curved rods (**vibrios**), short rods (**coccobacilli**) and flexible spirals (**spirochetes**). Examples of cell shapes are shown in Figures 3-64 through 3-71. In Figure 3-71, *Corynebacterium xerosis* illustrates **pleomorphism**. A variety of cell shapes—slender, ellipsoidal, or ovoid rods—may be seen in a given sample.

Cell arrangement, determined by the number of planes in which division occurs and whether the cells separate after division, is also useful in identifying bacteria. Spirilla rarely are seen as anything other than single cells, but cocci and bacilli do form multicellular associations. Because cocci exhibit the most variety in arrangements, they are used for illustration in Figure 3-72. If the two daughter cells remain attached after a coccus divides, a **diplococcus** is formed. The same process happens in bacilli that produce **diplobacilli**. If the cells continue to divide in the same plane and remain attached, they exhibit a **streptococcus** or **streptobacillus** arrangement.

If a second division occurs in a plane perpendicular to the first, a **tetrad** is formed. A third division plane perpendicular to the other two produces a cube-shaped arrangement of eight cells called a **sarcina**. Tetrads and sarcinae are seen only in cocci. If the division planes of a coccus are irregular, a cluster of cells is produced to form a **staphylococcus**. Figures 3-73 through 3-78 illustrate common cell arrangements.

Arrangement and morphology often are easier to see when the organisms are grown in a broth rather than a solid medium, or are observed from a direct smear. If you have difficulty identifying cell morphology or arrangement, consider transferring the organism to a broth culture and trying again.

One last bit of advice: Don't expect nature to conform perfectly to our categories of morphology and cell arrangement. These are convenient descriptive categories that will not be applied easily in all cases. When examining a slide, look for the most common morphology and most complex arrangement. Do not be afraid to report what you see. For instance, it's okay to say, "Cocci in singles, pairs and chains."

Cells are three-dimensional objects with a surface that contacts the environment and a volume made up of cytoplasm. The ability to transport nutrients into the cell and wastes out of the cell is proportional to the amount of surface area doing the transport. The demand for nutrients and production of wastes is proportional to a cell's volume. It's a mathematical fact of life that as an object gets bigger, its volume increases more rapidly than its surface area. Therefore, a cell can achieve a size at which its surface area is not adequate to supply the nutrient needs of its cytoplasm. That is, its **surface-to-volume ratio** is too small. At this point, a cell usually divides its volume to increase its surface area. This phenomenon is a major factor in limiting cell size and determining a cell's habitat.

Bacilli, cocci, and spirilla with the same volume have different amounts of surface area. A sphere has a lower surface-to-volume ratio than a bacillus or spirillum of the same volume. A streptococcus, however, would have approximately the same surface-to-volume ratio as a bacillus of the same volume. ✦

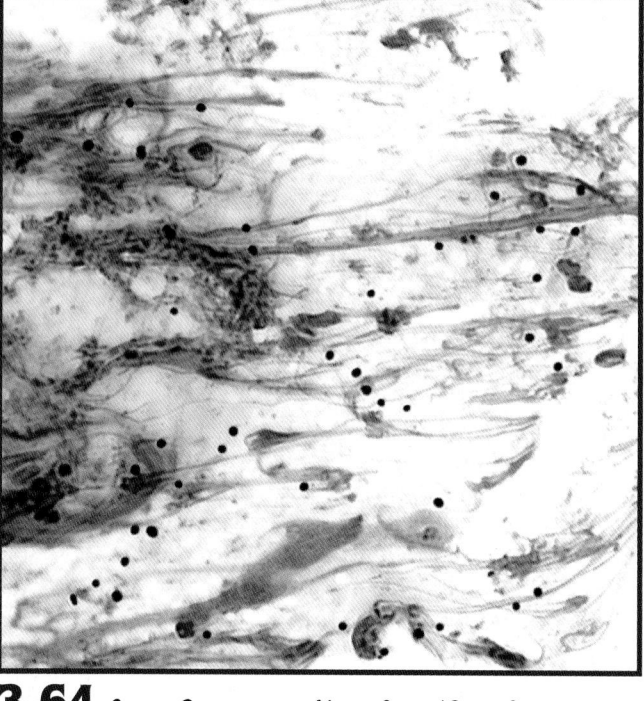

3-64 SINGLE COCCI FROM A NASAL SWAB (GRAM STAIN, X1000) ✦ This direct smear of a nasal swab illustrates unidentified cocci (dark circles) stained with crystal violet. The red background material is mostly mucus.

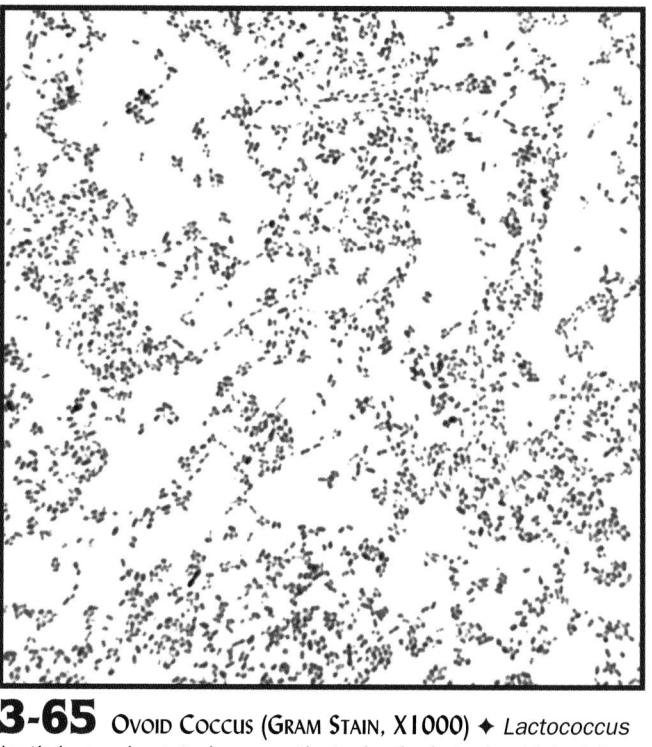

3-65 OVOID COCCUS (GRAM STAIN, X1000) ✦ *Lactococcus lactis* is an elongated coccus that a beginning microbiologist might confuse with a rod. Notice the slight elongation of the cells, and also that most cells are not more than twice as long as they are wide. *L. lactis* is found naturally in raw milk and milk products, but these cells were grown in culture.

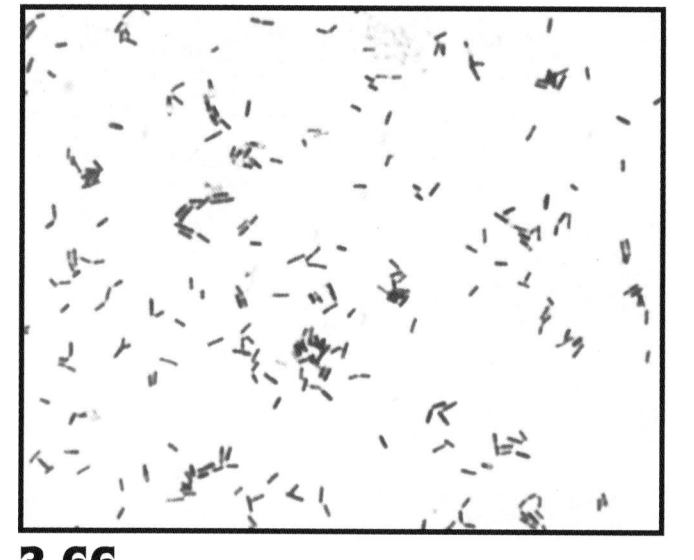

3-66 "TYPICAL" BACILLUS (CRYSTAL VIOLET STAIN, X1050) ✦ Notice the variability in rod length (because of different ages of the cells) in this stain of the soil organism *Bacillus subtilis* grown in culture. Also notice the squared ends on the cells, typical of the genus *Bacillus*.

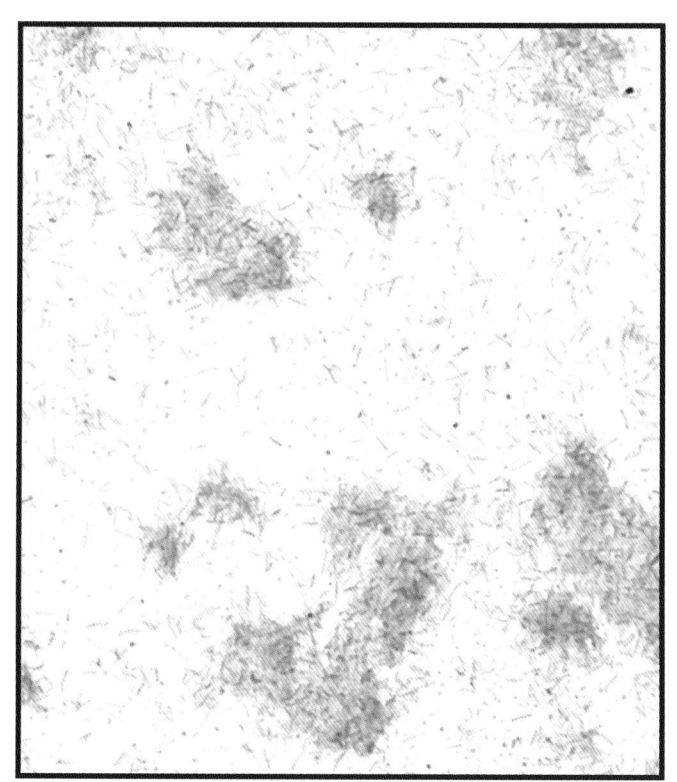

3-67 LONG, THIN BACILLUS (GRAM STAIN, X1000) ✦ The cells of *Aeromonas sobria*, a freshwater organism, are considerably longer than they are wide. These cells were grown in culture. Notice that a cell can be a bacillus without being in the genus *Bacillus*.

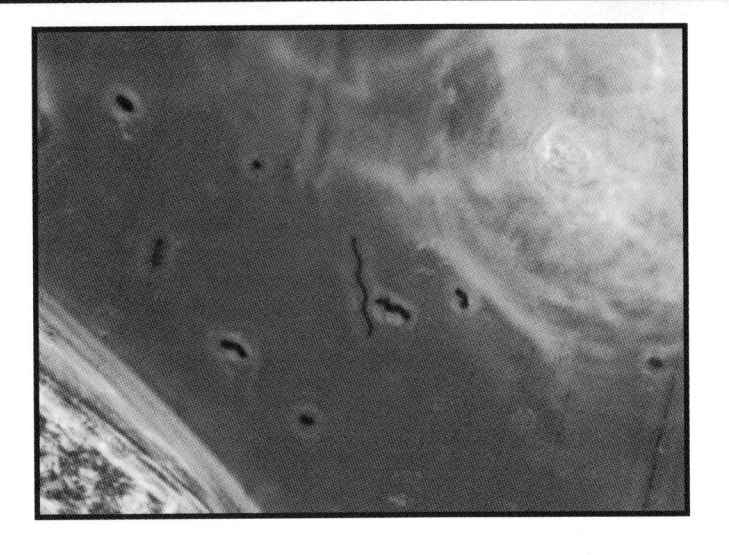

3-68 TWO DIFFERENT SPIRILLA (PHASE CONTRAST WET MOUNT, X1000) ✦ The two different spirilla are undoubtedly different species based on their different morphologies: one is long and slender with loose spirals, the other is shorter and fatter with tighter coils.

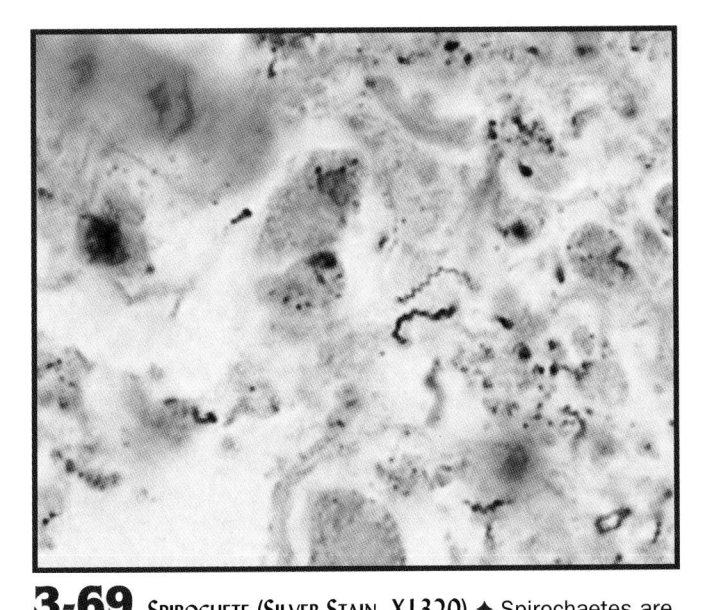

3-69 SPIROCHETE (SILVER STAIN, X1320) ✦ Spirochaetes are flexible, curved rods. Shown is *Treponema pallidum* in tissue stained with a silver stain that makes the cells appear black. *T. pallidum* is the causative agent of syphilis in humans and cannot be cultured.

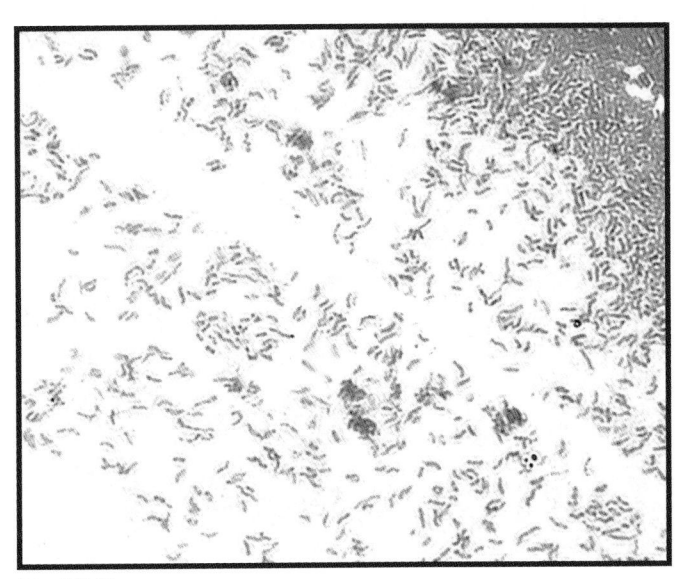

3-70 VIBRIO (GRAM STAIN, X1000) ✦ *Vibrio natriegens* grows naturally in salt marshes. This specimen was grown in culture.

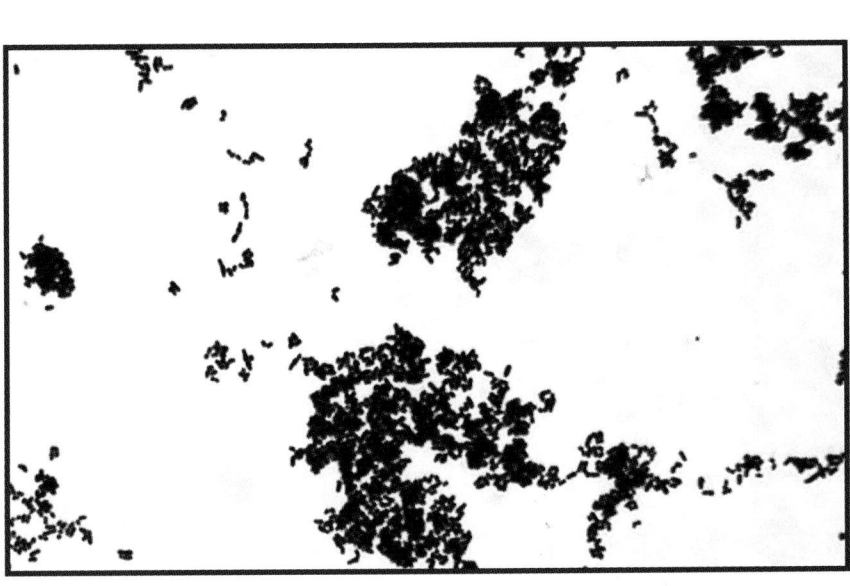

3-71 BACTERIAL PLEOMORPHISM (GRAM STAIN, X1000) ✦ Some organisms grow in a variety of shapes and are said to be **pleomorphic**. Notice that the rods of *Corynebacterium xerosis* range from almost spherical to many times longer than wide. This organism normally inhabits skin and mucous membranes and may be an opportunistic pathogen in compromised patients.

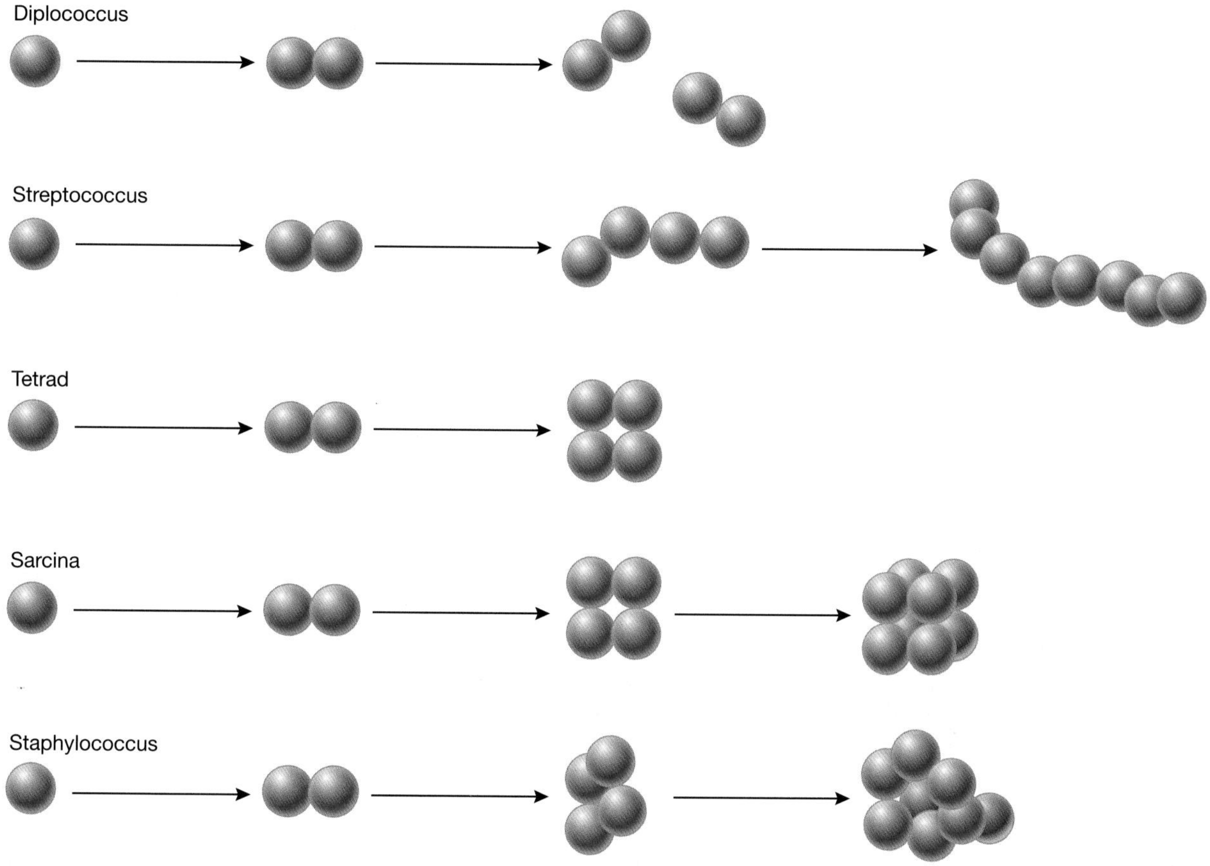

Diplococcus

Streptococcus

Tetrad

Sarcina

Staphylococcus

3-72 DIVISION PATTERNS AMONG COCCI ✦ Diplococci have a single division plane and the cells generally occur in pairs. Streptococci also have a single division plane, but the cells remain attached to form chains of variable length. If there are two perpendicular division planes, the cells form tetrads. Sarcinae have divided in three perpendicular planes to produce a regular cuboidal arrangement of cells. Staphylococci have divided in more than three planes to produce a characteristic grapelike cluster of cells. (**Note:** Rarely will a sample be composed of just one arrangement. Report what you see, and emphasize the most complex arrangement.)

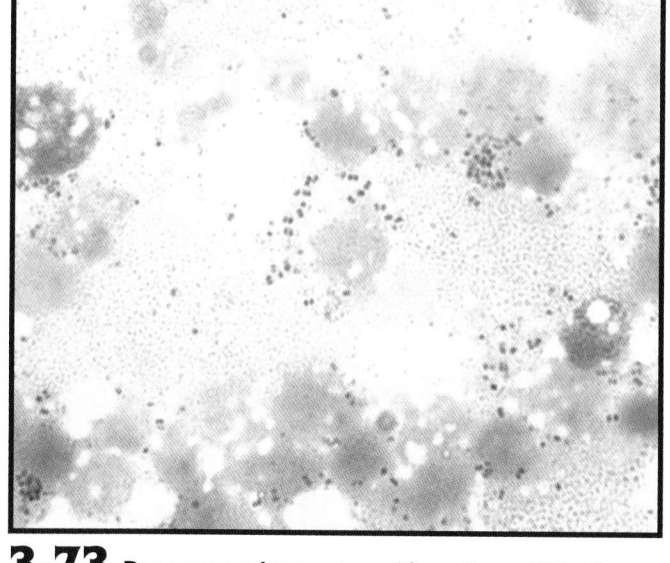

3-73 DIPLOCOCCUS ARRANGEMENT (GRAM STAIN, X1000) ✦ *Neisseria gonorrhoeae*, a diplococcus, causes gonorrhea in humans. Members of this genus produce diplococci with flattened adjacent sides.

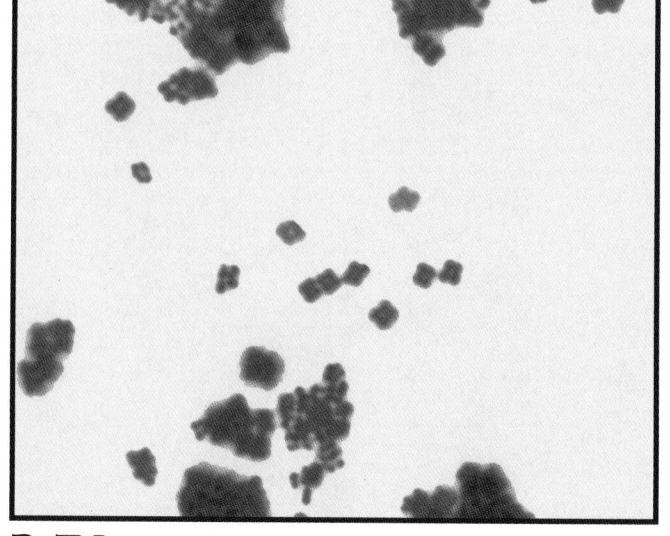

3-74 TETRAD ARRANGEMENT (GRAM STAIN, X1320) ✦ *Micrococcus roseus* grows in squared packets of cells, even when they are bunched together. The normal habitat for *Micrococcus* species is the skin, but the ones here were obtained from culture.

3-75 STREPTOCOCCUS ARRANGEMENT (GRAM STAIN, X1000) ✦ *Enterococcus faecium* is a streptococcus that inhabits the digestive tract of mammals. This specimen is from a broth culture (which enables the cells to form long chains) and was stained with crystal violet. Notice the slight elongation of these cells along the axis of the chain.

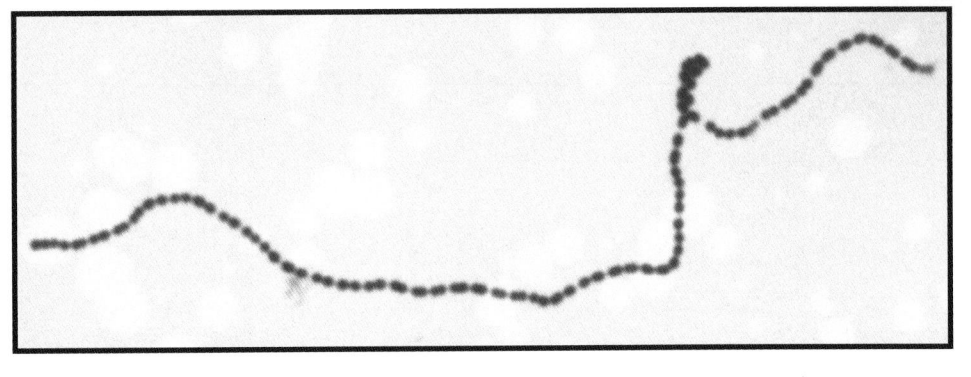

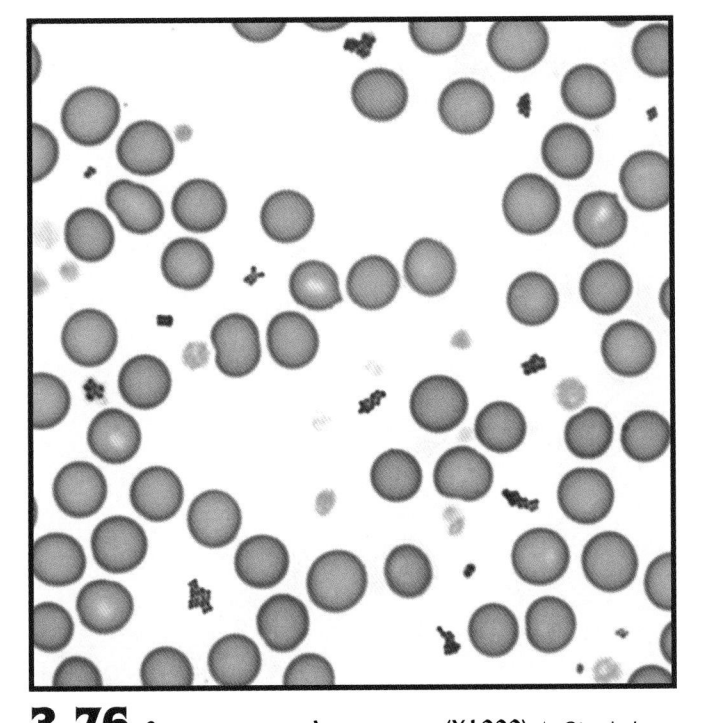

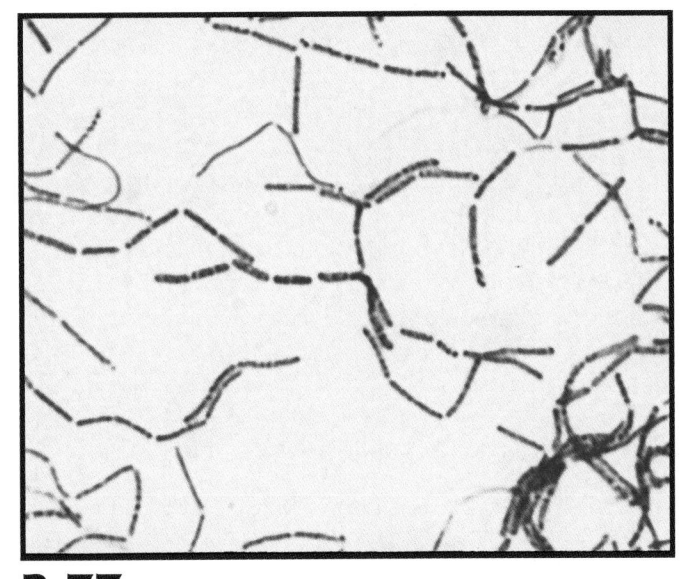

3-77 STREPTOBACILLUS ARRANGEMENT (CRYSTAL VIOLET STAIN, X1200) ✦ *Bacillus megaterium* is a streptobacillus. These cells were obtained from culture.

3-76 STAPHYLOCOCCUS ARRANGEMENT (X1000) ✦ *Staphylococcus aureus* is shown in a blood smear. Note the staphylococci interspersed between the erythrocytes. *S. aureus* is a common opportunistic pathogen of humans.

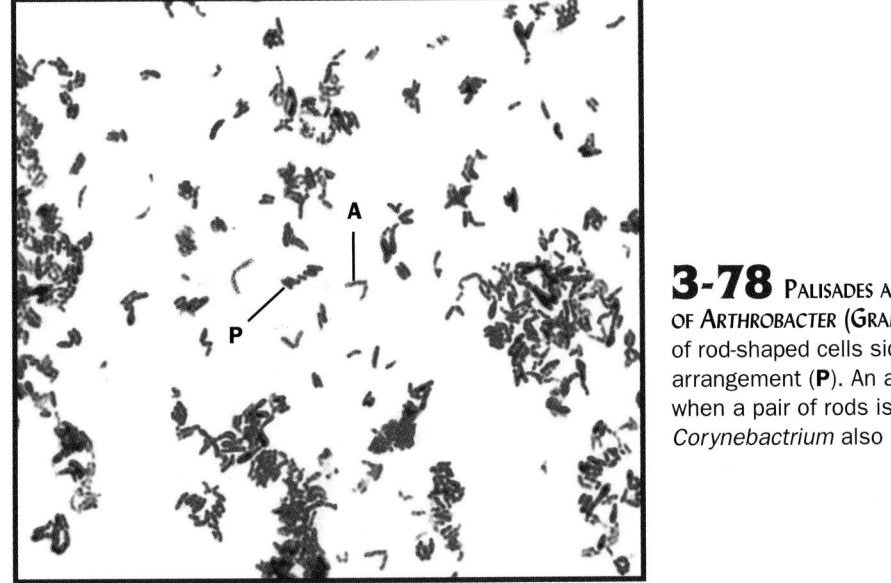

3-78 PALISADES AND ANGULAR ARRANGEMENTS OF ARTHROBACTER (GRAM STAIN, X1000) ✦ Stacking of rod-shaped cells side-by-side is the palisades arrangement (**P**). An angular arrangement (**A**) is when a pair of rods is bent where the cells join. *Corynebactrium* also shows these arrangements.

EXERCISE

3-5   Simple Stains

✦ Theory

Stains are solutions consisting of a solvent (usually water or ethanol) and a colored molecule (often a benzene derivative), the **chromogen**. The portion of the chromogen that gives it its color is the **chromophore**. A chromogen may have multiple chromophores, with each adding intensity to the color. The **auxochrome** is the charged portion of a chromogen and allows it to act as a dye through ionic or covalent bonds between the chromogen and the cell. **Basic stains**[1] (where the auxochrome becomes positively charged as a result of picking up a hydrogen ion or losing a hydroxide ion) are attracted to the negative charges on the surface of most bacterial cells. Thus, the cell becomes colored (Figure 3-79). Common basic stains include methylene blue, crystal violet and safranin. Examples of basic stains may be seen in Figures 3-66, 3-77, and 3-80.

Basic stains are applied to bacterial smears that have been **heat-fixed**. Heat-fixing kills the bacteria, makes them adhere to the slide, and coagulates cytoplasmic proteins to make them more visible. It also distorts the cells to some extent.

✦ Application

Because cytoplasm is transparent, cells usually are stained with a colored dye to make them more visible under the microscope. Then cell morphology, size, and arrangement can be determined. In a medical laboratory, these are usually determined with a Gram stain (Exercise 3-7), but you will be using simple stains as an introduction to these.

---

[1] Notice that the term "basic" means "alkaline," not "elementary;" however, coincidentally, *basic* stains can be used for *simple* staining procedures.

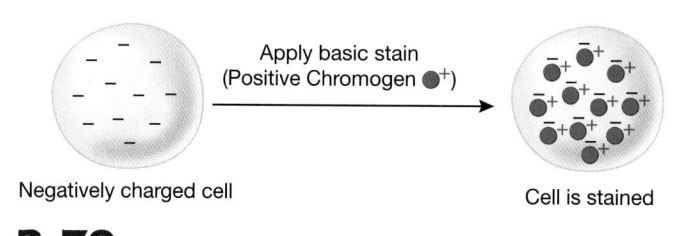

Negatively charged cell

Apply basic stain
(Positive Chromogen ●⁺)

Cell is stained

3-79 CHEMISTRY OF BASIC STAINS ✦ Basic stains have a positively charged chromogen (●⁺), which forms an ionic bond with the negatively charged bacterial cell, thus colorizing the cell.

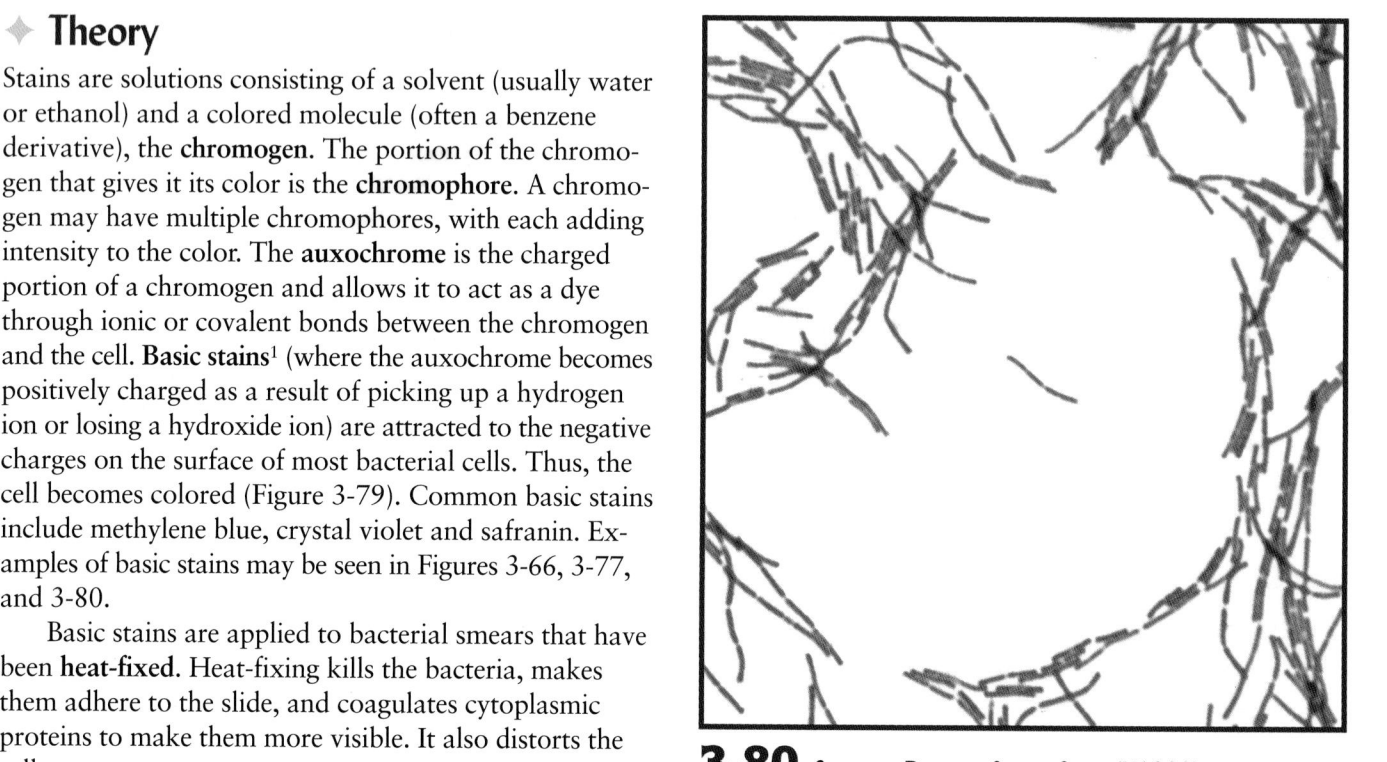

3-80 SAFRANIN DYE IN A SIMPLE STAIN (X1000) ✦ This is a simple stain using safranin, a basic stain. Notice that the stain is associated with the cells and not the background. The organism is Bacillus subtilis, the type species for the genus *Bacillus*, grown in culture.

✦ In This Exercise

Today you will learn how to prepare a bacterial smear (emulsion) and perform simple stains. Several different organisms will be supplied so you can begin to see the variety of cell morphologies and arrangements in the bacterial world. We suggest that you perform all the stains on one or two organisms (to get practice) and look at your lab partners' stains to see the variety of cell types. Be sure that you view all the available organisms.

✦ Materials

Per Group

✦ clean glass microscope slides
✦ methylene blue stain
✦ safranin stain
✦ crystal violet stain

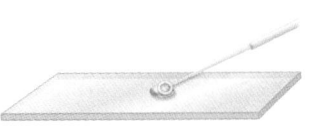

1. Place a small drop of water (not too much) on a clean slide using an inoculating loop. If you are staining from a broth culture, begin with Step 2.

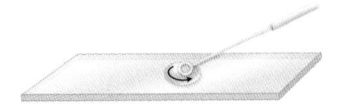

2. Aseptically add bacteria to the water. Mix in the bacteria and spread the drop out. Avoid spattering the emulsion as you mix. Flame your loop when done.

3. Allow the smear to air dry. If prepared correctly, the smear should be slightly cloudy.

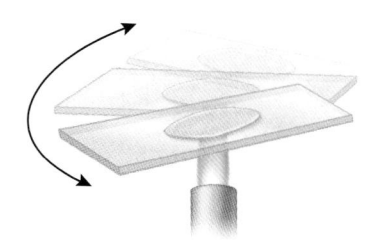

4. Using a slide holder, pass the smear through the upper part of a flame two or three times. This heat-fixes the preparation. Avoid overheating the slide as aerosols may be produced.

5. Allow the slide to cool, then continue with the staining protocol.

3-81 PROCEDURAL DIAGRAM: MAKING A BACTERIAL SMEAR (EMULSION) ✦ Preparation of uniform bacterial smears will make consistent staining results easier to obtain. Heat-fixing the smear kills the bacteria, makes them adhere to the slide, and coagulates their protein for better staining. Caution: Avoid producing aerosols. Do not spatter the smear as you mix it, do not blow on or wave the slide to speed-up air-drying, and do not overheat when heat-fixing.

✦ disposable gloves
✦ squirt bottle with water
✦ staining tray
✦ staining screen
✦ bibulous paper (or paper towels)
✦ slide holder
✦ compound microscope with oil objective lens and ocular micrometer
✦ immersion oil
✦ lens paper
✦ recommended organisms:
  ◆ *Bacillus cereus*
  ◆ *Micrococcus luteus*
  ◆ *Moraxella catarrhalis* (BSL-2)
  ◆ *Rhodospirillum rubrum*
  ◆ *Staphylococcus epidermidis*
  ◆ *Vibrio harveyi*

 **Procedure**

1 A bacterial smear (emulsion) is made prior to most staining procedures. Follow the Procedural Diagram in Figure 3-81 to prepare bacterial smears of each organism. If working in groups, each student should perform the various stains on one or two organisms, then observe each other's slides to see the variety of cell shapes and arrangements. Prepare several emulsions at once so they can be air drying at the same time. If staining more than one organism, you can make separate emulsions of each on the same slide.

2 Heat-fix each smear as described in Figure 3-81.

3 Following the basic staining procedure illustrated in the Procedural Diagram in Figure 3-82, prepare two slides with each stain using the following times:

crystal violet: stain for 30 to 60 seconds
safranin: stain for up to 1 minute
methylene blue: stain for 30 to 60 seconds

Be sure to wear gloves when staining. Record your actual staining times in the table provided in the Data Sheet so you can adjust for over-staining or understaining.

4 Using the oil immersion lens, observe each slide. Record your observations of cell morphology, arrangement, and size in the chart provided on the Data Sheet.

5 Dispose of the slides and used stain according to your laboratory's policy.

1. Begin with a heat-fixed emulsion (see Figure 3-81). More than one organism can be put on a slide.

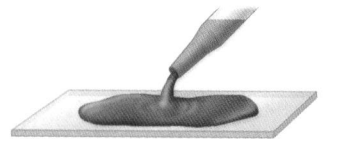

2. Cover the smear with stain. Use a staining tray to catch excess stain. Be sure to wear gloves.

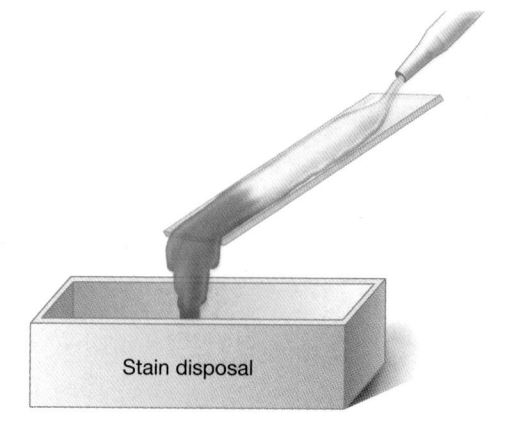

Stain disposal

3. Grasp the slide with a slide holder. Rinse the slide with water. Dispose of the excess stain according to your lab's practices.

Bibulous

4. Gently blot dry in a tablet of bibulous paper or paper towels. (Alternatively, a page from the tablet can be removed and used for blotting.) Do not rub. When dry, observe under oil immersion.

3-82 PROCEDURAL DIAGRAM: SIMPLE STAIN ✦ Staining times differ for each stain, but cell density of your smear also affects staining time. Strive for consistency in making your smears. Caution: Be sure to flame your loop after cell transfer and properly dispose of the slide when you are finished observing it.

References

Chapin, Kimberle. 1995. Chapter 4 in *Manual of Clinical Microbiology*, 6th ed., edited by Patrick R. Murray, Ellen Jo Baron, Michael A. Pfaller, Fred C. Tenover, and Robert H. Yolken. American Society for Microbiology, Washington, DC.

Chapin, Kimberle C., and Patrick R. Murray. 2003. Pages 257–259 in *Manual of Clinical Microbiology*, 8th ed., edited by Patrick R. Murray, Ellen Jo Baron, James H. Jorgensen, Michael A. Pfaller, and Robert H. Yolken. American Society for Microbiology, Washington, DC.

Forbes, Betty A., Daniel F. Sahm, and Alice. S. Weissfeld. 2002. Chapter 9 in *Bailey & Scott's Diagnostic Microbiology*, 11th ed. Mosby-Year Book, Inc. St. Louis.

Murray, R. G. E., Raymond N. Doetsch, and C. F. Robinow. 1994. Page 27 in *Methods for General and Molecular Bacteriology*, edited by Philipp Gerhardt, R. G. E. Murray, Willis A. Wood, and Noel R. Krieg. American Society for Microbiology, Washington, DC.

Norris, J. R., and Helen Swain. 1971. Chapter II in *Methods in Microbiology*, Vol. 5A, edited by J. R. Norris and D. W. Ribbons. Academic Press, Ltd., London.

Power, David A., and Peggy J. McCuen. 1988. Page 4 in *Manual of BBL™ Products and Laboratory Procedures*, 6th ed. Becton Dickinson Microbiology Systems, Cockeysville, MD.

EXERCISE

3-6

Negative Stains

✦ Theory

The negative staining technique uses a dye solution in which the chromogen is acidic and carries a negative charge. (An acidic chromogen gives up a hydrogen ion, which leaves it with a negative charge.) The negative charge on the bacterial surface repels the negatively charged chromogen, so the cell remains unstained against a colored background (Figure 3-83). A specimen stained with the acidic stain nigrosin is shown in Figure 3-84.

✦ Application

The negative staining technique is used to determine morphology and cellular arrangement in bacteria that are too delicate to withstand heat-fixing. A primary example is the spirochete *Treponema*, which is distorted by the heat-fixing of other staining techniques. Also, where determining the accurate size is crucial, a negative stain can be used because it produces minimal cell shrinkage.

✦ In This Exercise

Today you will perform negative stains on three different organisms. You also will have the opportunity to compare the size of *M. luteus* measured with the negative stain to its size as determined using a simple stain.

✦ Materials

Per Student Group

✦ nigrosin stain or eosin stain
✦ clean glass microscope slides
✦ disposable gloves

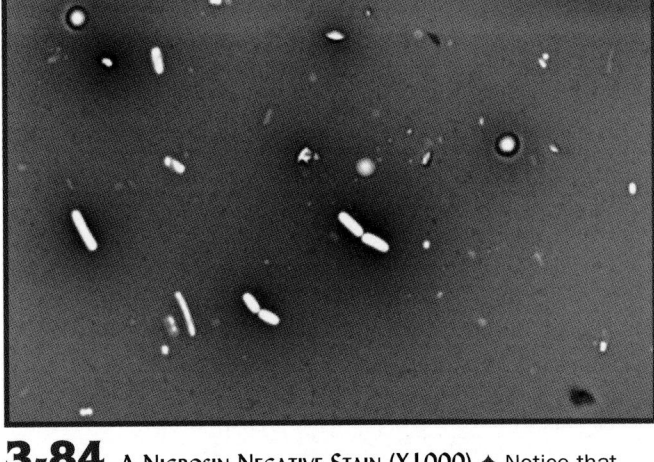

3-84 A NIGROSIN NEGATIVE STAIN (X1000) ✦ Notice that the *Bacillus megaterium* cells are unstained against a dark background. The circular objects are bubbles.

✦ compound microscope with oil objective lens and ocular micrometer
✦ immersion oil
✦ lens paper
✦ bibulous paper or paper towel
✦ recommended organisms:
  ✦ *Micrococcus luteus*
  ✦ *Bacillus cereus*
  ✦ *Rhodospirillum rubrum*

✦ Procedure

1 Follow the Procedural Diagram in Figure 3-85 to prepare a negative stain of each organism.

2 Dispose of the spreader slide in a disinfectant jar or sharps container immediately after use.

3 Observe using the oil immersion lens. Record your observations in the chart on the Data Sheet.

4 Dispose of the specimen slide in a disinfectant jar or sharps container after use.

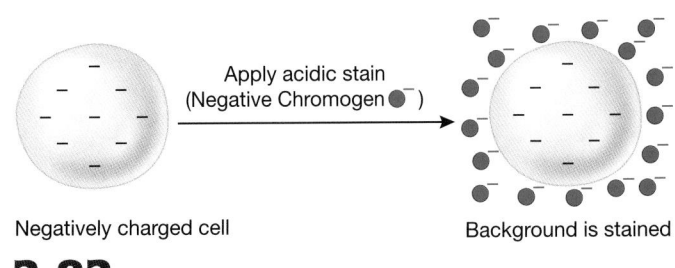

Apply acidic stain
(Negative Chromogen ●⁻)

Negatively charged cell   Background is stained

3-83 CHEMISTRY OF ACIDIC STAINS ✦ Acidic stains have a negatively charged chromogen (●⁻) that is repelled by negatively charged cells. Thus, the background is colored and the cell remains transparent.

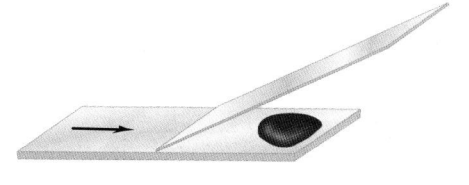

1. Begin with a drop of acidic stain at one end of a clean slide. Be sure to wear gloves.

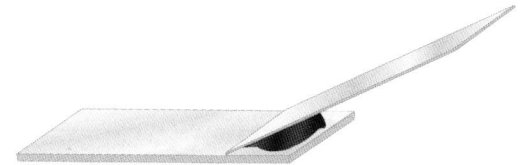

2. Aseptically add organisms and emulsify with a loop. Do not over-inoculate and avoid spattering the mixture. Sterilize the loop after emulsifying.

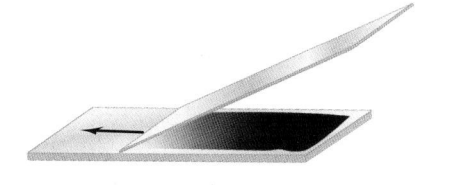

3. Take a second clean slide, place it on the surface of the first slide, and draw it back into the drop.

4. When the drop flows across the width of the spreader slide...

5. ...push the spreader slide to the other end. Dispose of the spreader slide in a jar of disinfectant or sharps container.

6. Air dry and observe under the microscope. Do NOT heat fix.

3-85 PROCEDURAL DIAGRAM: NEGATIVE STAIN ✦ Be sure to sterilize your loop after transfer, and to appropriately dispose of the spreader and specimen slides.

References

Claus, G. William. 1989. Chapter 5 in *Understanding Microbes—A Laboratory Textbook for Microbiology*. W. H. Freeman and Co., New York.

Murray, R. G. E., Raymond N. Doetsch, and C. F. Robinow. 1994. Page 27 in *Methods for General and Molecular Bacteriology*, edited by Philipp Gerhardt, R. G. E. Murray, Willis A. Wood, and Noel R. Krieg. American Society for Microbiology, Washington, DC.

Differential and Structural Stains

Differential stains allow a microbiologist to detect differences between organisms or differences between parts of the same organism. In practice, these are used much more frequently than simple stains because they not only allow determination of cell size, morphology, and arrangement (as with a simple stain) but information about other features as well.

The Gram stain is the most commonly used differential stain in bacteriology. Other differential stains are used for organisms not distinguishable by the Gram stain and for those that have other important cellular attributes, such as acid-fastness, a capsule, spores, or flagella. With the exception of the acid-fast stain, these other stains sometimes are referred to as **structural stains**. ✦

Gram Stain

✦ Theory

The Gram stain is a differential stain in which a **decolorization** step occurs between the application of two basic stains. The Gram stain has many variations, but they all work in basically the same way (Figure 3-86). The **primary stain** is crystal violet. Iodine is added as a **mordant** to enhance crystal violet staining by forming a **crystal violet–iodine complex**. Decolorization follows and is the most critical step in the procedure. Gram-negative cells are decolorized by the solution (of variable composition— generally alcohol or acetone) whereas Gram-positive cells are not. Gram-negative cells can thus be colorized by the **counterstain** safranin. Upon successful completion of a Gram stain, Gram-positive cells appear purple and Gram-negative cells appear reddish-pink (Figure 3-87).

Electron microscopy and other evidence indicate that the ability to resist decolorization or not is based on the different wall constructions of Gram-positive and Gram-negative cells. Gram-negative cell walls have a higher lipid content (because of the outer membrane) and a thinner peptidoglycan layer than Gram-positive cell walls (Figure 3-88). The alcohol/acetone in the decolorizer extracts the lipid, making the Gram-negative wall more porous and incapable of retaining the crystal violet– iodine complex, thereby decolorizing it. The thicker peptidoglycan and greater degree of cross-linking (because of teichoic acids) trap the crystal violet–iodine complex more effectively, making the Gram-positive wall less susceptible to decolorization.

Although some organisms give Gram-variable results, most variable results are a consequence of poor technique. The decolorization step is the most crucial and most likely source of Gram stain inconsistency. It is possible to **over-decolorize** by leaving the alcohol on too long and get reddish Gram-*positive* cells. It also is

Cells are transparent prior to staining.

Crystal violet stains Gram-positive and Gram-negative cells. Iodine is used as a mordant.

Decolorization with alcohol or acetone removes crystal violet from Gram-negative cells.

Safranin is used to counterstain Gram-negative cells.

3-86 GRAM STAIN ✦ After application of the primary stain (crystal violet), decolorization, and counterstaining with safranin, Gram-positive cells stain violet and Gram-negative cells stain pink/red. Notice that crystal violet and safranin are both basic stains, and that the decolorization step is what makes the Gram stain differential.

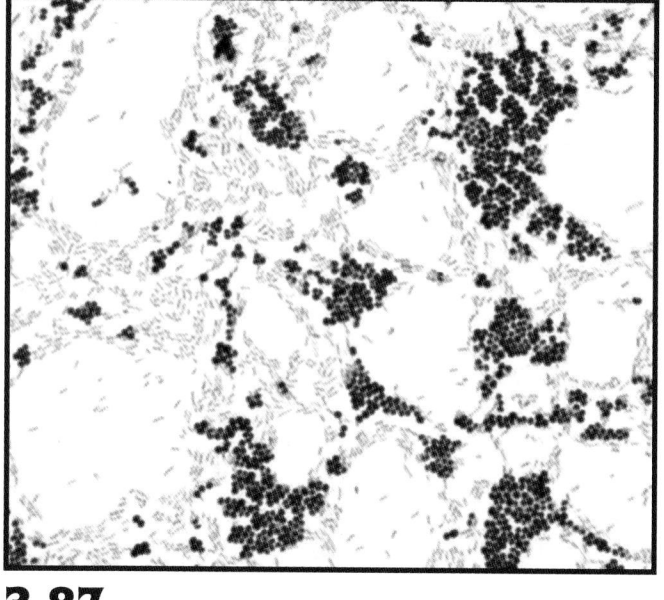

3-87 GRAM STAIN OF *STAPHYLOCOCCUS* (+) AND *PROTEUS* (−) (X1320) ✦ *Staphylococcus* has a staphylococcal arrangement, whereas *Proteus* is a bacillus.

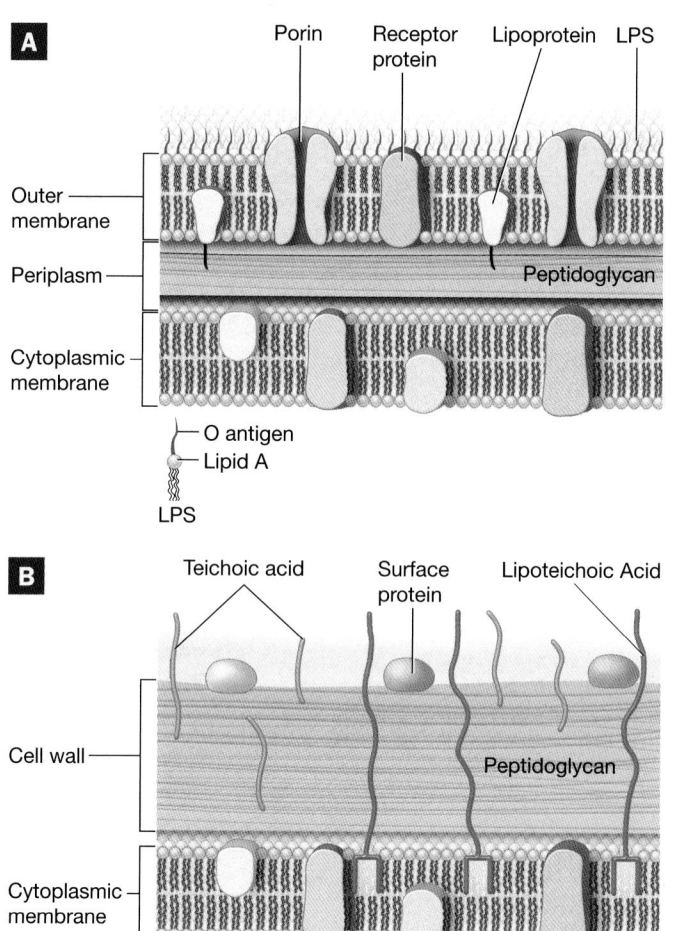

3-88 BACTERIAL CELL WALLS ✦ **A** The Gram-negative wall is composed of less peptidoglycan (as little as a single layer) and more lipid (due to the outer membrane) than the Gram-positive wall **B**.

possible to **under-decolorize** and produce purple Gram-*negative* cells. Neither of these situations changes the actual Gram reaction for the organism being stained. Rather, these are false results because of poor technique.

A second source of poor Gram stains is inconsistency in preparation of the emulsion. Remember, a good emulsion dries to a faint haze on the slide.

Until correct results are obtained consistently, it is recommended that control smears of Gram-positive and Gram-negative organisms be stained along with the organism in question (Figure 3-89). As an alternative control, a direct smear made from the gumline may be Gram-stained (Figure 3-90) with the expectation that both Gram-positive and Gram-negative organisms will be seen. Over-decolorized and under-decolorized gumline direct smears are shown for comparison (Figures 3-91 and 3-92). Positive controls also should be run when using new reagent batches.

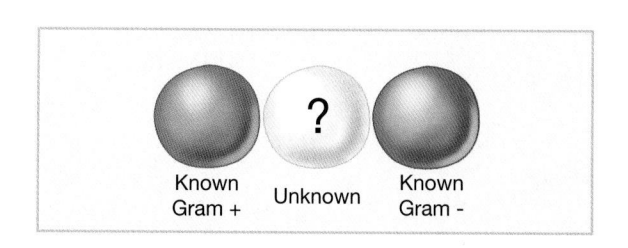

3-89 POSITIVE CONTROLS TO CHECK YOUR TECHNIQUE ✦ Staining known Gram-positive and Gram-negative organisms on either side of your unknown organism act as positive controls for your technique. Try to make the emulsions as close to one another as possible. Spreading them out across the slide makes it difficult to stain and decolorize them equally.

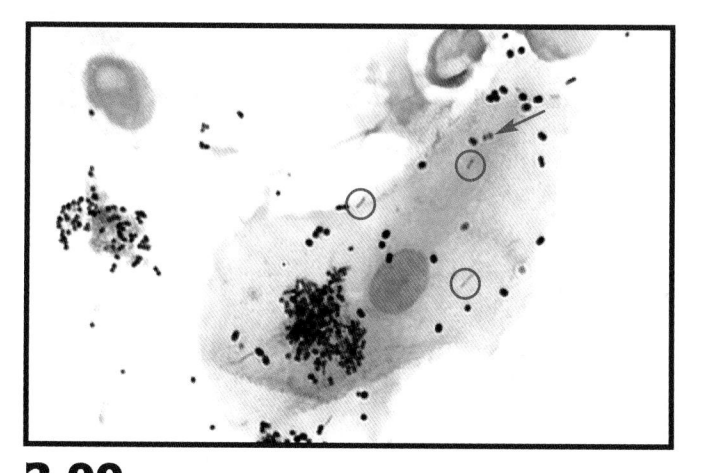

3-90 DIRECT SMEAR POSITIVE CONTROL (GRAM STAIN, X1000) ✦ A direct smear made from the gumline may also be used as a Gram stain control. Expect numerous Gram-positive bacteria (especially cocci) and some Gram-negative cells, including your own epithelial cells. In this slide, Gram-positive cocci predominate, but a few Gram-negative cells are visible, including Gram-negative rods (circled) and a Gram-negative diplococcus (arrow) on the surface of the epithelial cell.

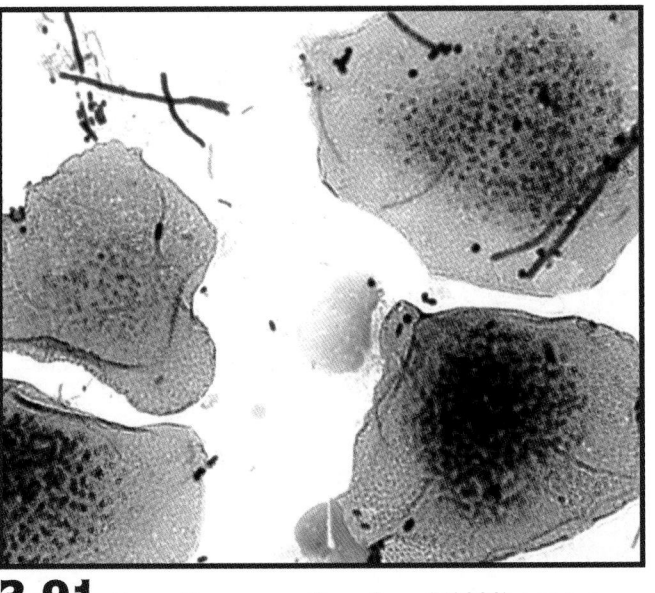

3-91 UNDER-DECOLORIZED GRAM STAIN (X1000) ✦ This is a direct smear from the gumline. Notice the purple patches of stain on the epithelial cells. Also notice the variable quality of this stain—the epithelial cell to the left of center is stained better than the others.

3-93 CRYSTAL VIOLET CRYSTALS (GRAM STAIN, X1000) ✦ If the staining solution is not adequately filtered or is old, crystal violet crystals may appear. Although they are pleasing to the eye, they obstruct your view of the specimen. Crystals from two different Gram stains are shown here: **A** a gumline direct smear; **B** *Micrococcus roseus* grown in culture.

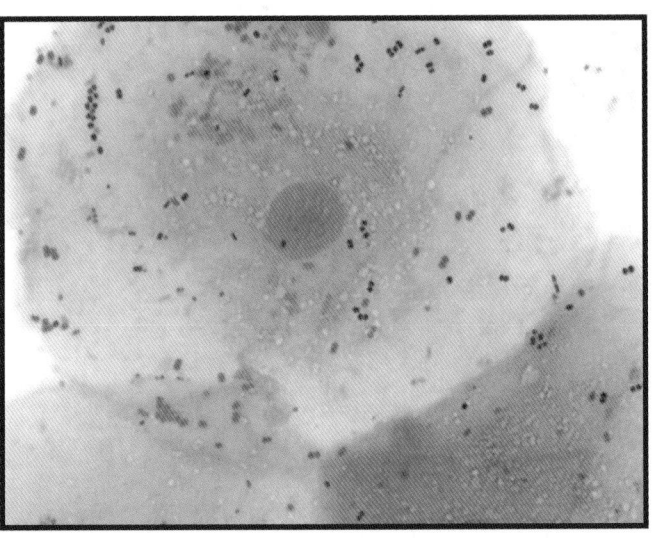

3-92 OVER-DECOLORIZED GRAM STAIN (X1000) ✦ This also is a direct smear from the gumline. Notice the virtual absence of any purple cells, a certain indication of over-decolorization.

Interpretation of Gram stains can be complicated by nonbacterial elements. For instance, stain crystals from an old or improperly made stain solution can disrupt the field (Figure 3-93) or stain precipitate may be mistakenly identified as bacteria (Figure 3-94).

Age of the culture also affects Gram stain consistency. Older Gram-positive cultures may lose their ability to resist decolorization and give an artifactual Gram-negative result. The genus *Bacillus* is notorious for this. *Staphylococcus* can also be a culprit. Cultures 24 hours old or less are best for this procedure.

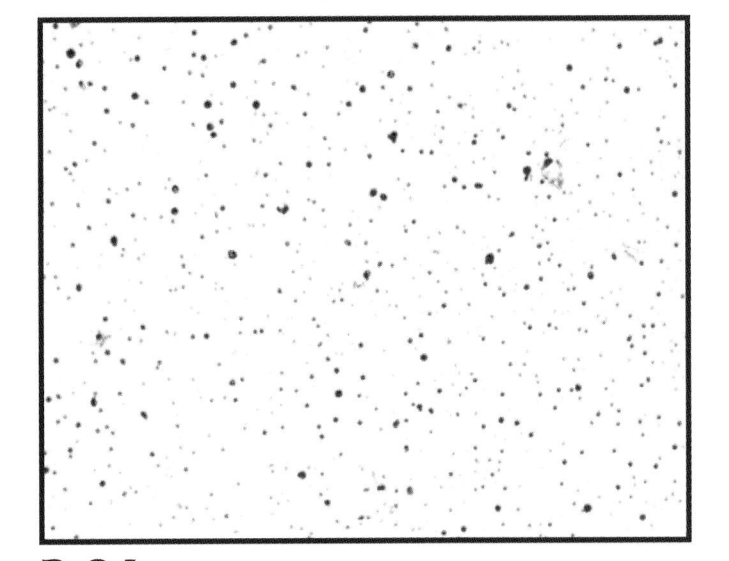

3-94 STAIN PRECIPITATE (GRAM STAIN, X1000) ✦ If the slide is not rinsed thoroughly or the stain is allowed to dry on the slide, spots of stain precipitate may form and may be confused with bacterial cells. Their variability in size is a clue that they are not bacteria.

✦ Application

The Gram stain, used to distinguish between Gram-positive and Gram-negative cells, is the most important and widely used microbiological differential stain. In addition to Gram reaction, this stain allows determination of cell morphology, size, and arrangement. It typically is the first differential test run on a specimen brought into the laboratory for identification. In some cases, a rapid, presumptive identification of the organism or elimination of a particular organism is possible.

✦ In This Exercise

The Gram stain is the single most important differential stain in bacteriology. Therefore, you will have to practice it and practice it some more to become proficient in its execution. The organisms should be used in the combinations given so you will have one Gram-positive and one Gram-negative on each slide.

✦ Materials

Per Student Group

- ✦ compound microscope with oil objective lens and ocular micrometer
- ✦ clean glass microscope slides
- ✦ sterile toothpick
- ✦ Gram stain solutions (commercial kits are available)
  - ✦ Gram crystal violet
  - ✦ Gram iodine
  - ✦ 95% ethanol (or ethanol/acetone solution)
  - ✦ Gram safranin
- ✦ squirt bottle with water
- ✦ bibulous paper or paper towels
- ✦ disposable gloves
- ✦ staining tray
- ✦ staining screen
- ✦ slide holder
- ✦ recommended organisms (overnight cultures grown on agar slants or turbid broth cultures):
  - ✦ *Staphylococcus epidermidis*
  - ✦ *Escherichia coli*
  - ✦ *Moraxella catarrhalis* (BSL-2)
  - ✦ *Corynebacterium xerosis*

 Procedure

1 Follow the procedure illustrated in Figure 3-81 to prepare and heat-fix smears of *Staphylococcus*

epidermidis and *Escherichia coli* immediately next to one another on the same clean glass slide. (If you make the emulsions at opposite ends of the slide, you may find it difficult to stain and decolorize each equally.) Strive to prepare smears of uniform thickness, as thick smears risk being under-decolorized.

2 Repeat step 1 for *Moraxella catarrhalis* and *Corynebacterium xerosis* on a second slide.

3 Because Gram stains require much practice, we recommend that you prepare several slides of each combination and let them air-dry simultaneously. Then they'll be ready when you need them.

4 Use the sterile toothpick to obtain a sample from your teeth at the gumline. (Do not draw blood! What you want is easily removed from your gingival pockets.) Transfer the sample to a drop of water on a clean glass slide, air-dry, and heat-fix.

5 Follow the basic staining procedure illustrated in Figure 3-95. We recommend staining the pure cultures first. After your technique is consistent, stain the oral sample. Be sure to wear gloves.

6 Observe using the oil immersion lens. Record your observations of cell morphology and arrangement, dimensions, and Gram reactions in the chart provided on the Data Sheet.

7 Dispose of the specimen slides in a jar of disinfectant or a sharps container after use.

References

Chapin, Kimberle C., and Patrick R. Murray. 2003. Pages 258–260 in *Manual of Clinical Microbiology*, 8th ed., edited by Patrick R. Murray, Ellen Jo Baron, James H. Jorgensen, Michael A. Pfaller, and Robert H. Yolken. American Society for Microbiology, Washington, DC.

Forbes, Betty A., Daniel F. Sahm, and Alice. S. Weissfeld. 2002. Chapter 9 in *Bailey & Scott's Diagnostic Microbiology*, 11th ed. Mosby-Year Book, St. Louis.

Koneman, Elmer W., Stephen D. Allen, William M. Janda, Paul C. Schreckenberger, and Washington C. Winn, Jr. 1997. Chapter 14 in *Color Atlas and Textbook of Diagnostic Microbiology*, 5th Ed. J. B. Lippincott Co., Philadelphia.

Murray, R. G. E., Raymond N. Doetsch, and C. F. Robinow. 1994. Pages 31 and 32 in *Methods for General and Molecular Bacteriology*, edited by Philipp Gerhardt, R. G. E. Murray, Willis A. Wood, and Noel R. Krieg. American Society for Microbiology, Washington, DC.

Norris, J. R., and Helen Swain. 1971. Chapter II in *Methods in Microbiology*, Vol 5A, edited by J. R. Norris and D. W. Ribbons. Academic Press, Ltd., London.

Power, David A., and Peggy J. McCuen. 1988. Page 261 in *Manual of BBL™ Products and Laboratory Procedures*, 6th ed. Becton Dickinson Microbiology Systems, Cockeysville, MD.

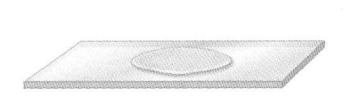

1. Begin with a heat-fixed emulsion.
(See Figure 3-81.)

2. Cover the smear with crystal violet stain for 1 minute.
Use a staining tray to catch excess stain. Be sure to
wear gloves.

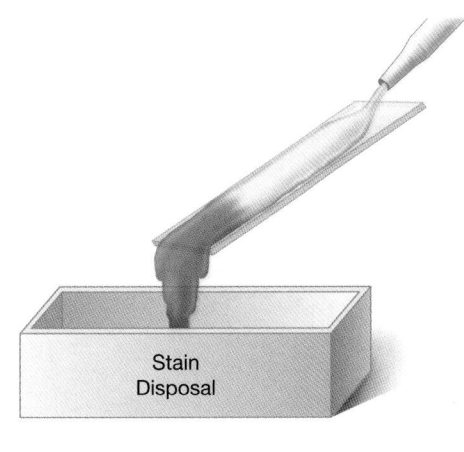

3. Grasp the slide with a slide holder.
Gently rinse the slide with distilled water.
Alternatively, you can tap the edge of the
slide to remove the excess stain and
eliminate the wash step.

4. Cover the smear with Iodine stain for 1 minute.
Use a staining tray to catch excess stain.

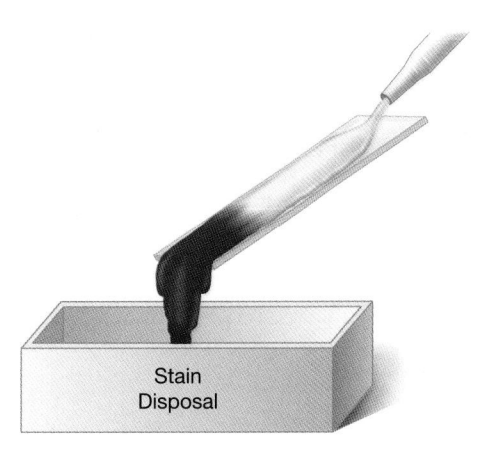

5. Grasp the slide with a slide holder.
Gently rinse the slide with distilled water.

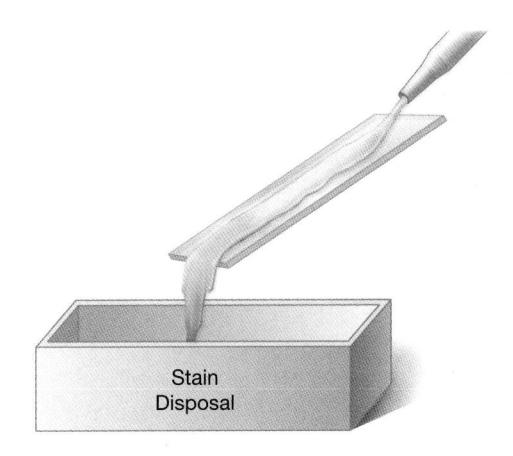

6. Decolorize with 95% ethanol or ethanol/acetone
by allowing it to trickle down the slide until the
run-off is clear.
Gently rinse the slide with distilled water.

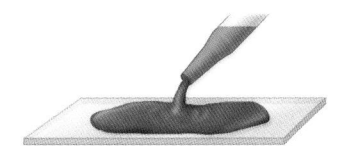

7. Counterstain with safranin stain for 1 minute.
Rinse with distilled water.

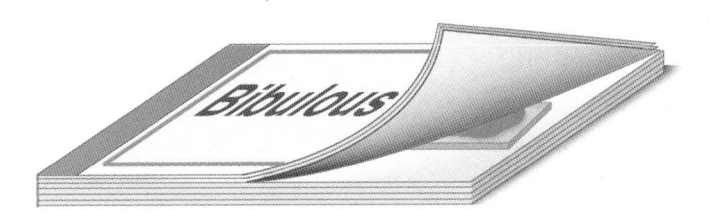

8. Gently blot dry in a tablet of bibulous paper or paper towels.
(Alternatively, a page from the tablet can be removed and used
for blotting.) Do not rub. When dry, observe under oil immersion.

3-95 PROCEDURAL DIAGRAM: GRAM STAIN ✦ Pay careful attention to the staining times. If your preparations do not give "correct" results, the most likely source of error is in the decolorization step. Adjust its timing accordingly on subsequent stains.

EXERCISE 3-8

Acid-Fast Stains

✦ Theory

The presence of mycolic acids in the cell walls of acid-fast organisms is the cytological basis for the acid-fast differential stain. Mycolic acid is a waxy substance that gives acid-fast cells a higher affinity for the primary stain and resistance to decolorization by an acid alcohol solution. A variety of acid-fast staining procedures are employed, two of which are the Ziehl-Neelsen (ZN) method and the Kinyoun (K) method. These differ primarily in that the ZN method uses heat as part of the staining process, whereas the K method is a "cold" stain. In both protocols the bacterial smear may be prepared in a drop of serum to help the "slippery" acid-fast cells adhere to the slide. The two methods provide comparable results.

The waxy wall of acid-fast cells repels typical aqueous stains. (As a result, most acid-fast positive organisms are only weakly Gram-positive.) In the ZN method (Figure 3-96), the phenolic compound carbolfuchsin is used as the primary stain because it is lipid-soluble and penetrates the waxy cell wall. Staining by carbolfuchsin is further enhanced by steam-heating the preparation to melt the wax and allow the stain to move into the cell. Acid alcohol is used to decolorize nonacid-fast cells; acid-fast cells resist this decolorization. A counterstain, such as methylene blue, then is applied. Acid-fast cells are reddish-purple; nonacid-fast cells are blue (Figure 3-97).

The Kinyoun method (Figure 3-98) uses a slightly more lipid-soluble and concentrated carbolfuchsin as the primary stain. These properties allow the stain to penetrate the acid-fast walls without the use of heat but make this method slightly less sensitive than the ZN method. Decolorization with acid alcohol is followed by a contrasting counterstain, such as brilliant green (Figure 3-99) or methylene blue.

✦ Application

The acid-fast stain is a differential stain used to detect cells capable of retaining a primary stain when treated with an acid alcohol. It is an important differential stain used to identify bacteria in the genus *Mycobacterium*, some of which are pathogens (*e.g.*, *M. leprae* and *M. tuberculosis*, causative agents of leprosy and tuberculosis, respectively). Members of the actinomycete genus *Nocardia* (*N. brasiliensis* and *N. asteroides* are

	Acid-Fast	Acid-Fast Negative
Cells prior to staining are transparent.		
After staining with carbolfuchsin, cells are reddish-purple. Steam heat enhances the entry of carbolfuchsin into cells.		
Decolorization with acid alcohol removes stain from acid-fast negative cells.		
Methylene blue is used to counterstain acid-fast negative cells.		

3-96 ZIEHL-NEELSEN ACID-FAST STAIN ✦ Acid-fast cells stain reddish-purple; acid-fast negative cells stain blue or the color of the counterstain if a different one is used.

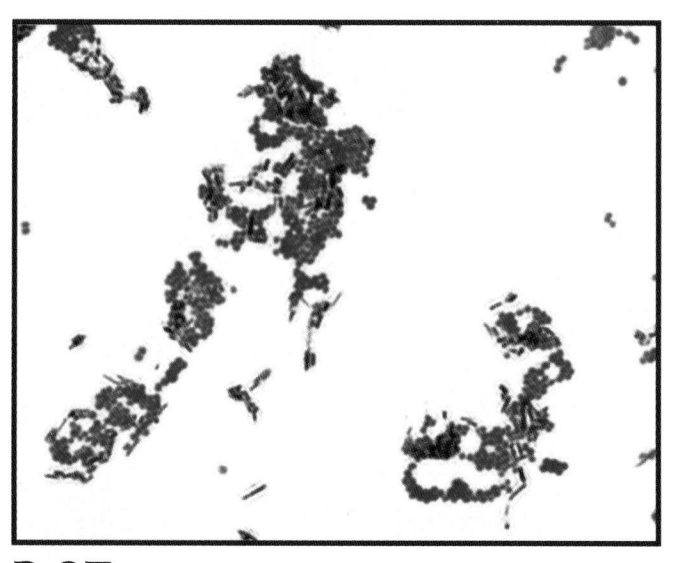

3-97 AN ACID-FAST STAIN USING THE ZN METHOD (X1000) ✦ Notice how most of the *Mycobacterium phlei* (AF+) cells are in clumps, an unusual state for most rods. They do this because their waxy cell walls make them sticky. A few individual cells are visible, however, and they clearly are rods. The *Staphylococcus epidermidis* cells (AF−) are also in clumps, but that is because they grow as grape-like clusters. Each cell's diameter is approximately 1μm. Compare this micrograph with Figure 3-99.

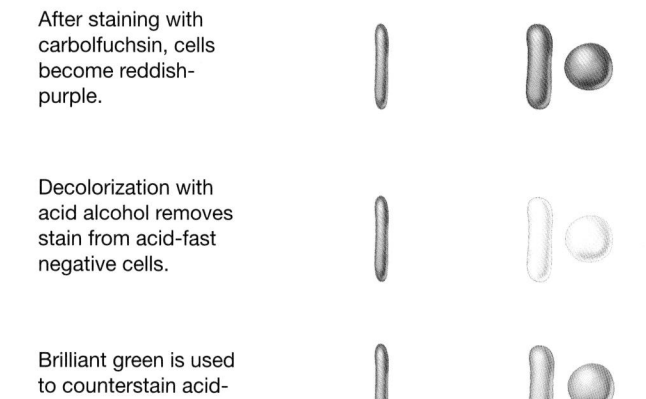

	Acid-Fast	Acid-Fast Negative
Cells prior to staining are transparent.		
After staining with carbolfuchsin, cells become reddish-purple.		
Decolorization with acid alcohol removes stain from acid-fast negative cells.		
Brilliant green is used to counterstain acid-fast negative cells.		

3-98 KINYOUN ACID-FAST STAIN ✦ Acid-fast cells stain reddish-purple; nonacid-fast cells stain green or the color of the counterstain if a different one is used.

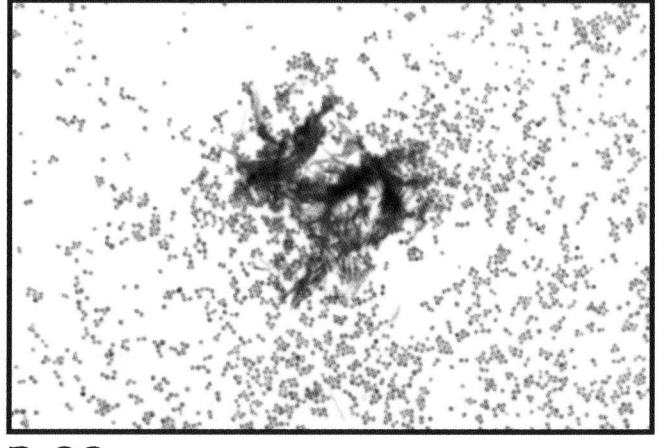

3-99 AN ACID-FAST STAIN USING THE KINYOUN METHOD (X1000) ✦ This is an acid-fast stain of *Mycobacterium smegmatis* (+) and *Staphylococcus epidermidis* (−) from cultures. Again, notice the clumping of the acid-fast organisms. Brilliant green dye commonly stains more of a gray color, but it still contrasts with the carbolfuchsin of acid-fast positive cells. Compare this micrograph with the one in Figure 3-97.

opportunistic pathogens) are partially acid-fast. Oocysts of coccidian parasites, such as *Cryptosporidium* and *Isospora*, are also acid-fast. Because so few organisms are acid-fast, the acid-fast stain is run only when infection by an acid-fast organism is suspected.

Acid-fast stains are useful in identifying **acid-fast bacilli (AFB)** and in rapid, preliminary diagnosis of tuberculosis (with greater than 90% predictive value from sputum samples). It also can be performed on patient samples to track the progress of antibiotic therapy and determine the degree of contagiousness. A prescribed number of microscopic fields is examined and the number of AFB is determined and reported using a standard scoring system.

✦ In This Exercise

Today you will perform an acid-fast stain designed primarily to identify members of the genus *Mycobacterium*. Because you know ahead of time which organisms should give a positive result and which should give a negative result, it is okay to mix them into a single emulsion.

✦ Materials

Per Student Group

- ✦ compound microscope with oil objective lens and ocular micrometer
- ✦ clean glass microscope slides
- ✦ staining tray
- ✦ staining screen
- ✦ bibulous paper or paper towel
- ✦ slide holder
- ✦ Ziehl-Neelsen Stains (complete kits are commercially available)
  - ◆ methylene blue stain
  - ◆ Ziehl's carbolfuchsin stain
  - ◆ acid alcohol (95% ethanol + 3% HCl)
- ✦ Kinyoun Stains (complete kits are commercially available)
  - ◆ Kinyoun carbolfuchsin
  - ◆ acid alcohol (95% ethanol + 3% HCl)
  - ◆ brilliant green stain
- ✦ sheep serum
- ✦ squirt bottle with water
- ✦ heating apparatus (steam or hotplate)—for ZN only
- ✦ nonsterile Petri dish for transporting slides
- ✦ disposable gloves
- ✦ lab coat
- ✦ eye goggles for ZN only
- ✦ recommended organisms:
  - ◆ *Mycobacterium phlei* (BSL-2)
  - ◆ *Staphylococcus epidermidis*

Procedure: Ziehl-Neelsen (ZN) Method

1 Prepare a smear of each organism on a clean glass slide as illustrated in Figure 3-81, substituting a drop

of sheep serum for the drop of water. Air-dry and then heat-fix the smears. *Note:* You may make two separate smears right next to one another on the slide or mix the two organisms in one smear.

2   Follow the staining protocol shown in the Procedural Diagram (Figure 3-100). Be sure to wear gloves and eye protection. Use a steaming apparatus (such as the one in Figure 3-101) to heat the slide. If the slide must be carried to and from the steaming apparatus,

put it in a covered Petri dish. Be sure to have adequate ventilation while staining.

3   Observe using the oil immersion lens. Record your observations of cell morphology and arrangement, dimensions, and acid-fast reaction in the chart on the Data Sheet.

4   When finished, dispose of slides in a disinfectant jar or sharps container.

1. Begin with a heat-fixed emulsion. (The emulsion can be prepared in a drop of sheep serum.) (See Figure 3-81.)

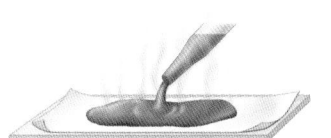

2. Cover the smear with a strip of bibulous paper. Apply ZN carbolfuchsin stain. Steam (as shown in Figure 3-101) for 5 minutes. Keep the paper moist with stain. Perform this step with adequate ventilation, eye protection, and gloves. Do not boil the stain.

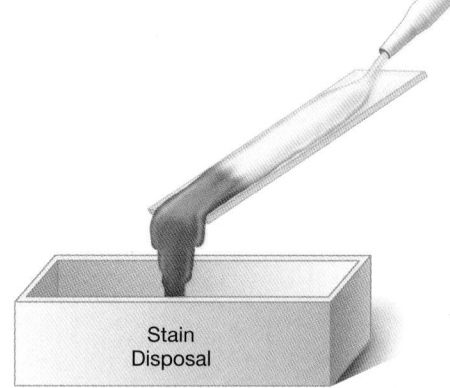

Stain Disposal

3. Grasp the slide with a slide holder. Remove the paper and dispose of it properly. Gently rinse the slide with distilled water.

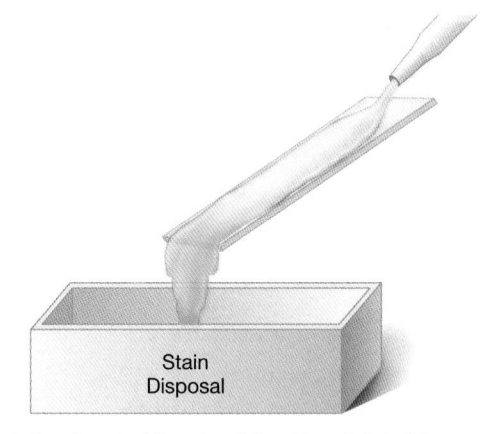

Stain Disposal

4. Continue holding the slide with a slide holder. Decolorize with acid-alcohol (CAUTION!) until the run-off is clear. Gently rinse the slide with distilled water.

5. Counterstain with methylene blue stain for 1 minute. Rinse with distilled water.

6. Gently blot dry in a tablet of bibulous paper or paper towels. (Alternatively, a page from the tablet can be removed and used for blotting.) Do not rub. When dry, observe under oil immersion.

3-100 PROCEDURAL DIAGRAM: ZN ACID-FAST STAIN ✦ Be sure to perform this stain in a fume hood or well-ventilated area. Carry the slide to and from the steaming apparatus in a covered Petri dish.

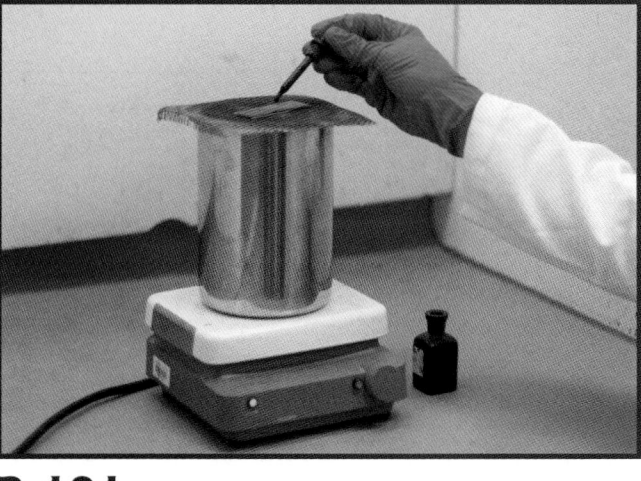

3-101 STEAMING THE SLIDE DURING THE ZIEHL-NEELSEN PROCEDURE ✦ Carefully steam the slide to melt the waxy wall so the carbolfuchsin can get into acid-fast cells. Do not boil the slide or let it dry out. Keep it moist with stain for the entire 5 minutes of steaming. Caution: This should be performed in a well-ventilated area with hand, clothing, and eye protection.

 ## Procedure: Kinyoun Method

1 Prepare a smear of each organism on a clean glass slide as illustrated in Figure 3-81, substituting a drop of sheep serum for the drop of water. Air-dry and then heat-fix the smears. *Note:* You may make two separate smears right next to one another on the slide or mix the two organisms in one smear.

2 Follow the staining protocol shown in the Procedural Diagram (Figure 3-102). Be sure to wear gloves and perform the stain with adequate ventilation.

3 Observe using the oil immersion lens. Record your observations of cell morphology and arrangement, dimensions, and acid-fast reaction on the Data Sheet.

4 When finished, dispose of slides in a disinfectant jar or a sharps container.

References

Chapin, Kimberle C. and Patrick R. Murray. 2003. Pages 259–261 in *Manual of Clinical Microbiology*, 8th ed., edited by Patrick R. Murray, Ellen Jo Baron, James H. Jorgensen, Michael A. Pfaller, and Robert H. Yolken. American Society for Microbiology, Washington, DC.

Doetsch, Raymond N. and C. F. Robinow. 1994. Page 32 in *Methods for General and Molecular Bacteriology*, edited by Philipp Gerhardt, R. G. E. Murray, Willis A. Wood, and Noel R. Krieg. American Society for Microbiology, Washington, DC.

Forbes, Betty A., Daniel F. Sahm, and Alice. S. Weissfeld. 2002. Chapter 9 in *Bailey & Scott's Diagnostic Microbiology*, 11th ed. Mosby-Year Book, St. Louis.

Norris, J. R., and Helen Swain. 1971. Chapter II in *Methods in Microbiology*, Vol. 5A, edited by J. R. Norris and D. W. Ribbons. Academic Press, Ltd., London.

Power, David A., and Peggy J. McCuen. 1988. Page 5 in *Manual of BBL™ Products and Laboratory Procedures*, 6th ed. Becton Dickinson Microbiology Systems, Cockeysville, MD.

1. Begin with a heat-fixed emulsion. (The emulsion can be prepared in a drop of sheep serum.) (See Figure 3-81.)

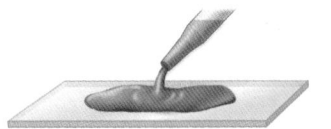

2. Apply Kinyoun carbolfuchsin stain for 5 minutes. Perform this step with adequate ventilation and wear gloves.

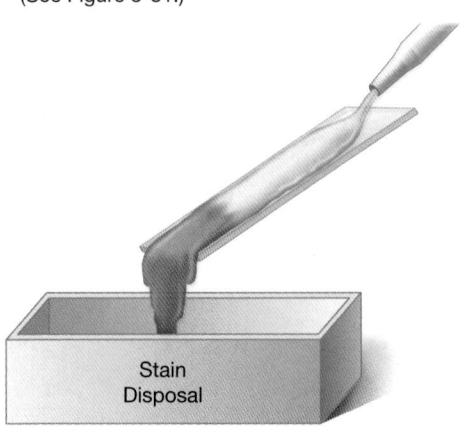

3. Grasp the slide with a slide holder. Gently rinse the slide with distilled water.

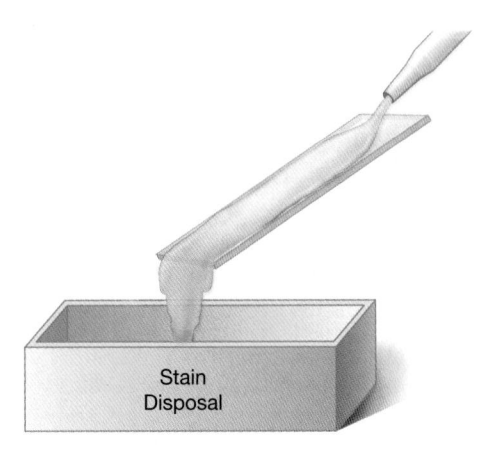

4. Continue holding the slide with a slide holder. Decolorize with acid-alcohol (CAUTION!) until the run-off is clear. Gently rinse the slide with distilled water.

5. Counterstain with brilliant green stain (or methylene blue stain) for 1 minute. Rinse with distilled water.

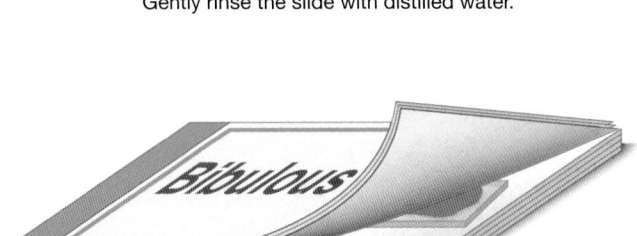

6. Gently blot dry in a tablet of bibulous paper or paper towels. (Alternatively, a page from the tablet can be removed and used for blotting.) Do not rub. When dry, observe under oil immersion.

3-102 PROCEDURAL DIAGRAM: KINYOUN ACID-FAST STAIN ✦

EXERCISE 3-9

Capsule Stain

✦ Theory

Capsules are composed of mucoid polysaccharides or polypeptides that repel most stains. The capsule stain technique takes advantage of this characteristic by staining *around* the cells. Typically an acidic stain such as Congo red or nigrosin, which stains the background, and a basic stain that colorizes the cell proper, are used. The capsule remains unstained and appears as a white halo between the cells and the colored background (Figure 3-103).

This technique begins as a negative stain; cells are spread in a film with an acidic stain and are not heat-fixed. Heat-fixing causes the cells to shrink, leaving an artifactual white halo around them that might be interpreted as a capsule. In place of heat-fixing, cells may be emulsified in a drop of serum to promote their adhering to the glass slide.

✦ Application

The capsule stain is a differential stain used to detect cells capable of producing an extracellular capsule. Capsule production increases virulence in some microbes (such as the anthrax bacillus *Bacillus anthracis* and the pneumococcus *Streptococcus pneumoniae*) by making them less vulnerable to phagocytosis.

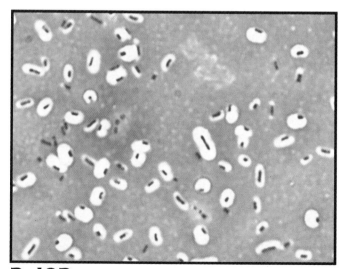

3-103 CAPSULE STAIN (X1000) ✦ The acidic stain colorizes the background while the basic stain colorizes the cell, leaving the capsules as unstained, white clearings around the cells. Notice the lack of uniform capsule size, and even the absence of a capsule in some cells.

✦ In This Exercise

The capsule stain allows you to visualize an extracellular capsule, if present. Be careful to distinguish between a tiny white halo (as a result of cell shrinkage) and a true capsule.

✦ Materials

Per Student
- ✦ clean glass slides
- ✦ sheep serum
- ✦ Maneval's stain
- ✦ congo red stain
- ✦ squirt bottle with water
- ✦ staining tray
- ✦ staining screen
- ✦ bibulous paper or paper towel
- ✦ slide holder
- ✦ disposable gloves
- ✦ sterile toothpicks
- ✦ compound microscope with oil objective lens and ocular micrometer
- ✦ immersion oil
- ✦ lens paper
- ✦ recommended organisms (18 to 24 hour skim milk or trypticase soy agar slant pure cultures)
  - ✦ *Klebsiella pneumoniae* (BSL-2)
  - ✦ *Aeromonas sobria*

Procedure

1 Follow the protocol in the Procedural Diagram (Figure 3-104) to make stains of the organisms supplied. Use a separate slide for each specimen. These specimens are not heat-fixed. Be sure to wear gloves.

2 Using a sterile toothpick, obtain a sample from below the gumline in your mouth. (Do not draw blood!) Mix the sample into a drop of water or serum on a slide, then perform a capsule stain on it.

3 Observe, using the oil immersion lens. Record your observations of cell morphology and arrangement,

1. Begin with a drop of serum at one end of a clean slide. Add a drop of Congo Red stain. Be sure to wear gloves.

2. Aseptically add organisms and emulsify with a loop. Do not over-inoculate and avoid spattering the mixture. Sterilize the loop after emulsifying.

3. Take a second clean slide, place it on the surface of the first slide, and draw it back into the drop.

4. When the drop flows across the width of the spreader slide...

5. ...push the spreader slide to the other end. Dispose of the spreader slide in a jar of disinfectant or sharps container.

6. Air dry and do NOT heat fix.

7. Flood the slide with Maneval's Stain for 1 minute. Rinse with water.

8. Gently blot dry in a tablet of bibulous paper or paper towels. (Alternatively, a page from the tablet can be removed and used for blotting.) Do not rub. When dry, observe under oil immersion.

3-104 PROCEDURAL DIAGRAM: CAPSULE STAIN ✦

cell dimensions, and presence or absence of a capsule in the chart on the Data Sheet.

4   Dispose of the specimen slides in a jar of disinfectant or sharps container after use.

References

Murray, R. G. E., Raymond N. Doetsch, and C. F. Robinow. 1994. Page 35 in *Methods for General and Molecular Bacteriology,* edited by Philipp Gerhardt, R. G. E. Murray, Willis A. Wood, and Noel R. Krieg. American Society for Microbiology, Washington, DC.

Norris, J. R., and Helen Swain. 1971. Chapter II in *Methods in Microbiology,* Vol 5A, edited by J. R. Norris and D. W. Ribbons. Academic Press, Ltd, London.

EXERCISE 3-10
Endospore Stain

✦ Theory

An **endospore** is a dormant form of the bacterium that allows it to survive poor environmental conditions. Spores are resistant to heat and chemicals because of a tough outer covering made of the protein **keratin**. The keratin also resists staining, so extreme measures must be taken to stain the spore. In the Schaeffer-Fulton method (Figure 3-105), a primary stain of malachite green is forced into the spore by steaming the bacterial emulsion. Alternatively, malachite green can be left on the slide for 15 minutes or more to stain the spores. Malachite green is water-soluble and has a low affinity for cellular material, so **vegetative cells** and **spore mother cells** can be decolorized with water and counterstained with safranin (Figure 3-106).

Spores may be located in the middle of the cell (**central**), at the end of the cell (**terminal**), or between the end and middle of the cell (**subterminal**). Spores also may be differentiated based on shape—either **spherical** or **elliptical** (**oval**)—and size relative to the cell (*i.e.*, whether they cause the cell to look swollen or not). These structural features are shown in Figure 3-107 and Figure 3-108.

	Spore producer	Spore nonproducer
Cells and spores prior to staining are transparent.		
After staining with Malachite green, cells and spores are green. Heat is used to force the stain into spores, if present.		
Decolorization with water removes stain from cells, but not spores.		
Safranin is used to counterstain cells.		

3-105 THE SCHAEFFER-FULTON SPORE STAIN ✦ Upon completion, spores are green, and vegetative and spore mother cells are red.

✦ Application

The spore stain is a differential stain used to detect the presence and location of spores in bacterial cells. Only a few genera produce spores. Among them are the genera *Bacillus* and *Clostridium*. Most members of *Bacillus* are soil, freshwater, or marine **saprophytes**, but a few are pathogens, such as *B. anthracis*, the causal agent of anthrax. Most members of *Clostridium* are soil or aquatic saprophytes or inhabitants of human intestines, but four pathogens are fairly well known: *C. tetani*,

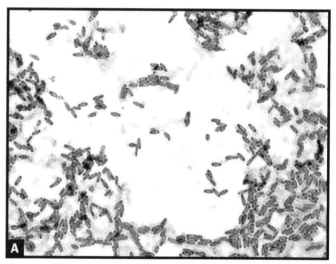

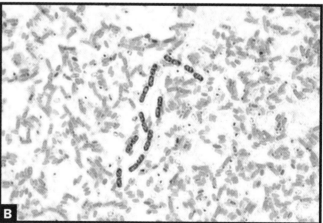

3-106 CULTURE AGE CAN AFFECT SPORULATION (X1000) ✦ Bacteria capable of producing spores don't do so uniformly during their culture's growth. Sporulation is done in response to nutrient depletion, and so is characteristic of older cultures. These two cultures illustrate different degrees of sporulation. **A** Most cells in this specimen contain spores; very few have been released. **B** This specimen consists mostly of released spores.

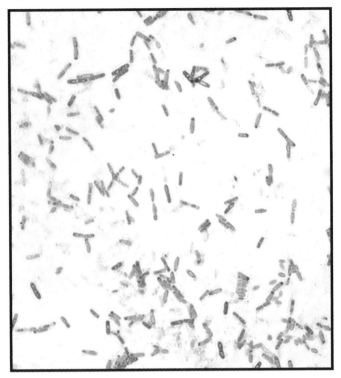

3-107 ELLIPTICAL SUBTERMINAL SPORES OF *BACILLUS COAGULANS*
(X1000) ✦ Notice that even though the majority of cells contain spores, some do not. The white spots in the cells without spores are probably lipid (poly-β-hydroxybutyrate) granules.

C. botulinum, *C. perfringens*, and *C. difficile*, which produce tetanus, botulism, gas gangrene, and pseudo-membranous colitis, respectively.

✦ In This Exercise

In the spore stain you will examine spores from different-aged *Bacillus* cultures. You will not have to stain each species. Divide the work within your lab group, but be sure to look at each others' slides. Also be sure to compare spore shape and location for each species.

✦ Materials

Per Student Group

- ✦ clean glass microscope slides
- ✦ malachite green stain
- ✦ safranin stain
- ✦ squirt bottle with water
- ✦ heating apparatus (steam apparatus or hot plate)
- ✦ bibulous paper or paper towel
- ✦ staining tray
- ✦ staining screen
- ✦ slide holder
- ✦ disposable gloves
- ✦ lab coat

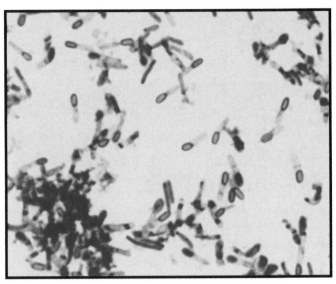

3-108 ELLIPTICAL TERMINAL SPORES (X1200) ✦ *Clostridium tetani* stained by a different spore stain protocol using carbol-fuchsin. Notice the swollen ends of the cells because of the spores.

- ✦ goggles
- ✦ compound microscope with oil objective lens and ocular micrometer
- ✦ immersion oil
- ✦ lens paper
- ✦ nonsterile Petri dish for transporting slides
- ✦ recommended organisms (per student group):
    - ✦ 48 hour and 5 day Sporulating Agar slant pure cultures of *Bacillus cereus*
    - ✦ 48 hour and 5 day Sporulating Agar slant pure cultures of *Bacillus coagulans*
    - ✦ 48 hour and 5 day Sporulating Agar slant pure cultures of *Bacillus megaterium*
    - ✦ 48 hour and 5 day Sporulating Agar slant pure cultures of *Bacillus subtilis*

Note: *Trypticase Soy Agar can be substituted for Sporulating Agar)*

✦ Medium Recipe

Sporulating Agar

✦ Pancreatic digest of gelatin	6.0 g
✦ Pancreatic digest of casein	4.0 g
✦ Yeast extract	3.0 g
✦ Beef extract	1.5 g
✦ Dextrose	1.0 g
✦ Agar	15.0 g
✦ Manganous sulfate	0.3 g
✦ Distilled or deionized water	1.0 L

Final ph 6.6 ± 0.2 at 25°C

Procedure

1 Prepare and heat-fix a smear of each species on the same slide as illustrated in Figure 3-81. Divide the work within your lab group. Minimally, each student should prepare a smear of the 48-hour and 5-day cultures of one species.

2 Follow the instructions in the Procedural Diagram in Figure 3-109. Use a steaming apparatus like the one shown in Figure 3-110. Be sure to have adequate ventilation, eye protection, and gloves. If you must carry the slide to and from the steaming apparatus, put it in a covered Petri dish.

3 Observe using the oil immersion lens. Record your observations of cell morphology and arrangement, cell dimensions, and spore presence, position and shape in the chart provided on the Data Sheet.

4 Dispose of specimen slides in a disinfectant jar or a sharps container.

1. Begin with a heat-fixed emulsion (See Figure 3-81).

2. Cover the smear with a strip of bibulous paper.
Apply malachite green stain.
Steam (as shown in Figure 3-110) for 5 to 10 minutes.
Keep the paper moist with stain.
Perform this step with adequate ventilation, gloves, and eye protection.

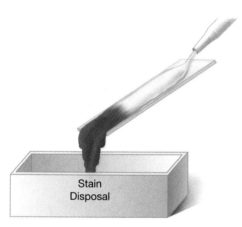

3. Grasp the slide with a slide holder.
Remove the paper and dispose of it properly.
Gently rinse the slide with water.

Stain Disposal

4. Counterstain with Safranin stain for 1 minute.
Rinse with water.

Bibulous

5. Gently blot dry in a tablet of bibulous paper or paper towels. (Alternatively, a page from the tablet can be removed and used for blotting.) Do not rub. When dry, observe under oil immersion.

3-109 PROCEDURAL DIAGRAM: SCHAEFFER-FULTON SPORE STAIN ✦ Steam the staining preparation; do not boil it. Be sure to perform this procedure with adequate ventilation (preferably a fume hood), eye protection, a lab coat, and gloves.

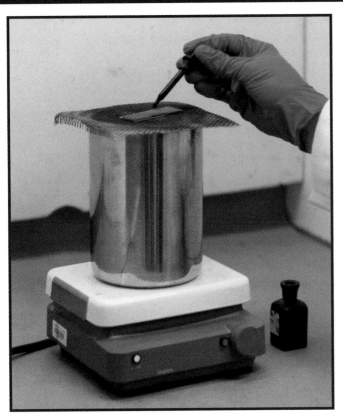

References

Claus, G. William. 1989. Chapter 9 in *Understanding Microbes—A Laboratory Textbook for Microbiology*. W. H. Freeman and Co., New York.

Murray, R. G. E., Raymond N. Doetsch, and C. F. Robinow. 1994. Page 34 in *Methods for General and Molecular Bacteriology*, edited by Philipp Gerhardt, R. G. E. Murray, Willis A. Wood, and Noel R. Krieg. American Society for Microbiology, Washington, DC.

3-110 STEAMING THE SLIDE DURING THE ENDOSPORE STAIN ✦
Carefully steam the slide to force the malachite green into the endospores. Do not boil the slide or let it dry out. Keep it moist with stain for up to 10 minutes of steaming (not 10 minutes on the apparatus—10 minutes of steaming!). ***Caution:*** This should be performed in a well-ventilated area (preferably a fume hood) with hand, clothing, and eye protection.

EXERCISE 3-11

Parasporal Crystal Stain

✦ Theory

Bacillus thuringiensis produces spores, as do all members of the genus *Bacillus*. But they also produce proteinaceous bodies near the spores called **parasporal crystals** (Figure 3-111). Parasporal crystals kill the larvae of various insect groups (especially Lepidopterans) and are often specific to hosts of different *B. thuringiensis* strains.[1] After ingestion of the crystal, it is activated in the larval gut by a protease enzyme. The result is cytolysis of larval cells and, presumably, a ready nutrient source for the endospore.

Some insecticides contain the **Bt toxin** (as it is commercially known). The genes encoding the toxin (most of which are on plasmids) have also been genetically engineered into crop plants as a natural alternative to insecticide use.

✦ Application

Production of parasporal crystals is a unique ability (among *Bacillus* species) of *Bacillus thuringiensis*. This stain is a means of rapid identification of the species.

✦ In This Exercise

You will perform a parasporal crystal stain on pure cultures of *Bacillus thuringiensis* and *B. subtilis*.

✦ Materials

Per Student Group

✦ one-week old sporulation medium or TSA cultures of
  ✦ *Bacillus thuringiensis*
  ✦ *Bacillus subtilis*
✦ clean microscope slide
✦ spore stain kit (optional)
✦ parasporal crystal stains
  ✦ methanol
  ✦ 0.5% basic fuchsin
✦ fume hood

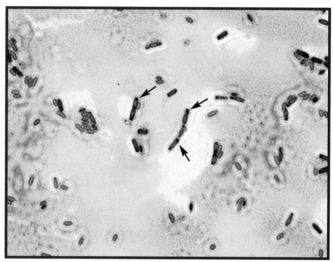

3-111 PARASPORAL CRYSTALS ✦ The dark objects (arrows) at the end of some cells are parasporal crystals. (X1000)

Procedure

1 If desired, perform a spore stain on the two specimens to verify that they are producing spores. If they aren't, keep incubating the cultures for a couple of days until they are.

2 Prepare a heat fixed emulsion of the two species on one slide (Figure 3-81).

3 Add methanol for 30 seconds, then pour it off.

4 Add the 0.5% basic fuchsin and gently heat until steam rises.

5 Remove the slide from the steam, wait 2 minutes, and repeat the steaming.

6 Allow the slide to cool, and then rinse with water.

7 Observe under oil immersion. If positive, darkly stained, angular crystals will be apparent (Figure 3-111).

References

Atlas, Ronald M. and Richard Bartha. 1998. *Microbial Ecology—Fundamentals and Applications,* 4th ed. Addison Wesley Longman, Inc., Menlo Park, CA.

Madigan, Michael T., John M. Martinko, Paul V. Dunlap, and David P. Clark. 2009. *Brock Biology of Microorganisms,* 12th ed. Pearson Benjamin Cummings, 1301 Sansome Street, San Francisco, CA 94111.

---

[1] *Paenibacillus popilliae* (formerly *B. popilliae*) also produces a larvicide that is fatal to scarab beetles. It does not have the breadth of insect hosts that the *B. thuringiensis* has.

Wet Mount and Hanging Drop Preparations

✦ Theory

A wet mount preparation is made by placing the specimen in a drop of water on a microscope slide and covering it with a cover glass. Because no stain is used and most cells are transparent, viewing is best done with little illumination as possible (Figure 3-112). Motility often can be observed at low or high dry magnification, but viewing must be done quickly because of drying of the preparation. As the water recedes, bacteria will appear to be herded across the field. This is not motility. You should look for independent darting of the cells.

A hanging drop preparation allows longer observation of the specimen because it doesn't dry out as quickly. A thin ring of petroleum jelly is applied around the well of a depression slide. A drop of water then is placed in the center of the cover glass and living microbes are transferred into it. A depression microscope slide is carefully placed over the cover glass in such a way that the drop is received into the depression and is undisturbed. The petroleum jelly causes the cover glass to stick to the slide.

The preparation then may be picked up, inverted so the cover glass is on top, and placed under the microscope for examination. As with the wet mount, viewing is best done with as little illumination as possible. The petroleum jelly forms an airtight seal that slows drying of the drop, allowing a long period for observation of cell size, shape, binary fission, and motility.

If these techniques are done to determine motility, the observer must be careful to distinguish between true motility and the **Brownian motion** created by collisions with water molecules. In the latter, cells will appear to vibrate in place. With true motility, cells will exhibit independent movement over greater distances.

✦ Application

Most bacterial microscopic preparations result in death of the microorganisms as a result of heat-fixing and staining. Simple **wet mounts** and the **hanging drop technique** allow observation of living cells to determine motility. They also are used to see natural cell size, arrangement, and shape. All of these characteristics may be useful in identification of a microbe.

✦ In This Exercise

Today you will have the opportunity to view living bacterial cells swimming. The hanging drop preparation allows longer viewing, whereas the simple wet mount can be used to view motility, and it is the starting point for a flagella stain (Exercise 3-13).

✦ Materials

Per Student
+ compound microscope with oil lens and ocular micrometer
+ depression slide and cover glass
+ clean microscope slides and cover glasses
+ petroleum jelly
+ toothpick
+ immersion oil
+ lens paper
+ disposable gloves
+ recommended organisms (overnight cultures grown in broth media):
  + *Aeromonas sobria*
  + *Staphylococcus epidermidis*

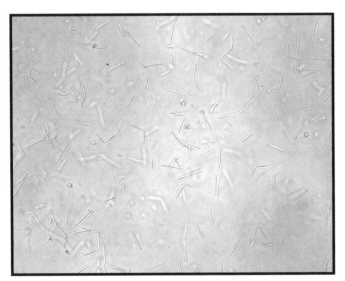

3-112 WET MOUNT (X1000) ✦ Shown is an unstained wet mount preparation of a motile Gram-negative rod. Because of the thickness of the water in the wet mount, cells show up in many different focal planes and are mostly out of focus. To get the best possible image, adjust the condenser height and reduce the light intensity with the iris diaphragm.

Procedure: Wet Mount Preparation

1 Place a loopful of water on a clean glass slide.

2 Add bacteria to the drop. Don't over-inoculate. Flame the loop after transfer.

3 Gently lower a cover glass with your loop supporting one side over the drop of water. Avoid trapping air bubbles.

4 Observe under high dry or oil immersion and record your results in the chart on the Data Sheet. Avoid hitting the cover slip with the oil objective.

5 Dispose of the slide and cover glass in a disinfectant jar or a sharps container when finished.

 ## Procedure: Hanging Drop Preparation

1 Follow the procedure illustrated in Figure 3-113 for each specimen.

2 Observe under high dry or oil immersion (if possible without hitting the cover slip with the objective lens) and record your results in the chart on the Data Sheet. Avoid hitting the cover slip with the oil objective.

3 When finished, remove the cover glass from the depression slide with an inoculating loop and soak both in a disinfectant jar for at least 15 minutes. Flame the loop. After soaking, rinse the slide with 95% ethanol to remove the petroleum jelly, then with water to remove the alcohol. Dry the slide for reuse.

References

Iino, Tetsuo, and Masatoshi Enomoto. 1969. Chapter IV in *Methods in Microbiology,* Vol 1, edited by J. R. Norris and D. W. Ribbins. Academic Press, Ltd., London.

Murray, R. G. E., Raymond N. Doetsch, and C. F. Robinow. 1994. Page 26 in *Methods for General and Molecular Bacteriology,* edited by Philipp Gerhardt, R. G. E. Murray, Willis A. Wood, and Noel R. Krieg. American Society for Microbiology, Washington, DC.

Quesnel, Louis B. 1969. Chapter X in *Methods in Microbiology,* Vol 1, edited by J. R. Norris and D. W. Ribbins. Academic Press, Ltd., London.

1. Apply a light ring of petroleum jelly around the well of a depression slide with a toothpick.

2. Apply a drop of water to a cover glass. If using a broth culture, omit the water. Do not use too much water.

3. Aseptically add a drop of bacteria to the water. Flame your loop after the transfer.

4. Invert the depression slide so the drop is centered in the well. Gently press until the petroleum jelly has created a seal between the slide and cover glass.

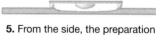

5. From the side, the preparation should look like this. Notice that the drop is "hanging," and is not in contact with the depression slide. Observe under high dry or oil immersion.

3-113 PROCEDURAL DIAGRAM: HANGING DROP PREPARATION ✦ The hanging drop method is used for long-term observation of a living specimen.

EXERCISE 3-13

Flagella Stain

✦ Theory

Bacterial flagella typically are too thin to be observed with the light microscope and ordinary stains. Various special flagella stains have been developed that use a **mordant** to assist in encrusting flagella with stain to a visible thickness. Most require experience and advanced techniques, and typically are not performed in beginning microbiology classes. The method provided in this exercise, though comparatively simple, still requires practice and a bit of good luck.

The number and arrangement of flagella may be observed with a flagella stain. A single flagellum is said to be **polar** and the cell has a **monotrichous** arrangement (Figure 3-114). Other arrangements (shown in Figures 3-115 through 3-117) include **amphitrichous**, with flagella at both ends of the cell; **lophotrichous**, with tufts of flagella at the end of the cell; and **peritrichous**, with flagella emerging from the entire cell surface.

The high magnifications in Figures 3-115 through Figure 3-117 are the result of enlarging photomicrographs shot at 1000X. No light microscope magnifies this high.

✦ Application

The flagella stain allows direct observation of flagella. The presence and arrangement of flagella may be useful in identifying bacterial species.

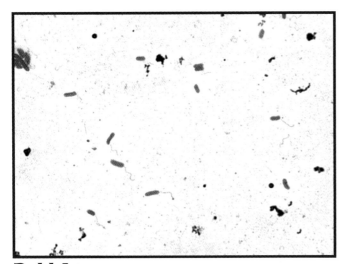

3-114 POLAR FLAGELLA (X1000) ✦ *Pseudomonas aeruginosa* is often suggested as a positive control for flagella stains. Notice the single flagellum emerging from the ends of many (but not all) cells. This is a result of the fragile nature of flagella, which can be broken from the cells during slide preparation.

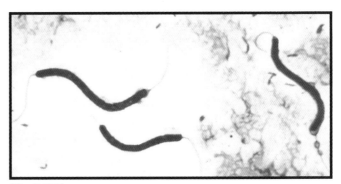

3-115 AMPHITRICHOUS FLAGELLA (X2200) ✦ *Spirillum volutans* has a flagellum at each end.

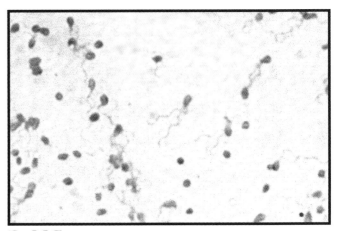

3-116 LOPHOTRICHOUS FLAGELLA (X2000) ✦ Several flagella emerge from one end of this *Pseudomonas* species. Not all cells have flagella because they were too delicate to stay intact during the staining procedure.

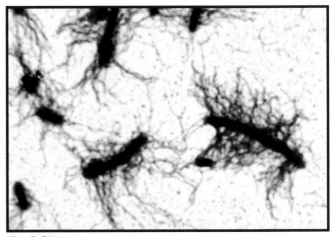

3-117 PERITRICHOUS FLAGELLA (X2640) ✦ This *Proteus* species has flagella emerging from the entire cell surface. It is an intestinal inhabitant of humans and animals.

✦ In This Exercise

You will begin this exercise with a simple wet mount preparation (Exercise 3-12). Once you have determined that the organism is motile, you will stain it to demonstrate its flagella. Be sure to note the flagellar arrangement.

✦ Materials

- ✦ compound microscope with ocular micrometer
- ✦ inoculating loop
- ✦ clean microscope slides and cover slips
- ✦ hypodermic syringe with 0.2 µm filter
- ✦ Ryu flagella stain
- ✦ disposable gloves
- ✦ commercially prepared slides of motile organisms showing flagella.
- ✦ recommended organisms (overnight cultures grown in broth media):
  - ✦ *Aeromonas sobria*
  - ✦ *Proteus vulgaris* (BSL-2)
  - ✦ *Staphylococcus epidermidis*

Procedure

1 Use a separate slide for each preparation. Make sure the slide is clean, and handle it by its edges to prevent oil from your hands soiling the surface.

2 Prepare a wet mount of the organism as directed in Exercise 3-12 with the following modification. Because bacterial flagella are extremely fragile, handle the preparations gently to minimize the inevitable damage. Place a drop of water on the slide. Then place the loop over the growth and don't move it; only make contact with the growth. Allow motile bacteria to swim in for about 30 seconds. Carefully remove the loop and place it in the drop of water on the slide, again allowing motile bacteria to swim into the drop. Gently put on a cover slip.

3 Observe the specimen under high dry power and note any motility. If the specimen is motile, continue with the procedure. If the specimen is nonmotile, observe the Brownian motion and then prepare a wet mount of a different organism.

4 If you see motility, allow the slide to dry for about 10–15 minutes. If practical to do so, leave the slide on the microscope stage so you will be able to find the cells again easily. Then apply the Ryu stain to the edge of the coverslip (Figures 3-118 and 3-119) using the syringe. Capillary action will draw the stain under. Be careful not to get stain on the microscope.

5 Allow the preparation to stain for a while. Keep it on the microscope. Check periodically for faint,

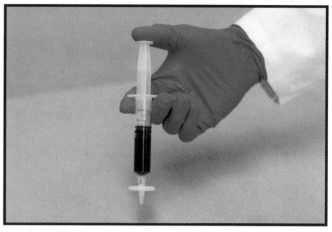

3-118 SYRINGE WITH RYU STAIN ✦ Apply the stain using a syringe fitted with a 0.2 µm filter. For safety, we strongly recommend not using a needle with the syringe, but care must be taken not to deliver too much stain.

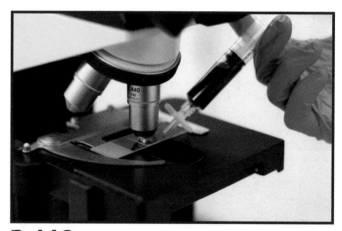

3-119 APPLYING THE STAIN ✦ Carefully place a couple of drops of stain next to the coverslip. Capillary action will draw the stain under. After about 10 minutes of staining, begin examining the slide for flagella. Be careful not to spill stain on the microscope.

stained threads emerging from the cells. Record your observations in the chart on the Data Sheet.

6 Dispose of your specimen slide in a disinfectant jar or sharps container after use.

7 Examine prepared slides of motile organisms and record your results on the Data Sheet.

References

Heimbrook, Margaret E., Wen Lan L. Wang, and Gail Campbell (1989). *Staining Bacterial Flagella Easily.* J. Clin. Microbiol. 27:2612–2615.

Iino, Tetsuo and Masatoshi Enomoto. 1971. Chapter IV in *Methods in Microbiology,* Vol 5A, edited by J. R. Norris and D. W. Ribbins. Academic Press, Ltd., London.

Murray, R. G. E., Raymond N. Doetsch and C. F. Robinow. 1994. Page 35 in *Methods for General and Molecular Bacteriology,* edited by Philipp Gerhardt, R. G. E. Murray, Willis A. Wood, and Noel R. Krieg. American Society for Microbiology, Washington, DC.

EXERCISE 3-14 Morphological Unknown

✦ Theory

In this exercise you will be given one pure bacterial culture selected from the organisms listed in Table 3-4. Your job will be to identify it using only the staining techniques covered in Section Three.

Your first task is to convert the information organized in Table 3-4 into flowchart form. A flowchart is simply a visual tool to illustrate the process of elimination that is the foundation of unknown identification. We have started it for you on the Data Sheet by giving you a few of the branches and listing appropriate organisms. You must complete it by adding necessary branches until you have shown a path to identify each of the organisms. The Procedure below contains a detailed explanation of the process.

Once you have designed your flowchart, you will run one stain at a time on your organism. As you match your staining results with those in the flowchart, you will follow a path to its identification.

A final differential stain will be performed as a **confirmatory test**. It will serve as further evidence that you have identified your unknown correctly.

✦ Application

Schemes employing differential tests are the main strategy for microbial identification.

✦ In This Exercise

You will practice the stains you have learned to this point, as well as familiarize yourself with the standard process of elimination used in bacterial identification.

✦ Materials

Per Student Group

✦ compound microscope with oil immersion lens
✦ immersion oil
✦ clean microscope slides
✦ clean cover glasses
✦ lens paper
✦ Gram stain kit
✦ acid-fast stain kit
✦ capsule stain kit
✦ spore stain kit
✦ Bunsen burner
✦ striker
✦ inoculating loop
✦ unknown organisms in numbered tubes (one per student). Fresh slant or broth cultures of [1]
  ◆ *Aeromonas sobria*
  ◆ *Bacillus subtilis*
  ◆ *Corynebacterium xerosis*
  ◆ *Klebsiella pneumoniae* (BSL-2)
  ◆ *Lactococcus lactis*
  ◆ *Micrococcus luteus*
  ◆ *Mycobacterium phlei* (BSL-2)
  ◆ *Moraxella catarrhalis* (BSL-2)
  ◆ *Rhodospirillum rubrum*
  ◆ *Shigella flexneri* (BSL-2)
  ◆ *Staphylococcus epidermidis*
     (**Note:** Your instructor will choose organisms appropriate to your specific microbiology course and facilities.)
✦ Gram-positive and Gram-negative control organisms (one set per group)—18–24 hour Trypticase Soy Agar slant cultures in tubes labeled with Gram reaction.
✦ A set of organisms (listed above) to be used for positive controls (per class)
✦ sterile Trypticase Soy Broth tubes (one per student)
✦ sterile Trypticase Soy Agar slants (one per student)

📖 Procedure

1 Using the information contained in Table 3-4, complete the flowchart on the Data Sheet. Each of the 11 organisms should occupy a solitary position at the end of a branch. Do not include stains in a branch that do not differentiate any organisms. Have your instructor check your flowchart.

2 Obtain one unknown slant culture. Record its number in the space labeled "Unknown Number" on the Data Sheet.

---

[1] Cultures should have abundant growth.

TABLE **3-4** TABLE OF RESULTS FOR ORGANISMS USED IN THE EXERCISE ✦ These results are typical for the species listed. Your specific strains may vary because of their genetics, their age, or the environment in which they are grown. It is important that you record *your* results in case strain variability leads to misidentification.

TABLE OF RESULTS

Organism	Gram Stain	Cell Morphology	Cell Arrangement[1]	Acid-Fast Stain	Motility (Wet Mount)	Capsule	Spore Stain[2]
Aeromonas sobria	−	rod	single cells	−	+	−	−
Bacillus subtilis	+	rod	usually single cells	−	+	−	+
Corynebacterium xerosis	+	rod	single cells or multiples in angular or palisade arrangement	−	−	−	−
Klebsiella pneumoniae	−	rod	usually single cells, sometimes pairs or short chains	−	−	+	−
Lactococcus lactis	+	ovoid cocci (appearing stretched in the direction of the chain or pair)	(pairs or) **chains**	−	−	+	−
Micrococcus luteus	+	coccus	(pairs or) **tetrads**	−	−	−	−
Mycobacterium phlei	weak +	rod	single cells or branched; sometimes in dense clusters	+	−	−	−
Moraxella catarrhalis	−	coccus	pairs with adjacent sides flattened	−	−	+	−
Rhodospirillum rubrum	−	spirillum or bent rod	single	−	+	−	−
Shigella flexneri	−	rod	single	−	−	−	−
Staphylococcus epidermidis	+	coccus	(singles, pairs, tetrads or) **clusters**	−	−	+	−

[1] It may be necessary to grow your unknown in broth culture to see the arrangement clearly.

[2] If spores are not readily visible, continue incubating your culture for another 24–48 hours and check again.

3 Perform a Gram stain of the organism. The stain should be run with known Gram-positive and Gram-negative controls to verify your technique (see Figure 3-89). In addition to providing information on Gram reaction, the Gram stain will allow observation of cell size, shape, and arrangement. Enter all these and the date in the appropriate boxes of the chart provided on the Data Sheet. (**Note:** Cell size for the unknown candidates is not given and should not be used to differentiate the species. To give you an idea of typical cell sizes, the cocci should be about 1 μm in diameter, and the rods 1 to 5 μm in length and about 1 μm in width.)

4 If you have cocci, or are unclear about cell arrangement, aseptically transfer your unknown to sterile Trypticase Soy Broth and examine it again in 24–48 hours using a crystal violet or carbolfuchsin simple stain. Cell arrangement often is easier to interpret using cells grown in broth.

5 Based on the Gram stain and cell morphology results, follow the appropriate branches in the flowchart until you reach the list of possible organisms that matches your unknown.

6 Perform the next stain and determine to which new branch of the flowchart your unknown belongs. Record the result and date of this stain in the chart provided on the Data Sheet.

7 Repeat Step 6 until you eliminate all but one organism in the flowchart—this is your unknown! (If you are not doing all the stains on the day the unknowns were handed out, be sure to inoculate appropriate media so you will have fresh cultures to work with. If you need to do a spore stain, incubate your original culture for another 24–48 hours, then perform the stain.)

8 After you identify your unknown, perform one more stain to confirm your result. This **confirmatory test** should be one you did not perform in the identification and should be added to your flowchart. The result of this test should agree with the predicted result (as given in Table 3-4). If your confirmatory test result doesn't match the expected result, repeat any suspect stains and find the source of error. (**Note:** The source of error may be your technique or bacterial strain variability. The results in Table 3-4 are typical for each organism, but some strains may vary. By rerunning suspect stain(s) *with controls*, you probably will be able to identify the source of error.)

9 Use a colored marker to highlight the path on the flowchart that leads to your unknown. Have your instructor check your work.

Selective Media

Individual microbial species in a **mixed culture** must be isolated and cultivated as **pure cultures** before they can be tested and properly identified. The most common means of isolating an organism from a mixed culture is to streak for isolation. Refer to Exercise 1-4 for a description of the streak-plate method of isolation.

The **media** introduced in this section are designed to enhance the isolation procedure by inhibiting growth of some organisms while encouraging the growth of others. Thus, they are referred to as **selective media**. Many types of selective media contain indicators to expose differences between organisms and are called **differential media**. The media illustrated in this section are used specifically to isolate pathogenic Gram-negative bacilli and Gram-positive cocci from human or environmental samples containing a mixture of microorganisms.

Clinical microbiologists, who are familiar with human pathogens and the types of infections they cause, choose selective media that will screen out normal flora that also are likely to be in a sample. Environmental microbiologists often choose selective and differential media to detect **coliform** bacteria. Coliforms are a subgroup of *Enterobacteriaceae* (*Escherichia coli* being the most prominent member) that produce gas from lactose fermentation. Fermentation will be discussed more fully in Section 5. Most coliforms are normal inhabitants of the human intestinal tract; therefore, their presence in the environment may be evidence of fecal contamination.

In the exercises that follow, you will examine some commonly used selective media for isolating Gram-positive cocci and Gram-negative rods. Most of the exercises contain two examples of growth—spot inoculations with pure cultures to magnify growth characteristics and streak-plates of mixed samples to illustrate actual colony morphology.

All of the exercises contain a Table of Results, identifying the various reactions produced on a specific medium. These include color results, interpretations of results, symbols used to quickly identify the various reactions, and presumptive identification of typical organisms encountered with these media. Presumptive identification is not final identification. It is an "educated guess" based on evidence provided by the selective/differential medium coupled with information about the origin of a sample. ✦

A Word About Selective Media

As you read the Theory of each of the following media and learn a little about how the tests actually perform on a biochemical level, you will see that each medium has six fundamental features. For a fuller understanding of the media and tests included in this section, watch for and identify these six features and make sure you can identify them. (**Note:** In the following, a distinction is made between nutritional components and substrates. Since, in the strictest sense all reactants in a chemical reaction are substrates, this is an artificial distinction. We separate the terms to illustrate the importance of the specific substance added to allow observations of biochemical distinctions to be made.)

1. Selective, Differential, Defined or Undefined—Which Is It?

The medium is any combination of the above. If it is selective, it encourages growth of some organisms and discourages growth of others. If it is differential, it allows us to distinguish between different microbes. If it is defined or "chemically defined", each of its chemical ingredients is known and in exactly what amounts. If it is undefined or "complex" it contains one or more ingredient made up of known ingredients, but of unknown composition, such as yeast extract, or beef extract, or digest of gelatin, *etc.*

2. Nutritional Components

Nutritional components are selected to obtain the optimum growth of the organisms being tested for or suspected of being in the sample. Although many ingredients are suitable for many different types of microorganisms, typically, slight variations are made to tailor the medium for a specified group. Yeast extract can be added for fastidious organisms requiring specific vitamins, or sodium chloride (although not really a nutritional component) can be added for organisms where osmotic equilibrium is critical. As you will see, the inhibitor is the major selective component added to the medium, but nutritional components can enhance the overall selectivity.

3. Inhibitors

Inhibitors make the medium selective. They are designed to exploit weaknesses in specific groups of organisms and thus prevent or inhibit their growth, while allowing other organisms to grow. Some inhibitors function by interrupting DNA synthesis or expression of a gene. Other inhibitors function at the enzymatic level, interfering with a critical reaction. Still other inhibitors interfere with membrane permeability, thus upsetting homeostasis and starting a cascade of catastrophic changes inside the cell.

4. Substrates

Substrates are almost always what make the medium differential. Differentiation and identification of organisms frequently relies on their differing abilities to perform a specific chemical reaction or set of reactions and to do them in a way that can be observed. This can all be staged in an artificial environment (differential medium) by providing the organisms all the required components for growth and by including at least one substrate that only one organism or group of organisms can utilize. Once the reaction has taken place, presumably, organisms that before the test were indistinguishable are now (with the help of indicators—see #5) easily differentiated. For example, if a fecal sample is being tested for the presence of a pathogen known not to ferment lactose, then lactose is an important substrate needed to identify and isolate *E. coli* and other coliforms, all of which ferment lactose and all of which are likely to be found in the sample.

5. Indicators

Indicators make a desired or expected reaction visible. An indicator is virtually always included in differential media because bacterial growth, by itself, is usually not enough to reliably differentiate between microbial groups. The medium must change fairly dramatically for us to be able to see it. Subtle changes can lead to false readings. The indicator frequently is a dye that changes color as pH changes. All dyes have a specific functional range and are chosen based on the starting pH of the medium, which in turn is based on the optimum pH range of the organisms being tested.

Indicators can also be chemicals that react with products of a reaction and produce a color change. For example, if the organism being tested reduces the medium substrate, an indicator that is oxidized (ferric sulfate in solution releases Fe^{3+}) can produce a pretty spectacular change. Oxidized iron is a common reactant used to form a visible precipitate.

6. Positive or Negative?

The last item covered is a description of the positive and negative reactions. We tell you what to look for. ✦

Selective Media for Isolation of Gram-Positive Cocci

Gram-negative organisms frequently inhibit growth of Gram-positive organisms when they are cultivated together. Therefore, when looking for staphylococci or streptococci in a clinical sample, it may be necessary to begin by streaking the unknown mixture onto a selective medium that inhibits Gram-negative growth. The four selective media introduced in this unit—Phenylethyl Alcohol Agar, Columbia CNA Agar, Bile Esculin Agar, and Mannitol Salt Agar—are all used to isolate streptococci, enterococci, or staphylococci in human samples. ✦

EXERCISE 4-1

Phenylethyl Alcohol Agar

✦ Theory

Phenylethyl Alcohol Agar (PEA) is an undefined, selective medium that encourages growth of Gram-positive organisms and inhibits growth of most Gram-negative organisms (Figure 4-1). It is not a differential medium because it does not distinguish between different organisms; it merely encourages or discourages growth. Digests of casein and soybean meal provide nutrition while sodium chloride provides a stable osmotic environment suitable for the addition of sheep blood if desired. Phenylethyl alcohol is the selective agent that inhibits Gram-negative organisms by breaking down their membrane permeability barrier, thus allowing influx of substances ordinarily blocked and leakage of large amounts of cellular potassium. This ultimately disrupts or halts DNA synthesis.

✦ Application

PEA is used to isolate staphylococci and streptococci (including enterococci and lactococci) from specimens containing mixtures of bacterial flora. Typically, it is used to screen out the common contaminants *Echerichia coli* and *Proteus* species. When prepared with 5% sheep blood, it is used for cultivation of Gram-positive anaerobes.

✦ In This Exercise

You will spot-inoculate one PEA plate and one Nutrient Agar (NA) plate with three test organisms. The NA plate will serve as a comparison for growth quality on the PEA plate.

✦ Materials

Per Student Group

✦ one PEA plate
✦ one NA plate
✦ fresh broth cultures of:
   ✦ *Escherichia coli*
   ✦ *Enterococcus faecalis* (BSL-2)
   ✦ *Staphylococcus aureus* (BSL-2)

4-1 NUTRIENT AGAR
VERSUS PHENYLETHYL ALCOHOL
AGAR ✦ NA is on the left; PEA
is on the right. Both plates were
inoculated with the same three
organisms—one Gram-positive coccus and two Gram-negative rods. All three organisms grow well on the NA, but only the Gram-positive organism (top) grows well on the PEA. Note the stunted growth of the Gram-negative rods on PEA.

TABLE **4-1** PEA Results and Interpretations

TABLE OF RESULTS		
Result	**Interpretation**	**Presumptive ID**
Poor growth or no growth (P)	Organism is inhibited by phenylethyl alcohol	Probable Gram-negative organism
Good growth (G)	Organism is not inhibited by phenylethyl alcohol	Probable *Staphylococcus*, *Streptococcus*, *Enterococcus*, or *Lactococcus*

✦ Medium Recipes

Phenylethyl Alcohol Agar

✦ Pancreatic digest of casein	15.0 g
✦ Papaic digest of soybean meal	5.0 g
✦ Sodium chloride	5.0 g
✦ β-Phenylethyl alcohol	2.5 g
✦ Agar	15.0 g
✦ Distilled or deionized water	1.0 L
pH 7.1–7.5 at 25°C	

Nutrient Agar

✦ Beef extract	3.0 g
✦ Peptone	5.0 g
✦ Agar	15.0 g
✦ Distilled or deionized water	1.0 L
pH 6.6 – 7.0 at 25°C	

Procedure

Lab One

1 Mix each culture well.

2 Using a permanent marker, divide the bottom of each plate into three sectors.

3 Label the plates with the organisms' names, your name, and the date.

4 Spot-inoculate the sectors on the PEA plate with the test organisms. Refer to Appendix B if necessary.

5 Repeat Step 4 with the Nutrient Agar plate.

6 Invert and incubate the plates at 35 ± 2°C for 24 to 48 hours.

Lab Two

1 Examine and compare the plates for color and quality of growth.

2 Record your results on the Data Sheet. Refer to Table 4-1 as needed.

References

Forbes, Betty A., Daniel F. Sahm, and Alice S. Weissfeld. 2002. Chapter 10 in *Bailey & Scott's Diagnostic Microbiology*, 11th ed. Mosby, St. Louis.

Zimbro, Mary Jo, and David A. Power. 2003. Page 443 in *Difco™ & BBL™ Manual—Manual of Microbiological Culture Media*. Becton, Dickinson and Co., Sparks, MD.

Columbia CNA With 5% Sheep Blood Agar

✦ Theory

Columbia CNA with 5% Sheep Blood Agar is an undefined, differential, and selective medium that allows growth of Gram-positive organisms (especially staphylococci, streptococci, and enterococci) and stops or inhibits growth of most Gram-negative organisms (Figure 4-2). Casein, digest of animal tissue, beef extract, yeast extract, corn starch, and sheep blood provide a range of carbon and energy sources to support a wide variety of organisms. In addition, sheep blood supplies the X factor (heme) and yeast extract provides B-vitamins. The antibiotics colistin and nalidixic acid (CNA) act as selective agents against Gram-negative organisms by affecting membrane integrity and interfering with DNA replication, respectively. They are particularly effective against *Klebsiella, Proteus,* and *Pseudomonas* species. Further, sheep blood makes possible differentiation of Gram-positive organisms based on hemolytic reaction.

✦ Application

Columbia CNA with 5% Sheep Blood Agar is used to isolate staphylococci, streptococci , and enterococci, primarily from clinical specimens.

✦ In This Exercise

You will spot-inoculate one Columbia CNA with 5% Sheep Blood Agar plate and one Nutrient Agar (NA) plate with four test organisms. The NA plate will serve as a comparison for growth quality on the Columbia CNA with 5% Sheep Blood Agar plate.

✦ Materials

Per Student Group

✦ one Columbia CNA with 5% Sheep Blood Agar plate
✦ one Nutrient Agar Plate
✦ fresh broth cultures of:
  ✦ *Escherichia coli* (or other Gram-negative, as available)
  ✦ *Streptococcus spp.*
  ✦ *Enterococcus spp.*
  ✦ *Staphylococcus spp.*

4-2 COLUMBIA CNA WITH 5% BLOOD AGAR ✦ The plate was inoculated with four organisms—two Gram-positive cocci, and two Gram-negative rods. Only the Gram-positive organisms (left and top quadrants) grow well on the Columbia CNA agar. The two Gram-negatives either didn't grow (bottom) or grew poorly (right). Further, the top Gram-positive is β-hemolytic, whereas the one on the left is nonhemolytic.

✦ Medium Recipes

Columbia CNA with 5% Sheep Blood Agar

✦ Pancreatic digest of casein	12.0 g
✦ Peptic digest of animal tissue	5.0 g
✦ Yeast extract	3.0 g
✦ Beef extract	3.0 g
✦ Corn starch	1.0 g
✦ Sodium chloride	5.0 g
✦ Colistin	10.0 mg
✦ Nalidixic acid	10.0 mg
✦ Agar	13.5 g
✦ Distilled or deionized water	1 L
pH 7.1–7.5 at 25°C	
✦ Sheep blood (defibrinated)	50 mL added to 950 mL of above

Nutrient Agar

✦ Beef extract	3.0 g
✦ Peptone	5.0 g
✦ Agar	15.0 g
✦ Distilled or deionized water	1.0 L
pH 6.6–7.0 at 25°C	

Procedure

Lab One

1 Mix each culture well.

2 Using a permanent marker, divide the bottom of each plate into four sectors.

3 Label the plates with the organisms' names, your name, and the date.

4 Spot-inoculate (Appendix B) the sectors on the CNA plate with the test organisms.

5 Repeat step 4 with the Nutrient Agar plate. Use the same position for each specimen.

6 Invert and incubate the plates at 35°C for 24 to 48 hours. If possible, incubate the plates in a candle jar or anaerobic jar with 3–5% CO_2.

Lab Two

1 Referring to Table 4-2, examine and compare the plates for quality of growth and hemolytic reaction. For more information on hemolysis, see Exercise 5-25.

2 Record your results in the space provided on the Data Sheet.

References

Chapin, Kimberle C., and Tsai-Ling Lauderdale. 2007. Chapter 21 in *Manual of Clinical Microbiology*, 8th Ed., edited by Patrick R. Murray, Ellen Jo Baron, James. H. Jorgensen, Marie Louise Landry, and Michael A. Pfaller. ASM Press, Washington, DC.

Forbes, Betty A., Daniel F. Sahm, and Alice S. Weissfeld. 2002. Pages 136 and 138 in *Bailey & Scott's Diagnostic Microbiology*, 11th Ed. Mosby, Inc., St. Louis, MO.

Zimbro, Mary Jo and David A. Power. 2003. Page 156 in *DIFCO & BBL Manual—Manual of Microbiological Culture Media*. Becton, Dickinson and Company, Sparks, MD.

TABLE **4-2**   Columbia CNA with 5% Sheep Blood Agar Results and Interpretations

TABLE OF RESULTS		
Result	**Interpretation**	**Presumptive ID**
Poor growth or no growth (P)	Organism is inhibited by colistin and nalidixic acid	Probable Gram-negative organism
Good growth (G) with clearing (β-hemolysis)	Organism is not inhibited by colistin and nalidixic acid and completely hemolyzes RBCs	Probable β-hemolytic *Staphylococcus*, *Streptococcus*, or *Enterococcus* (**Note:** Artifactual greening may occur with some β-hemolytic organisms)
Good growth (G) with greening of medium (α-hemolysis)	Organism is not inhibited by colistin and nalidixic acid and partially hemolyzes RBCs	Probable α-hemolytic *Staphylococcus*, *Streptococcus*, or *Enterococcus*
Good growth (G) with no change of medium's color (γ-hemolysis)	Organism is not inhibited by colistin and nalidixic acid and does not hemolyze RBCs	Probable γ-hemolytic *Staphylococcus*, *Streptococcus*, or *Enterococcus*

Bile Esculin Test

✦ Theory

Bile Esculin Agar is an undefined, selective and differential medium containing beef extract, digest of gelatin, esculin, oxgall (bile), and ferric citrate. Esculin extracted from the bark of the Horse Chestnut tree, is a glycoside composed of glucose and esculetin. Beef extract and gelatin provide nutrients and energy; bile is the selective agent added to separate the *Streptococcus bovis* group and enterococci from other streptococci. Ferric citrate is added as a source of oxidized iron to indicate a positive test.

Many bacteria can hydrolyze esculin under acidic conditions (Figure 4-3), and many bacteria, especially Gram-negative enterics, demonstrate tolerance to bile. However, among the streptococci, typically only enterococci and members of the *Streptococcus bovis* group (*S. equinus*, *S. gallolyticus*, *S. infantarius*, and *S. alactolyticus*) tolerate bile and hydrolyze esculin.

In this test, when esculin molecules are split esculetin reacts with the Fe^{3+} from the ferric citrate and forms a dark brown precipitate (Figure 4-4). This precipitate darkens the medium surrounding the growth. An organism that darkens the medium even slightly is Bile Esculin-positive (Figure 4-5). An organism that does not darken the medium is negative.

✦ Application

This test is used most commonly for presumptive identification of enterococci and members of the *Streptococcus bovis* group, all of which are positive.

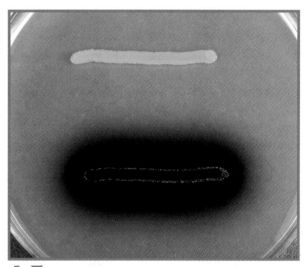

4-5 BILE ESCULIN TEST RESULTS ✦ This plate was inoculated with two Gram-positive cocci. The organism on the bottom is bile esculin-positive. The organism on the top is negative.

4-3 ACID HYDROLYSIS OF ESCULIN ✦ Many organisms can hydrolyze esculin, but the group D streptococci and enterococci are unique in their ability to do this in the presence of bile salts.

Esculin

Acid

β-D-Glucose   +   Esculetin

Glycolysis

4-4 BILE ESCULIN TEST INDICATOR REACTION ✦ Esculetin, produced during the hydrolysis of esculin, reacts with Fe^{3+} in the medium to produce a dark brown to black color in the medium.

Esculetin   + Fe^{3+} ⟶ Dark Brown Color

✦ In This Exercise

Today you will spot inoculate one Bile Esculin Agar plate with two test organisms and incubate it for up to 48 hours. Any blackening of the medium before the end of 48 hours can be recorded as positive. Negative results must incubate the full 48 hours before final determination is made. Use Table 4-3 as a guide.

✦ Materials

Per Student Group

✦ one Bile Esculin Agar plate
✦ fresh cultures of:
  ◆ *Lactococcus lactis*
  ◆ *Enterococcus faecalis* (BSL-2)

✦ Medium Recipe

Bile Esculin Agar

◆ Pancreatic Digest of Gelatin	5.0 g
◆ Beef Extract	3.0 g
◆ Oxgall	20.0 g
◆ Ferric citrate	0.5 g
◆ Esculin	1.0 g
◆ Agar	14.0 g
◆ Distilled or deionized water	1.0 L

pH 6.6–7.0 at 25°C

Procedure

Lab One

1 Obtain a Bile Esculin Agar plate.

2 Using a permanent marker, divide the bottom into two halves.

3 Label the plate with the organisms' names, your name, and the date.

4 Spot-inoculate each half with one test organism. Refer to Appendix B if necessary.

5 Invert and incubate the plate at 35 ± 2°C for 24 to 48 hours.

Lab Two

1 Examine the plate for any darkening of the medium.

2 Record your results on the Data Sheet.

References

Delost, Maria Dannessa. 1997. Page 132 in *Introduction to Diagnostic Microbiology*. Mosby, St. Louis.

Forbes, Betty A., Daniel F. Sahm, and Alice S. Weissfeld. 2002. Page 264 in *Bailey & Scott's Diagnostic Microbiology*, 11th ed. Mosby, St. Louis.

Lányi, B. 1987. Page 56 in *Methods in Microbiology*, Vol. 19, edited by R. R. Colwell and R. Grigorova. Academic Press, New York.

MacFaddin, Jean F. 2000. Page 8 in *Biochemical Tests for Identification of Medical Bacteria*, 3rd ed. Lippincott Williams & Wilkins, Philadelphia.

Zimbro, Mary Jo, and David A. Power, Eds. 2003. Page 76 in *Difco™ and BBL™ Manual—Manual of Microbiological Culture Media*. Becton Dickinson and Co., Sparks, MD.

TABLE **4-3** Bile Esculin Test Results and Interpretations

TABLE OF RESULTS		
Result	**Interpretation**	**Symbol**
Medium is darkened within 48 hours	Presumptive identification as a member of Group D *Streptococcus* or *Enterococcus*	+
No darkening of the medium after 48 hours	Presumptive determination as not a member of Group D *Streptococcus* or *Enterococcus*	−

EXERCISE 4-4

Mannitol Salts Agar

✦ Theory

Mannitol Salt Agar (MSA) contains the carbohydrate mannitol, 7.5% sodium chloride (NaCl), and the pH indicator phenol red. Phenol red is yellow below pH 6.8, red at pH 7.4 to 8.4, and pink at pH 8.4 and above. Mannitol provides the substrate for fermentation and makes the medium differential. Sodium chloride makes the medium selective because its concentration is high enough to dehydrate and kill most bacteria. Staphylococci thrive in the medium, largely because of their adaptation to salty habitats such as human skin. Phenol red indicates whether fermentation has taken place by changing color as the pH changes. (See Section 5, A Word About Biochemical Tests and Acid-Base Reactions.

Most staphylococci are able to grow on MSA, but do not ferment the mannitol, so the growth appears pink or red and the medium remains unchanged. *Staphylococcus aureus* ferments the mannitol, which produces acids and lowers the pH of the medium (Figure 4-6). The result is formation of bright yellow colonies usually surrounded by a yellow halo (Figure 4-7 and 4-8). For more information on fermentation, refer to Exercise 5-3 and Appendix A.

✦ Application

Mannitol Salt Agar is used for isolation and differentiation of *Staphylococcus aureus*.

✦ In This Exercise

Today you will spot inoculate one MSA plate and one Nutrient Agar (NA) plate with three test organisms. The NA will serve as a comparison for growth quality on the MSA plate.

✦ Materials

Per Student Group

✦ one MSA plate

✦ one NA plate

✦ fresh broth cultures of:
  ✦ *Staphylococcus aureus* (BSL-2)
  ✦ *Staphylococcus epidermidis*
  ✦ *Escherichia coli*

✦ Medium Recipes

Mannitol Salt Agar

✦ Beef extract	1.0 g
✦ Peptone	10.0 g
✦ Sodium chloride	75.0 g
✦ D-Mannitol	10.0 g
✦ Phenol red	0.025 g
✦ Agar	15.0 g
✦ Distilled or deionized water	1.0 L

pH 7.2–7.6 at 25°C

Nutrient Agar

✦ Beef extract	3.0 g
✦ Peptone	5.0 g
✦ Agar	15.0 g
✦ Distilled or deionized water	1.0 L

pH 6.6–7.0 at 25°C

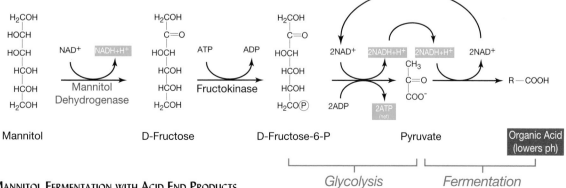

4-6 MANNITOL FERMENTATION WITH ACID END PRODUCTS

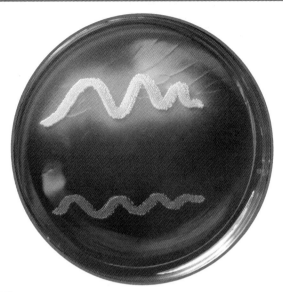

4-7 MANNITOL SALT AGAR ✦ This MSA was inoculated with two of the organisms used in today's lab. Both grew well, but only the top one fermented mannitol and produced acid end products. This is evidenced by the yellow growth and halo surrounding it. Compare this plate with the one in Figure 4-8 and see if you can identify the organisms.

4-8 MANNITOL SALT AGAR STREAKED FOR ISOLATION ✦ This MSA was inoculated with the same two organisms as in Figure 4-7. The solid growth in the first three quadrants is still a mixture of the two organisms, so disregard the color in this region. Note the small red colonies and larger yellow colonies in the fourth streak.

TABLE **4-4** Mannitol Salt Agar Results and Interpretations

TABLE OF RESULTS

Result	Interpretation	Presumptive ID
Poor growth or no growth (P)	Organism is inhibited by NaCl	Not *Staphylococcus*
Good growth (G)	Organism is not inhibited by NaCl	*Staphylococcus*
Yellow growth or halo (Y)	Organism produces acid from mannitol fermentation	Possible pathogenic *Staphylococcus aureus*
Red growth (no halo) (R)	Organism does not ferment mannitol. No reaction	*Staphylococcus* other than *S. aureus*

 Procedure

Lab One

1 Mix each culture well.

2 Using a permanent marker, divide the bottom of each plate into three sectors.

3 Label the plates with the organisms' names, your name, and the date.

4 Spot-inoculate the sectors on the Mannitol Salt Agar plate with the test organisms. Refer to Appendix B if necessary.

5 Repeat Step 4 with the Nutrient Agar plate.

6 Invert and incubate the plates at $35 \pm 2°C$ for 24 to 48 hours.

Lab Two

1 Examine and compare the plates for color and quality of growth.

2 Record your results on the Data Sheet.

References

Delost, Maria Dannessa. 1997. Page 112 in *Introduction to Diagnostic Microbiology, a Text and Workbook*. Mosby, St. Louis.

Forbes, Betty A., Daniel F. Sahm, and Alice S. Weissfeld. 2002. Chapter 19 in *Bailey & Scott's Diagnostic Microbiology*, 11th ed. Mosby, St. Louis.

Zimbro, Mary Jo, and David A. Power. 2003. Page 349 in *Difco™ & BBL™ Manual—Manual of Microbiological Culture Media*. Becton, Dickinson and Co., Sparks, MD.

Selective Media for Isolation of Gram-negative Rods

Members of the family *Enterobacteriaceae*—the enteric "gut" bacteria—are commonly found in clinical samples. Depending on the circumstances, some organisms in a mixed sample are contaminants and relatively benign, others are potentially harmful and must be isolated. The media in this unit are designed to isolate and differentiate these organisms from each and to discourage growth of other organisms.

The four examples selected for this unit are MacConkey Agar, Eosin Methylene Blue (EMB) Agar, Hektoen Enteric (HE) Agar, and Xylose Lysine Desoxycholate (XLD) Agar. MacConkey Agar and EMB Agar are selective for Gram-negative organisms and both contain indicators to differentiate lactose fermenters from lactose nonfermenters. EMB Agar is commonly used to test for the presence of coliforms in environmental samples. As mentioned in the introduction to this section, the presence of coliforms in the environment suggests fecal contamination and the possible presence of pathogens. HE Agar differentiates *Salmonella* and *Shigella* from each other and from other enterics based on their ability to overcome the inhibitory effects of bile, reduce sulfur to H_2S, and ferment lactose, sucrose, or salicin. XLD Agar favors growth of *Salmonella, Shigella* or *Providencia* based on their ability to overcome the inhibitory effects of desoxycholate, and differentiates them based on their ability to reduce sulfur to H_2S, decarboxylate the amino acid lysine, and ferment xylose or lactose. (For more information about fermentation or decarboxylation reactions, refer to Section 5. Also see, Appendix A, and "A Word About Biochemical Tests and Acid-Base Reactions.") ✦

EXERCISE 4-5

MacConkey Agar

✦ Theory

MacConkey Agar is a selective and differential medium containing lactose, bile salts, neutral red, and crystal violet. Bile salts and crystal violet inhibit growth of Gram-positive bacteria. Neutral red dye is a pH indicator that is colorless above a pH of 6.8 and red at a pH less than 6.8. Acid accumulating from lactose fermentation turns the dye red. Lactose fermenters turn a shade of red on MacConkey Agar, whereas lactose nonfermenters retain their normal color or the color of the medium (Figure 4-9 and Figure 4-10). Formulations without crystal violet allow growth of *Enterococcus* and some species of *Staphylococcus*, which ferment the lactose and appear pink on the medium.

✦ Application

MacConkey Agar is used to isolate and differentiate members of the *Enterobacteriaceae* based on the ability to ferment lactose. Variations on the standard medium include MacConkey Agar w/o CV (without crystal violet) to allow growth of Gram-positive cocci, or MacConkey Agar CS to control swarming bacteria (*Proteus*) that interfere with other results.

✦ In This Exercise

You will spot-inoculate one MacConkey Agar plate and one Nutrient Agar (NA) plate with three test organisms. The NA plate will serve as a comparison for growth quality on the MacConkey Agar plate.

✦ Materials

Per Student Group

✦ one MacConkey Agar plate

✦ one Nutrient Agar plate

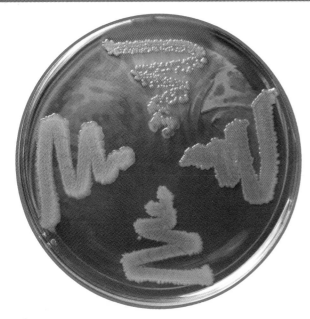

4-9 MACCONKEY AGAR ✦ This MacConkey Agar was inoculated with four members of *Enterobacteriaceae*, all of which show abundant growth. The top organism and the one on the right are lactose fermenters, as evidenced by the pink color. Note the bile precipitate around the top organism. The organisms on the bottom and on the left produced no color, so they do not appear to be lactose fermenters.

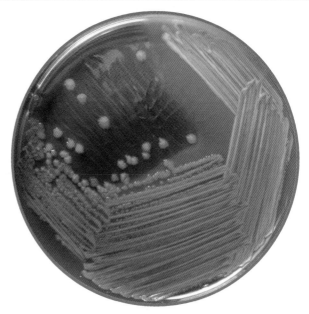

4-10 MACCONKEY AGAR STREAKED FOR ISOLATION ✦ This MacConkey Agar was inoculated with two enteric organisms similar to those selected for today's exercise. Both grew well, but only one fermented the lactose. The solid growth in the first three quadrants is a mixture of the two organisms, so the color produced in that region is not determinative. Note the colors of the individual colonies.

✦ fresh broth cultures of:
- *Enterococcus faecalis* (BSL-2)
- *Escherichia coli*
- *Salmonella typhimurium* (BSL-2)

✦ Medium Recipes

MacConkey Agar
- Pancreatic digest of gelatin   17.0 g
- Pancreatic digest of casein   1.5 g
- Peptic digest of animal tissue   1.5 g
- Lactose   10.0 g
- Bile salts   1.5 g
- Sodium chloride   5.0 g
- Neutral red   0.03 g
- Crystal violet   0.001
- Agar   13.5 g
- Distilled or deionized water   1.0 L
    pH 6.9–7.3 at 25°C

Nutrient Agar
- Beef extract   3.0 g
- Peptone   5.0 g
- Agar   15.0 g
- Distilled or deionized water   1.0 L
    pH 6.6–7.0 at 25°C

Procedure

Lab One

1. Mix each culture well.

2. Using a permanent marker, divide the bottom of each plate into three sectors.

3. Label the plates with the organisms' names, your name, and the date.

4. Spot-inoculate the sectors on the MacConkey Agar plate with the test organisms. Refer to Appendix B if necessary.

5. Repeat Step 4 with the Nutrient Agar plate.

6. Invert and incubate the plates at 35 ± 2°C for 24 to 48 hours.

Lab Two

1. Examine and compare the plates for color and for quality of growth.

2. Refer to Table 4-5 when recording your results on the Data Sheet.

TABLE **4-5** MacConkey Agar Results and Interpretations

TABLE OF RESULTS

Result	Interpretation	Presumptive ID
Poor growth or no growth (P)	Organism is inhibited by crystal violet and/or bile	Gram-positive
Good growth (G)	Organism is not inhibited by crystal violet or bile	Gram-negative
Pink to red growth with or without bile precipitate (R)	Organism produces acid from lactose fermentation	Probable coliform
Growth is "colorless" (not red or pink) (C)	Organism does not ferment lactose. No reaction	Noncoliform

References

Forbes, Betty A., Daniel F. Sahm, and Alice S. Weissfeld. 2002. Chapter 10 in *Bailey & Scott's Diagnostic Microbiology,* 11th ed. Mosby-Yearbook, St. Louis.

Winn, Washington, C., *et al.* 2006. *Koneman's Color Atlas and Textbook of Diagnostic Microbiology,* 6th ed. Lippincott Williams & Wilkins, Baltimore.

Zimbro, Mary Jo, and David A. Power. 2003. Page 334 in *Difco™ & BBL™ Manual—Manual of Microbiological Culture Media.* Becton, Dickinson and Co., Sparks, MD.

EXERCISE 4-6

Eosin Methylene Blue Agar

✦ Theory

Eosin Methylene Blue (EMB) Agar is a complex (chemically undefined), selective, and differential medium. It contains peptone, lactose, sucrose, and the dyes eosin Y and methylene blue. The peptone provides a complex mixture of carbon, nitrogen and other nutritional components. The sugars are included to encourage growth of enteric bacteria and to differentiate them based on color reactions created when combined with the dyes. Lactose supports coliforms such as *Escherichia coli* and sucrose supports pathogens such as *Proteus* or *Salmonella* species (Figure 4-11).

The purpose of the dyes is twofold, 1) they inhibit the growth of Gram-positive organisms and 2) they react with vigorous lactose fermenters and (in the acidic environment) turn the growth dark purple or black. This dark growth is typical of *Escherichia coli* and is usually accompanied by a green metallic sheen (Figures 4-12 and 4-13). Other less aggressive lactose fermenters such as *Enterobacter* or *Klebsiella* species produce colonies that can range from pink to dark purple on the medium. Nonfermenters and sucrose fermenters typically retain their normal color or take on the coloration of the medium.

✦ Application

EMB Agar is used for the isolation of fecal coliforms. It can be streaked for isolation or used in the Membrane Filter Technique as discussed in Exercise 8-12.

✦ In This Exercise

You will spot-inoculate one EMB Agar plate and one Nutrient Agar (NA) plate with four test organisms. The NA plate will serve as a comparison for growth quality on the EMB Agar plate.

✦ Materials

Per Student Group

✦ one EMB plate
✦ one NA plate

✦ fresh broth cultures of:
  ◆ *Enterobacter aerogenes*
  ◆ *Enterococcus faecalis*
  ◆ *Escherichia coli*
  ◆ *Salmonella typhimurium*

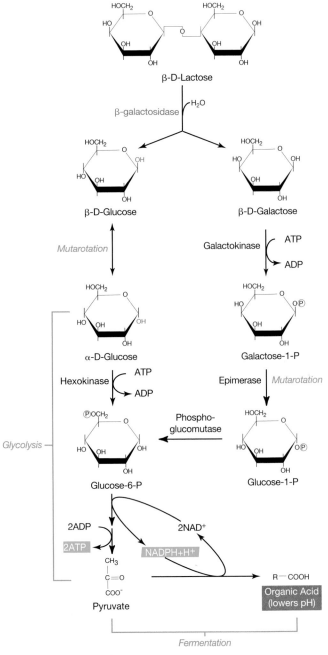

4-11 LACTOSE FERMENTATION WITH ACID END PRODUCTS

✦ Medium Recipes

Eosin Methylene Blue Agar

✦ Peptone	10.0 g
✦ Lactose	5.0 g
✦ Sucrose	5.0 g
✦ Dipotassium phosphate	2.0 g
✦ Agar	13.5 g
✦ Eosin Y	0.4 g
✦ Methylene blue	0.065 g
✦ Distilled or deionized water	1.0 L

pH 6.9–7.3 at 25°C

Nutrient Agar

✦ Beef extract	3.0 g
✦ Peptone	5.0 g
✦ Agar	15.0 g
✦ Distilled or deionized water	1.0 L

pH 6.6–7.0 at 25°C

4-12 EOSIN METHYLENE BLUE AGAR ✦ This EMB Agar was inoculated with (clockwise from the top) two coliforms, a Gram-negative noncoliform, and a Gram-positive organism. Note the characteristic green metallic sheen of the coliform at the top and the pink coloration of the one at the right. Both organisms are lactose fermenters; the difference in color results from the degree of acid production. The organism on the bottom is a nonfermenter, as indicated by the lack of color. Growth of the Gram-positive organism on the left was inhibited by eosin Y and methylene blue.

 ## Procedure

Lab One

1 Mix each culture well.

2 Using a permanent marker, divide the bottom of each plate into four sectors.

3 Label the plates with the organisms' names, your name, and the date.

4 Spot-inoculate the four sectors on the EMB plate with the test organisms. Refer to Appendix B if necessary.

5 Repeat step 4 with the Nutrient Agar plate.

6 Invert and incubate the plates at 35 ± 2°C for 24 to 48 hours.

Lab Two

1 Examine and compare the plates for color and quality of growth.

2 Record your results on the Data Sheet.

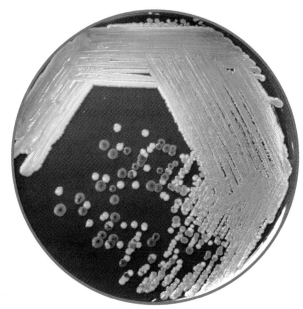

4-13 EOSIN METHYLENE BLUE AGAR STREAKED FOR ISOLATION ✦ This EMB agar was inoculated with two of the Gram-negative rods shown in Figure 4-12. One is a coliform; the other is not. Color produced by mixtures of bacteria is not determinative, so disregard the first three quadrants and pay attention only to the individual colonies. The coliform that produced a green metallic sheen on the plate shown in Figure 4-12 produced pink colonies with dark centers in this photo. The noncoliform colonies are beige.

References

Forbes, Betty A., Daniel F. Sahm, and Alice S. Weissfeld. 2002. Chapter 10 in *Bailey & Scott's Diagnostic Microbiology,* 11th ed. Mosby, St. Louis.

Winn, Washington, C., *et al.* 2006. *Koneman's Color Atlas and Textbook of Diagnostic Microbiology,* 6th ed. Lippincott Williams & Wilkins, Baltimore.

Zimbro, Mary Jo, and David A. Power. 2003. Page 218 in *Difco™ & BBL™ Manual—Manual of Microbiological Culture Media.* Becton, Dickinson and Co., Sparks, MD.

TABLE **4-6**   EMB Results and Interpretations

TABLE OF RESULTS

Result	Interpretation	Presumptive ID
Poor growth or no growth (P)	Organism is inhibited by eosin and methylene blue	Gram-positive
Good growth (G)	Organism is not inhibited by eosin and methylene blue	Gram-negative
Growth is pink and mucoid (Pi)	Organism ferments lactose with little acid production	Possible coliform
Growth is "dark" (purple to black, with or without green metallic sheen) (D)	Organism ferments lactose and/or sucrose with acid production	Probable coliform
Growth is "colorless" (no pink, purple, or metallic sheen) (C)	Organism does not ferment lactose or sucrose. No reaction	Noncoliform

Hektoen Enteric Agar

✦ Theory

Hektoen Enteric (HE) Agar is a complex (chemically un-defined), moderately selective, and differential medium designed to isolate *Salmonella* and *Shigella* species from other enterics. The test is based on the ability to ferment lactose, sucrose, or salicin, and to reduce sulfur to hydrogen sulfide gas (H_2S). Sodium thiosulfate is included as the source of oxidized sulfur. Ferric ammonium citrate is included as a source of oxidized iron to react with any sulfur that becomes reduced (H_2S) to form the black precipitate ferrous sulfide (FeS). Bile salts are included to prevent or inhibit growth of Gram-positive organisms. The bile salts also have a moderate inhibitory effect on enterics, so relatively high concentrations of animal tissue and yeast extract are included to offset this situation. Bromthymol blue and acid fuchsin dyes are added to indicate pH changes.

Differentiation is possible as a result of the various colors produced in the colonies and in the agar. Enterics that produce acid from fermentation will produce yellow to salmon-pink colonies. Neither *Salmonella* nor *Shigella* species ferment any of the sugars; instead they break down the animal tissue, which raises the pH of the medium slightly and gives the colonies a blue-green color. Additionally, *Salmonella* species reduce sulfur to H_2S, so the colonies formed also contain FeS, which makes them partially or completely black. Refer to Figures 4-14 and 4-15.

✦ Application

HE Agar is used to isolate and differentiate *Salmonella* and *Shigella* species from other Gram-negative enteric organisms.

✦ In This Exercise

You will streak-inoculate one Hektoen Enteric Agar plate and one Nutrient Agar plate with four test organisms. The Nutrient Agar will serve as a comparison for growth quality on the Hektoen Enteric Agar plate.

✦ Materials

Per Student Group

✦ one HE Agar plate
✦ one NA plate

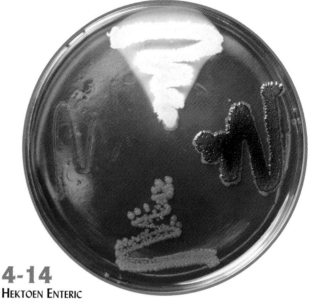

4-14

**HEKTOEN ENTERIC
AGAR** ✦ This HE Agar was inoculated with (clockwise from top), a Gram-negative lactose fermenter (indicated by the yellow growth), two lactose nonfermenters, and a Gram-positive organism. Note the blue-green growth of the lactose nonfermenters. Note also the black precipitate in the growth on the right, indicative of a reaction between ferric ammonium citrate and H_2S produced from sulfur reduction. The Gram-positive organism on the left was severely inhibited by the bile salts.

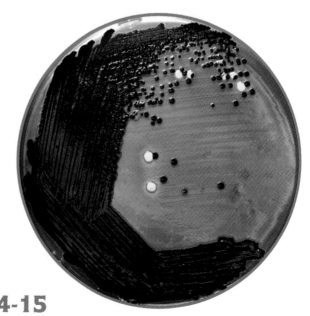

4-15

HEKTOEN ENTERIC AGAR STREAKED FOR ISOLATION ✦ This HE agar was streaked with two of the organisms chosen for today's exercise. One is a lactose fermenter, evidenced by the yellow growth; the other is a sulfur reducer, as indicated by the black precipitate in the colonies. Can you identify these two organisms?

✦ fresh broth cultures of:
- *Enterococcus faecalis*
- *Escherichia coli*
- *Salmonella typhimurium* (BSL-2)
- *Shigella flexneri* (BSL-2)

✦ Medium Recipes

Hektoen Enteric Agar

◆ Yeast extract	3.0 g
◆ Peptic digest of animal tissue	12.0 g
◆ Lactose	12.0 g
◆ Sucrose	12.0 g
◆ Salicin	2.0 g
◆ Bile salts	9.0 g
◆ Sodium chloride	5.0 g
◆ Sodium thiosulfate	5.0 g
◆ Ferric ammonium citrate	1.5 g
◆ Bromthymol blue	0.064 g
◆ Acid fuchsin	0.1 g
◆ Agar	13.5 g
◆ Distilled or deionized water	1.0 L

pH 7.4–7.8 at 25°C

Nutrient Agar

◆ Beef extract	3.0 g
◆ Peptone	5.0 g
◆ Agar	15.0 g
◆ Distilled or deionized water	1.0 L

pH 6.6–7.0 at 25°C

 Procedure

Lab One

1. Mix each culture well.

2. Using a permanent marker, divide the bottom of each plate into four sectors.

3. Label the plates with the organisms' names, your name, and the date.

4. Spot-inoculate the four sectors on the HE plate with the test organisms. Refer to Appendix B if necessary.

5. Repeat Step 4 with the Nutrient Agar plate.

6. Invert and incubate the plates aerobically at $35 \pm 2°C$ for 48 hours.

Lab Two

1. Examine and compare the plates for color and quality of growth.

2. Record your results on the Data Sheet.

References

Forbes, Betty A., Daniel F. Sahm, and Alice S. Weissfeld. 2002. Chapter 10 in *Bailey & Scott's Diagnostic Microbiology*, 11th ed. Mosby-Yearbook, St. Louis.

Winn, Washington, C., et al. 2006. *Koneman's Color Atlas and Textbook of Diagnostic Microbiology*, 6th ed. Lippincott Williams & Wilkins, Baltimore.

Zimbro, Mary Jo, and David A. Power. 2003. Page 265 in *Difco™ & BBL™ Manual—Manual of Microbiological Culture Media*. Becton, Dickinson and Co., Sparks, MD.

TABLE **4-7**   Hektoen Enteric Agar Results and Interpretations

TABLE OF RESULTS

Result	Interpretation	Presumptive ID
Poor growth or no growth (P)	Organism is inhibited by bile and/or one of the dyes included	Gram-positive
Good growth (G)	Organism is not inhibited by bile or any of the dyes included	Gram-negative
Pink to orange growth (Pi)	Organism produces acid from lactose fermentation	Not *Shigella* or *Salmonella*
Blue-green growth with black precipitate (Bppt)	Organism does not ferment lactose, but reduces sulfur to hydrogen sulfide (H_2S)	Possible *Salmonella*
Blue-green growth without black precipitate (B)	Organism does not ferment lactose or reduce sulfur. No reaction	Possible *Shigella* or *Salmonella*

EXERCISE
4-8

Xylose Lysine Desoxycholate Agar

✦ Theory

Xylose Lysine Desoxycholate (XLD) Agar is a selective and differential medium containing sodium desoxycholate, xylose, lactose, sucrose, lysine, sodium thiosulfate, phenol red, and ferric ammonium citrate. Desoxycholate is a bile salt that inhibits growth of Gram-positive organisms. Xylose, lactose, and sucrose are sugars provided for fermentation and lysine, an amino acid, is provided for decarboxylation (removal of the carboxyl group– COOH, see Exercise 5-10). Sodium thiosulfate provides oxidized sulfur for organisms capable of reducing it to hydrogen sulfide gas (H_2S). Ferric ammonium citrate performs as an indicator because it releases ferric ions (Fe^{3+}) into the medium that will react with H_2S and form ferrous sulfide (FeS), a black precipitate. Phenol red, which is yellow when acidic and red or pink when alkaline, is added as a pH indicator.

Organisms that ferment xylose will acidify the medium and produce yellow colonies (see A Word About Biochemical Tests and Acid-Base Reactions). Organisms able to decarboxylate lysine will release CO_2 into the medium, raise the pH, and produce red colonies. Organisms able to reduce sulfur will produce H_2S, which will precipitate with Fe^{3+} and form black colonies.

Shigella and *Providencia* species, which do not ferment xylose but decarboxylate lysine, form red colonies on the medium. *Salmonella* species reduce sulfur, ferment xylose, and decarboxylate lysine. The fermentation acidifies the medium, but as the xylose is in relatively short supply the lysine decarboxylation raises the pH and produces red colonies with black centers. Other enterics (*Escherichia, Klebsiella, Enterobacter*) that would ordinarily deplete the small amount of xylose in the medium and then raise the pH by decarboxylating the lysine, are discouraged from doing so by the ample supply of the fermentable sugars, lactose, and sucrose. These organisms typically produce yellow colonies on the medium (Figure 4-16).

✦ Application

XLD Agar is a selective and differential medium used to isolate and identify *Shigella* and *Providencia* from stool samples.

4-16 XYLOSE LYSINE DESOXYCHOLATE AGAR ✦ This XLD agar was inoculated with (clockwise from top) three members of *Enterobacteriaceae*—two sulfur-reducing lactose nonfermenters and one lactose fermenter. Note the black precipitate in the growth on top resulting from the reaction of ferric ammonium citrate with the reduced sulfur (H_2S). Note also the yellow color of the growth on the left as a result of acid production from lactose fermentation. The organism on the right is a nonfermenting Gram-negative rod and typically is a sulfur reducer. Lack of black precipitate suggests that it either has lost this ability or is a contaminant.

✦ In This Exercise

You will spot inoculate one XLD Agar plate and one Nutrient Agar (NA) plate with five test organisms. The NA plate will serve as a comparison for growth quality on the XLD Agar plate.

✦ Materials

Per Student Group

✦ one XLD Agar plate
✦ one NA plate
✦ fresh broth cultures of:
   ✦ *Enterococcus faecalis* (BSL-2)
   ✦ *Escherichia coli*
   ✦ *Providencia stuartii*
   ✦ *Salmonella typhimurium* (BSL-2)

◆ Medium Recipe

Xylose Lysine Desoxycholate Agar

◆ Xylose	3.5 g
◆ L-Lysine	5.0 g
◆ Lactose	7.5 g
◆ Sucrose	7.5 g
◆ Sodium chloride	5.0 g
◆ Yeast extract	3.0 g
◆ Phenol red	0.08 g
◆ Sodium desoxycholate	2.5 g
◆ Sodium thiosulfate	6.8 g
◆ Ferric ammonium citrate	0.8 g
◆ Agar	13.5 g
◆ Distilled or deionized water	1.0 L

pH 7.3–7.7 at 25°C

Nutrient Agar

◆ Beef extract	3.0 g
◆ Peptone	5.0 g
◆ Agar	15.0 g
◆ Distilled or deionized water	1.0 L

pH 6.6–7.0 at 25°C

 ## Procedure

Lab One

1 Mix each culture well.

2 Using a permanent marker, divide the bottom of each plate into four sectors.

3 Label the plates with the organisms' names, your name, and the date.

4 Spot inoculate the sectors on the XLD Agar plate with the test organisms. Refer to Appendix B if necessary.

5 Repeat Step 4 with the Nutrient Agar plate.

6 Invert and incubate the plates at $35 \pm 2°C$ for 18 to 24 hours. (Be sure to remove the plates from the incubator at 24 hours. Reversions from fermentation to decarboxylation, due to exhaustion of sugar in the medium, with the resulting yellow-to-red color shift may lead to false identification as *Shigella* or *Providencia*.)

Lab Two

1 Examine and compare the plates for color and quality of growth.

2 Record your results on the Data Sheet.

References

Forbes, Betty A., Daniel F. Sahm, and Alice S. Weissfeld. 2002. Chapter 10 in *Bailey & Scott's Diagnostic Microbiology,* 11th ed. Mosby-Yearbook, St. Louis.

Zimbro, Mary Jo, and David A. Power. 2003. Page 625 in *Difco™ & BBL™ Manual—Manual of Microbiological Culture Media.* Becton, Dickinson and Co., Sparks, MD.

TABLE **4-8** XLD Results and Interpretations

TABLE OF RESULTS

Result	Interpretation	Presumptive ID
Poor growth (P)	Organism is inhibited by desoxycholate	Gram-positive
Good growth (G)	Organism is not inhibited by desoxycholate	Gram-negative
Growth is yellow (Y)	Organism produces acid from xylose fermentation (A)	Not *Shigella* or *Providencia*
Red growth with black center (RB)	Organism reduces sulfur to hydrogen sulfide (H_2S)	Not *Shigella* or *Providencia* (probable *Salmonella*)
Red growth without black center (R)	Organism does not ferment xylose or ferments xylose slowly; alkaline products from lyzine decarboxylation (K)	Probable *Shigella* or *Providencia*

EXERCISE 5-2

Oxidation–Fermentation Test

✦ Theory

The Oxidation–Fermentation (O–F) Test is designed to differentiate bacteria on the basis of fermentative or oxidative metabolism of carbohydrates. In oxidation pathways a carbohydrate is directly oxidized to pyruvate and is further converted to CO_2 and energy by way of the Krebs cycle and the electron transport chain (ETC). As mentioned in the introduction, an inorganic molecule such as oxygen is required to act as the final electron acceptor. Fermentation also converts carbohydrates to pyruvate but uses it to produce one or more acids (as well as other compounds). Consequently, fermenters identified by this test acidify O–F medium to a greater extent than do oxidizers.

Hugh and Leifson's O–F medium includes a high sugar-to-peptone ratio to reduce the possibility that alkaline products from peptone utilization will neutralize weak acids produced by oxidation of the carbohydrate. Bromthymol blue dye, which is yellow at pH 6.0 and green at pH 7.1, is added as a pH indicator. A low agar concentration makes it a semi-solid medium that allows determination of motility.

The medium is prepared with glucose, lactose, sucrose, maltose, mannitol, or xylose and is not slanted. Two tubes of the specific sugar medium are stab-inoculated several times with the test organism. After inoculation, one tube is sealed with a layer of sterile mineral oil to promote anaerobic growth and fermentation (Figure 5-4). The other tube is left unsealed to allow aerobic growth and oxidation. (*Note:* Tubes of O–F medium are heated in boiling water and then cooled prior to inoculation. This removes free oxygen from the medium and ensures an anaerobic environment in all tubes. The tubes covered with oil will remain anaerobic, whereas the uncovered medium quickly will become aerobic as oxygen diffuses back in.)

Organisms that are able to ferment the carbohydrate or ferment *and* oxidize the carbohydrate will turn the sealed and unsealed media yellow throughout. Organisms that are able to oxidize only will turn the unsealed medium yellow (or partially yellow) and leave the sealed medium green or blue. Slow or weak fermenters will turn both tubes slightly yellow at the top. Organisms that are not able to metabolize the sugar will either produce no color change or turn the medium blue because of alkaline products from amino acid degradation. The

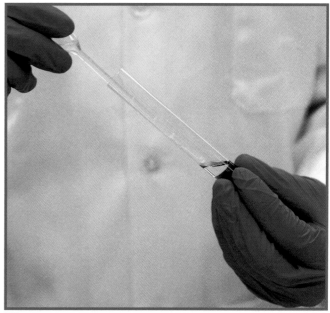

5-4 ADDING THE MINERAL OIL LAYER ✦ Tip the tube slightly to one side, and gently add 3–4 mm of sterile mineral oil. Be sure to use a sterile pipette for each tube.

results are summarized in Table 5-1 and shown in Figure 5-5.

✦ Application

The O–F Test is used to differentiate bacteria based on their ability to oxidize or ferment specific sugars. It allows presumptive separation of the fermentative *Enterobacteriaceae* from the oxidative *Pseudomonas* and *Bordetella*, and the nonreactive *Alcaligenes* and *Moraxella*.

✦ In This Exercise

Eight O–F media tubes will be used for this exercise—six for inoculation (two for each organism) and two for controls. One of each pair will receive a mineral oil overlay to maintain an anaerobic environment and force fermentation if the organism is capable of doing so. You will not do a motility determination, as motility is covered in Exercise 5-28. For best results, use a heavy inoculum and several stabs for each tube. Use only sterile pipettes to add the mineral oil.

TABLE **5-1** O–F Medium Results and Interpretations

TABLE OF RESULTS			
Sealed	**Unsealed**	**Interpretation**	**Symbol**
Green or blue	Any amount of yellow	Oxidation	O
Yellow throughout	Yellow throughout	Oxidation and fermentation or fermentation only	O–F or F
Slightly yellow at the top	Slightly yellow at the top	Oxidation and slow fermentation or slow fermentation only	O–F or F
Green or blue	Green or blue	No sugar metabolism; organism is nonsaccharolytic	N

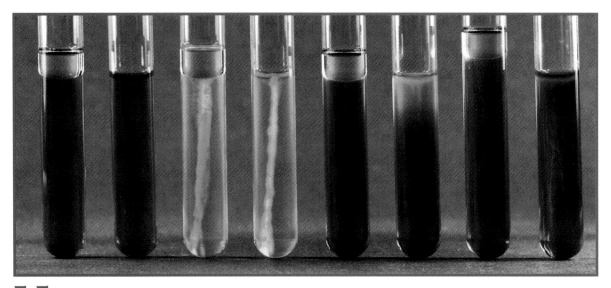

5-5 OXIDATION–FERMENTATION TEST ✦ These pairs of tubes represent three possible results in the oxidation–fermentation (O–F) test. Each pair contains one tube sealed with an overlay of mineral oil and one unsealed tube. The mineral oil creates an environment unsuitable for oxidation because it prevents the diffusion of oxygen from the air into the medium. The result is that an organism capable of fermentation will turn both tubes yellow, whereas an organism capable only of oxidizing glucose will turn only the oxygen-containing portion of the unsealed medium yellow. An organism incapable of utilizing glucose by any means either will not change the color of the medium or will turn it blue-green as a result of alkaline products from protein degradation. Reading from left to right, the first pair of tubes on the left are uninoculated controls for color comparison. The second pair of tubes were inoculated with an organism capable of both oxidative and fermentative utilization of glucose (O–F). Unfortunately, this determination cannot be made simply by visual examination, as the results of a fermentative organism (F) look exactly the same as an organism capable of both oxidation and fermentation (O–F). Therefore, when both tubes are yellow, the organism is assumed to be *either* (F) *or* (O–F). The third pair of tubes were inoculated with a glucose nonfermenter. This organism is capable only of oxidation. Note the yellowing only of the oxygenated portion of the unsealed tube. The fourth pair of tubes (right) were inoculated with an organism incapable of utilizing glucose. Note the blue color in the oxygenated portion of the unsealed tube suggesting that the organism is both nonsaccharolytic (N) and a strict aerobe.

✦ Materials

Per Student Group

+ eight O–F glucose tubes
+ sterile mineral oil
+ sterile transfer pipettes
+ fresh agar slants of:
  ◆ *Escherichia coli*
  ◆ *Pseudomonas aeruginosa* (BSL-2)
  ◆ *Alcaligenes faecalis* (BSL-2)

✦ Medium Recipe

Hugh and Leifson's O–F Medium with Glucose

◆ Pancreatic digest of casein	2.0 g
◆ Sodium chloride	5.0 g
◆ Dipotassium phosphate	0.3 g
◆ Agar	2.5 g
◆ Bromthymol blue	0.08 g
◆ Glucose	10.0 g
◆ Distilled or deionized water	1.0 L

pH = 6.6–7.0 at 25°C

Procedure

Lab One

1 Obtain eight O–F tubes. Label six (in three pairs) with the names of the organisms, your name, and the date. Label the last pair "control."

2 Stab-inoculate each pair with the appropriate test organism. Stab several times to a depth of about 1 cm from the bottom of the agar. Do not inoculate the controls.

3 Overlay one of each pair of tubes (including the controls) with 3–4 mm of sterile mineral oil (see Figure 5-4).

4 Incubate all tubes at 35 ± 2°C for 48 hours.

Lab Two

1 Examine the tubes for color changes. Be sure to compare them to the controls before making a determination.

2 Record your results in the chart provided on the Data Sheet.

References

Collins, C. H., Patricia M. Lyne, and J. M. Grange. 1995. Page 112 in *Collins and Lyne's Microbiological Methods*, 7th ed. Butterworth-Heinemann, United Kingdom.

Delost, Maria Dannessa. 1997. Pages 218–219 in *Introduction to Diagnostic Microbiology*. Mosby, St. Louis.

Forbes, Betty A., Daniel F. Sahm, and Alice S. Weissfeld. 2002. Pages 154–155 in *Bailey & Scott's Diagnostic Microbiology,* 11th ed. Mosby, St. Louis.

MacFaddin, Jean F. 2000. Page 379 in *Biochemical Tests for Identification of Medical Bacteria*, 3rd ed. Lippincott Williams & Wilkins, Philadelphia.

Smibert, Robert M., and Noel R. Krieg. 1994. Page 625 in *Methods for General and Molecular Bacteriology*, edited by Philipp Gerhardt, R. G. E. Murray, Willis A. Wood, and Noel R. Krieg, American Society for Microbiology, Washington, DC.

Zimbro, Mary Jo, and David A. Power, Eds. 2003. Page 410 in *Difco™ and BBL™ Manual—Manual of Microbiological Culture Media*. Becton Dickinson and Co., Sparks, MD.

Fermentation Tests

As defined at the beginning of this section, carbohydrate fermentation is the metabolic process by which an organic molecule acts as an electron donor (becoming oxidized in the process) and one or more of its organic products act as the final electron acceptor (FEA). In actuality, the term "carbohydrate fermentation" is used rather broadly to include hydrolysis of disaccharides prior to the fermentation reaction. Thus, a "lactose fermenter" is an organism that splits the disaccharide lactose into the monosaccharides glucose and galactose and then ferments the monosaccharides. In this section you will see the term "fermenter" frequently. Unless it is expressly used otherwise, this term should be assumed to include the initial hydrolysis and/or conversion reactions.

Fermentation of glucose begins with the production of pyruvate. Although some organisms use alternative pathways, most bacteria accomplish this by glycolysis. The end products of pyruvate fermentation include a variety of organic acids, alcohols, and hydrogen or carbon dioxide gas. The specific end products depend on the specific organism and the substrate fermented. Refer to Figure 5-6 and Appendix A.

In this unit you will perform tests using two differential fermentation media—Phenol Red (PR) Broth and MR-VP Medium. Phenol Red Broth is a general-purpose fermentation medium. Typically, it contains any one of several carbohydrates (*e.g.,* glucose, lactose, sucrose) and a pH indicator to detect acid formation. MR-VP broth is a dual-purpose medium that tests an organism's ability to follow either (or both) of two specific fermentation pathways. The methyl red (MR) test detects what is called a **mixed acid fermentation**. The Voges-Proskauer (VP) test identifies bacteria that are able to produce acetoin as part of a **2,3-butanediol fermentation**. See Appendix A for more information about fermentation. ✦

EXERCISE 5-3

Phenol Red Broth

✦ Theory

Phenol Red (PR) Broth is a differential test medium prepared as a base to which a carbohydrate is added. Included in the base medium are peptone and the pH indicator phenol red. Phenol red is yellow below pH 6.8, pink to magenta above pH 7.4, and red in between. During preparation the pH is adjusted to approximately 7.3 so it appears red. Finally, an inverted Durham tube is added to each tube as an indicator of gas production.

Acid production from fermentation of the carbohydrate lowers the pH below the neutral range of the indicator and turns the medium yellow (Figure 5-7). Deamination of peptone amino acids produces ammonia (NH_3), which raises the pH and turns the broth pink. Gas production, also from fermentation, is indicated by a bubble or pocket in the Durham tube where the broth has been displaced.

✦ Application

PR broth is used to differentiate members of *Enterobacteriaceae* and to distinguish them from other Gram-negative rods.

✦ In This Exercise

Today you will inoculate PR broths with four organisms to determine their fermentation characteristics. Use Table 5-2 as a guide when interpreting and recording your results. Be sure to use an uninoculated control of each medium for color comparison.

✦ Materials

Per Student Group
✦ five PR Glucose Broths with Durham tubes

✦ five PR Lactose Broths with Durham tubes
✦ five PR Sucrose Broths with Durham tubes
✦ five PR Base Broths with Durham tubes
✦ fresh cultures of:
  ◆ *Escherichia coli*
  ◆ *Pseudomonas aeruginosa* (BSL-2)
  ◆ *Proteus vulgaris* (BSL-2)
  ◆ *Enterococcus faecalis* (BSL-2)

✦ Medium Recipe

PR (Carbohydrate) Broth[1]
  ◆ Pancreatic digest of casein   10.0 g
  ◆ Sodium chloride   5.0 g

_____

[1] This formulation without carbohydrate is PR Base Broth.

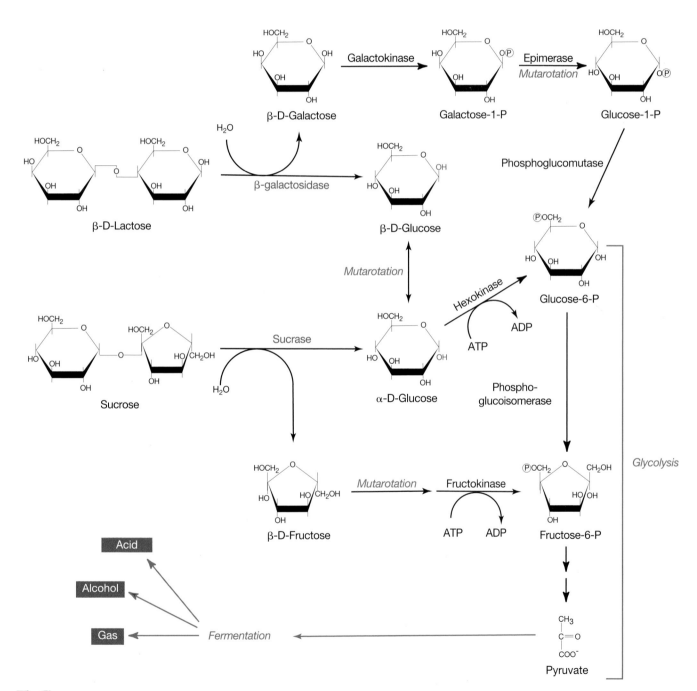

5-6 FERMENTATION OF THE DISACCHARIDES LACTOSE AND SUCROSE ✦ Notice that fermentation of the disaccharides lactose and sucrose relies on enzymes to hydrolyze them into two monosaccharides. Once the monosaccharides are formed, they are fermented via glycolysis, just as glucose is.

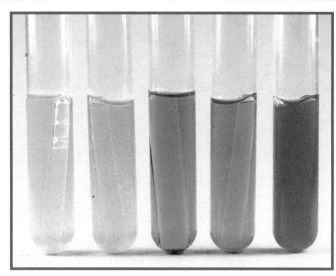

5-7 PR GLUCOSE BROTH RESULTS ✦ From left to right, these five PR Glucose Broths demonstrate acid and gas production (A/G), acid production without gas (A/−), uninoculated control tube for comparison, no reaction (−/−), and alkaline condition as a result of peptone degradation (K).

♦ Carbohydrate
    (glucose, lactose, sucrose)   5.0 g
♦ Phenol red   0.018 g
♦ Distilled or deionized water   1.0 L
        pH = 7.1–7.5 at 25°C

 Procedure

Lab One

1 Obtain five of each PR broth. Label each with the name of the organism, your name, the medium name, and the date. Label the fifth tube of each medium "control."

2 Inoculate one of each broth with each test organism.

3 Incubate all of the tubes at 35 ± 2°C for 48 hours.

Lab Two

Using the uninoculated controls for color comparison and Table 5-2 as a guide, examine all of the tubes and enter your results in the chart provided on the Data Sheet. When indicating the various reactions, use the standard symbols as shown, with the acid reading first, followed by a slash and then the gas reading. K indicates alkalinity and –/– symbolizes no reaction. *Note:* Do not try to read these results without control tubes for comparison.

References

Lányi, B. 1987. Page 44 in *Methods in Microbiology*, Vol. 19, edited by R. R. Colwell and R. Grigorova. Academic Press, New York.

MacFaddin, Jean F. 2000. Page 57 in *Biochemical Tests for Identification of Medical Bacteria*, 3rd ed. Lippincott Williams & Wilkins, Philadelphia.

Zimbro, Mary Jo and David A. Power, editors. 2003. Page 440 in *Difco™ and BBL™ Manual—Manual of Microbiological Culture Media*. Becton Dickinson and Co., Sparks, MD.

TABLE **5-2** PR Broth Results and Interpretations

TABLE OF RESULTS		
Result	**Interpretation**	**Symbol**
Yellow broth, bubble in tube	Fermentation with acid and gas end products	A/G
Yellow broth, no bubble in tube	Fermentation with acid end products; no gas produced	A/ −
Red broth, no bubble in tube	No fermentation	− / −
Pink broth, no bubble in tube	Degradation of peptone; alkaline end products	K

Methyl Red and Voges-Proskauer Tests

✦ Theory

Methyl Red and Voges-Proskauer (MR-VP) Broth is a combination medium used for both Methyl Red (MR) *and* Voges-Proskauer (VP) tests. It is a simple solution containing only peptone, glucose, and a phosphate buffer. The peptone and glucose provide protein and fermentable carbohydrate, respectively, and the potassium phosphate resists pH changes in the medium.

The MR test is designed to detect organisms capable of performing a **mixed acid fermentation**, which overcomes the phosphate buffer in the medium and lowers the pH (Figure 5-8 and Figure 5-9). The acids produced by these organisms tend to be stable, whereas acids produced by other organisms tend to be unstable and subsequently are converted to more neutral products.

Mixed acid fermentation is verified by the addition of methyl red indicator dye following incubation. Methyl red is red at pH 4.4 and yellow at pH 6.2. Between these two pH values, it is various shades of orange. Red color is the only true indication of a positive result. Orange is negative or inconclusive. Yellow is negative (Figure 5-10).

The Voges-Proskauer test was designed for organisms that are able to ferment glucose, but quickly convert their acid products to acetoin and 2,3-butanediol (Figure 5-9 and Figure 5-11). Adding VP reagents to the medium oxidizes the **acetoin** to **diacetyl**, which in turn reacts with **guanidine nuclei** from peptone to produce a red color (Figure 5-12). A positive VP result, therefore, is red. No color change (or development of copper color) after the addition of reagents is negative. The copper color is a result of interactions between the reagents and should not be confused with the true red color of

a positive result (Figure 5-13). Use of positive and negative controls for comparison is usually recommended.

After incubation, two 1 mL volumes are transferred from the MR-VP broth to separate test tubes. Methyl red indicator reagent is added to one tube, and VP reagents are added to the other. Color changes then are observed and documented. Refer to the procedural diagram in Figure 5-14.

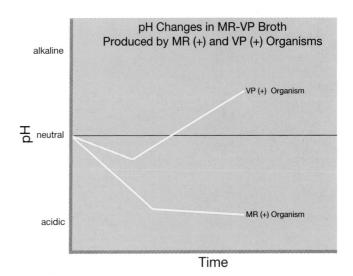

5-9 pH Changes in MR-VP Broth ✦ The MR test identifies organisms that perform a mixed acid fermentation and produce stable acid end products. MR (+) organisms lower the broth's pH permanently. The VP test is used to identify organisms that perform a 2,3-butanediol fermentation. VP (+) organisms initially may produce acid and temporarily lower the pH, but because the 2,3-butanediol fermentation end products are neutral, the pH at completion of the test is near neutral.

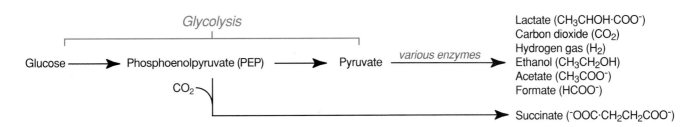

5-8 Mixed Acid Fermentation of E. coli ✦ *E. coli* is a representative Methyl Red positive organism and is recommended as a positive control for the test. Its mixed acid fermentation produces the end products listed in order of abundance. Most of the formate is converted to H_2 and CO_2 gases. **Note:** The amount of succinate falls between acetate and formate but is derived from PEP, not pyruvate. *Salmonella* and *Shigella* also are Methyl Red-positive.

✦ Application

The Methyl Red and Voges-Proskauer tests are components of the *IMViC* battery of tests (Indole, Methyl red, Voges-Proskauer, and Citrate) used to distinguish between members of the family *Enterobacteriaceae* and differentiate them from other Gram-negative rods. For more information, refer to Appendix A.

5-10 METHYL RED TEST ✦ The tube on the left is MR-positive. The tube on the right is MR-negative.

✦ In This Exercise

Today you will perform Methyl Red and Voges-Proskauer tests on *Escherichia coli* and *Enterobacter aerogenes*. After a 48-hour incubation period, you will observe the results of a mixed-acid fermentation and a 2,3-butanediol fermentation. The exercise involves transfer of a portion of broth containing living organisms and the measured addition of reagents to complete the observable reactions. Please use care in your transfers and in your measurements. For help with the procedure, refer to Figure 5-14. For help interpreting your results, refer to Figures 5-10 and 5-13 and Tables 5-3 and 5-4.

✦ Materials

Per Student Group

✦ three MR-VP broths
✦ Methyl Red reagent
✦ VP reagents A and B
✦ six nonsterile test tubes
✦ three nonsterile 1 mL pipettes

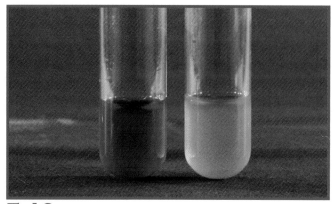

5-11 2,3-BUTANEDIOL FERMENTATION ✦ Acetoin is an intermediate in this fermentation. Reduction of acetoin by NADH leads to the end product 2,3-butanediol. Oxidation of acetoin produces diacetyl, as in the indicator reaction for the VP test (see Figure 5-9).

5-12 INDICATOR REACTION OF VOGES-PROSKAUER TEST ✦ Reagents A and B are added to VP broth after 48 hours' incubation. These reagents react with acetoin and oxidize it to diacetyl, which in turn reacts with guanidine (from the peptone in the medium) to produce a red color.

5-13 THE VOGES-PROSKAUER TEST ✦ The tube on the left is VP-negative. The tube on the right is VP-positive. The copper color at the top of the VP-negative tube is the result of the reaction of KOH and α-naphthol and should not be confused with a positive result.

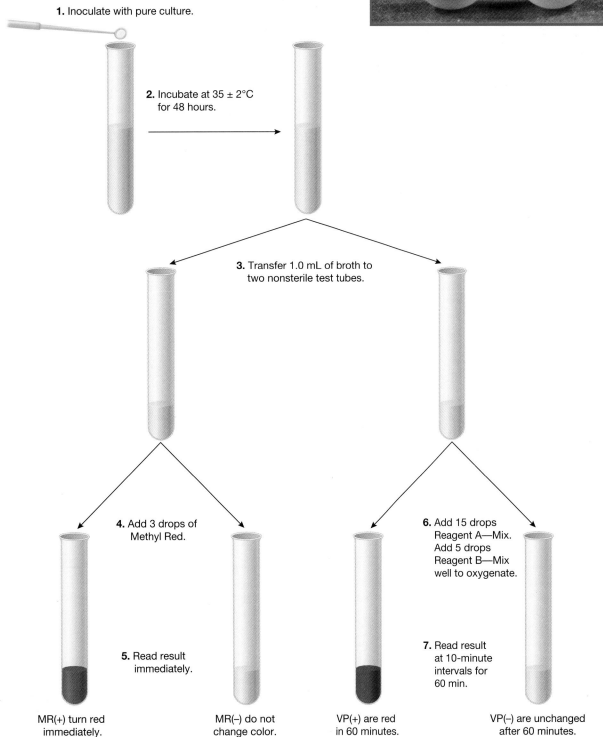

1. Inoculate with pure culture.

2. Incubate at 35 ± 2°C for 48 hours.

3. Transfer 1.0 mL of broth to two nonsterile test tubes.

4. Add 3 drops of Methyl Red.

5. Read result immediately.

6. Add 15 drops Reagent A—Mix. Add 5 drops Reagent B—Mix well to oxygenate.

7. Read result at 10-minute intervals for 60 min.

MR(+) turn red immediately.

MR(–) do not change color.

VP(+) are red in 60 minutes.

VP(–) are unchanged after 60 minutes.

5-14 PROCEDURAL DIAGRAM FOR THE METHYL RED AND VOGES-PROSKAUER TESTS ✦

◆ fresh cultures of:
  ◆ *Escherichia coli*
  ◆ *Enterobacter aerogenes*

✦ Medium and Reagent Recipes

MR-VP Broth

◆ Buffered peptone	7.0 g
◆ Dipotassium phosphate	5.0 g
◆ Dextrose (glucose)	5.0 g
◆ Distilled or deionized water	1.0 L

      pH 6.7–7.1 at 25°C

Methyl Red Reagent

◆ Methyl Red dye	0.1 g
◆ Ethanol	300.0 mL
◆ Distilled water to bring volume to	500.0 mL

VP Reagent A

◆ α-naphthol	5.0 g
◆ Absolute alcohol to bring volume to	100.0 mL

VP Reagent B

◆ Potassium hydroxide	40.0 g
◆ Distilled water to bring volume to	100.0 mL

Procedure

Lab One

1 Obtain three MR-VP broths. Label two of them with the names of the organisms, your name, and the date. Label the third tube "control."

2 Inoculate two broths with the test cultures. Do not inoculate the control.

3 Incubate all tubes at $35 \pm 2°C$ for 48 hours.

Lab Two

1 Mix each broth well. Transfer 1.0 mL from each broth to two nonsterile tubes. The pipettes should be sterile, but the tubes need not be. Refer to the procedural diagram in Figure 5-14.

2 For the three pairs of tubes, add the reagents as follows.

Tube #1: Methyl Red Test

a. Add three drops of Methyl Red reagent. Observe for red color formation immediately.

b. Using Table 5-3 as a guide, record your results in the chart provided on the Data Sheet.

Tube #2: VP Test

a. Add 15 drops (0.6 mL) of VP reagent A. Mix well to oxygenate the medium.

b. Add 5 drops (0.2 mL) of VP reagent B. Mix well again to oxygenate the medium.

c. Place the tubes in a test tube rack and observe for red color formation at regular intervals starting at 10 minutes for up to 1 hour. Mix again periodically to oxygenate the medium.

d. Using Table 5-4 as a guide, record your results in the chart provided on the Data Sheet. Sometimes it is difficult to differentiate weak positive reactions from negative reactions. VP (−) reactions often produce a copper color. Weak VP (+) reactions produce a pink color. The strongest color change should be at the surface.

References

Chapin, Kimberle C., and Tsai-Ling Lauderdale. 2003. Page 361 in *Manual of Clinical Microbiology,* 8th ed. Edited by Patrick R. Murray, Ellen Jo Baron, James H. Jorgensen, Michael A. Pfaller, and Robert H. Yolken. ASM Press, American Society for Microbiology, Washington, DC.

Delost, Maria Dannessa. 1997. Pages 187–188 in *Introduction to Diagnostic Microbiology.* Mosby, St. Louis.

Forbes, Betty A., Daniel F. Sahm, and Alice S. Weissfeld. 2002. Page 275 in *Bailey & Scott's Diagnostic Microbiology,* 11th ed. Mosby, St. Louis.

MacFaddin, Jean F. 2000. Pages 321 and 439 in *Biochemical Tests for Identification of Medical Bacteria,* 3rd ed. Williams & Wilkins, Baltimore.

Zimbro, Mary Jo, and David A. Power, Eds. 2003. Page 330 in *Difco™ and BBL™ Manual—Manual of Microbiological Culture Media.* Becton Dickinson and Co., Sparks, MD.

TABLE **5-3** Methyl Red Test Results and Interpretations

TABLE OF RESULTS		
Result	**Interpretation**	**Symbol**
Red	Mixed acid fermentation	+
No color change	No mixed acid fermentation	−

TABLE **5-4** Voges-Proskauer Test Results and Interpretations

TABLE OF RESULTS		
Result	**Interpretation**	**Symbol**
Red	2,3-butanediol fermentation (acetoin produced)	+
No color change	No 2,3-butanediol fermentation (acetoin is not produced)	−

Tests Identifying Microbial Ability to Respire

In this unit we will examine several techniques designed to differentiate bacteria based on their ability to respire. As mentioned earlier in this section, respiration is the conversion of glucose to energy in the form of ATP by way of glycolysis, the Krebs cycle, and oxidative phosphorylation in the electron transport chain (ETC). In respiration, the final electron acceptor at the end of the ETC is always an inorganic molecule.

Tests that identify an organism as an aerobic or an anaerobic respirer generally are designed to detect specific products of (or constituent enzymes used in) the reduction of the final (or terminal) electron acceptor. Aerobic respirers reduce oxygen to water or other compounds. Anaerobic respirers reduce other inorganic molecules such as nitrate or sulfate. Nitrate is reduced to nitrogen gas (N_2) or other nitrogenous compounds such as nitrite (NO_2), and sulfate is reduced to hydrogen sulfide gas (H_2S). Metabolic pathways are described in detail in Appendix A.

The exercises and organisms chosen for this section directly or indirectly demonstrate the presence of an ETC. The catalase test detects organismal ability to produce catalase—an enzyme that detoxifies the cell by converting hydrogen peroxide produced in the ETC to water and molecular oxygen. The oxidase test identifies the presence of cytochrome c oxidase in the ETC. The nitrate reduction test examines bacterial ability to transfer electrons to nitrate at the end of the ETC and thus respire anaerobically. For more information on biochemical tests that demonstrate anaerobic respiration, refer to Exercise 5-20, SIM Medium, and Exercise 5-21, Triple Sugar Iron Agar/Kligler Iron Agar. ✦

EXERCISE 5-5

Catalase Test

✦ Theory

The electron transport chains of aerobic and facultatively anaerobic bacteria are composed of molecules capable of accepting and donating electrons as conditions dictate. As such, these molecules alternate between the oxidized and reduced forms, passing electrons down the chain to the final electron acceptor. Energy lost by electrons in this sequential transfer is used to perform oxidative phosphorylation (*i.e.*, produce ATP from ADP).

One carrier molecule in the ETC called **flavoprotein** can bypass the next carrier in the chain and transfer electrons directly to oxygen (Figure 5-15). This alternative pathway produces two highly potent cellular toxins—hydrogen peroxide (H_2O_2) and superoxide radical (O_2^-).

Organisms that produce these toxins also produce enzymes capable of breaking them down. **Superoxide dismutase** catalyzes conversion of superoxide radicals (the more lethal of the two compounds) to hydrogen peroxide. **Catalase** converts hydrogen peroxide into water and gaseous oxygen (Figure 5-16).

$$\underset{\substack{\text{Reduced} \\ \text{Flavoprotein}}}{FPH_2} + O_2 \longrightarrow \underset{\substack{\text{Oxidized} \\ \text{Flavoprotein}}}{FP} + \underset{\substack{\text{Hydrogen} \\ \text{Peroxide}}}{H_2O_2}$$

$$2H^+ + \underset{\substack{\text{Superoxide} \\ \text{Radical}}}{2 O_2^-} \xrightarrow{\substack{\text{Superoxide} \\ \text{dismutase}}} \underset{\substack{\text{Hydrogen} \\ \text{Peroxide}}}{H_2O_2} + O_2$$

5-15 MICROBIAL PRODUCTION OF H_2O_2 ✦ Hydrogen peroxide may be formed through the transfer of electrons from reduced flavoprotein to oxygen or from the action of superoxide dismutase.

$$\underset{\substack{\text{Hydrogen} \\ \text{Peroxide}}}{2H_2O_2} \xrightarrow{\text{Catalase}} 2H_2O + O_2\,(g)$$

5-16 CATALASE MEDIATED CONVERSION OF H_2O_2 ✦ Catalase is an enzyme of aerobes, microaerophiles, and facultative anaerobes that converts hydrogen peroxide to water and oxygen gas.

Bacteria that produce catalase can be detected easily using typical store-grade hydrogen peroxide. When hydrogen peroxide is added to a catalase-positive culture, oxygen gas bubbles form immediately (Figure 5-17 and Figure 5-18). If no bubbles appear, the organism is catalase-negative. This test can be performed on a microscope slide or by adding hydrogen peroxide directly to the bacterial growth.

✦ Application

This test is used to identify organisms that produce the enzyme catalase. It is used most commonly to differentiate members of the catalase-positive *Micrococcaceae* from the catalase-negative *Streptococcaceae*. Variations on this test also may be used in identification of *Myco-bacterium* species.

✦ In This Exercise

You will perform the Catalase Test. The materials provided will enable you to conduct both the slide test and the slant test. The slide test has to be done first because the tube test requires the addition of hydrogen

5-17 CATALASE SLIDE TEST ✦ Visible bubble production indicates a positive result in the catalase slide test. A catalase-positive organism is on the left. A catalase-negative organism is on the right.

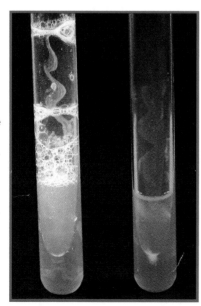

5-18 CATALASE TUBE TEST ✦ The catalase test also may be performed on an agar slant. A catalase-positive organism is on the left. A catalase-negative organism is on the right.

peroxide directly to the slant, after which the culture will be unusable.

✦ Materials

Per Student Group

- ✦ hydrogen peroxide (3% solution)
- ✦ transfer pipettes
- ✦ microscope slides
- ✦ one Nutrient Agar slant
- ✦ fresh Nutrient Agar cultures of:
  - ✦ *Staphylococcus epidermidis*
  - ✦ *Enterococcus faecalis* (BSL-2)

✦ Procedure

Slide Test

1 Transfer a large amount of growth to a microscope slide. (Be sure to perform this test in the proper order. Placing the metal loop into H_2O_2 could catalyze a false-positive reaction.)

2 Aseptically place one or two drops of hydrogen peroxide directly onto the bacteria and immediately observe for the formation of bubbles. (When running this test on an unknown, a positive control should be run simultaneously to verify the quality of the peroxide.)

3 If you get a negative result, observe the slide, using the scanning objective on your microscope.

4 Record your results in the chart provided on the Data Sheet.

Slant Test

1 Add approximately 1 mL of hydrogen peroxide to each of the three slants (one uninoculated control and two cultures). Do this *one slant at a time*, observing and recording the results as you go.

2 Record your results in the chart provided on the Data Sheet.

TABLE **5-5** Catalase Test Results and Interpretations

T A B L E O F R E S U L T S		
Result	Interpretation	Symbol
Bubbles	Catalase is present	+
No bubbles	Catalase is absent	−

References

Collins, C. H., Patricia M. Lyne, and J. M. Grange. 1995. Page 110 in *Collins and Lyne's Microbiological Methods,* 7th ed. Butterworth-Heinemann, Oxford, United Kingdom.

Forbes, Betty A., Daniel F. Sahm, and Alice S. Weissfeld. 2002. Pages 151 and 166 in *Bailey & Scott's Diagnostic Microbiology,* 11th ed. Mosby, St. Louis.

Lányi, B. 1987. Page 20 in *Methods in Microbiology,* Vol. 19, edited by R. R. Colwell and R. Grigorova. Academic Press, New York.

MacFaddin, Jean F. 2000. Page 78 in *Biochemical Tests for Identification of Medical Bacteria,* 3rd ed. Williams & Wilkins, Baltimore.

Smibert, Robert M., and Noel R. Krieg. 1994. Page 614 in *Methods for General and Molecular Bacteriology*, edited by Philipp Gerhardt, R. G. E. Murray, Willis A. Wood, and Noel R. Krieg. American Society for Microbiology, Washington, DC.

EXERCISE 5-6

Oxidase Test

✦ Theory

In the introduction to this section, we discussed the major metabolic reactions and talked a little bit about coenzymes. In Exercise 5-1 Reduction Potential we talked about energy transformations. This test, the oxidase test, is one requiring an understanding of those reactions.

Consider the life of a glucose molecule entering the cell. It is first split (oxidized) in glycolysis where it is converted to two molecules of pyruvate and reduces two NAD (coenzyme) molecules to NADH ($+H^+$). Then each of the pyruvate molecules is oxidized and converted to a two-carbon molecule called acetyl–CoA and one molecule of CO_2, which reduces another NAD to NADH. Then the Krebs cycle finishes the oxidation by producing two more molecules of CO_2 (per acetyl–CoA) and reduces three more NADs and one FAD to $FADH_2$.

As you can see, by this time the cell is becoming quite full of reduced coenzymes. Therefore, in order to continue oxidizing glucose, these coenzymes must be converted back to the oxidized state. This is the job of the electron transport chain (Figure 5-19).

Many aerobes, microaerophiles, facultative anaerobes, and even some anaerobes have ETCs. The functions of the ETC are to 1) transport electrons down a chain of molecules with increasingly positive reduction potentials (Exercise 5-1) to the terminal electron acceptor ($\frac{1}{2}O_2$, NO_3^{2-}, SO_4^{3-}) and 2) generate a proton motive force by pumping H^+ out of the cell thus creating an ionic imbalance that will drive the production of ATP by way of membrane ATPases. The protons pumped out of the cell come from the hydrogen atoms whose electrons are being transferred down the chain. Because only alternating ETC molecules are able to carry associated protons along with their electrons, the positively charged ions are expelled from the cell. **Flavoproteins, iron-sulfur proteins,** and **cytochromes** are important ETC molecules unable to donate protons.

There are many different types of electron transport chains, but all share the characteristics listed above. Some organisms use more than one type of ETC depending on the availability of oxygen or other preferred terminal electron acceptor. *Escherichia coli*, for example, has two pathways for respiring aerobically and at least one for respiring anaerobically. Many bacteria have ETCs resembling mitochondrial ETCs in eukaryotes. These chains contain a series of four large enzymes broadly named Complexes I, II, III, and IV, each of which contains several molecules jointly able to transfer electrons and use the free energy released in the reactions. The last enzyme in the chain, Complex IV, is called cytochrome c oxidase because it makes the final electron transfer of the chain from cytochrome c, residing in the periplasm, to oxygen inside the cell.

The oxidase test is designed to identify the presence of cytochrome c oxidase. It is able to do this because cytochrome c oxidase has the unique ability to not only oxidize cytochrome c, but to catalyze the *reduction* of cytochrome c by a **chromogenic reducing agent** called tetramethyl-*p*-phenylenediamine. Chromogenic reducing agents are chemicals that develop color as they become oxidized (Figure 5-20).

In the oxidase test, the reducing reagent is added directly to bacterial growth on solid media (Figure 5-21), or (more conveniently) a bacterial colony is transferred to paper saturated with the reagent (Figure 5-22). A dramatic color change occurs within seconds if the reducing agent becomes oxidized, thus indicating that cytochrome c oxidase is present. Lack of color change within the allotted time means that cytochrome c oxidase is not present and signifies a negative result (Table 5-6).

✦ Application

This test is used to identify bacteria containing the respiratory enzyme cytochrome c oxidase. Among its many uses is the presumptive identification of the oxidase-positive *Neisseria*. It also can be useful in differentiating the oxidase-negative *Enterobacteriaceae* from the oxidase-positive *Pseudomonadaceae*.

✦ In This Exercise

You will be performing the oxidase test (likely) using one of several commercially available systems. Your instructor will have chosen the system that is best for your laboratory. Generally, test systems produce a color change from colorless to blue/purple. The system shown in Figure 5-22 is BBL™ *Dry*Slide™, which calls for a transfer of inoculum to the slide and a reading within 20 seconds. Reagents used for this test are unstable and may oxidize independently shortly after they become moist. Therefore, it is important to take your readings within the recommended time.

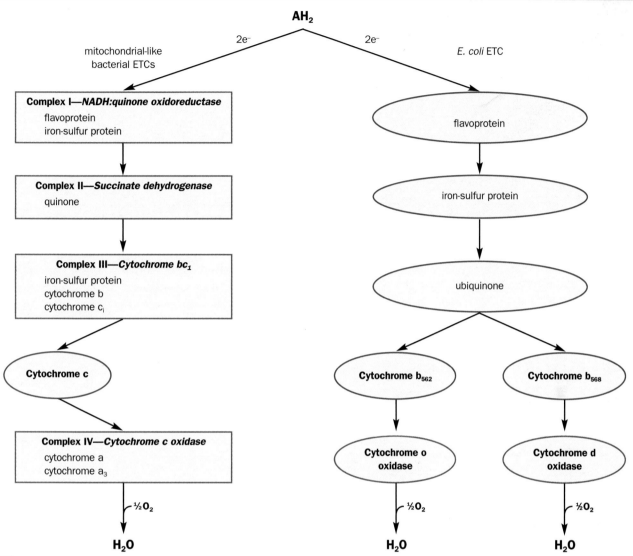

5-19 AEROBIC ELECTRON TRANSPORT CHAINS (ETCs) ✦ There are a variety of different bacterial aerobic electron transport chains. All begin with flavoproteins that pick up electrons from coenzymes, such as NADH or FADH$_2$ (represented by AH$_2$). Bacteria with chains that resemble the mitochondrial ETC in eukaryotes contain cytochrome c oxidase (Complex IV), which transfers electrons from cytochrome c to oxygen. These organisms give a positive result for the oxidase test. Other bacteria, such as members of the *Enterobacteriaceae*, are capable of aerobic respiration, but have a different terminal oxidase system and give a negative result for the oxidase test. The two paths shown on the right in the diagram are both found in *E. coli*. The amount of available O$_2$ determines which pathway is most active. Enzymatic complexes and oxidases are outlined in red.

Tetramethyl-*p*-phenylenediamine$_{red}$
(colorless)

Tetramethyl-*p*-phenylenediamine$_{ox}$
(deep purple/blue)

5-20 CHEMISTRY OF THE OXIDASE REACTION ✦ The oxidase enzyme shown is not involved directly in the indicator reaction as shown. Rather, it removes electrons from cytochrome c, making it available to react with the phenylenediamine reagent.

5-21 OXIDASE TEST ON BACTERIAL GROWTH ✦ A few drops of reagent on oxidase-positive bacteria will produce a purple-blue color immediately.

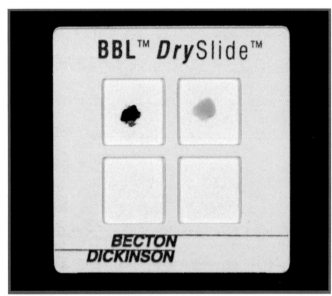

5-22 OXIDASE SLIDE TEST ✦ Positive results with this test should appear within 20 seconds. The dark blue is a positive result. No color change is a negative result. The organism in the top right panel demonstrates a negative result (not blue) but appears yellow because it is pigmented. (BBL™ DrySlide™ systems available from Becton Dickinson, Sparks, MD.)

TABLE **5-6** Oxidase Test Results and Interpretations

TABLE OF RESULTS		
Result	**Interpretation**	**Symbol**
Dark blue within 20 seconds	Cytochrome c oxidase is present	+
No color change to blue within 20 seconds	Cytochrome c oxidase is not present	−

✦ Materials

Per Student Group

✦ sterile cotton-tipped applicator or plastic loop
✦ sterile water
✦ fresh slant cultures of:
  ◆ *Escherichia coli*
  ◆ *Moraxella catarrhalis* (BSL-2)

Procedure

1 Following your instructor's guidelines, transfer some of the culture to the reagent slide. Get a visible mass of growth on the wooden end of a sterile cotton applicator or plastic loop, and roll it on the reagent slide.

2 Observe for color change within 20 seconds.

3 Repeat steps one and two for the second organism.

4 Enter your results in the chart provided on the Data Sheet.

5 When performing this test on an unknown, run a positive control simultaneously to verify the quality of the test system.

References

BBL™ DrySlide™ package insert.

Collins, C. H., Patricia M. Lyne, and J. M. Grange. 1995. Page 116 in Collins and Lyne's *Microbiological Methods,* 7th ed. Butterworth-Heinemann, Oxford, United Kingdom.

Forbes, Betty A., Daniel F. Sahm, and Alice S. Weissfeld. 2002. Pages 151–152 in *Bailey & Scott's Diagnostic Microbiology,* 11th ed. Mosby, St. Louis.

Lányi, B. 1987. Page 18 in *Methods in Microbiology,* Vol. 19, edited by R. R. Colwell and R. Grigorova. Academic Press, New York.

MacFaddin, Jean F. 2000. Page 368 in *Biochemical Tests for Identification of Medical Bacteria,* 3rd ed. Lippincott Williams & Wilkins, Philadelphia.

Smibert, Robert M., and Noel R. Krieg. 1994. Page 625 in *Methods for General and Molecular Bacteriology,* edited by Philipp Gerhardt, R. G. E. Murray, Willis A. Wood, and Noel R. Krieg. American Society for Microbiology, Washington, DC.

White, David, 2000. Chapter 4 in *The Physiology and Biochemistry of Prokaryotes,* 2nd ed. Oxford University Press, New York.

EXERCISE 5-7

Nitrate Reduction Test

✦ Theory

As mentioned in the introduction to this unit, anaerobic respiration involves the reduction of (*i.e.*, transfer of electrons to) an inorganic molecule other than oxygen. As the name implies, nitrate reduction is one such example. Many Gram-negative bacteria (including most *Enterobacteriaceae*) contain the enzyme **nitrate reductase** and perform a single-step reduction of nitrate to nitrite ($NO_3 \rightarrow NO_2$). Other bacteria, in a multi-step process known as **denitrification**, are capable of enzymatically converting nitrate to molecular nitrogen (N_2). Some products of nitrate reduction are shown in Figure 5-23.

Nitrate broth is an undefined medium of beef extract, peptone, and potassium nitrate (KNO_3). An inverted Durham tube is placed in each broth to trap a portion of any gas produced. In contrast to many differential media, no color indicators are included. The color reactions obtained in Nitrate Broth take place as a result of reactions between metabolic products and reagents added after incubation (Figure 5-24).

Before a broth can be tested for nitrate reductase activity (nitrate reduction to nitrite), it must be examined for evidence of denitrification. This is simply a visual inspection for the presence of gas in the Durham tube (Figure 5-25). If the Durham tube contains gas and the organism is known not to be a fermenter (as evidenced by a fermentation test), the test is complete. Denitrification has taken place. Gas produced in a nitrate reduction test by an organism capable of fermenting is not determinative because the source of the gas is unknown. For more information on fermentation, see Exercises 5-2 and 5-3.

$$NO_3 \xrightarrow{2e^-} NO_2 \begin{cases} \xrightarrow{6e^-} NH_4 \text{ ammonium (-3)} \\ \xrightarrow{e^-} NO \xrightarrow{1e^-} \tfrac{1}{2} N_2O \xrightarrow{2e^-} \tfrac{1}{2} N_2 \end{cases}$$

nitrate (+5) nitrite (+3)

NO nitric oxide (+2)   ½ N₂O nitrous oxide (+1)   ½ N₂ molecular nitrogen (0)

5-23 **POSSIBLE END PRODUCTS OF NITRATE REDUCTION** ✦ Nitrate reduction is complex. Many different organisms under many different circumstances perform nitrate reduction with many different outcomes. Members of the *Enterobacteriaceae* simply reduce NO_3 to NO_2. Other bacteria, functionally known as "denitrifiers," reduce NO_3 all the way to N_2 via the intermediates shown, and are important ecologically in the nitrogen cycle. Both of these are anaerobic respiration pathways (also known as "nitrate respiration"). Other organisms are capable of assimilatory nitrate reduction, in which NO_3 is reduced to NH_4, which can be used in amino acid synthesis. The oxidation state of nitrogen in each compound is shown in parentheses.

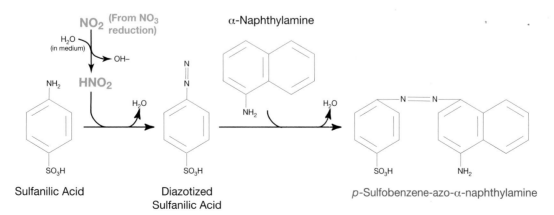

Sulfanilic Acid   Diazotized Sulfanilic Acid   *p*-Sulfobenzene-azo-α-naphthylamine

5-24 **INDICATOR REACTION** ✦ If nitrate is reduced to nitrite, nitrous acid (HNO_2) will form in the medium. Nitrous acid then reacts with sulfanilic acid to form diazotized sulfanilic acid, which reacts with the α-naphthylamine to form *p*-sulfobenzene-azo-α-naphthylamine, which is red. Thus, a red color indicates the presence of nitrite and is considered a positive result for nitrate reduction to nitrite.

5-25 INCUBATED NITRATE BROTH BEFORE THE ADDITION OF
REAGENTS ✦ Numbered 1 through 5 from left to right, these are
Nitrate Broths immediately after incubation before the addition
of reagents. Tube 2 is an uninoculated control used for color
comparison. Note the gas produced by tube 5. It is a known
nonfermenter and, therefore, will receive no reagents. The gas
produced is an indication of denitrification and a positive result.
Tubes 1 through 4 will receive reagents. See Figure 5-26.

If there is no visual evidence of denitrification,
sulfanilic acid (nitrate reagent A) and naphthylamine
(nitrate reagent B) are added to the medium to test for
nitrate reduction to nitrite. If present, nitrite will form
nitrous acid (HNO_2) in the aqueous medium. Nitrous acid
reacts with the added reagents to produce a red, water-
soluble compound (Figure 5-26). Therefore, formation
of red color after the addition of reagents indicates that
the organism reduced nitrate to nitrite. If no color change
takes place with the addition of reagents, the nitrate either
was not reduced or was reduced to one of the other nitrogenous compounds shown in Figure 5-23. Because it
is visually impossible to tell the difference between these
two occurrences, another test must be performed.

In this stage of the test, a small amount of powdered
zinc is added to the broth to catalyze the reduction of
any nitrate (which still may be present as KNO_3) to
nitrite. If nitrate is present at the time zinc is added, it
will be converted immediately to nitrite, and the above-
described reaction between nitrous acid and reagents
will follow and turn the medium red. In this instance,
the red color indicates that nitrate was *not* reduced by
the organism (Figure 5-27). No color change after the
addition of zinc indicates that the organism reduced the
nitrate to NH_3, NO, N_2O, or some other nongaseous
nitrogenous compound.

✦ In This Exercise

After inoculation and incubation of the Nitrate Broths,
you first will inspect the Durham tubes included in the

5-26 INCUBATED NITRATE BROTH AFTER ADDITION OF REAGENTS ✦
After the addition of reagents, tube 1 shows a positive result.
Tube 3 and tube 4 are inconclusive because they show no color
change. Zinc dust must be added to tubes 2 (control), 3, and 4
to verify the presence or absence of nitrate. See Figure 5-27.

5-27 INCUBATED NITRATE BROTH AFTER ADDITION OF REAGENTS
AND ZINC ✦ Finally, a pinch of zinc is added to tubes 2, 3, and 4
because they have been colorless up to this point. Tube 2 (the
control tube) and tube 3 turned red. This is a negative result
because it indicates that nitrate is still present in the tube.
Tube 4 did not change color, which indicates that the nitrate
was reduced by the organism beyond nitrite to some other
nitrogenous compound. This is a positive result.

media for bubbles—an indication of gas production
(Figure 5-25). One or more of the organisms selected
for this exercise are fermenters, and possible producers
of gas other than molecular nitrogen; therefore, you will
proceed with adding reagents to all tubes. Record any
gas observed in the Durham tubes and see if there is a
correlation between that and the color results after the
reagents and zinc dust have been added. Follow the
procedure carefully, and use Table 5-7 as a guide when
interpreting the results.

✦ Application

Virtually all members of *Enterobacteriaceae* perform a one-step reduction of nitrate to nitrite. The nitrate test differentiates them from Gram-negative rods that either do not reduce nitrate or reduce it beyond nitrite to N_2 or other compounds.

✦ Materials

Per Student Group

+ four Nitrate Broths
+ nitrate test reagents A and B
+ zinc powder
+ fresh cultures of:
  + *Erwinia amylovora*
    (Alternative: *Enterococcus faecalis*)
  + *Escherichia coli*
  + *Pseudomonas aeruginosa* (BSL-2)

✦ Medium and Reagent Recipes

Nitrate Broth

◆ Beef extract	3.0 g
◆ Peptone	5.0 g
◆ Potassium nitrate	1.0 g
◆ Distilled or deionized water	1.0 L
pH 6.8–7.2 at 25°C	

Reagent A

◆ N,N-Dimethyl-1-naphthylamine	0.6 g
◆ 5N Acetic acid (30%)	100.0 mL

Reagent B

◆ Sulfanilic acid	0.8 g
◆ 5N Acetic acid (30%)	100.0

Procedure

Lab One

1 Obtain four Nitrate Broths. Label three of them with the names of the organisms, your name, and the date. Label the fourth tube "control."

2 Inoculate three broths with the test organisms. Do not inoculate the control.

3 Incubate all tubes at $35 \pm 2°C$ for 24 to 48 hours.

Lab Two

1 Examine each tube for evidence of gas production. Record your results in the chart provided on the Data Sheet. Refer to Table 5-7 when making your interpretations. (Be methodical when recording your results. Red color can have opposite interpretations depending on where you are in the procedure.)

2 Using Figure 5-28 as a guide, proceed to the addition of reagents to *all* tubes as follows:

   a. Add eight drops each (approximately 0.5 mL) of reagent A and reagent B to each tube. Mix well, and let the tubes stand undisturbed for 10 minutes. Using the control as a comparison, record your test results in the chart provided in the Data Sheet. Set aside any test broths that are positive.

   b. Where appropriate, add a pinch of zinc dust. (Only a small amount is needed. Dip a wooden applicator into the zinc dust and transfer only the amount that clings to the wood.) Let the tubes stand for 10 minutes. Record your results in the chart provided on the Data Sheet.

TABLE **5-7** Nitrate Test Results and Interpretations

TABLE OF RESULTS		
Result	**Interpretation**	**Symbol**
Gas (Nonfermenter)	Denitrification—production of nitrogen gas ($NO_3 \rightarrow NO_2 \rightarrow N_2$)	+
Gas (Fermenter, or status is unknown)	Source of gas is unknown; requires addition of reagents	
Red color (after addition of reagents A and B)	Nitrate reduction to nitrite ($NO_3 \rightarrow NO_2$)	+
No color (after the addition of reagents)	Incomplete test; requires the addition of zinc dust	
No color change (after addition of zinc)	Nitrate reduction to nongaseous nitrogenous compounds ($NO_3 \rightarrow NO_2 \rightarrow$ nongaseous nitrogenous products)	+
Red color (after addition of zinc dust)	No nitrate reduction	−

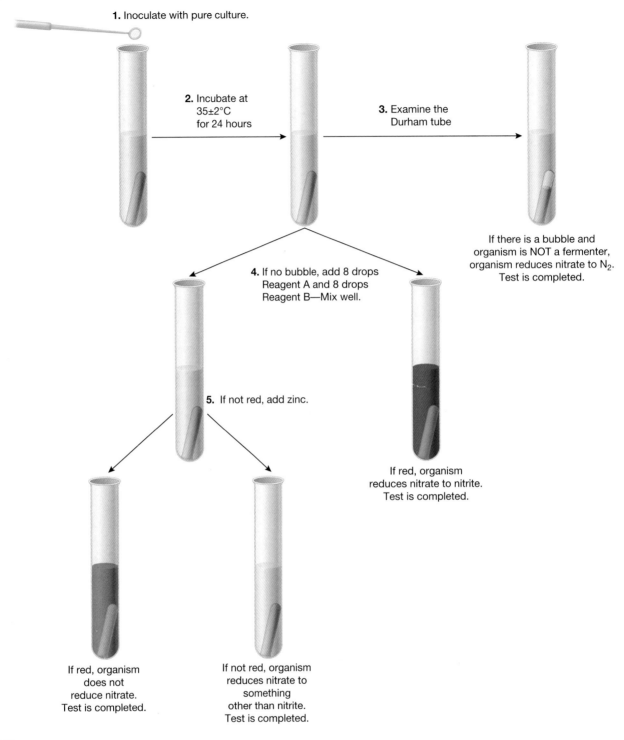

1. Inoculate with pure culture.

2. Incubate at 35±2°C for 24 hours

3. Examine the Durham tube

If there is a bubble and organism is NOT a fermenter, organism reduces nitrate to N₂. Test is completed.

4. If no bubble, add 8 drops Reagent A and 8 drops Reagent B—Mix well.

5. If not red, add zinc.

If red, organism reduces nitrate to nitrite. Test is completed.

If red, organism does not reduce nitrate. Test is completed.

If not red, organism reduces nitrate to something other than nitrite. Test is completed.

5-28 PROCEDURAL DIAGRAM FOR THE NITRATE REDUCTION TEST ✦

References

Atlas, Ronald M., and Richard Bartha. 1998. Pages 423–425 in *Microbial Ecology—Fundamentals and Applications*, 4th ed. Benjamin/Cummings Science Publishing, Menlo Park, CA.

Forbes, Betty A., Daniel F. Sahm, and Alice S. Weissfeld. 2002. Page 277 in *Bailey & Scott's Diagnostic Microbiology*, 11th ed. Mosby, St. Louis.

Lányi, B. 1987. Page 21 in *Methods in Microbiology*, Vol. 19, edited by R. R. Colwell and R. Grigorova. Academic Press, New York.

MacFaddin, Jean F. 2000. Page 348 in *Biochemical Tests for Identification of Medical Bacteria*, 3rd ed. Williams & Wilkins, Philadelphia.

Moat, Albert G., John W. Foster, and Michael P. Spector. Chapter 14 in *Microbial Physiology*, 4th ed. Wiley-Liss, New York.

Zimbro, Mary Jo, and David A. Power, Eds. 2003. Page 400 in *Difco™ and BBL™ Manual—Manual of Microbiological Culture Media*. Becton Dickinson and Co., Sparks, MD.

Nutrient Utilization Tests

In this unit you will perform tests using two examples of utilization media—Simmons Citrate Medium and Malonate Broth. Utilization media are highly defined formulations designed to differentiate organisms based on their ability to grow when an essential nutrient (*e.g.,* carbon or nitrogen) is strictly limited. For example, citrate medium contains sodium citrate as the only carbon-containing compound and ammonium ion as the only nitrogen source. Malonate broth contains three sources of carbon but prevents utilization of all but one by competitive inhibition of a specific enzyme. ✦

EXERCISE 5-8

Citrate Test

✦ Theory

In many bacteria, citrate (citric acid) produced as acetyl coenzyme A (from the oxidation of pyruvate or the β-oxidation of fatty acids) reacts with oxaloacetate at the entry to the Krebs cycle. Citrate then is converted back to oxaloacetate through a complex series of re-actions, which begins the cycle anew. For more infor-mation on the Krebs cycle and fatty acid metabolism, refer to Appendix A and Figure 5-54.

In a medium containing citrate as the only available carbon source, bacteria that possess **citrate-permease** can transport the molecules into the cell and enzymatically convert it to pyruvate. Pyruvate then can be converted to a variety of products, depending on the pH of the environment (Figure 5-29).

Simmons Citrate Agar is a defined medium that contains sodium citrate as the sole carbon source and ammonium phosphate as the sole nitrogen source. Bromthymol blue dye, which is green at pH 6.9 and blue at pH 7.6, is added as an indicator. Bacteria that survive in the medium and utilize the citrate also convert the ammonium phosphate to ammonia (NH_3) and am-monium hydroxide (NH_4OH), both of which tend to alkalinize the agar. As the pH goes up, the medium changes from green to blue (Figure 5-30). Thus, conver-sion of the medium to blue is a positive citrate test result (Table 5-8).

Occasionally a citrate-positive organism will grow on a Simmons Citrate slant without producing a change in color. In most cases, this is because of incomplete in-cubation. In the absence of color change, growth on the

5-29 CITRATE CHEMISTRY ✦ In the presence of citrate-permease enzyme, citrate enters the cell and is converted to pyruvate. The pyruvate then is converted to a variety of products depending on the pH of the environment.

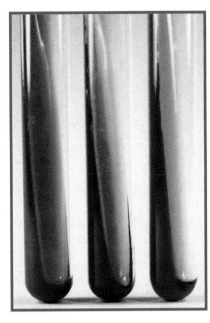

5-30 CITRATE TEST RESULTS ✦ These Simmons Citrate slants were inoculated with a citrate-positive (+) organism on the left and a citrate-negative (−) organism in the center. The slant on the right is uninoculated.

TABLE **5-8** Citrate Test Results and Interpretations

TABLE OF RESULTS

Result	Interpretation	Symbol
Blue (even a small amount)	Citrate is utilized	+
No color change; growth	Citrate is utilized	+
No color change; no growth	Citrate is not utilized	−

slant indicates that citrate is being utilized and is evidence of a positive reaction. To avoid confusion between actual growth and heavy inoculum, which may appear as growth, citrate slants typically are inoculated lightly with an inoculating needle rather than a loop.

✦ Application

The citrate utilization test is used to determine the ability of an organism to use citrate as its sole source of carbon. Citrate utilization is one part of a test series referred to as the IMViC (Indole, Methyl Red, Voges-Proskauer and Citrate tests) that distinguishes between members of the family *Enterobacteriaceae* and differentiate them from other Gram-negative rods.

✦ In This Exercise

You will be inoculating Simmons Citrate slants with a citrate-positive and a citrate-negative organism. To avoid confusing growth on the slant with a heavy inoculum, be sure to inoculate using a needle instead of a loop.

✦ Materials

Per Student Group
✦ three Simmons Citrate slants
✦ fresh slant cultures of:
  ✦ *Enterobacter aerogenes*
  ✦ *Escherichia coli*

✦ Medium Recipe

Simmons Citrate Agar
✦ Ammonium dihydrogen phosphate	1.0 g
✦ Dipotassium phosphate	1.0 g
✦ Sodium chloride	5.0 g
✦ Sodium citrate	2.0 g
✦ Magnesium sulfate	0.2 g
✦ Agar	15.0 g
✦ Bromthymol blue	0.08 g
✦ Distilled or deionized water	1.0 L

pH 6.7–7.1 at 25°C

 ## Procedure

Lab One

1 Obtain three Simmons Citrate tubes. Label two with the names of the organisms, your name, and the date. Label the third tube "control."

2 Using an inoculating needle and *light inoculum*, streak the slants with the test organisms. Do not inoculate the control.

3 Incubate all tubes at 35 ± 2°C for 48 hours.

Lab Two

1 Observe the tubes for color changes and/or growth.

2 Record your results in the chart provided on the Data Sheet.

References

Collins, C. H., Patricia M. Lyne, and J. M. Grange. 1995. Page 111 in *Collins and Lyne's Microbiological Methods*, 7th ed. Butterworth-Heinemann, Oxford, United Kingdom.

Forbes, Betty A., Daniel F. Sahm, and Alice S. Weissfeld. 2002. Page 266 in *Bailey & Scott's Diagnostic Microbiology*, 11th ed. Mosby, St. Louis.

MacFaddin, Jean F. 2000. Page 98 in *Biochemical Tests for Identification of Medical Bacteria*, 3rd ed. Lippincott Williams & Wilkins, Philadelphia.

Smibert, Robert M., and Noel R. Krieg. 1994. Page 614 in *Methods for General and Molecular Bacteriology*, edited by Philipp Gerhardt, R. G. E. Murray, Willis A. Wood, and Noel R. Krieg, American Society for Microbiology, Washington, DC.

Zimbro, Mary Jo, and David A. Power, Eds. 2003. Page 514 in *Difco™ and BBL™ Manual—Manual of Microbiological Culture Media*. Becton Dickinson and Co., Sparks, MD.

EXERCISE 5-9

Malonate Test

✦ Theory

One of the many enzymatic reactions of the Krebs cycle, as illustrated in Appendix A, is the oxidation of succinate to fumarate. In the reaction, which requires the enzyme **succinate dehydrogenase**, the coenzyme FAD is reduced to $FADH_2$. Refer to the upper reaction in Figure 5-31.

Malonate (malonic acid), which can be added to growth media, is similar enough to succinate to replace it as the substrate in the reaction (Figure 5-31, lower reaction). This **competitive inhibition** of succinate dehydrogenase, in combination with the subsequent buildup of succinate in the cell, shuts down the Krebs cycle and will kill the organism unless it can ferment or utilize malonate as its sole remaining carbon source.

Malonate Broth is the medium used to make this determination. It contains a high concentration of sodium malonate, yeast extract, and a very small amount of glucose to promote growth of organisms that otherwise are slow to respond. Buffers are added to stabilize the medium at pH 6.7. Bromthymol blue dye, which is green in uninoculated media, is added to indicate any shift in pH. If an organism cannot utilize malonate but manages to ferment a small amount of glucose, it may turn the medium slightly yellow or produce no color change at all. These are negative results. If the organism utilizes malonate, it will alkalinize the medium and change the indicator from green to deep blue (Figure 5-32). Deep blue is positive (Table 5-9).

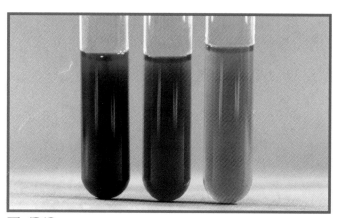

5-32 MALONATE TEST RESULTS ✦ Malonate broths illustrating malonate-positive (+) on the left, an uninoculated control in the center, and malonate-negative (−) on the right.

TABLE **5-9** Malonate Test Results and Interpretations

TABLE OF RESULTS		
Result	**Interpretation**	**Symbol**
Dark blue	Malonate is utilized	+
No color change; or slightly yellow	Malonate is not utilized	−

✦ Application

The Malonate Test was designed originally to differentiate between *Escherichia*, which will not grow in the medium, and *Enterobacter*. Its use as a differential medium has now broadened to include other members of *Enterobacteriaceae*.

✦ In This Exercise

Today, you will inoculate two Malonate Broths and observe for color changes. Blue color is the only positive for this test; all others are negative.

✦ Materials

Per Student Group

✦ three Malonate Broths

✦ fresh cultures of:
  ✦ *Enterobacter aerogenes*
  ✦ *Escherichia coli*

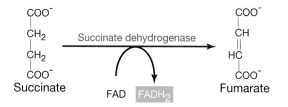

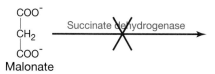

5-31 COMPETITIVE INHIBITION OF SUCCINATE DEHYDROGENASE BY MALONATE ✦ The attachment of malonate to the enzyme succinate dehydrogenase prevents the attachment of succinate and, thus, the conversion of succinate to fumarate.

◆ Medium Recipe

Malonate Broth

◆ Yeast extract	1.0 g
◆ Ammonium sulfate	2.0 g
◆ Dipotassium phosphate	0.6 g
◆ Monopotassium phosphate	0.4 g
◆ Sodium chloride	2.0 g
◆ Sodium malonate	3.0 g
◆ Glucose	0.25 g
◆ Bromthymol blue	0.025 g
◆ Distilled or deionized water	1.0 L

pH 6.5–6.9 at 25°C

Procedure

Lab One

1. Obtain three Malonate Broths. Label two of them with the names of the organisms, your name, and the date. Label the third tube "control."

2. Inoculate two broths with the test organisms. Do not inoculate the control.

3. Incubate all tubes at $35 \pm 2°C$ for 48 hours.

Lab Two

1. Observe the tubes for color changes.

2. Record your results in the chart provided on the Data Sheet.

References

MacFaddin, Jean F. 2000. Page 310 in *Biochemical Tests for Identification of Medical Bacteria*, 3rd ed. Lippincott Williams & Wilkins, Philadelphia.

Zimbro, Mary Jo, and David A. Power, Eds. 2003. Page 343 in *Difco™ and BBL™ Manual—Manual of Microbiological Culture Media*. Becton Dickinson and Co., Sparks, MD.

Decarboxylation and Deamination Tests

Decarboxylation and deamination tests were designed to differentiate members of *Enterobacteriaceae* and to distinguish them from other Gram-negative rods. Most members of *Enterobacteriaceae* produce one or more enzymes that are necessary to break down amino acids. Enzymes that catalyze the removal of an amino acid's carboxyl group (COOH) are called decarboxylases. Enzymes that catalyze the removal of an amino acid's amine group (NH_2) are called deaminases.

Each decarboxylase and deaminase is specific to a particular substrate. Decarboxylases catalyze reactions that produce alkaline products. Thus, we are able to identify the ability of an organism to produce a specific decarboxylase by preparing base medium, adding a known amino acid, and including a pH indicator to mark the shift to alkalinity. Differentiation of an organism in deamination media employs the same principle of substrate exclusivity by including a single known amino acid but requires the addition of a chemical reagent to produce a readable result.

In the next two exercises you will be introduced to the most common of both types of tests. In Exercise 5-10 you will test bacterial ability to decarboxylate the amino acids arginine, lysine, and ornithine. In Exercise 5-11 you will test bacterial ability to deaminate the amino acid phenylalanine. ✦

EXERCISE 5-10 Decarboxylation Tests

✦ Theory

Møller's Decarboxylase Base Medium contains peptone, glucose, the pH indicator bromcresol purple, and the **coenzyme** pyridoxal phosphate. Bromcresol purple is purple at pH 6.8 and above, and yellow below pH 5.2. Base medium can be used with one of a number of specific amino acid substrates, depending on the decarboxylase to be identified.

After inoculation, an overlay of mineral oil is used to seal the medium from external oxygen and promote fermentation (Figure 5-33). Glucose fermentation in the anaerobic medium initially turns it yellow because of the accumulation of acid end products. The low pH and presence of the specific amino acid induces **decarboxylase-positive** organisms to produce the enzyme.

Decarboxylation of the amino acid results in accumulation of alkaline end products that turn the medium purple (Figure 5-34 through Figure 5-37). If the organism is a glucose fermenter (as are all of the *Enterobacteriaceae*) but does not produce the appropriate decarboxylase, the medium will turn yellow and remain so. If the organism does not ferment glucose, the medium will exhibit no color change. Purple color is the only positive result; all others are negative.

✦ Application

Decarboxylase media can include any one of several amino acids. Typically, these media are used to differentiate organisms in the family *Enterobacteriaceae* and to distinguish them from other Gram-negative rods.

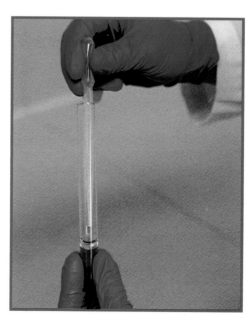

5-33
ADDING MINERAL OIL TO THE TUBES ✦

✦ In This Exercise

You will perform decarboxylation tests on three organisms in Møller's arginine, lysine, ornithine, and base decarboxylase media. This procedure requires the aseptic addition of mineral oil. Be careful not to introduce contaminants while adding the oil. Refer to Figure 5-38 and Table 5-10 when interpreting your results.

✦ Materials

Per Student Group

- ✦ four Lysine Decarboxylase Broths
- ✦ four Ornithine Decarboxylase Broths
- ✦ four Arginine Decarboxylase Broths
- ✦ four Decarboxylase Base Media
- ✦ sterile mineral oil
- ✦ sterile transfer pipettes
- ✦ recommended organisms:
  - ◆ *Salmonella typhimurium* (BSL-2)
  - ◆ *Enterobacter aerogenes*
  - ◆ *Proteus vulgaris* (BSL-2)

5-34 AMINO ACID DECARBOXYLATION ✦ Removal of an amino acid's carboxyl group results in the formation of an amine and carbon dioxide.

5-35 LYSINE DECARBOXYLATION ✦ Decarboxylation of the amino acid lysine produces cadaverine and CO_2.

5-36 ORNITHINE DECARBOXYLATION ✦ Decarboxylation of the amino acid ornithine produces putrescine and CO_2.

5-38 DECARBOXYLATION TEST RESULTS ✦ A lysine decarboxylase-positive (+) organism is on the left, an uninoculated control is in the center, and a lysine decarboxylase negative (−) is on the right. The color results shown here are representative of arginine and ornithine decarboxylase media as well.

5-37 ARGININE DECARBOXYLATION ✦ Decarboxylation of the amino acid arginine produces the amine agmatine. Members of *Enterobacteriaceae* are capable of degrading agmatine into putrescine and urea. Those strains with urease can break down the urea further into ammonia and carbon dioxide. Thus, the end products of arginine catabolism are carbon dioxide, putrescine and urea, *or* (in the presence of urease) carbon dioxide, putrescine, and ammonia.

TABLE **5-10** Decarboxylase Test Results and Interpretations

TABLE OF RESULTS

Result	Interpretation	Symbol
No color change	No decarboxylation	−
Yellow	Fermentation; no decarboxylation	−
Purple (may be slight)	Decarboxylation; organism produces the specific decarboxylase enzyme	+

✦ Medium Recipe

Decarboxylase Medium (Møller)

♦ Peptone	5.0 g
♦ Beef extract	5.0 g
♦ Glucose (dextrose)	0.5 g
♦ Bromcresol purple	0.01 g
♦ Cresol red	0.005 g
♦ Pyridoxal	0.005 g
♦ L-Lysine, L-Ornithine, or L-Arginine	10.0 g
♦ Distilled or deionized water	1.0 L

pH 5.8–6.2 at 25°C

 Procedure

Lab One

1 Obtain four of each decarboxylase broth. Label three of each with the name of the organism, your name, and the date. Label the fourth of each broth "control."

2 Inoculate the media with the test organisms. Do not inoculate the control.

3 Overlay all tubes with 3 to 4 mm sterile mineral oil (Figure 5-33).

4 Incubate all tubes aerobically at 35 ± 2°C for one week.

Lab Two

1 Remove the tubes from the incubator and examine them for color changes.

2 Record your results in the chart provided on the Data Sheet.

References

Collins, C. H., Patricia M. Lyne, and J. M. Grange. 1995. Page 111 in *Collins and Lyne's Microbiological Methods,* 7th ed. Butterworth-Heinemann, Oxford, United Kingdom.

DIFCO Laboratories. 1984. Page 268 in *DIFCO Manual,* 10th ed. DIFCO Laboratories, Detroit.

Forbes, Betty A., Daniel F. Sahm, and Alice S. Weissfeld. 2002. Page 267 in *Bailey & Scott's Diagnostic Microbiology,* 11th ed. Mosby, St. Louis.

Lányi, B. 1987. Page 29 in *Methods in Microbiology,* Vol. 19, edited by R. R. Colwell and R. Grigorova. Academic Press, New York.

MacFaddin, Jean F. 2000. Page 120 in *Biochemical Tests for Identification of Medical Bacteria,* 3rd ed. Lippincott Williams & Wilkins, Philadelphia.

EXERCISE
5-11

Phenylalanine Deaminase Test

✦ Theory

Organisms that produce phenylalanine deaminase can be identified by their ability to remove the amine group (NH_2) from the amino acid phenylalanine. The reaction, as shown in Figure 5-39, splits a water molecule and produces ammonia (NH_3) and phenylpyruvic acid. Deaminase activity, therefore, is evidenced by the presence of phenylpyruvic acid.

Phenylalanine Agar provides a rich source of phenylalanine. A reagent containing ferric chloride ($FeCl_3$) is added to the medium after incubation. The normally colorless phenylpyruvic acid reacts with the ferric chloride and turns a dark green color almost immediately (Figure 5-40). Formation of green color indicates the presence of phenylpyruvic acid and, hence, the presence of phenylalanine deaminase. Yellow is negative (Figure 5-41).

✦ Application

This medium is used to differentiate the genera *Morganella*, *Proteus*, and *Providencia* from other members of the *Enterobacteriaceae*.

✦ In This Exercise

You will inoculate two phenylalanine agar slants with test organisms. After incubation you will add 12% $FeCl_3$ (Phenylalanine Deaminase Test Reagent) and watch for a

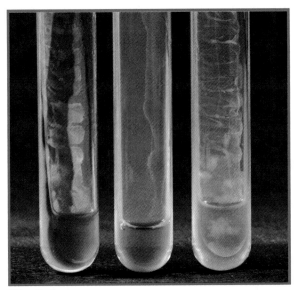

5-41 PHENYLALANINE DEAMINASE TEST ✦ Note the color produced by the stream of ferric chloride in each tube. A phenylalanine deaminase-positive (+) organism is on the left, an uninoculated control is in the middle, and a phenylalanine deaminase-negative (−) organism is on the right.

color change. Refer to Figure 5-41 and Table 5-11 when making your interpretations and recording your results.

✦ Materials

Per Student Group

✦ three Phenylalanine Agar slants

✦ 12% ferric chloride solution

✦ fresh slant cultures of:
  ◆ *Escherichia coli*
  ◆ *Proteus vulgaris* (BSL-2)

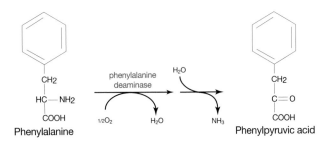

5-39 DEAMINATION OF PHENYLALANINE ✦ Deamination is the removal of an amine (NH_2) from an amino acid.

Phenylpyruvic Acid + $FeCl_3$ ⟶ Green Color

5-40 INDICATOR REACTION ✦ Phenylpyruvic acid produced by phenylalanine deamination reacts with $FeCl_3$ to produce a green color and indicates a positive phenylalanine test. Be sure to read the result immediately, as the color may fade.

TABLE **5-11** Phenylalanine Test Results and Interpretations

TABLE OF RESULTS		
Result	**Interpretation**	**Symbol**
Green color	Phenylalanine deaminase is present	+
No color change	Phenylalanine deaminase is absent	−

✦ Medium and Reagent Recipes

Phenylalanine Agar

◆ DL-Phenylalanine	2.0 g
◆ Yeast extract	3.0 g
◆ Sodium chloride	5.0 g
◆ Sodium phosphate	1.0 g
◆ Agar	12.0 g
◆ Distilled or deionized water	1.0 L

pH 7.1–7.5 at 25°C

Phenylalanine Deaminase Test Reagent

◆ Ferric chloride	12.0 g
◆ Concentrated hydrochloric acid	2.5 mL
◆ Deionized water (to bring the total to)	100.0 mL

 # Procedure

Lab One

1 Obtain three Phenylalanine Agar slants. Label two slants with the name of the organism, your name, and the date. Label the third slant "control."

2 Streak two slants with heavy inocula of the test organisms. Do not inoculate the control.

3 Incubate all slants aerobically at 35 ± 2°C for 18 to 24 hours.

Lab Two

1 Add a few drops of 12% ferric chloride solution to each tube and observe for color change. (**Note:** This color may fade quickly, so read and record your results immediately.)

2 Record your results in the chart provided on the Data Sheet.

References

Forbes, Betty A., Daniel F. Sahm, and Alice S. Weissfeld. 2002. Page 280 in *Bailey & Scott's Diagnostic Microbiology,* 11th ed. Mosby, St. Louis.

Lányi, B. 1987. Page 28 in *Methods in Microbiology,* Vol. 19, edited by R. R. Colwell and R. Grigorova. Academic Press, New York.

MacFaddin, Jean F. 2000. Page 388 in *Biochemical Tests for Identification of Medical Bacteria,* 3rd ed. Lippincott Williams & Wilkins, Philadelphia.

Zimbro, Mary Jo, and David A. Power, Eds. 2003. Page 442 in *Difco™ and BBL™ Manual—Manual of Microbiological Culture Media.* Becton Dickinson and Co., Sparks, MD.

Tests Detecting Hydrolytic Enzymes

Reactions that use water to split complex molecules are called **hydrolysis** (or hydrolytic) reactions. The enzymes required for these reactions are called hydrolytic enzymes. When the enzyme catalyzes its reaction inside the cell, it is referred to as **intracellular.** Enzymes secreted from the organism to catalyze reactions outside the cell are called **extracellular** enzymes, or **exoenzymes**. In this unit you will be performing tests that identify the actions of both intracellular and extracellular types of enzymes.

The hydrolytic enzymes detected by the Urease Test are intracellular and produce identifiable color changes in the medium. Nutrient Gelatin detects gelatinase—an exoenzyme that liquefies the solid medium. DNase Agar, Milk Agar, Tributyrin Agar, and Starch Agar are plated media that identify extracellular enzymes capable of diffusing into the medium and producing a distinguishable halo of clearing around the bacterial growth. ✦

EXERCISE 5-12   Starch Hydrolysis

✦ Theory

Starch is a polysaccharide made up of α-D-glucose subunits. It exists in two forms—linear (amylose) and branched (amylopectin)—usually as a mixture with the branched configuration being predominant. The α-D-glucose molecules in both amylose and amylopectin are bonded by 1,4-α-glycosidic (acetal) linkages (Figure 5-42). The two forms differ in that amylopectin contains polysaccharide side chains connected to approximately every 30th glucose in the main chain. These side chains are identical to the main chain except that the number 1 carbon of the first glucose in the side chain is bonded to carbon number 6 of the main chain glucose. The bond, therefore, is a 1,6-α-glycosidic linkage.

Starch is too large to pass through the bacterial cell membrane. Therefore, to be of metabolic value to the bacteria, it first must be split into smaller fragments or individual glucose molecules. Organisms that produce and secrete the extracellular enzymes α-**amylase** and **oligo-1,6-glucosidase** are able to hydrolyze starch by breaking the glycosidic linkages between sugar subunits. As shown in Figure 5-42, the end result of these reactions is the complete hydrolysis of the polysaccharide to its individual α-glucose subunits.

Starch agar is a simple plated medium of beef extract, soluble starch, and agar. When organisms that produce α-amylase and oligo-1,6-glucosidase are cultivated on starch agar, they hydrolyze the starch in the area surrounding their growth. Because both starch and its sugar subunits are virtually invisible in the medium, the reagent iodine is used to detect the presence or absence of starch in the vicinity around the bacterial growth. Iodine reacts with starch and produces a blue or dark brown color; therefore, any microbial starch hydrolysis will be revealed as a clear zone surrounding the growth (Figure 5-43 and Table 5-12).

✦ Application

Starch agar originally was designed for cultivating *Neisseria*. It no longer is used for this, but with pH indicators, it is used to isolate and presumptively identify *Gardnerella vaginalis*. It aids in differentiating members of the genera *Corynebacterium*, *Clostridium*, *Bacillus*, *Bacteroides*, *Fusobacterium*, and *Enterococcus*, most of which have (+) and (−) species.

✦ In This Exercise

You will inoculate a Starch Agar plate with two organisms. After incubation you will add a few drops of iodine to reveal clear areas surrounding the growth. Iodine will not color the agar where the growth is, only the area surrounding the growth. Before you add the iodine, make a note of the growth patterns on the plate so you don't confuse thinning growth at the edges with the halo produced by starch hydrolysis.

α-Amylose
[1,4-α-glucosidic (acetal) linkages]

α-D-Glucose
(many)

Amylopectin
[1,4-α-glucosidic (acetal) linkages and 1,6-α-glucosidic (acetal) branch linkages]

α-D-Glucose
(many)

5-42 STARCH HYDROLYSIS BY α-AMYLASE AND OLIGO-1,6-GLUCOSIDASE ✦ The enzymes α-amylase and oligo-1,6-glucosidase hydrolyze starch by breaking glycosidic linkages, resulting in release of the component glucose subunits.

5-43 STARCH HYDROLYSIS TEST ✦ This is a Starch Agar plate with iodine added to detect amylase activity. The organism above shows no clearing and is negative (−). The organism below is surrounded by a halo of clearing and is positive (+).

TABLE **5-12** Amylase Test Results and Interpretations

T A B L E O F R E S U L T S		
Result	**Interpretation**	**Symbol**
Clearing around growth	Amylase is present	+
No clearing around growth	No amylase is present	−

✦ Materials

Per Student Group

✦ one Starch Agar plate

✦ Gram iodine (from your Gram stain kit)

✦ recommended organisms:

  ♦ *Bacillus cereus*

  ♦ *Escherichia coli*

◆ Medium Recipe

Starch Agar

- Beef extract   3.0 g
- Soluble starch   10.0 g
- Agar   12.0 g
- Distilled or deionized water   1.0 L

       pH 7.3–7.7 at 25°C

Procedure

Lab One

1. Using a marking pen, divide the Starch Agar plate into three equal sectors. Be sure to mark on the bottom of the plate.

2. Label the plate with the names of the organisms, your name, and the date.

3. Spot-inoculate two sectors with the test organisms.

4. Invert the plate and incubate it aerobically at 35 ± 2°C for 48 hours.

Lab Two

1. Remove the plate from the incubator, and, before adding the iodine, note the location and appearance of the growth.

2. Cover the growth and surrounding areas with Gram iodine. Immediately examine the areas surrounding the growth for clearing. (Usually the growth on the agar prevents contact between the starch and the iodine so no color reaction takes place at that point. Beginning students sometimes look at this lack of color change and judge it incorrectly as a positive result. Therefore, when examining the agar for clearing, look for a halo *around* the growth, not at the growth itself.)

3. Record your results in the chart provided on the Data Sheet.

References

Collins, C. H., Patricia M. Lyne, and J. M. Grange. 1995. Page 117 in *Collins and Lyne's Microbiological Methods*, 7th ed. Butterworth-Heinemann, Oxford, United Kingdom.

Lányi, B. 1987. Page 55 in *Methods in Microbiology*, Vol. 19, edited by R. R. Colwell and R. Grigorova. Academic Press, New York.

MacFaddin, Jean F. 2000. Page 412 in *Biochemical Tests for Identification of Medical Bacteria*, 2nd ed. Lippincott Williams & Wilkins, Philadelphia.

Smibert, Robert M., and Noel R. Krieg. 1994. Page 630 in *Methods for General and Molecular Bacteriology*, edited by Philipp Gerhardt, R. G. E. Murray, Willis A. Wood, and Noel R. Krieg. American Society for Microbiology, Washington, DC.

Zimbro, Mary Jo, and David A. Power, Eds. 2003. *Difco™ and BBL™ Manual—Manual of Microbiological Culture Media*. Becton Dickinson and Co., Sparks, MD.

EXERCISE 5-13 Urea Hydrolysis

✦ Theory

Urea is a product of decarboxylation of certain amino acids. It can be hydrolyzed to ammonia and carbon dioxide by bacteria containing the enzyme **urease**. Many enteric bacteria (and a few others) possess the ability to metabolize urea, but only members of *Proteus, Morganella,* and *Providencia* are considered rapid urease-positive organisms.

Urea Agar was formulated to differentiate rapid urease-positive bacteria from slower urease-positive and urease-negative bacteria. It contains urea, peptone, potassium phosphate, glucose, and phenol red. Peptone and glucose provide essential nutrients for a broad range of bacteria. Potassium phosphate is a mild buffer used to resist alkalinization of the medium from peptone metabolism. Phenol red, which is yellow or orange below pH 8.4 and red or pink above, is included as an indicator.

Urea hydrolysis (Figure 5-44) to ammonia by urease-positive organisms will overcome the buffer in the medium and change it from orange to pink. The agar must be examined daily during incubation. Rapid urease-positive organisms will turn the entire slant pink within 24 hours. Weak positives may take several days (Table 5-13). Urease-negative organisms either produce no color change in the medium or turn it yellow from acid products (Figure 5-45).

Urea broth differs from urea agar in two important ways. First, its only nutrient source is a trace (0.0001%) of yeast extract. Second, it contains buffers strong enough to inhibit alkalinization of the medium by all but the rapid urease-positive organisms mentioned above. Phenol red, which is yellow or orange below pH 8.4 and red or pink above, is included to expose any increase in pH. Pink color in the medium in less than 24 hours indicates a rapid urease-positive organism. Orange or yellow is negative (Figure 5-46 and Table 5-14).

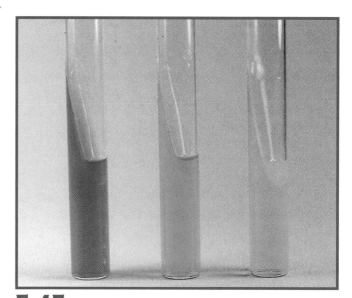

5-45 UREASE AGAR TEST RESULTS ✦ Urease agar tubes after a 24-hour incubation illustrate a rapid urea splitter (+) on the left and a urease-negative organism on the right. An uninoculated control is in the center.

$$\underset{\text{Urea}}{\overset{\displaystyle H_2N}{\underset{\displaystyle H_2N}{\Big\rangle}} C=O} \xrightarrow[\text{Urease}]{H_2O} 2NH_3 + CO_2$$

5-44 UREA HYDROLYSIS ✦ Urea hydrolysis produces ammonia, which raises the pH in the medium and turns the pH indicator pink.

TABLE **5-13** Urease Agar Test Results and Interpretations

TABLE OF RESULTS			
	Result		
24 hours	**24 hours to 6 days**	**Interpretation**	**Symbol**
All pink		Rapid urea hydrolysis; strong urease production	+
Partially pink	All pink or partially pink	Slow urea hydrolysis; weak urease production	w+
Orange or yellow	All pink or partially pink	Slow urea hydrolysis; weak urease production	w+
Orange or yellow	Orange or yellow	No urea hydrolysis; urease is absent	−

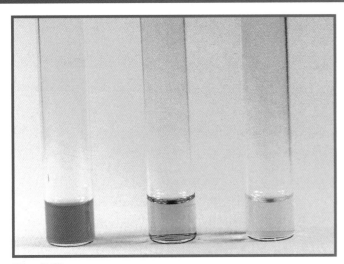

5-46 UREASE BROTH TEST RESULTS ✦ Urease broth tubes after a 24-hour incubation illustrate a urease-positive (+) organism on the left, an uninoculated control in the middle, and urease-negative (−) organism on the right.

✦ Application

This test is used to differentiate organisms based on their ability to hydrolyze urea with the enzyme urease. Urinary tract pathogens from the genus *Proteus* may be distinguished from other enteric bacteria by their rapid urease activity.

✦ In This Exercise

You will perform both forms of the test to determine urease activity of three organisms. Urease slants may require incubation up to 6 days, but broths should be read at 24 hours. If your labs are scheduled more than 24 hours apart, arrange to have someone place your broths in a refrigerator until you can examine them.

✦ Materials

Per Student Group

✦ four Urea Broths
✦ four Urea Agar slants

✦ fresh slant cultures of:
  ✦ *Escherichia coli*
  ✦ *Proteus vulgaris* (BSL-2)
  ✦ *Klebsiella pneumoniae* (BSL-2)

✦ Medium Recipes

Christensen's Urea Agar

✦ Peptone	1.0 g
✦ Dextrose (glucose)	1.0 g
✦ Sodium chloride	5.0 g
✦ Potassium phosphate, monobasic	2.0 g
✦ Urea	20.0 g
✦ Agar	15.0 g
✦ Phenol red	0.012 g
✦ Distilled or deionized water	1.0 L

pH 6.8 at 25°C

Rustigian and Stuart's Urea Broth

✦ Yeast extract	0.1 g
✦ Potassium phosphate, monobasic	9.1 g
✦ Potassium phosphate, dibasic	9.5 g
✦ Urea	20.0 g
✦ Phenol red	0.01 g
✦ Distilled or deionized water	1.0 L

pH 6.8 at 25°C

Procedure

Lab One

1 Obtain four tubes of each medium. Label three with the names of the organisms, your name, and the date. Label the fourth tube of each medium "control."

2 Inoculate three broths with heavy inocula from the test organisms. Do not inoculate the control.

3 Streak-inoculate three slants with the test organisms, covering the entire agar surface with a heavy inoculum. Do not stab the agar butt. Do not inoculate the control.

4 Incubate all tubes aerobically at 35 ± 2°C for 24 hours. (If your labs are scheduled more than 1 day

TABLE **5-14** Urease Broth Test Results and Interpretations

TABLE OF RESULTS		
Result	**Interpretation**	**Symbol**
Pink	Rapid urea hydrolysis; strong urease production	+
Orange or yellow	No urea hydrolysis; organism does not produce urease or cannot live in broth	−

apart, try to read these media at 24 hours or arrange to have someone else read them for you. Positive results can be removed and placed in the refrigerator for later examination.)

Lab Two

1 Remove all tubes from the incubator and examine them for color changes. Record all broth results and any rapid positive and slow positive agar test results in the charts provided on the Data Sheet. Discard all broth tubes and positive agar tubes in an appropriate autoclave container.

2 Return any negative agar tubes to the incubator. Inspect them daily for pink color formation for up to 6 days. (Negative tubes must not be recorded as negative until they have incubated the full 6 days.)

3 Enter your agar results daily in the chart provided. Refer to Tables 5-13 and 5-14 when interpreting your results.

References

Collins, C. H., Patricia M. Lyne, and J. M. Grange. 1995. Page 117 in *Collins and Lyne's Microbiological Methods,* 7th ed. Butterworth-Heinemann, Oxford, United Kingdom.

Delost, Maria Dannessa. 1997. Page 196 in *Introduction to Diagnostic Microbiology.* Mosby, St. Louis.

DIFCO Laboratories. 1984. Page 1040 in *DIFCO Manual,* 10th ed. DIFCO Laboratories, Detroit.

Forbes, Betty A., Daniel F. Sahm, and Alice S. Weissfeld. 2002. Page 283 in *Bailey & Scott's Diagnostic Microbiology,* 11th ed. Mosby, St. Louis.

Lányi, B. 1987. Page 24 in *Methods in Microbiology,* Vol. 19, edited by R. R. Colwell and R. Grigorova. Academic Press, New York.

MacFaddin, Jean F. 2000. Page 298 in *Biochemical Tests for Identification of Medical Bacteria,* 3rd ed. Lippincott Williams & Wilkins, Philadelphia.

Smibert, Robert M., and Noel R. Krieg. 1994. Page 630 in *Methods for General and Molecular Bacteriology,* edited by Philipp Gerhardt, R. G. E. Murray, Willis A. Wood, and Noel R. Krieg. American Society for Microbiology, Washington, DC.

Zimbro, Mary Jo and David A. Power, editors. 2003. Page 606 in *Difco™ and BBL™ Manual—Manual of Microbiological Culture Media.* Becton Dickinson and Co., Sparks, MD.

EXERCISE 5-14

Casein Hydrolysis Test

✦ Theory

Many bacteria require proteins as a source of amino acids and other components for synthetic processes. Some bacteria have the ability to produce and secrete enzymes into the environment that catalyze the breakdown of large proteins to smaller peptides or individual amino acids, thereby enabling their uptake across the membrane (Figure 5-47).

Casease is an enzyme that some bacteria produce to hydrolyze the milk protein **casein**, the molecule that gives milk its white color. When broken down into smaller fragments, the ordinarily white casein loses its opacity and becomes clear.

The presence of casease can be detected easily with the test medium Milk Agar (Figure 5-48). Milk Agar is an undefined medium containing pancreatic digest of casein, yeast extract, dextrose, and powdered milk. When plated Milk Agar is inoculated with a casease-positive organism, secreted casease will diffuse into the medium around the colonies and create a zone of clearing where the casein has been hydrolyzed. Casease-negative organisms do not secrete casease and, thus, do not produce clear zones around the growth.

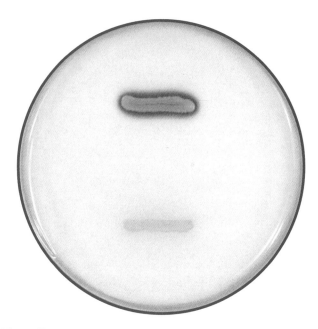

5-48 CASEIN HYDROLYSIS TEST RESULTS ✦ This Milk Agar plate was inoculated with a casease-positive organism above and a casease-negative organism below.

✦ Application

The casein hydrolysis test is used for the cultivation and differentiation of bacteria that produce the enzyme casease.

✦ In This Exercise

You will inoculate a single Milk Agar plate with a casease-positive and a casease-negative organism. Use Table 5-15 as a guide when making your interpretations.

✦ Materials

Per Student Group
✦ one Milk Agar plate

protein (n amino acids)

$$H_2N-CH(R)-C(O)-NH-CH(R)-C(O)----NH-CH(R)-C(O)-OH$$

casease   ← H₂O

amino acid   **protein (n-1 amino acids)**

$$H_2N-CH(R)-C(O)-OH + H_2N-CH(R)-C(O)----NH-CH(R)-C(O)-OH$$

5-47 CASEIN HYDROLYSIS ✦ Hydrolysis of any protein occurs by breaking peptide bonds (red arrow) between adjacent amino acids to produce short peptides or individual amino acids.

TABLE **5-15** Casease Test Results and Interpretations

TABLE OF RESULTS		
Result	**Interpretation**	**Symbol**
Clearing in agar	Casease is present	+
No clearing in agar	Casease is absent	−

✦ fresh cultures of:
- *Bacillus cereus*
- *Escherichia coli*

✦ Medium Recipe

Skim Milk Agar

◆ Pancreatic digest of casein	5.0 g
◆ Yeast extract	2.5 g
◆ Powdered nonfat milk	100.0 g
◆ Glucose	1.0 g
◆ Agar	15.0 g
◆ Distilled or deionized water	1.1 L

pH 6.9–7.1 at 25°C

Procedure

Lab One

1 Using a marking pen, divide the plate into three equal sectors. Be sure to mark on the bottom of the plate.

2 Label the plate with the names of the organisms, your name, and the date.

3 Spot-inoculate two sectors with the test organisms and leave the third sector uninoculated as a control.

4 Invert the plate and incubate it aerobically at 35 ± 2°C for 24 hours.

Lab Two

1 Examine the plates for clearing around the bacterial growth.

2 Record your results in the chart provided on the Data Sheet.

References

Atlas, Ronald M. 2004. Page 1563 in *Handbook of Microbiological Media,* 3rd ed. CRC Press LLC, Boca Raton, FL.

Holt, John G., Ed. 1994. *Bergey's Manual of Determinative Bacteriology,* 9th ed. Williams and Wilkins, Baltimore.

Smibert, Robert M., and Noel R. Krieg. 1994. Page 613 in *Methods for General and Molecular Bacteriology,* edited by Philipp Gerhardt, R. G. E. Murray, Willis A. Wood, and Noel R. Krieg. American Society for Microbiology, Washington, DC.

Zimbro, Mary Jo, and David A. Power, Eds. 2003. Page 364 in *Difco™ and BBL™ Manual—Manual of Microbiological Culture Media.* Becton Dickinson and Co., Sparks, MD.

5-15 Gelatin Hydrolysis Test

✦ Theory

Gelatin is a protein derived from collagen—a component of vertebrate connective tissue. **Gelatinases** comprise a family of extracellular enzymes produced and secreted by some microorganisms to hydrolyze gelatin. Subsequently, the cell can take up individual amino acids and use them for metabolic purposes. Bacterial hydrolysis of gelatin occurs in two sequential reactions, as shown in Figure 5-49.

The presence of gelatinases can be detected using Nutrient Gelatin, a simple test medium composed of gelatin, peptone, and beef extract. Nutrient Gelatin differs from most other solid media in that the solidifying agent (gelatin) is also the substrate for enzymatic activity. Consequently, when a tube of Nutrient Gelatin is stab-inoculated with a gelatinase-positive organism, secreted gelatinase (or gelatinases) will liquefy the medium. Gelatinase-negative organisms do not secrete the enzyme and do not liquefy the medium (Figure 5-50 and Figure 5-51). A 7-day incubation period is usually sufficient to see liquefaction of the medium. However, gelatinase activity is very slow in some organisms. All tubes still

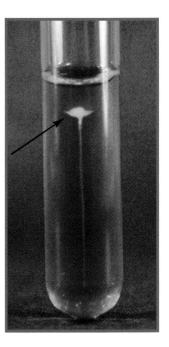

5-51 CRATERIFORM LIQUEFACTION ✦ This form of liquefaction also may be of diagnostic use because not all gelatinase-positive microbes liquefy the gelatin completely. Shown here is an organism liquefying the gelatin in the shape of a crater (arrow).

Gelatin $\xrightarrow[\text{Gelatinase}]{\text{H}_2\text{O}}$ Polypeptides $\xrightarrow[\text{Gelatinase}]{\text{H}_2\text{O}}$ Amino Acids

5-49 GELATIN HYDROLYSIS ✦ Gelatin is hydrolyzed by the gelatinase family of enzymes.

negative after 7 days should be incubated an additional 7 days.

A slight disadvantage of Nutrient Gelatin is that it melts at 28°C (82°F). Therefore, inoculated stabs are typically incubated at 25°C along with an uninoculated control to verify that any liquefaction is not temperature-related.

✦ Application

This test is used to determine the ability of a microbe to produce gelatinases. *Staphylococcus aureus*, which is gelatinase-positive, can be differentiated from *S. epidermidis*. *Serratia* and *Proteus* species are positive members of *Enterobacteriaceae*, whereas most others in the family are negative. *Bacillus anthracis*, *B. cereus*, and several other members of the genus are gelatinase-positive, as are *Clostridium tetani* and *C. perfringens*.

✦ In This Exercise

You will inoculate two Nutrient Gelatin tubes and incubate them for up to a week. Gelatin liquefaction is a positive result but is difficult to differentiate from gelatin that has melted as a result of temperature. Therefore, be sure to incubate a control tube along with the

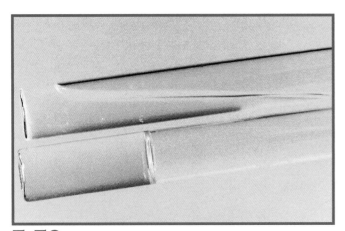

5-50 NUTRIENT GELATIN STABS ✦ A gelatinase-positive (+) organism is above, and a gelatinase–negative (−) organism is below.

TABLE **5-16** Gelatinase Test Results and Interpretations

TABLE OF RESULTS		
Result	**Interpretation**	**Symbol**
Gelatin is liquid (control is solid)	Gelatinase is present	+
Gelatin is solid	No gelatinase is present	−

others. If the control melts as a result of temperature, refrigerate all tubes until it resolidifies. Use Table 5-16 as a guide to interpretation.

✦ Materials

Per Student Group

✦ three Nutrient Gelatin stabs
✦ ice bath
✦ fresh cultures of:
  ◆ *Bacillus subtilis*
  ◆ *Escherichia coli*

✦ Medium Recipe

Nutrient Gelatin

◆ Beef Extract	3.0 g
◆ Peptone	5.0 g
◆ Gelatin	120.0 g
◆ Distilled or deionized water	1.0 L

pH 6.6–7.0 at 25°C

Procedure

Lab One

1 Obtain three Nutrient Gelatin stabs. Label each of two tubes with the name of the organism, your name, and the date. Label the third tube "control."

2 Stab-inoculate two tubes with heavy inocula of the test organisms. Do not inoculate the control.

3 Incubate all tubes at 25°C for up to 1 week.

Lab Two

1 Examine the control tube. If the gelatin is solid, the test can be read. If it is liquefied, place all tubes in the ice bath until the control has resolidified.

2 When the control has solidified, examine the inoculated media for gelatin liquefaction.

3 Record your results in the chart provided on the Data Sheet.

References

Collins, C. H., Patricia M. Lyne, and J. M. Grange. 1995. Page 112 in *Collins and Lyne's Microbiological Methods,* 7th ed. Butterworth-Heinemann, Oxford, United Kingdom.

Forbes, Betty A., Daniel F. Sahm, and Alice S. Weissfeld. 2002. Page 271 in *Bailey & Scott's Diagnostic Microbiology,* 11th ed. Mosby, St. Louis.

Lányi, B. 1987. Page 44 in *Methods in Microbiology,* Vol. 19, edited by R. R. Colwell and R. Grigorova. Academic Press, New York.

MacFaddin, Jean F. 2000. Page 128 in *Biochemical Tests for Identification of Medical Bacteria,* 3rd ed. Lippincott Williams & Wilkins, Philadelphia.

Smibert, Robert M., and Noel R. Krieg. 1994. Page 617 in *Methods for General and Molecular Bacteriology,* edited by Philipp Gerhardt, R. G. E. Murray, Willis A. Wood, and Noel R. Krieg. American Society for Microbiology, Washington, DC.

Zimbro, Mary Jo, and David A. Power, Eds. 2003. Page 408 in *Difco™ and BBL™ Manual—Manual of Microbiological Culture Media.* Becton Dickinson and Co., Sparks, MD.

EXERCISE 5-16 DNA Hydrolysis Test

✦ Theory

An enzyme that catalyzes the depolymerization of DNA into small fragments is called a **deoxyribonuclease**, or **DNase** (Figure 5-52). Ability to produce this enzyme can be determined by culturing and observing an organism on a DNase Test Agar plate.

DNase Test Agar contains an emulsion of DNA, peptides as a nutrient source, and methyl green dye. The dye and polymerized DNA form a complex that gives the agar a blue-green color. Bacterial colonies that secrete DNase will hydrolyze DNA in the medium into smaller fragments unbound from the methyl green dye. This results in clearing around the growth (Figure 5-53).

✦ Application

DNase Test Agar is used to distinguish *Serratia* species from *Enterobacter* species, *Moraxella catarrhalis* from *Neisseria* species, and *Staphylococcus aureus* from other *Staphylococcus* species.

✦ In This Exercise

You will inoculate one DNase plate with two of the organisms the test was designed to differentiate. For best results, limit incubation time to 24 hours. Depending on the age of the plates and the length of incubation, the clearing around a DNase-positive organism may be

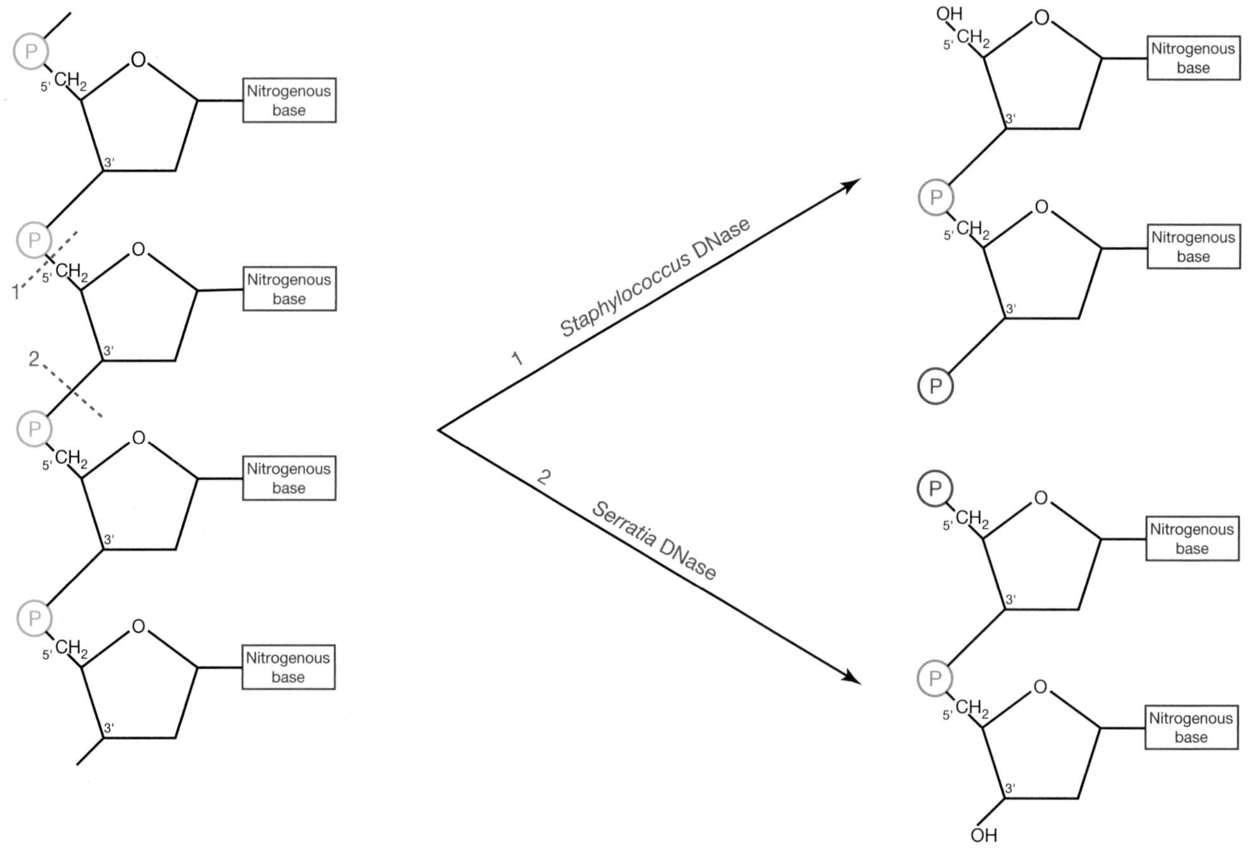

5-52 TWO PATTERNS OF DNA HYDROLYSIS ✦ DNase from *Staphylococcus* hydrolyzes DNA at the bond between the 5'-carbon and the phosphate (illustrated by line 1), thereby producing fragments with a free 3'-phosphate (shown in red on the upper fragment). Most fragments are one or two nucleotides long. A dinucleotide is shown here. *Serratia* DNase cleaves the bond between the phosphate and the 3'-carbon (illustrated by line 2) and produces fragments with free 5'-phosphates (shown in red on the lower fragment). Most fragments are two to four nucleotides in length. A dinucleotide is shown here.

If no clear zones ~~~~
(Figure 5-55). Ta~~~~
interpretations.

✦ Applicati~~~~

The lipase test is ~~~~
bacteria, especial~~~~
of other lipid sub~~~~
soybean oil, are ~~~~
among members ~~~~
Staphylococcus, ~~~~
demonstrate lipo~~~~

✦ In This E~~~~

You will inoculat~~~~
ganisms, one of ~~~~
the growth.

✦ Material~~~~

Per Student Gr~~~~

✦ one Tributyri~~~~

✦ fresh cultures ~~~~
  ✦ *Bacillus sub~~~~*
  ✦ *Escherichia~~~~*

5-55 LIPID ~~~~
plate was inocula~~~~
and a lipase-nega~~~~

5-53 DNA HYDROLYSIS TEST RESULTS ✦ DNase agar plate is inoculated with a DNase-positive (+) organism below and a DNase-negative (−) organism above.

subtle and difficult to see. When reading the plate, it might be helpful to look through it while holding it several inches above a white piece of paper. Use Table 5-17 to assist in interpretation.

✦ Materials

Per Student Group

✦ one DNase Test Agar plate

✦ fresh cultures of:
  ✦ *Staphylococcus aureus* (BSL-2)
  ✦ *Staphylococcus epidermidis*

✦ Medium Recipe

DNase Test Agar with Methyl Green

✦ Tryptose	20.0 g
✦ Deoxyribonucleic acid	2.0 g
✦ Sodium chloride	5.0 g
✦ Agar	15.0 g
✦ Methyl green	0.05 g
✦ Distilled or deionized water	1.0 L

pH 7.1–7.4 at 25°C

TABLE **5-17** DNA Hydrolysis Test Results and Interpretations

TABLE OF RESULTS		
Result	**Interpretation**	**Symbol**
Clearing in agar around growth	DNAse is present	+
No clearing in agar around growth	DNAse is absent	−

Procedure

Lab One

1 Using a marking pen, divide the DNase Test Agar plate into three equal sectors. Be sure to mark the bottom of the plate.

2 Label the plate with the names of the organisms, your name, and the date.

3 Spot-inoculate two sectors with the test organisms and leave the third sector as a control.

4 Invert the plate and incubate it aerobically at $35 \pm 2°C$ for 24 hours. (If your labs are scheduled more than 24 hours apart, arrange to have someone place your plates in a refrigerator until you can examine them.)

Lab Two

1 Examine the plates for clearing around the bacterial growth.

2 Record your results in the table provided on the Data Sheet.

References

Collins, C. H., Patricia M. Lyne, and J. M. Grange. 1995. Page 114 in *Collins and Lyne's Microbiological Methods,* 7th ed. Butterworth-Heinemann, Oxford, United Kingdom.

Delost, Maria Dannessa. 1997. Page 111 in *Introduction to Diagnostic Microbiology.* Mosby, St. Louis.

Forbes, Betty A., Daniel F. Sahm, and Alice S. Weissfeld. 2002. Page 268 in *Bailey & Scott's Diagnostic Microbiology,* 11th ed. Mosby, St. Louis.

Lányi, B. 1987. Page 33 in *Methods in Microbiology,* Vol. 19, edited by R. R. Colwell and R. Grigorova. Academic Press, New York.

MacFaddin, Jean F. 2000. Page 136 in *Biochemical Tests for Identification of Medical Bacteria,* 3rd ed. Lippincott Williams & Wilkins, Philadelphia.

Zimbro, Mary Jo, and David A. Power, Eds. 2003. Page 170 in *Difco™ and BBL™ Manual—Manual of Microbiological Culture Media.* Becton Dickinson and Co., Sparks, MD.

EXERCISE
5-17

EXERCISE 5-18 ONPG Test

✦ Theory

Lipid is the word fats. The enzyme Bacteria can be d produce and secr fats can be used the most commo because it is the s fats and oils.

Simple fats a **glycerols** (Figure glycerol and thre many biochemic cell, so a lipase i

✦ Theory

For bacteria to ferment lactose, they must possess two enzymes: β-**galactoside permease**, a membrane-bound transport protein; and β-**galactosidase**, an intracellular enzyme that hydrolyzes the disaccharide into β-glucose and β-galactose (Figure 5-56).

Bacteria possessing both enzymes are active β-lactose fermenters. Bacteria that cannot produce β-galactosidase, even if their membranes contain β-galactoside permease, cannot ferment β-lactose. Bacteria that possess β-galactosidase but no β-galactoside permease may mutate and, over a period of days or weeks, begin to produce the permease. Distinguishing these **late lactose fermenters**

from nonfermenters is made possible by *o*-nitrophenyl-β-D-galactopyranoside (ONPG).

When ONPG is made available to bacteria, it freely enters the cells without the aid of a permease. Because of its similarity to β-lactose, it then can become the substrate for any β-galactosidase present. In the reaction that occurs, ONPG is hydrolyzed to β-galactose and *o*-nitrophenol (ONP), which is yellow (Figures 5-57 and 5-58, and Table 5-19).

It should be noted that β-galactosidase is an inducible enzyme; it is produced in response to the presence of an appropriate substrate. Therefore, organisms to be tested with ONPG are typically prepared by growing them overnight in a lactose-rich medium, such as Kligler Iron

5-56 HYDROLYSIS OF LACTOSE BY β-GALACTOSIDASE ✦ β-galactosidase is an inducible enzyme in many bacteria; that is, it is produced only in the presence of the inducer lactose. Its activity is to hydrolyze the disaccharide lactose into its component monosaccharides.

Glycolysis

5-54 LIPID M
three fatty acid ch
(glycerol) and the

5-57 CONVERSION OF ONPG TO β-GALACTOSE AND *o*-NITROPHENOL BY β-GALACTOSIDASE ✦ ONPG is similar enough to lactose that β-galactosidase will catalyze its hydrolysis into β-galactose and *o*-nitrophenol. The latter compound is yellow and is indicative of a positive ONPG test.

5-58 THE ONPG TEST ✦ ONPG-positive (+) is on the left, and ONPG-negative (−) is on the right.

TABLE 5-19 ONPG Test Results and Interpretations

TABLE OF RESULTS		
Result	Interpretation	Symbol
Yellow color formation	Organism produces β-galactosidase	+
No color change	Organism does not produce β-galactosidase	−

Agar or Triple Sugar Iron Agar, to ensure that they will be actively producing β-galactosidase, given that capability.

✦ Application

The ONPG test is used to differentiate late lactose fermenters from lactose nonfermenters in the family *Enterobacteriaceae*.

✦ In This Exercise

You will be performing the ONPG test using Key Scientific Products' K1490 ONPG WEETABS. If your instructor uses a different version of the test, he/she will give you instructions to follow based on the manufacturer's recommendations. The organisms chosen for this exercise are two of the enterics that this test was designed to differentiate.

✦ Materials

Per Student Group

✦ Key Scientific Products' K1490 ONPG WEETABS
✦ sterile distilled or deionized water

✦ sterile transfer pipettes
✦ fresh Kligler Iron Agar or Triple Sugar Iron Agar cultures of:
  ✦ *Escherichia coli*
  ✦ *Proteus vulgaris* (BSL-2)

Procedure

1 Dispense 0.5 mL volumes of sterile water into three WEETAB tubes, and allow the tablets to dissolve.

2 Transfer a heavy inoculum of the test organisms to two tubes. Do not inoculate the third tube.

3 Incubate all tubes at 35 ± 2°C for 2 hours.

4 Remove all tubes from the incubator and examine them for color changes. Record your results on the Data Sheet.

References

DIFCO Laboratories. 1984. Page 294 in *DIFCO Manual*, 10th ed., DIFCO Laboratories, Detroit.

Forbes, Betty A., Sahm, Daniel F., and Alice S. Weissfeld. 1998. Page 442 in *Bailey & Scott's Diagnostic Microbiology*, 10th ed. Mosby, St. Louis.

Key Scientific Products' K1490 ONPG WEETABS Package Insert. Key Scientific Products. Round Rock, TX.

MacFaddin, Jean F. 2000. Page 160 in *Biochemical Tests for Identification of Medical Bacteria*, 2nd ed. Lippincott Williams & Wilkins, Philadelphia.

EXERCISE 5-19

PYR Test

✦ Theory

Group A streptococci (*S. pyogenes*) and enterococci produce the enzyme L-pyrrolidonyl arylamidase. This enzyme hydrolyzes the amide pyroglutamyl-β-naphthylamide to produce L-pyrrolidone and β-naphthylamine, all of which are colorless. β-naphthylamine will react with *p*-dimethylaminocinnamaldehyde and form a red precipitate (Figure 5-59).

PYR may be performed as an 18-hour agar test, a four-hour broth test or, as used in this example, a rapid disc test. In each case the medium (or disc) contains pyroglutamyl-β-naphthylamide to which is added a heavy inoculum of the test organism. After the appropriate incubation or waiting period, a 0.01% *p*-dimethylaminocinnamaldehyde solution is added. Formation of a deep red color within a few minutes is interpreted as PYR-positive. Yellow or orange is PYR-negative (Figure 5-60).

✦ Application

The PYR test is designed for presumptive identification of group A streptococci (*S. pyogenes*) and enterococci by determining the presence of the enzyme L-pyrrolidonyl arylamidase.

✦ In This Exercise

You will inoculate two discs with test organisms to determine which one contains the enzyme L-pyrrolidonyl arylamidase.

✦ Materials

Per Student Group

✦ one empty Petri dish
✦ PYR discs
✦ fresh slant cultures of:
  ✦ *Enterococcus faecalis* (BSL-2)
  ✦ *Staphylococcus epidermidis*

📖 Procedure

1 Obtain a Petri dish, the PYR discs, and two sterile pipettes. (A clean microscope slide can be used in place of the Petri dish.)

2 Open the Petri dish and mark two spots where you will place the PYR discs. Label each spot with the name of its respective organism.

3 Place two PYR discs inside on the appropriate spots.

4 With a loop or other appropriate inoculating device, inoculate each disc with a large amount of its respective organism.

5 Add a drop of the *p*-dimethylaminocinnamaldehyde solution to each disc.

6 If the disc turns red within 5 minutes, it is positive. Orange or no color change are negative results.

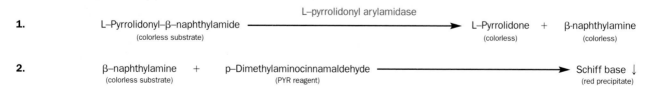

1.   L–Pyrrolidonyl–β–naphthylamide —— L-pyrrolidonyl arylamidase ——→ L-Pyrrolidone + β-naphthylamine
      (colorless substrate)   (colorless)   (colorless)

2.   β–naphthylamine + p–Dimethylaminocinnamaldehyde ————————→ Schiff base ↓
      (colorless substrate)   (PYR reagent)   (red precipitate)

5-59 PYR CHEMISTRY ✦ Reaction 1 shows the hydrolytic splitting of the substrate L–pyrrolidonyl–β–naphthylamide included in the medium. In this form of the test the filter-paper discs were saturated with substrate. Both products of the reaction are colorless. Reaction 2 illustrates the reaction of the PYR reagent and β–naphthylamine. In the reaction, the amino group (H₂N) of β–naphthylamine reacts with the aldehyde group (–CH=O) of the reagent and produces a Schiff base (–CH=N–), which is red.

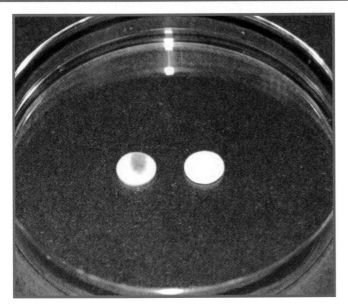

References

Claus, G. William. 1989. Chapter 14 in *Understanding Microbes—A Laboratory Textbook for Microbiology*. W.H. Freeman and Company, New York, NY.

Collins, C. H., Patricia M. Lyne and J. M. Grange. 1995. Chapter 6 in *Collins and Lyne's Microbiological Methods, 7th Ed.* Butterworth-Heineman, Oxford.

Forbes, Betty A., Daniel F. Sahm, and Alice S. Weissfeld. 2002. Chapter 10 in *Bailey & Scott's Diagnostic Microbiology, 11th Ed.* Mosby-Yearbook, St. Louis, MO.

Koneman, Elmer W., Stephen D. Allen, William M. Janda, Paul C. Schreckenberger and Washington C. Winn, Jr. 1997. Chapter 2 in *Color Atlas and Textbook of Diagnostic Microbiology, 5th Ed.* J. B. Lippincott Company, Philadelphia, PA.

5-60 PYR DISK TEST ✦ The disk on the left was inoculated with a PYR-positive organism; the disk on the right was inoculated with a PYR–negative organism.

TABLE **5-20** PYR Test Results and Interpretations

TABLE OF RESULTS		
Result	**Interpretation**	**Symbol**
Red color formation	Organism produces L-pyrrolidonyl arylamidase	+
Orange color or no color change	Organism des not produce L-pyrrolidonyl arylamidase	−

Combination Differential Media

Combination differential media combine components of several compatible tests into one medium, thus saving critical diagnostic time and money. Typically, they include core tests to differentiate members of specific bacterial groups. SIM medium, for example, tests for sulfur reduction, indole production, and motility—important characteristics of members of *Enterobacteriaceae*.

As with most biochemical tests, combination media can be used as a follow-up to selective media. Selective media promote isolation of the organisms of interest and help determine the sequence of tests to follow. MacConkey Agar (Ex. 4-5), for example, may be streaked initially to encourage and isolate Gram-negative organisms, followed by Kligler Iron Agar (KIA). KIA tests for glucose and lactose fermentation and for sulfur reduction. ✦

EXERCISE 5-20

SIM Medium

✦ Theory

SIM medium is used for determination of three bacterial activities: sulfur reduction, indole production from tryptophan, and motility. The semisolid medium includes casein and animal tissue as sources of amino acids, an iron-containing compound, and sulfur in the form of sodium thiosulfate.

Sulfur reduction to H_2S can be accomplished by bacteria in two different ways, depending on the enzymes present.

1. The enzyme **cysteine desulfurase** catalyzes the putrefaction of the amino acid cysteine to pyruvate (Figure 5-61).

2. The enzyme **thiosulfate reductase** catalyzes the reduction of sulfur (in the form of sulfate) at the end of the anaerobic respiratory electron transport chain (Figure 5-62).

Both systems produce hydrogen sulfide (H_2S) gas. When either reaction occurs in SIM medium, the H_2S that is produced combines with iron, in the form of ferrous ammonium sulfate, to form ferric sulfide (FeS), a black precipitate (Figure 5-63). Any blackening of the medium is an indication of sulfur reduction and a positive test. No blackening of the medium indicates no sulfur reduction and a negative reaction (Figure 5-64).

Indole production in the medium is made possible by the presence of tryptophan (contained in casein and animal protein). Bacteria possessing the enzyme **tryptophanase** can hydrolyze tryptophan to pyruvate, ammonia (by deamination), and indole (Figure 5-65).

The hydrolysis of tryptophan in SIM medium can be detected by the addition of Kovacs' reagent after a period of incubation. Kovacs' reagent contains **dimethylaminobenzaldehyde** (DMABA) and HCl dissolved in amyl alcohol. When a few drops of Kovacs' reagent are

5-61 PUTREFACTION OF CYSTEINE ✦ Putrefaction involving cysteine desulfurase produces H_2S. This reaction is a mechanism for getting energy out of the amino acid cysteine.

$$3S_2O_3^= + 4H^+ + 4e^- \xrightarrow{\text{Thiosulfate reductase}} 2SO_3^= + 2H_2S \ (g)$$

5-62 REDUCTION OF THIOSULFATE ✦ Anaerobic respiration with thiosulfate as the final electron acceptor also produces H_2S.

$$H_2S + FeSO_4 \longrightarrow H_2SO_4 + FeS \ (g)$$

5-63 INDICATOR REACTION ✦ Hydrogen sulfide, a colorless gas, can be detected when it reacts with ferrous sulfate in the medium to produce the black precipitate ferric sulfide.

5-64 SULFUR REDUCTION IN SIM MEDIUM ✦ The organism on the left is H_2S-negative. The organism on the right is H_2S-positive.

added to the tube, DMABA reacts with any indole present and produces a quinoidal compound that turns the reagent layer red (Figure 5-66 and Figure 5-67). The formation of red color in the reagent layer indicates a positive reaction and the presence of tryptophanase. No red color is indole-negative.

Determination of motility in SIM medium is made possible by the reduced agar concentration and the method of inoculation. The medium is inoculated with a single stab from an inoculating needle. Motile organisms are able to move about in the semisolid medium and can be detected by the radiating growth pattern extending outward in all directions from the central stab line. Growth that radiates in *all* directions and appears slightly fuzzy is an indication of motility (Figure 5-68). This should not be confused with the (seemingly) spreading growth produced by lateral movement of the inoculating needle when stabbing.

✦ Application

SIM medium is used to identify bacteria that are capable of producing indole, using the enzyme tryptophanase. The Indole Test is one component of the IMViC battery of tests (Indole, Methyl red, Voges-Proskauer, and Citrate) used to differentiate the *Enterobacteriaceae*. SIM medium also is used to differentiate sulfur-reducing members of

Tryptophan $+ H_2O \xrightarrow{\text{tryptophanase}}$ Indole $+ NH_3 +$ Pyruvate $\rightarrow$ Fermentation / Respiration

5-65 TRYPTOPHAN CATABOLISM IN INDOLE-POSITIVE ORGANISMS ✦ This reaction is a mechanism for getting energy out of the amino acid tryptophan.

2 Indole + p-Dimethylaminobenzaldehyde $N(CH_3)_2$ + HCl + Amyl Alcohol $\longrightarrow$ Rosindole dye (cherry red)

Kovacs' Reagent

5-66 INDOLE REACTION WITH KOVACS' REAGENT ✦ If present, indole reacts with the DMABA in Kovacs' reagent to produce a red color.

5-67 INDOLE TEST RESULTS ✦ These SIM tubes were inoculated with an indole-negative (−) organism on the left and an indole-positive (+) organism on the right.

5-68 MOTILITY IN SIM ✦ These SIM tubes were inoculated with a motile organism on the left and a nonmotile organism on the right.

Enterobacteriaceae, especially members of the genera *Salmonella, Francisella,* and *Proteus* from the negative *Morganella morganii* and *Providencia rettgeri*. In addition to the first two functions of SIM, motility is an important differential characteristic of *Enterobacteriaceae*.

✦ In This Exercise

You will stab-inoculate SIM medium with three organisms demonstrating a variety of results. When reading your test results, make the motility and H_2S determinations before adding the indole reagent. Use Tables 5-21, 5-22 and 5-23, and Figures 5-64, 5-67, and 5-68 as guides when making your interpretations.

✦ Materials

Per Student Group

✦ four SIM tubes

✦ Kovac's reagent

✦ fresh cultures of:
  ✦ *Escherichia coli*
  ✦ *Salmonella typhimurium* (BSL-2)
  ✦ *Klebsiella pneumoniae* (BSL-2)

✦ Medium and Reagent Recipes

SIM (Sulfur-Indole-Motility) Medium

✦ Pancreatic digest of casein	20.0 g
✦ Peptic digest of animal tissue	6.1 g
✦ Ferrous ammonium sulfate	0.2 g
✦ Sodium thiosulfate	0.2 g
✦ Agar	3.5 g
✦ Distilled or deionized water	1.0 L
pH 7.1–7.4 at 25°C	

TABLE **5-22** Indole Production Results and Interpretations

TABLE OF RESULTS		
Result	**Interpretation**	**Symbol**
Red in the alcohol layer of Kovac's reagent	Tryptophan is broken down into indole and pyruvate	+
Reagent color is unchanged	Tryptophan is not broken down into indole and pyruvate	−

TABLE **5-21** Sulfur Reduction Results and Interpretations

TABLE OF RESULTS		
Result	**Interpretation**	**Symbol**
Black in the medium	Sulfur reduction (H_2S production)	+
No black in the medium	Sulfur is not reduced	−

TABLE **5-23** Motility Results and Interpretations

TABLE OF RESULTS		
Result	**Interpretation**	**Symbol**
Growth radiating outward from stab line	Motility	+
No radiating growth	Nonmotile	−

Kovacs' Reagent

- ◆ Amyl alcohol   75.0 mL
- ◆ Hydrochloric acid, concentrated   25.0 mL
- ◆ *p*-dimethylaminobenzaldehyde   5.0 g

 Procedure

Lab One

1. Obtain four SIM tubes. Label three with the names of the organisms, your name, and the date. Label one tube "control."

2. Stab-inoculate three tubes with the test organisms. Insert the needle into the agar to within 1 cm of the bottom of the tube. Be careful to remove the needle along the original stab line. Do not inoculate the control.

3. Incubate all tubes aerobically at $35 \pm 2°C$ for 24 to 48 hours.

Lab Two

1. Examine the tubes for spreading from the stab line *and* formation of black precipitate in the medium. Record any H_2S production and/or motility in the chart provided on the Data Sheet.

2. Add Kovacs' reagent to each tube (to a depth of 2–3 mm). After several minutes, observe for the formation of red color in the reagent layer.

3. Record your results in the chart on the Data Sheet.

References

Delost, Maria Dannessa. 1997. Page 186 in *Introduction to Diagnostic Microbiology*. Mosby, St. Louis.

MacFaddin, Jean F. 1980. Page 162 in *Biochemical Tests for Identification of Medical Bacteria*, 2nd ed. Williams & Wilkins, Baltimore.

Zimbro, Mary Jo, and David A. Power, Eds. 2003. Page 490 in *Difco™ and BBL™ Manual—Manual of Microbiological Culture Media*. Becton Dickinson and Co., Sparks, MD.

Triple Sugar Iron Agar / Kligler Iron Agar[1]

✦ Theory

Triple Sugar Iron Agar (TSIA) is a rich medium designed to differentiate bacteria on the basis of glucose fermentation, lactose fermentation, sucrose fermentation, and sulfur reduction. In addition to the three carbohydrates, it includes animal proteins as sources of carbon and nitrogen, and both ferrous sulfate and sodium thiosulfate as sources of oxidized sulfur. Phenol red is the pH indicator, and the iron in the ferrous sulfate is the hydrogen sulfide indicator.

The medium is prepared as a shallow agar slant with a deep butt, thereby providing both aerobic and anaerobic growth environments. It is inoculated by a stab in the agar butt followed by a fishtail streak of the slant. The incubation period is 18 to 24 hours for carbohydrate fermentation and up to 48 hours for hydrogen sulfide reactions. Many reactions in various combinations are possible (Figure 5-69 and Table 5-24).

When TSIA is inoculated with a glucose-only fermenter, acid products lower the pH and turn the entire medium yellow within a few hours. Because glucose is in short supply (0.1%), it will be exhausted within about 12 hours. As the glucose diminishes, the organisms located in the aerobic region (slant) will begin to break down available amino acids, producing NH_3 and raising the pH. This process, which takes 18 to 24 hours to complete, is called a **reversion** and only occurs in the slant because of the anaerobic conditions in the butt. Thus, a TSIA with a red slant and yellow butt after a 24-hour incubation period indicates that the organism ferments glucose but not lactose.

Organisms that are able to ferment glucose *and* lactose *and/or* sucrose also turn the medium yellow throughout. However, because the lactose and sucrose concentrations are ten times higher than that of glucose, resulting in greater acid production, both slant and butt will remain yellow after 24 hours. Therefore, a TSIA

with a yellow slant and butt at 24 hours indicates that the organism ferments glucose and one or both of the other sugars. Gas produced by fermentation of any of the carbohydrates will appear as fissures in the medium or will lift the agar off the bottom of the tube.

Hydrogen sulfide (H_2S) may be produced by the reduction of thiosulfate in the medium or by the breakdown of cysteine in the peptone. Ferrous sulfate reacts with the H_2S to form a black precipitate, usually seen in the butt. Acid conditions must exist for thiosulfate reduction; therefore, black precipitate in the medium is an indication of sulfur reduction *and* fermentation. If the black precipitate obscures the color of the butt, the color of the slant determines which carbohydrates have been fermented (*i.e.*, red slant = glucose fermentation, yellow slant = glucose and lactose fermentation).

An organism that does not ferment any of the carbohydrates but utilizes peptone and amino acids will alkalinize the medium and turn it red. If the organism can use the peptone aerobically and anaerobically, both the slant and butt will appear red. An obligate aerobe will turn only the slant red. (**Note:** The difference between

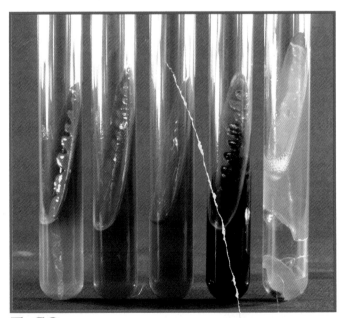

5-69 **KLIGLER IRON AGAR SLANTS** ✦ These KIA slants from left to right illustrate: alkaline slant/acid butt (K/A); alkaline slant/no change in the butt (K/NC); uninoculated control; alkaline slant/acid butt, hydrogen sulfide present (K/A, H₂S); acid slant/acid butt, gas (A/A, G); and. Refer to Table 5-24 for interpretations.

---

[1] This exercise is written for either Triple Sugar Iron Agar (TSIA) or Kligler Iron Agar (KIA). In form and function the two media are virtually identical except that KIA does not include sucrose. The results of both media look exactly the same. In results indicating fermentation of more than one carbohydrate, TSIA is reported as *glucose and lactose and/or sucrose fermentation*, KIA (because it contains no sucrose) is reported as *glucose and lactose fermentation*. To avoid the awkwardness of repeatedly including both names, we refer only to TSIA. If you are using KIA, simply substitute the name where applicable.

a red butt and a butt unchanged by the organism may be subtle; therefore, comparison with an uninoculated control is always recommended.)

Not surprisingly, timing is critical when reading results. An early reading could reveal yellow throughout the medium, leading you to believe that the organism ferments more than one sugar when it simply has not yet exhausted the glucose. A reading after all of the sugars have been depleted could reveal a yellow butt and red slant leading you to falsely believe that the organism is a glucose-only fermenter. The timing for interpreting sulphur reduction is not as critical, so tubes that have been interpreted for carbohydrate fermentation can be re-incubated for 24 hours before final H_2S determination is made. Refer to Table 5-24 for information on the correct symbols and method of reporting the various reactions.

✦ Application

TSIA and KIA are primarily used to differentiate members of *Enterobacteriaceae* and to distinguish them from other Gram-negative rods such as *Pseudomonas aeruginosa*.

✦ In This Exercise

Today, you will inoculate four TSIA or KIA slants. Use a large inoculum and try not to introduce excessive air when stabbing the agar butt. Also be sure to remove all tubes from the incubator after no more than 24 hours to take fermentation readings.

✦ Materials

Per Student Group
✦ five TSIA or KIA slants
✦ recommended organisms (grown on solid media):
  ◆ *Pseudomonas aeruginosa* (BSL-2)
  ◆ *Escherichia coli*
  ◆ *Morganella morganii* (BSL-2)
  ◆ *Salmonella typhimurium* (BSL-2)

✦ Media Recipes

Triple Sugar Iron Agar
  ◆ Pancreatic Digest of Casein   10.0 g
  ◆ Peptic Digest of Animal Tissue   10.0 g
  ◆ Dextrose (glucose)   1.0 g

TABLE **5-24** TSIA/KIA Results and Interpretations

TABLE OF RESULTS		
Result	**Interpretation**	**Symbol**
Yellow slant/yellow butt—KIA	Glucose and lactose fermentation with acid accumulation in slant and butt.	A/A
Yellow slant/yellow butt—TSIA	Glucose and lactose and/or sucrose fermentation with acid accumulation in slant and butt.	A/A
Red slant/yellow butt	Glucose fermentation with acid production. Proteins catabolized aerobically (in the slant) with alkaline products (reversion).	K/A
Red slant/red butt	No fermentation. Peptone catabolized aerobically and anaerobically with alkaline products. Not from *Enterobacteriaceae*.	K/K
Red slant/no change in the butt	No fermentation. Peptone catabolized aerobically with alkaline products. Not from *Enterobacteriaceae*.	K/NC
No change in slant / no change in butt	Organism is growing slowly or not at all. Not from *Enterobacteriaceae*.	NC/NC
Black precipitate in the agar	Sulfur reduction. (An acid condition, from fermentation of glucose or lactose, exists in the butt even if the yellow color is obscured by the black precipitate.)	H_2S
Cracks in or lifting of agar	Gas production.	

- ◆ Lactose   10.0 g
- ◆ Sucrose   1.0 g
- ◆ Ferrous Ammonium Sulfate   0.5 g
- ◆ Sodium chloride   5.0 g
- ◆ Sodium thiosulfate   0.2 g
- ◆ Agar   13.0 g
- ◆ Phenol red   0.025 g
- ◆ Distilled or deionized water   1.0 L
     pH 7.1–7.5 at 25°C

Kligler Iron Agar

- ◆ Pancreatic Digest of Casein   10.0 g
- ◆ Peptic Digest of Animal Tissue   10.0 g
- ◆ Lactose   10.0 g
- ◆ Dextrose (glucose)   1.0 g
- ◆ Ferric Ammonium Citrate   0.5 g
- ◆ Sodium chloride   5.0 g
- ◆ Sodium thiosulfate   0.5 g
- ◆ Agar   15.0 g
- ◆ Phenol red   0.025 g
- ◆ Distilled or deionized water   1.0 L
     pH 7.2 – 7.6 at 25°C

 Procedure

Lab One

1 Obtain five slants. Label four of the slants with the names of the organisms, your name, and the date. Label the fifth slant "control."

2 Inoculate four slants with the test organisms. Using a heavy inoculum, stab the agar butt and then streak the slant. Do not inoculate the control.

3 Incubate all slants aerobically at $35 \pm 2°C$ for 24 hours.

Lab Two

Examine the tubes for characteristic color changes and gas production. Use Table 5-24 as a guide while recording your results on the Data Sheet. The proper format for recording results is: slant reaction/butt reaction, gas production, hydrogen sulfide production. For example, a tube showing yellow slant, yellow butt, fissures in or lifting of the agar, and black precipitate would be recorded as: A/A, G, H_2S.

References

Delost, Maria Dannessa. 1997. Pages 184–185 in *Introduction to Diagnostic Microbiology*. Mosby, St. Louis.

Forbes, Betty A., Daniel F. Sahm, and Alice S. Weissfeld. 2002. Page 282 in *Bailey & Scott's Diagnostic Microbiology*, 11th ed. Mosby, St. Louis.

MacFaddin, Jean F. 2000. Page 239 in *Biochemical Tests for Identification of Medical Bacteria*, 3rd ed. Lippincott Williams & Wilkins, Philadelphia.

Zimbro, Mary Jo, and David A. Power, editors. 2003. Page 283 in *Difco™ and BBL™ Manual—Manual of Microbiological Culture Media*. Becton Dickinson and Co., Sparks, MD.

5-22 Lysine Iron Agar

✦ Theory

Lysine Iron Agar (LIA) is a combination medium that detects bacterial ability to decarboxylate or deaminate lysine and to reduce sulfur. It contains peptone and yeast extract to support growth, the amino acid lysine for deamination and decarboxylation reactions, and sodium thiosulfate—a source of reducible sulfur. A small amount of glucose (0.1%) is included as a fermentable carbohydrate. Ferric ammonium citrate is included as a sulfur reduction indicator and bromcresol purple is the pH indicator. Bromcresol purple is purple at pH 6.8 and yellow at or below pH 5.2.

LIA is prepared as a slant with a deep butt. This results in an aerobic zone in the slant and an anaerobic zone in the butt. After it is inoculated with two stabs of the butt and a fishtail streak of the slant, the tube is tightly capped and incubated for 18 to 24 hours.

If the medium has been inoculated with a lysine decarboxylase-positive organism, acid production from glucose fermentation will induce production of decarboxylase enzymes. The acidic pH will turn the medium yellow, but subsequent decarboxylation of the lysine will alkalinize the agar and return it to purple. Purple color throughout indicates lysine decarboxylation. Purple color in the slant with a yellow (acidic) butt indicates glucose fermentation, but no lysine decarboxylation took place; degradation of peptone alkalinized the slant.

If the organism produces lysine deaminase, the resulting deamination reaction will produce compounds that react with the ferric ammonium citrate and produce a red color. Deamination reactions require the presence of oxygen. Therefore, any evidence of deamination will be seen only in the slant. A red slant with yellow (acidic) butt indicates lysine deamination.

Hydrogen sulfide (H_2S) is produced in Lysine Iron Agar by the anaerobic reduction of thiosulfate. Ferric ions in the medium react with the H_2S to form a black precipitate in the butt. Refer to Figure 5-70 and Table 5-25 for the various reactions and symbols used to record them.

✦ Application

LIA is used to differentiate enterics based on their ability to decarboxylate or deaminate lysine and produce hydrogen sulfide (H_2S). LIA also is used in combination with Triple Sugar Iron Agar to identify members of *Salmonella* and *Shigella*.

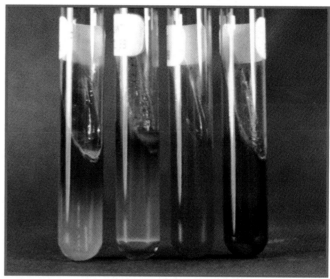

5-70 LIA RESULTS ✦ Lysine Iron Agar tubes illustrating from left to right: (R/A); (K/A, H_2S, [note the small amount of black precipitate near the middle and the gas production from glucose fermentation at the base]); uninoculated control; and (K/K [obscured by the black precipitate], H_2S).

TABLE **5-25** LIA Results and Interpretations

TABLE	OF	RESULTS
Result	**Interpretation**	**Symbol**
Purple slant/purple butt	Lysine deaminase negative; Lysine decarboxylase positive	K/K
Purple slant/yellow butt	Lysine deaminase negative; Lysine decarboxylase negative; Glucose fermentation	K/A
Red slant/yellow butt	Lysine deaminase positive; Lysine decarboxylase negative; Glucose fermentation	R/A
Black precipitate	Sulfur reduction	H_2S

✦ In This Exercise

You will be inoculating 3 LIA slants. Use heavy inocula and tighten the caps before incubating, to maintain anaerobic conditions in the butt. Because the decarboxylase/H_2S–positive organisms being differentiated produce results rapidly, incubation time will be 18 to 24 hours. If your labs are scheduled more than 24 hours apart, have someone remove the tubes from the incubator at the appropriate time and refrigerate them until time to record the results. When entering your data, be sure to write the slant information first, followed by the butt and hydrogen sulfide results (example: K/K, H_2S). Refer to Table 5-25 and Figure 5-70 when making your interpretations.

✦ Materials

Per Student Group

✦ four LIA slants

✦ fresh cultures on solid media:
  ◆ *Proteus mirabilis* (BSL-2)
  ◆ *Escherichia coli*
  ◆ *Salmonella typhimurium* (BSL-2)

✦ Medium Recipe

Lysine Iron Agar

◆ Peptone	5.0 g
◆ Yeast Extract	3.0 g
◆ Dextrose (glucose)	1.0 g
◆ L-Lysine hydrochloride	10.0 g
◆ Ferric ammonium citrate	0.5 g
◆ Sodium thiosulfate	0.04 g
◆ Bromcresol purple	0.02 g
◆ Agar	15.0 g

pH 6.5–6.9 at 25°C

Procedure

Lab One

1 Obtain four LIA slants. Label three slants with the names of the organisms, your name, and the date. Label the fourth slant "control."

2 Inoculate three slants with the test organisms. Using heavy inocula, stab the agar butt twice and then streak the slant. Do not inoculate the control.

3 Incubate all slants with the caps tightened at 35 ± 2°C for 18 to 24 hours.

Lab Two

1 Examine the tubes for characteristic color changes.

2 Record your results in the chart provided on the Data Sheet.

References

Delost, Maria Dannessa. 1997. Page 194 in *Introduction to Diagnostic Microbiology*. Mosby, St. Louis.

Forbes, Betty A., Daniel F. Sahm, and Alice S. Weissfeld. 2002. Page 274 in *Bailey & Scott's Diagnostic Microbiology*, 11th ed. Mosby, St. Louis.

Zimbro, Mary Jo, and David A. Power, editors. 2003. *Difco™ and BBL™ Manual—Manual of Microbiological Culture Media*. Becton Dickinson and Co., Sparks, MD.

EXERCISE 5-23

Litmus Milk Medium

✦ Theory

Litmus Milk is an undefined medium consisting of skim milk and the pH indicator azolitmin. Skim milk provides nutrients for growth, lactose for fermentation, and protein in the form of casein. Azolitmin (litmus) is pink at pH 4.5 and blue at pH 8.3. Between these extremes it is light purple.

Four basic reactions occur in Litmus Milk: lactose fermentation, reduction of litmus, casein coagulation, and casein hydrolysis. In combination these reactions yield a variety of results, each of which can be used to differentiate bacteria. Several possible combinations are described in Table 5-26.

Lactose fermentation acidifies the medium and turns the litmus pink (Figure 5-71, second tube from the right). This acid reaction typically begins with the splitting of the disaccharide into the monosaccharides glucose and galactose by the enzyme β-galactosidase. Refer to Figure 5-72 and Appendix A as necessary. Accumulating acid may cause the casein to precipitate and form an acid clot (Figures 5-73 and 5-74). Acid clots solidify the medium and can appear pink or white with a pink band at the top (Figure 5-71, second tube from the left) depending on the oxidation-reduction status of litmus. Reduced

litmus is white; oxidized litmus is purple. Acid clots can be dissolved in alkaline conditions. Fissures or cracks in the clot are evidence of gas production (Figure 5-75). Heavy gas production that breaks up the clot is called stormy fermentation.

In addition to being a pH indicator, litmus is an E_h (oxidation-reduction) indicator. As mentioned above, reduced litmus is white. If litmus is reduced during lactose

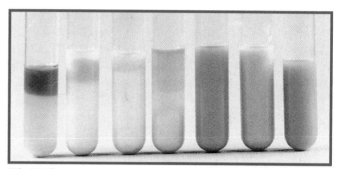

5-71 REACTIONS IN LITMUS MILK ✦ These Litmus Milk tubes illustrate, from left to right: digestion, alkaline reaction (DK); acid clot formation, reduction of litmus (ACR); acid clot formation, reduction of litmus, gas production from fermentation (ACRG—note small gas fissure in clot); curd formation, reduction of litmus (CR); uninoculated control; acid reaction (A); and alkaline reaction (K).

TABLE **5-26** Litmus Milk Results and Interpretations*

TABLE OF RESULTS		
Result	**Interpretation**	**Symbol**
Pink color	Acid reaction	A
Pink and solid (white in the lower portion if the litmus is reduced); clot not movable	Acid clot	AC
Fissures in clot	Gas	G
Clot broken apart	Stormy fermentation	S
White color (lower portion of medium)	Reduction of litmus	R
Semisolid and not pink; clear to gray fluid at top	Curd	C
Clarification of medium; loss of "body"	Digestion	D
Blue medium or blue band at top	Alkaline reaction	K
No change	None of the above reactions	NC

*These results may appear together in a variety of combinations.

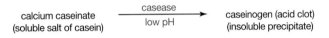

5-72 LACTOSE HYDROLYSIS ✦ Lactose hydrolysis requires the enzyme β-galactosidase and produces glucose and galactose—two fermentable sugars.

5-73 ACID CLOT FORMATION ✦ An acid clot is the result of casease catalyzing the formation of caseinogen, an insoluble precipitate, under acidic conditions.

ammonia (NH_3), which raises the pH of the medium and turns the litmus blue. Formation of a blue or purple ring at the top of the clear fluid or bluing of the entire medium indicates an alkaline reaction.

✦ Application

Litmus Milk is used primarily to differentiate members within the genus *Clostridium*. It differentiates *Enterobacteriaceae* from other Gram negative bacilli based on the ability of enterics to reduce litmus. Litmus Milk also is used to cultivate and maintain cultures of lactic acid bacteria.

5-74 ACID CLOT ✦ An acid clot appears in the top tube; an uninoculated control is below. Note the reduced litmus (white) at the bottom of the clot.

5-75 GAS PRODUCTION FROM FERMENTATION ✦ This organism has produced gas fissures in the clot.

fermentation, it will turn the medium white in the lower portion of the tube where the reduction rate is greatest.

Some bacteria produce proteolytic enzymes such as rennin, pepsin, or chymotrypsin that coagulate casein and produce a curd (Figure 5-71, center tube, and Figure 5-76). A curd differs from an acid clot in that it will not dissolve in alkaline conditions and tends to retract from the sides of the tube, revealing a straw colored fluid called whey.

Certain enzymes can digest both acid clots and curds. A digestion reaction leaves only a clear to brownish fluid behind (Figure 5-71, tube on the left). Bacteria that are able only to partially digest the casein typically produce

5-76 CURD FORMATION ✦ Rennin converts casein to paracasein to form a curd.

✦ In This Exercise

You will observe several different reactions in Litmus Milk. Most tubes will illustrate more than one reaction, so refer to Figure 5-71 and Table 5-26 when making your interpretations.

✦ Materials

Per Student Group

✦ six Litmus Milk tubes
✦ fresh cultures of:
  ◆ *Alcaligenes faecalis*
  ◆ *Pseudomonas aeruginosa* (BSL-2)
  ◆ *Klebsiella pneumoniae* (BSL-2)
  ◆ *Lactococcus lactis*
  ◆ *Enterococcus faecium* (BSL-2)

✦ Medium Recipe

◆ Litmus Milk Medium
◆ Skim milk .... 100.0 g
◆ Azolitmin .... 0.5 g
◆ Sodium sulfite .... 0.5 g
◆ Distilled or deionized water ... 1.0 L
      pH 6.3–6.7 at 25°C

Procedure

Lab One

1 Obtain six Litmus Milk tubes. Label five tubes with the names of the organisms, your name, and the date. Label the sixth tube "control."

2 Inoculate five tubes with the test cultures. Do not inoculate the control.

3 Incubate all tubes aerobically at 35 ± 2°C for 7 to 14 days.

Lab Two

1 Examine the tubes for color changes, gas production, and clot formation. Be sure to compare all tubes to the control, and refer to Table 5-26 when making your interpretations.

2 Record your results in the chart provided on the Data Sheet.

References

Forbes, Betty A., Daniel F. Sahm, Alice S. Weissfeld. 2002. Pages 272–273 in *Bailey & Scott's Diagnostic Microbiology,* 11th ed. Mosby, St. Louis.

MacFaddin, Jean F. 2000. Page 294 in *Biochemical Tests for Identification of Medical Bacteria,* 3rd ed. Lippincott Williams & Wilkins, Philadelphia.

Zimbro, Mary Jo, and David A. Power, editors. 2003. Page 309 in *Difco™ and BBL™ Manual—Manual of Microbiological Culture Media.* Becton Dickinson and Co., Sparks, MD.

Antibacterial Susceptibility Testing

Antibacterial susceptibility testing can be used to help identify an organism or simply select an appropriate antibacterial agent to be used to fight it. All antibacterial susceptibility tests included in this unit apply the **disc-diffusion method**.

In the disc-diffusion method, a commercially prepared filter paper disc impregnated with a specific concentration of antibiotic is placed on an agar plate inoculated to produce a lawn of bacterial growth. As the antibacterial agent diffuses from the disc into the agar, it creates a concentration gradient, becoming less concentrated the farther it travels. As bacterial growth develops on the agar surface, it will not grow in the area where it is susceptible to the agent. This area is called the **zone of inhibition**. The size of the zone relates to the **minimum inhibitory concentration (MIC)** of the antibacterial agent that is effective against the test organism.

The National Committee for Clinical Laboratory Standards (NCCLS) periodically publishes a Zone Diameter Interpretive Chart to be used in the testing and determinations of bacterial susceptibility or resistance to various antibacterial agents. The chart lists the concentrations of all antibacterial agents used and the zone diameters, in millimeters, used to make the determinations. For example, a 10 USP unit penicillin disc used to test susceptible staphylococci will produce a zone 29 mm or greater. ✦

Bacitracin, Novobiocin, and Optochin Susceptibility Tests

✦ Theory[1]

Bacitracin, produced by *Bacillus licheniformis*, is a powerful peptide antibiotic that inhibits bacterial cell wall synthesis (Figure 5-77). Thus, it is effective only on bacteria that have cell walls and are in the process of growing. In this test, any zone of clearing 10 mm or greater around the disk is interpreted as bacitracin susceptibility (Figure 5-78). For more information on antimicrobial susceptibility, refer to Exercise 7-3, Antimicrobial Susceptibility Test.

Novobiocin is an antibiotic, produced by *Streptomyces niveus*. It interferes with ATPase activity and ultimately the production of ATP. In this test, a 5 μg disc should produce a zone of clearing 16 mm or more to be considered novobiocin susceptible (Figure 5-79).

Optochin is an antibiotic that interferes with ATPase activity and ATP production in susceptible bacteria. A 6 mm disc (containing 5μg of optochin) should produce

a zone of inhibition 14 mm or more to be considered optochin susceptible (Figure 5-80).

✦ Application

The bacitracin test is used to differentiate and presumptively identify β-hemolytic group A streptococci (*Streptococcus pyogenes*) from other β-hemolytic streptococci. It also differentiates the genus *Staphylococcus* (resistant) from the susceptible *Micrococcus* and *Stomatococcus*.

The novobiocin test is used to differentiate coagulase-negative staphylococci. Most frequently it is used to presumptively identify the novobiocin-resistant *Staphylococcus saprophyticus*.

$$CH_3-\overset{\overset{\displaystyle CH_3}{|}}{C}=CH-CH_2-[CH_2-\overset{\overset{\displaystyle CH_3}{|}}{C}=CH-CH_2]_9-CH_2-\overset{\overset{\displaystyle CH_3}{|}}{C}=CH-CH_2-O-\overset{\overset{\displaystyle O}{\|}}{\underset{\underset{\displaystyle O^-}{|}}{P}}-O^-$$

5-77 UNDECAPRENYL PHOSPHATE ✦ Undecaprenyl phosphate is involved in transporting peptidoglycan subunits across the cell membrane during cell wall synthesis. Bacitracin interferes with its release from the peptidoglycan subunit. It is a C_{55} molecule derived from 11 isoprene subunits plus a phosphate.

---

[1] Because of the repetitive nature of antibacterial susceptibility tests, information that applies to this and other tests appears only once in the introduction to this unit. Include that information as part of Theory in this exercise.

5-78 BACITRACIN SUSCEPTIBILITY ON A SHEEP BLOOD AGAR PLATE ✦ The organism above has no clear zone and is resistant (R); the organism below has a clear zone larger than 10 mm and is susceptible (S).

5-79 NOVOBIOCIN DISK TEST ✦ A novobiocin-resistant (R) organism is on the left; a susceptible (S) organism is on the right.

The optochin test is used to presumptively differentiate *Streptococcus pneumoniae* from other α-hemolytic streptococci.

✦ In This Exercise

You will inoculate from one to three Blood Agar plates (depending on the tests chosen by your instructor) to produce confluent growth of the test organisms. You will then place specific antibiotic discs in the center of the inoculum. Following incubation, you will measure the clear zones surrounding the discs to determine susceptibility or resistance of the test organisms to the antibiotics.

✦ Materials

Per Student Group

✦ one Blood Agar plate per test (commercial preparation of TSA containing 5% sheep blood)

✦ sterile cotton applicators

✦ beaker of alcohol with forceps

Bacitracin Test

✦ 0.04 unit bacitracin discs

✦ fresh broth cultures of:
  ✦ *Staphylococcus aureus* (BSL-2)
  ✦ *Micrococcus luteus*

5-80 OPTOCHIN SUSCEPTIBILITY TEST ✦ The zone of inhibition surrounding the disk indicates susceptibility to optochin and presumptive identification of *S. pneumoniae*.

Novobiocin Test

✦ 5 μg novobiocin discs

✦ fresh broth cultures of:
  ✦ *Staphylococcus epidermidis*
  ✦ *Staphylococcus saprophyticus*

Optochin Test

✦ 5 µg (6 mm) optochin discs

✦ fresh broth cultures of:

    ◆ *Streptococcus pneumoniae* (BSL-2)

    ◆ *Streptococcus agalactiae* (BSL-2)

 Procedure (Per Test)

Lab One

1. Obtain one Blood Agar plate for each test you are performing, and draw a line on the bottom dividing the plate into two halves. Using a sterile cotton applicator, inoculate half of the plate with one of the two test organisms. (Make the inoculum as light as possible by wiping and twisting the wet cotton swab on the inside of the culture tube before removing it.) Inoculate the plate by making a single streak nearly halfway across its diameter. Turn the plate 90° and spread the organism evenly to produce a bacterial lawn covering nearly half the agar surface. Refer to the photo in Figure 5-78.

2. Being careful not to mix the cultures, repeat the process on the other half of the plate using the second test organism. Allow the broth to be absorbed by the agar for 5 minutes before proceeding to step 3.

3. Sterilize the forceps by placing them in the Bunsen burner flame long enough to ignite the alcohol. (**Note:** Do not hold the forceps in the flame.) Once the alcohol has burned off, use the forceps to place an antibacterial disc in the center of the half containing organism 1. Gently tap the disk into place to ensure that it makes full contact with the agar surface. Return the forceps to the alcohol.

4. Repeat step 3, placing a disc on the other half of the plate—the area containing organism 2. Tap the disk into place and return the forceps to the alcohol.

5. Invert the plate, label it appropriately, and incubate it at $35 \pm 2°C$ for 24 to 48 hours.

Lab Two

1. Remove the plate from the incubator and examine it for clearing around the disks.

2. Record your results in the chart on the Data Sheet.

TABLE **5-27** Bacitracin Test Results and Interpretations

TABLE OF RESULTS		
Result	**Interpretation**	**Symbol**
Zone of clearing 10 mm or greater	Organism is sensitive to bacitracin	+
Zone of clearing less than 10 mm	Organism is resistant to bacitracin	−

TABLE **5-28** Novobiocin Test Results and Interpretations

TABLE OF RESULTS		
Result	**Interpretation**	**Symbol**
Zone of clearing 16 mm or greater	Organism is sensitive to novobiocin	+
Zone of clearing less than 16 mm	Organism is resistant to novobiocin	−

TABLE **5-29** Optochin Test Results and Interpretations

TABLE OF RESULTS		
Result	**Interpretation**	**Symbol**
Zone of clearing 14 mm or greater	Organism is sensitive to optochin	+
Zone of clearing less than 14 mm	Organism is resistant to optochin	−

References

Baron, Ellen Jo, Lance R. Peterson, and Sydney M. Finegold. 1994. Page 329 in *Bailey & Scott's Diagnostic Microbiology*, 9th ed. Mosby–Year Book, St. Louis.

Delost, Maria Dannessa. 1997. Page 107 in *Introduction to Diagnostic Microbiology*. Mosby, St. Louis.

DIFCO Laboratories. 1984. Page 292 in *DIFCO Manual*, 10th ed. DIFCO Laboratories, Detroit.

Forbes, Betty A., Daniel F. Sahm, and Alice S. Weissfeld. 2002. Page 290 in *Bailey & Scott's Diagnostic Microbiology*, 11th ed. Mosby, St. Louis.

Winn, Washington C., *et al.* 2006. Pages 645 and 1472 in *Koneman's Color Atlas and Textbook of Diagnostic Microbiology*, 6th ed. Lippincott Williams & Wilkins, Baltimore.

MacFaddin, Jean F. 2000. Page 3 in *Biochemical Tests for Identification of Medical Bacteria*, 3rd ed. Lippincott Williams & Wilkins, Philadelphia.

Other Differential Tests

This unit includes tests that do not fit elsewhere but are important to consider. Blood Agar is especially useful for detecting hemolytic ability of Gram-positive cocci—typically *Streptococcus* species. It also is used as a general-purpose growth medium appropriate for fastidious and non-fastidious microorganisms alike. The coagulase tests are commonly used to presumptively identify pathogenic *Staphylococcus* species. Motility Agar is used to detect bacterial motility, especially in differentiating *Enterobacteriaceae* and other Gram-negative rods. ✦

EXERCISE 5-25 Blood Agar

✦ Theory

Several species of Gram-positive cocci produce exotoxins called **hemolysins**, which are able to destroy red blood cells (RBCs) and hemoglobin. Blood Agar, sometimes called Sheep Blood Agar because it includes 5% sheep blood in a Tryptic Soy Agar base, allows differentiation of bacteria based on their ability to hemolyze RBCs.

The three major types of hemolysis are β-hemolysis, α-hemolysis, and γ-hemolysis. β-hemolysis, the complete destruction of RBCs and hemoglobin, results in a clearing of the medium around the colonies (Figure 5-81). α-hemolysis is the partial destruction of RBCs and produces a greenish discoloration of the agar around the colonies (Figure 5-82). Actually, γ-hemolysis is non-hemolysis and appears as simple growth with no change to the medium (Figure 5-83).

Hemolysins produced by streptococci are called **streptolysins**. They come in two forms—type O and type S. **Streptolysin O** is oxygen-labile and expresses maximal activity under anaerobic conditions. **Streptolysin S** is

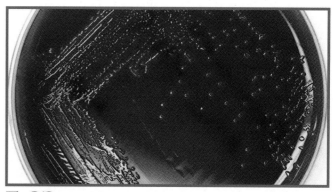

5-82 α-HEMOLYSIS ✦ This is a streak plate of *Streptococcus pneumoniae* demonstrating α-hemolysis. The greenish zone around the colonies results from incomplete lysis of red blood cells.

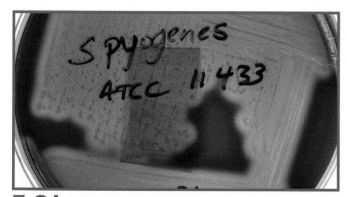

5-81 β-HEMOLYSIS ✦ *Streptococcus pyogenes* demonstrates β-hemolysis. The clearing around the growth is a result of complete lysis of red blood cells. This photograph was taken with transmitted light.

5-83 γ-HEMOLYSIS ✦ This streak plate of *Staphylococcus epidermidis* on a Sheep Blood Agar illustrates no hemolysis.

oxygen-stable but expresses itself optimally under anaerobic conditions as well. The easiest method of providing an environment favorable for streptolysins on Blood Agar is what is called the **streak–stab technique.** In this procedure the Blood Agar plate is streaked for isolation and then stabbed with a loop. The stabs encourage streptolysin activity because of the reduced oxygen concentration of the subsurface environment (Figure 5-84).

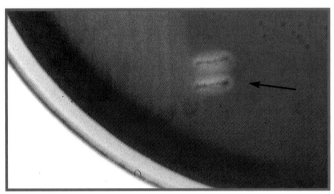

5-84 AEROBIC VERSUS ANAEROBIC HEMOLYSIS ✦ An unidentified throat culture isolate demonstrates α-hemolysis when growing on the surface, but β-hemolysis beneath the surface surrounding the stabs (arrow). This results from production of an oxygen-labile hemolysin.

✦ Application

Blood Agar is used for isolation and cultivation of many types of fastidious bacteria. It also is used to differentiate bacteria based on their hemolytic characteristics, especially within the genera *Streptococcus*, *Enterococcus*, and *Aerococcus*.

✦ In This Exercise

You will not be provided with organisms to inoculate your plate. Instead, your partner will take a swab of your throat. Then you will inoculate the Blood Agar plate in one quadrant to begin a streak for isolation. Isolation of the different colonies is the only way to observe the different forms of hemolysis properly. (**Note:** When you

streak your plate, do only the first streak with the swab, and do streaks two, three, and four with your inoculating loop. If you are not sure how to do this, refer to Exercise 1-4. Because of the potential for cultivating a pathogen, tape down the plate lid immediately after streaking. Upon removing your plate from the incubator, do not open it until your instructor says it is safe to do so.)

✦ Materials

Per Student Group

✦ one Sheep Blood Agar plate (commercially available— TSA containing 5% sheep blood)
✦ sterile cotton swabs
✦ sterile tongue depressors

✦ Medium Recipe

5% Sheep Blood Agar
(TSA + 5% Sheep Blood)

✦ Infusion from beef heart (solids)	2.0 g
✦ Pancreatic digest of casein	13.0 g
✦ Sodium chloride	5.0 g
✦ Yeast extract	5.0 g
✦ Agar	15.0 g
✦ Defibrinated sheep blood	50.0 mL
✦ Distilled or deionized water	1.0 L

pH 7.1–7.5 at 25°C

Procedure

Lab One

1 Have your lab partner obtain a culture from your throat. Follow the procedure in Appendix B.

2 Immediately transfer the specimen to a Blood Agar plate. Use the swab to begin a streak for isolation. Refer to Exercise 1-4 if necessary.

3 Dispose of the swab in a container designated for autoclaving.

TABLE **5-30** Blood Agar Results and Interpretations

TABLE OF RESULTS		
Result	Interpretation	Symbol
Clearing around growth	Organism hemolyzes RBCs completely	β-hemolysis
Greening around growth	Organism hemolyzes RBCs partially	α-hemolysis
No change in the medium	Organism does not hemolyze RBCs	no (γ) hemolysis

4 Finish the isolation procedure by streaking quadrants 2, 3, and 4 with your loop.

5 After completing the streak, use your loop to stab the agar in two or three places in the first streak pattern, and then in two or three places not previously inoculated.

6 Label the plate with your name, the specimen source ("throat culture"), and the date.

7 Tape down the lid to prevent it from opening accidentally. Invert and incubate the plate aerobically at $35 \pm 2°C$ for 24 hours.

Lab Two

1 After incubation, do not open your plate until your instructor has seen it and given permission to do so.

2 Using transmitted light and referring to Table 5-30, observe for color changes and clearing around the colonies. This can be done using a colony counter or by holding the plate up to a light. Record your results in the chart on the Data Sheet.

References

Delost, Maria Dannessa. 1997. Page 103 in *Introduction to Diagnostic Microbiology*. Mosby, Inc. St. Louis.

Forbes, Betty A., Daniel F. Sahm, and Alice S. Weissfeld. 2002. Page 16 in *Bailey & Scott's Diagnostic Microbiology*, 11th ed. Mosby, St. Louis.

Krieg, Noel R. 1994. Page 619 in *Methods for General and Molecular Bacteriology*, edited by Philipp Gerhardt, R. G. E. Murray, Willis A. Wood, and Noel R. Krieg. American Society for Microbiology, Washington, DC.

Power, David A., and Peggy J. McCuen. 1988. Page 115 in *Manual of BBL™ Products and Laboratory Procedures*, 6th ed. Becton Dickinson Microbiology Systems, Cockeysville, MD.

Winn, Washington C., *et al.* 2006. Chapter 13 in *Koneman's Color Atlas and Textbook of Diagnostic Microbiology*, 6th ed. Lippincott Williams & Wilkins, Baltimore.

Zimbro, Mary Jo, and David A. Power, Eds. 2003. *Difco™ and BBL™ Manual—Manual of Microbiological Culture Media*. Becton Dickinson and Co., Sparks, MD.

EXERCISE 5-26

CAMP Test

✦ Theory

Group B *Streptococcus agalactiae* produces the CAMP factor—a hemolytic protein that acts synergistically with the β-hemolysin of *Staphylococcus aureus* subsp. *aureus*. When streaked perpendicularly to an *S. aureus* subsp. *aureus* streak on blood agar (Figure 5-85), an arrowhead-shaped zone of hemolysis forms and is a positive result.

✦ Application

The CAMP test (an acronym of the developers of the test—Christie, Atkins, and Munch-Peterson) is used to differentiate Group B *Streptococcus agalactiae* (+) from other *Streptococcus* species (−).

✦ In This Exercise

You will inoculate a Blood Agar plate as shown in Figure 5-86 with *Streptococcus agalactiae* and *Staphylococcus aureus*. Following incubation, you will examine your plate for the characteristic arrowhead pattern of clearing.

5-86 CAMP TEST INOCULATION ✦ Two inoculations are made. First *Staphylococcus aureus* subsp. *aureus* is streaked along one edge of a fresh Blood Agar plate (I). Then the isolate (when testing an unknown organism) is inoculated densely in the other half of the plate opposite *S. aureus* (II). Finally, a single streak is made from inside Streak II toward, but not touching, *S. aureus* (III).

✦ Materials

✦ two Blood Agar plates (commercial preparation of TSA containing 5% sheep blood)
✦ sterile cotton applicators
✦ fresh broth cultures of:
  ✦ *Staphylococcus aureus* subsp. *aureus* (BSL-2)
  ✦ *Streptococcus agalactiae* (BSL-2)
  ✦ *Streptococcus pneumoniae* (BSL-2)

Procedure

Lab One

1 Obtain two Blood Agar plates

2 Using a cotton swab, inoculate both plates as shown in Figure 5-86, each with a single streak of *S. aureus* along one edge (Streak I).

3 Inoculate each plate with a dense smear of one of the other two organisms across from the *S. aureus* streak, as shown in Streak II. Finish each plate with a single streak (as in Streak III) from the *Streptococcus*

5-85 POSITIVE CAMP TEST RESULTS ✦ Note the arrowhead zone of clearing in the region where the CAMP factor of *Streptococcus agalactiae* acts synergistically with the β-hemolysin of *Staphylococcus aureus* subsp. *aureus*.

species toward, but not touching, the *S. aureus*. ***Note:*** It works best and saves material to do this third streak with the same swab used in Streak II before moving on to the other plate. Repeat procedure in the second plate with *S. aureus* and *S. pneumoniae*.

4 Label the plates with your name, the date, and the names of the organisms.

5 Invert the plates and incubate them at $35 \pm 2°C$ for 24 hours.

Lab Two

1 Remove the plates from the incubator and observe for the characteristic arrowhead pattern.

2 Discard the plates in an appropriate biohazard or autoclave container.

TABLE **5-31** CAMP Test Results and Interpretations

TABLE OF RESULTS		
Result	**Interpretation**	**Symbol**
Arrowhead pattern	Organism produces hemolytic CAMP protein. Presumptively identified as *S. agalactiae*	+
No arrowhead pattern	Organism does not produce hemolytic CAMP protein. Presumptively identified as no *S. agalactiae*	−

EXERCISE 5-27 Coagulase Tests

✦ Theory

Staphylococcus aureus is an opportunistic pathogen that can be highly resistant to both the normal immune response and antimicrobial agents. Its resistance results, in part, from the production of a coagulase enzyme. Coagulase works in conjunction with normal plasma components to form protective fibrin barriers around individual bacterial cells or groups of cells, shielding them from phagocytosis and other types of attack.

Coagulase enzymes occur in two forms—**bound coagulase** and **free coagulase**. Bound coagulase, also called **clumping factor**, is attached to the bacterial cell wall and reacts directly with fibrinogen in plasma. The fibrinogen then precipitates, causing the cells to clump together in a visible mass. Free coagulase is an extracellular enzyme (released from the cell) that reacts with a plasma component called coagulase-reacting factor (CRF). The resulting reaction is similar to the conversion of prothrombin and fibrinogen in the normal clotting mechanism.

Two forms of the Coagulase Test have been devised to detect the enzymes: the Tube Test and the Slide Test. The Tube Test detects the presence of either bound or free coagulase, and the Slide Test detects only bound coagulase. Both tests utilize rabbit plasma treated with anticoagulant to interrupt the normal clotting mechanisms.

The Tube Test is performed by adding the test organism to rabbit plasma in a test tube. Coagulation of the plasma (including any thickening or formation of fibrin threads) within 24 hours indicates a positive reaction (Figure 5-87). The plasma typically is examined for clotting (without shaking) periodically for about 4 hours. After 4 hours coagulase-negative tubes can be incubated overnight, but no more than a total of 24 hours, because coagulation can take place early and revert to liquid within 24 hours.

In the Slide Test, bacteria are transferred to a slide containing a small amount of plasma. Agglutination of the cells on the slide within 1 to 2 minutes indicates the presence of bound coagulase (Figure 5-88). Equivocal or negative Slide Test results typically are given the Tube Test for confirmation.

✦ Application

The Coagulase Test typically is used to differentiate *Staphylococcus aureus* from other Gram-positive cocci.

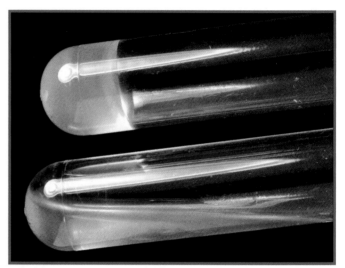

5-87 COAGULASE TUBE TEST ✦ These coagulase tubes illustrate a coagulase-negative (−) organism, below, and a coagulase-positive (+) organism, above. The Tube Test identifies both bound and free coagulase enzymes. Coagulase increases bacterial resistance to phagocytosis and antibodies by surrounding infecting organisms with a clot.

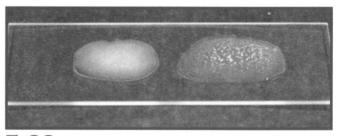

5-88 COAGULASE SLIDE TEST (CLUMPING FACTOR) ✦ This slide illustrates a coagulase-positive (+) organism on the right and a coagulase-negative (−) organism on the left. Agglutination of the coagulase plasma is indicative of a positive result for bound coagulase.

✦ In This Exercise

You will perform both Tube and Slide Coagulase Tests. Immediately enter your Slide Test results on the Data Sheet. Read the Tube Test the following day.

✦ Materials

Per Student Group

✦ three sterile rabbit plasma tubes (0.5 mL in 12 mm × 75 mm test tubes)

✦ sterile 1 mL pipettes

✦ sterile saline (0.9% NaCl)

✦ microscope slides

✦ fresh slant cultures of:

  ◆ *Staphylococcus aureus* (BSL-2)

  ◆ *Staphylococcus epidermidis*

 Procedure

Lab One: Tube Test

1 Obtain three coagulase tubes. Label two tubes with the names of the organisms, your name, and the date. Label the third tube "control."

2 Inoculate two tubes with the test organisms. Mix the contents by gently rolling the tube between your hands. Do not inoculate the control.

3 Incubate all tubes at 35 ± 2°C for up to 24 hours, checking for coagulation periodically for the first 4 hours.

Lab One: Slide Test (Clumping Factor)

1 Obtain two microscope slides, and divide them in half with a marking pen. Label the sides A and B.

2 Place a drop of sterile saline on side A and a drop of coagulase plasma on side B of each slide.

3 Transfer a loopful of *S. aureus* to each half of one slide, making sure to completely emulsify the bacteria in the solutions. Observe for agglutination within 2 minutes. Clumping after 2 minutes is not a positive result.

4 Repeat step 3 using the other slide and *S. epidermidis*.

5 Record your results in the chart on the Data Sheet. Refer to Figure 5-88 and Table 5-32 when making your interpretations. Confirm any negative results by comparing with the completed Tube Test in 24 hours.

Lab Two

1 Remove all tubes from the incubator no later than 24 hours after inoculation. Examine for clotting of the plasma.

2 Record your results on the Data Sheet. Refer to Figure 5-87 and Table 5-33 when making your interpretations.

TABLE **5-32** Coagulase Slide Test Results and Interpretations

TABLE OF RESULTS		
Result	**Interpretation**	**Symbol**
Clumping of cells	Plasma has been coagulated	+
No clumping of cells	Plasma has not been coagulated	−

TABLE **5-33** Coagulase Tube Test Results and Interpretations

TABLE OF RESULTS		
Result	**Interpretation**	**Symbol**
Medium is solid	Plasma has been coagulated	+
Medium is liquid	Plasma has not been coagulated	−

References

Collins, C. H., Patricia M. Lyne, and J. M. Grange. 1995. Page 111 in *Collins and Lyne's Microbiological Methods*, 7th ed. Butterworth-Heinemann, Oxford, United Kingdom.

Delost, Maria Dannessa. 1997. Pages 98–99 in *Introduction to Diagnostic Microbiology*. Mosby, St. Louis.

DIFCO Laboratories. 1984. Page 232 in *DIFCO Manual*, 10th ed. DIFCO Laboratories, Detroit.

Forbes, Betty A., Daniel F. Sahm, and Alice S. Weissfeld. 2002. Pages 266–267 in *Bailey & Scott's Diagnostic Microbiology*, 11th ed. Mosby, St. Louis.

Holt, John G., Ed. 1994. *Bergey's Manual of Determinative Bacteriology*, 9th ed. Williams and Wilkins, Baltimore.

Lányi, B. 1987. Page 62 in *Methods in Microbiology*, Vol. 19, edited by R. R. Colwell and R. Grigorova. Academic Press, New York.

MacFaddin, Jean F. 2000. Page 105 in *Biochemical Tests for Identification of Medical Bacteria*, 3rd ed. Lippincott Williams & Wilkins, Philadelphia.

EXERCISE 5-28   Motility Test

✦ Theory

Motility Test Medium is a semisolid medium designed to detect bacterial motility. Its agar concentration is reduced from the typical 1.5% to 0.4%—just enough to maintain its form while allowing movement of motile bacteria. It is inoculated by stabbing with a straight transfer needle. Motility is detectable as diffuse growth radiating from the central stab line.

A tetrazolium salt (TTC) sometimes is added to the medium to make interpretation easier. TTC is used by the bacteria as an electron acceptor. In its oxidized form, TTC is colorless and soluble; when reduced it is red and insoluble (Figure 5-89). A positive result for motility is indicated when the red (reduced) TTC is seen radiating outward from the central stab. A negative result shows red only along the stab line (Figure 5-90).

✦ Application

This test is used to detect bacterial motility. Motility is an important differential characteristic of *Enterobacteriaceae*.

✦ In This Exercise

You will inoculate two Motility Stabs with an inoculating needle. Straighten the needle before you stab the medium. It is important, also, to stab straight into the medium and remove the needle along the same line. Lateral movement of the needle will make interpretation more difficult.

✦ Materials

Per Student Group
✦ three Motility Test Media stabs
✦ fresh cultures of:
  ◆ *Enterobacter aerogenes*
  ◆ *Klebsiella pneumoniae* (BSL-2)

✦ Medium Recipe

Motility Test Medium
  ◆ Beef extract   3.0 g
  ◆ Pancreatic digest of gelatin   10.0 g
  ◆ Sodium chloride   5.0 g

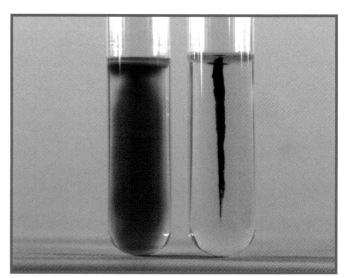

5-90 MOTILITY TEST RESULTS ✦ These Motility Test Media were inoculated with a motile (+) organism on the left and a nonmotile (−) organism on the right.

2,3,5-Triphenyltetrazolium chloride oxidized (TTC$_{OX}$)
colorless and soluble

reductase 2H$^+$

Formazan reduced
red color and insoluble

+ HCl

5-89 REDUCTION OF TTC ✦ Reduction of 2,3,5-Triphenyltetrazolium chloride by metabolizing bacteria results in its conversion from colorless and soluble to the red and insoluble compound formazan. The location of growing bacteria can be determined easily by the location of the formazan in the medium.

◆ Agar   4.0 g
◆ Triphenyltetrazolium chloride (TTC)   0.05 g
◆ Distilled or deionized water   1.0 L
    pH 7.1–7.4 at 25°C

 Procedure

Lab One

1 Obtain three motility stabs. Label two tubes, each
   with the name of the organism, your name, and the
   date. Label the third tube "control."

2 Stab-inoculate two tubes with the test organisms.
   (Motility can be obscured by the careless stabbing
   technique. Try to avoid lateral movement when
   performing this stab.) Do not inoculate the control.

3 Incubate the tubes aerobically at 35 ± 2°C for 24 to
   48 hours.

Lab Two

1 Examine the growth pattern for characteristic
   spreading from the stab line. Growth will appear
   red because of the additive in the medium.

2 Record your results in the chart provided on the
   Data Sheet.

TABLE **5-34** Motility Test Results and Interpretations

TABLE OF RESULTS		
Result	**Interpretation**	**Symbol**
Red diffuse growth radiating outward from the stab line	The organism is motile	+
Red growth only along the stab line	The organism is nonmotile	−

References

Forbes, Betty A., Daniel F. Sahm, and Alice S. Weissfeld. 2002. Page 276 in *Bailey & Scott's Diagnostic Microbiology*, 11th ed. Mosby, St. Louis.

MacFaddin, Jean F. 2000. Page 327 in *Biochemical Tests for Identification of Medical Bacteria*, 3rd ed. Lippincott Williams & Wilkins, Philadelphia.

Zimbro, Mary Jo, and David A. Power, Eds. 2003. Page 374 in *Difco™ and BBL™ Manual—Manual of Microbiological Culture Media*. Becton Dickinson and Co., Sparks, MD.

Multiple Test Systems

Multiple test systems are systems designed to run an entire battery of tests simultaneously. They employ the same biochemical principles discussed earlier in this section and are read essentially in the same manner. All conditions possible with standard tubed media also can be achieved with multiple test media, including aerobic or anaerobic growth conditions and the addition of reagents for confirmatory testing.

The savings in time and money alone make these systems enormously valuable. Even more important, what might take days or weeks to do with media preparation and individual tests, multiple test systems can do in as little as 18 hours. In addition, the media chosen for this unit come with booklets for fast and easy identification. The API 20 E system even offers software and a phone number for help with difficult identifications!

These exercises will give you an opportunity to have fun while using some of the skills you have learned. The organisms chosen for these tests will be provided as unknowns for you to identify. Alternatively, your instructor may allow you to isolate an environmental sample to identify. Finally, the system you use may be used in conjunction with your bacterial unknown project or with one of the identification exercises in Section 7. ✦

EXERCISE 5-29

API 20 E Identification System for *Enterobacteriaceae* and Other Gram-negative Rods

✦ Theory

The API 20 E system is a plastic strip of 20 microtubes and cupules, partially filled with different dehydrated substrates. Bacterial suspension is added to the microtubes, rehydrating the media and inoculating them at the same time. As with the other biochemical tests in this section, color changes take place in the tubes either during incubation or after addition of reagents. These color changes reveal the presence or absence of chemical action and, thus, a positive or negative result (Figure 5-91).

After incubation, spontaneous reactions—those that do not require addition of reagents—are evaluated first. Then tests that require addition of reagents are performed and evaluated. Finally, the results are entered on the Result Sheet (Figure 5-92). An oxidase test is performed separately and constitutes the 21st test.

As shown in Figure 5-92, the Result Sheet divides the tests into groups of three, with the members of a group having numerical values of 1, 2, or 4, respectively.

These numbers are assigned for positive (+) results only. Negative (−) results are not counted. The values for positive results in each group are added together to produce a number from 0 to 7, which is entered in the oval below the three tests. The totals from each group are combined sequentially to produce a seven-digit code, which can then be interpreted on the Analytical Profile Index[*] (Figure 5-93).

In rare instances, information from the 21 tests (and the 7-digit code) is not discriminatory enough to identify an organism. When this occurs, the organism is grown and examined on MacConkey Agar and supplemental tests are performed for nitrate reduction, oxidation/reduction of glucose, and motility. The results are entered separately in the supplemental spaces on the Result Sheet and used for final identification.

---

[*]The index is now available only as a subscription through BioMérieux-USA.com.

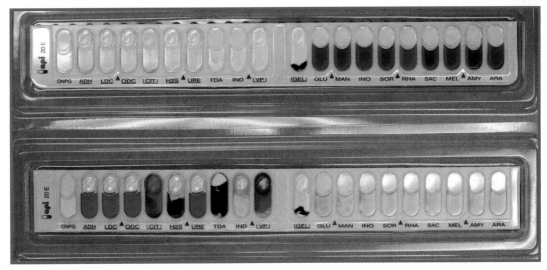

5-91 API 20 E

TEST STRIPS ✦ The top strip is uninoculated. The bottom strip (with the exception of GEL) illustrates all positive results.

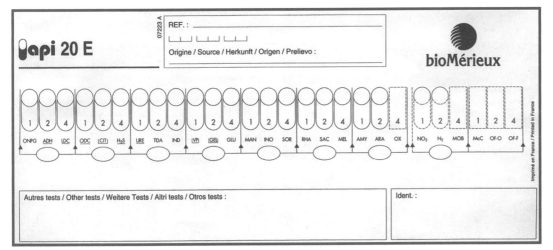

5-92 API 20 E RESULT SHEET ✦ The API 20 E Result Sheet divides the tests into groups of three, assigning each member of a group a numerical value of 1, 2, or 4. The values for positive results in each group are added together to produce a single-digit number. This number, when combined with the numbers from other groups, produces a seven-digit code that then is interpreted using the Analytical Profile Index supplied by the company (visit http//:www.BioMerieux-USA.com).

✦ Application

The API 20 E multitest system (available from Bio-Mérieux, Inc.) is used clinically for the rapid identification of *Enterobacteriaceae* (more than 5,500 strains) and other Gram-negative rods (more than 2,300 strains).

✦ In This Exercise

Today you will use a multitest system to run an entire battery of tests designed to identify a member of the *Enterobacteriaceae*.

✦ Materials

✦ API 20 E identification system for *Enterobacteriaceae* and other Gram-negative rods

✦ API Analytical Profile Index or online subscription

✦ sterile suspension medium

✦ ferric chloride reagent (see Exercise 5-11)

✦ Kovacs' reagent (see Exercise 5-20)

✦ potassium hydroxide reagent (see Exercise 5-4)

✦ alpha-naphthol reagent (see Exercise 5-4)

✦ sulfanilic acid reagent (see Exercise 5-7)

✦ N,N-Dimethyl-1-naphthylamine reagent (see Exercise 5-7)

✦ oxidase test slides (see Exercise 5-6)

✦ hydrogen peroxide

✦ zinc dust

✦ sterile mineral oil

✦ sterile Pasteur pipettes

```
7 046 124     TRES BONNE IDENTIFICATION        VERY GOOD IDENTIFICATION       SEHR GUTE IDENTIFIZIERUNG       7 046 124
              A L'ESPECE                        TO THE SPECIES LEVEL           AUF SPEZIES EBENE
                                                          I                    GLUCOSEg  ESC (HYD.)  O/129 R    RN/NR
   Aer.hydrophila gr. 1  %id=56.1 T=0.77 (LDC  25%)(ANY  75%) IAeromonas caviae      -        +         +         +
                                         (ARA  75%)           IAeromonas hydrophila  +        +         +        86%
   Aer.hydrophila gr. 2  %id=43.1 T=0.73 (CIT  80%)(VP  80%) IVibrio fluvialis      0%       NT         -        NT
                                         (ANY  75%)           IAeromonas sobria      +        -         +         -
   -POSSIBILITE DE Vibrio fluvialis          -POSSIBILITY OF Vibrio fluvialis     -MOEGLICHKEIT VON Vibrio fluvialis
-----------------------------------------------------------------------------------------------------------------------
7 046 125     EXCELLENTE IDENTIFICATION        EXCELLENT IDENTIFICATION       AUSGEZEICHNETE IDENTIFIZIERUNG 7 046 125
              A L'ESPECE                        TO THE SPECIES LEVEL           AUF SPEZIES EBENE
                                                          I                    GLUCOSEg  ESC (HYD.)  O/129 R    RN/NR
   Aer.hydrophila gr. 1  %id=56.5 T=0.84 (LDC  25%)(ARA  75%) IAeromonas caviae      -        +         +         +
   Aer.hydrophila gr. 2  %id=43.4 T=0.80 (CIT  80%)(VP  80%) IAeromonas hydrophila  +        +         +        86%
                                                             IVibrio fluvialis      0%       NT         -        NT
                                                             IAeromonas sobria      +        -         +         -
   -POSSIBILITE DE Vibrio fluvialis          -POSSIBILITY OF Vibrio fluvialis     -MOEGLICHKEIT VON Vibrio fluvialis
-----------------------------------------------------------------------------------------------------------------------
7 046 126     BONNE IDENTIFICATION             GOOD IDENTIFICATION            GUTE IDENTIFIZIERUNG           7 046 126
                                                          I                    GLUCOSEg  ESC (HYD.)  O/129 R    RN/NR
   Aer.hydrophila gr. 1  %id=97.9 T=0.84 (LDC  25%)(ANY  75%) IAeromonas caviae      -        +         +         +
                                                             IAeromonas hydrophila  +        +         +        86%
                                                             IVibrio fluvialis      0%       NT         -        NT
                                                             IAeromonas sobria      +        -         +         -
   -POSSIBILITE DE Vibrio fluvialis          -POSSIBILITY OF Vibrio fluvialis     -MOEGLICHKEIT VON Vibrio fluvialis
```

5-93 A PORTION OF THE ANALYTICAL PROFILE INDEX (API) ✦ Identification is made by locating the seven-digit number in the Analytical Profile Index. This information is also available through online subscription.

✦ distilled water

✦ agar plates containing unidentified bacterial colonies

 ## Procedure

Lab One

1 Perform a Gram stain on your unknown organism to confirm that it is a Gram-negative rod.

2 Open an incubation box and distribute about 5 mL of distilled water into the honeycombed wells of the tray to create a humid atmosphere (Figure 5-94).

3 Record the unknown number on the elongated flap of the tray.

4 Remove the strip from its packaging and place it in the tray.

5 Perform the oxidase test on a colony identical to the colony that will be tested. If you need help with this, refer to Exercise 5-6. Record the result on the score sheet as the 21st test.

6 Using a sterile Pasteur pipette, remove a single, well-isolated colony from the plate (Figure 5-95) and fully emulsify it in a tube of Suspension Medium or sterile 0.85% saline (Figure 5-96). Try not to pick up any agar.

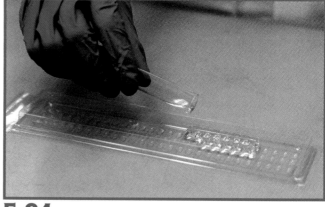

5-94 ADD WATER TO HONEYCOMBED WELLS ✦ The purpose of the water is to humidify the strip during incubation. Spread the water as uniformly as possible in the honeycomb, but it is not necessary to add identical amounts to each well.

7 Using the same pipette, fill both tube and cupule of test CIT, VP, and GEL with bacterial suspension (Figure 5-97). Fill only the tubes (not the cupules) of all other tests.

8 Overlay the ADH, LDC, ODC, H₂S, and URE microtubes by filling the cupules with sterile mineral oil.

9 Close the incubation box and incubate at $35 \pm 2°C$ for 24 hours. (*Note:* If the tests cannot be read at 24 hours, have someone place the incubation box in a refrigerator until the tests can be read.)

Lab Two

1 Refer to Table 5-35 (Table 2 on the package insert) as you examine the test results on the strip.

2 Record all spontaneous reactions on the Result Sheet. (Spontaneous reactions are reactions completed without the addition of reagents.)

3 Add 1 drop of ferric chloride to the TDA microtube (Figure 5-98). A dark brown color indicates a positive reaction to be recorded on the Result Sheet.

4 Add 1 drop of potassium hydroxide reagent and 1 drop of alpha-naphthol reagent to the VP microtube. A pink or red color indicates a positive reaction to be recorded on the Result Sheet. If a slightly pink color appears in 10 to 12 minutes, the reaction should be considered negative.

5 Add 2 drops of sulfanilic acid reagent and 2 drops of N,N-dimethyl-1-naphthylamine reagent to the GLU microtube. Wait 2 to 3 minutes. A red color indicates a positive reaction ($NO_3 \rightarrow NO_2$). Enter this on the

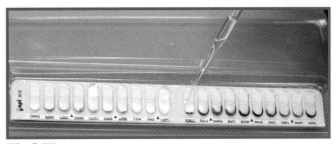

5-97 INOCULATE THE TUBES ✦ Inoculate the tubes by placing the pipette at the side of the tube and gently filling them with the suspension. If necessary to avoid creating bubbles, tilt the strip slightly. Tap gently with your index finger if necessary to remove bubbles.

5-98 ADD REAGENTS ✦ Follow the instructions in the text.

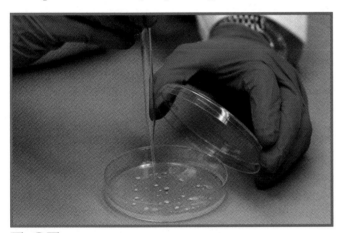

5-95 PICK A COLONY ✦ Being careful not to disturb the agar surface, pick up one isolated colony with a Pasteur pipette.

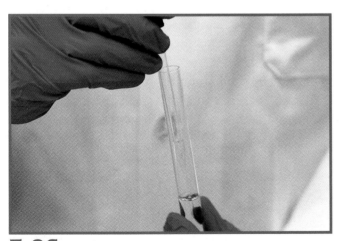

5-96 EMULSIFY THE COLONY ✦ Emulsify the colony in 5 mL Suspension Medium or sterile 0.85% saline. Do this by filling and emptying the pipette several times until uniform turbidity is achieved in the medium.

Result Sheet as positive. A yellow color at this point is inconclusive because the nitrate may have been reduced to nitrogen gas ($NO_3 \rightarrow N_2$, sometimes evidenced by gas bubbles). If the substrate in the tube remains yellow after adding reagents, add 2 to 3 mg of zinc dust to the tube. A yellow tube after 5 minutes is positive for N_2 and is recorded on the Result Sheet as positive. If the test turns pink-red after the addition of zinc, this is a negative reaction; the nitrates *still present* in the tube have been reduced by the zinc.

6 Add 1 drop of Kovacs' reagent to the IND microtube. Wait 2 minutes. A red ring indicates a positive reaction to be recorded on the Result Sheet.

7 Add the positive results within each group on the Result Sheet, using the values given, and enter the number in the circle below each group's results (Figure 5-92).

8 Locate the seven-digit code in the Analytical Profile Index, and identify your organism. Enter the name on the Result Sheet and tape it on the Data Sheet. Answer the questions.

References

API 20 E Identification System for *Enterobacteriaceae* and Other Gram-Negative Rods package insert.

API 20 E Analytical Profile Index.

TABLE **5-35** API 20 E Results and Interpretations

TABLE OF RESULTS

#	Test	Substrate/Activity	Result	Interpretation	Symbol
1	ONPG	O-nitrophenyl-β-D-galactopyranoside	Yellow[1]	Organism produces β-galactosidase	+
			Colorless	Organism does not produce β-galactosidase	−
2	ADH	Arginine	Red/Orange	Organism produces arginine dehydrolase	+
			Yellow	Organism does not produce arginine dehydrolase	−
3	LDC	Lysine	Red/Orange[2]	Organism produces lysine decarboxylase	+
			Yellow	Organism does not produce lysine decarboxylase	−
4	ODC	Ornithine	Red/Orange[2]	Organism produces ornithine decarboxylase	+
			Yellow	Organism does not produce ornithine decarboxylase	−
5	CIT	Sodium citrate	Blue-green/Blue[3]	Organism utilizes citrate as sole carbon source	+
			Pale green/Yellow	Organism does not utilize citrate	−
6	H$_2$S	Sodium thiosulfate	Black	Organism reduces sulfur	+
			Colorless/grayish	Organism does not reduce sulfur	−
7	URE	Urea	Red/Orange[2]	Organism produces urease	+
			Yellow	Organism does not produce urease	−
8	TDA	Tryptophan	Brown-red	Organism produces tryptophan deaminase	+
			Yellow	Organism does not produce tryptophan deaminase	−
9	IND	Tryptophane	Red ring	Organism produces indole	+
			Yellow	Organism does not produce indole	−
10	VP	Creatine Sodium pyruvate	Pink/red[4]	Organism produces acetoin	+
			Colorless	Organism does not produce acetoin	−
11	GEL	Kohn's charcoal gelatin	Diffusion of black pigment	Organism produces gelatinase	+
			No black pigment diffusion	Organism does not produce gelatinase	−
12	GLU	Glucose[5]	Yellow	Organism ferments glucose	+
			Blue/blue-green	Organism does not ferment glucose	−
13	MAN	Mannitol[5]	Yellow	Organism ferments mannitol	+
			Blue/blue-green	Organism does not ferment mannitol	−
14	INO	Inositol[5]	Yellow	Organism ferments inositol	+
			Blue/blue-green	Organism does not ferment inositol	−
15	SOR	Sorbitol[5]	Yellow	Organism ferments sorbitol	+
			Blue/blue-green	Organism does not ferment sorbitol	−

[1] A very pale yellow is also positive.
[2] Orange after 36 hours is negative.
[3] Reading made in the cupule.

[4] A slightly pink color after 5 minutes is negative.
[5] Fermentation begins in the lower portion of the tube; oxidation begins in the cupule.

TABLE **5-35** API 20 E Results and Interpretations (*continued*)

TABLE OF RESULTS

#	Test	Substrate/Activity	Result	Interpretation	Symbol
16	RHA	Rhamnose[5]	Yellow	Organism ferments rhamnose	+
			Blue/blue-green	Organism does not ferment rhamnose	−
17	SAC	Sucrose (saccharose)[5]	Yellow	Organism ferments sucrose	+
			Blue/blue-green	Organism does not ferment sucrose	−
18	MEL	Melibiose[5]	Yellow	Organism ferments melibiose	+
			Blue/blue-green	Organism does not ferment melibiose	−
19	AMY	Amygdalin[5]	Yellow	Organism ferments amygdalin	+
			Blue/blue-green	Organism does not ferment amygdalin	−
20	ARA	Arabinose[5]	Yellow	Organism ferments arabinose	+
			Blue/blue-green	Organism does not ferment arabinose	−
21	OX	Separate test done on paper test strip	Violet	Organism possesses cytochrome-oxidase	+
			Colorless	Organism does not possess cytochrome-oxidase	−
22	GLU (nitrate reduction)	Potassium nitrate	Red after addition of reagents	Organism reduces nitrate to nitrite	+
			Yellow after addition of reagents	Organism does not reduce nitrate to nitrite	−
			Yellow after addition of zinc	Organism reduces nitrate to N_2 gas	+
			Orange-red after addition of zinc	Organism does not reduce nitrate	−
23	MOB	Motility Medium or wet mount slide	Motility	Organism is motile	+
			Nonmotility	Organism is not motile	−
24	McC	MacConkey Medium	Growth	Organism is probably *Enterobacteriaceae*	+
			No growth	Organism is not *Enterobacteriaceae*	−
25	OF	Glucose	Yellow under mineral oil	Organism ferments glucose	+
			Green under mineral oil	Organism does not ferment glucose (not *Enterobacteriaceae*)	−
			Yellow without mineral oil	Organism either ferments glucose or utilizes it oxidatively	+
			Green without mineral oil	Organism does not utilize glucose (not *Enterobacteriaceae*)	−

EXERCISE 5-30   Enterotube® II

✦ Theory

The Enterotube® II is a multiple test system designed to identify enteric bacteria based on: glucose, adonitol, lactose, arabinose, sorbitol, and dulcitol fermentation, lysine and ornithine decarboxylation, sulfur reduction, indole production, acetoin production from glucose fermentation, phenylalanine deamination, urea hydrolysis, and citrate utilization.

The Enterotube® II, as diagramed in Figure 5-99, is a tube containing 12 individual chambers. Inside the tube, running lengthwise through its center, is a removable wire. After the end-caps are removed (aseptically), one end of the wire is touched to an isolated colony (on a streak plate) and drawn back through the tube to inoculate the media in each chamber (Figures 5-100 through 5-104).

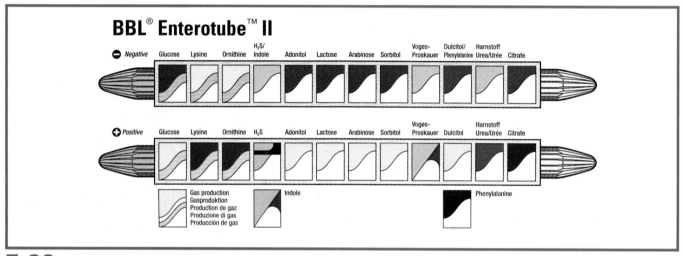

5-99   ENTEROTUBE® II TEST RESULT DIAGRAM ✦

Illustration courtesy Becton Dickinson and Co.

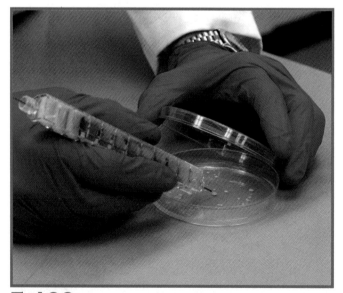

5-100   HARVEST GROWTH ✦ Using the sterile tip of the wire, aseptically remove growth from a colony on the agar surface. Do not dig into the agar. The inoculum should be large enough to be visible.

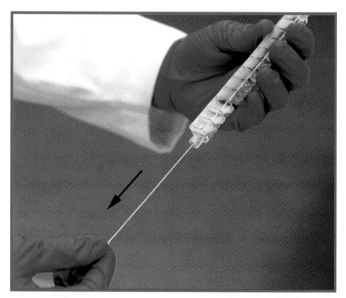

5-101   PULL WIRE THROUGH ✦ Loosen the wire by turning it slightly. While continuing to rotate it withdraw the wire until its tip is inside the last compartment (glucose).

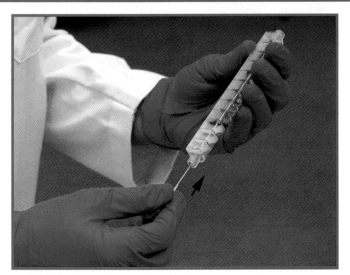

5-102 REINSERT WIRE ✦ Slide the wire back into the Enterotube® II until you can see the tip inside the citrate compartment. The notch in the wire should be lined up with the end of the tube nearest the glucose compartment.

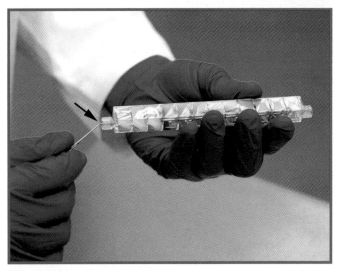

5-103 BREAK WIRE ✦ Bend the wire until it breaks off at the notch.

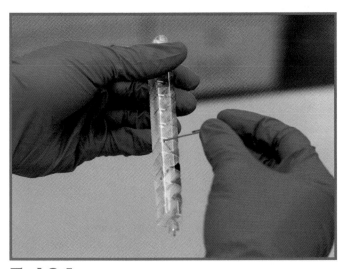

5-104 PUNCTURE AIR INLETS ✦ Using the broken wire, puncture the plastic covering the eight air inlets on the back side of the tube.

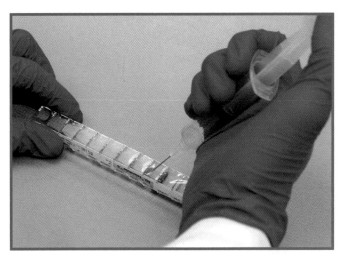

5-105 ADD REAGENTS AFTER INCUBATION ✦ Use a needle and syringe to add the reagents through the plastic film on the flat side of the tube. Do this with the tube in a test tube rack and the glucose end pointing down. The indole test should be done only after other tests have been read. Add the Voges-Proskauer reagents only if directed to do so by the CCIS.

After 18 to 24 hours' incubation, the results are interpreted, the indole test is performed (Figure 5-105), and the tube is scored on an Enterotube® II Results Sheet (Figure 5-106). As shown in the figure, the combination of positives entered on the score sheet results in a five-digit numeric code. This code is used for identification in the Enterotube® II Computer Coding and Identification System (CCIS) (Figure 5-107).

The CCIS is a master list of all enterics and their assigned numeric codes. In most cases, the five-digit number applies to a single organism, but when two or more species share the same code, a confirmatory VP test is performed to further differentiate the organisms.

An Enterotube® II before and after inoculation is shown in Figure 5-108.

✦ Application

The Enterotube® II is a multiple test system used for rapid identification of bacteria from the family *Enterobacteriaceae*.

✦ In This Exercise

You will use a multitest system to run an entire battery of tests designed to identify a member of the *Enterobacteriaceae*.

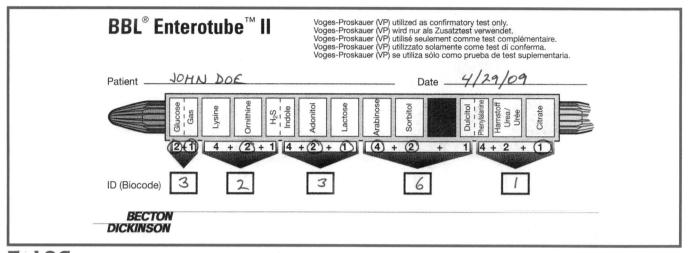

BBL® Enterotube™ II

Voges-Proskauer (VP) utilized as confirmatory test only.
Voges-Proskauer (VP) wird nur als Zusatztest verwendet.
Voges-Proskauer (VP) utilisé seulement comme test complémentaire.
Voges-Proskauer (VP) utilizzato solamente come test di conferma.
Voges-Proskauer (VP) se utiliza sólo como prueba de test suplementaria.

Patient __JOHN DOE__   Date __4/29/09__

BECTON
DICKINSON

5-106 BBL™ ENTEROTUBE® II RESULT SHEET ✦ This is the result sheet for the tube in Figure 5-108B. The ID value obtained was 32361, which identifies the organism as *Enterobacter aerogenes*.

ID VALUE	ORGANISM IDENTIFICATION	ATYPICAL TESTS	CONFIRMATORY TESTS				
			MAL	TRE			
20406	PROTEUS VULGARIS	H2S−	+	V(30%)			
	MORGANELLA MORGANII	ORN−	−	−			
	PROVIDENCIA STUARTII	PAD−	−	+			
20407	PROVIDENCIA STUARTII	NONE					
			SHA				
20410	*SHIGELLA SEROGROUPS A, B OR C	DUL+	+				
			SHA	ADA			
20420	*SHIGELLA SEROGROUPS A, B OR C	SOR+	+	−			
	ESCHERICHIA COLI (AD)	ARA−	−	+			
20425	PROVIDENCIA STUARTII	SOR+					
20427	PROVIDENCIA STUARTII	SOR+					
20430	ESCHERICHIA COLI (AD)	ARA−					
			SHA	ADA	MOT	VP	YEL
20440	*SHIGELLA SEROGROUPS A, B OR C	NONE	+	−	−	−	−
	ESCHERICHIA COLI (AD)	SOR−	−	+	−	−	−
	ENTEROBACTER AGGLOMERANS	IND+	−	−	+	V(70%)	+(75%)
20441	ENTEROBACTER AGGLOMERANS	IND+					

5-107 A SAMPLE OF THE ENTEROTUBE® II COMPUTER CODING AND IDENTIFICATION SYSTEM ✦ The five-digit numbers on the Enterotube® II Results Pad are matched against the database in the CCIS to get identification.

5-108 ENTEROTUBE® II TEST RESULTS ✦ **A** Uninoculated tube. **B** Tube inoculated with *Enterobacter aerogenes* after 24 hours incubation. This tube shows atypical negative results for lysine decarboxylase; after an additional 24 hours incubation the lysine medium turned purple (+).

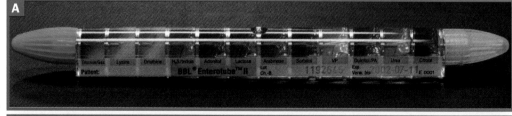

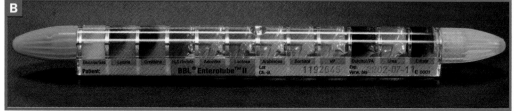

✦ Materials

Per Class

✦ Enterotube® II Computer Coding and Identification System (CCIS) identification booklet from Becton Dickinson Microbiology Systems

Per Student Group

✦ Becton Dickinson Microbiology Systems' Enterotube® II Identification System for *Enterobacteriaceae*

✦ Kovacs' reagent

✦ VP reagents (KOH and α-naphthol)

✦ needles and syringes or disposable pipettes for addition of reagents

✦ streak plate containing colonies of an unknown enteric (appropriate to facilities and course level)

✦ Reagent Recipes

Kovacs' Reagent

✦ Amyl alcohol	75.0 mL
✦ Hydrochloric acid, concentrated	25.0 mL
✦ *p*-dimethylaminobenzaldehyde	5.0 g

KOH Reagent

✦ Potassium hydroxide	20.0 g
✦ Distilled water to bring volume to	100.0 mL

α-Naphthol Reagent

✦ α-naphthol	5.0 g
✦ Absolute Ethanol to bring volume to	100.0 mL

Procedure

Lab One

1 Perform a Gram stain on your isolate to verify that it is a Gram-negative rod.

2 Place a paper towel on the tabletop and soak it with disinfectant.

3 Remove the blue cap and then the white cap from the Enterotube® II, being careful not to contaminate the sterile wire tip. Place the caps open end down on the paper towel.

4 Aseptically remove a large amount of growth from one of the plated colonies (Figure 5-100). Try not to remove any of the agar with it.

5 Grasp the looped end of the wire and, while turning, gently pull it back through all of the Enterotube® II compartments (Figure 5-101). You do not have to

completely remove the wire, but be careful to pull it back far enough to inoculate the last compartment.

6 Using the same turning motion as described above, slide the wire back into the tube (Figure 5-102). Push it in until the tip of the wire is inside the citrate compartment and the notch in the wire lines up with the opposite end of the tube.

7 Bend the wire at the notch until it breaks off (Figure 5-103).

8 Locate the air inlets on the side of the tube opposite the Enterotube® II label. Using the removed piece of inoculating wire, puncture the plastic membrane in the adonitol, lactose, arabinose, sorbitol, Voges-Proskauer, dulcitol/PA, urea, and citrate compartments (Figure 5-104). These openings will create the necessary aerobic conditions for growth. Be careful not to perforate the plastic film covering the flat side of the tube.

9 Discard the wire in disinfectant or a sharps container and replace the Enterotube® II caps.

10 Incubate the tube lying flat at 35 ± 2°C for 18 to 24 hours. (***Note:*** If the tube cannot be read at 24 hours, have someone place it in a refrigerator until it can be read.)

Lab Two

1 Examine the tube and circle the appropriate positive results on the Enterotube® II Results Pad.

2 *After* circling all positive test results on the Enterotube® II Results Pad, place the tube in a rack with the glucose compartment on the bottom. Puncture the plastic membrane of the H₂S/Indole compartment and (with a needle and syringe or disposable pipette) add one or two drops Kovacs' reagent to the compartment (Figure 5-105).

3 Observe the compartment for the formation of a red color within 10 seconds. Record a positive result on the Enterotube® II Results Pad.

4 Add the numbers in each bracketed section on the Enterotube® II Results Pad to obtain a five-digit number. Find the number in the CCIS. If not directed to perform a VP test, record the name of your organism on the score sheet and skip to #7 below. If directed to run a confirmatory Voges-Proskauer test, proceed to #5.

5 Puncture the plastic membrane of the VP compartment and (with a needle and syringe or disposable pipette) add two drops of the KOH reagent and three drops of the α-Naphthol reagent. Observe

for the formation of red color within 20 minutes. Interpret the result using the CCIS.

6 Discard the Enterotube II® in an appropriate autoclave container.

7 Enter all results on the Data Sheet, and answer the questions.

References

Enterotube® II Identification System for *Enterobacteriaceae* package insert.

Enterotube® II Computer Coding and Identification System.

TABLE **5-36** Enterotube® II Results and Interpretations

TABLE OF RESULTS

Compartment	Test	Result	Interpretation	Symbol
1	Glucose	Red	Organism does not ferment glucose	−
		Yellow, wax not lifted	Organism ferments glucose to acid	+
		Yellow, wax lifted	Organism ferments glucose to acid and gas	+
2	Lysine	Purple	Organism decarboxylates lysine	+
		Yellow	Organism does not decarboxylate lysine	−
3	Ornithine	Purple	Organism decarboxylates ornithine	+
		Yellow	Organism does not decarboxylate ornithine	−
4	H₂S	Beige	Organism does not reduce sulfur	−
		Black	Organism reduces sulfur	+
	Indole	Red after Kovacs' reagent	Organism produces indole	+
		No red after Kovacs' reagent	Organism does not produce indole	−
5	Adonitol	Red	Organism does not ferment adonitol	−
		Yellow	Organism ferments adonitol	+
6	Lactose	Red	Organism does not ferment lactose	−
		Yellow	Organism ferments lactose	+
7	Arabinose	Red	Organism does not ferment arabinose	−
		Yellow	Organism ferments arabinose	+
8	Sorbitol	Red	Organism does not ferment sorbitol	−
		Yellow	Organism ferments sorbitol	+
9	VP	Colorless	Organism does not produce acetoin	−
		Red	Organism produces acetoin	+
10	Dulcitol	Green	Organism does not ferment dulcitol	−
		Yellow	Organism ferments dulcitol	+
	PA	Green	Organism does not deaminate phenylalanine	−
		Black-smokey-gray	Organism deaminates phenylalanine	+
11	Urea	Beige	Organism does not hydrolyze urea	−
		Red-purple	Organism hydrolyzes urea	+
12	Citrate	Green	Organism does not utilize citrate	−
		Blue	Organism utilizes citrate	+

Bacterial Unknowns Project

The exercise in this unit is actually a term project to be done when you have completed Sections One through Five. These sections prepare you with the skills needed to successfully do what professional laboratory microbiologists do daily—that is, *isolate* from mixed culture, *grow* in pure culture, and *identify* unknown species of bacteria. Scared? Don't worry. You will not be asked to perform any task not introduced in previous sections. If you have any uncertainty regarding performance of these activities, refer to the appropriate exercises.

Although this project may seem intimidating, it is manageable if you don't try to do too much too fast. Take your time and think about what you want to do next, based on earlier results. Use the information from the lessons learned in Exercise 3-14 and closely follow the procedures listed.

This exercise gives you a unique opportunity (and responsibility) to set your own course in a learning project. Take advantage of it, as there is great fun and satisfaction to be had. Our experience tells us that students traditionally do one of three things. They

1. treat the project as a fun puzzle to solve,

2. become completely stressed out and hate every minute of it,

3. get lost early on and, for whatever reason, don't ask for help until it is too late to receive meaningful assistance.

Many of you will be nervous, perhaps even confused at first. Your instructor understands this. It is normal and actually can help you to focus if you use it in a positive way. For your success and overall satisfaction with this project, *ask your instructor for help if you need it*. Happy hunting! ✦

EXERCISE

5-31 Bacterial Unknowns Project

✦ In This Exercise

You will be given a mixture of two bacteria in broth culture. Your first job is to get the bacteria isolated and growing in pure culture, using the techniques of Sections Two and Four. Next, you will employ Section Three techniques as you perform Gram stains on each to determine Gram reaction, as well as cell morphology and size. Then, results from the differential biochemical tests of Section Five will lead you to identification of your unknowns. And, of course, all work must be done safely and aseptically using the methods covered in Lab Safety and Section One.

As in Exercise 3-14 (the Morphological Unknown), you will be expected to keep accurate records of all activities, including mistakes and unexpected or equivocal results. You also will be expected to construct a flowchart for each to show the tests you ran and the organisms eliminated by each until you have eliminated all but your unknown. The flowchart is a visual presentation of your thought processes in solving this problem.

✦ Materials

Recommended organisms (to be mixed in pairs in a broth by the lab technician immediately prior to use—generally, a few drops of the Gram-negative added to an overnight culture of the Gram-positive provides an appropriate ratio for isolation and growth).

✦ **Gram-positives**
  ✦ *Bacillus cereus*
  ✦ *Bacillus coagulans*
  ✦ *Corynebacterium xerosis*

- ◆ *Enterococcus faecalis* (BSL-2)
- ◆ *Kocuria rosea*
- ◆ *Lactobacillus plantarum*
- ◆ *Lactococcus lactis*
- ◆ *Micrococcus luteus*
- ◆ *Mycobacterium smegmatis* (BSL-2)
- ◆ *Staphylococcus aureus* (BSL-2)
- ◆ *Staphylococcus epidermidis*

✦ **Gram-negatives**
- ◆ *Aeromonas hydrophila* (BSL-2)
- ◆ *Alcaligenes faecalis* (BSL-2)
- ◆ *Chromobacterium violaceum* (BSL-2)
- ◆ *Citrobacter amalonaticus*
- ◆ *Enterobacter aerogenes*
- ◆ *Erwinia amylovora*
- ◆ *Escherichia coli*
- ◆ *Hafnia alvei*
- ◆ *Moraxella catarrhalis* (BSL-2)
- ◆ *Morganella morganii* (BSL-2)
- ◆ *Neisseria sicca*
- ◆ *Proteus mirabilis* (BSL-2)
- ◆ *Pseudomonas aeruginosa* (BSL-2)

✦ Appropriate stains and biochemical media

Procedure

1 Preliminary duties

a. Your instructor will tell you which organisms will actually be used based on your lab's inventory. You also will be advised as to the optimum temperature for the specific strain your lab has of each organism.

b. Your instructor will tell you which media and stains will be available for testing.

c. Working as a class, you will determine the results of each organism for each test available. This information will provide a database of results that you can use to compare against your unknown's results. We recommend running and using these "class controls" in lieu of referring to results in *Bergey's Manual of Systematic Bacteriology* or some other standard reference, for the following reasons:

- ◆ As many as 10% of the strains of a species listed as positive or negative on a test give the opposite result.
- ◆ Many species have even higher strain variability.
- ◆ Not all test results are listed for all organisms.

d. Class control tests should be run for the standard times, unless the timing of class sessions makes this impractical. It is imperative that incubation times for tests on unknowns be exactly the same as was used for the controls.

e. You are to incubate your class controls at the optimum temperature for each species as given to you by your instructor.

f. Class control results will be collected, tabulated, and distributed to each student. Your instructor will provide details on this process.

2 Isolation of the Unknown

a. You will be given a broth containing a *fresh* mixture of two unknown bacteria selected from the list of possible organisms. Enter the number of your unknown on your Data Sheets.

b. Mix the broth, then streak for isolation on two agar plates. Your instructor will tell you what media are available for use. You may be supplied with an undefined medium, such as Trypticase Soy Agar, or a selective medium, such as Phenylethyl Alcohol Agar or Desoxycholate Agar. Enter all relevant information concerning your isolation procedures on the Data Sheets, including date, medium, source of bacteria, type of inoculation, and incubation temperature.

c. Incubate one agar plate at 25°C and the other at 35 ± 2°C for at least 24 hours.

d. After incubation, check for isolated colonies that have different morphologies. If you have isolation of both, go on to step 3a. If you do *not* have isolation of *either*, continue with step 2e. If you have isolation of only one, go to step 3a for the isolated species and step 2e for the one not isolated. Be sure to enter relevant information on the Data Sheets.

e. If you do not have isolation, follow the advice that best matches your situation. (Be sure to record everything you do in the isolation. Include the date, source of inoculum, medium to which it is being transferred, type of inoculation and incubation temperature.)

- ◆ Look for growth with different colony morphologies, even if they're not separate. If you see different growth, use a portion of each and streak more plates. Then incubate them for either a shorter time or at a suboptimal temperature, because they grew *too* well the first time.
- ◆ If you see only one type of growth, ask for a selective medium that favors growth of the one you're missing. (This will require a Gram stain

of the one you do have.) Streak the mixture and incubate again. Also, reincubate your original plate. Some species are slow growers, so their absence may be a result of a slow growth rate.

After incubation, observe the plate(s) and repeat or go to step 3a, whichever is appropriate. You also may consult with your instructor for guidance in particularly difficult situations.

3 Growing the Unknown in Pure Culture

a. Once you have isolation of an unknown, transfer a portion of its colony to an agar slant to produce a pure culture. Use the rest of the colony for a Gram stain. (If you don't have enough of the colony left for a Gram stain, do the Gram stain on your pure culture after incubation.)

b. After Gram staining, label your pure culture accordingly.

c. Enter the following information on your Data Sheet: optimum growth temperature, colony morphology, cell morphology and arrangement, and cell size. Also, complete the description of your isolation procedure by noting the source of the isolate, the medium to which you are transferring it, and the incubation temperature.

4 Identification of the Unknown

a. Follow this procedure for each unknown. Be sure to enter on the Data Sheet, the inoculation and reading dates for each test. Also include the test result and any comments about the test that explain any deviation from standard procedure (*e.g.*, reading tests before or after the accepted incubation time, running a test and not using it in the flowchart).

b. Construct a flowchart that divides all the organisms first by Gram reaction, then by cellular morphology. Use the flowchart in Figure 5-109 as a style guide even though the actual organisms you use will be different.

c. You will have two options for proceeding at this point. Your instructor will let you know which to use.

(1) Find the group of organisms on the flowchart that matches the results of your unknown. Then, look at your class controls results for *just those organisms*, and choose a test that will divide these organisms into at least two groups. (Also consider your ability to return and read the test after the appropriate incubation time. That is, don't inoculate a 48-hour

test on Thursday if you can't get into the lab on Saturday!) Continue the flowchart from the branch with the remaining candidates for your unknown.

(2) Rather than allowing you to choose just any test that works, you may be required to follow some form of standard approach to identification. Your instructor will provide you with relevant information based on the organism inventory you are working with.

d. Inoculate the test medium you have chosen, and incubate it for the appropriate time. Use the optimum temperature for growth as given by your instructor. *It is important to run your tests at that optimum temperature because this is the temperature at which the class controls were run.*

e. While the test is incubating, you should begin planning what your next test will be. A final decision about the next test cannot be made until you have results from the first, but you can decide which test to run if the result is positive, and which one to run if it is negative. This applies to all stages of the flowchart: Because you won't know if a subsequent test is relevant or not until you have results from the current one, *you should not inoculate a medium until you have those results.* It's really easy: Run one test at a time for each unknown. (**Note:** In a clinical situation, where rapid identification of a pathogen may save a patient's life, tests routinely are run concurrently. But remember that correct identification is only one objective of this project. More important is for you to demonstrate an understanding of the logic behind the process and execute it in the most efficient manner.)

f. Repeat the process of inoculating a medium, getting the results, and then choosing a subsequent test until you eliminate all but one organism. This *should* be your unknown. Then continue with step 4g.

g. When you have eliminated all but one organism, you will run one more test—the confirmatory test. This must be a test that has not been run previously on your organism. It's also nice (but not necessary) if the test you choose gives a positive result. (In general, we have more confidence in positive results than in negative results, because false positives are usually harder to get than false negatives.) The confirmatory test provides you with the unique opportunity to predict the result before you run the test. If it matches, you are more certain that you have correctly identified your unknown. Continue with step 4j. If it doesn't match, continue with step 4h.

h. If your confirmatory test doesn't match the result you expected for your unknown, check with your instructor to see where your organism was eliminated incorrectly. In most cases, it will be difficult at this point to determine what was responsible— you, the class controls, or the organism itself. Misidentification may be a result of one or any combination of factors:

♦ The test procedure could have been done incorrectly by you or the person responsible for running class controls on your organism.

♦ The test may have been interpreted incorrectly by you or the person responsible for running class controls on your organism.

♦ The inoculum in your test or the class controls might have been too small to give a positive result in the limited incubation time.

♦ The wrong organism could have been inoculated at the time of the test or the class controls (many look alike in a tube, and once the label goes on, as far as the microbiologist is concerned, that culture becomes the labeled organism regardless of what's really in there!)

♦ For whatever reason, the organism maybe didn't react "correctly" during your test or during the class controls.

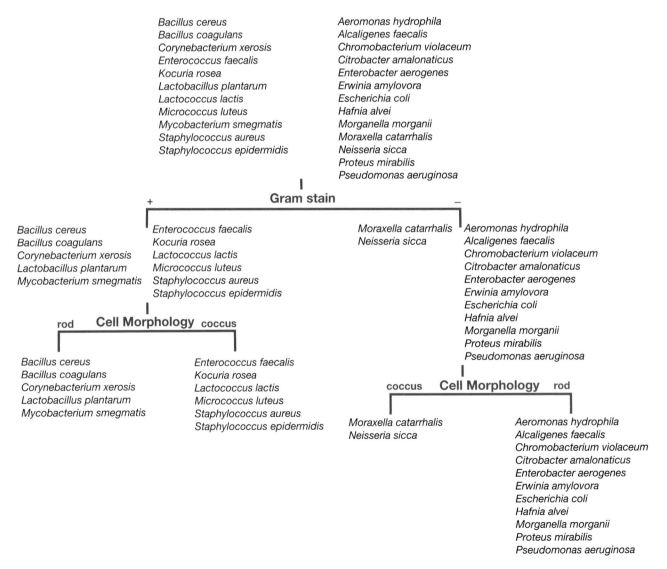

5-109 SAMPLE FLOW CHART ♦ Use this flow chart as a guide in constructing your own containing the organisms available in your laboratory.

i. Based on your instructor's advice, you will do one of the following:

- ◆ Rerun the test where you incorrectly eliminated your unknown (and perhaps rerun the test on the remaining organisms to check the class control results), or

- ◆ Rerun the test where you incorrectly eliminated your unknown without running the controls again, or

- ◆ Eliminate the problematic test from your flowchart, but use the other tests you've already done. If these don't allow you to identify your unknown, more tests will need to be run. Continue with step 4f.

(**Note:** There will undoubtedly be some "mistakes" uncovered in the class controls as they get repeated. These have to be reported to the class so classmates can incorporate the correct information into their flowcharts.)

j. When you have correctly identified your unknown, complete the Data Sheet, and turn it in. Your instructor will advise you as to the point value of each section, the grading scale, and any other items that are required.

Quantitative Techniques

Quantitative techniques in microbiology usually consist of counting cells or counting colonies formed by cells growing on agar plates. A typical result is a close estimate of the number of cells (or viruses) in one milliliter of sample—the population density or **cell density**.

Knowledge of microbial **population density** is an important part of many areas of microbiology. Food and environmental microbiologists use population density for food production as well as the detection of food or water contamination. Medical microbiologists manipulate population density for use in standardized tests. Many researchers use population density to measure the effect of varying nutritional or environmental conditions. Industrial microbiologists maintain microbial populations at optimum levels in large-scale fermenters for the manufacture of many products including enzymes, antibiotics, beer, and wine.

In this section we include five bacterial counting techniques and one exercise to measure a viral population. One technique involves directly counting cells, the other four involve plate counts. As you will see, all techniques involve dilutions and all require calculations to determine the *original* population density.

The standard dilution procedure used in microbial counts is the **serial dilution**. A serial dilution is a series of individual measured transfers, starting with an undiluted sample, one tube to the next, each reducing the cell density by a known amount, until it becomes dilute enough to be used in an appropriate counting procedure (Figure 6-1). The advantage of a serial dilution is that it not only reduces the cell density to a manageable size, but does so systematically so that each dilution tube contains a *known* volume of the undiluted original.

An important term you will see in the following exercises is **colony forming units**, or **CFU**. In plate counts, colonies are counted, but the count is usually recorded in CFU. CFU better describes the cell density, because colonies that grow on plates may start as single cells or as groups of cells, depending on the typical cellular arrangement of the organism. (**Note:** Viral samples are recorded in PFU [plaque forming units], which will be explained fully in Exercise 6-5.)

As you proceed through the exercises in this section, you will gain understanding and proficiency with serial dilutions. Because these exercises involve math, some of you will be a little (or a lot) intimidated. Take heart; these calculations are much simpler than they appear! If you are nervous about the math, we encourage you to see your instructor and to work as many practice problems as you can get. Before you know it, you will see these equations as a welcome solution to the very real problem of how to manage otherwise unmanageable numbers!

Most of the quantitative techniques in this section were originally designed for measurements and calculations in milliliters. Many school laboratories now are equipped with digital micropipettes that have the ability to deliver volumes as small as 1.0 microliter (1.0 μL = 0.001 mL). To accommodate schools with modest budgets and for ease of instruction, we have written the following exercises for measurements and calculations in milliliters. Alternative procedures, adapted for digital micropipettes, are included in Appendix F. ✦

Standard Plate Count (Viable Count)

✦ Theory

The standard plate count is a procedure that allows microbiologists to estimate the population density in a liquid sample by plating a very dilute portion of that sample and counting the number of colonies it produces.

The inoculum that is transferred to the plate contains a *known* proportion of the original sample because it is the product of a **serial dilution**.

As shown in Figure 6-1, a serial dilution is simply a series of controlled transfers down a line of **dilution blanks** (tubes containing a known volume of sterile

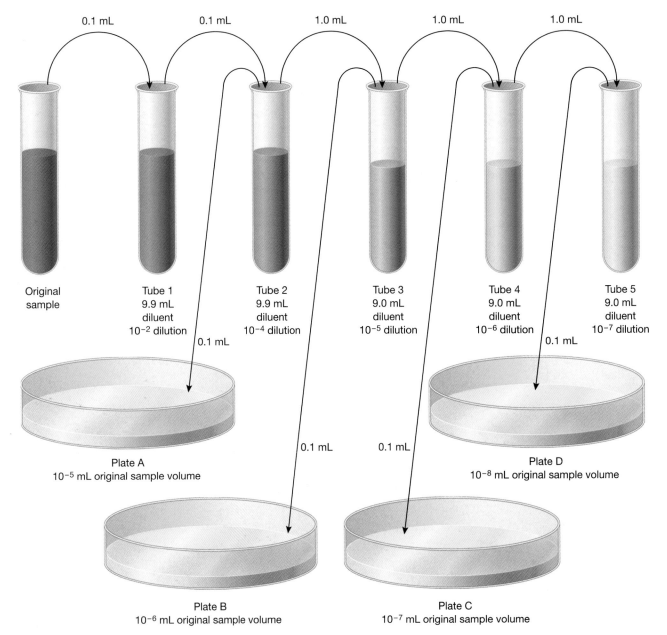

| 0.1 mL | 0.1 mL | 1.0 mL | 1.0 mL | 1.0 mL |

| Original sample | Tube 1 9.9 mL diluent 10^{-2} dilution | Tube 2 9.9 mL diluent 10^{-4} dilution | Tube 3 9.0 mL diluent 10^{-5} dilution | Tube 4 9.0 mL diluent 10^{-6} dilution | Tube 5 9.0 mL diluent 10^{-7} dilution |

0.1 mL

0.1 mL   0.1 mL

Plate A
10^{-5} mL original sample volume

Plate D
10^{-8} mL original sample volume

0.1 mL   0.1 mL

Plate B
10^{-6} mL original sample volume

Plate C
10^{-7} mL original sample volume

6-1   SERIAL DILUTION PROCEDURAL DIAGRAM ✦ This is an illustration of the dilution scheme outlined in the Procedure. The dilution assigned to each tube (written below the tube) represents the proportion of original sample inside that tube. For example, if the dilution is 10^{-4}, the proportion of original sample inside the tube would be 1/10000th of the total volume inside. When 0.1 mL of that solution is transferred to a plate, the volume of sample in the plate is 0.1 mL $\times$ 10^{-4} = 10^{-5} mL.

diluent—water, saline, or buffer). The series begins with a sample containing an unknown concentration of cells (density) and ends with a very dilute mixture containing only a few cells. Each dilution blank in the series receives a known volume from the mixture in the previous tube and delivers a known volume to the next, typically reducing the cell density to 1/10 or 1/100 at each step.

For example, if the original sample contains 1,000,000 cells/mL, following the first transfer the 1/100 dilution in dilution tube 1 would contain 10,000 cells/mL. In the second dilution (tube 2) the 1/100 dilution would reduce it further to 100 cells/mL. Because the cell density of the original sample is not known at this time, only the dilutions (without mL units) are recorded on the dilution tubes. By convention, dilutions are expressed in scientific notation. Therefore, a 1/10 dilution is written as 10^{-1} and a 1/100 dilution is written as 10^{-2}.

A small portion of appropriate dilutions (depending on the *estimated* cell density of the original sample) is then spread onto agar plates to produce at least one **countable plate**. A countable plate is one that contains between 30 and 300 colonies (Figure 6-2). A count lower than 30 colonies is considered statistically unreliable and greater than 300 is typically too many to be viewed as individual colonies.

In examining the procedural diagram, you can see that the first transfer in the series is a simple dilution, but that all successive transfers are compound dilutions.

Both types of dilutions can be calculated using the following formula,

$$V_1 D_1 = V_2 D_2$$

where V_1 and D_1 are the volume and dilution of the concentrated broth, respectively, while V_2 and D_2 are the volume and dilution of the completed dilution. Undiluted samples are always expressed as 1. Therefore, to calculate the dilution of a 1 mL sample transferred to 9 mL of diluent, the permuted formula would be used as follows.

$$D_2 = \frac{V_1 D_1}{V_2} = \frac{1.0\ \text{mL} \times 1}{10\ \text{mL}} = \frac{1}{10} = 10^{-1}$$

As mentioned above, compound dilutions are calculated using the same formula. However, because D_1, in compound dilutions, no longer represents undiluted sample, but rather a fraction of the original density, it must be represented as something less than 1 (*i.e.*, 10^{-1}, 10^{-2}, *etc.*) For example, if 1 mL of the 10^{-1} dilution from the last example were transferred to 9 mL of diluent, it would become a 10^{-2} dilution as follows.[1]

$$D_2 = \frac{V_1 D_1}{V_2} = \frac{1.0\ \text{mL} \times 10^{-1}}{10\ \text{mL}} = 10^{-1} \times 10^{-1} = 10^{-2}$$

Spreading a known volume of this dilution onto an agar plate and counting the colonies that develop would give you all the information you need to calculate the original cell density (OCD). Below is the basic formula for this calculation.

$$OCD = \frac{CFU}{D \times V}$$

CFU (colony forming units) is actually the number of colonies that develop on the plate. CFU is the preferred term because colonies could develop from single cells or from groups of cells, depending on the typical cellular arrangement of the organism. *D* is the dilution as written on the dilution tube from which the inoculum comes. *V* is the volume transferred to the plate. (**Note:** The volume is included in the formula because densities are expressed in CFU/mL, therefore a 0.1 mL inoculation (which would contain 1/10th as many cells as 1 mL) must be accounted for.)

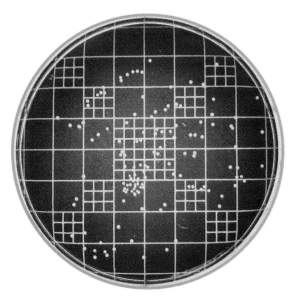

6-2 COUNTABLE PLATE ✦ A countable plate has between 30 and 300 colonies. Therefore, this plate with approximately 130 colonies is countable and can be used to calculate cell density in the original sample. Plates with fewer than 30 colonies are TFTC ("too few to count"). Plates with more than 300 colonies are TMTC ("too many to count").

[1] Permutations of this formula work with all necessary dilution calculations. For calculations involving unconventional volumes or dilutions, the formula is essential, but for simple ten-fold or hundred-fold dilutions like the ones described in this exercise, the final compounded dilution in a series can be calculated simply by multiplying each of the simple dilutions by each other. For example, a series of three 10^{-1} dilutions would yield a final dilution of 10^{-3} ($10^{-1} \times 10^{-1} \times 10^{-1} = 10^{-3}$). Three 10^{-2} dilutions would yield a final dilution of 10^{-6} ($10^{-2} \times 10^{-2} \times 10^{-2} = 10^{-6}$). We encourage you to use whatever means is best for you. In time you will be doing the calculations in your head.

As you can see in the formula, the volume of *original sample* being transferred to a plate is the product of the *volume transferred* and the *dilution* of the tube from which it came. Therefore 0.1 mL transferred from 10^{-2} dilution contains only 10^{-3} mL of the original sample (0.1 mL $\times$ 10^{-2}). The convention among microbiologists is to condense D and V in the formula into "Original sample volume" (expressed in mL). The formula thus becomes,

$$OCD = \frac{CFU}{\text{Original sample volume}}$$

The sample volume is written on the plate at the time of inoculation. Following a period of incubation, the plates are examined, colonies are counted on the countable plates, and calculation is a simple division problem. Suppose, for example, you inoculated a plate with 0.1 mL of a 10^{-5} dilution. This plate now contains 10^{-6} mL of original sample. Calculation would be as follows.

$$OCD = \frac{CFU}{\text{Sample volume}} = \frac{37\ CFU}{10^{-6}\ mL} = 3.7 \times 10^7\ CFU/mL$$

✦ Application

The viable count is one method of determining the density of a microbial population. It provides an estimate of actual *living* cells in the sample.

✦ In This Exercise

You will perform a dilution series and determine the population density of a broth culture of *Escherichia coli*. You will inoculate the plates using the **spread plate technique,** as illustrated in Exercise 1-5. As described in Figure 1-31, the inocula from the dilution tubes will be evenly dispersed over the agar surface with a bent glass rod. You will be sterilizing the glass rod between inoculations by immersing it in alcohol and igniting it. Be careful to organize your work area properly and *at all times keep the flame away from the alcohol beaker.*

✦ Materials

Per Student Group

- sterile 0.1 mL, 1.0 mL, and 10.0 mL pipettes
- mechanical pipettor
- five sterile dilution tubes
- flask of sterile normal saline
- eight Nutrient Agar plates
- beaker containing ethanol and a bent glass rod
- hand tally counter
- colony counter
- 24-hour broth culture of *Escherichia coli* (This culture will have between 10^7 and 10^{10} CFU/mL.)

 ## Procedure

Refer to the procedural diagram in Figure 6-1 as needed. Appendix F includes an alternative procedure for digital micropipettes using µL volumes.

Lab One

1 Obtain eight plates, organize them into four pairs, and label them A_1, A_2, B_1, B_2, *etc.*

2 Obtain five dilution tubes, and label them 1–5 respectively. Make sure they remain covered until needed.

3 Aseptically add 9.9 mL sterile water to dilution tubes 1 and 2. Cover when finished. Aseptically add 9.0 mL sterile water to dilution tubes 3, 4, and 5. Cover when finished.

4 Mix the broth culture, and aseptically transfer 0.1 mL to dilution tube 1. Mix well. This is a 10^{-2} dilution.

5 Aseptically transfer 0.1 mL from dilution tube 1 to dilution tube 2; mix well. This is 10^{-4}.

6 Aseptically transfer 1.0 mL from dilution tube 2 to dilution tube 3; mix well. This is 10^{-5}.

7 Aseptically transfer 1.0 mL from dilution tube 3 to dilution tube 4; mix well. This is 10^{-6}.

8 Aseptically transfer 1.0 mL from dilution tube 4 to dilution tube 5; mix well. This is 10^{-7}.

9 Aseptically transfer 0.1 mL from dilution tube 2 to plate A_1. Using the spread plate technique, disperse the sample evenly over the entire surface of the agar. Repeat the procedure with plate A_2 and label both plates "10^{-5} mL".

10 Following the same procedure, transfer 0.1 mL volumes from dilution tubes 3, 4, and 5 to plates B, C, and D, respectively. Label the plates accordingly. Allow the inocula to soak into the agar for a few minutes before continuing.

11 Invert the plates and incubate at 35°C for 24 to 48 hours.

Lab Two

1 After incubation, examine the plates and determine the countable pair—plates with 30 to 300 colonies. Only one pair of plates *should* be countable. The remainder should be "Too Few to Count" (TFTC) or "Too Numerous to Count" (TNTC).

2 Count the colonies on both plates, and calculate the average (Figure 6-3). Record these in the chart provided on the Data Sheet. (*Note:* In error, you may have more than one pair that is countable. Count *all* plates that have between 30–300 colonies for the practice, and try to identify which plate(s) you have the most confidence in. If *no* plates are in the 30–300 colony range, count the pair that is closest just for the practice and for purposes of the calculations.)

3 Using the formula provided in Data and Calculations on the Data Sheet, calculate the density of the original sample and record it in the space provided.

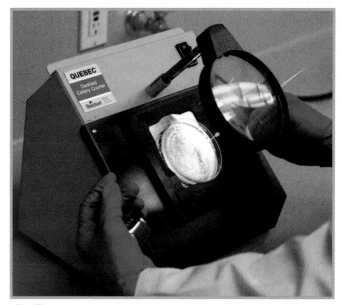

6-3 COUNTING BACTERIAL COLONIES ✦ Place the plate upside down on the colony counter. Turn on the light and adjust the magnifying glass until all the colonies are visible. Using the grid in the background as a guide, count colonies one section at a time. Mark each colony with a felt-tip marker as you record with a hand tally counter.

References

Collins, C. H., Patricia M. Lyne, and J. M. Grange. 1995. Page 149 in *Collins and Lyne's Microbiological Methods,* 7th ed. Butterworth-Heinemann, United Kingdom.

Koch, Arthur L. 1994. Page 254 in *Methods for General and Molecular Bacteriology,* edited by Philipp Gerhardt, R.G. E. Murray, Willis A. Wood, and Noel R. Krieg, American Society for Microbiology, Washington, DC.

Postgate, J. R. 1969. Page 611 in *Methods in Microbiology,* Vol. 1., edited by J. R. Norris and D. W. Ribbons. Academic Press, New York.

Urine Culture

✦ Theory

Urine culture is a semiquantitative CFU counting method that quickly produces countable plates without a serial dilution. The instrument used in this procedure is a volumetric loop, calibrated to hold 0.001 mL or 0.01 mL of sample. Urine culture procedures using volumetric loops are useful in situations where a rapid diagnosis is essential and approximations ($\pm 10^2$ CFU/mL) are sufficient to choose a course of action. Volumetric loops are useful in situations where population density is not likely to exceed 10^5 CFU/mL.

In this standard procedure, a loopful of urine is carefully transferred to a Blood Agar plate. The initial inoculation is a single streak across the diameter of the agar plate. Then the plate is turned 90° and (without flaming the loop) streaked again, this time across the original line in a zigzag pattern to evenly disperse the bacteria over the entire plate (Figures 6-4 and 6-5). Following a period of incubation, the resulting colonies are counted and population density, usually referred to as "original cell density," or OCD, is calculated.

OCD is recorded in "colony forming units," or CFU per milliliter (CFU/mL), as described in the introduction to this section. CFU/mL is determined by dividing the

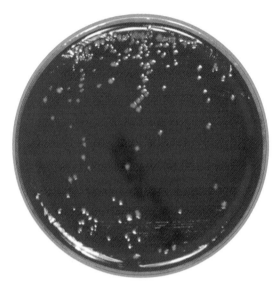

6-5 **URINE STREAK ON SHEEP BLOOD AGAR** ✦ This plate was inoculated with a 0.01 mL volumetric loop. The cell density can be determined by multiplying the number of colonies by 0.01.

number of colonies on the plate by the volume of the loop. For example, if 75 colonies are counted on a plate inoculated with a 0.001 mL loop, the calculation would be as follows:

$$OCD = \frac{CFU}{\text{loop volume}}$$

$$OCD = \frac{75 \text{ CFU}}{0.001 \text{ mL}}$$

$$OCD = 7.5 \times 10^4 \text{ CFU/mL}$$

✦ Application

Urine culture is a common method of detecting and quantifying urinary tract infections. It frequently is combined with selective media for specific identification of members of *Enterobacteriaceae* or *Streptococcus*.

✦ In This Exercise

You will estimate cell density in a urine sample using a volumetric loop and the above formula. Be sure to hold the loop vertically and transfer slowly.

6-4 **SEMIQUANTITATIVE STREAK METHOD** ✦ Streak 1 is a simple streak line across the diameter of the plate. Streak 2 is a tight streak across Streak 1 to cover the entire plate.

✦ Materials

Per Student Group

✦ one blood agar plate (TSA with 5% sheep blood)

✦ one sterile volumetric inoculating loop (either 0.01 mL or 0.001 mL)

✦ a fresh urine sample

 Procedure

Lab One

1 Holding the loop vertically, immerse it in the urine sample. Then carefully withdraw it to obtain the correct volume of urine. This loop is designed to fill to capacity in the vertical position. Do not tilt it until you get it in position over the plate.

2 Inoculate the blood agar by making a single streak across the diameter of the plate.

3 Turn the plate 90° and, without flaming the loop, streak the urine across the entire surface of the agar, as shown in Figure 6-4.

4 Invert, label, and incubate the plate for 24 hours at 35±2°C.

Lab Two

1 Remove the plate from the incubator and count the colonies. Also note any differing colony morphologies, which would indicate possible colonization by more than one species. Enter the data in the chart on the Data Sheet.

2 Calculate the original cell density of the sample using the formula on the Data Sheet. If two species are present, calculate each.

3 Enter the cell density(-ies) in the chart.

References

Forbes, Betty A., Daniel F. Sahm, and Alice S. Weissfeld. 2002. Pages 933–934 in *Bailey & Scott's Diagnostic Microbiology*, 11th ed. Mosby-Yearbook, St. Louis.

Winn, Washington C., *et al.* 2006. Page 85 in *Koneman's Color Atlas and Textbook of Diagnostic Microbiology*, 6th ed. Lippincott Williams & Wilkins, Baltimore.

Direct Count (Petroff-Hausser)

✦ Theory

Microbial direct counts, like plate counts, take a small portion of a sample and use the data gathered from it to calculate the overall population cell density. This is made possible with a device called a **Petroff-Hausser counting chamber.** The Petroff-Hausser counting chamber is very much like a microscope slide with a 0.02 mm deep chamber or "well" in the center containing an etched grid (Figure 6-6). The grid is one square millimeter and consists of 25 large squares, each of which contains 16 small squares, making a total of 400 small squares. Figures 6-7 and 6-8 illustrate the counting grid.

When the well is covered with a cover glass and filled with a suspension of cells, the volume of liquid above each small square is 5×10^{-8} mL. This may seem like an extremely small volume, but the space above each small square is large enough to hold about 50,000 average-size cocci! Fortunately, dilution procedures prevent this scenario from occurring and cell counting is easily done using the microscope.

As mentioned in the introduction to this section, cell density usually is referred to as "original cell density," or OCD, because most samples must be diluted before attempting a count. OCD is determined by counting the cells found in a predetermined group of small squares and dividing by the number of squares counted multiplied by the dilution[1] and the volume of sample above one small square.

The following is a standard formula for calculating original cell density in a direct count.

$$OCD = \frac{\text{Cells counted}}{(\text{Squares})(\text{Dilution})(\text{Volume})}$$

To maintain accuracy, some experts recommend a minimum overall count of 600 cells in one or more samples taken from a single population. Optimum density for counting is between 5 and 15 cells per small square.

---

[1] Dilutions are calculated using the following formula.

$$D_2 = \frac{V_1 D_1}{V_2}$$

D_2 is the new dilution to be determined. V_1 is the volume of sample being diluted. D_1 is the dilution of the sample before adding diluent (undiluted samples have a dilution factor of 1). V_2 is the new combined volume of sample and diluent after the dilution is completed.

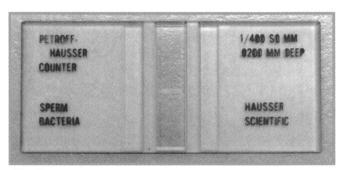

6-6 PETROFF-HAUSSER COUNTING CHAMBER ✦ The Petroff-Hausser counting chamber is a device used for the direct counting of bacterial cells. Bacterial suspension is drawn by capillary action from a pipette into the chamber enclosed by a coverslip. The cells are then counted against the grid of small squares in the center of the chamber.

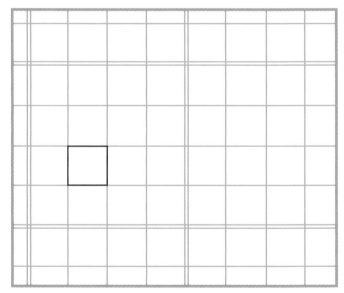

6-7 PETROFF-HAUSSER COUNTING CHAMBER GRID ✦ Shown is a portion of the Petroff-Hausser counting chamber grid. The smallest squares are the ones referred to in the formula. The volume above a small square is 5×10^{-8} mL. When cells land on a line, count them with the square below or to the right.

If, for example, 200 cells from a sample with a dilution factor of 10^{-2} were counted in 16 squares (remembering that the volume above a single small square is 5×10^{-8} mL), the cell density in the original sample would be calculated as follows:

$$OCD = \frac{\text{Cells counted}}{(\text{Squares})(\text{Dilution})(\text{Volume})}$$

6-8 ORGANISM IN A PETROFF-HAUSSER COUNTING CHAMBER ✦
This is a 10^{-4} dilution of *Vibrio natriegens* on the grid. Can you determine the original cell density?

$$OCD = \frac{200 \text{ cells}}{(16)(10^{-2})(5 \times 10^{-8} \text{ mL})}$$

$$OCD = 2.5 \times 10^{10} \text{ cells/mL}$$

✦ Application

The direct count method is used to determine bacterial cell density in a sample.

✦ In This Exercise

You will employ the direct count method to estimate cell density in a bacterial sample.

✦ Materials

Per Student Group

✦ Petroff-Hausser counting chamber with coverslip
✦ 1 mL pipettes with pipettor
✦ Pasteur pipettes with bulbs or disposable transfer pipettes
✦ hand counter
✦ staining agents A and B (Appendix G)
✦ test tubes
✦ overnight broth culture of:
   ✦ *Proteus vulgaris* (BSL-2)

 Procedure

1 Transfer 0.1 mL from the original 24-hour culture tube to a nonsterile test tube.

2 Add 0.4 mL Agent A and 0.5 mL Agent B; mix well. (This dilution may not be suitable in all situations. There should be 5 to 15 cells per small square for optimal results. Adjust the proportions of the broth culture and agents A and B if necessary to obtain a countable dilution, but try to keep the total solution volume at 1.0 mL for easier calculation of the dilution and cell density.)[2]

3 Place a coverslip on the Petroff-Hausser counting chamber and add a drop of the mixture at the edge of the coverslip. Capillary action will fill the chamber.

4 Place the counting chamber on the microscope and locate the grid in the center, using the low-power objective lens. Do not increase the magnification until you have found the grid on low power. (**Hint:** Close the iris diaphragm to bring the grid lines into view. Too much light will make them impossible to see.) Increase the magnification, focusing one objective at a time, until you have the cells and the grid in focus with the oil immersion lens. (In some cases, high dry is sufficient.)

5 Count the number of cells in at least 5 but no more than 16 small squares. Be consistent in your counting of cells that are on a line between squares. By convention, cells on lines belong to the square below or the square to the right.

6 Enter your data in the space provided on the Data Sheet and answer the questions.

References

Koch, Arthur L. 1994. Page 251 in *Methods for General and Molecular Bacteriology*, edited by Philipp Gerhardt, R. G. E. Murray, Willis A. Wood, and Noel R. Krieg. American Society for Microbiology, Washington, DC.

Postgate, J. R. 1969. Page 611 in *Methods in Microbiology*, Vol. 1, edited by J. R. Norris and D. W. Ribbons. Academic Press, Inc., New York.

---

[2] If the total solution volume results in something other than 1 mL, adjust your dilution as follows. For example, if you added 1.0 mL of sample to 0.4 mL Agent A and 1.0 mL Agent B, your dilution would be 1.0 mL in a total volume of 2.4 mL. In scientific notation, this dilution would be calculated as follows:

$$\frac{1.0 \text{ mL} \times 10^0}{2.4 \text{ mL} \times 10^0} = \frac{1.0 \text{ mL}}{2.4 \text{ mL}} \times 10^{0-0} = 0.42 = 4.2 \times 10^{-1}$$

Closed-System Growth

✦ Theory

A closed system—in this exercise, a broth culture in a flask—is one in which no nutrients are added beyond those present in the original medium and no wastes are removed. Bacteria grown in a closed system demonstrate four distinct growth phases: lag phase, exponential phase, stationary phase, and death phase.

The four phases together form a characteristic shape known as a microbial growth curve (Figure 6-9). **Lag phase,** the first phase, constitutes what might be called an adjustment period. Initially, there is no cell division. It is believed that microorganisms use this time to repair damaged cellular components and synthesize enzymes to begin using the resources of the new environment. The duration of lag phase can be quite variable and depends on many factors. Actively reproducing cells transferred to a medium identical to the one from which they are removed, will undergo a short lag period. Cells that are near death, have been refrigerated, or must adjust to a completely different medium will demonstrate a longer lag phase.

Eventually, as cells begin to use resources of the new environment and start reproducing, lag phase gradually gives way to the **exponential (or log) phase.** This phase is a time of maximum growth that is limited almost exclusively by the organism's reproductive potential. In a closed system, conditions usually are adjusted to be optimal for a specific organism; however, even under the best of conditions, the medium and other physical factors influence the growth rate slightly. Exponential growth is characterized by cellular division and doubling of the population size at regular intervals depending on the organism's generation time. Because not all cells are dividing at exactly the same time, exponential growth is marked by a smooth and dramatic upward sweep of growth, as shown in Figure 6-9.

As nutrients in the medium decrease and toxic waste products increase, the growth rate gradually declines to where the population's death rate is more or less equal to its reproductive rate. This leveling of growth is called the **stationary phase.** Stationary phase will last until the nutrients are depleted or the medium becomes toxic to the organism. **Death phase,** the final phase, is marked by the decline of the organism. As discussed in Exercise 6-6, microbial death is the reverse of microbial growth and is exponential. That is, a fixed proportion of the population will die in a given time (the time is specific to the organism), regardless of population size.

Several measurements are possible in a closed-system growth experiment. These measurements include the duration of each phase, the **mean growth rate constant, generation time,** and the organism's **minimum, maximum, and optimum (cardinal) temperatures.**

Mean growth rate constant (k) is calculated using the increase in population size between two points in time during exponential growth. The following is the standard formula:

$$k = \frac{\log N_2 - \log N_1}{(0.301)\, t}$$

where N_1 is the population size early in exponential phase, N_2 is the population size later in exponential phase, (t) is the elapsed time between the N_1 and N_2 readings, and 0.301 is the log of 2. Log 2 is used because bacterial reproduction typically occurs by binary fission; therefore, each reproductive cycle results in a doubling of the population size. Calculations of organisms using a different reproductive pattern require the log of a value appropriate to their typical number of offspring.

If population density calculation is not required, absorbance readings taken with the spectrophotometer early and late in the exponential phase (A_1 and A_2) can be substituted in the numerator as follows:

$$k = \frac{\log A_2 - \log A_1}{(0.301)\, t}$$

6-9 MICROBIAL GROWTH CURVE ✦ Note the different phases of growth.

For example, the mean growth rate constant of an organism whose absorbance readings increase from 0.195 to 0.815 in 2 hours would be:

$$k = \frac{\log 0.815 - \log 0.195}{(0.301)\, 2 \text{ hr}}$$

$$k = \frac{0.62}{0.602 \text{ hr}}$$

$$k = 1.03 \text{ generations/hr}$$

Whereas (k) is the number of generations per unit time, (g) is mean generation time—the amount of time required for a population to double. Mean generation time is simply the inverse of (k). It can be easily calculated using the following formula:

$$g = \frac{1}{k}$$

Taking the above example, the generation time of the organism would be:

$$g = \frac{1}{1.03 \text{ generations/hr}}$$

$$g = 0.97 \text{ hr/generation or 58 minutes/generation}$$

✦ Application

This exercise is designed to demonstrate the pattern of microbial growth in a closed system, expose you to the methods involved in determining growth, and demonstrate the calculation of mean growth rate and mean generation time.

✦ In This Exercise

You will use a spectrophotometer to measure the growth of *Vibrio natriegens* incubated at five different temperatures. The class will be split into five groups—each group incubating at a different temperature. Absorbance readings will be taken every 15 minutes. (As cell density increases, turbidity increases, and so does absorbance of the solution.) Using the data collected for your group, you will plot the growth curve either by computer or on the graph paper provided, with absorbance substituting for population size. You then will calculate the mean growth rate constant and generation time. Finally, using the class data, you will plot the mean growth rate constants over the range of temperatures to estimate the minimum, maximum, and optimum growth (cardinal) temperatures for the organism.

✦ Materials

One of Each Per Student Group

✦ five water baths set at 22°C, 25°C, 31°C, 37°C, and 40°C

✦ thermometers (to monitor actual temperatures in the ice and water baths)

✦ five sterile side-arm flasks containing 47 mL of sterile Brain Heart Infusion (BHI) Broth + 2% NaCl

✦ five sterile side-arm flasks containing 50 mL BHI + 2% NaCl to be used as controls

✦ spectrophotometers

✦ lab tissues

✦ sterile 5 mL pipettes and pipettor or micropipettes (10–100 µL and 100–1000 µL) with sterile tips

✦ Bunsen burner

✦ 20 mL Overnight broth culture of *Vibrio natriegens* in BHI + 2% NaCl (one per class)

✦ Medium Recipe

Brain Heart Infusion Agar

◆ Infusion from calf brains	200 g
◆ Infusion from beef hearts	250 g
◆ Peptone	10 g
◆ Dextrose	2 g
◆ Sodium chloride	5 g
◆ Disodium Phosphate	2.5 g
◆ Distilled or deionized water	1.0 L

pH 7.2–7.6 at 25°C

Procedure

1 Collect all necessary materials: a spectrophotometer, a sterile 5 mL pipette and mechanical pipettor (or micropipette and tips), the side-arm flasks containing 47 mL and 50 mL of broth, and lab tissues. Label the 50 mL flask "control."

2 Immediately label your 47 mL growth flask; place it into the appropriate water bath and allow 15 minutes for the temperature to equilibrate. Turn on the spectrophotometer and let it warm up for a few minutes.

3 If the spectrophotometer is digital, set it to "absorbance." Set the wavelength to 650 nm. (For instruction on spectrophotometer operation, refer to Appendix E.)

4 When all groups are ready to start (all groups *must* start within minutes of each other), mix the

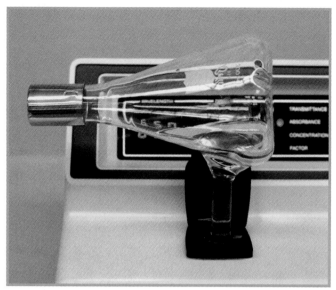

6-10 SIDE-ARM FLASK IN THE SPECTROPHOTOMETER ✦ Insert and remove the flask carefully to avoid spilling the broth. Place the flask in the spectrophotometer the same direction each time to avoid inconsistent readings caused by aberrations in the glass. Cover the port with your fingers to reduce light entering the machine.

V. natriegens culture and aseptically add 3.0 mL to your growth flask. Mix it, wipe it dry making sure the side arm is clean, and immediately take a turbidity reading (Figure 6-10). This time is T_0. (*Caution:* Be careful not to do this too quickly, as the side-arm flask is *very easy to spill* when it is tipped on its side for reading. Further, if it is pulled out on an angle, you risk breaking the side-arm and spilling the culture.)

5 Record the absorbance next to your temperature under T_0 in the chart provided on the Data Sheet. Remove the growth flask carefully and return it to the water bath.

6 Monitor the temperature in your water/ice bath frequently. It is more important to have a constant temperature than to have exactly the assigned temperature. Record your *actual* temperature in the chart on the Data Sheet.

7 Repeat the above procedure every 15 minutes (T_{15}, T_{30}, T_{45}, *etc.*), taking turbidity readings and recording them in the chart until you have completed and recorded the T_{240} reading. Make sure to place the flask in the spectrophotometer the same direction each time. Consistency in your readings will help compensate for irregularities in the glass.) Also be sure to cover the sample port with your fingers to reduce light from entering the machine.

References

Collins, C. H., Patricia M. Lyne, J. M. Grange. 1995. Page 149 in Collins and Lyne's *Microbiological Methods*, 7th ed. Butterworth-Heinemann, Oxford, United Kingdom.

Gerhardt, Philipp, et. al. 1994. Chapter 11 in *Methods for General and Molecular Bacteriology*. American Society for Microbiology, Washington DC.

Koch, Arthur L. 1994. Page 251 in *Methods for General and Molecular Bacteriology*, edited by Philipp Gerhardt, R. G. E. Murray, Willis A. Wood, and Noel R. Krieg. American Society for Microbiology, Washington, DC.

Postgate, J. R. 1969. Page 611 in *Methods in Microbiology*, Vol. 1, edited by J. R. Norris and D. W. Ribbons. Academic Press, New York.

Prescott, Lansing M., John P. Harley, and Donald A. Klein. 1999. Page 114 in *Microbiology*, 4th ed. WCB/McGraw-Hill, Boston.

allowing the smaller bacteriophages to diffuse short distances and infect surrounding cells. During incubation, the phage host produces a lawn of growth on the plate in which plaques appear where contiguous cells have been lysed by the virus (Figure 6-14).

The procedure for counting plaques is the same as that for the standard plate count. To be statistically reliable, countable plates must have between 30 and 300 plaques. Calculating phage titer (original phage density) uses the same formula as other plate counts except that PFU (plaque forming unit) instead of CFU (colony forming unit) becomes the numerator in the equation. Phage titer, therefore, is expressed in PFU/mL and the formula is written as follows:

$$\text{Phage titer} = \frac{\text{PFU}}{\text{Volume plated} \times \text{Dilution}}$$

As with the standard plate count (Exercise 6-1), it is customary to condense *volume plated* and *dilution* into *original sample volume*. The formula then becomes:

$$\text{Phage titer} = \frac{\text{PFU}}{\text{Original sample volume}}$$

The original sample volume is written on the plate at the time of inoculation. Following a period of incubation, the plates are examined, plaques are counted on the countable plates, and calculation is a simple division problem.

Suppose, for example, you inoculated a plate with 0.1 mL of a 10^{-4} dilution. (Remember, you are calculating the *phage density*; the *E. coli* has nothing to do with the calculations.) This plate now contains 10^{-5} mL of original phage sample. If you subsequently counted 45 plaques on the plate, calculation would be as follows.

$$\text{Original phage density} = \frac{45 \text{ PFU}}{10^{-5} \text{ mL}} = 4.5 \times 10^6 \text{ PFU/mL}$$

✦ Application

This technique is used to determine the concentration of viral particles in a sample. Samples taken over a period of time can be used to construct a viral growth curve.

✦ In This Exercise

You will be estimating the density (titer) of a T4 coliphage sample using a strain of *Escherichia coli* (*E. coli* B) as the host organism.

✦ Materials

Per Class

✦ 50°C hot water bath containing tubes of liquid soft agar (7 tubes per group)

Per Student Group

✦ sterile 0.1 mL, 1.0 mL, and 10.0 mL pipettes
✦ 7 sterile dilution tubes
✦ 7 sterile capped microtubes
✦ 7 Nutrient Agar plates
✦ 7 tubes containing 2.5 mL Soft Agar
✦ small tube of sterile normal saline
✦ 7 sterile transfer pipettes
✦ T4 coliphage
✦ 24-hour broth culture of *Escherichia coli* B (T-series phage host)

✦ Medium Recipe

Soft Agar

✦ Beef extract	3.0 g
✦ Peptone	5.0 g
✦ Sodium chloride	5.0 g
✦ Tryptone	2.5 g
✦ Yeast extract	2.5 g
✦ Agar	7.0 g
✦ Distilled or deionized water	1.0 L

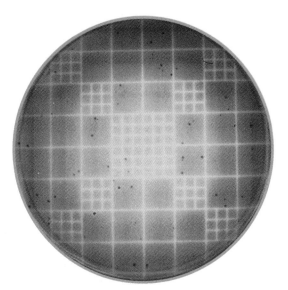

6-14 COUNTABLE PLATE ✦ This plaque assay plate has between 30 and 300 plaques; therefore, it is countable.

Procedure

Refer to the procedural diagram in Figure 6-12 as needed. Appendix F includes an alternative procedure for digital micropipettes and µL volumes.

Lab One

1. Obtain all materials except for the Soft Agar tubes. To keep the agar tubes liquefied, leave them in the water bath and take them out one at a time as needed.

2. Label seven tubes 1 through 7. Label the other seven tubes *E. coli* 1–7. Place all tubes in a rack, pairing like-numbered tubes.

3. Label the Nutrient Agar plates A through G. Place them in the 35°C incubator to warm them. Take them out one at a time as needed. This will keep the Soft Agar (added at step 14) from solidifying too quickly and result in a smoother agar surface.

4. Aseptically transfer 9.9 mL sterile normal saline to dilution tube 1.

5. Aseptically transfer 9.0 mL sterile normal saline to dilution tubes 2–7.

6. Mix the *E. coli* culture, and aseptically transfer 0.3 mL into each of the *E. coli* tubes.

7. Mix the T4 suspension, and aseptically transfer 0.1 mL to dilution tube 1. Mix well. This is a 10^{-2} dilution.

8. Aseptically transfer 1.0 mL from dilution tube 1 to dilution tube 2. Mix well. This is 10^{-3}.

9. Aseptically transfer 1.0 mL from dilution tube 2 to dilution tube 3. Mix well. This is 10^{-4}.

10. Continue in this manner through dilution tube 7. The dilution of tube 7 should be 10^{-8}.

11. Aseptically transfer 0.1 mL sterile normal saline to *E. coli* tube 1. This will be mixed with 2.5 mL Soft Agar and used to inoculate a control plate.

12. Aseptically transfer 0.1 mL from dilution tube 2 to its companion *E. coli* tube. Repeat this procedure with the remaining five tubes.

13. This is the beginning of the preadsorption period. Let all seven tubes stand undisturbed for 15 minutes.

14. Remove one Soft Agar tube from the hot water bath and, using a sterile transfer pipette, add the entire contents of *E. coli* tube 1. Mix well and immediately pour onto plate A. Gently tilt the plate back and forth until the soft agar mixture is spread evenly across the solid medium. Label the plate "Control."

15. Remove a second Soft Agar tube from the water bath and, using a sterile transfer pipette, add the entire contents of *E. coli* tube 2 . Mix well and immediately pour onto plate B. Tilt back and forth to cover the agar, and label it 10^{-4} mL.

16. Repeat this procedure with dilutions 10^{-4} thru 10^{-8} and plates C through G. Label the plates with the appropriate sample volume.

17. Allow the agar to solidify completely.

18. Invert the plates and incubate aerobically at $35 \pm 2°C$ for 24 to 48 hours.

Lab Two

1. After incubation, examine the control plate for growth and the absence of plaques.

2. Examine the remainder of your plates and determine which one is countable (30 to 300 plaques). Count the plaques and record the number in the chart provided on the Data Sheet. Record all others as either TNTC (too numerous to count) or TFTC (too few to count).

3. Using the sample volume on the countable plate and the formula provided on the Data Sheet, calculate the original phage titer. Record your results on the Data Sheet.

References

Collins, C. H., Patricia M. Lyne, J. M. Grange. 1995. Page 149 in *Collins and Lyne's Microbiological Methods*, 7th ed. Butterworth-Heinemann, United Kingdom.

DIFCO Laboratories. 1984. Page 619 in *DIFCO Manual*, 10th ed. DIFCO Laboratories, Detroit.

Province, David L., and Roy Curtiss III. 1994. Page 328 in *Methods for General and Molecular Bacteriology*, edited by Philipp Gerhardt, R. G. E. Murray, Willis A. Wood, and Noel R. Krieg. American Society for Microbiology, Washington, DC.

Thermal Death Time Versus Decimal Reduction Value

✦ Theory

The effects of excessive heat on bacteria range from disruption of cell membrane and protein function to complete combustion of the cell and all its components. Temperatures above 60°C are sufficient to kill most vegetative cells, whereas microbial spores may survive temperatures greater than 100°C.

The time needed to kill a specific number of microorganisms at a specific temperature is called **thermal death time (TDT)**. Because microbial death occurs exponentially—that is, the number of *dead* cells increases exponentially—TDT can be expressed in logarithmic terms. Figure 6-15 illustrates a typical death curve where the log of population size (in cells per milliliter) is plotted against time at a specific temperature. This hypothetical organism, in a suspension of 10^5 cells/mL, survives 25 minutes at 65°C.

One variation of TDT, commonly used by food processors, is the **decimal reduction value (D value)**. The D value is the time (in minutes at a specific temperature) needed to reduce the number of viable organisms in a given population by 90%—one logarithmic cycle. (In other words, a population that, upon heating, is reduced from a density of 1.2×10^7 to 1.2×10^6 cells/mL, converting to base 10 logarithmic terms, is reduced from $10^{7.08}$ to $10^{6.08}$ cells/mL—1 logarithmic cycle.) The temperature for the D value is customarily added as a subscript, (*e.g.,* D_{60}, D_{121}, *etc.*).

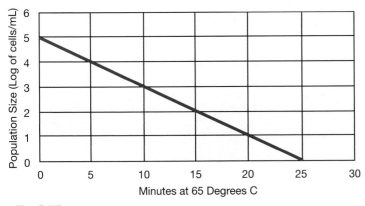

6-15 THERMAL DEATH TIME (TDT) CURVE ✦ With an original cell density of 10^5 cells/mL, this hypothetical population survives 25 minutes at 65°C. Notice that it takes 5 minutes to reduce the population size by a factor of 10, *regardless of the magnitude of the population*. What would be the TDT if the population had started at a density of 10^6 cells/mL?

D value is useful because it mathematically represents death rate as a constant for a species under particular circumstances and is independent of population size. An organism with a D_{60} value of 4 takes 4 minutes at 60°C to reduce a population by 90%, regardless of the starting number of cells. Mathematically, D value is defined as:

$$D_T = \frac{t}{\log_{10} x - \log_{10} y}$$

where

- T = temperature
- t = time in minutes
- x = cell density (cells per milliliter) before exposure to heat
- y = cell density after exposure to heat

For example, if a population of 1.7×10^5 cells heated at 60°C for 10 minutes was reduced to 2.6×10^3 cells, the D_{60} value would be 5.49 minutes.

$$D_{60} = \frac{10 \text{ min.}}{\log_{10} 1.7 \times 10^5 - \log_{10} 2.6 \times 10^3} = \frac{10 \text{ min.}}{5.23 \times 3.41} = 5.49 \text{ min.}$$

D value also can be plotted on a TDT standard curve, as shown in Figure 6-16. Either value (D value or TDT) can be determined from the other. A calculated D value requires two population density measurements—one before heating and one after heating. This produces two data points on the graph that can be used to construct a straight line TDT curve. Conversely, a TDT curve (determined by measuring the total time needed to kill the entire population) reveals the D value at the points (shown by arrows) where the curve intersects log values from the *y*-axis. The log values on the *y*-axis represent exponents for a base$_{10}$ logarithmic scale. In other words, 1 is interpreted as 10^1, 2 is 10^2, 3 is 10^3 cells per milliliter, *etc.*

✦ Application

This exercise is used to determine microorganismal thermal death time (TDT) and decimal reduction value at 60°C.

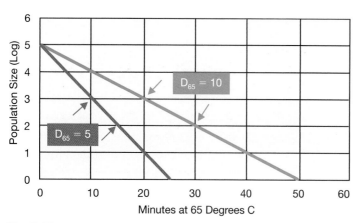

6-16 TDT CURVES OF TWO POPULATIONS ✦ Species A (purple) takes 5 minutes to reduce its population by 90% (10^3 to 10^2 cells) whereas species B (blue) takes 10 minutes. The graph can be constructed either by calculating the D value and extrapolating from two data points (shown by the arrows) or by plotting the TDT curve and determining D value from the points of intersection with log values. Both methods produce a straight line. As you can see, D value is independent of population size.

✦ In This Exercise

You will determine the D_{60} values of *Escherichia coli* and *Staphylococcus aureus* cultures using both methods described above. Your instructor will begin by adding bacteria to a broth maintained at 60°C. This is T_0—the beginning of the experiment. At this time and again at 10 minutes (T_{10}), a sample will be removed from the flask, diluted, and plated to perform viable counts to establish the density of the population.

In addition, beginning at T_1, samples will be removed from the flask at 1-minute intervals for 30 minutes (T_{30}). The samples will be used to inoculate Nutrient Broths. After a 48-hour incubation period, the Nutrient Broths will be examined for growth. The first broth in the sequence to produce no bacterial growth will establish the thermal death time of the population. The accumulated data then will be used to plot and calculate the D_{60} values of each organism.

In an attempt to divide the workload equitably and to allow time for all dilutions, transfers, and platings, the class will be divided into four groups, each of which will be further divided into a Plate Subgroup and a Broth Subgroup. Group One will be responsible for the T_0 platings of *E. coli*, and the inoculations of Nutrient Broths 1 through 15. Group 2 will be responsible for the T_{10} platings of *E. coli*, and the inoculations of Nutrient Broths 16 through 30. Group 3 will be responsible for the T_0 platings of *S. aureus*, and inoculations of Nutrient Broths 1 through 15. Group 4 will be responsible for the

T_{10} platings of *S. aureus*, and the inoculations of Nutrient Broths 16 through 30.

✦ Materials

Per Class

- ✦ stopwatch
- ✦ thermometers
- ✦ two flasks each containing 49.0 mL Nutrient Broth, maintained at 60°C in a water bath
- ✦ 24-hour broth cultures of:
  - ✦ *Escherichia coli*
  - ✦ *Staphylococcus aureus* (BSL-2)

Per Group

- ✦ 7 Nutrient Agar plates
- ✦ 16 Nutrient Broth tubes
- ✦ sterile 0.1 mL, 1.0 mL, 5.0 mL, and 10.0 mL pipettes
- ✦ sterile dilution tubes (large enough to hold 10.0 mL)
- ✦ sterile deionized water in a small flask or beaker
- ✦ beaker of alcohol and bent glass rod for spreading organisms

Procedure

[**Note:** This procedure uses the Spread Plate Technique described in Exercise 1-5. Refer to it as needed.] The Plate Subgroups will need at least two or three students for labeling tubes and plates, performing the serial dilution, and spreading plates. The Broth Subgroups will need one or two students to label and inoculate the 15 Nutrient Broths, keep track of time, and mix the broth prior to all transfers. Appendix F includes an alternate procedure for digital micropipettes.

Lab One: Group 1 (T_0)

Plate Subgroup

1 Obtain all plating materials. Place seven sterile test tubes in a rack. Label them 1 through 7. Add 9.0 mL of sterile water to tubes 2 through 7 and cover them until needed.

2 Label seven Nutrient Agar plates 1 through 7 respectively.

3 When your instructor adds 1.0 mL of the *E. coli* culture to the heated flask, he/she will start the stopwatch. This is T_0. Immediately swirl the flask to disperse the sample and remove 1.5 mL with your

pipette. If possible, do this step without removing the flask from the water bath. Begin the following serial dilution:

- Add the 1.5 mL of broth from the flask to tube 1.
- Transfer 1.0 mL from tube 1 to tube 2; mix. Discard the pipette. Use a clean pipette for each transfer from here forward.
- Transfer 1.0 mL from tube 2 to tube 3; mix.
- Transfer 1.0 mL from tube 3 to tube 4; mix.
- Transfer 1.0 mL from tube 4 to tube 5; mix.
- Transfer 1.0 mL from tube 5 to tube 6; mix.
- Transfer 1.0 mL from tube 6 to tube 7; mix.
- Transfer 0.1 mL from tube 1 to plate 1 and spread according to the spread-plate procedure.
- Transfer 0.1 mL from tube 2 to plate 2 and spread.
- Transfer 0.1 mL from tube 3 to plate 3 and spread.
- Transfer 0.1 mL from tube 4 to plate 4 and spread.
- Transfer 0.1 mL from tube 5 to plate 5 and spread.
- Transfer 0.1 mL from tube 6 to plate 6 and spread.
- Transfer 0.1 mL from tube 7 to plate 7 and spread.

4 Calculate the volume of original sample transferred to each plate (volume plated $\times$ dilution) and enter it in Chart 1 on the Data Sheet.

5 Tape the plates together, invert, and incubate them at $35 \pm 2°C$ for 48 hours.

Broth Subgroup

1 Obtain all broth transfer materials. Label 15 Nutrient Broth tubes T_1 through T_{15}. Label the 16th tube "Control."

2 When your instructor adds 1.0 mL of the *E. coli* culture to the heated flask, he/she will start the stopwatch. This is T_0. Slightly before T_1, gently swirl the flask to mix the broth. At exactly T_1, transfer 0.1 mL from the flask to the appropriately labeled Nutrient Broth tube. Discard the pipette. Use a clean pipette for each transfer.

3 Repeat this procedure at T_2 through T_{15}. Be sure to mix the broth before each transfer, and use a clean pipette.

4 Incubate all broths at $35 \pm 2°C$ for 48 hours.

---

* This procedure requires knowledge of dilutions and calculations, explained fully in Exercise 6-1, Standard Plate Count. If you haven't completed a standard plate count, or you have any doubt about your ability to complete this exercise, we suggest that you read Exercise 6-1 carefully and do some or all of the practice problems.

Lab One: Group 2 (T_{10})

Plate Subgroup

1 Obtain all plating materials. Place six sterile test tubes in a rack and label them 1 through 6. Add 9.0 mL of sterile water to tubes 2 through 6 and cover them until needed.

2 Label seven Nutrient Agar plates 1 through 7 respectively.

3 When your instructor adds 1.0 mL of the *E. coli* culture to the heated flask, he/she will start the stopwatch. This is T_0. At T_{10} remove 2.5 mL with your pipette. It may work best to wait for students in the broth group to withdraw their sample; then make your transfers as quickly as possible.

4 Begin the following serial dilution:

- Add the 2.5 mL of broth from the flask to tube 1.
- Transfer 1.0 mL from tube 1 to tube 2; mix. Discard the pipette. Use a clean pipette for each transfer from here forward.
- Transfer 1.0 mL from tube 2 to tube 3; mix.
- Transfer 1.0 mL from tube 3 to tube 4; mix.
- Transfer 1.0 mL from tube 4 to tube 5; mix.
- Transfer 1.0 mL from tube 5 to tube 6; mix.
- Transfer 1.0 mL from tube 1 to plate 1 and spread according to the spread plate procedure.
- Transfer 0.1 mL from tube 1 to plate 2 and spread.
- Transfer 0.1 mL from tube 2 to plate 3 and spread.
- Transfer 0.1 mL from tube 3 to plate 4 and spread.
- Transfer 0.1 mL from tube 4 to plate 5 and spread.
- Transfer 0.1 mL from tube 5 to plate 6 and spread.
- Transfer 0.1 mL from tube 6 to plate 7 and spread.

5 Calculate the volume of original sample transferred to each plate (volume plated $\times$ dilution) and enter it in Chart 1 on the Data Sheet.

6 Tape the plates together, invert, and incubate them at $35 \pm 2°C$ for 48 hours.

Broth Subgroup

1 Obtain all broth transfer materials. Label 15 Nutrient Broth tubes T_{16} through T_{30} respectively. Label the 16th tube "Control." Enter these numbers in the appropriate boxes in Chart 2 on the Data Sheet.

2 When your instructor adds 1.0 mL of the *E. coli* culture to the heated flask, he/she will start the stopwatch. This is T_0. Slightly before T_{16}, gently swirl the broth flask and at exactly T_{16}, transfer 0.1 mL from

the flask to the appropriately labeled Nutrient Broth tube. Discard the pipette tip.

3 Repeat this procedure at T_{17} through T_{30}. Be sure to mix the broth before each transfer, and use a clean pipette tip.

4 Incubate all broths at $35 \pm 2°C$ for 48 hours.

Lab One: Group 3 (T_0)

Plate Subgroup

Using the *S. aureus* culture, follow the Group 1 Plate Subgroup instructions.

Broth Subgroup

Using the *S. aureus* culture, follow the Group 1 Broth Subgroup instructions.

Lab One—Group 4 (T_{10})

Plate Subgroup

Using the *S. aureus* culture, follow the Group 2 Plate Subgroup instructions.

Broth Subgroup

Using the *S. aureus* culture, follow the Group 2 Broth Subgroup instructions.

Lab Two: All Groups

1 Remove all broths and plates from the incubator.

2 Pick the plate containing between 30 and 300 colonies. There should be only one. It is the countable plate. All others are either TFTC (too few to count) or TNTC (too numerous to count). Label them as such, and enter this information in Chart 1 on the Data Sheet.

3 If you haven't already done so, count the colonies on the countable plate and enter the number in Chart 1 on the Data Sheet.

4 Examine the broths for growth, comparing each one to the uninoculated control (tube #16). Any turbidity is read as positive for growth. Enter your results in Chart 2 on the Data Sheet. Circle the earliest time that no turbidity appears (*i.e.,* growth did not occur).

5 Calculate the population density of your group's organism for your specified time (T_0 or T_{10}), using the following formula:

$$\text{Cell density (CFU/mL)} = \frac{\text{CFU}}{\text{Original sample volume}}$$

(**Note:** "Original" in this formula refers to the *sample being heated,* not the culture tube used to inoculate it. As described in Exercise 6-1, calculate the *original sample volume* being transferred to a plate by multiplying the *dilution* of the sample by its *volume.*)

6 Enter your results in Chart 1 and again in Chart 3 on the Data Sheet. Also, be sure to convert your results to logarithmic form, and enter that in Chart 3 as well.

7 Collect the data from other groups and enter it in Chart 3 on the Data Sheet. Your instructor will likely provide a transparency or chalkboard space for class data.

8 Follow the directions on the Data Sheet to plot *and* calculate the D values of both organisms.

9 Answer the questions on the Data Sheet.

References

Ray, Bibek. 2001. Chapter 31 in *Fundamental Food Microbiology*, 2nd ed. CRC Press, Boca Raton.

National Canners Association Research Laboratories. 1968. *Laboratory Manual for Food Canners and Processors*, Vol. 1: *Microbiology and Processing.* AVI Publishing Co., Westport, CT.

Stumbo, Charles Raymond. 1973. Chapter 7 in *Thermobacteriology in Food Processing*, 2nd ed. Academic Press, New York.

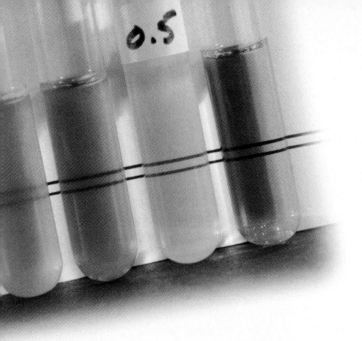

Medical Microbiology

The study and application of microbiological principles and those of medicine are inseparable. So, although this is not a medical microbiology manual, it nonetheless is devoted largely to the study of microorganisms and their relationship to human health. Our superficial treatment of medical microbiology in this section is not meant to represent a balanced body of information but, rather, is composed of exercises to augment those you have done already related to medical microbiology.

In this section you will perform a test to detect susceptibility to dental decay, do a quantitative test for an enzyme (lysozyme) that degrades Gram-positive cell walls, check the effectiveness of various antibiotics on sample bacteria, and demonstrate the formation of a medically important biofilm. These exercises are followed by two epidemiological exercises. In the first, you will use a standard resource—*Morbidity and Mortality Weekly Report* available from the Centers for Disease Control and Prevention (CDC) Web site—to follow the incidence of a disease of your choice over the course of two complete years. In the second exercise, you will simulate an epidemic outbreak in your class, determine the source of the outbreak, and perform epidemiological analyses of your class population.

The final three exercises in this section are a culmination of all you have done up to this point. You will be using skills developed in earlier exercises and apply them as a practicing microbiologist would. That is, you will be identifying unknown microbes from samples supplied to you by your instructor.

The opportunity to apply what you have learned to a real problem with practical applications is both challenging and exciting for most students. A word of caution, though. Your instructor will know what your unknown is and can tell you if your identification is correct. But out of the classroom and in a professional laboratory, no one can tell the microbiologist if he or she is correct or not. In fact, "identification" is more about finding out what the unknown *isn't,* rather than finding out with *certainty* what it is. There is an unspoken understanding that the unknown's identity is based on the best match between the unknown's test results and the accepted results for the same tests of the named species. But, the identification is provisional because other tests not run might lead to a different conclusion. Further, don't forget about those nasty false positives and false negatives. Now, with that in mind . . .

The three exercises employ traditional methods of microbial identification. That is, you will use staining properties, cell morphology, cell arrangement, and biochemical test results to identify an unknown organism assigned to you. In each exercise, a flowchart will direct you in your choice of tests. Matching your unknown's results to the flowchart will lead you, by the process of elimination, to a provisional identification. This methodology will be applied to common species of Gram-negative rods (Exercise 7-7), Gram-positive cocci (Exercise 7-8), and Gram-positive rods (Exercise 7-9).

You have already done most tests in the flowcharts. Some, such as fermentation of sugars, use the same PR medium as Exercise 5-2, just with a different sugar. These are read as in Exercise 5-2. The same applies to decarboxylase media (Exercise 5-10). ✦

EXERCISE

7-1

Snyder Test

✦ Theory

Snyder test medium is formulated to favor the growth of oral bacteria (Figure 7-1) and discourage the growth of other bacteria. This is accomplished by lowering the pH of the medium to 4.8. Glucose is added as a fermentable carbohydrate, and bromcresol green is the pH indicator. Lactobacilli and oral streptococci survive these harsh conditions, ferment the glucose, and lower the pH even further. The pH indicator, which is green at or above pH 4.8 and yellow below, turns yellow in the process. Development of yellow color in this medium, therefore, is evidence of fermentation and, further, is highly suggestive of the presence of dental decay-causing bacteria (Figure 7-2).

The medium is autoclaved for sterilization, cooled to just over 45°C, and maintained in a warm water bath until needed. The molten agar then is inoculated with a small amount of saliva, mixed well, and incubated for up to 72 hours. The agar tubes are checked at 24-hour intervals for any change in color. High susceptibility to dental caries is indicated if the medium turns yellow within 24 hours. Moderate and slight susceptibility are

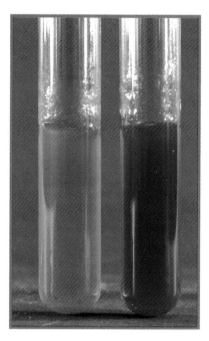

7-2 SNYDER TEST RESULTS ✦ A positive result is on the left, and a negative result is on the right.

indicated by a change within 48 and 72 hours, respectively. No change within 72 hours is considered a negative result. These results are summarized in Table 7-1.

✦ Application

The Snyder test is designed to measure susceptibility to dental caries (tooth decay), caused primarily by lactobacilli and oral streptococci.

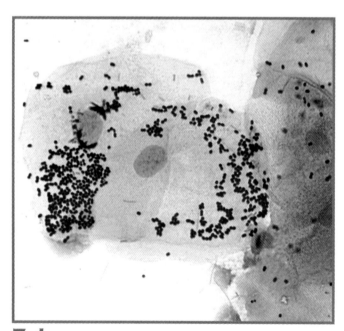

7-1 GRAM STAIN OF ORAL BACTERIA (X1000) ✦ Note the predominance of Gram-positive cocci. Also present are a few Gram-negative rods and very few Gram-positive rods (*Lactobacillus*). Note also the size difference between the large epithelial cells in the center and the bacteria.

TABLE **7-1** Snyder Test Results and Interpretations

TABLE OF RESULTS	
Result	**Interpretation**
Yellow at 24 hours	High susceptibility to dental caries
Yellow at 48 hours	Moderate susceptibility to dental caries
Yellow at 72 hours	Slight susceptibility to dental caries
Yellow at >72 hours	Negative

✦ In This Exercise

You will check your own oral microflora for its potential to produce dental caries.

✦ Materials

Per Student Group

✦ hot water bath set at 45°C

✦ small sterile beakers

✦ sterile 1 mL pipettes with bulbs

✦ two Snyder agar tubes

✦ Medium Recipe

Snyder Test Medium

◆ Pancreatic digest of casein	13.5 g
◆ Yeast extract	6.5 g
◆ Dextrose	20.0 g
◆ Sodium chloride	5.0 g
◆ Agar	16.0 g
◆ Bromcresol green	0.02 g
◆ Distilled or deionized water	1.0 L

pH 4.6–5.0 at 25°C

Procedure

Lab One

1 Collect a small sample of saliva (about 0.5 mL) in the sterile beaker.

2 Aseptically add 0.2 mL of the sample to a molten Snyder Agar tube (from the water bath), and roll it between your hands until the saliva is distributed uniformly throughout the agar.

3 Allow the agar to cool to room temperature. Do not slant.

4 Incubate with an uninoculated control at 35 ± 2°C for up to 72 hours.

Lab Two

1 Examine the tubes at 24-hour intervals for color changes.

2 Record your results in the chart provided on the Data Sheet.

Reference

Zimbro, Mary Jo, and David A. Power, Eds. 2003. Page 517 in *Difco™ and BBL™ Manual—Manual of Microbiological Culture Media.* Becton Dickinson and Co., Sparks, MD.

EXERCISE 7-2   Lysozyme Assay

✦ Theory

Lysozyme is an enzyme that occurs naturally in egg albumin, and normal body secretions such as tears, saliva, and urine. The enzyme provides limited protection from bacterial infection by breaking bonds in the cell wall's peptidoglycan. **Peptidoglycan** is made up of the alternating repeating glycan subunits *N*-acetylglucosamine (NAG) and *N*-acetylmuramic acid (NAM), cross-linked by peptides. Lysozyme functions by breaking the β-1,4 glycosidic linkages between the NAG and NAM subunits of the glycan polymers.

A lysozyme assay measures the ability of a sample to lyse cells of the substrate organism *Micrococcus lysodeikticus*. Cell lysis occurs as a result of damage caused by the lysozyme and the hypotonic diluent used to dilute the samples.

In this exercise you will measure the lysozyme concentrations in a variety of body fluids. To do this you will first construct a standard curve (see pages 8 and 9, and Figure 7-3) for comparison. First, you will mix diluted lysozyme samples of known concentration with an equal part of the bacterial substrate solution, then take an absorbance reading (with a spectrophotometer) of each after 20 minutes. You will then plot absorbance *vs.* lysozyme concentration to produce the standard curve. Dilutions of natural fluid samples can then be prepared, mixed with bacterial substrate solution and read for turbidity after 20 minutes. The light absorbance values of the samples can then be used to interpolate the lysozyme concentrations from the standard curve.

✦ Application

This exercise is used to estimate the relative concentrations of lysozyme in body fluids.

✦ In This Exercise

The work for this exercise will be divided among several groups. One group will perform the dilutions of known lysozyme concentrations for construction of the standard curve. Other groups will dilute and test samples of tears, saliva, urine and/or other body fluids as determined by your instructor. The data then will be shared among lab groups to complete the Data Sheet.

✦ Materials*

Per Student Group

- ✦ spectrophotometer with cuvettes
- ✦ timer
- ✦ micropipettes (100–1000 μL) with sterile tips
- ✦ 1 mL and 5 mL pipettes
- ✦ propipettes
- ✦ parafilm
- ✦ lysozyme buffer (20 mL for Group 1, 10 mL for other groups)
- ✦ lysozyme substrate—*Micrococcus lysodeikticus*—in solution measuring 10.0% transmittance (15 mL for Group 1, 5 mL for other groups)
- ✦ lysozyme dilutions shown in Table 7-2 (2.5 mL of each for Group 1)

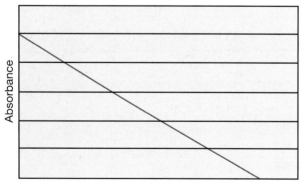

Standard Curve for Lysozyme Concentration and Absorbance After 20 Minutes Incubation

Absorbance (vertical axis) / Lysozyme Concentration (horizontal axis)

7-3 LYSOZYME STANDARD CURVE ✦ This is a typical standard curve illustrating an inverse correlation between light absorbance and lysozyme concentration. (See also pages 8 and 9 in the Introduction.) As the concentration of lysozyme increases and lysis of substrate cells increases, the solution in the tube absorbs less light. Lysozyme concentration in a sample can be estimated by finding the lysozyme concentration that corresponds to the absorbance value of the sample.

*Note to instructor: One liter of buffer is sufficient to make the lysozyme dilutions and substrate solution and for a class of 30 to 35 students.

TABLE **7-2** Lysozyme Dilutions

TABLE OF RESULTS		
Tube	Contains	In
A	0.15625 mg Lysozyme	100 mL Buffer
B	0.3125 mg Lysozyme	100 mL Buffer
C	0.625 mg Lysozyme	100 mL Buffer
D	1.25 mg Lysozyme	100 mL Buffer
E	2.5 mg Lysozyme	100 mL Buffer
F	5.0 mg Lysozyme	100 mL Buffer

Procedure

Group 1: Standard Curve

1 Turn on the spectrophotometer and allow it to warm up for a few minutes. Set the wavelength to 540 nm and, if it is digital, set it to absorbance. Be sure the filter is set properly for the wavelength used. (For directions on spectrophotometer use, see Appendix E.)

2 Obtain seven cuvettes. Transfer 2.0 mL of lysozyme substrate (*M. lysodeikticus*) solution to each of six cuvettes; transfer 4.0 mL straight lysozyme buffer to the seventh.

3 Blank the spectrophotometer with the cuvette containing the straight buffer. Continue to check the setting throughout the procedure and blank the machine as needed.

4 Transfer 2.0 mL of the 0.15625 mg/100 mL lysozyme dilution to a substrate-containing cuvette. Mark the time in the Standard Curve table on the Data Sheet. This is t_0 for this concentration. (**Note:** Time is critical for this part of the exercise. The absorbance of each lysozyme/substrate mixture must be read at *exactly 20 minutes*. Therefore, begin timing each transfer at the moment it is done.)

5 Continue transferring 2.0 mL each of the lysozyme dilutions to substrate-containing cuvettes, respectively. Mark each time on the Data Sheet.

6 Take the absorbance readings at t_{20} for each mixture. Record your results in the chart on the Data Sheet.

7 Share these data with the class.

Other Groups: Determination of Lysozyme in Body Fluid Samples

1 Turn on the spectrophotometer and allow it to warm up for a few minutes. Set the wavelength to 540 nm and, if it is digital, set it to absorbance. Be sure the filter is set properly for the wavelength used. (For directions on spectrophotometer use, see Appendix E.)

2 Obtain four cuvettes and all other necessary material for your body fluid sample.

3 Collect your assigned body fluid according to your teacher's instructions.

4 Add 0.2 mL (200 µL) undiluted sample to a cuvette. Add 3.8 mL lysozyme buffer. Mix well. Use this mixture to blank the spectrophotometer. Check this setting frequently throughout the exercise, and adjust as necessary.

5 Add 0.2 mL (200 µL) undiluted sample to a cuvette. Add 1.8 mL lysozyme buffer. Mix. This is a 10^{-1} dilution.

6 Transfer 0.2 mL from the 10^{-1} dilution to the third cuvette. Add 1.8 mL lysozyme buffer. Mix. This is a 10^{-2} dilution.

7 Add 1.8 mL lysozyme substrate to the 10^{-1} cuvette and mix well. Mark the time on the Data Sheet. This is t_0 for this mixture. (**Note:** Time is critical for this part of the exercise. The absorbance of each body fluid/substrate mixture must be read at *exactly 20 minutes*. Therefore, begin timing each transfer at the moment it is done.)

8 Add 2.0 mL lysozyme substrate to the 10^{-2} cuvette. Mix well. Mark the time on the Data Sheet. This is t_0 for this mixture.

9 Take the absorbance readings at t_{20} for each mixture. Record your results in the chart on the Data Sheet.

10 Share these t_{20} absorbance values of body fluids with the class.

References

DIFCO Laboratories. 1984. Page 515 in *DIFCO Manual*, 10th ed. DIFCO Laboratories, Detroit.

Sprott, G. Dennis, Susan F. Koval, and Carl A. Schnaitman. 1994. Page 78 in *Methods for General and Molecular Bacteriology*, edited by Philipp Gerhardt, R. G. E. Murray, Willis A. Wood, and Noel R. Krieg, American Society for Microbiology, Washington, DC.

Antimicrobial Susceptibility Test (Kirby-Bauer Method)

✦ Theory

Antibiotics are natural antimicrobial agents produced by microorganisms. One type of penicillin, for example, is produced by the mold *Penicillium notatum*. Today, because many agents that are used to treat bacterial infections are synthetic, the terms **antimicrobials** or **antimicrobics** are used to describe all substances used for this purpose.

The Kirby-Bauer test, also called the disk diffusion test, is a valuable standard tool for measuring the effectiveness of antimicrobics against pathogenic microorganisms. In the test, antimicrobic-impregnated paper disks are placed on a plate that is inoculated to form a bacterial lawn. The plates are incubated to allow growth of the bacteria and time for the agent to diffuse into the agar. As the drug moves through the agar, it establishes a concentration gradient. If the organism is susceptible to it, a clear zone will appear around the disk where growth has been inhibited (Figure 7-4).

The size of this **zone of inhibition** depends upon the sensitivity of the bacteria to the specific antimicrobial agent and the point at which the chemical's **minimum inhibitory concentration** (**MIC**) is reached. Some drugs kill the organism and are said to be **bactericidal**. Other drugs are **bacteriostatic**; they stop growth but don't kill the microbe.

All aspects of the Kirby-Bauer procedure are standardized to ensure reliable results. Therefore, care must be taken to adhere to these standards. Mueller-Hinton agar, which has a pH between 7.2 and 7.4, is poured to a depth of 4 mm in either 150 mm or 100 mm Petri dishes. The depth is important because of its effect upon the diffusion. Thick agar slows lateral diffusion and thus produces smaller zones than plates held to the 4 mm standard. Inoculation is made with a broth culture diluted to match a 0.5 McFarland turbidity standard (Figure 7-5).

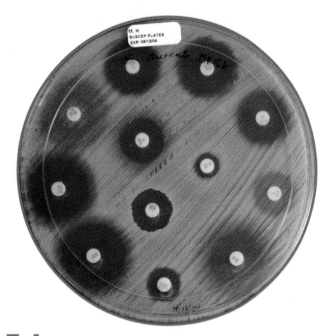

7-4 DISK DIFFUSION TEST OF METHICILLIN-RESISTANT STAPHYLO-COCCUS AUREUS (MRSA) ✦ This plate illustrates the effect of (clockwise from top outer right) Nitrofurantoin (F/M300) Norfloxacin (NOR 10), Oxacillin (OX 1), Sulfisoxazole (G 0.25), Ticarcillin (TIC 75), Trimethoprim-Sulfamethoxazole (SXT), Tetracycline (TE 30), Ceftizoxime (ZOX 30), Ciprofloxacin (CIP 5), and (inner circle from right) Penicillin (P 10), Vancomycin (VA 30), and Trimethoprim (TMP 5) on Methicillin-resistant *Staphylococcus aureus*.

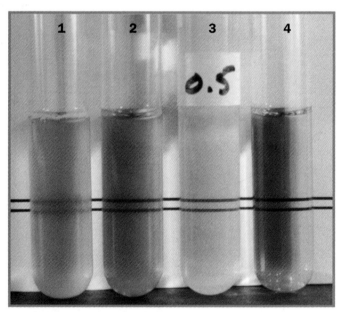

7-5 McFARLAND STANDARDS ✦ This is a comparison of a McFarland turbidity standard to three broths having varying degrees of turbidity. Each of the 11 McFarland standards (0.5 to 10) contains a specific percentage of precipitated barium sulfate to produce turbidity. In the Kirby-Bauer procedure, the test culture is diluted to match the 0.5 McFarland standard (roughly equivalent to 1.5×10^8 cells per mL) before inoculating the plate. Comparison is made visually by placing a card with sharp black lines behind the tubes. Tube 3 is the 0.5 McFarland standard. Notice that the turbidity of Tube 2 matches the McFarland standard exactly, whereas Tubes 1 and 4 are too turbid and too clear, respectively.

The disks, which contain a specified amount of the antimicrobial agent (printed on the disk) are dispensed onto the inoculated plate and incubated at $35 \pm 2°C$ (Figure 7-6). After 16 to 18 hours of incubation, the plates are removed and the clear zones are measured (Figure 7-7).

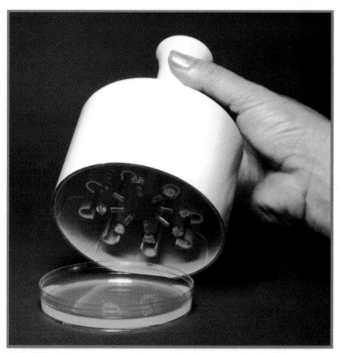

7-6 DISK DISPENSER ✦ This antibiotic disk dispenser is used to deposit disks uniformly on a Mueller-Hinton agar plate.

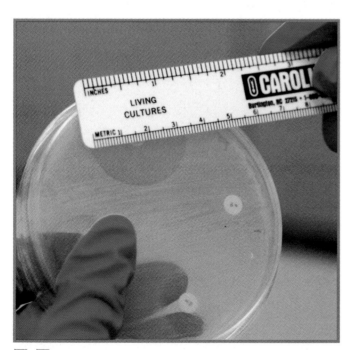

7-7 MEASURING THE ANTIMICROBIAL SUSCEPTIBILITY ZONES ✦ A metric ruler is used to measure the diameter of each clearing, in millimeters (mm).

✦ Application

Antimicrobial susceptibility testing is a standardized method that is used to measure the effectiveness of antibiotics and other chemotherapeutic agents on pathogenic microorganisms. In many cases, it is an essential tool in prescribing appropriate treatment.

✦ In This Exercise

You will test the susceptibility of *Escherichia coli* and *Staphylococcus aureus* strains to penicillin, chloramphenicol, trimethoprim, and ciprofloxacin. These antibiotics were chosen because they exhibit different modes of action on bacterial cells (Table 7-3).

✦ Materials

Per Student Group

+ two Mueller-Hinton agar plates
+ penicillin, chloramphenicol, trimethoprim, and ciprofloxacin antibiotic disks (and/or other disks as available)
+ antibiotic disk dispenser or forceps for placement of disks
+ small beaker of alcohol (for sterilizing forceps)
+ two sterile cotton swabs
+ one metric ruler
+ sterile saline (0.85%)
+ sterile transfer pipettes
+ one McFarland 0.5 standard with card
+ black, nonreflective poster board (8.5" ×11")
+ two Trypticase Soy Agar (or Nutrient Agar) plates (for optional procedure)
+ fresh broth cultures of:
  + *Escherichia coli*
  + *Staphylococcus aureus* (BSL-2)

✦ Medium Recipe

Mueller-Hinton II Agar

✦ Beef extract	2.0 g
✦ Acid hydrolysate of casein	17.5 g
✦ Starch	1.5 g
✦ Agar	17.0 g
✦ Distilled or deionized water	1.0 L

$pH = 7.2–7.4$ at 25°C

TABLE **7-3** Antibiotic Targets and Resistance Mechanisms ✦ Not all antibiotics affect cells in the same way. Some attack the bacterial cell wall, and others interfere with biosynthesis reactions. Resistance mechanisms can be broken down into three main categories: (a) altered target such that the antibiotic no longer can interact with the cellular process, (b) an alteration in how the drug is taken into the cell, and (c) enzymatic destruction of the drug.

Antibiotic	Cellular Target	Resistance Mechanism
Chloramphenicol	Prevents peptide bond formation during translation	1. Poor uptake of drug 2. Inactivation of drug
Ciprofloxacin	Interferes with DNA replication	1. Altered target 2. Poor uptake of drug
Trimethoprim	Inhibits purine and pyrimidine synthesis	1. Altered target
Penicillin	Inhibits cross-linking of the cell wall's peptidoglycan	One or more of: 1. Altered target 2. Poor uptake of drug 3. Production of β-lactamases

 Procedure

Lab One

1 Gently mix the *E. coli* culture and the McFarland standard until they reach their maximum turbidity.

2 Holding the culture and McFarland standard upright in front of you, place the card behind them so you can see the black line(s) through the liquid in the tubes. As you can see in Figure 7-5, the line becomes distorted by turbidity in the tubes. Use the black line to compare the turbidity level of the two tubes. Dilute the broth with sterile saline until it appears to have the same level of turbidity as the standard. If it is not turbid enough, incubate it until it reaches that level.

3 Repeat Steps 1 and 2 with the *S. aureus* culture.

4 Dip a sterile swab into the *E. coli* broth. As you remove it, press and rotate the cotton tip against the side of the tube to remove excess broth.

5 Inoculate a Mueller-Hinton plate with *E. coli* by streaking the entire surface of the agar three times with the swab. Your goal is confluent growth, so make the streaks right next to each other. When you have covered the surface, rotate the plate 1/3 turn and repeat the streaking of the inoculum already on the plate, using the same technique to produce confluent growth. Then rotate the plate another 1/3 turn and repeat.

6 Using a fresh sterile swab, inoculate the other plate with *S. aureus* in the same fashion to produce confluent growth.

7 Label the plates with the organisms' names, your name, and the date.

8 Apply the penicillin, chloramphenicol, ciprofloxacin, and trimethoprim disks to the agar surface of each plate. You can apply the disks either singly using sterile forceps, or with a dispenser (Figure 7-6). Be sure to space the disks sufficiently (4 to 5 cm) to prevent overlapping zones of inhibition. Also keep them away from the edge of the plate.

9 Press each disk gently with alcohol-flamed forceps so it makes full contact with the agar surface.

10 Invert the plates and incubate them aerobically at $35 \pm 2°C$ for 16 to 18 hours. Have a volunteer in the group remove and refrigerate the plates at the appropriate time.

Lab Two

1 Remove the plates from the incubator (or refrigerator). Hold the plate over the black, nonreflective posterboard and examine the plate with reflected light. The edge of a zone is where no growth is visible to the naked eye. Measure the diameter of each zone of inhibition in millimeters (Figure 7-7).

2 Using Table 7-4 and those provided with your antibiotic disks (if you used additional antibiotics), record your results in the table provided on the Data Sheet.

TABLE **7-4** Zone Diameter Interpretive Chart

Antibiotic	Organism(s)	Code	Disk Potency	Zone Diameter Interpretive Standards		
				Susceptible	Intermediate	Resistant
Chloramphenicol	Enterobacteriaceae and *Staphylococcus*	C 30	30 µg	≥18	13–17	≤12
Ciprofloxacin	Enterobacteriaceae and *Staphylococcus*	CIP 5	5 µg	≥21	16–20	≤15
Trimethoprim	Enterobacteriaceae and *Staphylococcus*	TMP 5	5 µg	≥16	11–15	≤10
Penicillin	*Staphylococcus*	P 10	10 U	≥29		≤28

This chart includes the antibiotics used in this exercise and contains data provided by the Clinical and Laboratory Standards Institute (CLSI). Permission to use portions (specifically Tables 2A and 2C) of M100-S19 (Performance Standards for Antimicrobial Susceptibility Testing; Nineteenth Informational Supplement) has been granted by CLSI. The interpretive data are valid only if the methodology in M02-A10 (Performance Standards for Antimicrobial Disk Susceptibility Tests—10th edition; Approved Standard) is followed. CLSI frequently updates the interpretive tables through new editions of the standard and supplements. Users should refer to the most recent editions. The current standard may be obtained from CLSI, 940 West Valley Road, Suite 1400, Wayne, PA 19087. Contact also may be made via the Web site (www.CLSI.org), email (customerservice@clsi.org) and by phone (1.877.447.1888).

Optional Procedure
Beginning with Lab Two

1 Obtain two TSA or NA plates. With your marking pen, divide the plates into four sectors (or more if you used more antibiotics).

2 Label each sector with an antibiotic. It is easiest if you label in the same order as these are found on the MH plates.

3 Label one plate *E. coli* and the other *S. aureus*.

4 Using a sterile loop for each transfer, obtain a sample from each antibiotic's zone of inhibition on the *E. coli* plate, and inoculate the corresponding sector on the TSA (or NA) plate. If there is no zone for a particular antibiotic, no transfer is necessary.

5 Repeat with the *S. aureus* MH plate.

6 Incubate the plates aerobically at $35 \pm 2°$ for 24–48 hours.

Lab Three

1 Remove the plates from the incubator and examine each sector for growth.

2 Record your observations and answer the questions on the Data Sheet.

References

Clinical Laboratory Standards Institute (CLSI). 2009. *Performance Standards for Antimicrobial Disk Susceptibility Tests; Approved Standard—10th Ed.* CLSI document M02-A10. Wayne, PA.

Collins, C. H., Patricia M. Lyne, and J. M. Grange. 1995. Page 128 in *Collins and Lyne's Microbiological Methods*, 7th ed. Butterworth-Heinemann, UK.

Ferraro, Mary Jane, and James H. Jorgensen. 2003. Chapter 15 in *Manual of Clinical Microbiology*, 8th ed., edited by Patrick R. Murray, Ellen Jo Baron, James. H. Jorgensen, Michael A. Pfaller, and Robert H. Yolken, ASM Press, Washington, DC.

Forbes, Betty A., Daniel F. Sahm, and Alice S. Weissfeld. 2002. Pages 236–240 in *Bailey & Scott's Diagnostic Microbiology*, 11th ed. Mosby-Yearbook, St. Louis, MO.

Koneman, Elmer W., Stephen D. Allen, William M. Janda, Paul C. Schreckenberger, and Washington C. Winn, Jr. 1997. Pages 818–822 in *Color Atlas and Textbook of Diagnostic Microbiology*, 5th ed. J.B. Lippincott Company, Philadelphia, PA.

Mims, Cedric, Hazel M. Dockrell, Richard V. Goering, Ivan Roitt, Derek Wakelin, and Mark Zuckerman. 2004. Chapter 33 in *Medical Microbiology*, 3rd ed. Mosby, Philadelphia, PA.

Zimbro, Mary Jo, and David A. Power, editors. 2003. Page 376 in *Difco™ and BBL™ Manual—Manual of Microbiological Culture Media*. Becton Dickinson and Company, Sparks, MD.

EXERCISE 7-4

Clinical Biofilms

✦ Theory

According to Elvers and Lappin-Scott (2004), a biofilm is a "Complex association or matrix of microorganisms and microbial products attached to a surface." Development begins with formation of a **conditioning film** composed of biomolecules and particles from the environment on the surface. This results in the surface becoming more hydrophilic and negatively charged. Following this comes attachment of **planktonic** microorganisms to the surface. The precise mechanism varies depending on the amount of fluid movement and other factors, however biofilm formation invariably depends upon the microbial community itself through secretion of an extracellular polysaccharide (**glycocalyx**). As more microbes become attached, they change the composition of the biofilm's chemistry, providing a suitable environment for still other microorganisms. Eventually, the biofilm gets thick enough that detachment, which can be the result of erosion, abrasion, or simple breakage, becomes a factor. At some point a steady state is reached in which addition to the biofilm is compensated by loss from the biofilm.

Biofilms occur in natural environments, but also are formed in industrial and medical settings. It is the medical with which this lab is concerned. Indwelling devices, such as needles and catheters, are common locations for biofilm development. Reduced susceptibility of the biofilm community to antimicrobics (by a factor of 100–1000 times compared to their planktonic counterparts!) with natural detachment make these biofilms problematic in the production of nosocomial infections.

✦ Application

Staphylococcus aureus and *S. epidermidis* are notorious for forming biofilms on invasive medical devices resulting in nosocomial infections.

✦ In This Exercise

You will grow a mixed culture of *Staphylococcus aureus* and *S. epidermidis* for one week in a test tube, stain it, and observe for evidence of a biofilm.

✦ Materials

Per Student Group

✦ overnight broth cultures of *Staphylococcus aureus* (BSL-2) and *S. epidermidis*
✦ two tubes of TSB enriched with 1% glucose
✦ phosphate buffered saline (PBS), pH 7.3
✦ 0.1% crystal violet
✦ deionized water bottle
✦ test tube rack
✦ receptacle for culture media disposal

✦ Medium, Stain, and Reagent Recipes

Tryptic Soy Broth Plus 1% Glucose

◆ Tryptic Soy Broth	1 L
◆ Glucose	10 g

Crystal Violet Stain (0.1%)

◆ Gram Crystal Violet	0.1 mL
◆ dH$_2$O	99.9 mL

10x Phosphate Buffered Saline (PBS)

◆ Na$_2$HPO$_4$, anhydrous, reagent grade	12.36 g
◆ NaH$_2$PO$_4$·H$_2$O, reagent grade	1.80 g
◆ NaCl, reagent grade	85.00 g

Dissolve ingredients in distilled water to a final volume of 1 L

Working Solution (0.01 M Phosphate, pH 7.6)

◆ Stock solution	100 mL
◆ dH$_2$O	900 mL

Procedure

Day 1

1 Heavily inoculate one TSB + 1% Glucose tube with both *Staphylococcus aureus* and *S. epidermidis*. Label this tube with the names of the organisms.

2 Label the second tube "control" and do not inoculate it.

3 Incubate the tubes for 24–48 hours at $35 \pm 2°C$.

Day 2

1 Decant the broth from each tube into the receptacle designated for disposal. Be sure not to spill or drip culture when pouring. If you do, clean it up with your lab's disinfectant and wash your hands with your lab's antiseptic.

2 Gently rinse each tube with PBS twice and pour it into the receptacle designated for disposal.

3 Air dry the tubes in an inverted position.

4 Stain both tubes with 0.1% crystal violet solution for ten minutes at room temperature. Be sure to add more stain than the original volume of broth in the tubes.

5 Gently wash the tubes with dH_2O to remove unbound crystal violet.

6 Air dry the tubes in an inverted position.

7 Compare your results with Figure 7-8. You are looking for a purple film adhering to the inside of the experimental tube. Ignore any dark ring at the surface. Record your observations and answer the questions on the Data Sheet.

7-8 A STAPHYLOCOCCAL BIOFILM ✦ The faint purple is the biofilm stained with dilute crystal violet. The tube on the right is from an uninoculated control.

References

Elvers, Karen T. and Hilary M. Lappin-Scott. 2004. *Biofilms and Biofouling*, Chapter 12 in *The Desk Encyclopedia of Microbiology*, Moselio Schaechter, Ed. Elsevier Academic Press, 525 B Street, Suite 1900, San Diego, CA 92101-4495, USA.

Hirshfield, Irvin N., Subit Barua, and Paramita Basu. 2009. *Overview of Biofilms and Some Key methods for Their Study*. Chapter 42 in *Practical Handbook of Microbiology*, 2nd ed., Edited by Emanuel Goldman and Lorrence H. Green. CRC Press, Taylor, and Francis Group, Boca Raton, FL.

EXERCISE 7-5

Morbidity and Mortality Weekly Report (MMWR) Assignment

✦ Theory

Epidemiology is the study of the causes, occurrence, transmission, distribution, and prevention of diseases in a population. The Centers for Disease Control and Prevention (CDC) in Atlanta, GA, is the national clearinghouse for epidemiological data. The CDC receives reports related to the occurrence of 26 notifiable diseases (Table 7-5) from the United States and its territories, and compiles the data into tabular form, available in the publication *Morbidity and Mortality Weekly Report* (MMWR).

Two important disease measures that **epidemiologists** collect are **morbidity** (sickness) and **mortality** (death). Morbidity relative to a specific disease is the number of susceptible people who have the disease within a defined population during a specific time period. It usually is expressed as a rate. Because population size fluctuates constantly, it is conventional to use the population size at the midpoint of the study period. Also, the units for the rate fraction are "cases per person" and usually are small decimal fractions. To make the calculated rate more "user-friendly," it is multiplied by some power of 10 ("K") to achieve a value that is a whole number. Thus, a morbidity rate of 0.00002 is multiplied by 100,000 (10^5) so it can be reported as 2 cases per 100,000 people rather than 0.00002 cases per person. Morbidity rate is calculated using the following equation:

$$\text{Morbidity Rate} = \frac{\text{number of existing cases in a time period}}{\text{size of at-risk population at midpoint of time period}} \times K$$

Mortality, also expressed as a rate, is the number of people who die from a specific disease out of the total population afflicted with that disease in a specified time period. It, too, is multiplied by a factor "K" so the rate can be reported as a whole number of cases. The equation is:

$$\text{Mortality Rate} = \frac{\text{number of deaths due to a disease in a time period}}{\text{number of people with that disease in the time period}} \times K$$

Minimally, an epidemiological study evaluates morbidity or mortality data in terms of **person** (age, sex, race, *etc.*), **place,** and **time.** Sophisticated analyses require training in biostatistics, but the simple epidemiological calculation you will be doing can be performed with little mathematical background. You will calculate **incidence rate,** which is the occurrence of new cases of a disease within a defined population during a specific period of time. As before, "K" is some power of 10 so the rate can be reported as a whole number of cases.

$$\text{Incidence Rate} = \frac{\text{number of new cases in a time period}}{\text{size of at-risk populations at midpoint of time period}} \times K$$

Because our focus is microbiology, we will deal only with **infectious diseases**—those caused by biological agents such as bacteria and viruses. **Noninfectious diseases,** such as stroke, heart disease, and emphysema, also are studied by epidemiologists but are not within the scope of microbiology.

✦ Application

An understanding of the causes and distribution of diseases in a population is useful to health-care providers in a couple of ways. First, awareness of what diseases are prevalent during a certain period of time aids in diagnosis. Second, an understanding of the disease, its causes, and transmission can be useful in implementing strategies for preventing it.

✦ In This Exercise

First you will choose a disease from the list of notifiable diseases in Table 7-5. Then you will collect data for the United States over the past two years. Once you have collected data, you will construct graphs illustrating the cumulative totals and incidence values for each year.

✦ Materials

✦ a computer with Internet access or access to printed copies of *Morbidity and Mortality Weekly Report.*

Procedure

1 Go to the CDC Web site http://www.cdc.gov. Then follow these links. (**Note:** Web sites often are revised, so if the site doesn't match this description exactly, it still is probably close. Improvise, and you'll find what you need. These links are current as of November 2009):

- ✦ Click on MMWR under "Publications" near the bottom of the home page.

- ✦ Click on "State Health Statistics" in the menu bar at the left. This will drop down a new menu.

- ✦ Click on "Morbidity Tables."

- ✦ Read the "Note" below the menu window on this page. It describes how the data in the tables are collected and why the numbers are provisional.

- ✦ Select *MMWR* Week 1 of the most recent complete *MMWR* year, and then click on "Submit."

- ✦ Table II is divided into 9 parts. Upon your first visit, examine the 9 parts and choose a disease you would like to work on.[1]

- ✦ On your Data Sheet, record the name of the disease you have selected and write the *cumulative* number of cases in the United States for Week 1 of the two years you are studying. For instance, if you are studying 2009 and 2008, you would take the numbers from the columns entitled *Cum 2009* and *Cum 2008*. Record these on the Data Sheet. Using the years 2008 and 2009 as examples, you will find that the 2008 number reported for a particular week in the *2009* table might differ from the reported number in the same week of the *2008* table. This is a result of corrections in reported 2008 numbers in the 2009 table. Don't fret. This is out of your control! Just record the numbers as given.

- ✦ Return to the page with the *MMWR* Week and *MMWR* Year and continue the process for each week through *MMWR* Week 52. (Alternatively, you can just change the week number in the URL and press return.[2]) Record the cumulative totals on the Data Sheet.

2 Answer the questions and complete the activities on the Data Sheet.

---

[1] Table I lists the occurrence of infrequently reported notifiable diseases (< 1000 cases in the previous year). Your instructor may allow you to choose a disease from this table in addition to Table II.

[2] Thanks to Joe Montes, Jr. for this shortcut tip!

TABLE **7-5** NOTIFIABLE DISEASES IN THE UNITED STATES AND ITS TERRITORIES POSTED IN TABLE II OF MMWR (AS OF NOVEMBER 2009.) ✦ Diseases highlighted in orange should not be selected because you won't have enough information to complete the calculations required for this assignment. Some of the more "interesting diseases" (*i.e.,* AIDS and tuberculosis) are listed in Table IV, but are reported only quarterly and so don't lend themselves to this assignment.

Chlamydia	Rabies, animal
Coccidioidomycosis	Rocky Mountain spotted fever
Cryptosporidiosis	Salmonellosis
Girardiasis	Shiga toxin-producing *E. coli* (STEC)
Gonorrhea	Shigellosis
Haemophilus influenzae, invasive, all ages and serotypes	Streptococcal disease, invasive, Group A
Hepatitis A	*Streptococcus pneumoniae,* invasive disease, age <5 years
Hepatitis B	*Streptococcus pneumoniae,* invasive disease, drug resistant, all ages
Legionellosis	*Streptococcus pneumoniae,* invasive disease, drug resistant, <5 years
Lyme Disease	Syphilis, primary and secondary
Malaria	Varicella (chickenpox)
Meningococcal diseases, invasive, all serogroups	West Nile Virus—neuroinvasive)
Pertussis	West Nile Virus—non-neuroinvasive)

Epidemic Simulation

✦ Theory

As mentioned in Exercise 7-5, epidemiology is the study of the causes, occurrence, transmission, distribution, and prevention of diseases in a population. Infectious diseases are transmitted by ingestion, inhalation, direct skin contact, open wounds or lesions in the skin, animal bites, direct blood-to-blood contact as in blood transfusions, and sexual contact. Infectious diseases can be transmitted by way of sick people or animals, healthy people or animals carrying the infectious organism or virus, water contaminated with human or animal feces, contaminated objects (**fomites**), aerosols, or biting insects (**vectors**).

When a disease is transmitted from an area such as the heating or cooling system of a building or from contaminated water that infects many people at once, it is called a **common source epidemic**. **Propagated transmission** is a disease transmitted from person to person. The first case of such a disease is called the **index case**. Determining the index case is the object of today's lab exercise.

A second objective of today's lab is to gather data to be used in performing simple epidemiological calculations. You will be calculating **incidence rate** and **prevalence** rate for the "disease" spreading through your class.

Incidence rate is the number of new cases of a disease reported in a defined population during a specific period of time. Incidence rate is calculated by the following equation:

$$\text{Incidence rate} = \frac{\substack{\text{number of new cases} \\ \text{in a time period}}}{\substack{\text{size of at-risk population} \\ \text{at midpoint of time period}}} \times K$$

Because population size fluctuates constantly, it is conventional to use the population size at the midpoint of the study period. Also, the usually small decimal fraction obtained for the rate is multiplied by some power of 10 (K in the equation) to get its value up to a number bigger than one. That is, an incidence rate of 0.00002 is multiplied by 100,000 (10^5) so it can be reported as two cases per 100,000 people, rather than 0.00002 cases per person.

You also will be calculating one form of prevalence rate called **point prevalence**, the number of cases of a disease at a specific point in time in a defined population. It is calculated by the following equation:

$$\text{Point prevalence} = \frac{\substack{\text{number of existing cases} \\ \text{at a point in time}}}{\text{size of at-risk population}} \times K$$

As with incidence rate, the calculated prevalence is multiplied by some power of 10 (K in the equation) to bring the number up to one or more.

✦ Application

Epidemiologists have the task of identifying the source of a disease and establishing the mode of its transmission. Further, epidemiologists characterize diseases quantitatively, using measures such as incidence and prevalence. Hopefully, this information will allow them to make recommendations for prevention.

✦ In This Exercise

One of you will become the "index case" in a simulation of an epidemic, and many of you who contact this person directly and indirectly will become "infected." Your job, besides having a little fun, will be to collect the data and determine which one of you is the index case. You will also use the data to calculate incidence and prevalence rates for your class over a hypothetical time period.

✦ Materials

Per Student

✦ examination glove (latex or synthetic)

✦ numbered Petri dish containing a piece of hard candy (one will be contaminated with *Serratia marcescens*[1] or other microbe as chosen by your instructor; the others will be moistened with water)[2]

✦ one Nutrient Agar plate

✦ two sterile cotton applicators and sterile saline

✦ marking pen

---

[1] *S. marcescens* was chosen because of its obvious color. Other, less distinctive organisms can be substituted, if desired.

[2] Petri dishes should be numbered according to the number of students in the class. There should be no gaps in the sequence.

Procedure

Lab One

1 Obtain a Petri dish containing candy, two sterile cotton applicators, a tube of sterile saline, and a Nutrient Agar plate. Record the number of your candy dish here and circle that number in the "case" column of the Data Sheet.

2 Mark a line on the bottom of the agar plate to divide it into two halves. Label the sides "1" and "2." Also record the number of your candy dish on your plate.

3 Open a package of sterile, cotton-tipped applicators so they can be easily removed.

4 Put a glove on your nondominant hand.

5 Using your ungloved hand, remove one cotton applicator, moisten it in the sterile saline, and sample the palm of the gloved hand. Be sure not to touch anything else with the palm of the glove. When finished, zigzag inoculate side 1 of the plate with the cotton applicator (Figure 1-27). Discard the applicator in an appropriate autoclave container.

6 Pick up the candy with your gloved hand and roll it around until a good amount of it is transferred to the glove. Drop the candy back into the Petri dish, close it with your ungloved hand, and touch nothing with your gloved hand. Just sit with your elbow on the table and your gloved hand in the air.

7 Using gloved hands only, student #1 rubs hands with student #2, making sure to transfer anything that may be on the palm of the glove to the palm of the other student's glove. Then #2 rubs gloved hands with #3, and #3 with #4, and so on until all

students have contacted gloved hands. Be careful not to snap the gloves when you separate, as this will produce aerosols. *Gently slide your hands apart. You may also need to hold on to the cuff when transferring and separating to prevent the glove from coming off your hand.*

8 When all students have rubbed hands, the last (highest numbered) student should rub hands with student #1.

9 With a second sterile applicator, sample the palm of your gloved hand as before and inoculate side 2 of the plate. Be sure to sample the area of the glove that contacted other students' gloves.

10 Remove your glove by inserting the thumb of your other hand under the cuff and rolling it off your hand, inverting it in the process. Dispose of it properly. Dispose of the applicator in an appropriate autoclave container.

11 Wash your hands thoroughly with antiseptic soap, if available.

12 Discard all materials (except your inoculated plate) in appropriate containers. Be sure to tape the lid on your candy dish.

13 Tape the lid on the inoculated plate.

14 Invert the inoculated plate and incubate it at 25°C for 24 to 48 hours.

Lab Two

1 Remove the plates from the incubator, and examine them for characteristic reddish growth of *Serratia marcescens*.

2 Enter your results on the Data Sheet and follow the instructions for establishing the index case, incidence rate, and point prevalence.

Identification of *Enterobacteriaceae*

✦ Theory

The *Enterobacteriaceae* (Figure 7-9) comprise a family of Gram-negative rods that mostly inhabit the intestinal tract. Major characteristics defining the group are:

- ✦ Growth on MacConkey agar
- ✦ Gram-negative rods
- ✦ Oxidase negative
- ✦ Acid production from glucose (with or without gas)
- ✦ Most are catalase-positive
- ✦ Most reduce NO_3 to NO_2
- ✦ Peritrichous flagella, if motile

Many **enterics** are pathogens (*e.g.*, *Escherichia coli* 0157:H7—bloody diarrhea; *Klebsiella pneumoniae*—pneumonia of various types; *Proteus mirabilis*—urinary tract infections; *Salmonella typhi*—typhoid fever; *Shigella dysenteriae*—bacillary dysentery; and *Yersinia pestis*—plague) but just as many are harmless gut **commensals** or **opportunists**. The organisms for which the identification flowcharts in this exercise work are listed in Figure 7-10, and some are commonly associated with human infections. Your instructor will choose enteric unknowns appropriate to your microbiology course and facilities.

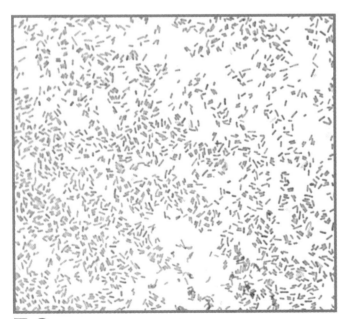

7-9 ESCHERICHIA COLI, A COMMON ENTERIC (X1000) ✦ All *Enterobacteriaceae* are Gram-negative, oxidase-negative rods. They also ferment glucose to acid end products.

✦ Application

Identification of enterics from human specimens using traditional methods requires a coordinated and integrated use of biochemical tests and stains. Although several multitest systems that utilize computer databases are available (see Exercises 5-29 and 5-30 for two examples), flowcharts are still a useful way to visualize the process of identification by elimination. It is striking how often a few test results, when taken in combination, are necessary to identify an organism.

✦ In This Exercise

This exercise will span several lab periods. You will be assigned an unknown enteric and run biochemical tests as directed by the flowcharts in Figures 7-11 through 7-14 to identify it.

List of *Enterobacteriaceae* in Identification Charts and IMViC Results

*Citrobacter freundii**
Edwardsiella tarda
Enterobacter aerogenes
Enterobacter cloacae
Escherichia coli
*Hafnia alvei**
*Klebsiella pneumoniae**
Morganella morganii
*Proteus mirabilis**

Proteus vulgaris
Providencia alcalifaciens
Providencia stuartii
Salmonella
Serratia liquefaciens
*Serratia marcescens**
*Shigella flexneri (Group B)**
Shigella sonnei (Group D)

*Organism shows up in more than one figure due to variable results of one or more IMViC tests

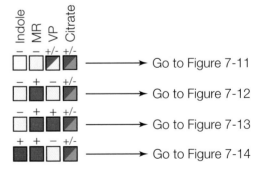

7-10 IMViC RESULTS OF ENTEROBACTERIACEAE ✦ The organisms listed are *Enterobacteriaceae* used in this Exercise. Match the IMViC results of your isolate with one of the four shown, then proceed with the flowchart in the figure indicated. Some organisms are listed in more than one flowchart because of variable results for a particular test. A box divided diagonally means either result for that test is a match. (Most results after Farmer *et al.*, 2007.)

TABLE **7-6** KEY TO ICONS USED IN THIS EXERCISE ✦ Colors are to remind you what results look like. They should not be used to compare with your results. **Note:** For the fermentation tests, "A+" means acid is produced; "A−" means no acid is produced.

+	−	
■	▨	Citrate Utilization
◧	▢	DNase
▨	▢	Glucose Fermentation-Gas
■	▢	H₂S Production
■	▢	Indole
■	▨	Lysine Decarboxylase
■	▢	Methyl Red Test
▨	■	Motility
■	▨	Ornithine Decarboxylase
▨	▢	Urease
▨	▢	Voges-Proskauer

A+	A−	
▢	■	Arabinose Fermentation
▨	■	Mannitol Fermentation
▢	■	Sucrose Fermentation
▢	■	Trehalose Fermentation
▢	▨	Xylose Fermentation

+	−	
◇	◈	Gelatinase

✦ Materials

Per Class

✦ Media listed in Table 7-6 for biochemical testing.

✦ Pure cultures of organisms listed in Figure 7-10 to be used for positive controls and unknowns. (All organisms are available from either Ward's or Carolina Biological Supply Company and should be chosen appropriate to the course level.)

Per Student

✦ two MacConkey Agar plates

✦ one unknown organism[1] from the list in Figure 7-10

✦ TSA slants (enough to keep the culture fresh over the time required for identification)

✦ Gram stain kit

✦ microscope slides

✦ compound microscope with oil objective lens and ocular micrometer

✦ immersion oil

✦ lens paper

Procedure

1 Obtain an unknown. Record its number on the Data Sheet.

---

[1] *Note to instructor:* This exercise can be combined with Exercises 7-8 or 7-9 by passing out a mixture of an enteric and a Gram-positive coccus or rod as unknowns. The student must isolate the two organisms from mixed culture and then proceed to identify the two independently. If this option is chosen, grow the two unknowns separately and then mix them immediately prior to handing them out in lab.

2 Streak the sample for isolation on two MacConkey Agar plates. Incubate one at 25°C and the other 35 ± 2°C for 24 hours or more. It is best to check your plate for isolation at 24 hours. If you have isolation, continue with Step 3. If you don't have time to continue with Step 3, refrigerate your plates until you do. If you don't see isolation, ask your instructor if you should re-streak for isolation, let the original plates continue to incubate, or do both. Record your isolation procedure as directed on the Data Sheet.

3 After incubation, record colony morphology (including medium) and optimum growth

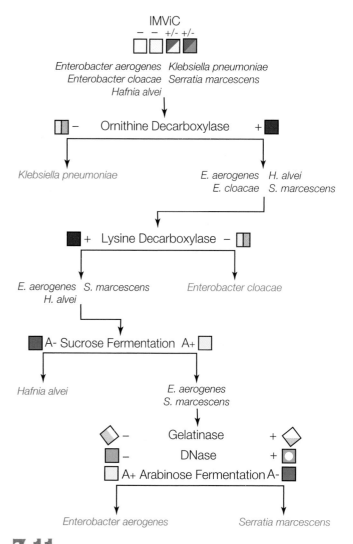

7-11 FLOWCHART FOR IDENTIFYING INDOLE-NEGATIVE, MR-NEGATIVE, VP-POSITIVE OR VP-NEGATIVE, CITRATE-POSITIVE OR -NEGATIVE ENTEROBACTERIACEAE ✦ See the appropriate exercises for instructions on how to run each test. Results are based on 48 hour incubation at 38°C. Colors in the icons are not intended to match media exactly. Use controls for this. (Most results after Farmer *et al.*, 2007.)

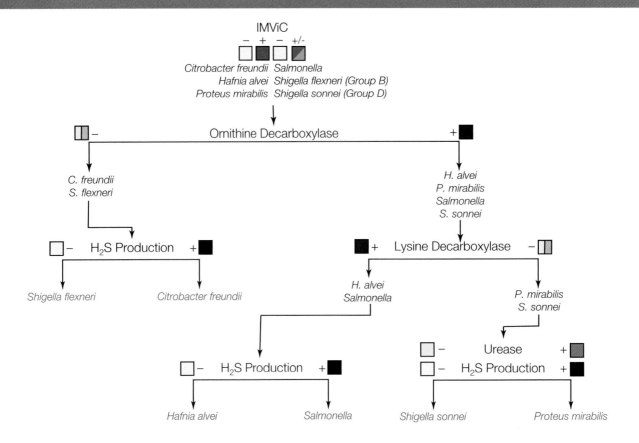

7-12 FLOWCHART FOR IDENTIFYING INDOLE-NEGATIVE, MR-POSITIVE, VP-NEGATIVE, CITRATE-POSITIVE OR -NEGATIVE ENTEROBACTERIACEAE ◆
See the appropriate exercises for instructions on how to run each test. Results are based on 48 hour incubation at 38°C. Colors in the icons are not intended to match media exactly. Use controls for this. (Most results after Farmer *et al.,* 2007.)

temperature under "preliminary observations" on the Data Sheet. Also include the result of the differential component of MacConkey agar as Test #1 under "Differential Tests." Continue to record your isolation procedure on the Data Sheet.

4 Perform a Gram stain on a portion of one well-isolated colony. Record its Gram reaction and cell morphology, arrangement and size on the Data Sheet under Preliminary Observations. If the isolate is a Gram-negative rod, transfer a portion of the *same* colony to a Trypticase Soy Agar slant (or another suitable growth medium as available in your lab) and incubate it at its optimum temperature. This is your pure culture to be used as a source of organisms for further testing. Complete the record of your isolation procedure on the Data Sheet.

5 Perform an oxidase test on what remains of the isolated colony, or wait until your pure culture has grown and do it then. Record the result as Test #2 under "Differential Tests" on the Data Sheet. If the isolate grew on MacConkey Agar and is an oxidase-negative, Gram-negative rod, it probably is a

member of the *Enterobacteriaceae*. At the discretion of your instructor, further tests chosen from the characteristics listed may be run to assure that it is an organism from *Enterobacteriaceae*.

6 You now will begin biochemical testing to identify your organism. The tests in the flowcharts were chosen because of their uniformity of results as published in standard microbiological references, so they give the best chance of correct identification. Be aware, however, that the symbol " + " indicates that 90% or more of the strains tested give a positive result. This means that as many as 10% of the strains tested give a negative result. The same is true of the symbol " − ." Bottom line: There are no guarantees that your particular lab strains will behave in the majority, and if not, you will misidentify your unknown. This issue, if relevant to your unknown, will be addressed in Step 10.

7 Perform the IMViC series of tests (Indole Production, Methyl Red Test, Voges-Proskauer Test, and Citrate Utilization) at your isolate's optimum temperature.

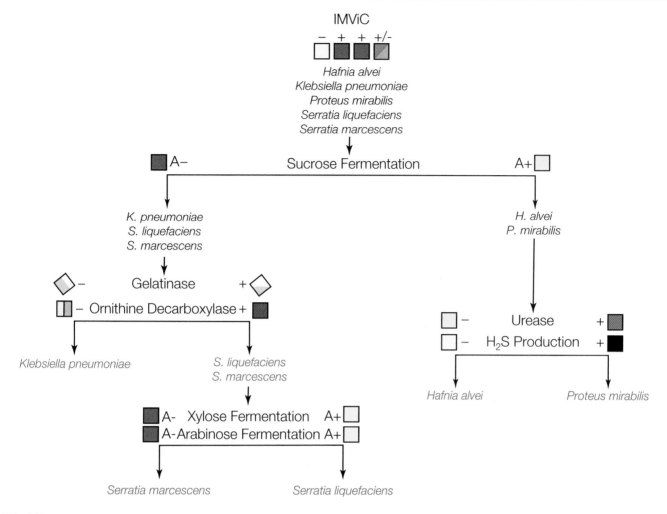

7-13 FLOWCHART FOR IDENTIFYING INDOLE-NEGATIVE, MR-POSITIVE, VP-POSITIVE, CITRATE-POSITIVE OR -NEGATIVE ENTEROBACTERIACEAE ✦
See the appropriate exercises for instructions on how to run each test. Results are based on 48 hour incubation at 38°C. Colors in the icons are not intended to match media exactly. Use controls for this. (Most results after Farmer *et al.*, 2007.)

♦ Record the relevant inoculation information and results on the Data Sheet under "Differential Tests."

♦ Use Test #3 for indole, Test #4 for MRVP, and Test #5 for citrate.

♦ Match your isolate's IMViC results with one shown in Figure 7-10, and proceed to the appropriate identification flowchart (Figures 7-11 through 7-14). Record on the Data Sheet the figure number of the flowchart you will use. (***Note:*** If a result is shown as +/− [as it is for all citrate tests], either result for that test is a match.)

8 Follow the tests in the appropriate flowchart to identify your unknown.[2] Use these guidelines.

♦ Do not run more than one test at a time unless your instructor tells you to do so.

♦ Where multiple tests are listed at a branch point, run only one, and then move to the next level in the flowchart as indicated by the result of that test.

♦ Continue to record the relevant inoculation information and results on the Data Sheet as you go. Keep accurate records of what you have done. Do *not* enter all your data at the end of the project.

9 When you have identified your organism, use *Bergey's Manual* or another standard reference to find one more test to run for confirmation. The confirmatory test doesn't have to separate the final organisms on the flowchart. It only has to be a test for which you know the result and that you haven't run already as part of the identification process. Record this test on the Data Sheet and identify it as the confirmatory test. Record the *expected* result on the Comments line. After incubation, note whether the organism's result matches or differs from the expected result.

---

[2] These flowcharts can be used only for the organisms listed in Figure 7-10.

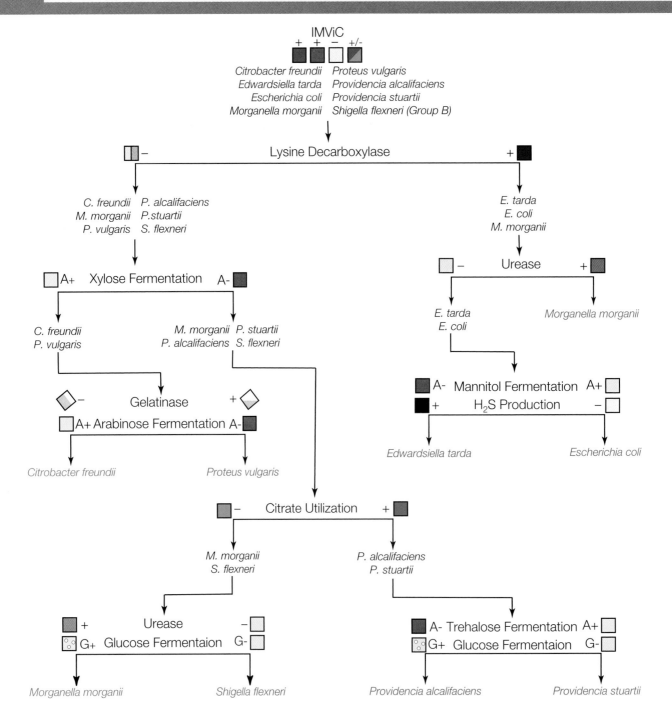

7-14 FLOWCHART FOR IDENTIFYING INDOLE-POSITIVE, MR-POSITIVE, VP-NEGATIVE, CITRATE-POSITIVE OR -NEGATIVE ENTEROBACTERIACEAE ✦
See the appropriate exercises for instructions on how to run each test. Results are based on 48 hour incubation at 38°C. Colors in the icons are not intended to match media exactly. Use controls for this. (Most results after Farmer *et al.*, 2007.)

10 After the confirmatory test, write the name of the organism on the back of the Data Sheet under "My unknown is" and check with your instructor to see if you are correct. If you are, congratulations! If you aren't, your instructor will advise you as to which test(s) gave "incorrect" results by writing the test name(s) in the "rerun" space next to your identification. These should be rerun with appropriate controls

from your school's inventory of organisms. Record the test(s) and control(s) on your Data Sheet. Checking the results with controls and your unknown will indicate if the "incorrect" test result was truly incorrect or if the strain of organism your school is using doesn't match the majority of strains for that test result.

References

Farmer III, J. J., K. D. Boatwright, and J. Michael Janda. 2007. Chapter 42 in *Manual of Clinical Microbiology*, 9th ed., edited by Patrick R. Murray, Ellen Jo Baron, James H. Jorgensen, Marie Louise Landry, and Michael A. Pfaller. ASM Press, Washington, DC.

Forbes, Betty A., Daniel F. Sahm, and Alice S. Weissfield. 2002. Chapter 25 in *Bailey & Scott's Diagnostic Microbiology*, 11th ed. Mosby, Inc., St. Louis, MO.

MacFaddin, Jean F. 2000. *Biochemical Tests for Identification of Medical Bacteria*, 3rd ed. Lippincott Williams & Wilkins, Philadelphia, PA.

Winn Jr., Washington, Stephen Allen, William Janda, Elmer Koneman, Gary Procop, Paul Schreckberger, and Gail Woods. 2006. Chapter 6 in *Koneman's Color Atlas and Textbook of Diagnostic Microbiology*, 6th ed., Lippincott Williams & Wilkins, Philadelphia, PA.

Identification of Gram-positive Cocci

✦ Theory

Gram-positive cocci are frequent isolates in a clinical setting because they are common inhabitants of skin and mucous membranes. Four main genera, briefly described below, will be used in this lab exercise.

Enterococcus

✦ Gram-positive cocci to elongated cocci in singles or short chains

✦ Catalase-negative

✦ Facultatively anaerobic

✦ Grows in 6.5% NaCl

✦ Grows in Bile Esculin

✦ Most are PYR-positive

✦ Key pathogen is *Enterococcus faecalis* (mostly nosocomial or opportunistic urinary tract infections, wound infections, and bacteremia in seriously ill elderly patients)

Micrococcus

✦ Gram-positive cocci in pairs, tetrads, or clusters

✦ Catalase-positive

✦ Strictly aerobic

✦ Oxidase-positive

✦ Does not produce acid from glucose

✦ Bacitracin-sensitive

✦ Key pathogen is *Micrococcus luteus* (opportunistic infections of immunocompromised patients)

Staphylococcus (Figure 7-15)

✦ Gram-positive cocci in singles, pairs, tetrads, or clusters

✦ Catalase-positive

✦ Facultatively anaerobic

✦ Oxidase negative

✦ Grows in 6.5% NaCl

✦ Most produce acid from glucose

✦ Most are resistant to bacitracin

✦ Key pathogen is *Staphylococcus aureus* (toxic shock syndrome and a variety of other skin and deep organ infections, including bacteremia)

Streptococcus (Figure 7-16)

✦ Gram-positive cocci to ovoid cocci in singles or short chains

✦ Catalase-negative

✦ Facultatively anaerobic

✦ Nutritionally fastidious

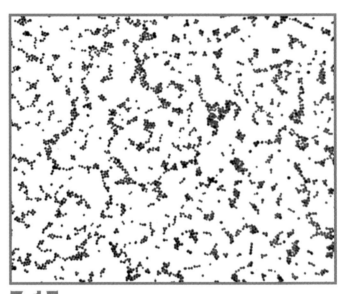

7-15 STAPHYLOCOCCUS AUREUS (X1000) ✦ This specimen grown in broth illustrates the grape-like clusters of cells characteristic of the genus. Specimens grown on solid media may not show the clusters as clearly.

7-16 STREPTOCOCCUS AGALACTIAE (X1000) ✦ This specimen grown in broth illustrates the streptococcal arrangement of cells characteristic of the genus. Specimens grown on solid media may not show the chains as clearly.

✦ Ferments glucose to acid, but not gas

✦ Key pathogens are *Streptococcus pyogenes* (strep throat, necrotizing fasciitis, scarlet fever) and *S. pneumoniae* (bacterial pneumonia, otitis media, and bacteremia)

One species of *Kocuria* is also included in the flow-charts. *Kocuria* species were once placed in *Micrococcus* but have been removed as a result of DNA and rRNA dissimilarities consistent with certain fundamental biochemical dissimilarities.

The specific organisms are listed in Table 7-7, as are the tests to be used in their identification. (**Note:** Your instructor will choose Gram-positive coccus unknowns appropriate to your microbiology course and facilities.)

✦ Application

Identification of Gram-positive cocci from human specimens requires a coordinated and integrated use of biochemical tests and stains. Although several serological tests allow rapid identification, flowcharts are still a useful way to visualize the process of identification by elimination.

✦ In This Exercise

This exercise will span several lab periods. You will be given an unknown Gram-positive coccus from the organisms listed in Table 7-7, then run biochemical tests as directed by the flowcharts in Figure 7-17 and Figure 7-18 to identify it.

✦ Materials

Per Class

✦ media listed in Table 7-7 for biochemical testing

✦ pure cultures of organisms listed in Table 7-7 to be used for positive controls and unknowns (appropriate to the course level)

✦ Todd-Hewitt Broth or Brain-Heart Infusion Broth

✦ candle jar set-up

Per Student

✦ two PEA, CNA, or 5% Sheep Blood Agar plates

✦ Gram stain kit

✦ acid-fast stain kit

✦ microscope slides

✦ 3% H_2O_2

✦ one unknown organism[1] from the list in Table 7-7

✦ compound microscope with oil objective lens and ocular micrometer

✦ lens paper

✦ immersion oil

TABLE **7-7** LIST OF ORGANISMS AND A KEY TO THE ICONS USED IN THIS EXERCISE
✦ The organisms listed are Gram-positive cocci available from Ward's or Carolina Biological Supply Companies. Icon colors are to remind you what the results look like. They should not be used to compare with your results. **Note:** For the fermentation tests, "A+" means acid is produced; "A−" means no acid is produced.

Gram-Positive Cocci and Identification Tests

Catalase-Positive (Figure 7-17)	Catalase-Negative (Figure 7-18)
Kocuria rosea (= *Micrococcus roseus*)	*Enterococcus faecalis*
Micrococcus luteus	*Streptococcus agalactiae*
Staphylococcus aureus	*Streptococcus equisimilis*
Staphylococcus epidermidis	*Streptococcus (bovis I) gallolyticus*
Staphylococcus saprophyticus	*Streptococcus mutans*
	Streptrococcus salivarius
	Streptococcus sanguis
	Streptococcus pneumoniae
	Streptococcus pyogenes

🖿 Procedure

1 Obtain an unknown. Record its number on the Data Sheet.

2 Streak the sample for isolation on two Sheep Blood Agar, PEA, or CNA plates. Incubate one in the candle jar and the

----

[1] **Note to instructor:** This exercise can be combined with Exercise 7-7 by passing out a mixture of a Gram-negative enteric and a Gram-positive coccus as unknowns. The student then must isolate the two organisms from a mixed culture, and then proceed to identify the two independently. If this option is chosen, grow the two unknowns separately, and then mix them immediately prior to handing them out in lab.

Gram-Positive, Catalase-Positive Cocci

7-17   IDENTIFICATION FLOWCHART FOR GRAM-POSITIVE, CATALASE-POSITIVE COCCI (USUALLY IN TETRADS OR CLUSTERS) ✦ See the appropriate exercises for instructions on how to run each test. Novobiocin is covered in this exercise. In some cases, final identification to species is difficult. Try to get that far, but you may reach a point where your lab strains can't be differentiated any further. (Compiled from sources listed in the References.)

other aerobically. Incubate both at 30–35°C. It is best to check your plate for isolation at 24 hours. If you have isolation, continue with Step 3. If you don't have time to continue with Step 3, refrigerate your plates until you do. If you don't see isolation, ask your instructor if you should restreak for isolation, let the original plates continue to incubate, or do both. Record your isolation procedure as directed on the Data Sheet.

3 After incubation, record colony morphology (including medium), optimum growth temperature, and CO_2 requirement under "Preliminary Results" on the Data Sheet. Also include the result of the differential component of blood agar as Test #1 if you used it. Continue to record your isolation procedure on the Data Sheet.

4 Perform a Gram stain on a portion of one well-isolated colony. Continue testing colonies until you find one that is a Gram-positive coccus.

    ✦ Record its Gram reaction and cell morphology, arrangement and size under "Preliminary Observations" on the Data Sheet.

    ✦ If the isolate is a Gram-positive coccus, transfer a portion of the *same* colony to a Todd-Hewitt broth (or another suitable growth medium as available in your lab) and incubate it at $35 \pm 2°C$. This is your pure culture to be used as a source of organisms for further testing. Complete the record of your isolation procedure on the Data Sheet.

5 Perform a catalase test on what remains of the isolated colony, or wait until your pure culture has grown and do it then. Record the result as Test #1 (or #2 if hemolysis reaction is already recorded) under Differential Tests on the Data Sheet. (**Note:** Be sure to use growth from the top of the colony if isolation was performed on a blood agar plate. This minimizes the possibility of a false positive from the catalase-positive erythrocytes in the medium.)

6 Use the chart in Figure 7-17 for identification of catalase-positive cocci (usually in tetrads or clusters). Use the chart in Figure 7-18 if the isolate is catalase-negative (usually with cocci in pairs or chains). Record the Figure number of the flowchart you will use on the Data Sheet. If your isolate is

Gram-Positive, Catalase-Negative Cocci

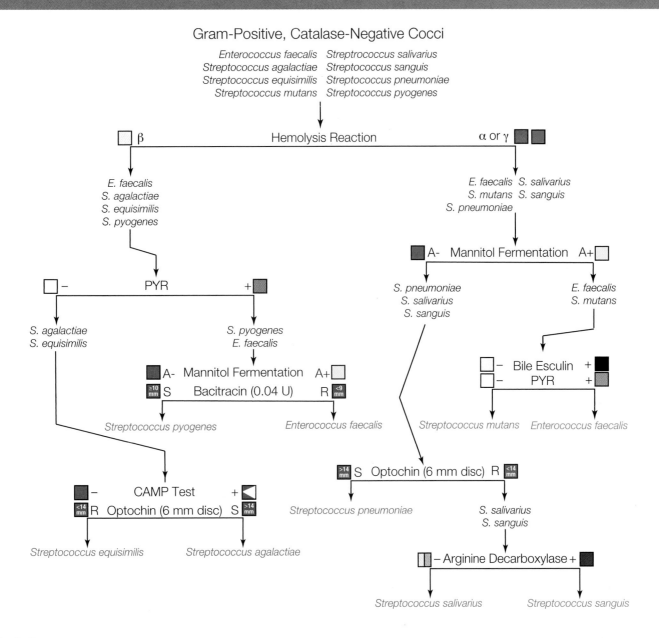

7-18 IDENTIFICATION FLOWCHART FOR GRAM-POSITIVE, CATALASE-NEGATIVE COCCI (USUALLY IN PAIRS OR CHAINS) ✦ See the appropriate exercises for instructions on how to run each test. In some cases, final identification to species is difficult. Try to get that far, but you may reach a point where your lab strains can't be differentiated any further.   (Compiled from sources listed in References.)

catalase-negative, ask your instructor if you would get better growth of your pure culture using Todd-Hewitt Broth or Brain-Heart Infusion Broth and incubating in a candle jar or CO_2 incubator.

7 Now you will begin biochemical testing to identify your organism. The tests in the flowcharts were chosen because of their uniformity of results as published in standard microbiological references, so they give the best chance for correct identification. Be aware, however, that the symbol " + " indicates that 90% or more of the strains tested give a positive result. This means that as many as 10% of the strains tested give a negative result. The same is true of the symbol " − ". Bottom line: There are no guarantees that your lab strains will behave in the majority, and if not, you will misidentify your unknown. This issue, if relevant to your unknown, will be addressed in Step 10.

8 Follow the tests in the appropriate flowchart to identify your unknown.[2] Use these guidelines.

- ♦ Do not run more than one test at a time unless your instructor tells you to do so.

- ♦ Where multiple tests are listed at a branch point, run only one, and then move to the next level in the flowchart as indicated by the result of that test.

- ♦ Continue to record the relevant inoculation information and results on the Data Sheet as you go. Keep accurate records of what you have done. *Do not* enter all your data at the end of the project.

9 When you have identified your organism, use *Bergey's Manual* or another standard reference to find one more test to run for confirmation. The confirmatory test doesn't have to separate the final organisms on the flowchart. It only has to be a test for which you know the result and that you haven't run already as part of the identification process. Record this test on the Data Sheet, and identify it as the confirmatory test. Record the *expected* result on the comments line. After incubation, note whether the organism's result matches or differs from the expected result.

10 After the confirmatory test, write the organism's name on the back of the Data Sheet under "My

Unknown is" and check with your instructor to see if you are correct. If you are, congratulations! If you aren't, your instructor will advise you as to which test(s) gave "incorrect" results by writing the test name(s) in the "rerun" space next to your identification. These should be rerun with appropriate controls from your school's inventory of organisms. Record the test(s) and control(s) on your Data Sheet. Checking the results with controls and your unknown will indicate if the "incorrect" test result was truly incorrect, or if the strain of organism your school is using doesn't match the majority of strains for that test result.

References

Bannerman, Tammy L., and Sharon J. Peacock. 2007. Chapter 28 in *Manual of Clinical Microbiology*, 8th ed., edited by Patrick R. Murray, Ellen Jo Baron, James. H. Jorgensen, Marie Louise Landry, and Michael A. Pfaller. ASM Press, Washington, DC.

Hardie, Jeremy M., Jiri Rotta, and J. Orvin Mundt. Section 12 (*Streptococcus, Enterococcus and Lactococcus*) in *Bergey's Manual of Systematic Bacteriology*, Vol. 2, edited by John G. Holt *et al.* Williams and Wilkins, Baltimore, MD.

Holt, John G., Noel R. Krieg, Peter H. A. Sneath, James T. Staley, and Stanley T. Williams. 1994. Group 17 (Gram-Positive Cocci) in *Bergey's Manual of Determinative Bacteriology*, 9th ed. Williams and Wilkins, Baltimore, MD.

Kloos, Wesley E., and Karl Heinz Schleifer. 1986. Section 12 (*Staphylococcus*) in *Bergey's Manual of Systematic Bacteriology*, Vol. 2, edited by John G. Holt *et al.* Williams and Wilkins, Baltimore, MD.

Kocur, Miloslav. 1986. Section 12 (*Micrococcus*) in *Bergey's Manual of Systematic Bacteriology*, Vol. 2, edited by John G. Holt *et al.* Williams and Wilkins, Baltimore, MD.

MacFaddin, Jean F. 2000. *Biochemical Tests for Identification of Medical Bacteria*, 3rd ed. Lippincott Williams & Wilkins, Philadelphia, PA.

Spellerberg, Barbara and Claudia Brandt. 2007. Chapter 29 in *Manual of Clinical Microbiology*, 8th ed., edited by Patrick R. Murray, Ellen Jo Baron, James. H. Jorgensen, Marie Louise Landry, and Michael A. Pfaller. ASM Press, Washington, DC.

Teixeira, Lúcia Martins, Maria da Glória Siqueira Carvalho, and Richard R. Facklam. 2007. Chapter 30 in *Manual of Clinical Microbiology*, 8th ed., edited by Patrick R. Murray, Ellen Jo Baron, James. H. Jorgensen, Marie Louise Landry, and Michael A. Pfaller. ASM Press, Washington, DC.

Winn Jr., Washington, Stephen Allen, William Janda, Elmer Koneman, Gary Procop, Paul Schreckberger, and Gail Woods. 2006. Chapters 12 and 13 in *Koneman's Color Atlas and Textbook of Diagnostic Microbiology*, 6th ed., Lippincott Williams & Wilkins, Philadelphia, PA.

---

[2] These flowcharts can be used only for the organisms listed in Table 7-7.

EXERCISE 7-9

Identification of Gram-positive Rods

✦ Theory

Several genera of Gram-positive rods have environmental, industrial, and clinical importance. Among these are *Bacillus, Clostridium, Corynebacterium, Lactobacillus* and *Mycobacterium*. Each is briefly characterized below.

Bacillus (Figure 7-19)

- ✦ A large, heterogeneous group
- ✦ Gram-positive rods (at least early in growth), in singles or chains
- ✦ Aerobic or facultatively anaerobic
- ✦ Produces endospores aerobically; spore shape and position are variable
- ✦ Most are catalase-positive
- ✦ Most are motile
- ✦ Most are soil saprophytes
- ✦ Key pathogens are *B. anthracis* (anthrax) and *B. cereus* (food poisoning and opportunistic infections)

Clostridium (Figure 7-20)

- ✦ Gram-positive rods (at least early in growth), in singles, pairs, or chains

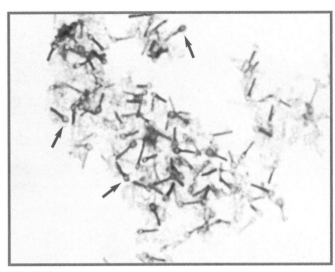

7-20 GRAM STAIN OF CLOSTRIDIUM TETANI (X1000) ✦ Note the round, terminal endospores (arrows) that distend the cells.

- ✦ Most are obligate anaerobes, but some are microaerophiles
- ✦ Produces endospores, but not aerobically; spore shape and position are variable, but usually distend the cell
- ✦ Most are catalase-negative
- ✦ Most are isolated from soil, sewage, or marine sediments
- ✦ Key pathogens are *Cl. tetani* (tetanus), *Cl. botulinum* (botulism), *Cl. perfringens* (food poisoning and gas gangrene), and *Cl. difficile* (pseudomembranous colitis)

Corynebacterium (Figure 7-21)

- ✦ Gram-positive rods, often club-shaped, in singles, pairs (V-forms), or arranged in stacks (palisades) or irregular clusters
- ✦ Metachromatic granules often present
- ✦ Facultatively anaerobic, but some are aerobes
- ✦ Catalase-positive
- ✦ No endospores
- ✦ Nonmotile
- ✦ Some are found in the environment, and many are in the normal flora of humans (skin and mucous membranes)
- ✦ Key pathogen is *C. diphtheriae* (diphtheria), although many species are opportunistic pathogens

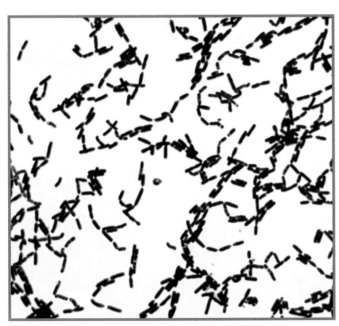

7-19 GRAM STAIN OF BACILLUS CEREUS (X1000) ✦ This particular specimen, obtained from culture, was not producing spores at the time of staining. Lesson: The absence of spores does not necessarily mean the inability to produce spores!

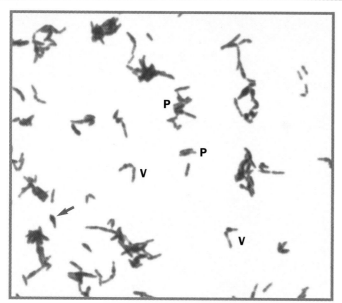

7-21 GRAM STAIN OF *CORYNEBACTERIUM DIPHTHERIAE* (X400) ✦
Note the club-shaped cells that might be mistaken for spore-producers (arrow). Cells of this genus often appear as "V-shaped" pairs (**V**) or palisades (**P**).

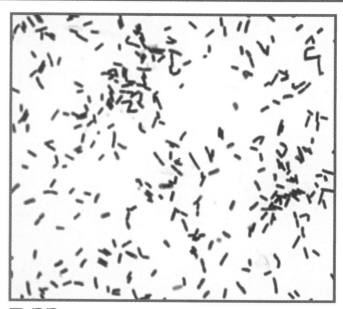

7-22 GRAM STAIN OF *LACTOBACILLUS SPP.* (1000X) ✦

Lactobacillus (Figure 7-22)

✦ Gram-positive rods, sometimes in chains

✦ Microaerophilic with fermentative metabolism; lactose is an abundant product

✦ Catalase-negative

✦ Oxidase-negative

✦ No endospores

✦ Nonmotile

✦ Isolated from a variety of foods (dairy, fish, grain, meat), and many are in normal flora (mouth, intestines, and vagina)

✦ No key pathogens and therefore not usually identified to species in clinical setting

Mycobacterium (Figure 7-23)

✦ Weakly Gram-positive curved or straight rods, sometimes branched or filamentous

✦ Acid-fast (generally in early stages of growth)

✦ Aerobic

✦ Catalase-positive

✦ Nonmotile

✦ No endospores

✦ Colony pigmentation with and without light is of taxonomic use

✦ Divided into slow growers (>7 days to form colonies on Löwenstein-Jensen medium) and rapid growers (<7 days to form colonies on L-J medium)

7-23
ACID-FAST STAIN OF *MYCOBACTERIUM TUBERCULOSIS* (X1000)
✦ Note the characteristic cord-like arrangement of cells.

✦ Many are aquatic saprophytes

✦ Key pathogens are *M. tuberculosis* (tuberculosis) and *M. leprae* (leprosy or Hansen's disease). Many rapidly growing species are opportunists.

✦ Application

Staining reactions and biochemical testing are used widely in bacterial identification.

✦ In This Exercise

This exercise will span several lab periods. You will combine staining and biochemical testing to identify one of several unknown organisms to species level. For obvious reasons, pathogens are not included, but the organisms chosen will give you an overview of the characteristics of each genus as well as practice in the identification process.

✦ Materials

Per Class

- ✦ media listed in Table 7-8 for bio-chemical testing

- ✦ pure cultures of organisms listed in Table 7-8 to be used for positive controls and unknowns (appropriate to the course level)

- ✦ anaerobic jars and GasPaks to accommodate one plate per student

- ✦ incubator or water bath set at 52°C

Per Student

- ✦ compound microscope with ocular micrometer and oil objective

- ✦ immersion oil

- ✦ clean glass slides

- ✦ Gram stain kit

- ✦ spore stain kit

- ✦ lens paper

- ✦ 3% H_2O_2

- ✦ one unknown organism in Thioglycollate Broth[1]

- ✦ two agar plates for streaking: instructor will choose from Trypticase Soy Agar, TSA plus 5% Sheep Blood, or Brain Heart Infusion Agar

- ✦ TSA slant or Thioglycollate Broth

TABLE **7-8** LIST OF ORGANISMS AND A KEY TO THE ICONS USED IN THIS EXERCISE ✦ The organisms listed are Gram-positive rods available from Ward's or Carolina Biological Supply Companies. Icon colors are to remind you what the results look like. They should not be used to compare your results. **Note:** For the fermentation tests, "A+" means acid is produced; "A−" means no acid is produced.

Gram-Positive Rods and Identification Tests

Spore-Positive
(Figure 7-24)

Bacillus cereus
Bacillus coagulans
Bacillus licheniformis
Bacillus megaterium
Bacillus mycoides
Bacillus subtilis
Bacillus thuringiensis
Clostridium acetobutylicum
Clostridium butyricum
Clostridium sporogenes

Spore-Negative
(Figure 7-25)

Corynebacterium pseudodiphtheriticum
Corynebacterium xerosis
Lactobacillus plantarum
Mycobacterium phlei
Mycobacterium smegmatis

Acid-Fast Reaction (24-Hour Culture)
Catalase
Citrate Utilization
Lipase
NO_3 reduced to NO_2
Growth at 52°C
Parasporal Crystal Stain
Spore Stain
Urease
VP

Dulcitol Fermentation
Glucose Fermentation
Mannitol Fermentation
Sucrose Fermentation

Gelatinase

Anaerobic Growth
(Thioglycollate)

✦ Procedure

1   Obtain an unknown and record its number on the Data Sheet.

2   Streak the sample for isolation on two plates (medium as provided by the instructor). Incubate both at 30°–35°C, one aerobically and the other in the Anaerobic Jar. It is best to check your plate for isolation at 24 hours. If you have isolation, continue with Step 3. If you don't have time to continue with Step 3, refrigerate your plates until you do. If you don't see isolation, ask your instructor if you should restreak for isolation, let the original plates continue

to incubate, or do both. Record your isolation procedure as directed on the Data Sheet.

3   After incubation, record colony morphology (including medium) under "Preliminary Observations" and note any differences between aerobic and anaerobic growth. Continue to record your isolation procedure on the Data Sheet.

4   Perform a Gram stain on a portion of one colony.

- ✦ Record its Gram reaction and cell morphology, arrangement, and size under "Preliminary Observations" on the Data Sheet.

- ✦ If the isolate is a Gram-positive rod, transfer a portion of the *same* colony to a Trypticase Soy Agar slant (or another suitable growth medium as available in your lab) or Thioglycollate Broth (if anaerobic) and incubate it at its optimum temperature. This is your pure culture to be used as a source of organisms for further testing. Complete the record of your isolation procedure on the Data Sheet. If available, also streak for isolation on Sporulation Agar to promote sporulation.

---

[1] **Note to instructor:** This exercise can be combined with Exercise 7-7 by passing out a mixture of a Gram-positive rod and an enteric. The student must isolate the two organisms from mixed culture and then proceed to identify the two independently. If this option is chosen, grow the two unknowns separately and then mix them immediately prior to handing them out in lab.

Gram-Positive, Spore-Forming Rods

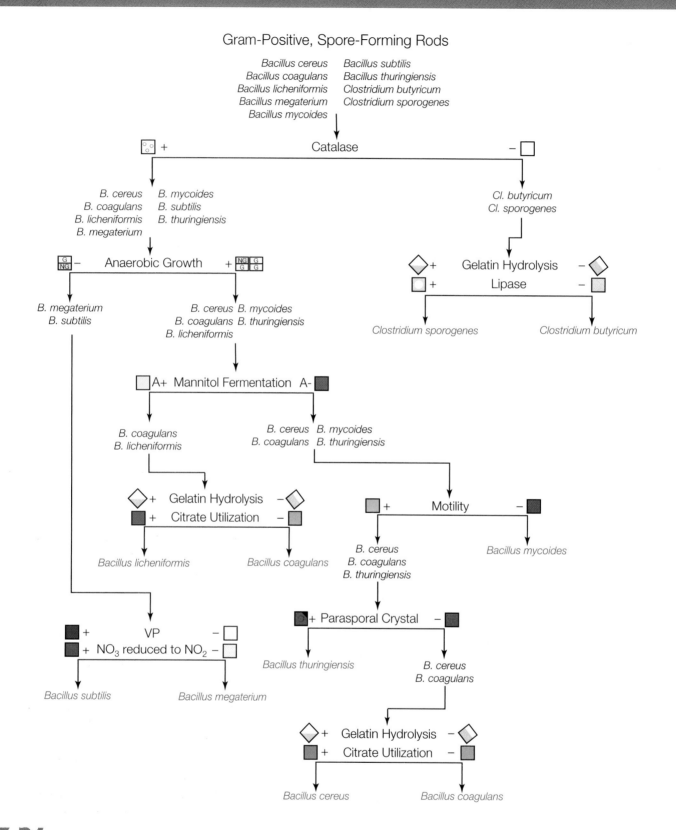

7-24 IDENTIFICATION FLOWCHART FOR SPORE-FORMING GRAM-POSITIVE RODS ✦ See the appropriate exercises for instructions on how to run each test.

(Compiled from sources listed in References)

Gram-Positive, Nonspore-Forming Rods

Corynebacterium pseudodiphtheriticum   *Mycobacterium phlei*
Corynebacterium xerosis   *Mycobacterium smegmatis*
Lactobacillus plantarum

7-25 IDENTIFICATION FLOWCHART FOR NONSPORE-FORMING GRAM-POSITIVE RODS ✦ See the appropriate exercises for instructions on how to run each test.

(Compiled from sources listed in References)

5 After incubation, perform a spore stain. Record the result as Test #1 on the Data Sheet. (Make sure you incubate long enough for sporulation to occur.)

6 Use the chart in Figure 7-24 if your isolate produces spores. If the organism does not produce spores, proceed to the identification chart in Figure 7-25. Be aware that a negative spore result may happen if a spore-former is not producing spores at the time of staining.

7 You now will begin biochemical testing to identify your organism. The tests in the flowcharts were chosen because of their uniformity of results, as published in standard microbiological references, so they give the best chance of correct identification. Be aware, however, that the symbol " + " indicates that 90% or more of the strains tested give a positive result. This means that as many as 10% of the strains tested give a negative result. The same is true of the symbol " − ." Bottom line: There are no guarantees that your particular lab strains will behave in the majority and if not, you will misidentify your unknown. This issue, if relevant to your unknown, will be addressed in Step 10.

8 Follow the tests in the appropriate flowchart to identify your unknown.[2] Use these guidelines.

♦ Do not run more than one test at a time unless your instructor tells you to do so.

♦ If your unknown is an obligate anaerobe, you will have to incubate media in an anaerobic jar. Coordinate your incubation with others who need the anaerobic jar to conserve GasPaks and incubator space.

♦ Where multiple tests are listed at a branch point, run only one, and then move to the next level in the flowchart as indicated by the result of that test.

♦ Continue to record the relevant information and results on the Data Sheet as you go. Keep accurate records of what you have done. Do *not* enter all of your data after you complete the project.

9 When you have identified your organism, use *Bergey's Manual* or another standard reference to find one more test to run for confirmation. The confirmatory test doesn't have to separate the final

---

[2] These flowcharts can be used only with the organisms listed in Table 7-8.

organisms on the flowchart. It only has to be a test for which you know the result and that you haven't run already as part of the identification process. Record this test on the Data Sheet under "My organism is" and identify it as the confirmatory test. Record the *expected* result on the comments line. After incubation, note whether the organism's result matches or differs from the expected result.

10 After the confirmation test, write the organism's name on the back of the Data Sheet under "My Unknown is" and check with your instructor to see if you are correct. If you are, congratulations! If you aren't, your instructor will advise you as to which test(s) gave "incorrect" results, by writing the test name(s) in the "rerun" space next to your identification. These should be rerun with appropriate controls from your school's inventory of organisms. Record the test(s) and control(s) on your Data Sheet. Checking the results with controls and your unknown will indicate if the "incorrect" test result was truly incorrect or if the strain of organism your school is using doesn't match the majority of strains for that test result.

References

Brown-Elliot, Barbara A., and Richard J. Wallace, Jr. 2007 Chapter 38 in *Manual of Clinical Microbiology*, 9th ed., edited by Patrick R. Murray, Ellen Jo Baron, James H. Jorgensen, Marie Louise Landry, and Michael A. Pfaller. ASM Press, Washington, DC.

Cato, Elizabeth P., W. Lance George, and Sydney M. Finegold. 1986. Section 13 (*Clostridium*) in *Bergey's Manual of Systematic Bacteriology*, Vol. 2, edited by John G. Holt. Williams and Wilkins, Baltimore, MD.

Claus, D and R. C. W. Berkeley. 1986. Section 13 (*Bacillus*) in *Bergey's Manual of Systematic Bacteriology*, Vol. 2, edited by John G. Holt. Williams and Wilkins, Baltimore, MD.

Collins, M. D., and C. S. Cummins. 1986. Section 15 (*Corynebacterium*) in *Bergey's Manual of Systematic Bacteriology*, Vol. 2, edited by John G. Holt. Williams and Wilkins, Baltimore, MD.

Funke, Guido, and Kathryn A. Bernard. 2007. Chapter 34 in *Manual of Clinical Microbiology*, 9th ed., edited by Patrick R. Murray, Ellen Jo Baron, James H. Jorgensen, Marie Louise Landry, and Michael A. Pfaller. ASM Press, Washington, DC.

Johnson, Eric A., Paula Summanen, and Sydney M. Finegold. 2007. Chapter 57 in *Manual of Clinical Microbiology*, 9th ed., edited by Patrick R. Murray, Ellen Jo Baron, James H. Jorgensen, Marie Louise Landry, and Michael A. Pfaller. ASM Press, Washington, DC.

Kandler, Otto, and Norbert Weiss. 1986. Section 14 (*Lactobacillus*) in *Bergey's Manual of Systematic Bacteriology*, Vol. 2, edited by John G. Holt. Williams and Wilkins, Baltimore, MD.

Könönen, Eija, and William G. Wade. 2007. Chapter 56 in *Manual of Clinical Microbiology*, 9th ed., edited by Patrick R. Murray, Ellen Jo Baron, James H. Jorgensen, Marie Louise Landry, and Michael A. Pfaller. ASM Press, Washington, DC.

Logan, Niall A, Tanja Popovic, and Alex Hoffmaster. 2007. Chapter 32 in *Manual of Clinical Microbiology*, 9th ed., edited by Patrick R. Murray, Ellen Jo Baron, James H. Jorgensen, Marie Louise Landry, and Michael A. Pfaller. ASM Press, Washington, DC.

MacFaddin, Jean F. 2000. *Biochemical Tests for Identification of Medical Bacteria*, 3rd ed. Lippincott Williams & Wilkins, Philadelphia, PA.

Wayne, Lawrence G., and George P. Kubrica. 1986. Section 15 (*Mycobacterium*) in *Bergey's Manual of Systematic Bacteriology*, Vol. 2, edited by John G. Holt. Williams and Wilkins, Baltimore, MD.

Winn Jr., Washington, Stephen Allen, William Janda, Elmer Koneman, Gary Procop, Paul Schreckberger, and Gail Woods. 2006. Chapters 14, 16, and 19 in *Koneman's Color Atlas and Textbook of Diagnostic Microbiology*, 6th ed., Lippincott Williams & Wilkins, Philadelphia, PA.

Environmental Microbiology

The preceding editions of this lab manual have emphasized basic techniques, molecular biology, and medical aspects of microbiology. Environmental Microbiology was limited to 4 exercises: Bioluminescence (Exercise 8-9), Soil Microbial Count (Exercise 8-11), Membrane Filter Technique (Exercise 8-12), and Multiple Tube Fermentation Method for Total Coliform Determination (Exercise 8-13). We have retained these four (though we have modified their sequence), and have added 9 new ones to round out this Section's offerings.

Section 8 begins with a fun exercise in which a mud culture is produced and then followed over the course of several weeks in the attempt to identify characteristic environmental microbes. This Winogradsky column will serve as a source of organisms for subsequent labs. A mature column will produce beautiful colors (by comparison) out of dull, old mud!

The next four exercises (8-2 through 8-5) illustrate various aspects of the nitrogen cycle. Not only is it one of the most important biogeochemical cycles, it is driven in large part by microbial activity.

Following the nitrogen cycle, are three exercises illustrating the sulfur cycle. It, too, relies heavily on microbial activity. It is in these exercises that you will harvest what you have grown in your Winogradsky column.

The only other new lab (Exercise 8-10) is a soil slide culture, in which you will place microscope slides into a beaker of soil and let them incubate at room temperature for several days. Then, with careful removal and a special stain, you will observe the spatial relationships of the soil microbes you have cultivated. ✦

Winogradsky Column

✦ Theory

The Winogradsky column bears the name of its developer, Sergei Winogradsky (1856–1953), a Russian microbiologist and pioneer in microbial ecology. He studied sulfur bacteria because of their ease of handling and cultivation, and then moved on to bacteria associated with the nitrogen cycle. One of his major discoveries was finding microorganisms (*Beggiatoa*) capable of the unheard of type of metabolism that came to be known as chemolithotrophic autotrophy (see below). Until he made his discovery, only photoautotrophs—those performing plant photosynthesis—were known to be autotrophs.

As a result of his work and the work of others, metabolic categories of microorganisms have been identified based on their carbon, energy, and electron sources. These are listed below. Note that in practice, terms are combined to describe the organism more fully.

Autotroph: an organism capable of obtaining all of its carbon from CO_2.

Heterotroph: an organism that can only get its carbon from organic molecules.

Chemotroph: an organism that gets its energy from the oxidation of chemicals.

Phototroph: an organism that gets its energy from light (hv).

Organotroph: an organism that gets its electrons from an organic molecule.

Lithotroph: an organism that gets its electrons from an inorganic molecule.

Winogradsky first used "his" column in the late 19th Century. It was (and is) used as a convenient laboratory source to supply for study a variety of **anaerobic, microaerophilic,** and **aerobic** bacteria, including purple nonsulfur bacteria, purple sulfur bacteria, green sulfur bacteria, chemoheterotrophs, and many others (Figure 8-1).

The basis for the Winogradsky column is threefold. The first two factors involve opposing gradients that impact the types of organisms that can grow. The first is the oxygen gradient, which gets more and more anaerobic toward the bottom. As a result, obligate aerobes, microaerophiles, facultative anaerobes, and obligate aerobes are found in different locations in the column.

The second is the H_2S gradient, which runs opposite in direction to the O_2 gradient. The third factor is the diffuse light shined upon the column. This promotes growth of phototrophic organisms at levels where they are adapted to the opposing gradients of O_2 and H_2S. These layers of phototrophs occur in natural ecosystems but are extremely thin because light doesn't penetrate mud sediments very far. But with the transparent column,

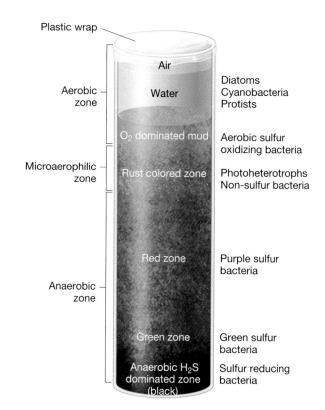

8-1 AN ARTIST'S RENDITION OF A WINOGRADSKY COLUMN ✦ What you put into a Winogradsky column dictates what you grow. Any well-constructed column has an oxygen gradient from top to bottom, with the aerobic zone penetrating perhaps only as much as 20% of the total depth. The remaining portion of the mud column becomes progressively more anaerobic. The different amounts of oxygen lead to layering of microbial communities adapted to that specific environment. This illustration is a generalized picture of the layering that you might see in a mature column. (The column often produces intermixed patches rather than distinct layers.) Starting at the top and working downward, the layers are: air, water (containing algae and cyanobacteria), aerobic mud (sulfur oxidizing bacteria), microaerophilic mud (nonsulfur photosynthetic bacteria), red/purple zone (purple sulfur photosynthetic bacteria), green zone (green sulfur photosynthetic bacteria), black anaerobic zone (sulfur reducing bacteria).

thicker layers develop, which are more easily sampled for cultures.

✦ Application

The Winogradsky column is a method for growing a variety of microbes with uniquely microbial metabolic abilities. Bacterial photoautotrophs, chemolithotrophs, and photoheterotrophs may be found in a mature column. And more "typical" chemoheterotrophs and photoauto-trophs also are likely to be found. A mature Winogradsky column is a good source for studying these organisms in the laboratory.

✦ In This Exercise

Using this procedure, you will produce an enrichment culture for growing aquatic sediment microorganisms with various metabolic abilities. In Exercises 8-6 through 8-8 you will be taking samples from the column for examination

✦ Materials

Per Student Group

✦ dirt or mud (approximately 250 g) from an aquatic source—to be collected by each student group and brought to lab

✦ one glass or plastic column, about 30 cm tall and 8 cm in diameter (available from biological supply houses)

✦ calcium carbonate—5 g

✦ calcium sulfate—5 g

✦ shredded newspaper (or other cheap paper)—10 g

✦ pan or bowl to hold the contents during mixing

✦ spoon or spatula for mixing

✦ dH$_2$O

✦ aluminum foil sheet

✦ plastic wrap or Parafilm

✦ rubberband

✦ light source (a north-facing window sill or lamp that produces "natural" light)

Procedure

Day 1

Note: This procedure is very forgiving with respect to amounts. Try to approximate the suggested amounts as much as you can, but don't fret if you are off by a little bit. This isn't brain surgery.

1 Remove large chunks, rocks, and debris from the mud/dirt sample.

2 To about 125 g mud in a mixing pan, add 5 g of calcium carbonate and 5 g of calcium sulfate. Add water (if it isn't already in the sample). Mix thoroughly.

3 Add the 10 g of shredded paper and mix again.

4 Add this enriched slurry to the column to a depth of 3 cm. Pack it firmly with a stirring rod so no air spaces remain. *Removing air spaces is probably the most important step.*

5 Add the remaining unenriched mud to a level about 5 cm from the top of the column and pack again.

6 Add water to a depth of 2–3 cm on top of the mud. Let it sit for about 30 minutes. If there is more water than you put in, remove the excess. If the water level has dropped, add more until you have the correct depth. There should be about 2–3 cm of air space above the water at this point.

7 Cover the top with foil or plastic wrap held on with a rubberband.

8 Incubate for 6 or 7 weeks in indirect sunlight at room temperature (Figure 8-2).

8-2 A Freshly Made Winogradsky Column ✦ The black layer comprising the majority of the column is the unenriched mud. The lighter gray area at the bottom contains mud, CaCO$_3$, CaSO$_4$, and shredded paper mixed into a slurry. Note the absence of air spaces.

9   Check the column weekly to watch its development (Figure 8-3).

10   Record your observations on the Data Sheet.

Weeks Later

Your instructor will tell you how you will use the column as a source of samples in Exercises 8-6 through 8-8.

References

Atlas, Ronald M., and Richard Bartha. 1998. *Microbial Ecology—Fundamentals and Applications,* 4th ed. Addison Wesley Longman, Inc., Menlo Park, CA.

Lens, Piet, Marcus Vallero, and Look Hulshoff Pol. 2004. *Sulfur Cycle,* Chapter 38 in *The Desk Encyclopedia of Microbiology,* Moselio Schaechter, Ed. Elsevier Academic Press, San Diego, CA.

Madigan, Michael T., John M. Martinko, Paul V. Dunlap, and David P. Clark. 2009. *Brock Biology of Microorganisms,* 12th ed. Pearson Benjamin Cummings, San Francisco, CA.

Perry, Jerome J., James T. Staley, and Stephen Lory. 2002. *Microbial Life.* Sinauer Associates, Publishers. Sunderland, MA.

Varnam, Alan H., and Malcom G. Evans. 2000. *Environmental Microbiology.* ASM Press, Washington, DC.

Willey, Joanne M., Linda M. Sherwood, and Christopher J. Woolverton. 2009. *Prescott's Principles of Microbiology.* McGraw-Hill Higher Education, Boston, MA.

8-3   A MATURE WINOGRADSKY COLUMN AT EIGHT WEEKS ✦
Notice the layers and colors! Also notice that the layers are not as well defined as in Figure 8-1. In fact, some look mixed (*e.g.,* the rust and red portions appear mixed in some regions). But the dark, anaerobic zone above the whitish layer at the bottom is well defined. The remainder is—pardon the expression—clear as mud.

Introduction to the Nitrogen Cycle

Biogeochemical cycles, such as the carbon cycle, the nitrogen cycle, and the sulfur cycle, are characterized by **environmental phases**, in which the element is not incorporated into an organism, and an **organismal phase**, in which it is. All are important, but because the nitrogen cycle has so many parts in which bacteria participate, it will be discussed here. It will be helpful for you to look at Figure 8-4 as you read the following.

Organisms require nitrogen as a component of the 20 amino acids, purine and pyrimidine nucleotides, and other compounds. These organic forms of nitrogen do not occur outside cells. Most of the air (approximately 80%) is nitrogen gas (N_2), but this is a form of nitrogen that is not usable by all organisms except **nitrogen-fixing bacteria**. Some nitrogen-fixing bacteria (*e.g., Rhizobium*) are symbionts of certain legumes and form nodules in their roots. Other nitrogen-fixing bacteria are free-living in the soil (*Azotobacter*) or water (*e.g., Nostoc*). The process of nitrogen fixation is highly endergonic, requiring about 16 moles of ATP per mole of N_2 reduced to ammonia (NH_3).

Because nitrogen fixation requires so much energy, the **nitrogenase** enzyme is not active if other forms of nitrogen are available. In addition, nitrogenase is inactivated by oxygen, so a variety of mechanisms have evolved to protect it from oxygen. You will examine N-fixation in Exercise 8-2.

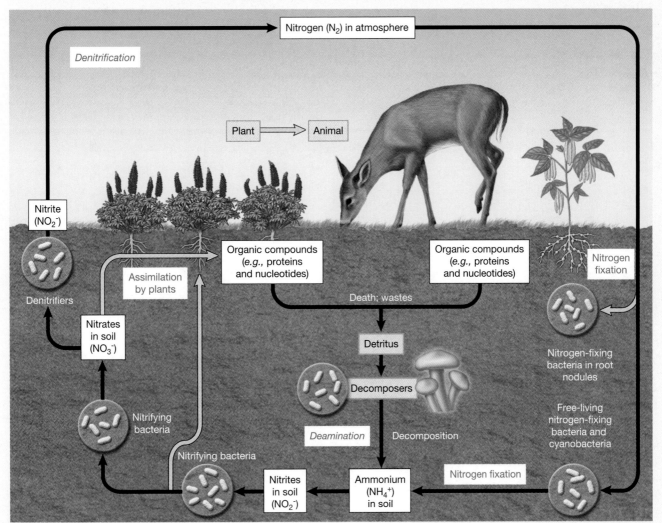

8-4 THE NITROGEN CYCLE ✦ The nitrogen cycle involves all forms of metabolism: aerobic respiration (chemoheterotrophic metabolism), anaerobic respiration, and chemolithotrophic metabolism, in addition to nitrogen fixation. And microbes are participants in all the steps.

Some **chemolithotrophic** bacteria are capable of **nitrification**, in which ammonium ion acts as an electron and energy source when it is oxidized to nitrite, and then to nitrate. Organisms such as *Nitrosomonas* oxidize ammonium to nitrite; then, other nitrifiying organisms, such as *Nitrobacter*, continue the oxidation of nitrite to nitrate. The same organisms do not do both. You will examine nitrification in Exercise 8-3.

Once nitrogen gas has been reduced to ammonium ion, it becomes available for **assimilation** into amino acids by green plants and other bacteria. It then enters the food chain as various forms of organic nitrogen, the only form of nitrogen animals can use. **Oxidative deamination** of amino acids during heterotrophic catabolism (including decomposition by bacteria and fungi) results in the formation of ammonium ion once again. You will examine this process of ammonification in Exercise 8-4.

Denitrifying bacteria use nitrate as the **final electron acceptor** of **anaerobic respiration**. The end products of nitrate reduction include nitrogen gas, nitrite, and ammonium ion. You will examine denitrification in Exercise 8-5. ✦

Nitrogen Fixation

✦ Theory

Fixation of atmospheric nitrogen (N₂) occurs through the following reaction:

$$N_2 + 8H^+ + 8e^- + 16ATP \longrightarrow 2NH_3 + H_2 + 16ADP + 16P_i$$

This is a highly endergonic process and is performed by only a few groups of bacteria. Among terrestrial bacteria, *Azotobacter* (a β-proteobacterium) and *Rhizobium* (an α-proteobacterium) usually are put forth as examples.

Azotobacter (Figure 8-5) is a common, free-living, aerobic Gram-negative rod that fixes N₂ in the soil. N-fixation is inhibited by oxygen because it inhibits the **nitrogenase** that is responsible for fixation. The ability of *Azotobacter* to fix nitrogen aerobically is attributable to a number of complex, species-specific mechanisms that are beyond the scope of this book. *Azotobacter* also has the ability to form resting vegetative cells called **cysts** when environmental conditions are not favorable (Figure 8-6). Though they perform somewhat the same function as bacterial endospores, they are different in that they are not as differentiated from the cell that formed them, nor are they as resistant to environmental factors such as desiccation and physical and chemical agents.

Rhizobium (Figure 8-7) is a mutualistic symbiont of leguminous plants (members of Fabaceae: the Pea family). They rely on a supply of carbon compounds as electron donors from the plant host and, in return, supply the plant with ammonium, which can be assimilated into organic nitrogen (amino acids, nucleotides, *etc.*). *Rhizobium* and other symbionts are the major N-fixers on land.

Rhizobium is a Gram-negative rod that enters the root and induces the formation of **root nodules** (Figure 8-8), tumors that are the location of N-fixation. Once in

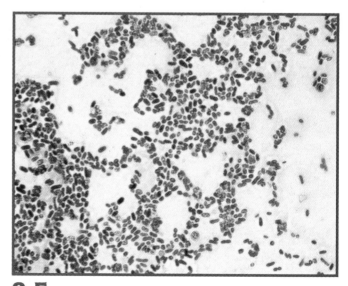

8-5 AZOTOBACTER GRAM STAIN ✦ Plump, Gram-negative rods characterize the free-living, nitrogen-fixing *Azotobacter*. These cells were grown in culture.

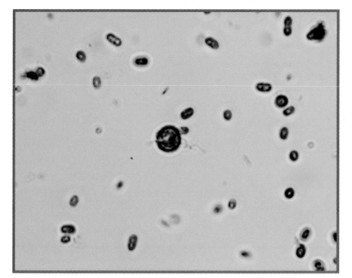

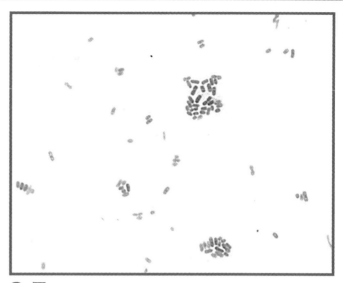

8-6 AZOTOBACTER CYST STAIN ✦ *Azotobacter* forms resting cells called "cysts" when the environment becomes unsuitable. They can be viewed with a wet mount stain.

8-7 RHIZOBIUM **Gram Stain** ✦ *Rhizobium* occurs naturally as a nitrogen-fixing symbiont of leguminous plants. This Gram stain was made from a culture.

8-8 CLOVER ROOT NODULES ✦ The roots of this small clover have been infected with *Rhizobium* that causes the formation of tumor-like root nodules (four are circled).

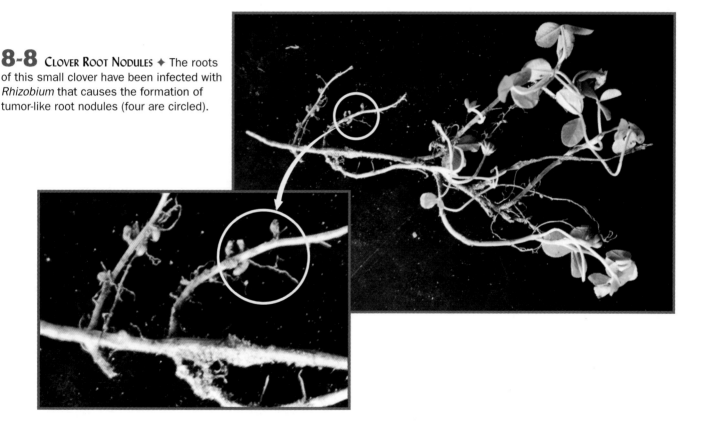

the plant, the cells become irregular in shape and are referred to as **bacteroids** (Figure 8-9). Their adaptation to fixing nitrogen aerobically is the presence of **leghemoglobin,** which, like our own hemoglobin, binds free oxygen. Interestingly, its production is induced through the combined efforts of the host plant and *Rhizobium*. Leghemoglobin is so effective at moderating O_2 concentrations in the nodule that the bound form is estimated to outnumber the free form 10,000:1!

In aquatic environments, Cyanobacteria are the main nitrogen fixers. An interesting example is the filamentous species *Anabaena* (Figure 8-10). Cyanobacteria perform photosynthesis using the same process as green plants. That is, they convert CO_2 and H_2O to carbohydrate and O_2. Because N-fixation is inhibited by oxygen, they restrict the process to thick-walled cells called **heterocysts,** which lack Photosystem II, the oxygen-producing component of photosynthesis (Figure 8-10). Further, they are

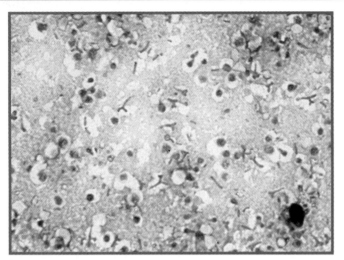

8-9 ROOT NODULE SECTION ✦ Symbiotic nitrogen-fixing bacteria, such as *Rhizobium*, induce tumor formation in the roots of certain legumes. Once in the nodule, infected cells become filled with differentiated forms of the bacterium called bacteroids. Uninfected cells are white in this preparation.

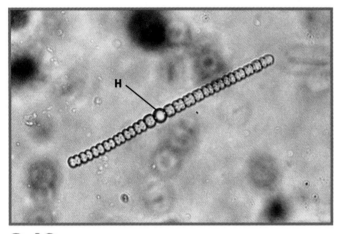

8-10 HETEROCYST IN ANABAENA ✦ *Anabaena* is a filamentous cyanobacterium. Specialized, thick-walled cells called hetero- cysts (**H**) are involved in nitrogen fixation. The oxygen-sensitive nitrogenase enzyme is protected from oxygen in these cells because they lack Photosystem II, which produces oxygen from water during photolysis, and the thick wall apparently limits oxygen diffusion from the environment.

thick-walled, which restricts entry of O_2 by diffusion. Heterocysts rely on physical connection with adjacent cells to supply reducing power for N-fixation and, in re- turn, supply the other cells of the filament with organic nitrogen compounds.

Nitrogen Free Broth is a selective medium lacking a nitrogen source other than the air. Therefore, only N-fixing organisms can survive in this broth. We will be testing for free-living N-fixers, such as *Azotobacter*. When *Azotobacter* is present, you will see a pellicle at the surface of the medium, and often gas bubbles (Figure

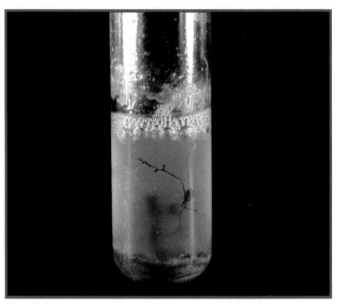

8-11 NITROGEN-FREE BROTH ✦ This nitrogen-free broth was inoculated with a soil sample. Because no nitrogen is present in the medium, only N-fixers can grow. Notice the bubbles and pellicle at the surface.

8-11). Whether you see a pellicle or not, N-fixers will produce ammonia and you will perform a qualitative test for its presence.

✦ Application

Soil fertility is determined largely by the rate and amount of N-fixation. This technique provides soil microbiologists with a mechanism to perform an assay for soil N-fixing organisms.

✦ In This Exercise

You will bring a soil sample to lab and incubate a portion of it in Nitrogen Free Broth. After incubation, you have several options.

1. To see that organisms can grow in the selective Nitrogen Free Broth—meaning that you have isolated a community of N-fixers. If so, you will perform a qualitative test for ammonia, the product of N-fixation.

2. To examine the culture for a pellicle, a sure sign that *Azotobacter* is present.

3. To perform a Gram stain on the broth.

4. To perform an *Azotobacter* cyst stain and a capsule stain on the pellicle.

Your choice(s) likely will be dictated by lab time and availability of resources.

✦ Materials

Per Student

- ✦ a capped 25 mL tube of soil from a location of your choice (Tubes should be handed out the previous lab.) This soil sample can be used for the other exercises relating to the Nitrogen Cycle (Exercises 8-3 through 8-5).
- ✦ two Nitrogen Free Broth tubes (5–8 mL)
- ✦ two clear plastic microtubes
- ✦ two 15 × 100 test tubes
- ✦ four transfer pipettes
- ✦ three microscope slides

Per Student Group

- ✦ API Freshwater Master Test Kit[1]
- ✦ microfuge
- ✦ Gram stain kit (optional)
- ✦ *Azotobacter* stain (optional)

✦ Medium and Stain Recipes

Nitrogen Free Broth

♦ Glucose	10 g
♦ $CaCO_3$	5 g
♦ $CaCl_2$	0.1 g
♦ KH_2PO_4	0.1 g
♦ K_2HPO_4	0.9 g
♦ $MgSO_4$	0.1 g
♦ $FeSO_4$	10 mg
♦ Na_2MoO_4	5 mg
♦ Distilled or deionized water	1 L

Azotobacter Cyst Stain

♦ Glacial Acetic Acid	8.5 mL
♦ Sodium Sulfate	3.25 g
♦ Neutral Red	200 mg
♦ Light Green	200 mg
♦ Ethanol	50 ml
♦ Distilled or deionized water	100 mL

Procedure

Day 1

1. Obtain two Nitrogen-Free Broth tubes.

2. Add a pinch of soil to one tube. Leave the other uninoculated.

3. Label and incubate both tubes for 7 days at 25°C or until a pellicle forms on the surface.

Day 2

1. Examine the inoculated tube for turbidity, which indicates you have isolated one or more N-fixing bacteria.

2. Examine the inoculated tube for a pellicle, which indicates that you likely have grown *Azotobacter* (Figure 8-11).

3. (*Optional*) If a pellicle is present, perform a:

   a. Gram stain (see Exercise 3-7 for the procedure)

   b. Capsule stain (see Exercise 3-9)

   c. *Azotobacter* cyst stain by making a wet mount in the stain solution.

4. Use a transfer pipette to dispense 1.0 mL of your culture into a clear plastic microtube. Close the cap tightly. Label it "NFB-Culture."

5. With a fresh transfer pipette, put an *equal* volume of uninoculated Nitrogen-Free Broth in the second clear plastic microtube. Close the cap tightly. Label it "NFB control."

6. Label both caps with your initials.

7. Spin both tubes for 2–5 minutes at high speed in a microfuge. Be sure to balance the microfuge by placing your tubes opposite one another.

8. With a transfer pipette, carefully draw the broth out of your sample without disturbing the pellet at the bottom, and transfer 0.6 mL to a clean glass test tube. Label this tube with your name and "NFB-Sample."

9. Using a fresh transfer pipette, remove 0.6 mL from the "NFB-Control" tube and transfer it to a clean glass test tube.

10. **Test for ammonia.** Test for ammonia using the reagents supplied.

    a. Add one drop of Ammonia Reagent #1 and one drop of Ammonia Reagent #2 to "NFB-Sample" and to "NFB-Control."

    b. Mix the tubes for 5 seconds.

    c. Watch for color to develop in 5 minutes (Figure 8-12A).

    d. Compare the color against the color card to determine the ammonia concentration in your soil sample (Figure 8-12B).

    e. Record your results on the Data Sheet.

---

[1] API Freshwater Test Kit is available from Mars Fishcare North America, Inc. PO Box 218, Chalfont, PA 18914. 1-800-847-0659 or www.aquariumpharm.com.

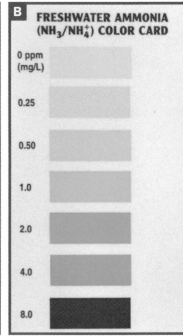

8-12 TEST FOR AMMONIA ✦
A Nitrogen fixation results in the production of ammonia. Shown are results of a quantitative test for ammonia, with negative on the left and positive on the right. **B** The color card matches test result colors with approximate ammonia concentrations. This sample has approximately 8 mg/L of ammonia. (**Note:** Use the actual color card, not this photograph, for comparison.)

References

Atlas, Ronald M. 2005. *Handbook of Media for Environmental Microbiology,* 2nd ed. CRC Press, an imprint of Taylor & Francis Group, LLC. 6000 Broken Sound Parkway NW, Boca Raton, FL.

Atlas, Ronald M., and Richard Bartha. 1998. *Microbial Ecology—Fundamentals and Applications,* 4th ed. Addison Wesley Longman, Inc., Menlo Park, CA.

Kennedy, Christina, Paul Rudnick, Melanie L. MacDonald, and Thoyd Melton. 2005. *Genus III. Azotobacter* Beijerinck 1901, 567[AL] in *Bergey's Manual of Systematic Bacteriology,* 2nd ed., Volume Two, *The Proteobacteria—Part B The Gammaproteobacteria.* Springer, 233 Spring Street, New York, NY.

Madigan, Michael T., John M. Martinko, Paul V. Dunlap, and David P. Clark. 2009. *Brock Biology of Microorganisms,* 12th ed. Pearson Benjamin Cummings, San Francisco, CA.

Microbiology Procedures: http://www.microbiologyprocedure.com/staining-methods-in-microbiology/stain-for-azotobacter-cysts.html

Varnam, Alan H. and Malcom G. Evans. 2000. *Environmental Microbiology.* ASM Press, Washington, DC.

Willey, Joanne M., Linda M. Sherwood, and Christopher J. Woolverton. 2009. *Prescott's Principles of Microbiology.* McGraw-Hill Higher Education, Boston, MA.

EXERCISE

8-3

Nitrification: The Production of Nitrate

✦ Theory

Nitrification is the process of making **nitrate.** This is strictly a microbial process performed by **nitrifying bacteria** and is a two-step, exergonic process involving two different groups of aerobic **chemolithotrophs** (which also are **autotrophs** because they use the energy released to fix carbon dioxide into organic carbon).

First, bacteria, such as *Nitrosomonas*, oxidize ammonium (NH_4^+) to form nitrite (NO_2), as follows:

$$NH_4^+ + 1\tfrac{1}{2}\, O_2 \longrightarrow NO_2^- + H_2O + 2H^+$$

The second step involves different chemolithotrophs, such as *Nitrobacter*, that oxidize nitrite to nitrate, as follows:

$$NO_2^- + \tfrac{1}{2}\, O_2 \longrightarrow NO_3^-$$

Clay particles in soil bind positively charged ions, whereas negatively charged ions are repelled and are more available for organismal use, as they are freely diffusible in the soil water. With that fact in mind, nitrification clearly is an important ecological process because it changes the charge on nitrogen from positive (in ammonium ion) to negative (nitrite and nitrate). Nitrate, along with ammonium, is a form of nitrogen that plants can absorb into their roots. Because of their solubility, however, nitrite and nitrate are readily leached from soil, making it less fertile.

✦ Application

This exercise shows a simple way to determine the presence of aerobic chemolithotrophic bacteria that (1) oxidize ammonia to nitrite, and (2) oxidize nitrite to nitrate.

✦ In This Exercise

You will bring a soil sample to class and check it for activity of microbial ammonia oxidizers that produce nitrite and nitrite oxidizers that produce nitrate.

✦ Materials

Per Student

✦ one capped 25 mL tube of soil from a location of your choice (Tubes should be handed out the previous lab.)

This soil sample can be used for the other exercises relating to the Nitrogen Cycle (Exercises 8-2, 8-4, and 8-5).

✦ two Ammonium Broth tubes (5-8 mL)
✦ five clear plastic microtubes
✦ five 15×100 glass test tubes
✦ six transfer pipettes

Per Student Group

✦ API Freshwater Master Test Kit[1]
✦ microfuge

✦ Medium Recipe

Ammonium Broth

✦ Ammonium sulfate	1.0 g
✦ MgSO$_4$	0.5 g
✦ NaCl	2.0 g
✦ K$_2$HPO$_4$	1.0 g
✦ CaCO$_3$	10.0 g
✦ FeCl$_3$	trace
✦ Distilled or deionized water	1 L

✦ Procedure

Day 1

1　Obtain two Ammonium Broth tubes.

2　Add a pinch of soil to one. Leave the other uninoculated.

3　Label and incubate both tubes for 7 days at 25°C.

Day 2

1　Examine the inoculated tube for turbidity, which indicates that you have isolated one or more ammonia-oxidizing bacteria. If there is none, you may wish to continue incubation for a few days longer.

2　Use a transfer pipette to dispense 1.5 mL of your culture into two clear plastic microtubes. Close the caps tightly. Label one "NO$_2$ Sample" and the other "NO$_3$ Sample."

---

[1] API Freshwater Test Kit is available from Mars Fishcare North America, Inc. PO Box 218, Chalfont, PA 18914. 1-800-847-0659 or www.aquariumpharm.com.

3 Transfer 1.5 mL of uninoculated Ammonium Broth to the other three clear plastic microtubes using fresh transfer pipettes. Label as follows: "AB-Control 1," "AB-Control 2," and "AB-Control 3."

4 Label the five caps with your initials.

5 Spin all tubes for 2–5 minutes at high speed in a microfuge. Be sure to balance the microfuge by placing your tubes opposite one another. You'll have to combine with another student to make the balancing even, or you can use water to make the sixth tube.

6 Carefully draw 1 mL of broth out of each sample tube without disturbing the pellet at the bottom, and transfer the broth to two clean glass test tubes. Label these tubes with your name and "NO$_2$ Sample" and "NO$_3$ Sample."

7 Transfer 1 mL from "AB-Control 1" to a clean glass test tube. Label it "Control 1." (You may use the same transfer pipette for Steps 8 and 9.)

8 Transfer 1 mL from "AB-Control 2" to a clean glass test tube. Label it "Control 2."

9 Transfer 1 mL from "AB-Control 3" to a clean glass test tube. Label it "Control 3."

10 **Test for nitrite.** Use the reagent supplied to perform the test for nitrite as follows:

   a. Add one drop of Nitrite Reagent to "NO$_2$-Sample" and "Control 1." Mix for 5 seconds.

   b. Watch for color to develop in 5 minutes (Figure 8-13A).

   c. Compare the color against the color card to determine the nitrite concentration in your soil sample (Figure 8-13B).

   d. Record your results on the Data Sheet.

11 **Test for nitrate.** Use the reagents supplied to perform the test for nitrate as follows:

   a. Add two drops of Nitrate Reagent #1 to "NO$_3$-Sample" and to "Control 2." Mix for 5 seconds.

   b. Shake Nitrate Reagent #2 for 30 seconds.

   c. Add two drops of Nitrate Reagent #2 to "NO$_3$-Sample" and to "Control 2." Mix vigorously for 1 minute.

   d. Watch for color to develop in 5 minutes (Figure 8-14A).

   e. Compare the color against the color card to determine the nitrite concentration in your soil sample (Figure 8-14B).

   f. Record your results on the Data Sheet.

12 **Test for ammonia.** Use the reagents supplied to test for ammonia as follows:

   a. Add one drop of Ammonia Reagent #1 and one drop of Ammonia Reagent #2 to "Control 3." Mix for 5 seconds.

   b. Color should develop in 5 minutes (Figure 8-15A).

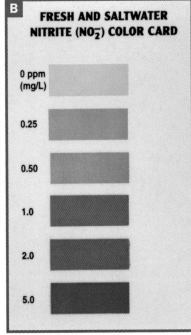

FRESH AND SALTWATER NITRITE (NO$_2^-$) COLOR CARD

0 ppm (mg/L)

0.25

0.50

1.0

2.0

5.0

8-13 TEST FOR NITRITE ✦ **A** Oxidation of ammonia can result in nitrite formation. Shown are results of a quantitative test for nitrite, with negative on the left and positive on the right. **B** The color comparison card is used to approximate nitrite concentration. This sample has about 1 mg/L of nitrite. (**Note:** Use the actual color card, not this photograph, for comparison.)

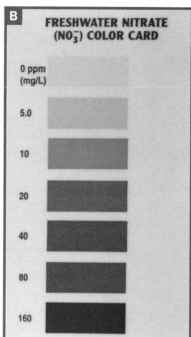

8-14 TEST FOR NITRATE ✦ **A** Ammonia oxidation can also produce nitrate. In this photo, two nitrate-positive tests are on either side of the negative control. **B** By comparing to the color card, the sample on the left has about 5 mg/L of nitrate and the one on the right has about 80 mg/L. (***Note:*** Use the actual color card, not this photograph, for comparison.)

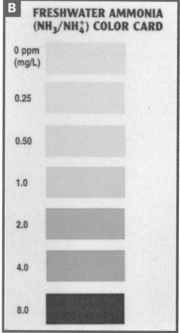

8-15 TEST FOR AMMONIA ✦ **A** The ammonia broth control (AB-Control #3) is also tested for ammonia after incubation. **B** Compare your tube against the color card, not this photograph.

c. Compare the color against the color card to determine the ammonia concentration in your soil sample (Figure 8-15B).

d. Record your results on the Data Sheet.

References

Atlas, Ronald, and Richard Bartha. 1998. *Microbial Ecology—Fundamentals and Applications*, 4th ed. Addison Wesley Longman, Inc., Menlo Park, CA.

Bartholomew, James W. (1967). *Laboratory Textbook and Exercises in Microbiology*. Wm. C. Brown Company Publishers, Dubuque, IA.

Varnam, Alan H., and Malcom G. Evans. 2000. *Environmental Microbiology*. ASM Press, Washington, DC.

White, David. 2000. *The Physiology and Biochemistry of Prokaryotes*, 2nd ed. Oxford University Press, New York and Oxford.

Willey, Joanne M., Linda M. Sherwood, and Christopher J. Woolverton. 2009. *Prescott's Principles of Microbiology*. McGraw-Hill Higher Education, Boston, MA.

EXERCISE 8-4

Ammonification

✦ Theory

In an ecosystem, much of the nitrogen is tied up in organic molecules. No organism lives forever, however, or is immune from producing wastes. **Ammonification** is a consequence of decomposition of organic nitrogen from dead animal and plant protein, and nitrogenous wastes (*e.g.*, urea) from animals, and is a byproduct of amino acid **deamination** (a step in amino acid catabolism) in cells. Many bacteria and fungi are capable of ammonification. The ammonia so produced can be recycled through uptake by plants, or it can be oxidized by ammonia oxidizing bacteria (see Exercise 8-3).

Deamination of an aminio acid occurs as follows:

$$\underset{\displaystyle H_2N-CH-COOH}{\overset{\displaystyle R}{|}} + H_2O \longrightarrow \underset{\displaystyle CH_2-COOH}{\overset{\displaystyle R}{|}} + NH_3$$

✦ Application

Ammonia is an important soil component because of its availability to plants as a source of nitrogen and their ability to incorporate it into organic nitrogen again. This test allows for rapid determination of ammonia production in soil.

✦ In This Exercise

You will bring a soil sample to class and check it for activity of ammonifying microbes that produce ammonia from organic nitrogen.

✦ Materials

Per Student

A capped 25 mL tube of soil from a location of your choice. (Tubes should be handed out during the previous lab.) This soil sample can be used for the other exercises relating to the Nitrogen Cycle (Exercises 8-2, 8-3, and 8-5).

✦ two Peptone Broth tubes (5–8 mL)

✦ two clear plastic microtubes

✦ four transfer pipettes

Per Student Group

✦ API Freshwater Master Test Kit[1]

✦ microfuge

✦ Medium Recipes

Peptone Water (Bartholomew)

✦ Peptone	40.0 g
✦ Distilled or deionized water	1 L

Procedure

Day 1

1 Obtain two Peptone Water tubes.

2 Add a pinch of soil to one tube. Leave the other uninoculated.

3 Label and incubate for 7 days at 25°C.

Day 2

1 Examine the inoculated tube for turbidity. You also may notice a distinctive odor. If there is none, you may wish to continue incubation for a few days longer.

2 Use a transfer pipette to dispense 1.5 mL of your Peptone Water culture into a clear plastic microtube. Close the cap tightly. Label it "PW-Sample."

3 Using a fresh transfer pipette, transfer 1.5 mL of uninoculated Peptone Water to the second clear plastic microtube. Label it "PW-Control 1."

4 Label the two caps with your initials.

5 Spin both tubes for 2–5 minutes at high speed in a microfuge. Be sure to balance the microfuge by placing your tubes opposite from one another.

---

[1] API Freshwater Test Kit is available from Mars Fishcare North America, Inc. PO Box 218, Chalfont, PA 18914. 1-800-847-0659 or www.aquariumpharm.com.

FRESHWATER AMMONIA (NH₃/NH₄⁺) COLOR CARD

- 0 ppm (mg/L)
- 0.25
- 0.50
- 1.0
- 2.0
- 4.0
- 8.0

8-16 TEST FOR AMMONIA ✦ **A** Ammonia is a product of decomposition of nitrogenous organic compounds. Shown are results of a quantitative test for ammonia, with negative on the left and positive on the right. **B** The color card matches test result colors with approximate ammonia concentrations. This sample has approximately 8 mg/L of ammonia. (***Note:*** Use the actual color card, not this photograph, for comparison.)

6 With a fresh transfer pipette, carefully draw 0.6 mL of broth out of your sample without disturbing the pellet at the bottom, and transfer to a clean glass test tube. Label this tube with your name and "PW-Sample."

7 With a fresh transfer pipette, carefully draw 1 mL of broth from the "PW-Control" and transfer to a clean glass test tube.

8 **Test for ammonia.** Use the reagents supplied to test for ammonia as follows:

a. Add one drop of Ammonia Reagent #1 and one drop of Ammonia Reagent #2 to "PW-Sample" and "PW-Control." Mix for 5 seconds.

b. Watch for the color to develop in 5 minutes (Figure 8-16A).

c. Compare the color against the color card to determine the ammonia concentration in your soil sample (Figure 8-16B).

d. Record your results on the Data Sheet.

References

Atlas, Ronald, and Richard Bartha. 1998. *Microbial Ecology—Fundamentals and Applications,* 4th ed. Addison Wesley Longman, Inc., Menlo Park, CA.

Bartholomew, James W. (1967). *Laboratory Textbook and Exercises in Microbiology.* Wm. C. Brown Company Publishers, Dubuque, IA.

White, David. 2000. *The Physiology and Biochemistry of Prokaryotes,* 2nd ed. Oxford University Press. New York and Oxford.

Willey, Joanne M., Linda M. Sherwood, and Christopher J. Woolverton. 2009. *Prescott's Principles of Microbiology.* McGraw-Hill Higher Education, Boston, MA.

EXERCISE 8-5 Denitrification: Nitrate Reduction

✦ Theory

Nitrate reduction is the result of **anaerobic respiration,** in which nitrate is used as the **final electron acceptor.** In some cases, nitrite is the end product, but in others nitrate is reduced to N_2 gas. We will be looking for the latter end product. Both processes are called **denitrification.** An example of this process is provided by *Pseudomonas denitrificans* and is shown below.

$$C_6H_{12}O_6 + 4NO_3 \longrightarrow 6CO_2 + 6H_2O + 2N_2$$

Notice that this reaction is very similar to the summary reaction for aerobic respiration. Because the final electron acceptor is $4NO_3$ and not O_2, however, the products are different. O_2 is not required for—in fact, it often inhibits—anaerobic respiration, so dentrification usually occurs in its absence.

Because nitrate is so soluble, it is leached from soils into aquatic environments and ultimately resides in the oceans. Without denitrification, all the nitrogen would end up in the ocean and be unavailable to terrestrial life.

✦ Application

This test allows for simple and fast determination of denitrification from soil samples.

✦ In This Exercise

You will bring a soil sample to class and check it for activity of denitrifying microbes that produce N_2 from nitrate.

✦ Materials

Per Student

- ✦ a capped 25 mL tube of soil from a location of your choice (Tubes should be handed out the previous lab.) This soil sample can be used for the other exercises relating to the Nitrogen Cycle (Exercises 8-2, 8-3, and 8-4).
- ✦ two Nitrate Broths with Durham tubes
- ✦ two clear plastic microtubes
- ✦ two 15 × 100 test tubes
- ✦ four transfer pipettes

Per Student Group

- ✦ API Freshwater Master Test Kit[1]
- ✦ microfuge

✦ Medium Recipe

Nitrate Broth

◆ Beef extract	3.0 g
◆ Peptone	5.0 g
◆ Potassium nitrate	1.0 g
◆ Distilled or deionized water	1.0 L
pH 6.8–7.2 at 25°C	

Procedure

Day 1

1 Obtain two Nitrate Broth tubes.

2 Add a pinch of soil to one medium. Label this tube "NB-Sample." Be sure the soil doesn't clog the opening of the Durham tube.

3 Do not inoculate the other tube and label it "NB-Control."

4 Incubate both tubes for 7 days at 25°C.

Day 2

1 Examine the tube for turbidity.

2 Compare the Durham tubes of the sample and the control.

3 If you see a bubble in the sample tube and not in the control, proceed (Figure 8-17). Otherwise, you may choose to let your culture incubate for a few more days.

4 Transfer 1.5 mL from the "NB-Sample" tube to a clean, clear plastic microtube. Label it "NO_3 Sample." Label the cap with your initials.

5 Repeat for the "NB-Control" tube with a clean transfer pipette.

---

[1] API Freshwater Test Kit is available from Mars Fishcare North America, Inc. PO Box 218, Chalfont, PA 18914. 1-800-847-0659 or www.aquariumpharm.com

8-17 NITRATE REDUCTION BROTH WITH A DURHAM TUBE ✦ The bubble in the Durham tube indicates that nitrate has been reduced to molecular nitrogen gas (N_2). This tube was inoculated with soil.

6 Spin both tubes in a microfuge for 2–5 minutes at high speed. Be sure to balance the microfuge by placing your tubes opposite one another.

7 With a clean transfer pipette, carefully draw 1 mL of broth out of the sample tube without disturbing the pellet at the bottom and transfer it to a clean glass test tube. Label the tube with your name and "NO_3-Sample."

8 With a clean transfer pipette, transfer 1 mL from "NB-Control" to a clean glass test tube. Label it as "NB-Control."

9 **Test for nitrate.** Use the reagents supplied to perform the test for nitrate as follows:

a. Add two drops of Nitrate Reagent #1 to "NO_3-Sample" and to "NB-Control." Mix for 5 seconds.

b. Shake Nitrate Reagent #2 for 30 seconds.

c. Add two drops of Nitrate Reagent #2 to "NO_3-Sample" and to "NB-Control." Mix vigorously for 1 minute.

d. Watch for color to develop in 5 minutes (Figure 8-18A).

e. Compare the color against the color card to determine the nitrite concentration in your soil sample (Figure 8-18B).

f. Record your results on the Data Sheet.

References

Atlas, Ronald and Richard Bartha. 1998. *Microbial Ecology—Fundamentals and Applications*, 4th ed. Addison Wesley Longman, Inc., Menlo Park, CA.

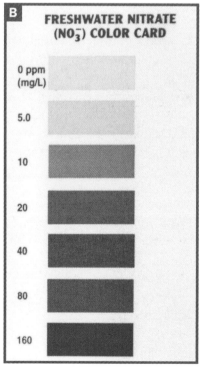

8-18 TEST FOR NITRATE ✦ **A** In this photo, two nitrate positive tests are on either side of the negative control. **B** Using this color card allows comparison to sample suspected of having nitrate. A comparison with Figure 8-18A shows that the sample on the left has about 5 mg/L of nitrate and the one on the right has about 80 mg. (**Note:** Use the actual color card, not this photograph, for comparison.)

A Word About Trophic Groups

Understanding how bacteria make their living (*i.e.* where they get their nutrition and/or, energy, whether they use organic or inorganic molecules for their purposes, where they live in relation to available oxygen, their optimum temperature, pH, and osmotic conditions) is fundamental to microbiology. In Section 2, we discussed major factors that affect growth, but we didn't talk about trophic groups. Trophic groups are categories designed by microbiologists to differentiate organisms based on their energy and carbon requirements.

As you may have already predicted, there are terms to describe the various categories. Undoubtedly, you have seen them. These are the words that always end with the suffix "-troph" or "-trophic," such as "autotroph," "heterotroph," or "phototroph."

The suffix, derived from the Greek word *trophe* literally means "nutrition," but is used, slightly more broadly, to describe how an organism obtains its nutrition (carbon) and energy. The terms all follow the same general format—energy source + carbon source + "-troph." Table 8-1 lists the sources of energy and nutrition and the root words used to describe them.

As shown in the table, an organism that gets its energy from sunlight and fixes CO_2 as a source of carbon would be called a "photoautotroph." An organism that gets energy from the sun, but uses organic molecules for its purposes is called a "photoheterotroph." An organism that gets energy from inorganic molecules and fixes CO_2 is a "chemolithoautotroph."

Unfortunately, strict use of the rules sometimes becomes a little cumbersome, so for example, you probably will never see the term, "chemoorganoheterotroph," although it clearly describes the sources of energy and carbon needed by an organism. The term is actually redundant because an organism that uses organic molecules for energy typically uses the carbon from the organic molecules as well. Therefore, you are more likely to see the term "chemoorganotroph."

Additionally, many descriptions do not include (or need) a comprehensive term such as "chemolithoheterotroph" if they are meant only to describe one aspect of an organism's metabolism. For example, if the intent is to describe only that the organism uses organics for its carbon needs, it may be called simply "heterotrophic." If the intent is to say that the organism uses chemicals for energy rather than light, it may be called "chemolithotrophic" or simply "chemotrophic."

In your reading, expect to see shortened versions of the terms and even inconsistencies in the definitions. It is difficult enough to learn and memorize the terms without being confused by the many variations possible. Therefore, in this manual we will try to stay consistent with the rules illustrated in the table.

Finally, when in doubt about the meaning of a term, consider the context in which it is being discussed and look for answers in supporting text. And if further reading doesn't clear up your uncertainty, be sure to ask your instructor. Everybody who has studied microbiology understands the possible inconsistencies and difficulties of these terms, so he or she will be happy to help. ✦

TABLE **8-1** Trophic Groups

TABLE OF RESULTS			
Energy Source	**Descriptive Root**	**Carbon Source**	**Descriptive Root**
Light	Photo-	CO_2	auto-
Organic chemical	Chemoorgano-	Organic chemicals	hetero-
Inorganic chemical	Chemolitho-		

Sulfur Cycle—Introduction

Sulfur is one of the most abundant elements on Earth. Having oxidation states from -2 to $+6$, it is able to form many different compounds usable by living things. Most of the sulfur compounds used by microorganisms are inorganic molecules, used strictly for energy or to be incorporated into organic molecules in biosynthetic processes. Table 8-2 summarizes some important sulfur compounds and their oxidation states.

TABLE **8-2** Sulfur Compounds And Sulfur Organisms That Use Them

Sulfur Compound	Chemical Formula	Oxidation State	Used By Oxidizers	Used By Reducers
Organic sulfur	R–SH	-2	X	X
Sulfide	H_2S, HS^-, S^{2-}	-2	X	
Elemental sulfur	S^0	0	X	X
Thiosulfate	$S_2O_3^{2-}$	$+2$ per S	X	X
Sulfur dioxide	SO_2	$+4$		
Sulfite	SO_3^{2-}	$+4$		X
Sulfate	SO_4^{2-}	$+6$		X

The sulfur microorganisms are a diverse group and include both Bacteria and Archaeans. They live in habitats as diverse as fresh water ponds, lakes, and rivers (especially where there is sewage contamination), water-saturated soils, salt-water lagoons, sulfur solfaturas as in Yellowstone National Park, and in and around deep ocean hydrothermal vents. This vast group includes, photoautotrophs, photoheterotrophs, chemolithoautotrophs, chemolithoheterotrophs, obligate aerobes, facultative anaerobes, and obligate anaerobes.

Many sulfur oxidizers and reducers live **syntrophically** in mutually dependent communities, in which sulfur is converted back and forth between reduced and oxidized forms. Conversely, sulfur oxidizers, living in and around hydrothermal vents, although still a complex community, have a never-ending source of reduced sulfur flowing up from the vents. These microbes, receiving no biologically reduced sulfur, thrive in the ecosystem and produce large living mats that cover surrounding surfaces.

The exercises in this study unit focus on sulfur bacteria, all of which fall into three major categories—photoautotrophs, chemolithoautotrophs, and the sulfur reducers. Table 8-3 lists the major groups of sulfur bacteria and some summary reactions they perform.

The photoautotrophs are anoxygenic photosynthesizers, that is, they perform a type of photosynthesis that does not produce oxygen. These organisms reside in the anoxic zone of a pond or other aquatic ecosystem close enough to the surface to use the sun's energy to fix carbon from CO_2. Rather than chloroplasts, as in green plants, anoxygenic phototrophs contain membrane-bound bacteriochlorophyll. In sulfur bacteria, bacteriochorophyll traps light energy and oxidizes H_2S or elemental sulfur and converts it to ATP, which ultimately is used to fix carbon from CO_2. This reaction is analogous to the oxygenic photosynthetic reactions by cyanobacteria and eukaryotes.

photosynthetic eukaryotes: $\qquad CO_2 + H_2O \longrightarrow [CH_2O] + O_2$

photosynthetic sulfur bacteria: $\qquad CO_2 + H_2S \longrightarrow [CH_2O] + S^0$

TABLE **8-3** Major Sulfur Bacteria Reactions

Microbial groups	Representative Organisms	Habitat	Reactions	Representative Summary Reactions
Photoautotrophs	*Chromatium, Chlorobium*	Anoxic	Anoxygenic photosynthesis	$H_2S + CO_2 \rightarrow S^0 + [CH_2O]$ $S^0 + CO_2 \rightarrow SO_4^{2-} + [CH_2O]$
Chemolithoautotrophs	*Beggiatoa, Macromonas, Thiobacillus, Thiobacterium*	Anoxic/oxic interface where H_2S and O_2 meet	Sulfur/sulfide/ thiosulfate oxidation	$HS^- + \frac{1}{2} O_2 + H^+ \rightarrow S^0 + H_2O$ $H_2S + 2\, O_2 \rightarrow SO_4^{2-} + 2H^+$ $S^0 + 1\frac{1}{2} O_2 + H_2O \rightarrow H_2SO_4$ $S_2O_3^{2-} + H_2O + 2\, O_2 \rightarrow 2\, SO_4^{2-} + 2H^+$
Sulfate/sulfur reducers	*Desulfovibrio, Desulfobulbus Desulfobacter, Desulfuromonas*	Either oxic or anoxic	Assimilatory sulfate reduction	$SO_4^{2-} \rightarrow S^{2-} + $ O-acetyl-L-serine $\rightarrow$ L- cysteine + acetate + H_2O
		Anoxic	Dissimilatory sulfate reduction	$SO_4^{2-} \rightarrow S^0$ $S^0 \rightarrow H_2S$
Many groups	Many organisms	Either oxic or anoxic	Desulfuration	Organic sulfur compounds $(R–SH) + H_2O \rightarrow R–OH + H_2S$

The chemolithoautotrophs are aerobic organisms that oxidize reduced sulfur compounds as an energy source and use it to fix carbon from CO_2. Because the most common form of reduced sulfur is sulfide gas (S^{2-}, HS^-, H_2S) produced in anoxic sediments, oxidation by these bacteria must occur as the gaseous sulfur rises and meets the oxic zone.

Sulfur reducers perform two important reduction reactions: dissimilatory and assimilatory sulfate reduction. Assimilatory sulfate reduction is the production of sulfide in the form of –SH groups needed for biosynthesis. Dissimilatory sulfate reduction is a purely energy releasing respiratory reaction. Finally, desulfuration (sulfur mineralization) is the reversal of assimilatory reduction and involves the release of H_2S to the environment. Figure 8-19 illustrates the major biogeochemical sulfur transformations. ✦

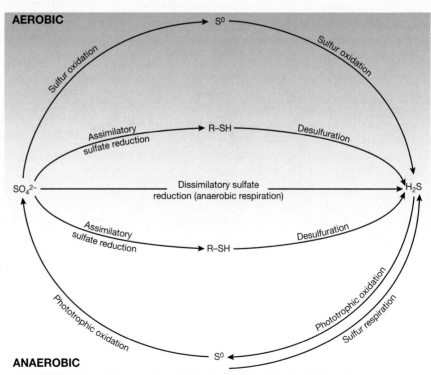

8-19 BIOGEOCHEMICAL SULFUR TRANSFORMATIONS ✦ These are the major sulfur transformations in the sulfur cycle. Refer to Tables 8-2 and 8-3 for details.

EXERCISE 8-6

Photosynthetic Sulfur Bacteria

✦ Theory

Both **purple sulfur bacteria** and **green sulfur bacteria**, although not closely related phylogenetically, appear in this exercise because they are both sulfur oxidizing anoxygenic phototrophs. In other words, they obtain energy from light as do green plants and cyanobacteria, but without producing O_2. They reside in the anoxic regions of lakes, ponds, and marine environments that contain reduced sulfur compounds and sufficient light to drive their photosystems. Purple sulfur bacteria include members of *Chromatium*, *Thiopedia*, *Thiospirillum*, and *Ectothiorhodospira*. Green sulfur bacteria include members of *Chlorobium*, *Chlorobaculum*, and *Prosthecochloris*. Both purple and green genera contain bacteriochlorophyll, which absorbs light in the range of 350–650 nm.

Purple and green sulfur bacteria each contain a photosystem that is similar to, but much simpler than that of green plants. Green plants contain two photosystems to oxidize water (H_2O) molecules; photosynthetic sulfur bacteria each contain one photosystem to oxidize sulfide (H_2S). Although the former is oxygenic and the latter is anoxygenic, the systems are analogous.

All photosystems do one critical thing; they use light energy to convert a pigment molecule from a low-energy oxidant (accepts electrons) to a high-energy reductant (donates electrons). As the pigment molecule becomes energized by light and gains reduction potential, it passes electrons down an electron transport chain (ETC), thus releasing free energy, creating a proton motive force in the membrane, and producing ATP. For more information on redox reactions, see Exercise 5-1 Reduction Potential.

In the green sulfur bacteria photosystem (Figure 8-20), the photo pigment P840 excites electrons (donated by H_2S or other reduced sulfur compounds to constituent cytochromes), which in turn are passed down an ETC to oxidized ferredoxin. The reduced ferredoxin then donates electrons to succinyl-CoA (a molecule in the citric acid cycle), which in turn combines with CO_2. With the help of ATP, the reduced succinyl-CoA eventually, through a series of redox reactions, is converted to citrate in what is called the **reverse citric acid cycle** (see Figure 8-20). Compare this reaction to the Krebs cycle in Appendix A.

The purple sulfur bacteria photosystem (Figure 8-21) does roughly the same thing as the green sulfur bacteria

photosystem with one important difference. In the purple sulfur bacteria photosystem, reduced coenzymes (NADPH) are required to transfer electrons to molecules in the Calvin Cycle to complete the synthesis of carbon compounds. However, the electrons needed to reduce NADP+ must come from quinone—a molecule in the ETC with a lower reduction potential (Exercise 5-1) than that of the coenzyme. Therefore, the cell must spend ATP energy to transfer those electrons from quinone to NADP+. This process is called **reverse electron flow** (see Figure 8-21).

The overall reaction of both green and purple sulfur bacteria could be summarized as follows:

$$CO_2 + H_2S \longrightarrow [CH_2O] + S^0$$

Note: This is simply an overview of the major components involved. As explained earlier, the CO_2 and H_2S do not actually interact in this scenario. The elemental sulfur is produced as it donates electrons to the pigment molecule. CO_2, with the help of the newly reduced coenzymes, is converted to sugars either in the reverse citric acid cycle (green bacteria) or in the Calvin cycle (purple bacteria). The elemental sulfur produced by the purple sulfur bacteria is stored inside the cell. The elemental sulfur produced by green sulfur bacteria is stored outside the cell.

✦ Application

This exercise is for observation of phototrophic sulfur bacteria.

✦ In This Exercise

The Winogradsky column (Figure 8-1) illustrates some of the important growth regions for sulfur bacteria. Today, you will be harvesting bacteria from the column and observing them on the microscope. Refer to the figure and the photomicrographs of sulfur bacteria as you do the exercise.

✦ Materials

✦ Bunsen burner

✦ Pasteur or other glass pipette (long enough to reach the red and green zone of the Winogradsky column) with bulb

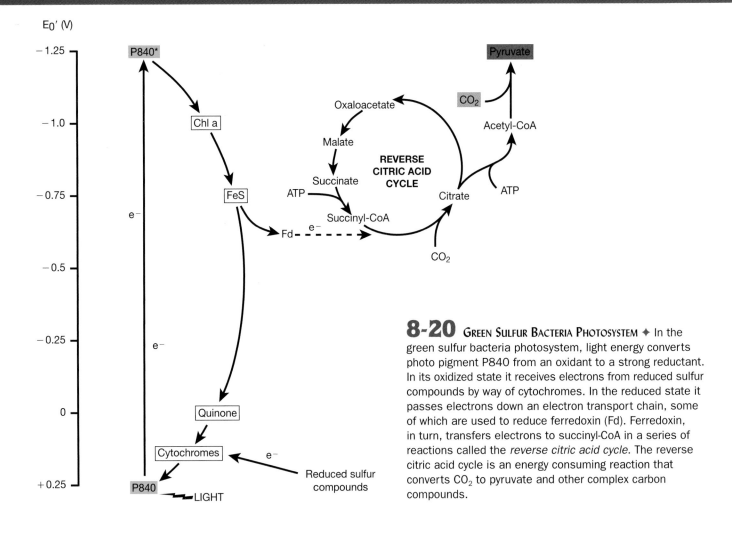

8-20 GREEN SULFUR BACTERIA PHOTOSYSTEM ✦ In the green sulfur bacteria photosystem, light energy converts photo pigment P840 from an oxidant to a strong reductant. In its oxidized state it receives electrons from reduced sulfur compounds by way of cytochromes. In the reduced state it passes electrons down an electron transport chain, some of which are used to reduce ferredoxin (Fd). Ferredoxin, in turn, transfers electrons to succinyl-CoA in a series of reactions called the *reverse citric acid cycle*. The reverse citric acid cycle is an energy consuming reaction that converts CO_2 to pyruvate and other complex carbon compounds.

✦ forceps
✦ microscope slides and coverslips
✦ Gram-stain materials

 Procedure

1   Holding the glass pipette in one hand and the forceps in the other, heat the bottom ½–1 inch of the pipette in the Bunsen burner until it becomes soft. Using the forceps, gently bend the soft glass tip approximately 45–90 degrees (1/8–1/4 turn). Do not clamp down on the glass with the forceps and do not attempt to bend it too far. Doing either will close the tip and make the pipette unusable.

2   Referring to Figure 8-1, examine your Winogradsky column. To find your way to the correct organisms, let your eyes follow the various regions downward from the top. The layers you will observe are air, water, and a mud layer immediately below the water. Slightly below this layer is the beginning of the microaerophilic zone which will be rust colored. This rust-colored zone will be inhabited by phototrophic, purple *non*-sulfur bacteria. Continue downward until you reach a red-violet region in the lower half of the column. This will be just above the green area very near the bottom of the column.

3   When you have found the red-violet area, tilt the column as far as you can without spilling the contents. Ease your bent pipette down the inside of the column until it reaches a spot with good color and remove a sample from next to the glass.

4   Place this sample on a microscope slide, add water as necessary, cover with a coverslip, and observe on the microscope. Watch for motility, and if you have a phase contrast scope, look for sulfur granules inside the cells. If you are doing a Gram-stain, look for Gram-negative cells.

5   Repeat the above process for the green region. Here you are looking for green sulfur bacteria, also Gram-negative, but with sulfur granules outside the cell.

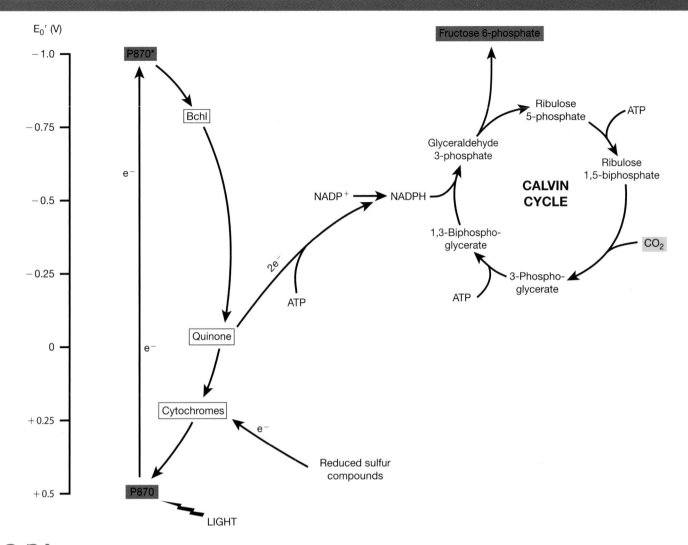

8-21 PURPLE SULFUR BACTERIA PHOTOSYSTEM ✦ In this photosystem light energy excites electrons in photo pigment P870 changing it from an oxidant to a strong reductant. In the oxidized state, it receives its electrons from reduced sulfur molecules, by way of cytochrome c_2. In the reduced state it transfers electrons down an electron transport chain, not unlike respiratory ETCs. The Calvin Cycle is used to reduce CO_2, but requires electrons from NADPH. Because quinone does not have enough reductive power to convert the $NADP^+$, energy is used to push electrons uphill (thermodynamically) in a process called *reverse electron flow*.

References

Atlas, Ronald M. and Richard Bartha. 1998. Chapter 11 in *Microbial Ecology: Fundamentals and Applications*, 4th Ed. Benjamin/Cummings Science Publishing. Menlo Park, CA.

Atlas, Ronald M. 2004. *Handbook of Microbiological Media*, 3rd Ed. CRC Press LLC. Boca Raton, Florida.

Holt, John G., Noel R. Krieg, Peter H. A. Sneath, James T. Staley, and Stanley T. Williams. 1994. Groups 7, 10, and 12 in *Bergey's Manual of Determinative Bacteriology*, 9th Ed. Lippincott Williams & Wilkins. Baltimore, MD.

Madigan, Michael T. and John M. Martinko. 2006. Chapters 12 and 17 in *Brock Biology of Microorganisms*, 11th Ed. Pearson Prentice Hall. Upper Saddle River, NJ.

Maier, Raina M., Ian L. Pepper, and Charles P. Gerba. 2009. Chapter 14 in *Environmental Microbiology*, 2nd Ed. Academic Press, subsidiary of Elsevier, Inc. Oxford, UK.

Radu Popa, multiple personal interviews, 6/09–10/09.

Chemolithotrophic Sulfur-Oxidizing Bacteria

✦ Theory

Chemolithotrophic sulfur oxidizing bacteria are organisms that get their energy from reduced sulfur compounds. The most common electron donors for these organisms are hydrogen sulfide (H_2S), elemental sulfur (S^0), and thiosulfate ($S_2O_3^{2-}$). The term "chemolithotroph" distinguishes these organisms from photosynthetic sulfur bacteria (phototrophs) that also oxidize reduced sulfur compounds (Exercise 8-6).

Chemolithotrophic sulfur oxidizers live in aquatic habitats both in the water and in upper sediment layers. As such, they range from strictly aerobic to microaerophilic to facultatively anaerobic. Very common among the sulfur oxidizing bacteria are the so-called **gradient bacteria**. These microaerophiles require both oxygen and H_2S, but only in a very narrow range of concentrations. They typically reside at the level where H_2S rising from sediments and O_2 diffusing down from the air create a critical composition of the two gases.

Gradient bacteria, such as *Beggiatoa*, *Thioplaca*, and *Thiothrix* are facultative chemolithotrophs. They usually obtain their energy from H_2S and fix carbon from CO_2, but are capable of obtaining energy and carbon from organic compounds as well. The chemolithotrophic energy reaction is as follows:

$$H_2S + \tfrac{1}{2}\,O_2 \longrightarrow S^0 + H_2O$$

These organisms typically store sulfur granules inside the cells (Figure 8-22).

Thiobacillus is another group of sulfur oxidizing bacteria. Several species, such as *T. thioparus* and *T. novellus* perform the same reaction as the gradient bacteria and also store sulfur granules inside the cell. Other, more acidophilic *Thiobacillus* species such as *T. thiooxidans* and *T. ferroxidans*, oxidize sulfur compounds and elemental sulfur all the way to sulfuric acid as follows:

$$S^0 + 1\tfrac{1}{2}\,O_2 + H_2O \longrightarrow H_2SO_4$$

Acidophilic *Thiobacillus* species are obligate chemolithotrophs and derive all of their energy and carbon from reduced sulfur compounds and carbon dioxide, respectively. Not surprisingly, they thrive in environments as acidic as pH 2.

Most *Thiobacillus* species are obligate aerobes, but one species *T. denitrificans*, is a facultative anaerobe

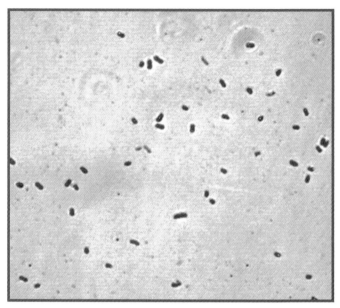

8-22 GRAM STAIN OF A SULFUR OXIDIZER ✦ This is a Gram stain of an unknown sulfur oxidizer. Note the clear spots indicating possible sulfur storage granules.

important to the sulfur *and* the nitrogen cycle. As shown in the following reaction, it oxidizes elemental sulfur by reducing nitrate. This not only produces sulfate (SO_4^{2-}) needed by sulfur reducers, but converts nitrate to nitrogen gas.

$$3S^0 + 4NO_3^- \longrightarrow 3SO_4^{2-} + 2N_2$$

Two media were chosen for this exercise—a modified *Thiobacillus* medium and sulfur microgradient tubes. The *Thiobacillus* medium contains reduced sulfur and all of the necessary trace minerals and vitamins needed to isolate chemolithotrophic sulfur oxidizers. The broth is incubated on a shaker to maximize the oxygen content and encourage aerobic growth.

Microgradient tubes are made with two layers of agar—a soft layer on top and a solid layer on the bottom (Figures 8-23 and 8-24). The bottom layer, called a "plug," contains concentrated hydrogen sulfide. The top layer is a soft mineral agar that provides necessary trace minerals and allows free movement of motile bacteria. The medium is first heated to remove the oxygen and cooled to room temperature. Then it is stab inoculated and capped loosely. As the oxygen diffuses downward into the medium and the sulfide diffuses upward from the plug, the organism establishes a population in the

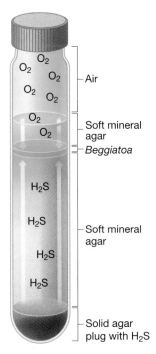

8-23 MICROGRADIENT TUBE DIAGRAM ✦ This is a diagram of a microgradient tube used to isolate *Beggiatoa*, a microaerophilic sulfur oxidizer. In the bottom is a solid agar "plug" containing sulfide (or other reduced sulfur compound). Above the plug is a deep soft mineral agar layer. The minerals provide necessary trace elements and help maintain the proper osmotic pressure and pH. Above the soft agar is air space. The sulfide from the plug diffuses upward through the soft agar and creates a gradient from high concentration at the bottom to low concentration somewhere above. The oxygen diffuses downward from the air space and creates its gradient from high concentration (~21%) at the top to low concentration below. The medium typically is stab inoculated with soil bacteria from a mixture of soil and water or buffer. The motile organisms move about freely and colonize the region that satisfies their needs.

area where optimal concentrations of both gases exist. This medium is best suited for gradient bacteria such as *Beggiatoa* and *Thiothrix*.

8-24 MICROGRADIENT TUBE ✦ This is a microgradient tube containing unknown soil bacteria. Microgradient tubes are commonly used to cultivate the sulfide oxidizer, *Beggiatoa*. Note the thin band of growth (arrow).

the very specific growth conditions required by gradient bacteria.

✦ Medium recipe

Modified *Thiobacillus* Medium
(See Appendix H for details)

✦ Sodium Thiosulfate	10.0 g
✦ K_2HPO_4	8.0 g
✦ KH_2PO_4	3.0 g
✦ NH_4Cl	0.4 g
✦ $MgSO_4 \cdot 7H_2O$	0.4 g
✦ Trace minerals solution (See Appendix G for details)	2.0 mL
✦ Vitamin B_{12}	20.0 mg
✦ Biotin	10.0 mg
✦ Distilled or deionized H_2O	to 1.0 L

pH 7.5 adjusted with KOH

Microgradient Tubes
(Simplified version. See Appendix H for details.)

Top Layer

✦ Sodium Thiosulfate	0.5 g
✦ Agar	2.5 g
✦ Trace mineral/salts solution	1.0 L

✦ Application

This exercise is for observation of sulfur oxidizing bacteria.

✦ In This Exercise

You will collect a sample of water and mud from a lake, pond, or other (indoor or outdoor) aquatic ecosystem and use your sample to cultivate aerobic and micro-aerophilic sulfur oxidizers in the media described above. You will be given an opportunity to observe and to stain organisms that develop in the broth medium. The solid microgradient tubes are included for you to observe

Bottom Layer (Plug)

- Na$_2$S   0.3 g
- Agar   15.0 g
- Trace mineral/salts solution   1.0 L

✦ Materials

Per Class

- ✦ orbital shaker

Per Group

- ✦ vortex mixer
- ✦ pH paper (pH range 2–14) (optional)
- ✦ Pasteur pipettes with bulbs
- ✦ microscope slides and coverslips
- ✦ Gram-stain materials
- ✦ four tubes of Modified *Thiobacillus* Medium
- ✦ three microgradient tubes
- ✦ sediment/water sample from an aquatic ecosystem (in a clean glass container)

Procedure

Lab One

1 Obtain all materials for your group and label your tubes with your name, date, sample source, *etc*. Label the tubes 1, 2, 3, and 4.

2 Shake your sediment sample well and transfer one drop to tube #1.

3 Vortex tube #1 well and, using a clean pipette, transfer one drop from it to tube #2.

4 Repeat this procedure with tubes #2 and #3. Do nothing to tube #4.

5 Using your inoculating loop, stab inoculate the microgradient tubes with a loopful of the (well-mixed) solution from tube #1. (**Note:** Stab the mineral agar portion only, and stop before you reach the sulfide agar plug in the bottom. Inoculate each tube individually, mixing well before each transfer.)

6 Place your four broths (along with all tubes from the class in a rack on an orbital shaker set at medium speed) in a 30°C incubator. Incubate for 24–48 hours, or longer if necessary to get turbidity.

7 Also incubate your microgradient tube at 30°C for 72 hours or until a thin line of microaerophilic growth appears as shown in Figure 8-24.

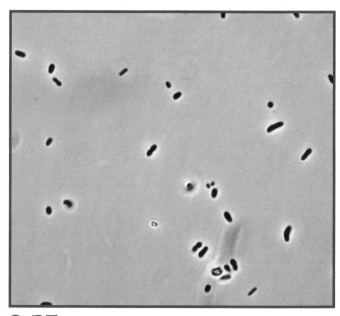

8-25 PHASE CONTRAST IMAGE OF A SULFUR OXIDIZER ✦ This is a phase-contrast photomicrograph of unknown sulfur oxidizers cultivated from an upper layer of pond sediment. See Figure 8-1.

Lab Two

1 Remove your broths from the incubator and examine them for turbidity.

2 Make wet mounts and Gram stains of organisms you find (Figures 8-22 and 8-25).

3 Compare and share your slides with other students.

4 Using a clean pipette, place a drop of each broth on a separate strip of pH paper.

5 Record your results on the Data Sheet and answer the questions.

6 Examine the gradient tubes for growth. Enter your observations on the Data Sheet.

References

Atlas, Ronald M. and Richard Bartha. 1998. Chapter 11 in *Microbial Ecology: Fundamentals and Applications*, 4th Ed. Benjamin/Cummings Science Publishing. Menlo Park, CA.

Atlas, Ronald M. 2004. *Handbook of Microbiological Media*, 3rd Ed. CRC Press LLC. Boca Raton, Florida.

Holt, John G., Noel R. Krieg, Peter H. A. Sneath, James T. Staley, and Stanley T. Williams. 1994. Groups 7, 10, and 12 in *Bergey's Manual of Determinative Bacteriology*, 9th Ed. Lippincott Williams & Wilkins. Baltimore, MD.

Madigan, Michael T. and John M. Martinko. 2006. Chapters 12 and 17 in *Brock Biology of Microorganisms*, 11th Ed. Pearson Prentice Hall. Upper Saddle River, NJ.

Maier, Raina M., Ian L. Pepper, and Charles P. Gerba. 2009. Chapter 14 in *Environmental Microbiology*, 2nd Ed. Academic Press, subsidiary of Elsevier, Inc. Oxford, UK.

Radu Popa, multiple personal interviews, 6/09–10/09.

Sulfur-Reducing Bacteria

✦ Theory

Sulfur reducing bacteria perform two important types of sulfur reduction, **assimilatory sulfate reduction** and **dissimilatory sulfate reduction**. Assimilatory sulfate reduction is the process of actively transporting sulfate (SO_4^{2-}) into the cell and converting it to sulfhydryl groups (R–SH) found in the amino acids cysteine and methionine. Because sulfur is an essential component needed for biosynthesis, many bacteria, fungi, and even green plants transport it into their cells. Figure 8-26 illustrates a typical five-step conversion of sulfate to a sulfhydryl group in the synthesis of cysteine. The overall reaction can be summarized as follows:

$$SO_4^{2-} + ATP \longrightarrow cysteine + H_2O.$$

Even though sulfide (S^{2-}) is the form needed by bacteria to generate sulfur-containing amino acids (reaction #5), sulfate (SO_4^{2-}) is the form actively transported and then converted inside the cell. This is because free sulfide reacts and precipitates with the iron in cytochromes,

$$H_2S + Fe^{2+} \longrightarrow H_2 + FeS$$

thus, severely restricting electron flow and ATP production. The S^{2-} produced (in reaction #4) is instantly incorporated into a molecule of serine to stabilize it and complete the conversion to cysteine.

Dissimilatory sulfate reduction is a type of anaerobic respiration done exclusively by sulfate reducing bacteria (SRB). Members of the SRB include *Desulfobacter*, *Desulfobulbus*, *Desulfococcus*, *Desulfotomaculum*, *Desulfonema*, *Desulfosarcina*, and *Desulfovibrio*. These organisms are part of an anaerobic bacterial consortium made up of fermenters, sulfate reducers, and methanogens that live in aquatic sediments. In the reaction, SO_4^{2-} is converted to H_2S, the reduced sulfur is immediately released back into the environment, and the freed electrons and energy are used to make ATP.

SRB are heterotrophs and must receive their carbon from organic molecules, but they can accept electrons from either inorganic or organic sources. Common electron donors for dissimilatory sulfate reduction are H_2 and a variety of low molecular weight organic molecules, such as acetate, alcohols, pyruvate, or lactate. Organic molecules are available in anaerobic habitats as products of fermentation and decomposition. Below are two common reactions. In each case, sulfide is released into the environment.

$$4H_2 + 2SO_4^{2-} \longrightarrow S^{2-} + 4H_2O$$

$$4CH_3OH + 3SO_4^{2-} \longrightarrow 4CO_2 + 3S^{2-} + 8H_2O$$

Some SRB can perform a **disproportionation reaction**. In this type of reaction, a sulfur compound (usually an intermediate such as thiosulfate, $S_2O_3^{2-}$ likely produced by a sulfur oxidizer) is split into two products, one of which is more oxidized than thiosulfate and one that is more reduced. A typical reaction would be:

$$S_2O_3^{2-} + H_2O \longrightarrow SO_4^{2-} + H_2S$$

Note: The oxidation states are $S_2O_3^{2-} = +2$ (per S), $SO_4^{2-} = +6$, $H_2S = -2$. Therefore, the sulfate is more oxidized than thiosulfate and the hydrogen sulfide is more reduced.

Desulfuromonas acetoxidans is one of at least two species known to reduce sulfur rather than sulfate. In

---

1 $SO_4^{2-} + ATP$ ——————$\xrightarrow{ATP\ sulfurylase}$—————→ $APS + PP_i$

2 $ATP + APS$ ——————$\xrightarrow{APS\ phosphokinase}$—————→ $PAPS + ADP$

3 $PAPS + thioredoxin_{red}$ ——————$\xrightarrow{PAPS\ reductase}$—————→ $SO_3^{2-} + PAP + thioredoxin_{ox}$

4 $SO_3^{2-} + 3NADPH$ ——————$\xrightarrow{NADPH-SO_3^{2-}\ reductase}$—————→ $S^{2-} + NADP$

5 $O-acetyl-L-serine + S^{2-}$ ——————$\xrightarrow{O-acetylserine\ sulfhydrylase}$—————→ $L-cysteine + acetate + H_2O$

8-26 ASSIMILATORY SULFATE REDUCTION ✦ This is an example of the multi-step conversion of sulfate to the amino acid cysteine. Assimilatory sulfate reduction is performed by many bacteria. It solves the problem of obtaining an essential form of sulfur without dealing with the very toxic, free hydrogen sulfide. (**Note:** One type of desulfuration, the release of sulfur from the cell, is the reverse of equation #5.)

this example of anaerobic respiration the organism oxidizes acetate and converts the S^0 to H_2S as follows:

$$CH_3COOH + 2H_2O + 4S^0 \longrightarrow 2CO_2 + 4H_2S$$

This reaction yields only a small amount of energy, but the organism survives for two important reasons, 1) it needs only the compounds shown in the formula for growth, and 2) it lives syntrophically in anoxic marine sediments with phototrophic green sulfur bacteria—*Chlorobium*. Each group provides an essential nutrient for the other. *Chlorobium* oxidizes the H_2S produced by *Desulfuromonas*; *Desulfuromonas* reduces the elemental sulfur released by *Chlorobium*. The acetate also comes from the green bacteria and other organics present in the environment.

Finally, there are a number of facultative anaerobes in the family *Enterobacteriaceae* that are able to reduce sulfur compounds. For more information on these organisms, see Exercises 5-20, 21, and 22.

The medium chosen for this exercise, Modified *Desulfovibrio* Medium, contains lactate as an electron donor and both sulfite and sulfate (from the ferrous ammonium sulfate) as oxidized sulfur. Additionally, the ferrous ammonium sulfate serves as an indicator of sulfite or sulfate reduction by precipitating with any sulfide produced. Trace minerals and yeast extract take care of necessary growth factors while L-cysteine is added as a reducing agent to reduce oxygen to water and maintain an anoxic environment. The agar also slows oxygen diffusion but more importantly, prevents generalized turbidity that would occur if these motile organisms were allowed to move freely. Finally, ammonium phosphate and dipotassium phosphate act as buffers to maintain a relatively constant pH.

✦ Application

This exercise is for observation of sulfate reducing bacteria.

✦ In This Exercise

You will collect a sample of sediment from a lake, pond, or other (indoor or outdoor) aquatic ecosystem and use your sample to cultivate sulfate-reducing bacteria (Figures 8-27 and 8-28). In this lab you will perform a dilution series to attempt to cultivate individual colonies. Your instructor may or may not have you calculate the original concentration of SRB in the sample. The medium you will use—modified *Desulfovibrio* Medium—is rich in sulfates and sulfites and encourages growth of organisms capable of reducing the oxidized sulfur.

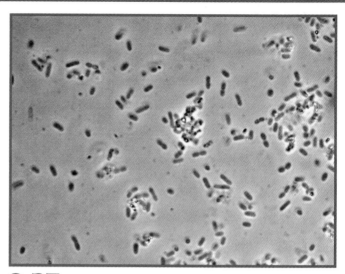

8-27 PHASE CONTRAST IMAGE OF A SULFUR REDUCER ✦ This is a phase-contrast photomicrograph of unknown sulfur reducers recovered from black (anoxic) pond sediment. See Figure 8-1.

✦ Medium recipe

Modified *Desulfovibrio* Medium

✦ DI H_2O	1.0 L
✦ Yeast extract	5.0 g
✦ Wolfe trace minerals (see Appendix H)	10.0 mL
✦ Ammonium phosphate, dibasic	0.3 g
✦ Sodium lactate	3.0 g
✦ Sodium sulfite	2.0 g
✦ Dipotassium phosphate	1.7 g
✦ L-Cysteine	0.5 g
✦ Ferrous ammonium sulfate	0.005g
✦ Agar	3.0 g

pH 7.5 ± 0.2 @ 25°C

✦ Materials

Lab One

Per Class

✦ 50°C water bath

Per Group

✦ vortex mixer

✦ digital pipettes and sterile tips to transfer 1 mL of solution

✦ six 9 mL dilution tubes

✦ seven Modified *Desulfovibrio* Medium tubes (labeled 1-7) tightly closed (retained in the water bath until needed)

✦ sediment sample from a pond, lake, or other aquatic ecosystem (in a clean glass container)

✦ tube of filter-sterilized 0.5% ferrous ammonium sulfate

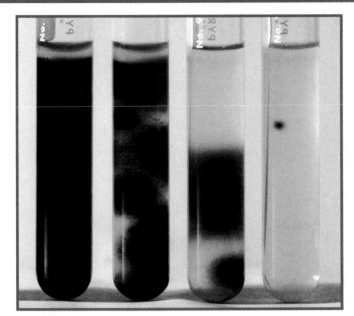

8-28 *DESULFOVIBRIO* MEDIUM WITH ANAEROBIC SULFUR REDUCERS ✦ This medium can be used solely for cultivation or for quantitative purposes. It is heated immediately prior to inoculation to drive off the dissolved oxygen and melt the agar. The entire medium does not remain anoxic, but when inoculated and gently mixed, the organisms will populate the regions where oxygen has not returned by diffusion.

Lab Two

Per Group

✦ microscope slides and coverslips

✦ Gram-stain materials

 Procedure

Lab One

1 Obtain all materials for your group and label the dilution tubes 1–6. Set them aside temporarily.

2 Removing one *Desulfovibrio* tube from the water bath at a time (to prevent having it resolidify prematurely), aseptically add 1 mL 0.5% filter-sterilized ferrous ammonium sulfate to the medium, mix well, and return it to the water bath. Do this with all seven tubes. (It helps to use a digital pipette and placing the tip just below the surface of the medium, forcefully eject the iron sulfate, then mix by rolling the tube between your hands. Use a different pipette for each transfer and do not shake the tube; this will introduce oxygen into the medium, which will inhibit growth of the desired organisms.

3 Start your dilution series (10^{-1} to 10^{-6}) by adding 1 g of sediment (1 mL if soupy) to dilution tube 1 (10^{-1}).

4 Vortex the tube very well and transfer 1 mL (while still homogeneous) to dilution tube 2 (10^{-2}).

5 Continue in this manner, transferring from tube 2 to 3, 3 to 4, *etc.*, until all dilution tubes have received an inoculum.

6 Again, removing one medium tube from the water bath at a time, and using a different sterile pipette for each transfer, inoculate six medium tubes with 1 mL of its corresponding dilution. BE SURE TO VORTEX EACH *DILUTION* VERY WELL IMMEDIATELY BEFORE TRANSFERRING IT TO GROWTH MEDIUM. DO NOT VORTEX OR SHAKE THE MEDIUM. As before when adding the ferrous ammonium sulfate solution, place the pipette tip just below the top surface of the medium and eject the inoculum forcefully. Cap tightly, and mix well by rolling the tube between your hands. Do not inoculate tube #7.

7 Allow the medium tubes to solidify upright in a test tube rack, then incubate them at 30°C for 48 hours to two weeks, checking daily for development of black colonies in the agar.

Lab Two

1 Remove your tubes from the incubator and examine them for black colonies in the agar.

2 Make wet mounts and Gram stains of organisms you find (Figure 8-27). The organisms can be retrieved using an inoculating loop, needle, wooden applicator, or small spatula.

3 Compare and share your slides with other students.

4 Record your results on the Data Sheet and answer the questions.

5 Be sure to dispose of all materials in appropriate autoclave containers.

References

Atlas, Ronald M. and Richard Bartha. 1998. Chapter 11 in *Microbial Ecology: Fundamentals and Applications*, 4th Ed. Benjamin/Cummings Science Publishing. Menlo Park, CA.

Atlas, Ronald M. 2004. *Handbook of Microbiological Media*, 3rd Ed. CRC Press LLC. Boca Raton, Florida.

Holt, John G., Noel R. Krieg, Peter H. A. Sneath, James T. Staley, and Stanley T. Williams. 1994. Groups 7, 10, and 12 in *Bergey's Manual of Determinative Bacteriology*, 9th Ed. Lippincott Williams & Wilkins. Baltimore, MD.

Madigan, Michael T. and John M. Martinko. 2006. Chapters 12 and 17 in *Brock Biology of Microorganisms*, 11th Ed. Pearson Prentice Hall. Upper Saddle River, NJ.

Maier, Raina M., Ian L. Pepper, and Charles P. Gerba. 2009. Chapter 14 in *Environmental Microbiology*, 2nd Ed. Academic Press, subsidiary of Elsevier, Inc. Oxford, UK.

Radu Popa, multiple personal interviews, 6/09–10/09.

Bioluminescence

✦ Theory

A few marine bacteria from genera *Vibrio* and *Photobacterium* are able to emit light by a process known as **bioluminescence**. Many of these organisms maintain mutualistic relationships with other marine life. For example, *Photobacterium* species living in the Flashlight Fish receive nutrients from the fish and in return provide a unique device for frightening would-be predators.

Bioluminescent bacteria are able to emit light because of an enzyme called **luciferase** (Figure 8-29). In the presence of oxygen and a long-chain aldehyde, luciferase catalyzes the oxidation of reduced flavin mononucleotide ($FMNH_2$). In the process, outer electrons surrounding FMN become excited. Light is emitted when the electronically excited FMN returns to its ground state (Figure 8-30).

It is estimated that a single *Vibrio* cell burns between 6000 and 60000 molecules of ATP per second emitting light. (ATP hydrolysis occurs in conjunction with synthesis of the aldehyde.) It also is known that their luminescence occurs only when a certain threshold population size is reached in a phenomenon called **quorum sensing**. This system is controlled by a genetically produced **autoinducer** that must be in sufficient concentration to trigger the reaction.

✦ In This Exercise

You will inoculate Seawater Complete Medium (SWC) with *Vibrio fischeri*, incubate it for a day or two, then observe for bioluminescence.

✦ Materials

Per Student Group

✦ overnight culture of *Vibrio fischeri*

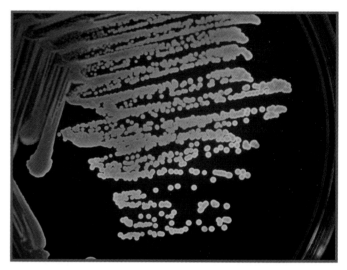

8-30 BIOLUMINESCENCE ON AN AGAR PLATE ✦ This is an unknown bioluminescent bacterium growing on SWC Agar.

Per Student

✦ one SWC plate

Medium Recipe

Seawater Complete Medium

✦ Pancreatic Digest of Casein	5.0 g
✦ Yeast Extract	3.0 g
✦ Glycerol	3.0 mL
✦ Seawater	750.0 mL

Procedure

Lab One

1 Obtain one SWC plate and aseptically inoculate it with *V. fischeri*. Any pattern will do.

2 Invert and incubate the plate at 25°C for 24 to 48 hours.

$$FMNH_2 + O_2 + R-\overset{\overset{\displaystyle O}{\|}}{C}H \xrightarrow{\text{Luciferase}} FMN + H_2O + R-\overset{\overset{\displaystyle O}{\|}}{C}-OH + \text{Light}$$

8-29 CHEMISTRY OF BIOLUMINESCENT BACTERIA ✦ The enzyme luciferase catalyzes the oxidation of reduced flavin mononucleotide in the presence of an aldehyde. During the reaction, an electron becomes excited. When it returns to its ground state, light is emitted.

Lab Two

Remove the plate from the incubator and examine it in a dark room for light emission. It may take awhile for your eyes to adjust to the dark and see the bioluminescence.

References

DIFCO Laboratories. 1984. *DIFCO Manual*, 10th ed. DIFCO Laboratories, Detroit.

Krieg, Noel R., and John G. Holt (Editor-in-Chief). 1984. Page 518 in *Bergey's Manual of Systematic Bacteriology*, Vol. 1. Lippincott Williams and Wilkins, Baltimore.

Power, David A., and Peggy J. McCuen. 1988. *Manual of BBL™ Products and Laboratory Procedures*, 6th ed. Becton Dickinson Microbiology Systems, Cockeysville, MD.

White, David. 2000. Pages 504–509 in *The Physiology and Biochemistry of Pro-karyotes*. Oxford University Press. New York.

EXERCISE 8-10 Soil Slide Culture

✦ Theory

Trophic levels, food webs, symbioses, and species diversity and abundance are all important aspects of community structure studied by ecologists. However, physical location and proximity to other community members is also important. On a macroscopic level, this is relatively easy to assess. On the other hand, physical associations between soil microbes are difficult to ascertain without disturbing those relationships. That is, damage is done when one digs up the soil.

The soil slide culture is an easy way to see those relationships *in situ*. The idea is a simple one: a microscope slide is buried in soil and then carefully removed after a week or so. The microbes adhering to the slide are stained and examined under the microscope in their natural locations relative to one another and to soil particles.

This is a qualitative procedure in that community members and associations can be identified, and an idea of diversity can be assessed, but actual numbers and percentages cannot.

✦ Application

The soil slide culture is used to view associations and diversity of soil microbes *in situ*.

✦ In This Exercise

Your group will bring a soil sample to class, prepare it, and then bury microscope slides in it. After a week, you will observe the slides under the microscope and see what organisms are present and identify associations between each other and with soil particles.

✦ Materials

Per Student Group

✦ soil sample (brought to class by a lab group member)

✦ one 600 mL (or other size depending on the size of each lab group) beaker

✦ one microscope slide per student

✦ parafilm or plastic wrap and a rubberband

✦ forceps (with tips capable of gripping a microscope slide by one end and lifting vertically)

✦ 40% v/v acetic acid

✦ gloves

✦ goggles

✦ fume hood

✦ 1% Rose Bengal stain[1]

✦ water bottle

✦ a beaker or other container in which to wash off the slides

✦ 2 Coplin jars—one with the 40% acetic acid, the other with Rose Bengal stain

✦ compound microscope

✦ Stain Recipe

Rose Bengal Stain (1%)

✦ Rose Bengal	0.1 g
✦ Distilled or deionized water	9.9 mL

📖 Procedure

Day 1

1 Label the beaker with your lab group's name/number and soil type.

2 Add soil to the beaker until it is half full.

3 Add water to the soil so it is damp, but not saturated.

4 Each student should label one slide with his/her name, and then place it vertically in the soil with approximately 2 cm extending above the surface as in the preparation on the left in Figure 8-31. The label should be on the part above the soil!

5 Cover the beaker with Parafilm (or plastic wrap held on with a rubberband).

6 Poke a few holes in the Parafilm to allow aeration of the soil, but not so many that the soil will dehydrate during incubation. The number of holes will be determined by the size of the beaker used. The finished product should look like the preparation on the right in Figure 8-31.

7 Incubate for one week at room temperature.

---

[1] Avavilable from Cole-Parmer, 625 East Bunker Court, Vernon Hills, IL 60061. http://www.coleparmer.com/catalog/product_index.asp?cls=37976.

8-31 PREPARED SOIL CULTURES ✦ The beaker on the left shows three microscope slides properly buried in the moist, but not saturated, soil. The beaker on the right is covered with Parafilm and is ready for incubation.

8-32 THE STAINING SET-UP IN A FUME HOOD ✦ Shown are the Coplin jars of acetic acid and Rose Bengal stain on either side of a rinse pan. Notice that the slides being "fixed" in acetic acid have the labels up. Coplin jars are covered with Parafilm when not in use. Gloves and goggles must be worn while staining.

Day 2

1 Remove the beaker from the incubator.

2 Remove your slide by tipping it *gently* to about 20 degrees so that the "upper" side is no longer in contact with the soil. This "upper" side is the one with your name. Then *gently* pull the slide out without rubbing or scraping the upper side.

3 *Gently* tap the slide on a paper towel on your bench top to remove chunks of soil. Thoroughly clean the lower surface with a damp paper towel. Let the slide dry at room temperature with the clean side down.

4 Your professor will have added 40% v/v acetic acid to one or more Coplin jars under a fume hood (Figure 8-32). Wearing gloves and goggles use forceps to place your slide in one of the slots of a Coplin jar.

Be sure to have your label at the top. Leave it in the acetic acid for 1 to 3 minutes.

5 Use forceps to remove your slide from the acetic acid. Then *gently* wash the acetic acid off your slide with water into the container provided.

6 Use forceps to place your slide in the Rose Bengal stain Coplin jar located under the fume hood. Leave the slide in the stain for 5 to 10 minutes.

7 Use forceps to remove your slide from the Rose Bengal stain and *gently* wash with water as in Step 5.

8 Air dry the slide on a paper towel at your desk.

9 Observe under oil immersion.

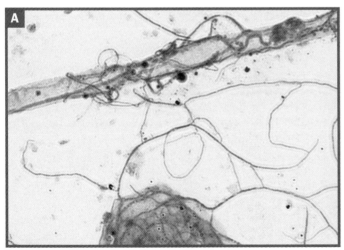
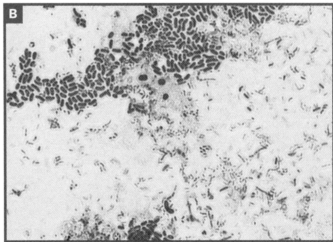

8-33 MICROGRAPHS OF SOIL SLIDE CULTURES ✦ **A** The thick filament is a fungal hypha. The thin, branching filaments are streptomycetes. **B** Soil bacteria dominate this field. The plump rods in the cluster are probably *Azotobacter*. Both micrographs are X1000.

10 Compare your slide to Figures 8-33A and 8-33B. Record your results on the Data Sheet and answer the questions.

References

Pepper, Ian and Charles Gerba. 2005. *Environmental Microbiology, A Laboratory Manual,* 2nd Ed. Elsevier Academic Press, Burlington, MA, USA.

Varnam, Alan H. and Malcom G. Evans. 2000. *Environmental Microbiology.* ASM Press, Washington, DC.

EXERCISE 8-11 Soil Microbial Count

✦ Theory

Ordinary soil is a storehouse of microorganisms, both beneficial and pathogenic. Actinobacteria living in soil are important organic matter decomposers and humus producers. Some fungi are plant and animal pathogens, but most are saprophytes that decompose dead or dying plant material. Bacteria are by far the most numerous of the three groups. Some have animal reservoirs and are simply lying dormant as spores; others are important environmental nutrient cyclers. See Exercises 8-1 through 8-5. The major groups of antibiotic producing genera—*Bacillus, Cephalosporium, Penicillium*, and *Streptomyces*—can be found in soil.

The media used for this exercise are specifically designed for the organisms to be isolated. Glycerol Yeast Extract Agar is designed for actinomycetes and doesn't offer enough nutritive value for typical bacteria or fungi. Nutrient Agar is designed for a broad range of bacteria, but not fungi. Sabouraud Dextrose Agar is designed for fungi but also will support bacteria; therefore, penicillin and streptomycin have been added to discourage bacterial growth.

✦ Application

This technique used to determine the relative concentrations of bacteria, actinomycetes, and fungi in soil. It is also a good exercise to demonstrate diversity in soil samples.

TABLE 8-4 Group Assignments

Group Number	Organism	Media (see materials)
1	Actinomycetes	GYEA
2	Actinomycetes	GYEA
3	Bacteria	NA
4	Bacteria	NA
5	Fungi	SDA
6	Fungi	SDA

✦ In This Exercise

You will perform a serial dilution of a soil sample and calculate the densities of three very prominent soil residents—bacteria, actinomycetes, and fungi. Colony counts will be performed on plates produced using the pour plate technique described in Exercise 6-5, Plaque Assay of Virus Titre. It is expected that the bacteria will outnumber the other groups and that the actinomycetes will outnumber the fungi. Therefore, the dilutions have been designed accordingly in order to produce countable plates for each group. To save time and material, the class will be divided into six groups. Refer to Table 8-4 for group assignments.

✦ Materials

Per Student Group

- ✦ soil sample
- ✦ water bath set at 45°C containing sterile molten:
  - ✦ Sabouraud Dextrose Agar
  - ✦ Nutrient Agar
  - ✦ Glycerol Yeast Extract Agar
- ✦ capped bottle containing 100 mL sterile water
- ✦ micropipettes (10–100 µL and 100–1000 µL) with sterile tips
- ✦ sterile microtubes
- ✦ flask of sterile water
- ✦ five Sterile Petri dishes
- ✦ hand tally counter
- ✦ colony counter

✦ Medium Recipes

Sabouraud Dextrose Agar (with antibiotics added to inhibit bacteria)

✦ Peptone	10.0 g
✦ Dextrose	40.0 g
✦ Agar	15.0 g
✦ Penicillin	20000.0 units
✦ Streptomycin	0.00004 g
✦ Distilled or deionized water	1.0 L

pH 5.4–5.8 at 25°C

Nutrient Agar

- ◆ Beef extract   3.0 g
- ◆ Peptone   5.0 g
- ◆ Agar   15.0 g
- ◆ Distilled or deionized water 1.0 L

  pH 6.6 ± 7.0 at 25°C

Glycerol Yeast Extract Agar

- ◆ Glycerol   5.0 mL
- ◆ Yeast extract   2.0 g
- ◆ Dipotassium phosphate   1.0 g
- ◆ Agar   15.0 g
- ◆ Distilled or deionized water 1.0 L

 Procedure

Note: This is a quantitative exercise employing serial dilutions, dilutions, and calculations that are explained fully in Exercise 6-1, Standard Plate Count. If you haven't completed a standard plate count, or you have any doubt about your ability to complete this exercise, we suggest that you read Exercise 6-1 carefully and do some or all of the practice problems.

Lab One: Groups 1 and 2

1 Obtain all materials including five Petri dishes. Label the plates GYEA—A through E, respectively. Plate E will be the control.

2 Obtain five microtubes and label them 1 through 5. These are your dilution tubes; make sure they remain covered until needed.

3 Aseptically add 900 μL (0.9 mL) sterile water to dilution tubes 2, 3, 4, and 5. Cover when finished.

4 Add 10 g of the soil sample to the 100 mL water bottle. Shake vigorously for several minutes. This is a 10^{-1} dilution. If you don't remember how to calculate dilutions, refer to Exercise 6-1.

5 Aseptically transfer 1000 μL of the suspended solution from the bottle to dilution tube 1. Remember— this is 10^{-1}.

6 Aseptically transfer 100 μL from dilution tube 1 to dilution tube 2; mix well. This is 10^{-2}.

7 Aseptically transfer 100 μL from dilution tube 2 to dilution tube 3; mix well. This is 10^{-3}.

8 Aseptically transfer 100 μL from dilution tube 3 to dilution tube 4; mix well. This is 10^{-4}.

9 Tube 5 will remain sterile and be used as a control.

10 Remove one molten GYEA tube from the water bath. Aseptically add 100 μL from dilution tube 1, mix well, and pour into plate A. This plate contains 10^{-2} mL original sample volume. Repeat the process using 100 μL from tubes 2 through 5 to plates B through E, respectively. Label the plates with the original sample volume. Label plate E "control."

11 Allow the plates time to cool and solidify. Invert and incubate them at 25°C for 2 to 7 days.

Lab One: Groups 3 and 4

1 Obtain all materials including five Petri dishes. Label the plates NA—A through E respectively. Plate E will be the control.

2 Obtain 5 microtubes and label them 1 through 5. These are your dilution tubes; make sure they remain covered until needed.

3 Aseptically add 900 μL (0.9 mL) sterile water to each dilution tube. Cover when finished.

4 Group 1 will have added 10 g of the soil sample to the 100 mL water bottle. (This is a 10^{-1} dilution. If you don't remember how to calculate dilution, refer to Exercise 6-1.)

5 Obtain the bottle from Group 1, mix the sample well, and aseptically transfer 100 μL of the suspended solution to dilution tube 1. Mix the contents of tube 1. This is 10^{-2}.

6 Aseptically transfer 100 μL from dilution tube 1 to dilution tube 2; mix well. This is 10^{-3}.

7 Aseptically transfer 100 μL from dilution tube 2 to dilution tube 3; mix well. This is 10^{-4}.

8 Aseptically transfer 100 μL from dilution tube 3 to dilution tube 4; mix well. This is 10^{-5}.

9 Tube 5 will remain sterile and be used as a control.

10 Remove one molten NA tube from the water bath. Aseptically add 100 μL from dilution tube 1, mix well, and pour into plate A. This plate contains 10^{-3} mL original sample volume. Repeat the process using 100 μL from tubes 2 through 5 to plates B through E, respectively. Label the plates with the original sample volume. Label plate E "control."

11 Allow the plates time to cool and solidify. Invert and incubate them at 25°C for 2 to 7 days.

Lab One: Groups 5 and 6

1 Obtain all materials including five Petri dishes. Label five plates SDA—A through E, respectively. Plate E will be the control.

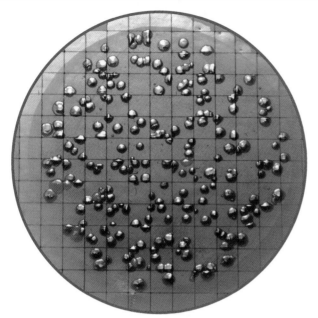

8-37 MEMBRANE FILTER APPARATUS ✦ Assemble the membrane filter apparatus as shown in this photograph. Use two suction flasks (as shown) to avoid getting water into the vacuum source. Secure the flasks on the table, as the tubing may make them top-heavy.

8-36 COLIFORM COLONIES ON A MEMBRANE FILTER ✦ Note the characteristic dark colonies with gold metallic sheen, indicating that this water sample is contaminated with fecal coliforms. Potable water has less than one coliform per 100 mL of sample tested.

✦ Application

The membrane filter technique is commonly used to identify the presence of fecal coliforms in water.

✦ In This Exercise

To determine total coliform population density, you will be using collected water samples and performing a membrane filter technique. Your results will be recorded in coliforms per 100 mL of sample.

✦ Materials

Per Student Group

✦ one m Endo Agar LES plate

✦ one sterile membrane filter (pore size 0.45 µm)

✦ sterile membrane filter suction apparatus (Figure 8-37)

✦ 100 mL water dilution bottle (to be distributed in the preceding lab)

✦ 100 mL water sample (obtained by student)

✦ 100 mL sterile water (for rinsing the apparatus)

✦ gloves

✦ household disinfectant and paper towels

✦ small beaker containing alcohol and forceps

✦ vacuum source (pump or aspirator)

✦ Medium Recipe

m Endo Agar LES

✦ Yeast extract	1.2 g
✦ Casitone	3.7 g
✦ Thiopeptone	3.7 g
✦ Tryptose	7.5 g
✦ Lactose	9.4 g
✦ Dipotassium phosphate	3.3 g
✦ Monopotassium phosphate	1.0 g
✦ Sodium chloride	3.7 g
✦ Sodium desoxycholate	0.1 g
✦ Sodium lauryl sulfate	0.05 g
✦ Sodium sulfite	1.6 g
✦ Basic fuchsin	0.8 g
✦ Agar	15.0 g
✦ Distilled or deionized water	1.0 L
pH 7.3–7.7 at 25°C	

Procedure

Prelab

1 Obtain a 100 mL water dilution bottle from your instructor.

2 Choose an environmental source to sample. (Your instructor may decide to approve your choice to avoid duplication among lab groups.)

3 Visit the environmental site as close to your lab period as possible. Bring your water dilution bottle, a pair of gloves, some household disinfectant, and paper towels. While wearing gloves, fill the bottle to the white line (100 mL) and replace the cap.

4 Wipe the outside of the bottle with disinfectant. Dispose of the towels and gloves in the trash.

5 Store the sample in the refrigerator until your lab period. If the sample must sit for a while before your lab, leave the cap loose to allow some aeration.

Lab One

1 Alcohol-flame the forceps (Figure 8-38) and place the membrane filter (grid facing up) on the lower half of the filter housing (Figure 8-39).

2 Clamp the two halves of the filter housing together and insert the filter housing into the suction flask as shown in Figure 8-40. (This assembly can be a little top-heavy; so have someone hold it or otherwise secure it to prevent tipping.)

3 Pour the appropriate volume of water sample into the funnel. (Refer to Table 8-5 for suggested sample volumes. If the sample size is smaller than 10.0 mL, add 10 to 20 mL of sterile water before filtering, to help distribute the cells evenly on the surface of the membrane filter.)

4 Turn on the suction pump (or aspirator), and filter the sample into the flask.

5 Before removing the membrane, and with the suction pump still running, rinse the sides of the funnel two or three times with 20 or 30 mL of sterile water. This will wash off any cells adhering to the funnel and reduce the likelihood of contaminating the next sample.

6 Sterilize the forceps again, and carefully transfer the filter to the Endo Agar plate, being careful not to fold it or create air pockets (Figure 8-41).

7 Wait a few minutes to allow the filter to adhere to the agar, then invert the plate and incubate it aerobically at $35 \pm 2°C$ for 22 to 24 hours.

Lab Two

1 Remove the plate and count the colonies on the membrane filter that are dark red, purple, have a black center, or produce a green or gold metallic sheen.

2 Record your data in the chart on the Data Sheet.

3 Calculate the coliform CFU per 100 milliliters using the following formula:

$$\frac{\text{Total coliforms}}{100 \text{ mL}} = \frac{\text{coliform colonies counted} \times 100}{\text{volume of original sample in mL}}$$

4 Record your results in the chart on the Data Sheet.

8-38 ALCOHOL-FLAMING THE FORCEPS ✦ Dip the forceps into the beaker of alcohol. Remove the forceps and quickly pass them through the Bunsen burner flame—just long enough to ignite the alcohol. **Be sure to keep the flame away from the beaker of alcohol.**

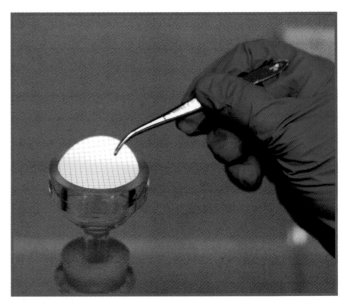

8-39 PLACING THE FILTER ON THE FILTER HOUSING ✦ Carefully place the filter on the bottom half of the filter housing. Clamp the filter funnel over the filter.

References

Collins, C. H., Patricia M. Lyne, and J. M. Grange. 1995. Page 270 in *Collins and Lyne's Microbiological Methods*, 7th ed. Butterworth-Heinemann, Oxford, United Kingdom.

Eaton, Andrew D. (AWWA), Lenore S. Clesceri (WEF), Eugene W. Rice (APHA), Arnold E. Greenberg (APHA), and Mary Ann H. Franson. 2005. Chapter 9 in *Standard Methods for the Examination of Water and Wastewater*, 21st Ed. American Public Health Association, American Water Works Association, Water Environment Federation. APHA Publication Office, Washington, DC.

Zimbro, Mary Jo, and David A. Power, Eds. 2003. Page 208 in *Difco™ and BBL™ Manual—Manual of Microbiological Culture Media*. Becton Dickinson and Co., Sparks, MD.

8-40 MEMBRANE FILTER ASSEMBLY ✦ The membrane filter assembly is made up of a two-piece funnel and clamp. The membrane filter is inserted between the two funnel halves, and the whole assembly is clamped together.

8-41 PLACING THE FILTER ON THE AGAR PLATE ✦ Using sterile forceps, carefully place the filter onto the agar surface with the grid facing up. Try not to allow any air pockets under the filter, because contact with the agar surface is essential for growth of bacteria. Allow a few minutes for the filter to adhere to the agar before inverting the plate.

TABLE **8-5** Suggested Sample Volumes for Membrane Filter Total Coliform Test

Water Source	Volume (●)To Be Filtered (mL)							
	100	50	10	1.0	0.1	0.01	0.001	0.0001
Drinking water	●							
Swimming pools	●							
Wells, springs	●	●	●					
Lakes, reservoirs	●	●	●					
Water supply intake			●	●	●			
Bathing beaches			●	●	●			
River water				●	●	●	●	
Chlorinated sewage				●	●	●		
Raw sewage					●	●	●	●

(Reprinted with permission from the American Public Health Association)

Multiple Tube Fermentation Method for Total Coliform Determination

✦ Theory

The **multiple tube fermentation method**, also called **most probable number**, or **MPN**, is a common means of calculating the number of coliforms present in 100 mL of a sample. The procedure determines both **total coliform** counts and *E. coli* counts.

The three media used in the procedure are: Lauryl Tryptose Broth (LTB), Brilliant Green Lactose Bile (BGLB) Broth, and EC (*E. coli*) Broth. LTB, which includes lactose and lauryl sulfate, is selective for the coliform group. Because it does not screen out all noncoliforms, it is used to *presumptively* determine the presence or absence of coliforms. BGLB broth, which includes lactose and 2% bile, inhibits noncoliforms and is used to *confirm* the presence of coliforms. EC broth, which includes lactose and bile salts, is selective for *E. coli* when incubated at $45.5°C$.

All broths are prepared in 10 mL volumes and contain an inverted Durham tube to trap any gas produced by fermentation. The LTB tubes are arranged in up to ten groups of five (see Figure 8-42, Procedural Diagram). Each tube in the first set of five receives 1.0 mL of the original sample. Each tube in the second group receives 1.0 mL of a 10^{-1} dilution. Each tube in group three receives 1.0 mL of 10^{-2}, *etc.*[1] (**Note:** The volume of broth in the tubes is not part of the calculation of dilution factor. Dilutions of the water sample are made using sterile water prior to inoculating the broths. One milliliter of diluted sample is added to each tube in its designated group. Refer to Exercise 6-1 for help with serial dilutions.)

After inoculation, the LTB tubes are incubated at $35 \pm 2°C$ for up to 48 hours, then examined for gas production (Figure 8-43). Any positive LTB tubes then are used to inoculate BGLB tubes. Each BGLB receives one or two loopfuls from its respective positive LTB tube. Again, the cultures are incubated 48 hours at $35 \pm 2°C$ and examined for gas production (Figure 8-44). Positive BGLB cultures then are transferred to EC broth and incubated at $45.5°C$ for 48 hours (Figure 8-45). After incubation the EC tubes with gas are counted. Calculation of BGLB MPN and EC MPN is based on the combinations of positive results in the BGLB and EC broths, respectively, using the following formula:

$$\text{MPN/100 mL} = \frac{100P}{\sqrt{V_n V_a}}$$

where

P = total number of positive results (BGLB or EC)

V_n = combined volume of sample in LTB tubes that produced negative results in BGLB or EC

V_a = combined volume of sample in all LTB tubes

It is customary to calculate and report *both* total coliform *and E. coli* densities. Total coliform MPN is calculated using BGLB broth results, and *E. coli* MPN is based on EC broth results.

Using the data from Table 8-6 and the formula above, the calculation would be as follows:

$$\text{MPN/100 mL} = \frac{100P}{\sqrt{V_n V_a}}$$

$$\text{MPN/100 mL} = \frac{100 \times 9}{\sqrt{0.24 \times 5.55}}$$

$$\text{MPN/100 mL} = 780$$

✦ Application

This standardized test is used to measure coliform density (cells/100 mL) in water. It may be used to calculate the density of all coliforms present (total coliforms) or to calculate the density of *Escherichia coli* specifically.

✦ In This Exercise

You will be using collected water samples to perform a multiple tube fermentation. The data collected then will be used to calculate a total coliform count and an *E. coli* count. Counts will be recorded in coliforms per 100 mL of sample and *E. coli* per 100 mL of sample, respectively.

✦ Materials

Per Student Group

✦ 15 Lauryl Tryptose Broth (LTB) tubes (containing 10 mL broth and an inverted Durham tube)

---

[1] This exercise has been simplified for ease of instruction. The number of groups, number of tubes in each group, dilutions necessary, volume of broth in each tube, and volumes of sample transferred vary significantly, depending on the source and expected use of the water being tested.

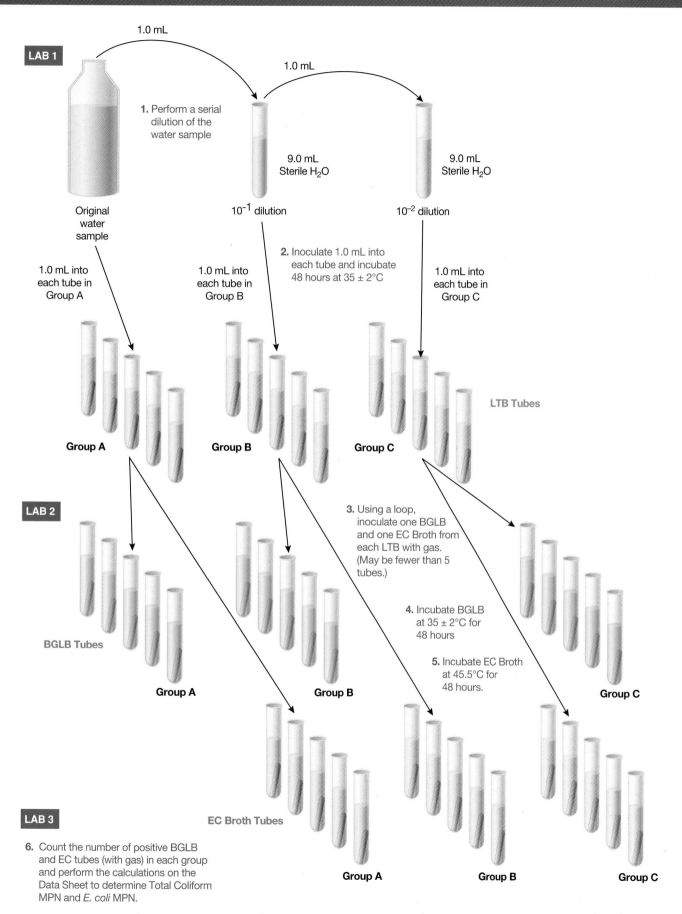

LAB 1

1.0 mL

1.0 mL

1. Perform a serial dilution of the water sample

9.0 mL Sterile H₂O

9.0 mL Sterile H₂O

Original water sample

10^{-1} dilution

10^{-2} dilution

2. Inoculate 1.0 mL into each tube and incubate 48 hours at 35 ± 2°C

1.0 mL into each tube in Group A

1.0 mL into each tube in Group B

1.0 mL into each tube in Group C

Group A

Group B

Group C

LTB Tubes

LAB 2

3. Using a loop, inoculate one BGLB and one EC Broth from each LTB with gas. (May be fewer than 5 tubes.)

4. Incubate BGLB at 35 ± 2°C for 48 hours

5. Incubate EC Broth at 45.5°C for 48 hours.

BGLB Tubes

Group A

Group B

Group C

LAB 3

EC Broth Tubes

6. Count the number of positive BGLB and EC tubes (with gas) in each group and perform the calculations on the Data Sheet to determine Total Coliform MPN and *E. coli* MPN.

Group A

Group B

Group C

8-42 PROCEDURAL DIAGRAM OF MULTIPLE TUBE FERMENTATION METHOD FOR DETERMINATION OF TOTAL COLIFORM ✦

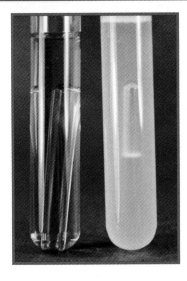

8-43 LTB TUBES ✦ The bubble in the Durham tube on the right is *presumptive* evidence of coliform contamination. The tube on the left is negative.

8-44 BGLB BROTH ✦ The bubble in the Durham tube on the right is seen as *confirmation* of coliform contamination. The tube on the left is negative.

8-45 EC BROTH ✦ The bubble in the Durham tube on the right is seen as *confirmation* of *E. coli* contamination. The tube on the left is negative.

✦ up to 15 Brilliant Green Lactose Bile (BGLB) broth tubes (These tubes are needed for Lab Two. The number of tubes required will be determined by the results of the LTB test.)

✦ up to 15 EC broth tubes (These tubes are needed for Lab Two. The number of tubes required will be determined by the results of the LTB test.)

✦ two 9.0 mL dilution tubes

✦ sterile 1.0 mL pipettes and pipettor

✦ water bath set at 45.5°C (Lab Two)

✦ test tube rack

✦ labeling tape

✦ 100 mL water dilution bottle (to be distributed in the preceding lab)

✦ 100 mL water sample (obtained by student)

✦ gloves

✦ household disinfectant and paper towels

✦ Sodium chloride	5.0 g
✦ Sodium lauryl sulfate	0.1 g
✦ Distilled or deionized water	1.0 L

pH 6.6–7.0 at 25°C

EC Broth

✦ Tryptose	20.0 g
✦ Lactose	5.0 g
✦ Dipotassium phosphate	4.0 g
✦ Monopotassium phosphate	1.5 g
✦ Sodium chloride	5.0 g
✦ Distilled or deionized water	1.0 L

pH 6.7–7.1 at 25°C

✦ Medium Recipes

Brilliant Green Lactose Bile Broth

✦ Peptone	10.0 g
✦ Lactose	10.0 g
✦ Oxgall	20.0 g
✦ Brilliant green dye	0.0133 g
✦ Distilled or deionized water	1.0 L

pH 7.0–7.4 at 25°C

Lauryl Tryptose Broth

✦ Tryptose	20.0 g
✦ Lactose	5.0 g
✦ Dipotassium phosphate	2.75 g
✦ Monopotassium phosphate	2.75 g

 ## Procedure

[Refer to the procedural diagram in Figure 8-42 as needed.]

Lab One

1 Mix the water sample. Then take a dilution by adding 1.0 mL of the water sample to one of the 9.0 mL dilution tubes. Mix well. This dilution is only 1/10 of the original sample; therefore, the dilution is 1/10 or 0.1 or 10^{-1} (preferred). For help with dilutions, refer to Exercise 6-1.

2 Make a second dilution by adding 1.0 mL of the 10^{-1} dilution to one of the 9.0 mL dilution tubes. Mix

well. This dilution contains 1/100 original sample and is 10^{-2}.

3 Arrange the 15 LTB tubes into three groups of five in a test tube rack. Label the groups A, B, and C, respectively.

4 Aseptically transfer 1.0 mL of the (undiluted) water sample to each LTB tube in the group labeled A. Mix well. This is undiluted sample, therefore it is a 10^0 dilution. (*Note:* The other dilutions are determined when transferred to the dilution tubes. The LTB is not part of the dilution.)

5 Add 1.0 mL of the 10^{-1} dilution to each of the LTB tubes in group B. Mix well.

6 Add 1.0 mL of the 10^{-2} dilution to each of the LTB tubes in the C group.

7 Incubate the LTB tubes at 35 to 37°C for 48 hours.

Lab Two

1 Remove the broths from the incubator and, one group at a time, examine the Durham tubes for accumulation of gas. Gas production is a positive result; absence of gas is negative if the broth is clear. (*Note:* If the LTB broth is turbid but contains no gas, the presence of coliforms is not yet ruled out. Therefore, record any turbidity as a positive result and include these tubes with those containing gas. Record positive and negative results in the chart provided on the Data Sheet. If necessary, refer to the example in Table 8-6.)

2 Using an inoculating loop, inoculate one BGLB broth with each positive LTB tube. (Make sure that each BGLB tube is *clearly labeled* A, B, or C according to the LTB tube from which it is inoculated.)

3 Inoculate EC broths with the positive LTB tubes in the same manner as the BGLB above. Again, be sure to clearly label all EC tubes appropriately.

TABLE **8-6** BGLB Test Results ✦ The results shown here are of a hypothetical BGLB test using three groups of five tubes (A, B, and C). (**Note:** On your data sheet, you are given two similar tables—one for BGLB results and one for EC broth results.) In this example, the original water sample was diluted to 10^0, 10^{-1}, and 10^{-2}. The three dilutions were used to inoculate the broths in groups A, B, and C, respectively. The first row contains the dilution of the inoculum used per group. The second row shows the actual amount of original sample that went into each broth. The third row contains the number of tubes in each group (five in this example). The fourth row shows the number of tubes from each group of five that showed evidence of gas production. This total (in red) inserts into the equation. The fifth row shows the number of tubes from each group that did *not* show evidence of gas production. The sixth row is used for calculating the "combined volume of sample in negative tubes" and refers to the inoculum that went into the LTB tubes that produced a negative result. This total (in red) inserts into the equation. The seventh row is used for calculating the "combined volume of sample in all tubes" and refers to the total volume of inoculum that went into all LTB tubes. This total (in red) inserts into the equation. As you can see, the undiluted inoculation (Group A) produced five positive results and zero negative results; the 10^{-1} dilution (Group B) produced three positive results and two negative; and the 10^{-2} dilution (Group C) produced one positive and four negative results. The total volume of original sample that went into LTB tubes was 5.55 mL, 0.24 mL of which produced no gas (shown in red in rows 7 and 6, respectively).

	Group	Group A	Group B	Group C	Totals (A+B+C)
1	Dilution (D)	10^0	10^{-1}	10^{-2}	NA
2	Portion of dilution added to each LTB tube that is original sample (1.0 mL × D)	1.0 mL	0.1 mL	0.01 mL	NA
3	# LTB tubes in group	5	5	5	NA
4	# positive (BGLB or EC) results	5	3	1	9
5	# negative (BGLB or EC) results	0	2	4	NA
6	Total volume of *original sample* in all LTB tubes that produced negative (BGLB or EC) results (D × 1.0 mL × # negative tubes)	0 mL	0.2 mL	0.04 mL	0.24 mL
7	Volume of *original sample* in all LTB tubes inoculated (D × 1.0 mL × # tubes)	5.0 mL	0.5 mL	0.05 mL	5.55 mL

4 Incubate the BGLB at 35 to 37°C for 48 hours. Incubate the EC tubes in the 45.5°C water bath for 48 hours.

Lab Three

1 Remove all tubes from the incubator and water bath and examine the Durham tubes for gas accumulation. (**Note:** At this point, turbidity does not count as positive; the only positive result is the presence of gas in the Durham tube.) Count the positive BGLB tubes and enter your results in the chart on the Data Sheet.

2 Using Table 8-6 as a guide, complete the BGLB chart on the Data Sheet.

3 Using the data in the BGLB chart, calculate the total coliform MPN using the formula on page 336. (This formula is used to calculate both Total Coliform MPN and *E. coli* MPN.)

4 Count the positive EC broth results in the same manner as the BGLB test. Record the results in the chart on the Data Sheet.

5 Determine the *E. coli* MPN in the same manner as described for total coliform count, and record your results in the chart on the Data Sheet.

Reference

Eaton, Andrew D. (AWWA), Lenore S. Clesceri (WEF), Eugene W. Rice (APHA), Arnold E. Greenberg (APHA), and Mary Ann H. Franson. 2005. Chapter 9 in *Standard Methods for the Examination of Water and Wastewater*, 21st Ed. 2005. American Public Health Association, American Water Works Association, Water Environment Federation. APHA Publication Office, Washington, DC.

Food Microbiology

Food microbiology is the study, utilization, and control of microorganisms in food. In Exercise 2-1 you learned that microorganisms exist virtually everywhere. Left undisturbed in their own habitat, most are beneficial in some way. Some microorganisms have been used successfully to produce our favorite foods, such as yogurt, wine, beer, sauerkraut, buttermilk, vinegar, bread, and cheeses. Generally speaking, however, the unintended introduction of microorganisms to our food (including otherwise beneficial microorganisms) is a serious health hazard. For example, *Escherichia coli*, one of the most common enterics in humans and other mammals, can cause mild to severe illness or even death if ingested.

It is not unusual to find unwanted microorganisms in unprocessed food. It is unavoidable and expected. This is why there are agencies such as the Food and Drug Administration (FDA) and the Centers for Disease Control and Prevention (CDC). These agencies were designed, in part, to protect consumers from the food-borne illnesses caused by improper management of food items. But these agencies exercise control only in the public arena; their influence over the practices of people in their homes typically is advisory only. This is why it is important to follow their recommendations and not only practice good personal hygiene, but to wash and cook food properly.

In the two exercises included in this unit, you will be performing a simple test of milk quality and then you will have a little fun by making yogurt. Don't forget to wash your hands before you start! ✦

EXERCISE 9-1

Methylene Blue Reductase Test

✦ Theory

Methylene blue dye is blue when oxidized and colorless when reduced. It can be reduced enzymatically either aerobically or anaerobically. In the aerobic electron transport system, methylene blue is reduced by cytochromes but immediately is returned to the oxidized state when it subsequently reduces oxygen. Anaerobically, the dye is in the reduced form, and in the absence of an oxidizing substance, remains colorless.

The reduction of methylene blue may be used as an indicator of milk quality. In the methylene blue reductase test, a small quantity of a dilute methylene blue solution is added to a sterilized test tube containing raw milk. The tube then is sealed tightly and incubated in a 35°C water bath. The time it takes the milk to turn from blue to white (because of methylene blue reduction) is a qualitative indicator of the number of microorganisms living in the milk (Figure 9-1). Good-quality milk takes longer than 6 hours to convert the methylene blue.

✦ Application

This test is helpful in differentiating the *enterococci* from other streptococci. It also tests for the presence of coliforms in raw milk.

✦ In This Exercise

You will test milk quality by measuring how long the indicator dye, methylene blue, takes to become oxidized by any bacterial contaminants present.

✦ Materials

Per Student Group

- ✦ milk samples (raw or processed; a variety is best)
- ✦ sterile screw-capped test tubes
- ✦ sterile 1 mL and 10 mL pipettes with mechanical pipettors
- ✦ hot water bath set at 35°C
- ✦ methylene blue reductase reagent
- ✦ overnight broth culture of *Escherichia coli*
- ✦ clock or wristwatch

Reagent

Methylene Blue Reductase Reagent

✦ Methylene blue dye	8.8 mg
✦ Distilled or deionized water	200.0 mL

Procedure

Lab One

1. Obtain sterile tubes for as many samples as you are testing, plus two more, to be used as positive and negative controls. Label them appropriately.

2. Using a sterile 10 mL pipette, aseptically add 10 mL milk to each tube.

3. Inoculate the tube marked "positive control" with 1 mL of *E. coli* culture.

4. Aseptically add 1.0 mL methylene blue solution to each test tube. Cap the tubes tightly and invert them several times to mix thoroughly.

5. Place the tube marked "negative control" in the refrigerator to prevent it from changing color.

6. Place all other tubes in the hot water bath, and note the time.

9-1 METHYLENE BLUE REDUCTASE TEST ✦ The tube on the left is a control to illustrate the original color of the oxidized medium. The tube on the right indicates bacterial reduction of methylene blue after 20 hours. The speed of reduction is related to the concentration of microorganisms present in the milk.

7 After 5 minutes, remove the tubes, invert them once to mix again, then return them to the water bath. Record the time in the chart in the Data Sheet under "Starting Time."

8 Using the control tubes for color comparison, check the tubes at 30-minute intervals, and record the time when each becomes white. Poor-quality milk takes less than 2 hours. Good-quality milk takes longer than 6 hours. If necessary, have someone check the tubes at 6 hours and record the results for you.

9 Using the chart provided, calculate the time it takes for each milk sample to become white.

References

Bailey, R. W., and E. G. Scott. 1966. Pages 114 and 306 in *Diagnostic Microbiology*, 2nd ed. Mosby, St. Louis.

Benathen, Isaiah. 1993. Page 132 in *Microbiology with Health Care Applications*. Star Publishing, Belmont, CA.

Power, David A., and Peggy J. McCuen. 1988. Page 62 in *Manual of BBL™ Products and Laboratory Procedures*, 6th ed. Becton Dickinson Microbiology Systems, Cockeysville, MD.

Richardson, Gary H., Ed. 1985. *Standard Methods for the Examination of Dairy Products*, 15th ed. American Public Health Association, Washington DC.

EXERCISE 9-2

Making Yogurt

✦ Theory

Several species of bacteria are used in the commercial production of yogurt. Most formulations include combinations of two or more species to synergistically enhance growth and to produce the optimum balance of flavor and acidity. One common pairing of organisms in commercial yogurt is that of *Lactobacillus delbrueckii* subsp. *bulgaricus* and *Streptococcus thermophilus*.

Yogurt gets its unique flavor from acetaldehyde, diacetyl, and acetate produced from fermentation of the milk sugar lactose. The proportions of products, and ultimately the flavor, in the yogurt depend upon the types of enzyme systems possessed by the species used. Both species mentioned above contain **constitutive** β-galactosidase systems that break down lactose and convert the glucose to lactate, formate, and acetate via pyruvate in the Embden-Meyerhof-Parnas (glycolysis) pathway. (See Appendix A.)

As you may remember, lactose is a disaccharide composed of glucose and galactose. *S. thermophilus* does not possess the enzymes needed to metabolize galactose, and *L. delbrueckii* preferentially metabolizes glucose. This results in an accumulation of galactose, which adds sweetness to the yogurt. Acetaldehyde is produced directly from pyruvate by *S. thermophilus* and through the conversion of proteolysis products threonine and glycine by *L. delbrueckii*. Some strains of *S. thermophilus* also produce glucose polymers, which give the yogurt a viscous consistency.

✦ Application

This exercise is designed to keep you away from the internet for a few minutes.

✦ In This Exercise

You will produce yogurt using a simple home recipe with a commercial yogurt as a starter. Read the label to see which microorganisms are included. We hope you enjoy it.

✦ Materials

Per Student Group

- ✦ whole, low-fat, or skim milk
- ✦ plain yogurt with active cultures (bring from home or supermarket)
- ✦ medium-size saucepan
- ✦ medium-size bowl
- ✦ wire whisk
- ✦ hotplate
- ✦ cooking thermometer
- ✦ measuring cup
- ✦ plastic wrap
- ✦ fresh fruit
- ✦ sugar (optional)
- ✦ pH meter or pH paper

Procedure

Lab One

1 Obtain all materials, and set them up in a clean work area.

2 While stirring, slowly heat 5 cups milk in the saucepan to 185°F. Remove the milk from the heat, and let it cool to 110°F.

3 Place 1/4 cup starter yogurt in the bowl. Slowly, stir in cooled milk, about 1/3 to 1/2 cup at a time, mixing after each addition until smooth.

4 Cover the bowl with plastic wrap, and puncture several times to allow gases and excess moisture to escape.

5 Label the bowl with your name, the date, and the cultures present in your yogurt starter.

6 Incubate 5–6 hours at 30°–35°C. Remove the bowl from the incubator at the correct time, and place it in the refrigerator.

Lab Two

1 Remove your yogurt from the refrigerator.

2 Perform a simple stain on a smear from your yogurt and examine on the microscope.

3 Compare flavor, consistency, and starter cultures with other groups in the lab. With a pH meter or pH paper, measure the pH of your yogurt. Record your results in the chart provided on the Data Sheet.

4 Serve with fresh fruit and enjoy!

References

Downes, Frances Pouch, and Keith Ito. 2001. Chapter 47 in *Compendium of Methods for the Microbiological Examination of Foods*, 4th ed. American Public Health Association, Washington, DC.

Ray, Bibek. 2001. Chapter 13 in *Fundamental Food Microbiology*, 2nd ed. CRC Press LLC, Boca Raton, FL.

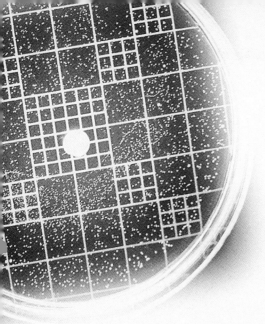

Microbial Genetics

In this unit you will be introduced to DNA experiments dealing with biotechnology (Exercises 10-1 through 10-4), mutation (Exercises 10-5 and 10-6), and molecular taxonomy (Exercise 10-7).

First you will perform a simple extraction of *E. coli* DNA. The second exercise provides an example of how DNA can be digested into fragments and how those fragments can be put to use. As such, it is an important test in the fight against cancer. The third exercise incorporates into one exercise bacterial transformation, use of antibiotic selective media, genetic regulation, and genetic engineering. You will be transferring a specific gene into *E. coli* cells, selecting for only those cells that have been transformed successfully, then manipulating the environment so the organisms produce the gene product only when you want them to. The fourth exercise illustrates a sophisticated means of copying DNA *in vitro*.

In the next two exercises you will examine mutations—alterations of DNA. The first of these examines the effects of ultraviolet (UV) radiation on bacteria. It illustrates some characteristics of a particular *mutagen*—how it causes damage, factors affecting its impact on the cell, and how bacteria are able to repair that damage. The second mutation exercise—the "Ames Test"—illustrates a simple method of screening substances (commercial products) for mutagenicity, and in turn, carcinogenicity!

In the last exercise, you will see how molecular differences in *E. coli* strains, as detected by different susceptibility to bacteriophage infection, can be used as an aid to identification. ✦

EXERCISE 10-1

Extraction of DNA from Bacterial Cells[1]

✦ Theory

DNA extraction from cells is surprisingly easy and occurs in three basic stages.

1. A detergent (*e.g.*, Sodium Dodecyl Sulfate—SDS) is used to lyse cells and release cellular contents, including DNA.

2. This is followed by a heating step (at approximately 65–70°C) that denatures protein (including DNases that would destroy the extracted DNA) and other cell components. Temperatures higher than 80°C will denature DNA, and this is undesirable. A protease also may be added to remove proteins. (Other techniques for purification may be used, but these will not be included in this exercise.)

3. Finally, the water-soluble DNA is precipitated in cold alcohol as a whitish, mucoid mass (Figure 10-1).

As an optional follow-up to extraction, an ultraviolet spectrophotometer (Figure 10-2) will be used to estimate DNA concentration in the sample by measuring absorbance at 260 nm, the optimum wavelength for absorption by DNA. An absorbance of A_{260nm} of 1 corresponds to 50 µg/mL of double-stranded DNA (dsDNA). Reading absorbance at 280 nm and calculating the following ratio can determine purity of the sample:

$$\frac{\text{Absorbance}_{260nm}}{\text{Absorbance}_{280nm}}$$

[1] *Note:* Thanks to the following individuals who offered helpful suggestions for this protocol: Dr. Melissa Scott of San Diego City College, Allison Shearer of Grossmont College, Donna Mapston and Dr. Ellen Potter of Scripps Institute for Biological Studies, and Dr. Sandra Slivka of Miramar College.

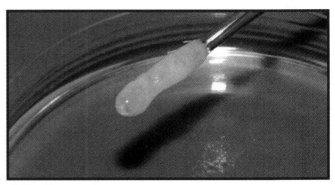

10-1 PRECIPITATED DNA ✦ This onion DNA has been spooled onto a glass rod.

If the sample is reasonably pure nucleic acid, the ratio will be about 1.8 (between 1.65 and 1.85). Because protein absorbs maximally at 280 nm, a ratio of less than 1.6 is likely because of protein contamination. If purity is crucial, the DNA extraction can be repeated. If the ratio is greater than 2.0, the sample is diluted and read again.

✦ Application

Extraction of DNA is a starting point for many lab procedures, including DNA sequencing and cloning.

✦ In This Exercise

You will extract DNA from *E. coli*. To improve yield, you first will concentrate an *E. coli* broth culture. The actual extraction involves cell lysis, denaturation of protein, and precipitation of the DNA in alcohol. Following extraction, an optional procedure may be used to determine DNA yield and the purity of your extract.

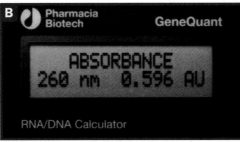

10-2 ULTRAVIOLET SPECTROPHOTOMETER ✦ **A** A UV spectrophotometer can be used to determine DNA concentration. A quartz cuvette is shown in the sample port. **B** This specimen has an A_{260nm} of 0.596. Because an A_{260nm} of 1.0 is equal to 50 µg/mL of dsDNA, this specimen has a concentration of 29.8 µg dsDNA/mL. Absorbance also can be used to determine purity of the sample. A relatively pure DNA sample will have an A_{260nm}/A_{280nm} value of approximately 1.8.

✦ Materials

Per Pair of Students

✦ overnight culture of *Escherichia coli* in Luria-Bertani broth (young cultures work best)

✦ water bath set to 65°C

✦ microtube floats

✦ ice bath (small cups with crushed ice work well)

✦ 300 µL 10% Sodium Dodecyl Sulfate (SDS)

✦ 50 µL 20 mg/mL Proteinase K solution (stored in freezer between uses) (**Note:** Meat tenderizer is an inexpensive substitute, though it may result in DNA hydrolysis if too much is used.)

✦ 300 µL 1.0 M Sodium Acetate solution (pH = 5.2)

✦ 3 mL 1X M Tris-Acetate-EDTA (TAE) buffer (dilute 10X TAE 9 + 1)

✦ 2 mL 90% Isopropanol (stored in a freezer or an ice bath)

✦ two calibrated disposable transfer pipettes

✦ 100–1000 µL digital pipettor and tips

✦ 25 mL centrifuge tube

✦ tabletop centrifuge

✦ minicentrifuge

✦ disposable inoculating loop

✦ two microtubes (at least 1.5 mL in volume)

✦ vortex mixer (optional)

✦ ultraviolet spectrophotometer (optional)

✦ two quartz cuvettes (optional)

✦ Medium Recipe

Luria-Bertani Broth

♦ Tryptone	10.0 g
♦ Yeast Extract	5.0 g
♦ NaCl	10.0 g
♦ Distilled or deionized water	1.0 L
pH 7.4 at 25°C	

Procedure

Refer to the procedural diagram in Figure 10-3 as you read and follow this protocol.

1 Obtain the *E. coli* culture. Mix the suspension until uniform turbidity is seen, then transfer 5 mL to a clean, nonsterile centrifuge tube.

2 Spin the 5 mL sample in the tabletop centrifuge slowly for 10 minutes to produce a cell pellet.

3 After spinning and without disturbing the pellet, remove 4.5 mL of the supernatant with a transfer pipette. Dispose of the supernatant in the original culture tube.

4 Transfer the remaining 0.5 mL to a nonsterile microtube.

5 Add 200 µL of 10% SDS to the *E. coli*.

6 Add 30 µL of 20 mg/mL Proteinase K solution or a "pinch" of meat tenderizer.

7 Close the cap and gently mix the tube for 5 minutes by tipping it upside down every few seconds.

8 Place the tube in a float and incubate in a 65°C water bath for 5 minutes.

9 Add 200 µL mL 1 M Sodium Acetate solution and mix gently.

10 Place the tube in the ice bath until it is at or below room temperature.

11 When cooled, squirt 400 µL mL of cold 95% isopropanol into the preparation. (If there are bubbles on the surface, remove them with a nonsterile transfer pipette prior to adding the isopropanol.)

12 Use a disposable loop to mix the preparation, moving in and out and turning occasionally. A glob of mucoid DNA will begin to appear as you mix and will adhere to the loop.

✦ Optional Procedure

(if your lab has an ultraviolet spectrophotometer)

Refer to the procedural diagram in Figure 10-3 as you read and complete the following protocol.

1 Remove the DNA from the original microtube and transfer it to a second microtube, using the loop.

2 Resuspend the DNA in 1000 µL isopropanol using a nonsterile transfer pipette, then spin it in a minicentrifuge for a few seconds. The DNA should be at the bottom of the tube.

3 With a nonsterile transfer pipette, remove the supernatant and allow the DNA pellet to air-dry.

4 Resuspend the dried DNA in 1000 µL of TAE buffer. Mix vigorously to dissolve the DNA in the TAE. This mixing may be done by hand, or a vortex mixer may be used.

5 Transfer the suspended DNA solution into a quartz cuvette.

6 Prepare a second cuvette containing 1000 µL of TAE as a blank.

DNA Extraction Procedural Diagram

1. Transfer 5 mL to a centrifuge tube

2. Centrifuge for 10 minutes

3. Remove 4.5 mL of supernatant with transfer pipette

24 hour
E. coli culture

Nonsterile centrifuge tube with 5 mL of E. coli culture

Cell pellet in centrifuge tube after spinning

0.5 mL
E. coli culture

4. Mix and transfer contents of centrifuge tube to a microtube

5. Add 200 μL SDS and 30 μL Proteinase K

6. Mix contents for 5 minutes by tipping every few seconds

0.5 mL (500 μL)
E. coli culture

65°C

7. Place tube in 65°C water bath for 5 minutes

BEP-MJPLeb. Inc

cooler   warmer   ON OFF

8. Add 200 μL 1 M Sodium Acetate and place in ice bath until room temperature

9. Squirt 400 μL cold isopropanol into the tube

10. Mix with a disposable loop—the DNA will be a thick, mucoid mass

Continue with optional procedure

10-3 PROCEDURAL DIAGRAM FOR BACTERIAL DNA EXTRACTION ✦

DNA Extraction—Optional Procedure

1. Transfer DNA to a second microtube

1000 µL isopropanol

2. Add 1000 µL isopropanol, then spin in a microcentrifuge for a few seconds

3. Remove the supernatant and allow pellet to air dry

4. Resuspend DNA with 1000 µL TAE buffer

5. Mix thoroughly to put DNA back into solution

6. Transfer to UV cuvette and take appropriate readings with a UV spectrophotometer

10-3 PROCEDURAL DIAGRAM FOR BACTERIAL DNA EXTRACTION (CONTINUED) ✦

7 Set the spectrophotometer to 260 nm wavelength. Follow the instructions for your UV spectrophotometer to check the absorbance of the extracted DNA, and record on the Data Sheet.

8 Set the spectrophotometer to 280 nm wavelength. Follow the instructions for your UV spectrophotometer to check the absorbance of the extracted DNA, and record on the Data Sheet.

9 Calculate the probable purity of your extracted DNA sample using the formula provided, and record on the Data Sheet.

10 If desired, absorbencies at other wavelengths may be taken to produce an absorption spectrum for DNA. Suggested wavelengths are: 220 nm, 240 nm, 300 nm, and 320 nm, in addition to the measure-ments for 260 and 280 nm taken above. Record these on the Data Sheet.

11 If step 10 was done, plot the absorption spectrum (Absorption *versus* Wavelength) of the DNA sample on the Data Sheet.

References

Bost, Rod. 1989. *Down and Dirty DNA Extraction.* Carolina Genes. North Carolina Biotechnology, Research Triangle Park, NC.

Davis, Leonard G., Mark D. Dibner, and James F. Battey. (1986). *Basic Methods in Molecular Biology.* Elsevier Science Publishing, New York.

Freifelder, David. 1982. Pages 504–505 in *Physical Biochemistry,* 2nd ed. W. H. Freeman and Co, New York.

Kreuzer, Helen, and Adrianne Massey. 2001. *Recombinant DNA and Bio-technology—A Guide For Teachers,* 2nd ed. ASM Press, Washington, DC.

Zyskind, Judith W., and Sanford I. Bernstein. 1992. *Recombinant DNA Laboratory Manual.* Academic Press, San Diego.

EXERCISE
10-2 Restriction Digest[1]

✦ Theory

Restriction enzymes are endonucleases contained in many bacteria and archea whose function appears to be both maintenance and protection against foreign DNA. Restriction enzymes function by bonding to the DNA and cutting out (cleaving) damaged portions or, in the case of invading viral DNA, disabling it. Each restriction enzyme has a different **recognition site** (short sequence) to which it attaches. Recognition sites are short DNA sequences, typically 4 to 8 nucleotides. Below is the recognition sequence of EcoRI[2], the restriction enzyme extracted from a strain of *Escherichia coli*.

$$5'-GAATTC-3'$$
$$3'-CTTAAG-5'$$

Cells are protected from damage from their own endonucleases by a process known as **methylation**, in which methyl groups ($R-CH_3$) attach to specific recognition sites rendering the DNA an unfit substrate for the enzyme. The following diagram illustrates a typical methylation by the enzyme *EcoRI methylase*. This methylation prevents attachment by the above-mentioned restriction enzyme *EcoRI*.

$$CH_3$$
$$|$$
$$5'-GAATTC-3'$$
$$3'-CTTAAG-5'$$
$$|$$
$$CH_3$$

Most recognition sites are palindromes. That is, both strands of DNA have the same sequence of nucleotides when read from the 5' (or 3') end. Each enzyme cuts the DNA at a specific location in the sequence. For example, *EcoRI* cleaves double stranded DNA as follows:

$$5'-GAATTC-3'$$
$$3'-CTTAAG-5'$$

Resulting in,

$$5'-G \qquad AATTC-3'$$
$$3'-CTTAA \qquad G-5'$$

Many cleavage patterns are possible. In fact, because some enzymes attack multiple variations of a sequence, many more fragment patterns are achievable than the restriction enzymes that perform them. For example, *Bgl*I (from *Bacillus globigii*) recognizes 625 variations of the following sequence:

$$5'-GCCNNNNNGGC-3'$$
$$3'-CGGNNNNNCCG-5'$$

where N can be A, G, C, or T. The enzyme reads the sequence flanking the variable segment, and will not attach if the nucleotides are not spaced as shown. Table 10-1 lists a few examples of restriction enzymes and cleavage patterns.

Agarose gel electrophoresis[3] is frequently used in conjunction with restriction enzymes to separate DNA fragments of different sizes. Many variables can affect migration patterns of DNA fragments, relating to shape, size, and a tendency of some fragments to form attachments to other molecules. Generally speaking, however, small DNA fragments (fewer base pairs) will migrate faster and farther than large fragments (Figure 10-4 and Figure 2 on the Data Sheet). Simply looking at the bands will not tell you the size of the fragments, but running a control sample of known DNA standards in the gel allows construction of a standard curve that can be used to determine unknown fragment sizes.

Because one of the three samples used in this exercise is bacteriophage Lambda DNA, special care will be required to achieve a reliable band pattern in the gel. The phage genome is double-stranded and linear with a 12 base-pair single-stranded tail at the 5' terminus of each strand. These strands are complementary to each other, so when the genome is released inside a host cell the complementary strands link to form a circular molecule. In restriction digestion reactions the tendency of

[1] This procedure is adapted, with permission, from Edvotek kit #213 Cleavage of DNA with Restriction Enzymes, for which the company holds copyrights. All materials and equipment required for this exercise can be purchased at Edvotek, Inc., PO Box 341232, Bethesda, MD, 20827, USA. email: info@edvotek.com www.edvotek.com

[2] Restriction enzymes get their names from a combination of genus, species, strain (when applicable), and a Roman numeral indicating position in order of identification. For example *EcoRI* comes from *Escherichia coli*, strain **RY13**, first restriction enzyme identified from this organism (**I**).

[3] This exercise includes agarose gel electrophoresis. Depending on the time available in your lab, your instructor may or may not include gel preparation as part of the exercise. Additionally, because gel preparation, electrophoresis, and gel staining are included in more than one exercise, only instructions for running the electrophoresis are included in this procedure. A short introduction to gel electrophoresis, agarose gel preparation, and various staining procedures appear in Appendix G.

TABLE **10-1** Restriction Enzymes and Their Recognition Sequences

EcoRI	*Escherichia coli*	5'–GAATTC–3' 3'–CTTAAG–5'	5'–G AATTC–3' 3'–CTTAA G–5'
TaqI	*Thermus aquaticus*	5'–TCGA–3' 3'–AGCT–5'	5'–T CGA–3' 3'–AGC T–5'
BamHI	*Bacillus amyloliquefaciens*	5'–GGATCC–3' 3'–CCTAGG–5'	5'–G GATCC–3' 3'–CCTAG G–5'
BglII	*Bacillus globigii*	5'–AGATCT–3' 3'–TCTAGA–5'	5'–A GATCT–3' 3'–TCTAG A–5'
HindII	*Haemophilus influenzae*	5'–GTPyPuAC–3' 3'–CAPuPyTG–5'	5'–GTPy PuAC–3' 3'–CAPu PyTG–5'
Sau3A	*Staphylococcus aureus*	5'–GATC–3' 3'–CTAG–5'	5'– GATC–3' 3'–CTAG –5'
SmaI	*Serratia marcescens*	5'–CCCGGG–3' 3'–GGGCCC–5'	5'–CCC GGG–3' 3'–GGG CCC–5'

Pu = Purine (G or A), Py = Pyrimidine (C or T)

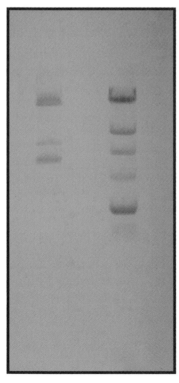

10-4 RESTRICTION DIGEST GEL ✦ This gel contains two lanes from a restriction digest. The bands on the right are from the DNA standard fragments (Markers). The bands on the left are from a DNA sample cleaved by one enzyme. In this photo the wells are at the top and migration was downward.

these cohesive ends or **cos sites** to form bonds can reduce the number of fragments desired. Therefore, a short period of heating must be done immediately prior to electrophoresis to minimize this activity.

✦ Application

Restriction enzymes are used extensively in genetic engineering.

✦ In This Exercise

You will be given three samples of DNA and two restriction enzymes (*Eco*RI and *Bam*HI) with which to run a restriction digestion. DNA 1 and 2 are plasmid DNA. DNA 3 is Lambda phage linear DNA. When your digestion is completed, you will run your samples, along with DNA standards (controls), in agarose electrophoresis and use the band migration distance to determine the DNA fragment sizes.

✦ Materials

Per Class

✦ one 37°C water bath

✦ one 65°C water bath

✦ microcentrifuge

Per Student Group

+ ice bath
+ pipettes and sterile tips to handle volumes from 5 μL to 35 μL
+ five reaction tubes
+ tubes containing:
  + DNA 1 (on ice)
  + DNA 2 (on ice)
  + DNA 3 (on ice)
  + *Eco*RI restriction enzyme (on ice)
  + *Bam*HI restriction enzyme (on ice)
  + Buffer
  + 10× gel loading solution
  + enzyme grade ultrapure water

Procedure

(Overall time ~ 2 to 4 hours)

Preparation and Digestion Reaction (~ 1 hour: 30 minutes)

Follow the procedural diagram in Figure 10-5.

1 Using the table on the Data Sheet, plan your digestion reactions. (**Note:** Although you can use any combination of restriction enzymes and DNA samples, it is best not to make it too complicated. Too many small fragments may be difficult to resolve as distinct bands. Large fragments will produce fewer bands, but will contain more biomass and stain more heavily.)

2 Calculate the amount of water required in each reaction tube and enter the data in Table 1 on the data sheet.

3 Obtain all materials including the number of reaction tubes (up to five) you have chosen to use. Place the reaction tubes in a rack along with the DNA standard marker tube. Label the tubes according to your plan entered in the table.

4 Using the plan you set up in the table, add water, buffer, and DNA to the reaction tubes. ALWAYS use a clean pipette for each transfer.

5 Add the appropriate enzyme to each tube and cap tightly. Tap the tubes or vortex them GENTLY to mix the ingredients.

6 Place the tubes in a microcentrifuge for 20 seconds to force the mixture to the bottom of the tube. Be sure to balance the centrifuge before starting the spin cycle.

7 Place the tubes in the 37°C water bath and incubate them for 30 to 60 minutes.

8 When the incubation is complete, remove the tubes and add the 5 μL of 10× gel loading solution to each one. Cap and mix gently.

Gel Electrophoresis (20 minutes to 2 hours, depending on gel size and voltage used)

1 Properly orient and place the gel bed in the electrophoresis chamber. Remember, the DNA placed in the wells will "run toward red"; therefore, the wells must always be positioned on the negative end opposite the positive "red" terminal.

2 Add ~300 mL buffer to completely cover the gel.

3 Prior to loading the gel, place the reaction tubes in the 65°C water bath for two minutes.

4 Allow the tubes to cool for 2 or 3 minutes and then load 35 μL of each sample into its appropriate well.

5 Immediately plug the red and black leads into the electrophoresis apparatus and power supply, turn on the power supply, and set the voltage to the level established by your instructor.

6 Check to see that it is working properly by looking for bubbles rising from the electrodes.

7 When the dye has migrated approximately 4 centimeters from the well, turn off the power supply, remove the cover, disconnect the cables, and go to Appendix G, Module B—Gel Staining.

8 When finished, record your data and follow the instructions on the data sheet.

References

Edvotek, Inc. 2001. *EDVO-Kit #213 Cleavage of DNA with Restriction Enzymes* Instruction Booklet. Edvotek, Inc., PO Box 341232, Bethesda, MD, 20827, USA.

Madigan, Michael T. and John M. Martinko. 2006. Chapter 7 in *Brock Biology of Microorganisms*, 11th Ed. Pearson Prentice Hall. Upper Saddle River, NJ.

Nelson, David L. and Michael M. Cox. 2008. Chapters 8 and 9 in *Lehninger: Principles of Biohemistry*, 5th Ed. W. H. Freeman and Company, New York, NY.

Roberts, Richard J. January 1980. *Restriction and modifications enzymes and their recognition sequences. Nucleic Acids Research*, Vol 8 Number 1, p. 197. Oxford University Press. Oxford, UK.

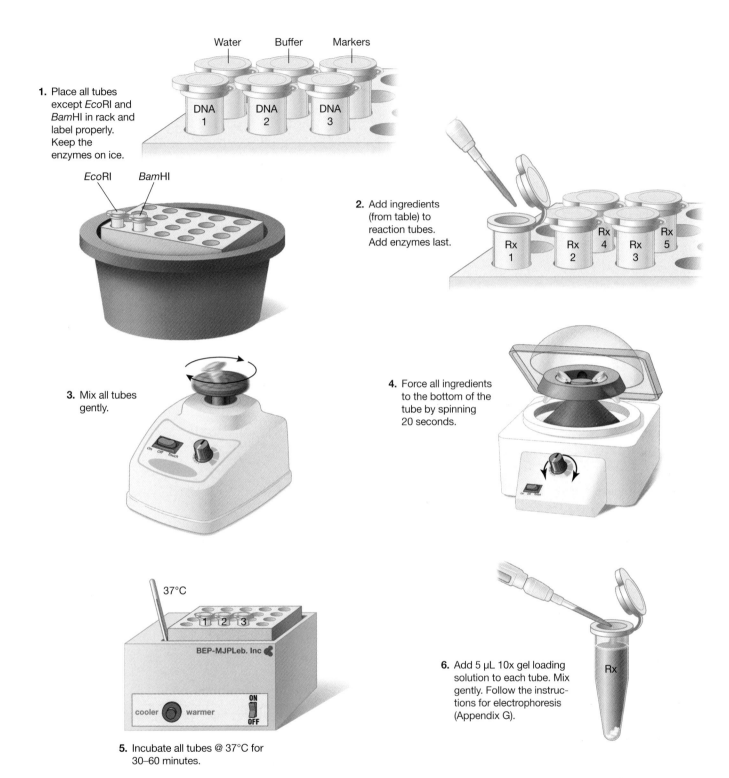

1. Place all tubes except *Eco*RI and *Bam*HI in rack and label properly. Keep the enzymes on ice.

Water Buffer Markers

DNA 1 DNA 2 DNA 3

*Eco*RI *Bam*HI

2. Add ingredients (from table) to reaction tubes. Add enzymes last.

Rx 1 Rx 2 Rx 4 Rx 3 Rx 5

3. Mix all tubes gently.

4. Force all ingredients to the bottom of the tube by spinning 20 seconds.

37°C

BEP-MJPLeb. Inc

cooler warmer ON OFF

5. Incubate all tubes @ 37°C for 30–60 minutes.

6. Add 5 µL 10x gel loading solution to each tube. Mix gently. Follow the instructions for electrophoresis (Appendix G).

Rx

10-5 PROCEDURAL DIAGRAM FOR RESTRICTION DIGEST ✦

Bacterial Transformation: the pGLO™ System

✦ Theory

This exercise utilizes a kit produced by Bio-Rad Laboratories[1] that efficiently illustrates the following principles of microbial genetics:

✦ bacterial transformation,

✦ use of an antibiotic selective medium to identify transformed cells, and

✦ the operon as a mechanism of microbial genetic regulation.

Bacterial **transformation** is the process by which **competent** bacterial cells pick up DNA from the environment and make use of the genes it carries. It was first demonstrated in a strain of pneumococcus in 1928 by Fred Griffith and since has been found to occur naturally in only certain genera. Modern techniques, however, have

[1] Catalog Number 166-0003-EDU. Bio-Rad Laboratories Main Office 2000 Alfred Nobel Drive, Hercules, CA 94547; www.bio-rad.com, 1-800-424-6723.

allowed biologists to make most cells (including eukaryotic cells) artificially competent, and this has made transformation a useful tool in genetic engineering. You will be using a CaCl$_2$ Transforming Solution and heat shock to make your *Escherichia coli* cells competent.

Green Fluorescent Protein (GFP) is responsible for **bioluminescence** in the jellyfish *Aequorea victoria*. The GFP gene was isolated and altered so the GFP in this experiment fluoresces more than the natural version. You will be introducing the GFP gene into competent *E. coli* cells and will be using the ability of the cell to fluoresce as visual evidence of successful transformation and subsequent gene expression.

Operons are structural and functional genetic units of prokaryotes. Each operon minimally includes a **promoter site** (for binding **RNA polymerase**) and two or more **structural genes** coding for enzymes in the same metabolic pathway. In this exercise you will be using part of the arabinose operon.

The complete arabinose operon (Figure 10-6) consists of the promoter (P$_{BAD}$) and three structural genes

10-6 THE ARABINOSE OPERON—NORMAL AND GENETICALLY ENGINEERED ✦ Operons are prokaryotic structural and functional genetic units: They carry genes for enzymes in the same metabolic pathway, and they are regulated together. Operons with genes for catabolic enzymes are transcribed only when the specific substrate is present. In this case, the substrate is the sugar arabinose. **A** The arabinose operon consists of a promoter (P$_{BAD}$) and three structural genes (*ara*B, *ara*A, and *ara*D). The DNA binding protein (*ara*C) attaches to the promoter and acts like a switch. Without arabinose present, RNA polymerase is unable to bind to the promoter and begin transcription of the genes. **B** and **C** Arabinose binds to a receptor on *ara*C and causes it to change to a shape that allows binding of RNA Polymerase to the promoter. **D** Transcription of the structural genes occurs, the enzymes are produced, and arabinose is catabolized. **E** The pGLO™ plasmid has been engineered to contain the arabinose promoter and the gene for Green Fluorescent Protein (GFP) instead of the genes for arabinose catabolism. If arabinose is present, the GFP gene will be transcribed. In addition, the pGLO™ plasmid has a replication origin and genes for antibiotic resistance and the DNA binding protein.

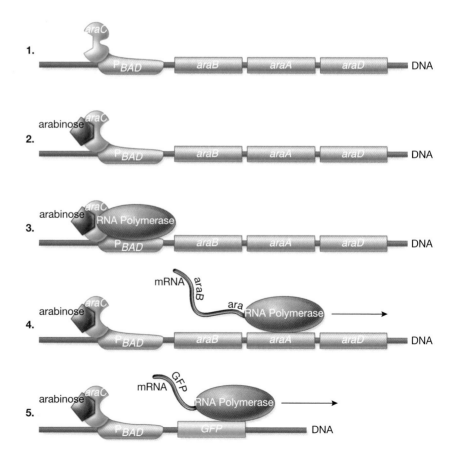

(*ara*B, *ara*A, and *ara*D) that code for enzymes used in arabinose digestion. *In vivo,* the arabinose enzymes are needed only when arabinose is present. After all, there is no point in the cell expending a lot of energy making the enzymes if the substrate is not there. A DNA binding protein called *ara*C attaches to the promoter of the arabinose operon and acts like a switch.

When arabinose is not present, *ara*C prevents RNA polymerase from binding to the promoter, so transcription can't occur—the switch is "off." When arabinose is present, it binds to *ara*C and changes its shape so RNA polymerase *can* bind to the promoter and transcribe the genes—the switch is "on." Then this sequence of events occurs: The genes are transcribed, the enzymes are synthesized, they catalyze their reactions, and eventually the arabinose is consumed. Now, without arabinose *ara*C returns to its "off" shape and the genes are no longer transcribed. In this exercise you will be using the regulatory portion of the arabinose operon (the arabinose promoter and the *ara*C gene), but the structural genes have been replaced with the GFP gene.

All that remains is a means of carrying the GFP gene into the cell and replicating it—a **vector**. In this exercise you will be using a **plasmid** as a vector: the pGLO™ plasmid. Plasmids are small, naturally occurring, circular DNA molecules that possess only a few genes and replicate independent of the chromosome (because they have their own replication origin). Although they are non-essential, they often carry genes that are beneficial to the bacterium, such as antibiotic resistance. The antibiotic resistance gene (*bla*) used in this experiment produces an enzyme called β-lactamase, which hydrolyzes certain antibiotics, including penicillin and ampicillin.

The bottom line is this: The pGLO™ plasmid used in this experiment has been genetically engineered to contain the arabinose promoter (P_{BAD}), the gene for *ara*C, an antibiotic-resistance gene (*bla*), the gene for Green Fluorescent Protein (GFP), and a replication origin (*ori*).

As you interpret the results of your experiment, you will see how these components come together and provide you with information about what is happening at the molecular level in your *E. coli* culture. Their functions are summarized in Table 10-2.

✦ Application

Introduction of foreign DNA into a cell, identification of transformed cells, and regulation of expression of the introduced genes are skills used in genetic engineering.

TABLE **10-2** Cast of Characters and a Legend of Abbreviations

Name	Symbol	Function in This Experiment
Green Fluorescent Protein	GFP	Used an indicator of gene transcription in this experiment.
Plasmid	pGLO™	Used as a vector to introduce the GFP gene into recipient *E. coli* cells.
Arabinose promoter	P_{BAD}	It is the attachment site for RNA polymerase during transcription of the GFP gene on the pGLO™ plasmid. Normally, it is the attachment site for RNA polymerase during transcription of the *ara*B, *ara*A, and *ara*D genes.
DNA binding protein	*ara*C	It is a regulatory molecule for the arabinose promoter. In the presence of arabinose, *ara*C has a shape that allows RNA polymerase to bind to the promoter. Without arabinose, *ara*C's shape prevents RNA polymerase from binding to the promoter.
Antibiotic resistance gene	*bla*	Produces β-lactamase, an enzyme that hydrolyzes certain antibiotics, including ampicillin. The gene provides a means of differentiating cells that were transformed and those that were not.
Replication origin	*ori*	Necessary for DNA replication. The pGLO™ plasmid is capable of replicating inside the cell because it has this. Because of replication, copies of the plasmid can be distributed to the descendants of the original, transformed *E. coli* cell.

✦ In This Exercise

You will first make *E. coli* cells competent. Then, you will transform the competent *E. coli* cells with a plasmid containing the gene for Green Fluorescent Protein. This technique of introducing a plasmid into cells is done routinely in genetic engineering protocols. You also will use arabinose as a genetic switch to regulate expression of the GFP gene by *E. coli*.

✦ Materials

Per Student Group

One Bio-Rad pGLO Transformation kit contains enough material for eight student workstations. Each workstation requires the following:

- ✦ one Luria-Bertani (LB) agar plate of *E. coli*
- ✦ one sterile LB plate
- ✦ two sterile LB + ampicillin plates
- ✦ one sterile LB + ampicillin + arabinose plate
- ✦ one vial of $CaCl_2$ transformation solution
- ✦ LB Broth
- ✦ seven disposable inoculating loops
- ✦ five disposable calibrated transfer pipettes
- ✦ one foam microtube holder/float
- ✦ one container of crushed ice
- ✦ one marking pen

Per Class

In addition, a class supply of the following is required:

- ✦ hydrated pGLO™ plasmid
- ✦ 42°C water bath and thermometer
- ✦ long-wave UV lamp

Procedure

Lab One

Refer to the procedural diagram in Figure 10-7 as you read and perform the following procedure.

1. Obtain two closed microtubes. Label them with your group's name, then label one "+DNA" and the other "−DNA." Put both tubes in the microtube holder/float.

2. Using a sterile calibrated transfer pipette, dispense 250 µL of $CaCl_2$ Transformation Solution into each tube. Close the caps. (The calibration marks are shown in Figure 10-7.)

3. Return the two tubes to the microtube holder/float and place them in the ice bath.

4. With a sterile loop, transfer one entire *E. coli* colony into the +DNA tube. Agitate the loop until all the growth is off of it and the cells are dispersed uniformly in the Transformation Solution. Close the lid and properly dispose of the loop.

5. Repeat Step 4 with a sterile loop and the −DNA tube.

6. Hold the UV light next to the vial of pGLO plasmid solution. Record your observation on the Data Sheet.

7. Using a sterile loop, remove a loopful of pGLO plasmid DNA solution. Be sure there is a film across the loop. Then transfer the loopful of plasmid solution to the +DNA tube and mix.

8. Leave the tubes on ice. Make sure they are far enough down in the microtube holder/float that they make good contact with the ice. Leave them on ice for 10 minutes.

9. As the tubes are cooling on ice for 10 minutes, label the four LB agar plates.
   - ◆ Label one LB/amp plate: +DNA
   - ◆ Label the LB/amp/ara plate: +DNA
   - ◆ Label the other LB/amp plate: −DNA
   - ◆ Label the LB plate: −DNA

10. After 10 minutes on ice, transfer the microtube holder/float (still containing both tubes) to the 42°C water bath for exactly 50 seconds. This transfer from ice to warm water must be done rapidly. Also, make sure the tubes make good contact with the water.

11. After 50 seconds, quickly place both tubes back in the ice for 2 minutes. This process of heat shock makes the cell membranes more permeable to DNA. (Timing is critical. According to the kit's manufacturer, 50 seconds is optimal. No heat shock results in a 90% reduction in transformants, whereas a 90 second heat shock yields about half the transformants.)

12. After 2 minutes, remove the microtube holder/float from the ice and place it on the table.

13. Using a sterile calibrated transfer pipette, add 250 µL of LB broth to the +DNA tube. Repeat with another sterile pipette and the −DNA tube. Properly dispose of both pipettes.

14. Incubate both tubes for 10 minutes at room temperature. Then mix the tubes by tapping them with your fingers.

15 Using a different sterile calibrated transfer pipette for each, inoculate the "LB/amp/+DNA" plate and the "LB/amp/ara/+DNA" plate with 100 μL from the +DNA tube.

16 Using a different sterile calibrated transfer pipette for each, inoculate the "LB/amp/−DNA" plate and the "LB/−DNA" plate with 100 μL from the −DNA tube.

17 Using a different sterile loop for each, spread the inoculum over the surface of all four plates to get confluent growth. Use the "face" of the loop for this, not its edge. Properly dispose of the loops.

18 Tape the plates together in a stack so they face the same direction. Then label them with your group name and the date, and incubate them for 24 hours at 37°C in an inverted position.

Lab Two

1 Retrieve your plates. Observe them in ambient room light, then in the dark with UV illumination.

1. Label one microtube "+DNA" and the other "−DNA."

2. Place the tubes in the foam microtube rack.

250 μL Transforming solution

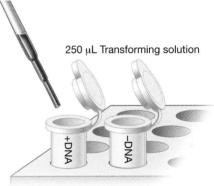

3. Use a sterile calibrated transfer pipette to dispense 250 μL Transformation solution into each tube. Close the caps.

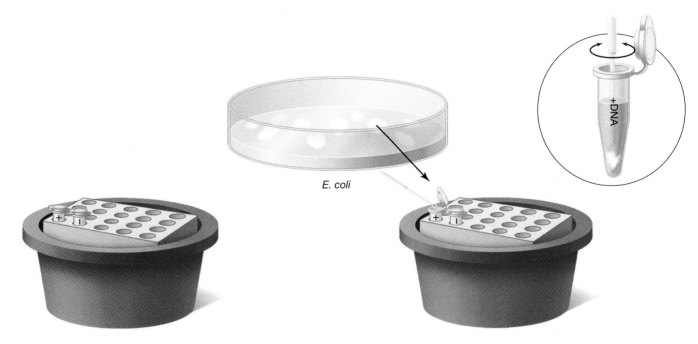

E. coli

4. Place the microtube rack on the ice bath. Push the tubes down into the ice.

5. Transfer one E. coli colony to each tube using a different sterile loop.
Agitate the loop to remove all growth (see detail). Close the caps and properly dispose of the loops.

10-7 PROCEDURAL DIAGRAM ✦ Be sure to dispose of all pipettes and loops properly. (Continued on next page.)

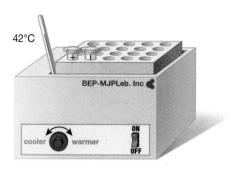

42°C

BEP-MJPLeb. Inc

cooler ⟷ warmer   ON OFF

6. Transfer one loopful of pGLO™ plasmid DNA to the "+DNA" tube only. Continue icing both tubes for 10 minutes.

7. Quickly transfer the entire microtube rack to the 42°C water bath for exactly 50 seconds. Make sure the tubes contact the water.

8. Quickly place the microtube rack back on ice for two minutes.

250 µL Luria-Bertani Broth

9. Place the microtube rack on the table.

10. Add 250 µL LB broth to each tube with a different sterile transfer pipette.

11. Incubate the tubes for 10 minutes at room temperature.

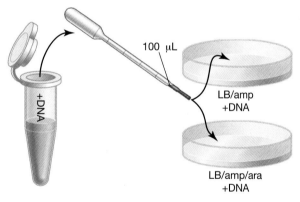

100 µL

LB/amp +DNA

LB/amp/ara +DNA

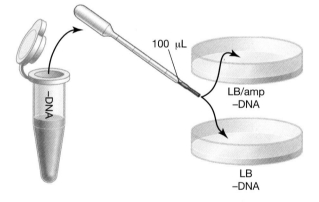

100 µL

LB/amp –DNA

LB –DNA

12. Using a different pipette for each, transfer 100 µL of +DNA to an LB/amp and an LB/amp/ara plate. Using a different sterile loop for each, spread the inoculum for confluent growth. Use the "face" of the loop for this, not its edge. Incubate for 24 hours at 37°C.

13. Using a different pipette for each, transfer 100 µL of –DNA to an LB/amp and an LB plate. Using a different sterile loop for each, spread the inoculum for confluent growth. Use the "face" of the loop for this, not its edge. Incubate for 24 hours at 37°C.

10-7 PROCEDURAL DIAGRAM (CONTINUED) ✦ Be sure to dispose of all pipettes and loops properly.

Record your observations on the Data Sheet, and answer the questions.

2 When finished, properly dispose of all plates.

Reference

Bio-Rad Laboratories. Instruction pamphlet for the *Bacterial Transformation—The pGLO™ System* kit (Catalog Number 166-0003-EDU). Bio-Rad Laboratories, Hercules, CA.

EXERCISE 10-4

Polymerase Chain Reaction[1]

✦ Theory

The **polymerase chain reaction (PCR)** is a technique conceived by the 1983 Chemistry Nobel Laureate Kary Mullis. This process is designed to make multiple copies of (amplify) a desired gene or other short DNA fragment. In the process, the double-stranded DNA sequence to be amplified (the **template**) is separated with heat and then replicated using free nucleotides, two commercially prepared **primers**, and a polymerase. The primers (short nucleic acid molecules) are selected to be complementary to positions on opposite strands of the DNA molecule. The points at which they attach flank the area to be replicated. The free nucleotides are thus able to attach to complementary bases on the template and make the desired copy (Figure 10-8).

The PCR process involves 20 to 40 cycles of three different activities: **denaturation** (DNA strand separation), **annealing** (of primers), and **extension** (DNA replication). In denaturation, the temperature of the reaction tube is raised to 92–96°C for one to a few minutes to separate the DNA strands. Following this, the temperature is lowered to 45–65°C, which allows the primers to anneal to their complementary sequences on opposite template strands. For the extension phase, the temperature is raised to 72°C, for a few seconds to a few minutes, to allow the polymerase to synthesize DNA complementary to the template (elongate primer).

This completes the first cycle of amplification. The process then repeats itself, but this time there are four template strands to replicate and the result is eight strands of DNA (four dsDNA molecules). By continuing the process and using the products of one cycle as templates for the next cycle, the number of DNA target strands doubles each cycle and can be amplified a million-fold, all in a few hours. (For 20 cycles, each strand will be replicated $2^{20} = 1,048,576$ times!) As mentioned above, two primers are used in the reaction, but they must be added to the PCR reaction mixture in great excess to flood the mixture and promote primer-DNA annealing and discourage DNA-DNA re-annealing. Other essential

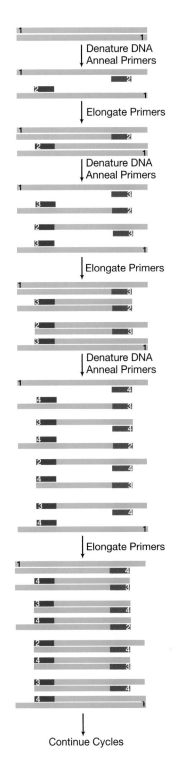

10-8 SCHEMATIC DIAGRAM OF PCR ✦ Each PCR cycle doubles the amount of DNA in the sample. The numbers on the strands give generations of DNA. In this example, three replication cycles are shown, in which 2 original strands are amplified to a total of 16 strands.

---

[1] This exercise is adapted, with permission, from Edvotek® Kit #330—The Molecular Biology of DNA Amplification by Polymerase Chain Reaction, for which the company holds copyrights. All necessary materials and equipment, including the EdvoCycler™, are available from Edvotek, Inc., PO Box 341232, Bethesda, MD, 20827; email: info@edvotek.com www.edvotek.com.

components of the mixture include sufficient quantities of free deoxyribonucleotides (dATP, dCTP, dTTP, and dGTP) needed for extension, buffers to maintain the appropriate pH, magnesiuim required by the polymerase, and other salts to create the proper osmotic balance. Because of the high temperatures required for the denaturation phase, the thermotolerant *Taq*DNA polymerase, isolated from the thermophile, *Thermus aquaticus*, is used to catalyze the extension.

Depending on the objectives of the experiment, the PCR product may be used in agarose or polyacrylamide electrophoresis (Figure 10-9). The isolated DNA in the gel can then be sequenced or, as in this exercise, be used to determine the size (in base pairs) of the target DNA (Figure 1 on the Data Sheet). For more about gel electrophoresis, turn to Appendix G, page 459.

---

[2] This exercise includes agarose gel electrophoresis. Depending on the time available in your lab, your instructor may or may not include gel preparation as part of the exercise. In addition, because gel preparation, electrophoresis, and gel staining are included in more than one exercise, only the electrophoresis instructions are included in this procedure. A short introduction to gel electrophoresis, agarose gel preparation, and various staining procedures appear in Appendix G.

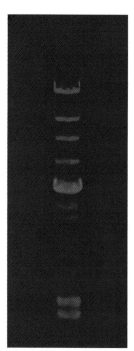
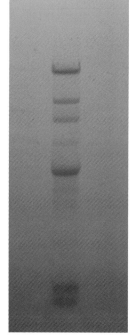

10-9 TWO STAINED GELS ✦ The gel on the left was stained with Edvotek Instastain® EtBr, as viewed with UV light. The gel on the right was stained with Edvotek Instastain® Blue and destained with water. It is viewed on a light box. Both gels show only the bands formed by standard DNA fragments. The relative position of the bands produced by your PCR product can be used to calculate the size of the fragments in base pairs. See Figure 1 on the Data Sheet and Figure G-1 in Appendix G for more information.

✦ Application

The polymerase chain reaction (PCR) is a relatively simple and convenient method of amplifying DNA. Once multiple copies of the DNA are made, they can be used in various ways ranging from research to diagnostics and forensics.

✦ In This Exercise

You will be programming a thermal cycler, preparing a PCR reaction mixture, and running the PCR. Following the PCR exercise, you will have three different samples of DNA—one that has not undergone PCR, one that has undergone 15 PCR cycles, and one that has undergone 30 PCR cycles.

✦ Materials

Per Class

✦ one thermal cycler

Per Student Group

✦ container of ice
✦ microcentrifuge
✦ pipettes and sterile tips to handle volumes from 2 μL to 35 μL
✦ tip disposal container
✦ three 0.5 mL tubes
✦ one PCR reaction tube
✦ containers with:
     ◆ PCR pellet
     ◆ 10× gel loading solution
     ◆ primer mix
     ◆ enzyme grade ultrapure water

Prelab Procedure
(~20 minutes)

| Program your thermal cycler for 30 cycles as shown in the table below. Depending on your thermal cycler, you will either set the final extension at 5 minutes or add a 5-minute post-extension period. Your instructor will show you how this is done.

Activity	Temperature	Time
Denaturing	94°C	45 seconds
Annealing	45°C	45 seconds
Extension	72°C	45 seconds
Final extension	72°C	5 minutes
Hold	4°C	indefinite

2 Clear your counter top of all unnecessary items and disinfect it.

3 Wash and dry your hands and put on appropriate gloves.

 Procedure

(Overall time ~2 to 4 hours)

Preparation of PCR Reaction Mixture (~30 minutes)

Follow the PCR procedural diagram in Figure 10-10.

1 Obtain all materials for your group and label three 0.5 mL tubes "0", "15", and "30."

2 Label the PCR reaction tube with your name or group name.

3 Using a different pipette tip for each transfer, add the following to tube 0:
  - 10 µL enzyme grade ultrapure water
  - 2 µL DNA template for amplification
  - 2 µL primer mix
  - 5 µL 10× gel loading solution

4 Place this mixture on ice until time for electrophoresis.

5 Transfer the PCR Reaction Pellet™ to the PCR reaction tube.

6 Tap the PCR tube on the counter top a couple of times to make sure the pellet is on the bottom.

7 Add the following:
  - 7 µL enzyme grade ultrapure water
  - 10 µL DNA template for amplification
  - 10 µL primer mix

8 Mix the contents of the tube by gently vortexing it or tapping it on the counter top. Immediately centrifuge it for 10–20 seconds to collect all of the contents at the bottom. (**Note:** Always make sure the tubes are firmly closed and the centrifuge contents are properly balanced before starting a spin cycle.)

PCR (~1 hour: 30 minutes)

1 If your thermal cycler has a heated cover, place your PCR tube inside as shown by your instructor. If your thermal cycler does not have a heated cover, add a wax bead to the PCR tube and then place it inside the cycler.

2 After completing 15 cycles, pause the thermal cycler, remove the PCR tube, and transfer 7 µL of the reaction mixture to tube 15.

(**Note:** If you have added wax to your reaction tube, place it on ice until the wax has solidified. Gently break through with a clean pipette tip. Once the wax has been broken, replace the tip with a new sterile one and transfer 7 mL to tube 15.)

3 Return the PCR tube to the thermal cycler and continue the program.

4 Add 5 µL of 10× Gel Loading solution and 8 µL of enzyme grade water to tube 15. Mix gently and store on ice until time for electrophoresis.

5 After the 30th cycle has completed (including the 5 minute elongation), transfer 7 µL of the reaction mixture to tube 30.

6 Add 5 µL of 10× Gel Loading solution and 8 µL of enzyme grade water to tube 30. Mix gently and store the tube on ice until time for electrophoresis.

Gel Electrophoresis (20 minutes to 2 hours, depending on gel size and voltage used)

Follow the Gel Electrophoresis procedural diagram in Figure G-1 in Appendix G.

1 Properly orient and place the gel bed in the electrophoresis chamber. Remember—the DNA placed in the wells will "run toward red"; therefore, the wells must always be positioned on the negative end opposite the positive "red" terminal.

2 Add ~300 mL buffer to completely cover the gel.

3 Load 20 µL of each sample in its appropriate well. See the chart below:

Gel Loading Instructions		
Lane	**Tube**	**Contents**
1	Marker	Standard DNA Fragments
2	0	Control sample, 0 cycles
3	15	Reaction sample after 15 cycles
4	30	Reaction sample after 30 cycles

4 Properly place the cover on the electrophoresis apparatus and plug the red and black leads into the terminals.

5 Plug the red and black leads into the like-colored terminals of the power supply.

6 Turn on the power supply and set the voltage to the level established by your instructor.

7 Check to see that it is working properly by looking for bubbles rising from the electrodes.

8 After electrophoresis is completed, turn off the power supply, remove the cover, and disconnect the cables.

9 To stain your gel, go to Appendix G, Module B— Gel Staining.

10 When finished, record your data and follow the instructions on the Data Sheet.

References

Edvotek, Inc. 2008. *EDVO-Kit #330 The Molecular Biology of DNA Amplification by Polymerase Chain Reaction* Instruction Booklet. Edvotek, Inc., PO Box 341232, Bethesda, MD, 20827, USA.

Madigan, Michael T. and John M. Martinko. 2006. Chapter 7 in *Brock Biology of Microorganisms,* 11th Ed. Pearson Prentice Hall. Upper Saddle River, NJ.

Nelson, David L. and Michael M. Cox. 2008. Chapters 8 and 9 in *Lehninger: Principles of Biohemistry,* 5th Ed. W. H. Freeman and Company, New York, NY.

1. Label three 0.5 mL tubes and one PCR reaction tube as shown.

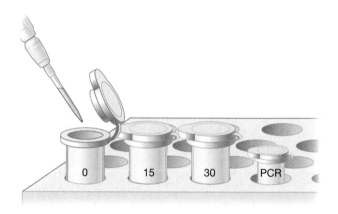

2. Add 10 μL H_2O, 2 μL DNA template, 2 μL primer mix, and 5 μL 10x gel loading solution to tube 0.

3. Place tube "0" on ice.

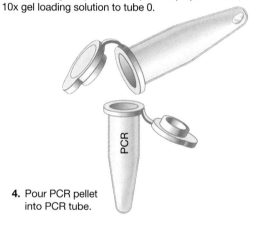

4. Pour PCR pellet into PCR tube.

5. Add 7 μL H_2O, 10 μL DNA template, and 10 μL primer mix.

6. Mix by tilting, tapping or vortexing gently.

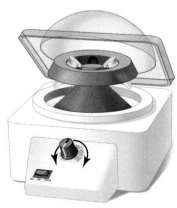

7. Centrifuge for 10–20 seconds.

10-10 PCR PROCEDURAL DIAGRAM ✦

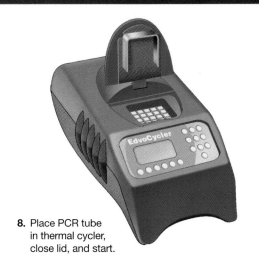

8. Place PCR tube in thermal cycler, close lid, and start.

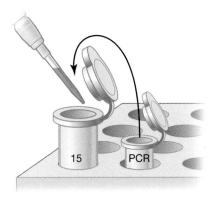

9. After 15 cycles, pause thermal cycler and transfer 7 µL to tube 15.

10. Place tube 15 on ice.

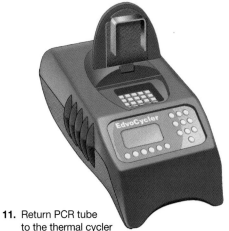

11. Return PCR tube to the thermal cycler and continue program.

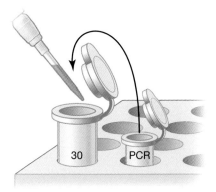

12. After completing the last cycle, transfer 7 µL to tube 30.

13. Place tube 30 on ice.

10-10 PCR PROCEDURAL DIAGRAM (CONTINUED) ✦

Ultraviolet Radiation Damage and Repair

✦ Theory

Ultraviolet radiation is part of the electromagnetic spectrum, but with shorter, higher energy wavelengths than visible light. Prolonged exposure can be lethal to cells because when DNA absorbs UV radiation at 254 nm, the energy is used to form new covalent bonds between adjacent pyrimidines: cytosine-cytosine, cytosine-thymine, or thymine-thymine. Collectively, these are known as **pyrimidine dimers,** with thymine dimers being the most common. These dimers distort the DNA molecule and interfere with DNA replication and transcription (Figure 10-11).

Many bacteria have mechanisms to repair such DNA damage. *Escherichia coli* performs **light repair** or **photoreactivation,** in which the repair enzyme, **DNA photolyase,** is activated by visible light (340–400 nm) and simply monomerizes the dimer by reversing the original reaction.

A second *E. coli* repair mechanism, **excision repair** or **dark repair,** involves a number of enzymes (Figure 10-12). The thymine dimer distorts the sugar-phosphate backbone of the strand. This is detected by an **endonuclease** (UvrABC) that breaks two bonds—eight nucleotides in the 5' direction from the dimer, and the other four nucleotides in the 3' direction. A **helicase** (UvrD) removes the 13-nucleotide fragment (including the dimer), leaving single-stranded DNA. **DNA polymerase I** inserts the

appropriate complementary nucleotides in a 5' to 3' direction to make the molecule double-stranded again. Finally, **DNA ligase** closes the gap between the last nucleotide of the new segment and the first nucleotide of the

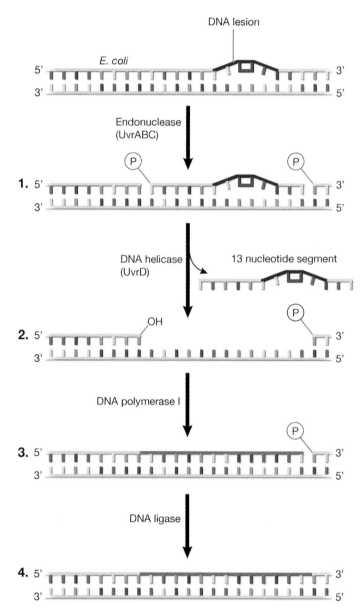

10-12 EXCISION OR DARK REPAIR IN *E. COLI* ✦ In the repair process, four enzymes are used: (1) An endonuclease (UvrABC) to break two covalent bonds in the sugar-phosphate backbone of the damaged strand, (2) a helicase (UvrD) to remove the nucleotides in the damaged segment, (3) DNA polymerase I to synthesize a new strand, and (4) DNA ligase to form a covalent bond between the new and the original strands.

10-11 A THYMINE DIMER IN ONE STRAND OF DNA ✦ Thymine dimers form when DNA absorbs UV radiation with a wavelength of 254 nm. The energy is used to form two new covalent bonds between the thymines, resulting in distortion of the DNA strand. The enzyme DNA photolyase can break this bond to return the DNA to its normal shape and function. If it doesn't, and excision repair fails, the distortion interferes with DNA replication and transcription.

old DNA, and the repair is complete. Both mechanisms are capable of repairing a small amount of damage, but long and/or intense exposures to UV produce more damage than the cell can repair, making UV radiation lethal.

✦ Application

Because ultraviolet radiation has a lethal effect on bacterial cells, it can be used in decontamination. Its use is limited, however, because it penetrates materials such as glass and plastic poorly. In addition, bacterial cells have mechanisms to repair UV damage.

✦ In This Exercise

The lethal effects of ultraviolet radiation, its ability to penetrate various objects, and the cells' ability to repair UV damage will all be demonstrated. Do not be misled by the apparent simplicity of the experimental design—there's a lot going on!

✦ Materials

Per Student Group

✦ seven Nutrient Agar plates

✦ disinfectant

✦ posterboard masks with 1" to 2" cutouts (Figure 10-13)

✦ UV light (shielded for eye protection)

✦ sterile cotton applicators

✦ 24-hour trypticase soy broth culture of *Serratia marcescens*

 Procedure

Lab One

Refer to the Procedural Diagram in Figure 10-14 as you read and follow this procedure:[1]

1 Using a sterile cotton applicator, streak a Nutrient Agar plate to form a bacterial lawn over the entire surface by using a tight pattern of streaks. Rotate the plate one-third of a turn and repeat, then rotate it another one-third of a turn and streak one last time. Repeat this process for all remaining plates.

2 Number the plates 1, 2, 3, 4, 5, 6, and 7.

3 Remove the lid from plate 1 and set it on a disinfectant-soaked towel. Place the plate under the UV light and cover it with a mask.

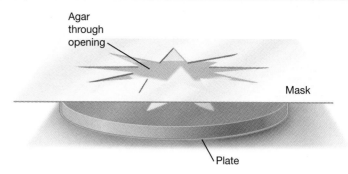

10-13 POSTERBOARD MASK ✦ This is an example of a mask placed over a Petri dish. The cutout may be any shape but should leave the outer 25% of the plate masked.

4 Turn the UV lamp on, but do not look at it. After 30 seconds, turn off the UV lamp, remove the mask, and replace its lid. If space permits, you may combine this step with Step 5.

5 Repeat the process for plate 2. If space permits, you may combine this step with Step 4.

6 Repeat the process for plate 3, but leave the UV lamp on for 3 minutes. If space permits, you may combine this step with Steps 7 and 8.

7 Irradiate plate 4 for 3 minutes, but leave the lid on and cover with the mask.

8 Repeat Step 7 for plate 5. If space permits, you may combine this step with Step 7.

9 Do not irradiate plates 6 and 7.

10 Incubate plates 1, 3, 4, and 6 for 24 to 48 hours at room temperature in an inverted position where they can receive natural light (*e.g.*, a windowsill) for 24–48 hours. Do not stack them.

11 Wrap plates 2, 5, and 7 in aluminum foil, invert them, and place with the others for 24–48 hours.

Lab Two

1 Remove the plates and examine the growth patterns.

2 Record your results on the Data Sheet.

References

Moat, Albert G., John W. Foster, and Michael P. Spector. 2002. Chapter 3 in *Microbial Physiology,* 4th ed. Wiley-Liss, New York.

Nelson, David L., and Michael M. Cox. 2005. Chapter 25 in *Lehninger's Principles of Biochemistry.* W. H. Freeman and Co., New York.

White, David. 2000. Chapter 19 in *The Physiology and Biochemistry of Prokaryotes.* Oxford University Press, Inc. New York.

[1] Thanks to Roberta Pettriess of Wichita State University for her helpful suggestions to improve this exercise.

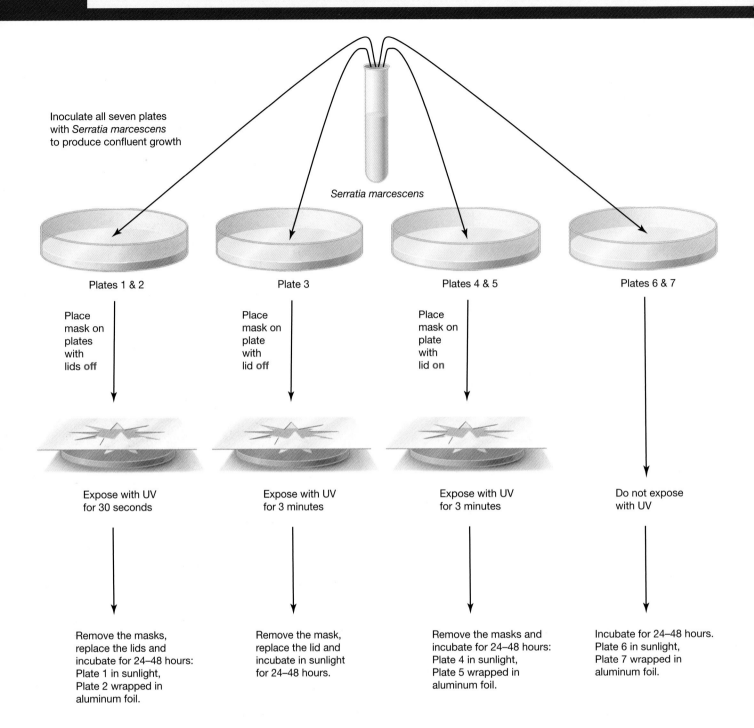

Inoculate all seven plates with *Serratia marcescens* to produce confluent growth

Serratia marcescens

Plates 1 & 2

Plate 3

Plates 4 & 5

Plates 6 & 7

Place mask on plates with lids **off**

Place mask on plate with lid **off**

Place mask on plate with lid **on**

Expose with UV for 30 seconds

Expose with UV for 3 minutes

Expose with UV for 3 minutes

Do not expose with UV

Remove the masks, replace the lids and incubate for 24–48 hours: Plate 1 in sunlight, Plate 2 wrapped in aluminum foil.

Remove the mask, replace the lid and incubate in sunlight for 24–48 hours.

Remove the masks and incubate for 24–48 hours: Plate 4 in sunlight, Plate 5 wrapped in aluminum foil.

Incubate for 24–48 hours. Plate 6 in sunlight, Plate 7 wrapped in aluminum foil.

10-14 PROCEDURAL DIAGRAM ✦ Inoculate and expose the plates as directed. Be sure to shield the UV light source adequately. Do not look at the light.

EXERCISE
10-6 Ames Test

✦ Theory

Bacteria that are able to enzymatically synthesize a particular necessary biochemical (such as an amino acid) are called **prototrophs** for that biochemical. If the bacterial strain can make the amino acid histidine, for instance, it is classified as a histidine prototroph. Bacteria that must be supplied with that biochemical (*e.g.*, histidine) are called (histidine) **auxotrophs.** Auxotrophs typically are created from prototrophs when a mutation occurs in the gene coding for an enzyme used in the pathway of the biochemical's synthesis.

The Ames Test employs mutant strains of *Salmonella typhimurium* that have lost their ability to synthesize histidine. One strain of histidine auxotrophs possesses a **frameshift mutation;** they are missing one or have an extra nucleotide in the DNA sequence that otherwise would code for an enzyme necessary for histidine production. Other strains are **substitution mutants,** in which one nucleotide in the histidine gene has been replaced, resulting in a faulty gene product.

The Ames Test determines the ability of chemical agents to cause a reversal—a **back mutation**—of these auxotrophs to the original prototrophic state. In this test, histidine auxotrophs are spread onto **minimal agar plates** that contain all nutrients for growth but only a trace of histidine.

When a filter paper disk saturated with a suspected mutagen is placed in the middle of the minimal agar plate, the substance will diffuse outward into the medium. If it is mutagenic, it will cause back mutation in some cells (converting them to histidine prototrophs) that freely grow into full-size colonies. (**Note:** Histidine initially is included in the medium to allow the auxotrophs to grow for several generations and expose them to the effects of the mutagen. The unmutated auxotrophs typically exhibit faint growth but do not develop into full-size colonies because of the rapid exhaustion of the histidine.) A sample plate is shown in Figure 10-15.

Several variations of the Ames test are possible. This exercise uses two minimal agar plates and two **complete agar plates.** Complete agar contains all of the nutrients necessary for growth of *Salmonella.* All four plates are inoculated with histidine auxotrophs. One minimal agar plate and one complete agar plate each receive a filter paper disk saturated with the test substance. The second minimal agar plate and complete agar plate each receive

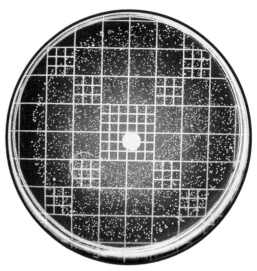

10-15 TESTING FOR MUTAGENICITY ✦ In this example, a *Salmonella* histidine auxotroph was grown on histidine minimal agar while exposed to a suspected mutagen. The colonies were derived from cells that underwent back mutations. It is not known from this plate alone, however, whether they are a result of the effects of the test substance, so it must be compared to a control minimal agar plate in which a nonmutagenic substance (DMSO) is substituted for the test substance.

a filter paper disk saturated with Dimethyl Sulfoxide, DMSO (a substance known to be nonmutagenic).

The minimal agar plate containing the test substance determines mutagenicity of the test substance. Only prototrophs will grow on minimal agar, so recovery of any colonies on the minimal agar inoculated with auxotrophs is indicative of back mutation. Depending on the strain used, the type of mutation (either frameshift or base substitution) also can be determined. The minimal agar plate with DMSO serves as a control for the minimal agar/test substance plate by measuring **spontaneous back mutations** (natural mutations that occur without the presence of a mutagen).

The purpose of the complete agar plate containing the test substance is a control to evaluate toxicity of the test substance. Creation of a **zone of inhibition** around the disk indicates toxicity (Figure 10-16). The more toxic the substance is, the larger the zone will be. If the substance is toxic, there may be no indication of mutagenicity because the cells are killed before they can back-mutate. Finally, the complete agar plate containing DMSO serves as a control for comparison to the growth and zone of inhibition on the complete agar/test substance plate.

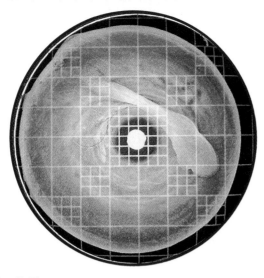

10-16 TESTING FOR TOXICITY ✦ Shown is a *Salmonella* histidine auxotroph grown on complete agar (containing histidine) and exposed to the test substance. The zone of inhibition around the paper disk indicates toxicity of the substance. (Note that a zone of inhibition is visible on the plate shown in Figure 10-15, but it is not as well defined as on complete agar because the growth is not as dense.)

✦ Application

Many substances that are mutagenic to bacteria are also carcinogenic to higher animals. The Ames Test is a rapid, inexpensive means of using specific bacteria to evaluate the mutagenic properties of potential carcinogens. Many variations of the basic Ames Test are used in specific applications.

✦ In This Exercise

You will test a household substance for its mutagenic properties. If you want the best chance of a positive result, check below your sink and in the garage for products that have many organic chemicals in the ingredient list. When handling these materials, be sure to wear gloves. Use caution in the laboratory if the material is flammable.

✦ Materials

Per Group

✦ four Minimal Medium (MM) plates

✦ four Complete Medium (CM) plates

   (***Note:*** This exercise can be run with two plates of each medium if replicates are not desired.)

✦ centrifuge

✦ two sterile centrifuge tubes

✦ small beaker containing alcohol and forceps

✦ bottle of 1× Vogel-Bonner salts (To make 1× Vogel-Bonner salts, add 1.0 mL 50× Vogel-Bonner solution to 49 mL water.) (Appendix A)

✦ eight sterile filter disks made with a paper punch

✦ two sterile Petri dishes (for soaking filter paper disks)

✦ two sterile transfer pipettes

✦ 100 µL–1000 µL digital pipettors and tips

✦ container for disposal of supernatant (to be autoclaved)

✦ DMSO

✦ test substance (any substance that has possible mutagenic properties and **does not contain histidine or protein**)

✦ hand tally counter

✦ broth culture[1] of *Salmonella typhimurium TA 1535*

✦ broth culture[1] of *Salmonella typhimurium TA 1538*

Procedure

Day One

Follow the procedural diagram in Figure 10-17.

1 Soak four filter paper disks in DMSO and four filter paper disks in the test substance.

2 Pipette 10.0 mL TA 1535 into a sterile centrifuge tube. Do the same with TA 1538, then label the tubes.

3 Centrifuge the tubes on high speed for 10 minutes. Be sure the centrifuge is balanced.

4 Being careful not to disturb the cell pellet at the bottom, decant the supernatant from each centrifuge tube using a different sterile transfer pipette.

5 Resuspend the cell pellets by adding 1.0 mL sterile 1× Vogel-Bonner salts to each tube and mixing well.

6 Using the spread plate technique (see Exercise 1-5), inoculate two MM plates and two CM plates with 100 µL of the resuspended TA 1535. Do the same with TA 1538.

---

[1] The cultures used for this exercise must be prepared as follows:

1. 24 hours before the test, inoculate two 10.0 mL broth tubes with TA 1535 and TA 1538. Incubate at 35±2°C together with two sterile 90.0 mL broths (in a 100 mL diluent bottle).

2. 5½ hours before the test, pour the TA 1538 culture into one of the sterile 90.0 mL broths and return it to the incubator until time for the exercise.

3. 4 hours before the test, pour the TA 1535 culture into the other 90.0 mL sterile broth and return it to the incubator until time for the exercise.

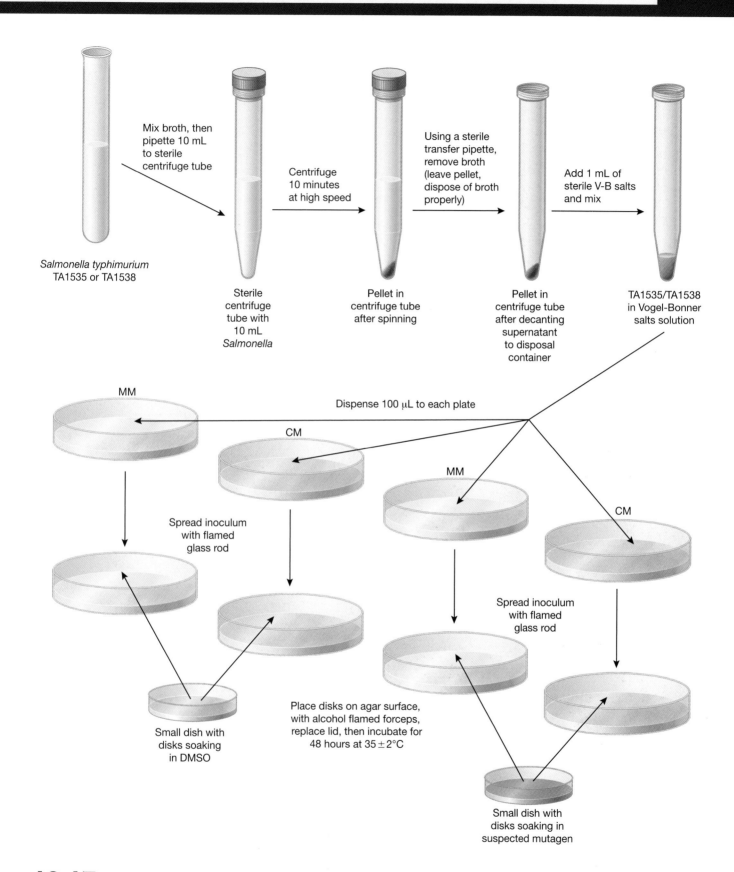

10-17 PROCEDURAL DIAGRAM ✦ Be sure to dispose of all pipettes, broth, and tubes properly as you perform the experiment. After 48 hours of incubation, use a metric ruler to measure the diameter of the cleared zone on all CM plates. Use a colony counter to count the back-mutant colonies on all the MM plates. (**Note:** These will be fairly large. Do not count the tiny "hazy" growth over the plate's surface.) Mark each colony with a toothpick (to avoid counting it more than once) as you keep track of the number using a hand tally counter.

7 Flame the forceps by passing them through the Bunsen burner flame, and allowing the alcohol to burn off.

8 Using the flamed forceps, place the DMSO and test disks in the centers of the plates. Gently tap down the disks with the forceps to prevent them falling off when the plates are inverted.

9 Incubate the plates aerobically at $35 \pm 2°C$ for 48 hours.

Day Two

1 Measure the diameters (in millimeters) of the zones of inhibition on the CM plates.

2 Count the colonies on the MM plates and compare. Count only the large colonies, not the "hazy" background growth. (These are the colonies produced by auxotrophs that did not back-mutate to prototrophs and only grew until the histidine in the minimal medium was exhausted.)

3 Record your observations for the Ames Test plates on the Data Sheet.

References

Eisenstadt, Bruce, C. Carlton, and Barbara J. Brown. 1994. Page 311 in *Methods for General and Molecular Bacteriology*, edited by Philipp Gerhardt, R. G. E. Murray, Willis A. Wood, and Noel R. Krieg, American Society for Microbiology, Washington, DC.

Maron, D. M. and B. N. Ames. 1983. *Mutation Research,* 113:173–215.

Nelson, David L., and Michael M. Cox. 2005. Chapter 25 in *Lehninger's Principles of Biochemistry.* W. H. Freeman and Co., New York.

EXERCISE 10-7

Phage Typing of *E. coli* Strains

✦ Theory

Identification of organisms is a primary function of microbiologists. The best identification methods involve utilizing genetically based differences to distinguish between organisms at every level: Domain through Species. In microbiology classical identification methods include cell morphology, staining properties, biochemical testing and serological reactions, all of which are still being used. Newer methods include comparisons of nucleic acids, such as hybridization and sequencing. **Phage typing** allows microbiologists to differentiate between strains of bacterial species based on susceptibility to infection by different viruses.

Viruses that infect bacteria are called **bacteriophages** or simply just **phages**. The infection process (see Figure 6-12) begins with attachment by the phage to a receptor on the host cell. The phage then injects its nucleic acid into the host cytoplasm and overtakes its metabolism; the host is now under viral control and becomes a factory for making a hundred or more copies of the virus. When completed, the host cell bursts and releases the viral progeny, which now infect other cells to repeat the replicative cycle. Depending on the virus, one round of replication can take as little as 25 minutes at 37°C!

The ability of a virus to infect a host has to do with its ability to attach to *specific* host receptors. Because receptor structure is genetically determined, different genetic strains of the bacterial species can be identified by their susceptibility to a particular virus. Challenging the same host with different viruses (which use different receptors) produces a composite picture of viral susceptibility. Different composite pictures indicate different strains of the *host* and provide a means of characterizing host strains based on their viral susceptibilities. This characterization can be used as a basis for identifying fresh isolates down to the level of "strain."

One way of determining host susceptibility to a particular phage is to inoculate a plate with the host so it produces confluent growth (called a "lawn"). Phage can be introduced at the same time and the plate incubated for 24–48 hours. Clear zones in the lawn are evidence that the virus has gone through multiple replicative cycles and destroyed all the host cells in that region. If no clearings are seen, then the bacterial strain is not a suitable host for that virus (Figure 10-18).

10-18 ZONE OF CLEARING AND NO CLEARING ✦ This plate was inoculated to produce a lawn of host bacteria. Two different viruses were spotted onto the plate, which was then incubated for 48 hours. Notice the clearing on the right and absence of clearing on the left. (For convenience, the places where virus was applied are circled.)

✦ Application

Phage typing is used to characterize and identify strains within a bacterial species.

✦ In This Exercise

Today you will be phage typing three strains of *Escherichia coli* (wild type, *B*, and *C*) by inoculating agar plates with each strain and with two *E. coli* bacteriophages (T$_2$ and φχ*174*).

✦ Materials

Per Student Group

✦ overnight broth cultures of
  ✦ *E. coli* (wild type)
  ✦ *E. coli* B
  ✦ *E. coli* C
✦ three Nutrient Agar or Trypticase Agar Plates
✦ vials of
  ✦ T$_2$ Phage
  ✦ Phage φχ*174*
✦ twelve sterile cotton swabs
✦ marking pen
✦ one vial of sterile distilled water

Procedure

Lab One

1 With your marking pen, divide each agar plate into three sectors. Label one sector "T_2," a second sector "φχ," and the third sector "dH_2O" (Figure 10-19).

2 Then, label one plate "Wild Type," a second plate "*E. coli* B," and the third plate "*E. coli* C."

3 Take the "Wild Type" plate and, with a sterile swab, streak it with "wild type *E. coli*" to produce a lawn. Your goal is confluent growth, so make the streaks right next to each other. When you have covered the surface, rotate the plate 1/3 turn and repeat the streaking of the inoculum already on the plate using the same technique to produce confluent growth. Then, rotate the plate another 1/3 turn and repeat. *Note:* failure to produce a lawn of confluent growth may mean your results are compromised. Take this inoculation process seriously.

4 Repeat Step 3 with the "*E. coli* B" plate and *E. coli* B culture.

5 Repeat Step 3 with the "*E. coli* C" plate and *E. coli* C culture.

6 With the wooden end of a sterile swab, transfer a *small* drop (which may mean really loading up the stick with fluid) of T_2 phage to the appropriate sector on the "Wild Type" *E. coli* plate (Figure 10-20). DO NOT touch the agar with the stick. Simply lower

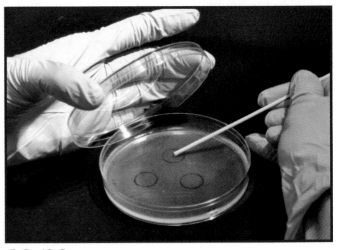

10-20 APPLYING THE PHAGE ✦ Use the wooden end of a cotton applicator to put a drop of the appropriate phage on each plate sector. Do not touch the agar with the stick or use it more than once.

it until the fluid on the end of the stick contacts the agar, then lift the stick. If you approach the agar from an angle the fluid will form a drop on the lower side, which may make it easier for you. Dispose of the stick in a sharps container.

7 Repeat Step 6 using phage φχ174 on the "Wild Type" *E. coli* plate.

8 Repeat Step 6 using sterile distilled water on the "Wild Type" *E. coli* plate.

9 Repeat Steps 6 through 8 using the *E. coli* B plate.

10 Repeat Steps 6 through 8 using the *E. coli* C plate.

11 Allow the phage/water drops to soak for about 10 minutes.

12 Tape the plates and incubate them inverted at $35 \pm 2°C$ for 24–48 hours.

Lab Two

1 Examine the plates for zones of clearing where the phage/sterile water was applied.

2 Record your results and answer the questions on the Data Sheet.

References

Acheson, Nicholas H. 2006. *Fundamentals of Virology*. John Wiley and Sons, Hoboken, NJ.

Rees, Catherine E. D., and Martin J. Loessner. 2009. *Phage Identification of Bacteria*. Chapter 8 in *Practical Handbook of Microbiology*, 2nd. Ed., Edited by Emanuel Goldman and Lorrence H. Green. CRC Press, Taylor and Francis Group, Boca Raton, FL 33487-2742.

Shors, Teri. 2009. *Understanding Viruses*. Jones and Bartlett Publishers, Inc. Sudbury, MA.

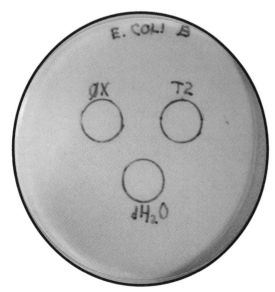

10-19 LABELING THE PLATE ✦ On each plate's base, divide the agar into three sectors with your marking pen. Label the sectors "T_2," "φχ174," and "dH_2O." Also label each plate with one of the *E. coli* strains.

Hematology and Serology

This section deals with blood cells and practical applications of the immune system. In Exercise 11-1 you will have the opportunity to perform a differential blood cell count and observe many of the cells involved in the body's defense system. Exercises 11-2 through 11-7 allow you to perform serological tests that are used to detect the presence of antigens or antibodies in a sample. Exercises 11-2 and 11-3 introduce the basic concept and utility of antigen/antibody specificity, which is at the heart of all serological reactions. They also introduce the necessity of an "indicator reaction" to show that antigen and antibody have reacted. These use precipitation, a type of test that is rarely used any more, but is a comfortable way of introducing these concepts. Exercises 11-4 through 11-6 introduce serological tests in which the indicator is agglutination. Exercise 11-7 illustrates a modern technique, the ELISA test. Further, this test illustrates how serological reactions can be used to quantify antigen or antibody in a sample. ✦

EXERCISE 11-1

Differential Blood Cell Count

✦ Theory

Leukocytes (white blood cells, or WBCs) are divided into two groups: **granulocytes** (which have prominent cytoplasmic granules) and **agranulocytes** (which lack these granules). There are three basic types of granulocytes: **neutrophils, basophils,** and **eosinophils.** The two types of agranulocytes are **monocytes** and **lymphocytes.**

Neutrophils (Figure 11-1A) are the most abundant WBCs in blood. They leave the blood and enter tissues to phagocytize foreign material. An increase in neutrophils in the blood is indicative of a systemic bacterial infection. Mature neutrophils sometimes are referred to as **segs** because their nucleus usually is segmented into two to five lobes. Because of the variation in nuclear appearance, they also are called **polymorphonuclear neutrophils (PMNs).** Immature neutrophils lack this nuclear segmentation and are referred to as **bands** (Figure 11-1B).

This distinction is useful because a patient with an active infection increases neutrophil production, which creates a higher percentage of the band (immature) type. Neutrophils are 12–15 μm in diameter—about twice the size of an erythrocyte (RBC).[1] Their cytoplasmic granules are neutral-staining and thus do not have the intense color of other granulocytes when prepared with Wright's or Giemsa stain.

Eosinophils are phagocytic, and their numbers increase during allergic reactions and parasitic infections (Figure 11-2). They are 12–15 μm in diameter (about twice the size of an RBC) and generally have two lobes

in their nucleus. Their cytoplasmic granules stain red in typical preparations.

Basophils (Figure 11-3) are the least abundant WBCs in normal blood. They are structurally similar to tissue mast cells and produce some of the same chemicals (histamine and heparin) but are derived from different stem cells in bone marrow. They are 12–15 μm in diameter. The nucleus usually is obscured by the dark-staining cytoplasmic granules, but it either has two lobes or is unlobed.

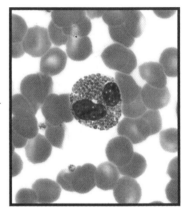

11-2 EOSINOPHIL ✦
These granulocytes are relatively rare and are about twice the size of red blood cells. Their cytoplasmic granules stain red, and their nucleus usually has two lobes. This specimen was prepared with Wright's stain and was magnified X1000.

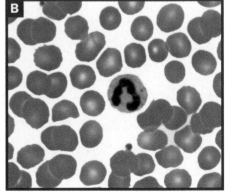

11-3 BASOPHIL ✦
Basophils comprise only about 1% of all white blood cells. They are slightly larger than red blood cells and have dark purple cytoplasmic granules that obscure the nucleus. This micrograph was prepared with Wright's stain and magnified X1000.

---

[1] It is convenient to discuss leukocyte size in terms of erythrocyte size because RBCs are so uniform in diameter. In an isotonic solution, erythrocytes are 7.5 μm in diameter.

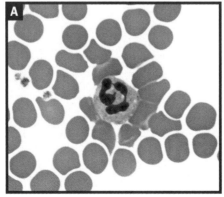

11-1 NEUTROPHIL ✦ **A** The segmented nucleus of this cell identifies it as a mature neutrophil (seg). About 30% of neutrophils in blood samples from females demonstrate a "drumstick" protruding from the nucleus, as in this specimen. This is the region of the inactive X chromosome. **B** This is an immature band neutrophil with an unsegmented nucleus. Both specimens were prepared with Wright's stain and were magnified X1000.

Agranulocytes include monocytes and lymphocytes. Monocytes (Figure 11-4) are the blood form of **macrophages**. They are the largest of the leukocytes, being two to three times the size of RBCs (12–20 μm). Their nucleus is horseshoe-shaped, and the cytoplasm lacks prominent granules (but may appear finely granular).

Lymphocytes (Figure 11-5A) are cells of the immune system. Two functional types of lymphocytes are the **T-cell**, involved in cell-mediated immunity, and the **B-cell**, which converts to a **plasma cell** when activated, and produces antibodies. The nucleus usually is spherical and takes up most of the cell. Lymphocytes are approximately the same size as RBCs or up to twice their size. The larger ones form a third functional group of lymphocytes, the **null cell**, many of which are **natural killer (NK) cells** that kill foreign or infected cells without antigen–antibody interaction (Figure 11-5B).

In a differential white cell count, a sample of blood is observed under the microscope, and at least 100 WBCs are counted and tallied (this task is automated now). Approximate normal percentages for each leukocyte are as follows and as summarized in Table 11-1.

neutrophils (mostly segs) 55%–65%

lymphocytes 25%–33%

monocytes 3%–7%

eosinophils 1%–3%

basophils 0.5%–1%.

✦ Application

A differential blood cell count is done to determine approximate numbers of the various leukocytes in blood. Excess or deficiency of all or a specific group is indicative of certain disease states. Even though differential counts are automated now, it is good training to perform one "the old-fashioned way" using a blood smear and a microscope to get an idea of the principle behind the technique.

✦ In This Exercise

You will be examining prepared blood smears and doing a differential count of white blood cells. As an optional activity, you may look at smears of abnormal blood and compare the differential count to normal blood.

✦ Materials

Per Student Group

✦ commercially prepared human blood smear slides (Wright's or Giemsa stain)

✦ (optional) commercially prepared abnormal human blood smear slides (*e.g.*, infectious mononucleosis, eosinophilia, or neutrophilia)

📼 Procedure

1 Obtain a blood smear slide and locate a field where the cells are spaced far enough apart to allow easy counting. (The cells should be fairly dense on the slide, but not overlapping.)

2 Using the oil immersion lens, scan the slide using the pattern shown in Figure 11-6. Be careful not to overlap fields when scanning the specimen. Choose a "landmark" blood cell at the right side of the field, and move the slide horizontally until that cell disappears off the left side. Avoid diagonal movement

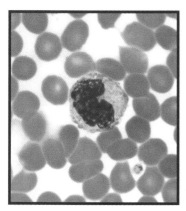

11-4 MONOCYTE ✦ Monocytes are the blood form of macrophages. They are about twice the size of red blood cells and have a round or indented nucleus. (Wright's stain, X1000.)

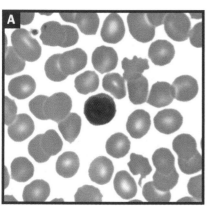
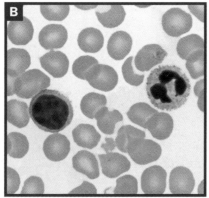

11-5 LYMPHOCYTE ✦ Lymphocytes are common in the blood, comprising up to 33% of all WBCs. Most are about the size of red blood cells and have only a thin halo of cytoplasm encircling their round nucleus. They belong to functional groups called B-cells and T-cells (which are morphologically indistinguishable). Micrograph **A** is a small lymphocyte and was prepared with Wright's stain. Some lymphocytes are larger, as in micrograph **B**. These are natural killer (NK) cells or some other type of null cell. Also visible is a neutrophil. Both micrographs are X1000.

TABLE **11-1** Typical Features of Human Leukocytes in Blood

TABLE OF RESULTS

Cell	Abundance in Blood (%)	Diameter (μm)	Nucleus	Cytoplasmic Granules (Wright's or Giemsa Stain)	Functions
Granulocyte					
Neutrophil	55–65	12–15	2–5 lobes	Present, but stain poorly; contain antimicrobial chemicals	Phagocytosis and digestion of (usually) bacteria
Eosinophil	1–3	12–15	2 lobes	Present and stain red; contain antimicrobial chemicals and histaminase	Present in inflammatory reactions and immune response against some multicellular parasites (worms)
Basophil	0.5–1	12–15	Unlobed or 2 lobes	Present and stain dark purple; contain histamine and other chemicals	Participate in inflammatory response
Agranulocyte					
Lymphocyte	25–33	7–18 (rare)	Spherical (leaving little visible cytoplasm)	Absent	Active in specific acquired immunity (as T and B cells)
Monocyte	3–7	12–20	Horseshoe-shaped (cytoplasm is prominent)	Absent	Phagocytosis (as macrophages)

11-6 FOLLOWING A SYSTEMATIC PATH ✦ A systematic scanning path is used to avoid wandering around the slide and perhaps counting some cells more than once. Remember that a microscope image is inverted. If you want the image to move left, you must move the slide to the right.

References

Brown, Barbara A. 1993. *Hematology—Principles and Procedures*, 6th ed. Lea and Febiger, Philadelphia.

Diggs, L. W., Dorothy Sturm, and Ann Bell. 1978. *The Morphology of Human Blood Cells*, 4th ed. Abbott Laboratories, North Chicago, IL.

Junqueira, L. Carlos, and Jose Carneiro. 2003. *Basic Histology, Text and Atlas*, 10th ed. Lange Medical Books, McGraw Hill, New York.

of the slide. As you scan, use the mechanical stage knobs separately to move the slide up and back or to the right and left in straight lines.

3 Make a tally mark in the appropriate box in the chart on the Data Sheet for the first 100 leukocytes you see.

4 Calculate percentages and compare your results with the accepted normal values.

5 Repeat with a pathological blood smear (if available).

Simple Serological Reactions

Antigen–antibody reactions are highly specific and occur *in vitro* as well as *in vivo*. Serology is the discipline that exploits this specificity as an *in vitro* diagnostic tool. Two simple serological reactions —agglutination and precipitation—are used in the following five exercises because they result in the formation of complexes that can be viewed with the naked eye and without sophisticated equipment.

The precipitin ring test (Exercise 11-2) and radial immunodiffusion (Exercise 11-3) illustrate precipitation. Both may be used to identify antigens or antibodies in a sample; the latter also is used to compare antigens in more than one sample. The slide agglutination test (Exercise 11-4) can be an important diagnostic (and highly specific) tool for the identification of organisms. It is especially useful for serotyping large genera such as *Salmonella*. Hemagglutination, a type of agglutination reaction, detects specific antigens on red blood cells, and is the standard test for determining blood type (Exercise 11-5). Other hemagglutination tests are used for diagnosing infections, such as the one in Exercise 11-6, which is used to diagnose infectious mononucleosis. The most sophisticated serological test covered in this section is in Exercise 11-7. It is a quantitative ELISA designed to detect antibody in a sample. ✦

EXERCISE 11-2 Precipitin Ring Test

✦ Theory

Soluble antigens may combine with **homologous antibodies** to produce a visible **precipitate**. Precipitate formation thus serves as evidence of antigen–antibody reaction and is considered to be a positive result.

Precipitation is produced because each antibody has (at least) two **antigen binding sites** and many antigens have multiple **epitopes** (sites for antibody binding). This results in the formation of a complex lattice of antibodies and antigens and produces the visible precipitate—a positive result. As shown in Figure 11-7, if either antibody or antigen is found in a concentration that is too high relative to the other, no visible precipitate will be formed even though both are present. **Optimum proportions** of antibody and antigen are necessary to form precipitate, and they occur in the **zone of equivalence**.

Several styles of precipitation tests are used. The **precipitin ring test** is performed in a small test tube or a capillary tube. Antiserum (containing antibodies homologous to the antigen being looked for) is placed in the bottom of the tube. The sample with the suspected antigen is layered on the surface of the antiserum in such a way that the two solutions have a sharp interface. As the two fluids diffuse into each other, precipitation occurs where optimum proportions of antibody and antigen are

found (Figure 11-8). This test also may be run to test for antibody in a sample.

✦ Application

Precipitation reactions can be used to detect the presence of either antigen or antibody in a sample. They have mostly been replaced by more sensitive serological techniques for diagnosis but are still useful as a simple demonstration of serological reactions.

✦ In This Exercise

You will perform a simple serological test to illustrate homology between equine albumin and equine albumin antiserum.

✦ Materials

Per Student Group

✦ two clean 6 × 50 mm Durham tubes

✦ equine serum (containing equine albumin)

✦ equine albumin antiserum (containing equine albumin antibodies)

✦ 0.9% saline solution

✦ three small transfer pipettes with a fine tip

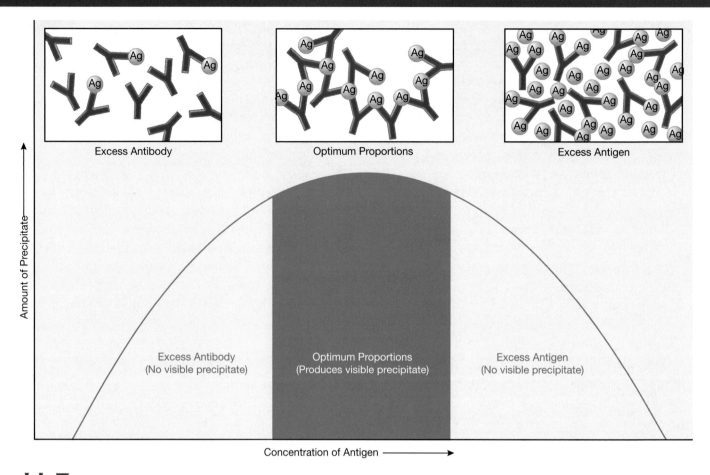

11-7 PRECIPITATION REACTIONS ✦ Precipitation occurs between soluble antigens and homologous antibodies where they are found in optimal proportions to produce a cross-linked lattice. Excess antigen or excess antibody prevents substantial cross-linking, so no lattice is formed and no visible precipitate is seen—even though both antigen and antibody are present. In this graph, antibody concentration is kept constant as antigen concentration is adjusted.

11-8 POSITIVE PRECIPITIN RING TEST ✦ A sample of antigen has been layered over an antiserum. The white precipitation ring has formed at the site of optimum proportions of antibodies and antigens.

Procedure

1 With a transfer pipette, carefully add equine antiserum to both Durham tubes. Fill from the bottom of the tube until it is about 1/3 full.

2 Mark one tube "A." Use a fresh transfer pipette to add the equine serum in such a way that a sharp and distinct second layer is formed without any mixing of the two solutions. It is critical not to allow any mixing. Success usually can be achieved by allowing the serum to trickle slowly down the inside of the glass (Figure 11-9).

3 Mark the second tube "B." Add the 0.9% saline in the same way you added equine serum to tube A. Use a fresh transfer pipette.

4 Incubate both at $35 \pm 2°C$ undisturbed for 1 hour. (Often, a positive result is seen sooner.)

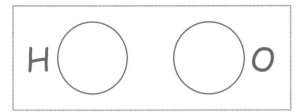

11-15 PREPARE THE SLIDE ✦ With your marking pen, draw two dime-sized circles on the slide. Label one circle "H" and the other "O." The circles will be where you check for the presence of *Salmonella* H and O antigens, respectively.

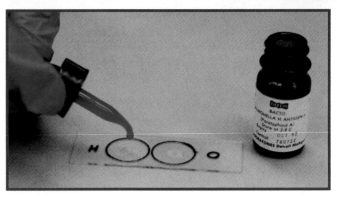

11-16 ADDING THE ANTIGEN ✦ Carefully add the H and O antigens to the drops of Anti-H antiserum already on the slide. Do not touch the dropper to the antiserum.

 ## Procedure

1 Using a marking pen, draw two circles approximately the size of a dime on a microscope slide (Figure 11-15). Label one "O" and the other "H."

2 Place a drop of *Salmonella* anti-H antiserum in each circle.

3 Place a drop of *Salmonella* O antigen in the "O" circle and a drop of *Salmonella* H antigen in the "H" circle (Figure 11-16). Be careful not to touch the dropper to the antiserum already on the slide.

4 Using a *different* toothpick for each circle, mix until each of the antigens are completely emulsified with the antiserum. Do not over-mix. Discard the toothpicks in a sharps container.

5 Gently rock this slide for a few minutes and observe for agglutination. Record your observations on the Data Sheet. (When rocking, the fluid should not spread. If it does, you are rocking too hard.)

✦ Alternative Test Procedure

Your instructor will cover the labels of the two *Salmonella* antigen bottles. You will use this procedure to identify which one contains the *Salmonella* H antigen.

1 Using a marking pen, draw two circles approximately the size of a dime on a microscope slide (Figure 11-15).

2 Place a drop of one *Salmonella* unknown antigen in one circle and a drop of the other *Salmonella* unknown antigen in the other circle (Figure 11-16).

3 Place a drop of *Salmonella* anti-H antiserum in each circle. Be careful not to touch the dropper to the antigen solutions already on the slide.

4 Using a *different* toothpick for each circle, mix until each of the antigens is completely emulsified with the antiserum. Do not *over*-mix. Discard the toothpicks in a biohazard container.

5 Gently rock this slide for a few minutes, then observe for agglutination. Record your observations on the Data Sheet.

References

Bopp, Cheryl A., Frances W. Brenner, Joy G. Wells, and Nancy A. Strockbine. 1999. Pages 467–471 in *Manual of Clinical Microbiology*, 7th ed., edited by Patrick R. Murray, Ellen Jo Baron, Michael A. Pfaller, Fred C. Tenover, and Robert H. Yolken. American Society for Microbiology, Washington, DC.

Collins, C. H., Patricia M. Lyne, J. M. Grange. 1995. Page 118 in *Collins and Lyne's Microbiological Methods*, 7th ed. Butterworth-Heinemann, Oxford, United Kingdom.

Constantine, Niel T., and Dolores P. Lana. 2003. Pages 222–223 in *Manual of Clinical Microbiology*, 8th ed., edited by Patrick R. Murray, Ellen Jo Baron, James H. Jorgensen, Michael A. Pfaller, and Robert H. Yolken. American Society for Microbiology, Washington, DC.

Forbes, Betty A., Daniel F. Sahm, and Alice S. Weissfeld. 2002. Pages 206–207 in *Bailey & Scott's Diagnostic Microbiology*, 11th ed. Mosby, St. Louis.

Lam, Joseph S., and Lucy M. Mutharia. 1994. Page 120 in *Methods for General and Molecular Bacteriology*, edited by Philipp Gerhardt, R. G. E. Murray, Willis A. Wood, and Noel R. Krieg. American Society for Microbiology, Washington, DC.

EXERCISE

11-5 Blood Typing

✦ Theory

Hemagglutination is a general term applied to any agglutination test in which clumping of red blood cells indicates a positive reaction. Blood tests as well as a number of indirect diagnostic serological tests are hemagglutinations.

The most common form of blood typing detects the presence and absence of **A** and/or **B antigens** on the surface of red blood cells. A person's ABO blood type is genetically determined. An individual with type A blood has RBCs with the A antigen and produces anti-B antibodies. Conversely, an individual with type B blood has RBCs with the B antigen and produces anti-A antibodies. People with type AB blood have *both* A and B antigens

on their RBCs and lack anti-A and anti-B antibodies. Type O individuals lack A and B antigens but produce *both* anti-A and anti-B antibodies.

ABO blood type is ascertained by adding a patient's blood separately to anti-A and anti-B antiserum and observing any signs of agglutination (Table 11-2 and Figure 11-17). Agglutination with anti-A antiserum indicates the presence of the A antigen and type A blood. Agglutination with anti-B antiserum indicates the presence of the B antigen and type B blood. If both agglutinate, the individual has type AB blood; lack of agglutination occurs in individuals with type O blood.

A similar test is used to determine the presence or absence of the **Rh factor** (antigen). If clumping of the

TABLE **11-2** Interpreting Blood Types

Reaction With				
Anti-A Antiserum	**Anti-B Antiserum**	**Anti-Rh Antiserum**	**Interpretation**	**Symbol**
Agglutination	No Agglutination	Agglutination	A antigen present B antigen absent Rh antigen present	A+
Agglutination	No Agglutination	No Agglutination	A antigen present B antigen absent Rh antigen absent	A−
No Agglutination	Agglutination	Agglutination	A antigen absent B antigen present Rh antigen present	B+
No Agglutination	Agglutination	No Agglutination	A antigen absent B antigen present Rh antigen absent	B−
Agglutination	Agglutination	Agglutination	A and B antigens present Rh antigen present	AB+
Agglutination	Agglutination	No Agglutination	A and B antigens present Rh antigen absent	AB−
No Agglutination	No Agglutination	Agglutination	A and B antigens absent Rh antigen present	O+
No Agglutination	No Agglutination	No Agglutination	A and B antigens absent Rh antigen absent	O−

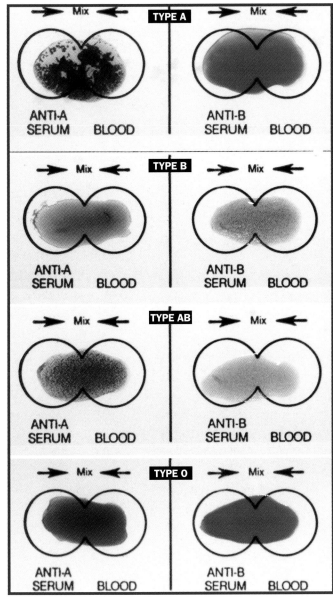

11-17 ABO BLOOD GROUPS ✦ Blood typing relies on agglutination of RBCs by Anti-A and/or Anti-B antisera. The blood types are as shown.

patient's blood occurs when mixed with anti-Rh antiserum (also known as anti-D), the patient is Rh positive.

✦ Application

Blood typing is a simple example of an agglutination test using blood. Clinically, it is used for cross-matching donor and recipient blood prior to transfusion.

✦ In This Exercise

You will determine your blood type with respect to two different markers: the ABO group and the Rh factor. Then, you will compile class data and compare it to blood type frequencies in the United States (Table 11-3).

✦ Materials

Per Student Group

- ✦ blood typing anti-A antiserum
- ✦ blood typing anti-B antiserum
- ✦ blood typing anti-Rh (anti-D) antiserum
- ✦ two microscope slides
- ✦ three toothpicks
- ✦ marking pen
- ✦ sterile lancets
- ✦ alcohol wipes
- ✦ small adhesive bandages
- ✦ sharps container
- ✦ disposable gloves

Procedure

1 With your marking pen, draw two circles on one microscope slide. Label one circle "A" and the other "B" (Figure 11-18A).

2 Draw a single circle with your marking pen in the center of a second microscope slide. Label it "Rh" (Figure 11-18B).

3 Place a drop of anti-A antiserum in the "A" circle.

4 Place a drop of anti-B antiserum in the "B" circle.

5 On the second microscope slide, place a drop of anti-Rh antiserum.

6 Clean the tip of your finger with an alcohol wipe. Let the alcohol dry.

7 Open a lancet package and remove the lancet, being careful not to touch the tip before you use it.

8 Shake your hand and "milk" blood down to the end of the finger you are going to prick.

9 Prick the end of your finger and immediately place a drop of blood beside each drop of antiserum. Do not touch the antisera with your finger. It's okay to have someone else prick your finger, but make sure he or she wears protective gloves.

10 Discard the lancet in the sharps container.

11 Put an adhesive bandage on your wound.

12 Using a circular motion, mix each set of drops with a toothpick. Be sure to *use a different toothpick* for each antiserum. Dispose of the toothpick in the sharps container.

TABLE **11-3** Approximate Percentages of Blood Types in the United States

Blood Type	White	African American	Hispanic	Asian
O+	37	47	53	39
O−	8	4	4	1
A+	33	24	29	27
A−	7	2	2	0.5
B+	9	18	9	25
B−	2	1	1	0.4
AB+	3	4	2	7
AB−	1	0.3	0.1	0.1

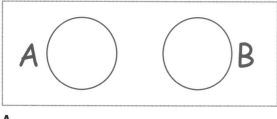

A

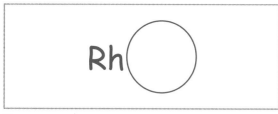

B

13 Gently rock the slides back and forth for a few minutes or until agglutination occurs.

14 After the agglutination reaction is complete, record the results in the table provided on the Data Sheet. Compare your results with the possible results in Table 11-2 to determine your blood type.

15 Collect class data and record these in the table provided on the Data Sheet. Compare class data with U. S. population data in Table 11-3.

Reference

American Red Cross. http://www.redcrossblood.org

11-18 PREPARE THE SLIDES ✦ **A** Draw two circles on a slide and label them "A" and "B" for the type of antiserum each will receive. **B** Draw one circle on a second slide and label it "Rh."

EXERCISE 11-6

Mononucleosis Hemagglutination Test

✦ Theory

Infectious mononucleosus is a viral disease caused by Epstein-Barr virus (EBV) infection of **B-lymphocytes** (the white blood cells that produce antibodies.) EBV stimulates B-cell division—it is a **mitogen**—and limits B-cell death, so more B-cells carry the virus. One consequence is the production of **heterophile antibodies**, which react only weakly with antigen and lack the specificity usually associated with antigen-antibody interaction. Symptoms of mononucleosis include fatigue, swollen lymph glands (lymphadenopathy), fever, sore throat, and enlargement of the liver and spleen. It is rarely fatal in otherwise healthy patients. Transmission is through saliva—hence the popular name, "kissing disease."

Hemagglutination tests use clumping of red blood cells as an indicator of antigen-antibody reaction. In this version of test, the heterophile antibodies (belonging to the IgM group of immunoglobins) show reactivity with an antigen on ovine (sheep), equine (horse), and bovine (cow) red blood cells, but not to a guinea pig kidney antigen. A positive test for heterophile antibodies can usually be obtained within a week of infection and may last months.

The specifics of the *Seradyn Color Slide II Mononucleosis Test* are these: Reagent A contains guinea pig kidney antigen, and Reagent B contains preserved horse red blood cells. When mixed with the heterophile antibodies from an infected patient, other antibodies that might cause agglutination (and a false positive result) of equine RBCs are removed because of their reactivity with the guinea pig antigen. The heterophile antibodies are nonreactive with the guinea pig kidney antigens and are free to agglutinate the horse RBCs.

✦ Application

This hemagglutination kit is used in medical laboratories to quickly demonstrate the presence of heterophile antibodies present in patients with infectious mononucleosis.

✦ In This Exercise

You will use the Seradyn Color Slide® II kit to illustrate positive and negative results for heterophile antibodies associated with infectious mononucleosis. In a medical lab this involves handling of human serum, fingertip blood, or plasma. To avoid this, you will be performing this test using positive and negative controls.

✦ Materials

One Seradyn Color Slide® Mononcleosis Test Kit[1] will provide all the materials needed as long as students share the reagents. Inasmuch as each step can be done in a relatively short time, this is a practical approach. An alternative is to make this lab a demonstration by the instructor.

Per Student Group

✦ one bottle of Reagent A (guinea pig kidney antigen)

✦ one bottle of Reagent B (preserved horse erythrocytes)

✦ one bottle of Patient 1 or Patient 2[2]

✦ one pipette from the kit or one transfer pipette

✦ two toothpicks if using a transfer pipette

✦ one Seradyn Color Slide II Mononucleosis Test Card[3]

✦ one pair of gloves per person actually handling the reagents and/or card

▱ Procedure

1 Obtain the materials listed. All reagents should be at room temperature.

2 Put on gloves (any person who will be handling the reagents and/or card), and be sure that all reagents are at room temperature.

3 Clear the dropper in Reagent A by squeezing the bulb. Replace the cap with while the dropper is empty, and then screw the cap on tightly. Mix the solution by inverting it repeatedly just prior to use. Be sure to mix thoroughly, as a heterogeneous mixture won't react properly.

---

[1] The kit may be ordered from Remel, 12076 Santa Fe Drive, Lenexa, KS 66215.

[2] Patient 1 sample is prepared from either the positive control serum or the negative control serum. Only the instructor or lab technician will know the actual identity until the test is run. The sample from Patient 2 is prepared from the other control.

[3] The card has six circles that can be separated along the perforations. Each student group will need one.

4 Holding the dropper of Reagent A, vertically dispense one free-falling drop on the left side of the circle on the card.

5 Clear the dropper and mix Reagent B as in Step 3.

6 Holding the dropper of Reagent B, vertically dispense one free-falling drop on the right side of the circle on the card.

7 Using the kit's pipette (or a transfer pipette), squeeze the bulb and insert the tip into the "patient's sample" you were given while still holding pressure on the bulb. Release pressure on the bulb to draw the sample up into the pipette. Then gently squeeze one free falling drop into Reagent A. If using a toothpick in Step 8, dispose of the pipette in an autoclave receptacle.

8 With the scoop end of the pipette (or a toothpick), gently, but thoroughly, mix the "patient's sample" with Reagent A; 10 to 15 circular strokes will be enough. Be careful not to mix with the drop of Reagent B.

9 After thorough mixing of Reagent A with the "patient's sample," gently begin mixing into Reagent B. This mixture should fill the entire circle. Dispose of the pipette or toothpick in an appropriate autoclave receptacle.

10 Gently begin rocking the card back and forth for 1 minute (approximately one "rock" every 4 seconds).

11 Read the card immediately. An obvious granular agglutinate indicates a positive result. A negative result will remain homogeneous (Figure 11-19).

12 Dispose of the card and gloves in an autoclave receptacle.

13 Record your results on the Data Sheet, and answer the questions.

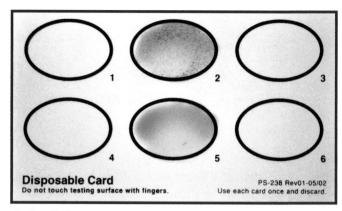

Disposable Card
Do not touch testing surface with fingers.

PS-238 Rev01-05/02
Use each card once and discard.

11-19 HEMAGGLUTINATION RESULTS ✦ The top circle is a positive result for hemagglutination; the lower one is negative. Note the finely granular agglutinate in the positive.

References

Murray, Patrick R., Ken S. Rosenthal, George S. Kobayashi, and Michael A. Pfaller. 2002. *Medical Microbiology*, 4th ed. Mosby, Inc. St. Louis, MO.

Seradyn Package Insert for the *Color Slide II Mononucleosis Test Kit*. August 2005.

EXERCISE
11-7

Quantitative Indirect ELISA[1]

✦ Theory

ELISA is an acronym for *Enzyme Linked Immunosorbant Assay*. As with other serological tests, ELISAs can be used to detect antigen in a sample or antibody in a sample. All rely on a secondary antibody with an attached, or **conjugated**, enzyme as an indicator of antigen–antibody reaction.

An indirect ELISA detects the presence of antibody in a sample. Follow along with Figure 11-20 as the general process is described. In this version of ELISA, antigen specific to the antibody being looked for is non-specifically coated onto the surface of microtiter plate wells (Fig. 11-21). Unbound antigen is then washed away with a buffer. The sample being assayed is added to the well(s), and if the antibody is present, it will react specifically with the antigen coating the wells; if none is present, no reaction occurs. To minimize the possibility of nonspecific binding between the sample's components (including other antibodies) and the well itself, a **blocking agent** (such as gelatin or other protein) is used to coat the well where antigens have not.

The enzyme-linked (**secondary** or **conjugate**) antibody is an **immunoglobin antibody**—its antigen is actually an antibody! When added to the well, it binds to the antibody in the sample, if present. After allowing time for the secondary antibody to react with the antigen (the antibody from the sample), the well is washed again with buffer to remove any unbound secondary antibody.

The specific enzyme chosen for use in an ELISA is only important in that it catalyzes a reaction that can be detected easily. A commonly used enzyme is **horseradish peroxidase**. Peroxidase catalyzes the conversion of hydrogen peroxide to water and oxygen ($2H_2O_2 \longrightarrow O_2 + H_2O$.) Hydrogen peroxide can be associated with a **chromogenic cosubstrate**. In this Exercise, azino-di-ethylbenz-thiazoline sulfonate—ABTS—is used, which turns a green color when it becomes oxidized during peroxidase activity, indicating a positive result; a negative result shows no color change (Figure 11-21).

✦ Application

ELISA may be used to detect the presence and amount of either antigen or antibody in a sample. The indirect ELISA is used to screen patients for the presence of HIV antibodies, rubella virus antibodies, and others. The direct ELISA may be used to detect hormones (such as HCG in some pregnancy tests and LH in ovulation tests), drugs, and viral and bacterial antigens.

If samples of known antibody concentration are available, a **standard curve** of degree of reaction with the secondary antibody can be constructed. Patient samples can then be compared to the standard curve to get an estimate of antibody concentration in their sample. This can be useful in tracking the progress of an immune response (and in turn, progress of a disease or infection) to a specific antigen over time and treatment. In this lab, only degree of reaction to different antigen concentrations is demonstrated.

✦ In This Exercise

You will perform a quantitative indirect ELISA, which allows determination of antibody titer in a patient's sample.

✦ Materials

✦ 1 Edvotek Quantitative ELISA Kit (EDVO-Kit #278)

Per Student Group
(The kit is designed for 6 groups)

✦ microtiter plate (6 × 8 wells)

✦ one permanent marker

✦ nine large transfer pipettes (labeled "PBS," "Ag1," "Ag2," "Block," "Ab1," "Ab2," "2°Ab," "Substrate," and "Stop")

✦ twelve small transfer pipettes (labeled "Ag1," "Ag2," and two each of "AB," "D," "F," "CE," and "GH")

✦ 50 mL Phosphate Buffered Saline (PBS)

✦ 1.5 mL Ag1 in microtube

✦ 1.5 mL Ag2 in microtube

✦ 3 mL Blocking Agent in larger microtube

✦ 0.5 mL 1:100 Ab1 in a microtube

✦ 0.5 mL 1:100 Ab2 in a microtube

✦ 3.0 mL Diluted Secondary (2°) Ab

[1] This procedure is adapted, with permission, from Edvotek kit #278, for which the company holds copyrights. The kit comes with an instruction manual for set-up by instructors. All materials and equipment required for this exercise can be purchased at Edvotek, Inc., PO Box 341232, Bethesda, MD, 20827, USA. email: info@edvotek.com www.edvotek.com

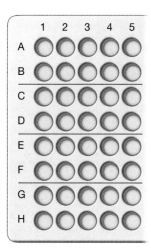

Step 1: Labeling the Microtiter Plate:
Using a permanent marker, label the columns 1 through 5 (Column 6 won't be used). Then, label the rows A through H. Draw lines between rows B and C, D and E, and F and G. Finally, label the plate with your group's name or number.

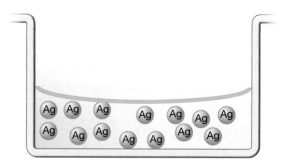

Step 2: Addition of Antigen:
Add one drop from the large pipette (or 50 μL) labeled "Ag 1" of Antigen 1 (Ag1) to Rows A, B, C, and D. Repeat with Antigen 2 (Ag2) in Wells E, F, G, and H using the large pipette (or 50 μL) labeled "Ag 2." Incubate at room temperature for 5 minutes.

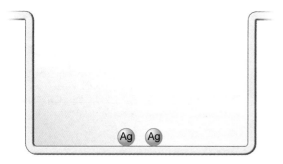

Step 3: Removal of Antigen:
Use the small pipette labeled "Ag1" to carefully remove the Ag1 solution from Wells A through D. Repeat with the small pipette labeled "Ag2" for wells E through H.

Step 4: PBS Wash:
Add 200 μL or 5 drops of PBS from the large pipette labeled "PBS" to each well (don't overfill). After filling all 40 wells, carefully remove PBS with the small pipettes as in Step 3.

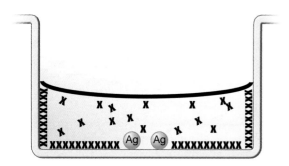

Step 5: Addition of Blocking Agent:
Add 50 μL or one drop from the large pipette labeled "Block" of **blocking agent** (x) to each well. Incubate for 10 minutes at 37°C. Then, remove the blocking agent with the small pipettes as in Step 3.

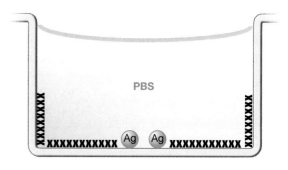

Step 6: PBS Wash:
Add 200 μL or 5 drops of PBS from the large pipette labeled "PBS" to each well (don't overfill). After filling all 40 wells, carefully remove PBS with the small pipettes as in Step 3. Discard the pipettes when finished.

Optional Stopping Point. Cover the plate with Parafilm and refrigerate until the next lab period.

11-20 PROCEDURAL DIAGRAM FOR THE QUANTITATIVE ELISA ✦ Pay special attention to volumes, solutions, and pipettes used in each step. Also, take care not to mix solutions from neighboring wells by overfilling.

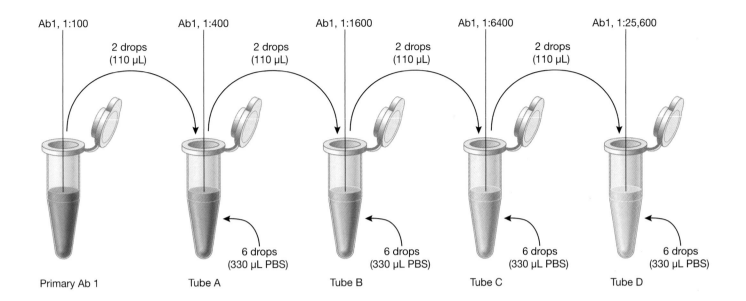

Ab1, 1:100   Ab1, 1:400   Ab1, 1:1600   Ab1, 1:6400   Ab1, 1:25,600

2 drops (110 µL) — repeated across each transfer

6 drops (330 µL PBS) — added to each tube

Primary Ab 1   Tube A   Tube B   Tube C   Tube D

Step 7: Preparation for Serial Dilution of Primary Antibody 1:

Obtain the 1:100 dilution of Primary Ab1 from your instructor. Label four tubes Tube A (Ab1, 1:400); Tube B (Ab1, 1:1600); Tube C (Ab1, 1:6400); and Tube D (Ab1 1:25,600). Add 330 µL of PBS or 6 drops from the large pipettes to Tubes A through D.

Step 8: Serial Dilution of Primary Antibody 1:

Add 2 drops or 110 µL of Primary Ab1, 1:100 to Tube A (labeled "Ab1, 1:400"). Mix by pipetting up and down, then cap the tube and continue mixing by inverting the tube several times. *It is important to mix well.* Transfer 110 µL or two drops from Tube A (labeled Ab1, 1:400) to Tube B (labeled "Ab1, 1:1600"). Mix well as before. Repeat transfers and mixing from Tube B to Tube C (labeled "Ab1, 1:6400") and Tube C to Tube D (labeled 1:25,600).

Step 9: Preparation for Serial Dilution of Primary Antibody 2:

Repeat Step 7, but use Ab2 and label the tubes E through H.

Step 10: Serial Dilution of Primary Antibody 2:

Repeat Step 8, but use Tubes E through H.

11-20 PROCEDURAL DIAGRAM FOR THE QUANTITATIVE ELISA *(CONTINUED)* ✦ Pay special attention to volumes, solutions, and pipettes used in each step. Also, take care not to mix solutions from neighboring wells by overfilling.

Step 11. Addition of Diluted Primary Ab1 and Ab2 to the Microtiter Plate:
Add PBS and Ab1 and Ab2 to the wells as follows (and as shown in the figure to the right). Use different large pipettes (labeled "PBA," "Ab1," or "Ab2") or pipette tips for each solution. You will be adding 50 μL or one drop of solution to each well.

Addition of PBS and Ab1: Be sure to use different pipettes or tips for PBS and Ab1.
- Add 50 μL or one drop of PBS to each well in Rows C and E.
- Add 50 μL or one drop from Tube D to Wells A5, B5, and F5.
- Add 50 μL or one drop from Tube C to Wells A4, B4, and F4.
- Add 50 μL or one drop from Tube B to Wells A3, B3, and F3.
- Add 50 μL or one drop from Tube A to Wells A2, B2, and F2.
- Add 50 μL or one drop from Ab1 Tube 1:100 (the one you got from your instructor) to Wells A1, B1, and F1.

Addition of Ab2: Be sure to use a fresh pipette or tip for Ab2.
- Add 50 μL or one drop from Tube H to Wells D5, G5, and H5.
- Add 50 μL or one drop from Tube G to Wells D4, G4, and H4.
- Add 50 μL or one drop from Tube F to Wells D3, G3, and H3.
- Add 50 μL or one drop from Tube E to Wells D2, G2, and H2.
- Add 50 μL or one drop from Ab2 Tube 1:100 (the one you got from your instructor) to Wells D1, G1, and H1.

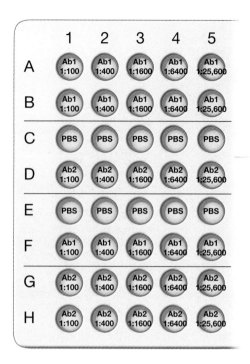

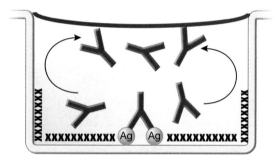

What's Happening in Step 11:
Addition of antibody to each well provides the opportunity for Ag-Ab binding. If the *homologous* Ag is in the well, the Ab will bind to it specifically (*i.e.*, Ag1/Ab1 or Ag2/Ab2). Antibodies are prevented from nonspecifically binding to the wells' walls by the blocking agent.

Step 12. Removal of Antibodies:
Label five small pipettes "AB," "D," "F," "CE," and "GH." Carefully remove the antibody solutions from each row using the appropriate pipettes. Be sure to go from the most dilute to the most concentrated antibody solutions. (For instance, remove A5, B5, A4, B4, *etc.* or C5, C4, C3, *etc.*, depending on the row contents.)

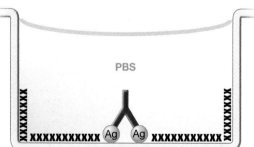

Step 13. PBS Wash:
Add 200 μL or 5 drops of PBS from the large pipette labeled "PBS" to each well (don't overfill). After filling all 40 wells, carefully remove PBS with the small pipette labeled "CE."

11-20 PROCEDURAL DIAGRAM FOR THE QUANTITATIVE ELISA *(CONTINUED)* ✦ Pay special attention to volumes, solutions, and pipettes used in each step. Also, take care not to mix solutions from neighboring wells by overfilling.

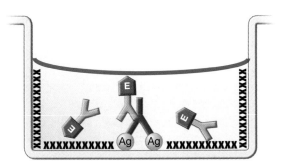

Step 14. Addition of Secondary Antibody:
Add one drop of Secondary Ab from a fresh large pipette (or 50 µL with a fresh tip) to all 40 wells.

Step 15. Removal of Secondary Antibodies:
Use the second set of five small pipettes labeled "AB," "D," "F," "CE," "GH." Carefully remove the antibody solutions from each row using the appropriate pipettes. Be sure to go from the most dilute to the most concentrated antibody solutions. (For instance, remove A5, B5, A4, B4, *etc.* or C5, C4, C3, *etc.*, depending on the row contents.)

Step 16. PBS Wash:
Add 200 µL or 5 drops of PBS from the large pipette labeled "PBS" to each well (don't overfill). After filling all 40 wells, carefully remove PBS with the small pipettes labeled "Ab1" and "Ab2" as in Step 3.

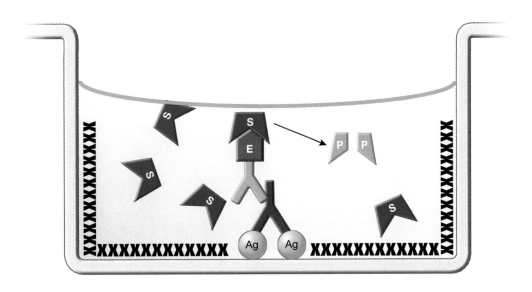

Step 17. Addition of Substrate:
Add one drop of Substrate from a fresh large pipette (or 50 µL with a fresh tip) to each of the 40 wells. Add from most dilute to most concentrated as before (that is, from Column 5 to Column 1). Incubate at room temperature until the color develops (see Figure 11-21). It should only take a few minutes. *Note:* Color doesn't usually develop in Column 5. When the colors have developed and show sufficient contrast, add 1 drop or 20 µL of Stop Solution *beginning with the most concentrated solution and working to the right to the least concentrated.*

11-20 PROCEDURAL DIAGRAM FOR THE QUANTITATIVE ELISA (CONTINUED) ✦ Pay special attention to volumes, solutions, and pipettes used in each step. Also, take care not to mix solutions from neighboring wells by overfilling.

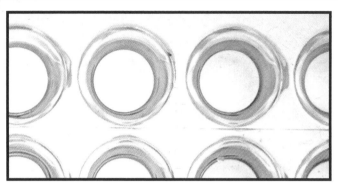

11-21 CHROMOGENIC **ELISA** ✦ A positive result in this ELISA produces a green color; no reaction is colorless. Intermediate intensities represent different degrees of reaction because of different concentrations of primary antibody in the sample.

✦ 2.5 mL Substrate (refrigerated until use)
✦ 1.5 mL Stop Solution in a larger test tube (warmed to 37° just prior to use)
✦ 25 mL disposal beaker
✦ 10-100 μL digital pipettor (optional)
✦ digital pipettor tips (optional)
✦ tip disposal container (optional)
✦ gloves
✦ goggles

 Procedure

Lab One

Refer to the Procedural Diagram in Figure 11-20 as you read and perform this procedure. If you are using a digital pipettor, dispose of tips in the container designated for them. Liquids should be disposed of in the disposal beaker.

1 Obtain a 6 × 8 well microtiter plate. With a permanent marker, label the columns 1 through 5 (Column 6 won't be used) and the rows A through H. (See Step 1 in Figure 11-20.) Then, label the large pipettes and the small pipettes as listed in the Materials section above.

2 With the Ag1 large transfer pipette or a digital pipettor, add one drop or 50 μL of Ag1 to Rows A, B, C, and D. Repeat with Ag2 using the large transfer pipette labeled Ag2 and Rows E, F, G, and H. *Be sure not to mix the pipettes with the wrong antigen.* Incubate for 5 minutes at room temperature.

3 Use the small pipettes labeled Ag1 and Ag2 to *gently* remove the antigen solutions from the wells. *Be sure not to mix the pipettes with the wrong antigen.*

4 With the large pipette labeled "PBS," add 5 drops of PBS (or, 200 μL with a digital pipettor) to each well, beginning with Row A and finishing with Row H. *Don't overfill the wells and allow solutions from different wells to mix.* This wash step is a bath not a shower—add the PBS gently! After all wells have been filled, use the small pipette labeled "Ag1" to gently remove the PBS from wells in Rows A through D. Repeat with the small pipette labeled "Ag2" for Rows E through H. Dispose of the PBS in the disposal beaker.

5 With the large pipette labeled "Block" gently add 1 drop of Blocking Agent (or 50 μL with a digital pipettor) to each well. Incubate at 37°C for 10 minutes. Then, carefully remove the Blocking Agent with the small pipettes as in Step 3 above. Dispose of the Blocking Agent in the disposal beaker.

6 Perform a PBS wash as in Step 4. Remove the PBS using the small pipettes labeled "Ag1" and "Ag2." When finished, throw away the two pipettes in a Biohazard Bag.

> **This represents an optional stopping point. If you will continue during the next lab period, cover the microtiter plate with Parafilm and refrigerate. Otherwise, continue with #7.**

7 Follow the dilution scheme in the Procedural Diagram to dilute Primary Antibody 1. Add 6 drops (330 μL) of PBS to 4 microtubes. Label as shown in the Procedural Diagram.

8 Perform the dilution scheme illustrated in the Procedural Diagram by transferring 2 drops (110 μL) of Primary Antibody 1 to Tube A. Mix well, and then transfer 2 drops (110 μL) from Tube A to Tube B. Mix well, and transfer 2 drops (110 μL) from Tube B to Tube C. Mix well, then transfer 2 drops (110 μL) from Tube C to Tube D.

9 Repeat Step 7 with Antibody 2 and 4 new microtubes.

10 Repeat Step 8 with Antibody 2.

11 Follow the instructions in the Procedural Diagram, Step 11, to add the diluted Primary Ab1 and Ab2, and PBS to the proper microtiter plate rows. Be careful to use the correct pipettes or tips. It seems imposing, but as you proceed, the loading pattern should become clear.

12 Remove the antibody solutions as follows: Use the small transfer pipettes labeled "AB," "D," "F,"

"CE," and "GH" to gently remove the antibody and PBS solutions. The letters indicate the rows for which each pipette is to be used. Remove the contents of those rows beginning with the *most dilute* and continue to the *most concentrated* (that is, from right to left). If two rows are the same, do the same dilution for both rows before moving to the next dilution. (That is, using Pipette AB, remove the antibody solution in the following order: A5, B5, A4, B4, A3, B3, A2, B2, A1, and B1.) Be sure not to mix pipettes or solutions.

13 Wash the 40 wells as before, using 5 drops of PBS (or 200 μL). Remove the PBS with the small pipette labeled "CE."

14 Using the large pipette labeled "2° Ab," add one drop (50 μL) of secondary antibody to all 40 wells.

15 Remove the secondary antibody with the second set of small pipettes labeled "AB," "D," "F," "CE," and "GH" to gently remove the antibody. Again, remove the most diluted samples first and work toward the left to the most concentrated.

16 Use the large pipette to wash each well with 5 drops (or 200 μL) of PBS as before. Use the two small pipettes labeled "Ab1" and "Ab2" to remove the PBS.

17 Using the large pipette labeled "Substrate," add 1 drop (or 50 μL) of Substrate Solution to each well. Begin with Column 5 and work back to Column 1. Incubate at room temperature for a minimum of 5 minutes.

18 When the colors have developed (often, no color develops in Column 5), add one drop (or 20 μL) of Stop Solution to each well from Column 1 to Column 5 (this time, you are moving left to right).

19 Examine the wells for color intensity, record your results on the Data Sheet, and answer the questions.

References

EDVOTEK Kit #278 Instruction Manual, Quantitative ELISA Laboratory Activity. 1-800-EDVOTEK and www.edvotek.com.

Lam, Joseph S., and Lucy M. Mutharia (1994). Antigen-Antibody Reactions, Chapter 5 in *Methods for General and Molecular Bacteriology*, edited by Philipp Gerhardt, Editor-in-Chief, R.G.E. Murray, Willis A. Wood, and Noel. R. Krieg. ASM Press, Washington, DC.

Madigan, Michael T., John M. Martinko, Paul V. Dunlop, and David P. Clark (2009). Pages 922–926 in *Brock's Biology of Microorganisms*, 12th ed. Pearson/Benjamin Cummings, San Francisco, CA.

Shores, Teri (2009). Pages 102–103 in *Understanding Viruses*. Jones and Bartlett Publishers, LLC, Sudbury, MA.

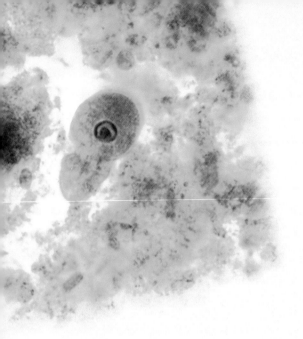

Eukaryotic Microbes

Microbiology in college courses is usually dominated by bacteriology, but the discipline also includes eukaryotic microscopic organisms. You have had the opportunity to look at some eukaryotic microorganisms in Exercises 3-3 and 3-4. In this section, you will examine some representative fungi (Exercise 12-1) and their culture (Exercise 12-2). These are followed by a survey of medically important Protozoans (Exercise 12-3). Exercise 12-4 introduces you to parasitic helminths (worms) because their identification is often the responsibility of microbiologists who examine patient samples. Thus, these parasites have entered the domain of microbiology. ✦

The Fungi—Common Yeasts and Molds

Members of the Kingdom Fungi are nonmotile eukaryotes. Their cell wall is usually made of the polysaccharide **chitin**, not cellulose as in plants. Unlike animals (that ingest, then digest their food), fungi are **absorptive heterotrophs:** They secrete exoenzymes into the environment, then absorb the digested nutrients. Most are **saprophytes** that decompose dead organic matter, but some are **parasites** of plants, animals, or humans.

Fungi are informally divided into unicellular **yeasts** (Figure 12-1) and filamentous **molds** (Figure 12-2) based on their overall appearance. **Dimorphic fungi** have both mold and yeast life-cycle stages. Filamentous fungi that produce fleshy reproductive structures—mushrooms, puffballs, and shelf fungi—are referred to as **macrofungi** (Figure 12-3), even though the majority of the fungus is filamentous and hidden underground or within decaying matter.

Individual fungal filaments are called **hyphae**, and collectively they form a **mycelium**. The hyphae are darkly pigmented in **dematiaceous fungi** and unpigmented in **hyaline** or **moniliaceous fungi** (Figure 12-4). Hyphae may be **septate**, in which walls separate adjacent cells, or **nonseptate** if walls are absent (Figure 12-5).

Fungal life cycles usually are complex, involving both sexual and asexual forms of reproduction. Gametes are produced by **gametangia**, and spores are produced by a variety of **sporangia**. Typically, the only diploid cell in the fungal life cycle is the zygote, which undergoes meiosis to produce haploid spores that are characteristic of the fungal group (see the following paragraph). Various asexual spores also may be produced during the life cycle of many fungi. If they form at the ends of hyphae, they are called **conidia**. Other asexual spores are **blastospores**, which are produced by budding, and **arthrospores**, which are produced when a hypha breaks. **Chlamydospores** (**chlamydoconidia**) are formed at the end of some hyphae and are a resting stage.

Formal taxonomic categories are based primarily on the pattern of sexual spore production and the presence of crosswalls in the hyphae. Members of the **Division Glomeromycota** (formerly **Class Zygomycetes**) are terrestrial, have nonseptate hyphae, and produce nonmotile **sporangiospores** and **zygospores**. Members of the **Division Ascomycota** (formerly **Class Ascomycetes**) produce a sac (an **ascus**) in which the zygote undergoes meiosis to produce haploid **ascospores**. Ascomycete hyphae are septate. This group also includes yeasts. Members of the **Division Basidiomycota** (formerly **Class Basidiomycetes**) have septate hyphae and during sexual reproduction produce a **basidium** that undergoes meiosis to produce four **basidiospores** attached to its surface.

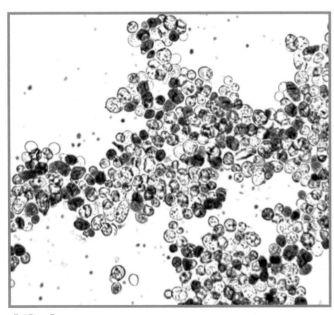

12-1 YEAST ✦ This is a wet mount of the Brewer's yeast, *Saccharomyces cerevisiae* stained with Methylene Blue. (X600)

12-2 MOLD ✦ Molds grow as fuzzy colonies. Their spores are abundant in the environment and frequently show up as contaminants on agar plates.

12-3 MACROFUNGUS ✦ This mushroom is the fruiting body from a mycelium that is busily engaged in decomposing the redwood log below it.

12-4 HYALINE HYPHAE ✦ This tangled mass of hyphae belongs to the bread mold, *Rhizopus*. They are hyaline because they lack pigment.

Following is a brief survey of fungi that are likely to be encountered in an introductory microbiology class and selected medically important fungi. Because of the medical emphasis of this exercise, the less formal groupings of "yeast" and "mold" are used.

Yeasts of Medical or Economic Importance

Candida albicans

Candida albicans (Figure 12-6) is part of the normal respiratory, gastrointestinal and female urogenital tract floras. Under the proper circumstances, it may flourish and produce pathological conditions, such as thrush in

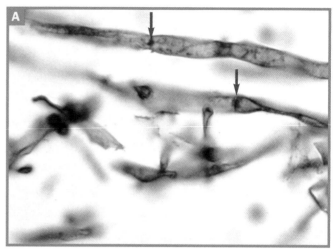

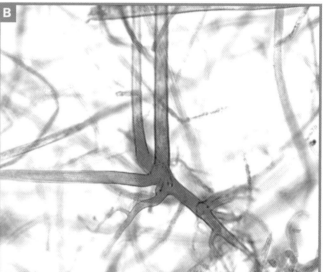

12-5 SEPTATE AND NONSEPTATE HYPHAE ✦ **A** The hyphae of *Aspergillus* are septate. Note the crosswalls (designated by arrows) dividing cells. (X1000). **B** *Rhizopus* provides an example of nonseptate hyphae. (X200)

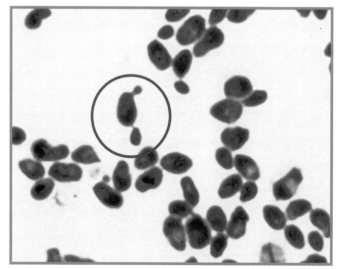

12-6 CANDIDA ALBICANS VEGETATIVE CELLS (X2640) ✦ Note the oval shape and nuclei. The circled cell is reproducing by forming a bud.

the oral cavity, vulvovaginitis of the female genitals, and cutaneous candidiasis of the skin. Systemic candidiasis may follow infection of the lungs, bronchi, or kidneys. Entry into the blood may result in endocarditis. Individuals most susceptible to *Candida* infections are diabetics, those with immunodeficiency (*e.g.,* AIDS), catheterized patients, and individuals taking antimicrobial medications. Budding results in chains of cells called **pseudohyphae,** which produce clusters of round, asexual blastoconidia at the cell junctions. Large, round, thick-walled **chlamydospores** form at the ends of pseudohyphae.

Saccharomyces cerevisiae

Saccharomyces cerevisiae is an ascomycete used in the production of bread, wine, and beer but is not an important human pathogen. It does not form a mycelium but, rather, produces a colony similar to bacteria (Figure 12-7). The vegetative cells (**blastoconidia**) are generally oval to round in shape, and asexual reproduction occurs by budding (Figure 12-8). Short **pseudohyphae** are

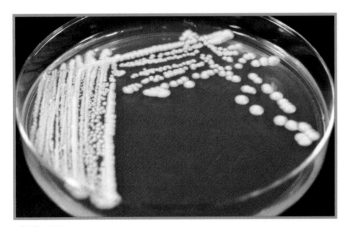

12-7 SACCHAROMYCES CEREVISIAE COLONY ✦ Note that the appearance is similar to a typical bacterial colony, not "fuzzy" like mold colonies.

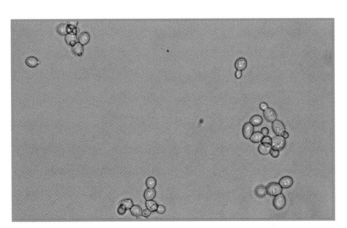

12-8 SACCHAROMYCES CEREVISIAE ✦ This bright-field micrograph is of an unstained wet mount of the yeast *Saccharomyces cerevisiae.* The cells are oval with dimensions of 3–8 × 5–10 μm. Short chains of cells (pseudohyphae) are visible in this field.

sometimes produced when the budding cells fail to separate. Meiosis produces one to four ascospores within the vegetative cell, which acts as the ascus. Ascospores may fuse to form another generation of diploid vegetative cells or they may be released to produce a population of haploid cells that are indistinguishable from diploid cells. Haploid cells of opposite mating types may also combine to create a diploid cell.

Molds of Medical or Economic Importance

Rhizopus

Rhizopus species are fast-growing molds that produce white or grayish, cottony growth. The mycelium becomes darker with age as sporangia are produced, giving it a "salt and pepper" appearance (Figure 12-9). Microscopically, *Rhizopus* species produce broad (10 μm), hyaline, and usually nonseptate surface and aerial hyphae. Anchoring **rhizoids** (Figure 12-10) are produced where the surface hyphae (**stolons**) join the bases of the long, unbranched **sporangiophores.**

The *Rhizopus* life cycle (Figure 12-11) has both sexual and asexual phases. Asexual **sporangiospores** are

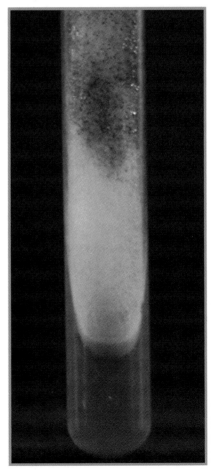

12-9 RHIZOPUS STOLONIFER ✦ Black asexual sporangia of this bread mold have begun to form, giving the growth a "salt and pepper" appearance.

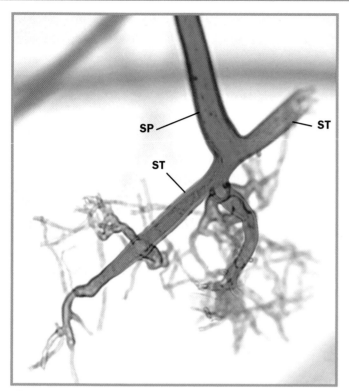

12-10 **RHIZOPUS RHIZOIDS (X200)** ✦ Anchoring rhizoids form at the junction of each sporangiophore (**SP**) and the stolon (**ST**). Note the absence of the septa.

produced by large, circular sporangia (Figure 12-12) borne at the ends of long, nonseptate, elevated sporangiophores. A hemispherical **columella** supports the sporangium. The spores develop into hyphae that are identical to those that produced them. On occasion, sexual reproduction occurs when hyphae of different mating types (designated + and − strains) make contact.

Initially, **progametangia** (Figure 12-13) extend from each hypha. Upon contact, a septum separates the end of each progametangium into a gamete (Figure 12-14). The walls between the two gametangia dissolve, and a thick-walled **zygospore** develops (Figures 12-15 and 12-16). Fusion of nuclei occurs within the zygospore and produces one or more diploid nuclei, or **zygotes**. After a dormant period, meiosis of the zygotes occurs. The zygospore then germinates and produces a sporangium similar to asexual sporangia. Haploid spores are released, develop into new hyphae, and the life cycle is completed.

Rhizopus species are common contaminants. *R. stolonifer* is the common bread mold. *R. oryzae* and *R. arrhizus* are responsible for producing zygomycosis, a condition found most frequently in diabetics and immunocompromised patients. Inhalation of spores may lead to hypersensitivity reactions in the respiratory

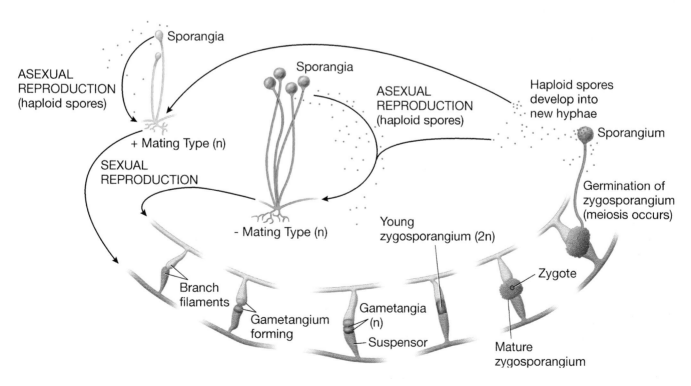

12-11 **RHIZOPUS LIFE CYCLE** ✦ Please refer to the text for details.

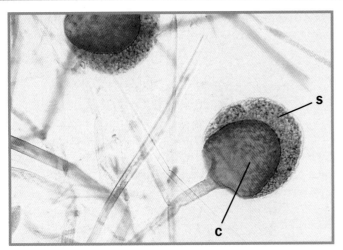

12-12 RHIZOPUS SPORANGIOPHORES (X264) ✦ The sporangium is found at the end of a long, unbranched, and nonseptate sporangiophore. The haploid asexual sporangiospores (**S**) cover the surface of the columella (**C**), which has a flattened base.

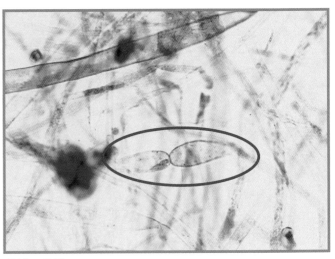

12-13 RHIZOPUS PROGAMETANGIA (X264) ✦ Progametangia from different hyphae are shown in the center of the field. Contact between the progametangia results in each forming a gamete.

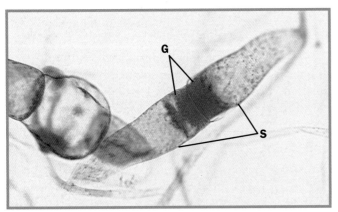

12-14 RHIZOPUS GAMETANGIA AND SUSPENSORS (X264) ✦ Gametangia (**G**) and suspensors (**S**) are shown in the center of the field. Gametangia contain haploid nuclei from each mating type.

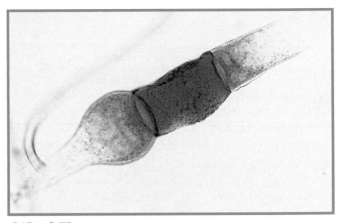

12-15 YOUNG RHIZOPUS ZYGOSPORE (X264) ✦ The zygospore forms when the cytoplasm from the two mating strains fuse (plasmogamy).

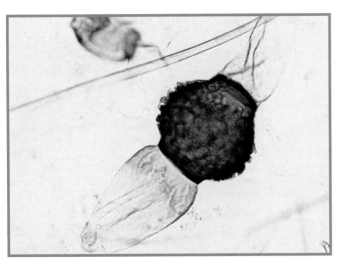

12-16 MATURE RHIZOPUS ZYGOSPORE (X264) ✦ Haploid nuclei from each strain fuse within the zygospore (**karyogamy**) to produce many diploid nuclei. Meiosis occurs to produce numerous haploid spores.

system. Entry into the blood leads to rapid spreading of the organism, occlusion of blood vessels, and necrosis of tissues.

Aspergillus

The genus *Aspergillus* is characterized by green to yellow or brown granular colonies with a white edge (Figure 12-17). One species, *A. niger*, produces distinctive black colonies. Vegetative hyphae are hyaline (unpigmented) and septate. The *Aspergillus* fruiting body is distinctive, with chains of conidia arising from one (uniseriate) or two (biseriate) rows of **phialides** attached to a swollen **vesicle** at the end of an unbranched conidiophore (Figure 12-18). The conidiophore grows from a foot cell in the vegetative hypha (Figure 12-19). Fruiting body structure and size, and conidia color are useful in species identification.

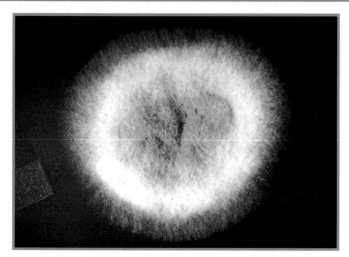

12-17 ASPERGILLUS FUMIGATUS COLONY ON SABOURAUD DEXTROSE AGAR ✦ Note the rugose topography and green, granular appearance with a white margin. The reverse is white. Although a common cause of aspergillosis, normally healthy people are not at great risk from *A. fumigatus* infection.

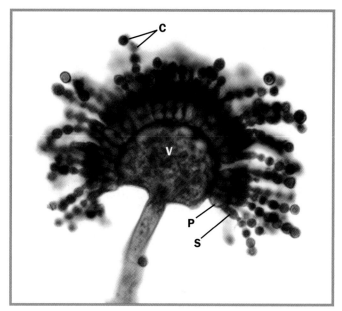

12-18 ASPERGILLUS CONIDIAL HEADS ✦ Shown is a section of an *Aspergillus niger* conidiophore (X1000). The conidia (**C**), primary (**P**) and secondary (**S**) phialides, and vesicle (**V**) are visible. This is a biseriate conidium.

A. fumigatus and other species are opportunistic pathogens that cause aspergillosis, an umbrella term covering many diseases. One form of pulmonary aspergillosis (referred to as fungus ball) involves colonization of the bronchial tree or tissues damaged by tuberculosis. Allergic aspergillosis may occur in individuals who are in frequent contact with the spores and become sensitized to them. Subsequent contact produces symptoms similar to asthma. Invasive aspergillosis is the most severe form. It results in necrotizing pneumonia and may spread to other organs.

Some species of *Aspergillus* are of commercial importance. Fermentation of soybeans by *A. oryzae* produces soy paste. Soy sauce is produced by fermenting soybeans with a mixture of *A. oryzae* and *A. soyae*. *Aspergillus* is also used in commercial production of citric acid.

Penicillium

Members of the genus *Penicillium* produce distinctive green, powdery, radially furrowed colonies with a white apron (Figure 12-20) and light colored reverse surface. The hyphae are septate and thin. Distinctive *Penicillium* fruiting bodies, consisting of **metulae, phialides** and chains of spherical conidia, are located at the ends of branched or unbranched **conidiophores** (Figure 12-21). Although not an important feature in laboratory identification, sexual reproduction results in the formation of ascospores within an ascus.

Penicillium is best known for its production of the antibiotic penicillin, but it is also a common contaminant. One pathogen, *P. marneffei*, is endemic to Asia and is responsible for disseminated opportunistic infections of

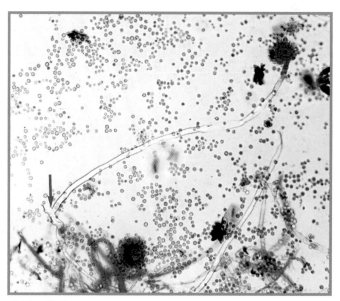

12-19 ASPERGILLUS FOOT CELL (LACTOPHENOL COTTON BLUE STAIN, X100) ✦ Sporangiophores emerge from a foot cell with the shape of an inverted "T" (arrow). This is a wet mount of *A. flavus*. Notice all the conidia in the field.

the lungs, liver, and skin in immunosuppressed and immunocompromised patients. It is thermally dimorphic, producing a typical velvety colony with a distinctive red pigment at 25°C but converting to a yeast form at 35°C. Other species of *Penicillium* are of commercial importance for fermentations used in cheese production. Examples include *P. roquefortii* (Roquefort cheese) and *P. camembertii* (Camembert and Brie cheeses).

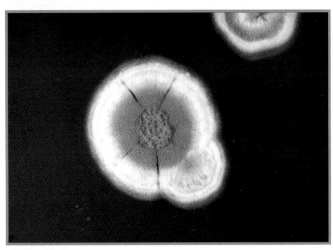

12-20 PENICILLIUM NOTATUM COLONY ON SABOURAUD DEXTROSE AGAr ✦ The green, granular surface with radial furrows and a white apron are typical of the genus.

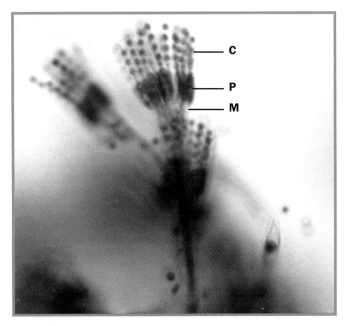

12-21 PENICILLIUM CONIDIOPHORE (X1000) ✦ *Penicillium* species produce a characteristic brush-shaped conidiophore (penicillus). Metulae (**M**), phialides (**P**), and chains of spherical conidia (**C**) are visible.

✦ Materials

Per Student Group

✦ agar slant of *Saccharomyces cerevisiae*

✦ Potato Dextrose Agar or Sabouraud Dextrose Agar plate culture of *Aspergillus spp.* (with lid taped on)

✦ Potato Dextrose Agar or Sabouraud Dextrose Agar plate culture of *Penicillium spp.* (with lid taped on)

✦ Potato Dextrose Agar or Sabouraud Dextrose Agar plate culture of *Rhizopus spp.* (with lid taped on)

✦ Gram's iodine stain or Methylene Blue

✦ dissecting microscope

✦ prepared slides of:
  ✦ *Aspergillus spp.* conidiophore
  ✦ *Candida albicans*
  ✦ *Penicillium spp.* conidiophore
  ✦ *Rhizopus spp.* sporangia
  ✦ *Rhizopus spp.* gametangia

Procedure

Yeasts

1 Using an inoculating loop and aseptic technique, make a wet mount slide of *Saccharomyces cerevisiae* as illustrated in Figure 3-19, and stain with iodine or Methylene Blue. Observe under high-dry and oil immersion. Identify vegetative cells and budding cells. Sketch representative cells in the space provided on the Data Sheet.

2 Observe prepared slides of *Candida albicans*. Identify vegetative cells and budding cells. Sketch representative cells in the space provided on the Data Sheet.

Molds

1 Obtain the plate culture of *Rhizopus. Do not remove the lid. Uncovering the organism will spread spores and contaminate the laboratory.*

  a. Examine the colony morphology, and sketch a representative colony in the space provided on the Data Sheet. Record the **color** on both the front (obverse) and reverse surfaces. Also record the **colony texture** as glabrous (leathery), velvety, yeast-like, cottony, or granular (powdery), and the **colony topography** as flat, rugose (with radial grooves), folded, crateriform, verrucose (warty, rough) or cerebriform (brain-like).

  b. Examine the colony under the dissecting microscope and identify hyphae, rhizoids, and sporangia. Sketch and label representative structures in the space provided on the Data Sheet.

2 Examine prepared slides of *Rhizopus* sporangia using medium and high-dry powers. Identify the following: sporangiophores, sporangia, and spores. Sketch and label representative structures in the space provided on the Data Sheet.

3 Examine prepared slides of *Rhizopus* gametangia using medium and high dry power. Identify the following: progametangia, gametangia, young zygosporangia, mature zygosporangia. Sketch and

label representative structures in the space provided on the Data Sheet.

4   Obtain the plate culture of *Penicillium. Do not remove the lid from the plate or you will spread spores and contaminate the laboratory.*

a. Examine the colony morphology and sketch a representative colony in the space provided on the Data Sheet. Record the *color* on both the front (obverse) and reverse surfaces. Also record the ***colony texture*** as glabrous (leathery), velvety, yeast-like, cottony, or granular (powdery), and the ***colony topography*** as flat, rugose (with radial grooves), folded, crateriform, verrucose (warty, rough) or cerebriform (brain-like).

b. Examine the colony under the dissecting microscope. Identify hyphae and conidia. Sketch and label representative structures in the space provided on the Data Sheet.

5   Observe prepared slides of *Penicillium* conidiophores. Identify the following: hyphae, conidiophores, and chains of conidia. Sketch and label representative structures in the space provided on the Data Sheet.

6   Obtain the plate culture of *Aspergillus. Do not remove the lid from the plate or you will spread spores and contaminate the laboratory.*

a. Examine the colony morphology and sketch a representative colony in the space provided on the Data Sheet. Record the *color* on both the front (obverse) and reverse surfaces. Also record the ***colony texture*** as glabrous (leathery), velvety, yeast-like, cottony, or granular (powdery), and the ***colony topography*** as flat, rugose (with radial grooves), folded, crateriform, verrucose (warty, rough), or cerebriform (brainlike).

b. Examine the colony under the dissecting microscope. Identify hyphae and conidia. Sketch and label representative structures in the space provided on the Data Sheet.

7   Observe prepared slides of *Aspergillus* conidiophores. Identify hyphae, conidiophores, and conidia. Sketch and label representative structures in the space provided on the Data Sheet.

References

Collins, C. H., Patricia M. Lyne, and J. M. Grange. 1995. Chapter 51 in *Collins and Lyne's Microbiological Methods,* 7th ed. Butterworth-Heineman, Oxford, United Kingdom.

Fisher, Fran, and Norma B. Cook. 1998. Chapter 2 in *Fundamentals of Diagnostic Mycology.* W. B. Saunders, Philadelphia.

Forbes, Betty A., Daniel F. Sahm, and Alice. S. Weissfeld. 2002. Chapter 53 in *Bailey & Scott's Diagnostic Microbiology,* 11th ed. Mosby-Year Book, St. Louis.

Koneman, Elmer W., Stephen D. Allen, William M. Janda, Paul C. Schreckenberger, and Washington C. Winn, Jr. 1997. Chapter 19 in *Color Atlas and Textbook of Diagnostic Microbiology,* 5th ed. J. B. Lippincott, Philadelphia.

12-2 Fungal Slide Culture

✦ Theory

The background material concerning fungi is contained in Exercise 12-1.

✦ Application

Classical fungal taxonomy has relied on the appearance and arrangement of sexual spores, as well as certain features of the hyphae. Because sporangia are so delicate, it is difficult to transfer a portion of mycelium to make a wet mount showing an intact sporangium with its spores. (For an example of what these wet mounts frequently look like, see Figure 12-19 and notice all the scattered, detached spores!) By growing the fungus on a microscope slide, intact sporangia can be observed without disturbing them.

✦ In This Exercise

You will grow a mold in a slide culture and observe its sexual spores.

✦ Materials

Per Class

✦ Fume hood or cell culture hood
✦ Aerosol disinfectant in hood

Per Student Group

✦ Sabouraud Agar slant or plate culture of mature (sporing) *Penicillium spp.* (Figures 12-20 and 12-22A) (This can be located in the fume or cell culture hood and one culture should serve the entire class.)
✦ Sabouraud Agar slant or plate culture of mature (sporing) *Aspergillus spp.* (Figures 12-19 and 12-22B) (This can be located in the fume or cell culture hood and one culture should serve the entire class.)
✦ 5 mL tube of sterile, molten Sabouraud Agar (stored in a 50°C water bath until needed)
✦ 50 mL beaker with alcohol
✦ 50 mL flask of sterile water
✦ Bunsen burner
✦ petroleum jelly
✦ sterile toothpicks or wooden applicators
✦ jar of disinfectant large enough to immerse the microscope slides when finished

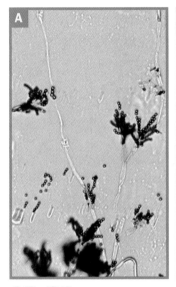

12-22 SLIDE CULTURES ILLUSTRATING CONIDIOPHORES ✦
A *Penicillium.* B *Aspergillus.*

Per Student

✦ sterile Petri dish
✦ glass rod bent to about 30 degrees in a way it will fit into the Petri dish (a bendable drinking straw will also work)
✦ filter paper cut to fit inside the Petri dish
✦ sterile transfer pipettes
✦ depression microscope slide
✦ cover slips
✦ forceps
✦ inoculating (or dissecting) needle
✦ scalpel
✦ gloves

✎ Procedure

Lab One

1. Place a folded paper towel on your desk. It should be about the size of a magazine. Put all the working materials on the towel: Petri dish, glass rod, depression slide, cover slip, forceps, inoculating needle, and sterile transfer pipettes.

2. Place the filter paper inside the Petri dish. Set the glass rod on top of the filter paper (Figure 12-23).

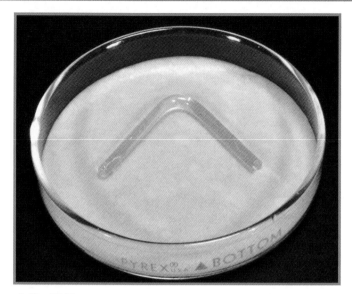

12-23 THE FUNGAL SLIDE CULTURE PRELIMINARY SET UP ✦ Shown are the Petri dish, bent glass rod, and moistened filter paper.

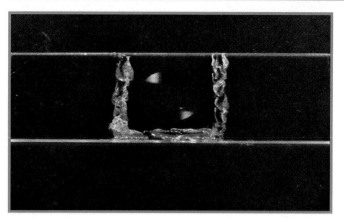

12-24 PETROLEUM JELLY SEAL ✦ Use a wooden applicator or toothpick to make a bead of petroleum jelly about the size of the cover slip on three sides of the slide's depression. This is easier to do if the petroleum jelly is slightly warm. In this photo, blue food coloring has been added to the petroleum jelly so it is easier to see.

12-25 AGAR IN THE WELL ✦ Shown here is the solidified agar in the well and covered with the cover slip.

3 Saturate (but don't flood) the filter paper with sterile distilled water using a sterile transfer pipette. Replace the lid.

4 Using a sterile transfer pipette, rinse the depression slide with alcohol. Tap the excess alcohol off on the paper towel. Then, holding the slide with a forceps pass it *through* the flame and let the alcohol burn off. Do NOT perform this step over the paper towel or the alcohol because a drop of flaming alcohol can start a fire. After it cools, place it on the paper towel with the depression upwards.

5 Repeat Step 4 with the cover slip. Be sure not to heat the cover slip too much or it will melt or crack.

6 Using the toothpick or wooden applicator, apply a bead of petroleum jelly around the two sides and bottom of the depression on the slide (Figure 12-24). This is made easier if the petroleum jelly is slightly warm. Make sure it is applied so its size matches the cover slip's size.

7 Dip the forceps in alcohol, then pass them through the flame. Let the alcohol burn off and set them aside on the paper towel. Do NOT perform this step over the paper towel or the alcohol because a drop of flaming alcohol can start a fire.

8 Using a sterile transfer pipette, place a drop of molten Sabouraud Agar in the depression of the slide. Immediately place the cover slip over the agar using the sterile forceps. Gently press it down to seal the edges. Allow the agar to solidify (Figure 12-25).

9 Once the agar has solidified, *gently* slide the cover slip away from the unsealed edge so about two-thirds of the agar is exposed.

10 Dip the scalpel blade in alcohol, then pass the blade through the flame. Let the alcohol burn off. Be sure not to catch the scalpel's handle on fire! Also, do NOT perform this step over the paper towel or the alcohol because a drop of flaming alcohol can start a fire.

11 Gently make a straight cut across the diameter of the agar with the scalpel. Remove the top (uncovered) half.

12 Place the slide on the glass rod inside the Petri dish and replace the lid. Then, take your slide, inoculating needle, forceps, and fungal culture (if located at your desk) to the fume or cell culture hood designated for fungal transfers.

13 Flame your inoculating needle (with alcohol if you are using a dissecting needle—but don't burn the handle!) and touch it to the surface of the fungal growth you are transferring (either *Penicillium* or *Aspergillus*.) Then, gently transfer some fungal growth to the center of the cut edge of agar (Figure 12-26). Flame the inoculating needle as before.

14 With your sterile forceps, gently push the cover slip back over the agar to completely cover it. Do not separate the cover slip from the agar.

15 Replace the cover of the Petri dish and incubate at 25°C for 7 days or until spores are visible. The plate should be placed right side up in the incubator.

16 Spray aerosol disinfectant in the air and surface of the fume or cell culture hood. Make sure the flame is off.

Lab Two

1 Remove the slide from the Petri dish and examine it under low and high dry power.

2 Sketch your observations on the Data Sheet.

3 Examine another student's preparation of the fungus you did not grow and sketch your observations on the Data Sheet.

4 Place your slide culture in a jar of disinfectant.

5 Answer the questions on the Data Sheet.

12-26 INOCULATION OF THE AGAR ✦ With the cover slip pulled down out of the way and the agar cut across its diameter, a small amount of fungal mycelium can be transferred to the center of the agar's cut edge. It is difficult to see in the photo, but the tip of the needle is at the agar's edge. After the transfer, the cover slip should be gently pushed back over the entire well and the plate incubated at 25°C for 7 days or until spores are visible.

Reference

Larone, Davise H. 2002. *Medically Important Fungi—A Guide to Identification*, 4th Ed., ASM Press, Washington, D.C.

Examination of Common Protozoans of Clinical Importance

Protozoans are unicellular eukaryotic heterotrophic microorganisms. A typical life cycle includes a vegetative **trophozoite** stage and a resting **cyst** stage. Some have additional stages, making their life cycles more complex.

As discussed in Exercise 3-3, protozoans are classified as follows:

Phylum Sarcomastigophora (including Subphylum Mastigophora [the flagellates] and Subphylum Sarcodina [the amoebas]),
Phylum Ciliophora (the ciliates), and
Phylum Apicomplexa (sporozoans and others).

Following is a survey of some commonly encountered protozoans of clinical importance. You will be examining these on prepared slides because most are pathogens that are not handled appropriately in a beginning microbiology laboratory.

Amoeboid Protozoans Found in Clinical Specimens

Entamoeba histolytica

Entamoeba histolytica is the causative agent of amoebic dysentery (amebiasis), a disease most common in areas with poor sanitation. Identification is made by finding either trophozoites (Figure 12-27) or cysts (Figure 12-28) in a stool sample. The diagnostic features of each are described in the captions.

Infection occurs when a human host ingests cysts, either through fecal–oral contact or, more typically, contaminated food or water. Cysts (but not trophozoites) are able to withstand the acidic environment of the stomach. Upon entering the less acidic small intestine, the cysts undergo **excystation**. Mitosis produces eight small trophozoites from each cyst.

The trophozoites parasitize the mucosa and submucosa of the colon, causing ulcerations. They feed on red blood cells and bacteria. The extent of damage determines whether the disease is acute, chronic, or asymptomatic. In the most severe cases, infection may extend to other organs, especially the liver, lungs, or brain. Among the symptoms of amoebic dysentery are abdominal pain, diarrhea, blood and mucus in feces, nausea, vomiting, and hepatitis.

Developing cysts undergo mitosis to produce mature quadranucleate cysts, which are shed in the feces and are

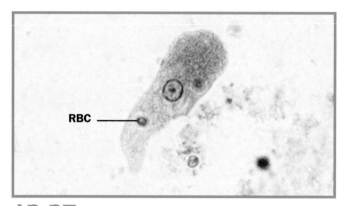

12-27 ENTAMOEBA HISTOLYTICA TROPHOZOITE (X800, IRON HEMATOXYLIN STAIN) ✦ Trophozoites range in size from 12 to 60 µm. Notice the small, central karyosome, the beaded chromatin at the margin of the nucleus, the ingested red blood cells, and the finely granular cytoplasm. Compare with an *Entamoeba coli* trophozoite in Figure 12-29.

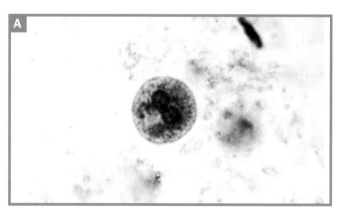

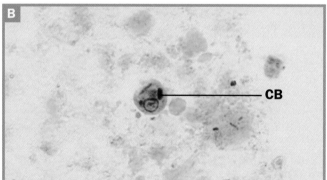

12-28 ENTAMOEBA HISTOLYTICA CYSTS ✦ A Cysts are spherical with a diameter of 10 to 20 µm. Two of the four nuclei are visible; other nuclear characteristics are as in the trophozoite. Compare with an *Entamoeba coli* cyst in Figure 12-30 (X1320, Iron Hematoxylin Stain). B *E. histolytica* cyst (X1200, Trichrome Stain) with cytoplasmic chromatoidal bars (**CB**). These are found in approximately 10% of the cysts, have blunt ends, and are composed of ribonucleoprotein.

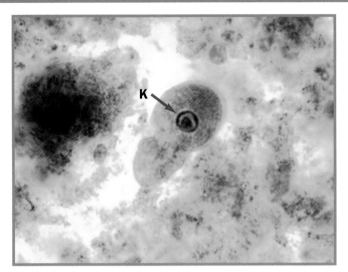

12-29 ENTAMOEBA COLI TROPHOZOITE (X1000, TRICHROME STAIN) ✦ Trophozoites range in size from 15 to 50 μm. Notice the relatively large and eccentrically positioned karyosome (**K**), the unclumped chromatin at the periphery of the nucleus, and the vacuolated cytoplasm lacking ingested red blood cells. The usually nonpathogenic *E. coli* must be distinguished from the potentially pathogenic *E. histolytica*, so compare with Figure 12-27.

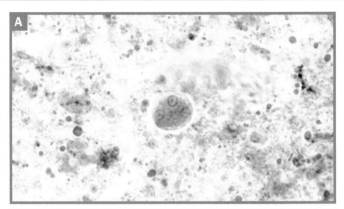

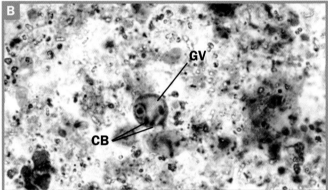

12-30 ENTAMOEBA COLI CYSTS ✦ Cysts typically are spherical and are between 10 and 35 μm in diameter. They contain 8, or sometimes 16 nuclei. This makes differentiation from *E. histolytica* cysts simpler, as they never have more than 4 nuclei. **A** Five nuclei are visible in this specimen (X1000, trichrome stain). **B** Chromatoidal bars (**CB**) and a large glycogen vacuole (**GV**) characteristic of immature cysts are visible in this specimen (X1000, Trichrome Stain).

infective. They also may persist in the original host, resulting in an **asymptomatic carrier**—a major source of contamination and infection.

Another member of the genus *Entamoeba* deserves mention here. *Entamoeba coli* is a fairly common, nonpathogenic intestinal commensal that must be differentiated from *E. histolytica* in stool samples. Its characteristic features are given in the captions to Figures 12-29 and 12-30.

Ciliate Protozoan Found in Clinical Specimens

Balantidium coli

Balantidium coli (Figures 12-31 and 12-32) is the causative agent of balantidiasis and exists in two forms: a vegetative trophozoite and a cyst. Laboratory diagnosis is made by identifying either the cyst or the trophozoite, with the latter being more commonly found.

The trophozoite is highly motile because of the cilia and has a macronucleus and a micronucleus. Cysts in sewage-contaminated water are the infective form. Trophozoites may cause ulcerations of the colon mucosa, but not to the extent produced by *Entamoeba histolytica*. Symptoms of acute infection are bloody and mucoid feces. Diarrhea alternating with constipation may occur in chronic infections. Most infections probably are asymptomatic.

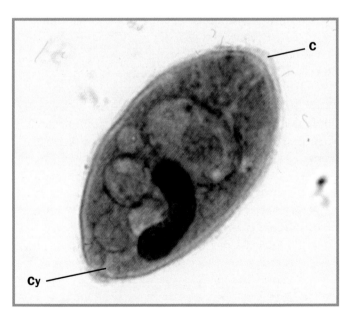

12-31 BALANTIDIUM COLI TROPHOZOITE (X800) ✦ Trophozoites are oval in shape with dimensions of 50 to 100 μm long by 40 to 70 μm wide. Cilia (**C**) cover the cell surface. Internally, the macronucleus is prominent; the adjacent micronucleus is not. An anterior cytostome (**Cy**) is usually visible.

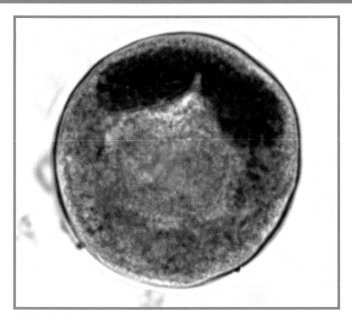

12-32 *BALANTIDIUM COLI* CYST (X1000) ✦ Cysts usually are spherical and have a diameter in the range of 50 to 75 μm. There is a cyst wall, and the cilia are absent. As in the trophozoite, the macronucleus is prominent, but the micronucleus may not be.

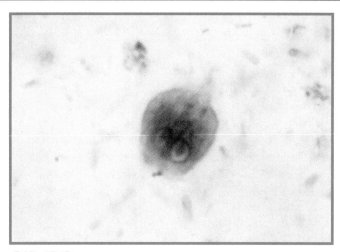

12-33 *GIARDIA LAMBLIA* TROPHOZOITE (X1320, IRON HEMATOXYLIN STAIN) ✦ Trophozoites have a long, tapering posterior end and range in size from 9 to 21 μm by 5 to 15 μm. There are two nuclei with small karyosomes. Two median bodies are visible, but the four pairs of flagella are not.

Flagellate Protozoans Found in Clinical Specimens

Giardia lamblia

Giardiasis is caused by *Giardia lamblia* (also known as *Giardia intestinalis)*, a flagellate protozoan. It is seen most frequently in the duodenum as a heart-shaped vegetative trophozoite (Figure 12-33) with four pairs of flagella and a sucking disc that allows it to resist gut peristalsis. Multinucleate cysts lacking flagella (Figure 12-34) are formed as the organism passes through the colon. Cysts are shed in the feces and may produce infection of a new host upon ingestion. Transmission typically involves fecally contaminated water or food, but direct fecal–oral contact transmission is also possible.

The organism attaches to epithelial cells but does not penetrate to deeper tissues. Most infections are asymptomatic. Chronic diarrhea, dehydration, abdominal pain, and other symptoms may occur if the infection produces a population large enough to involve a significant surface area of the small intestine. Diagnosis is made by identifying trophozoites or cysts in stool specimens.

Trichomonas vaginalis

Trichomonas vaginalis (Figure 12-35) is the causative agent of trichomoniasis (vulvovaginitis) in humans. It has four anterior flagella and an **undulating membrane**.

Trichomoniasis may affect both sexes but is more common in females. *T. vaginalis* causes inflammation

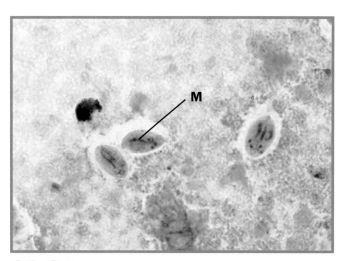

12-34 *GIARDIA LAMBLIA* CYSTS (X1000, TRICHROME STAIN) ✦ Giardia cysts are smaller than trophozoites (8 to 12 μm by 7 to 10 μm), but the four nuclei with eccentric karyosomes and the median bodies (**M**) are still visible.

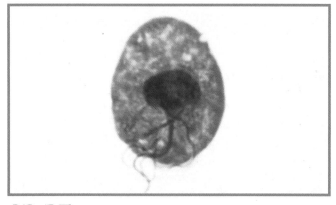

12-35 *TRICHOMONAS VAGINALIS* (X2027) ✦ The trophozoite is the only stage of the *Trichomonas* life cycle. Several flagella are visible.

of genitourinary mucosal surfaces—typically the vagina, vulva, and cervix in females and the urethra, prostate, and seminal vesicles in males. Most infections are asymptomatic or mild. Some erosion of surface tissues and a discharge may be associated with infection. The degree of infection is affected by host factors, especially the bacterial flora present and the pH of the mucosal surfaces. Transmission typically is by sexual intercourse.

The morphologically similar nonpathogenic *Trichomonas tenax* and *T. hominis* are residents of the oral cavity and intestines, respectively.

Trypanosoma brucei

Trypanosoma brucei (Figure 12-36) is a species of flagellated protozoans divided into subspecies: *T. brucei brucei* (which is nonpathogenic), and *T. brucei gambiense* and *T. brucei rhodesiense*, which produce African trypanosomiasis, also known as African sleeping sickness. The organisms are very similar morphologically but differ in geographic range and disease progress. West African trypanosomiasis (caused by *T. brucei gambiense*) is generally a mild, chronic disease that may last for years, whereas East African trypanosomiasis (caused by *T. brucei rhodesiense*) is more acute and results in death within a year. Modern molecular methods that compare proteins, RNA, and DNA are used to differentiate between them.

Trypanosomes have a complex life cycle. One stage of the life cycle, the **epimastigote**, multiplies in an intermediate host, the tsetse fly (genus *Glossina*). The infective **trypomastigote** stage then is transmitted to the human host through tsetse fly bites. Once introduced, trypomastigotes multiply and produce a chancre at the site of the bite. They enter the lymphatic system and spread through the blood, and ultimately to the heart and brain.

Immune response to the pathogen is hampered by the trypanosome's ability to change surface antigens faster than the immune system can produce appropriate antibodies. This antigenic variation also makes development of a vaccine unlikely. Diagnosis is made from clinical symptoms and identification of the trypomastigote in patient specimens (*e.g.,* blood, CSF, and chancre aspirate). An ELISA and an indirect agglutination test also have been developed to detect trypanosome antigens in patient samples.

Progressive symptoms include headache, fever, and anemia, followed by symptoms characteristic of the infected sites. The symptoms of sleeping sickness—sleepiness, emaciation, and unconsciousness—begin when the central nervous system becomes infected. Depending on the infecting strain, the disease may last for months or years, but the mortality rate is high. Death results from heart failure, meningitis, or severe debility of some other organ(s).

The infective cycle is complete when an infected individual (humans, cattle, and some wild animals are reservoirs) is bitten by a tsetse fly, which ingests the organism during its blood meal. It becomes infective for its lifespan.

Sporozoan Protozoans Found in Clinical Specimens

Plasmodium spp.

Plasmodia are sporozoan parasites with a complex life cycle, part of which is in various vertebrate tissues while the other part involves an insect. In humans, the tissues are the liver and red blood cells, and the insect vector is the female *Anopheles* mosquito. A generalized life cycle is shown in Figure 12-37. Representative life cycle stages for the various species are shown in Figures 12-38 to 12-40.

Four species of *Plasmodium* cause malaria in humans:

P. vivax (benign tertian malaria),
P. malariae (quartan malaria),
P. falciparum (malignant tertian malaria), and
P. ovale (ovale malaria).

The life cycles are similar for each species, as is the progress of the disease, so *P. falciparum* will be discussed as an example, with unique aspects compared to the others.

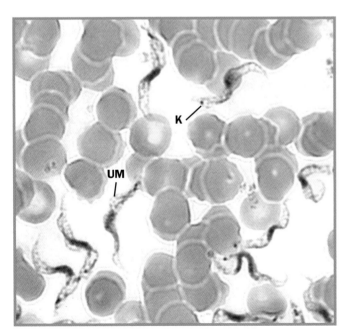

12-36 TRYPANOSOMA BRUCEI TRYPOMASTIGOTES IN A BLOOD SMEAR (X1500) ✦ The central nucleus, posterior kinetoplast (**K**), and undulating membrane (**UM**) are visible.

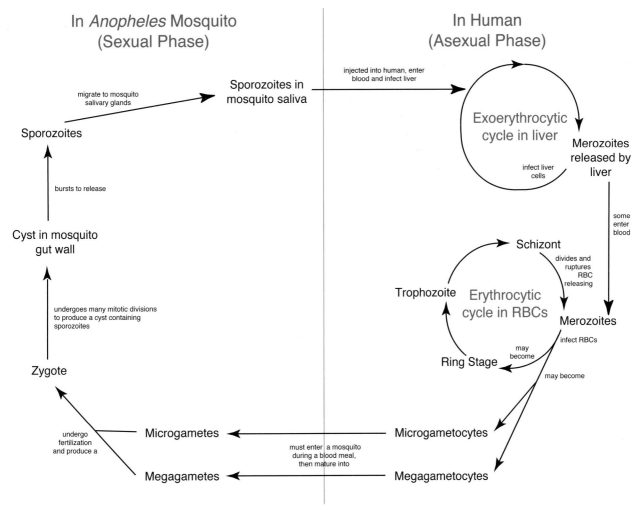

In *Anopheles* Mosquito (Sexual Phase)

In Human (Asexual Phase)

Sporozoites in mosquito saliva

migrate to mosquito salivary glands

Sporozoites

bursts to release

Cyst in mosquito gut wall

undergoes many mitotic divisions to produce a cyst containing sporozoites

Zygote

undergo fertilization and produce a

Microgametes

Megagametes

injected into human, enter blood and infect liver

Exoerythrocytic cycle in liver

infect liver cells

Merozoites released by liver

some enter blood

Schizont

divides and ruptures RBC releasing

Trophozoite   Erythrocytic cycle in RBCs

Ring Stage

may become

Merozoites

infect RBCs

may become

must enter a mosquito during a blood meal, then mature into

Microgametocytes

Megagametocytes

12-37 PLASMODIUM LIFE CYCLE ✦ Please see the text for details.

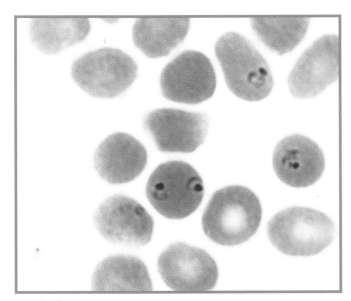

12-38 PLASMODIUM *FALCIPARUM* DOUBLE INFECTION OF A RED BLOOD CELL (X2640) ✦ Double infections are commonly seen in *P. falciparum* infections. A single infection is seen at the right. Young trophozoites are said to be in the "ring stage."

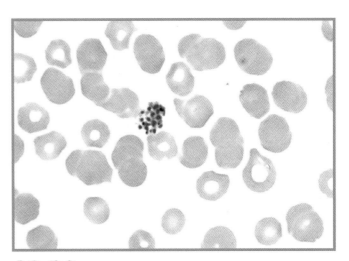

12-39 A MATURE PLASMODIUM *VIVAX* SCHIZONT COMPOSED OF APPROXIMATELY 16 MEROZOITES (X1200) ✦ More than 12 merozoites distinguish *P. vivax* from *P. malariae* and *P. ovale*, both of which typically have eight, but up to 12 merozoites. *P. falciparum* may have up to 24 merozoites, but they typically are not seen in peripheral blood smears and so are not confused with *P. vivax*.

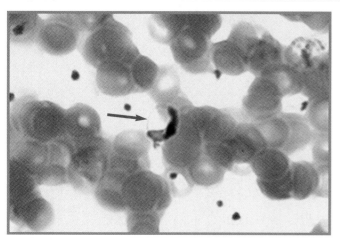

12-40 PLASMODIUM FALCIPARUM GAMETOCYTE IN AN ERYTHROCYTE (X1000) ✦ Differentiation between microgametocytes and megagametocytes is difficult in this species. The erythrocyte membrane is visible around the gametocyte (arrow).

12-41 TOXOPLASMA GONDII TROPHOZOITES (X1000) ✦ Notice the bow-shaped cells and the prominent nuclei.

The **sporozoite** stage of the pathogen is introduced into a human host during a bite from an infected female *Anopheles* mosquito. Sporozoites then infect liver cells and produce the asexual **merozoite** stage. Merozoites are released from lysed liver cells, enter the blood, and infect erythrocytes. (Reinfection of the liver occurs at this stage in all except *P. falciparum* infections.) Once in RBCs, merozoites enter a cyclic pattern of reproduction in which more merozoites are released from the red cells synchronously every 48 hours (hence *tertian*—every third day—malaria).

These events are tied to the symptoms of malaria. A chill, nausea, vomiting, and headache are symptoms that correspond to rupture of the erythrocytes. A spiking fever ensues and is followed by a period of sweating, after which the exhausted patient falls asleep. During this latter phase, the parasites reinfect the red cells and the cycle repeats.

The sexual phase of the life cycle begins when certain merozoites enter erythrocytes and differentiate into male or female **gametocytes**. The sexual phase of the life cycle continues when ingested by a female *Anopheles* mosquito during a blood meal. Fertilization occurs, and the zygote eventually develops into a cyst within the gut wall of the mosquito. After many divisions, the cyst releases sporozoites, some of which enter the mosquito's salivary glands ready to be transmitted back to the human host.

Most malarial infections are cleared eventually, but not before the patient has developed anemia and has suffered permanent damage to the spleen and liver. The most severe infections involve *P. falciparum*. Erythrocytes infected by *P. falciparum* develop abnormal projections that cause them to adhere to the lining of small blood vessels. This can lead to obstruction of the vessels, thrombosis, or local ischemia, which account for many of the fatal complications of this type of malaria—including liver, kidney, and brain damage.

Toxoplasma gondii

Like other sporozoans, the *Toxoplasma gondii* (Figure 12-41) life cycle has sexual and asexual phases. The sexual phase occurs in the lining of cat intestines where **oocysts** are produced and shed in the feces. Each oocyst undergoes division and contains eight **sporozoites**. If ingested by another cat, the sexual cycle may be repeated as the sporozoites produce **gametocytes,** which in turn produce gametes. If ingested by another animal host (including humans) the oocyst germinates in the duodenum and releases the sporozoites. Sporozoites enter the blood and infect other tissues, where they become trophozoites, which continue to divide and spread the infection to lymph nodes and other parts of the reticuloendothelial system. Trophozoites ingested by a cat eating an infected animal develop into gametocytes in the cat's intestines. Gametes are formed, fertilization produces an oocyst, and the life cycle is completed.

Infection via ingestion of the oocyst typically is not serious. The infected person may notice fatigue or muscle aches. The more serious form of the disease involves infection of a fetus across the placenta from an infected mother. This type of infection may result in stillbirth, or liver damage and brain damage. AIDS patients may incur fatal complications from infection.

✦ Materials

Per Student Group
✦ Prepared slides of:
- *Entamoeba histolytica* trophozoite and cyst
- *Entamoeba coli* trophozoite and cyst
- *Balantidium coli* trophozoite and cyst
- *Giardia lamblia* trophozoite and cyst
- *Trichomonas vaginalis* trophozoite
- *Trypanosoma spp.*
- *Plasmodium spp.*
- *Toxoplasma gondii* trophozoite

Procedure

1. Obtain prepared slides of the protozoan pathogens and observe them under appropriate magnification. You should observe the assigned structures on each organism. (Many of these slides are made from patient samples, so there will be a lot of other material on the slide besides the desired organism. You must search carefully and with patience.)

2. Sketch and label the assigned components on the Data Sheet.

Entamoeba histolytica
- Trophozoite
  - pseudopods
  - nucleus with small, central karyosome and beaded nucleus
  - ingested erythrocytes
- Cyst
  - multiple nuclei (up to four) with karyosomes and chromatin as in the trophozoite
  - cytoplasmic chromatoidal bars (maybe)

Entamoeba coli
- Trophozoite
  - same as *E. histolytica* except with eccentric karyosome and unclumped chromatin
- Cyst
  - up to eight nuclei (more than four is enough to distinguish it from *E. histolytica*) that are the same as in the trophozoite
  - cytoplasmic chromatoidal bars (maybe)

Balantidium coli
- Trophozoite
  - elongated shape
  - cilia

- macronucleus
- micronucleus (maybe)
- Cyst
  - spherical shape with multiple nuclei

Giardia lamblia
- Trophozoite
  - oval shape
  - flagella (four pairs)
  - nuclei (two)
  - median bodies (two)
- Cyst
  - multiple nuclei (four)
  - median bodies (four)

Trichomonas vaginalis trophozoite
- nucleus
- flagella (four)

Trypanosoma spp.
- nucleus
- flagellum
- undulating membrane
- kinetoplasm

Plasmodium spp.
- ring stage
- mature trophozoite
- schizont
- male gametocyte
- female gametocyte

Toxoplasma gondii
- trophozoite
- bow-shaped cells
- nucleus

References

Ash, Lawrence R., and Thomas C. Orihel. 1991. Chapter 14 in *Parasites: A Guide to Laboratory Procedures and Identification*. American Society for Clinical Pathology (ASCP) Press, Chicago.

Forbes, Betty A., Daniel F. Sahm, and Alice. S. Weissfeld. 2002. Chapter 52 (pages 650–687) in *Bailey & Scott's Diagnostic Microbiology*, 11th ed. Mosby-Year Book, St. Louis.

Garcia, Lynne Shore. 2001. Chapters 2, 3, 5, 7, and 9 in *Diagnostic Medical Parasitology*, 4th ed. ASM Press, Washington, DC.

Koneman, Elmer W., Stephen D. Allen, William M. Janda, Paul C. Schreckenberger, and Washington C. Winn, Jr. 1997. Chapter 20 in *Color Atlas and Textbook of Diagnostic Microbiology*, 5th ed. J. B. Lippincott, Philadelphia.

Lee, John J., Seymour H. Hutner, and Eugene C. Bovee. 1985. *Illustrated Guide to the Protozoa*. Society of Protozoologists, Lawrence, KS.

Markell, Edward K., Marietta Voge and David T. John. 1992. *Medical Parasitology*, 7th ed. W. B. Saunders, Philadelphia.

12-4 Parasitic Helminths

A study of helminths is appropriate to the microbiology lab because clinical specimens may contain microscopic evidence of helminth infection. The three major groups of parasitic worms encountered in lab situations are trematodes (flukes), cestodes (tapeworms), and nematodes (round worms). Life cycles of the parasitic worms are often complex, sometimes involving several hosts, and are beyond the scope of this book. Emphasis here is on a brief background and clinically important diagnostic features of selected worms.

Trematode Parasites Found in Clinical Specimens

Clonorchis (Opisthorchis) sinensis

Clonorchis sinensis is the Oriental liver fluke (Figure 12-42) and causes clonorchiasis, a liver disease. It is a common parasite of people living in Japan, Korea, Vietnam, China, and Taiwan and is becoming more common in the United States with the influx of Southeast Asian immigrants. Infection typically occurs when undercooked infected fish is ingested. The adults migrate to the liver bile ducts and begin laying eggs in approximately one month. Degree of damage to the bile duct epithelium and surrounding liver tissue is due to the number of worms and the duration of infection. Diagnosis is made by identifying the characteristic eggs in feces (Figure 12-43).

Paragonimus westermani

Paragonimus westermani (Figure 12-44) is a lung fluke and one of several species to cause paragonimiasis, a

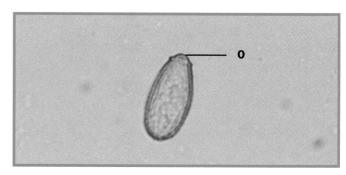

12-43 CLONORCHIS SINENSIS EGG IN A FECAL SPECIMEN (X1000, D'ANTONI'S IODINE STAIN) ✦ The eggs are thick-shelled and between 27 and 35 µm long. There is a distinctive operculum (**O**) (positioned to give the appearance of shoulders) and often a knob at the aboperccular end (not visible in this specimen).

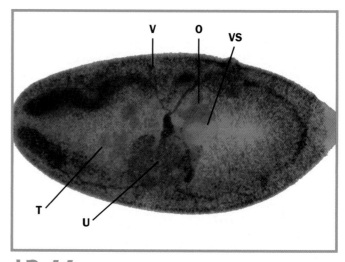

12-44 PARAGONIMUS WESTERMANI ADULT ✦ Adults are up to 1.2 cm long, 0.6 cm wide and 0.5 cm thick. Visible in this specimen are the ventral sucker (**VS**), ovary (**O**), uterus (**U**), testis (**T**), and vitellaria (**V**).

disease mostly found in Asia, Africa, and South America. *P. westermani* is primarily a parasite of carnivores, but humans (omnivores) may get infected when eating undercooked crabs or crayfish infected with the cysts. Ingested juveniles excyst in the duodenum and travel to the abdominal wall. After several days, they resume their journey and find their way to the bronchioles where the adults mature. Eggs (Figure 12-45) are released in approximately two to three months and are diagnostic of infection. They may be recovered in sputum, lung fluids, or feces. Consequences of lung infection are a local inflammatory

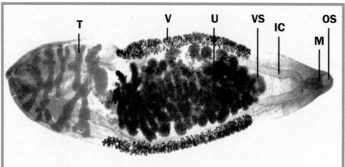

12-42 CLONORCHIS SINENSIS ADULT (X5) ✦ Adults range in size from 1 cm to 2.5 cm. Visible in this specimen are the oral sucker (**OS**), mouth (**M**), intestinal ceca (**IC**), ventral sucker (**VS**), uterus (**U**), vitellaria (**V**), and testis (**T**).

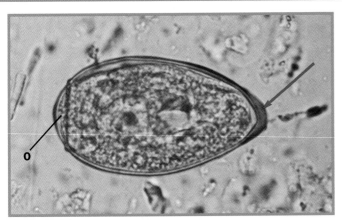

12-45 PARAGONIMUS WESTERMANI EGG IN A FECAL SPECIMEN (X1000, D'ANTONI'S IODINE STAIN) ✦ *Paragonimus westermani* eggs are ovoid and range in size from 80 to 120 μm long by 45 to 70 μm wide. They have an operculum (**O**) and the shell is especially thick at the aapercular end (arrow). They are unembryonated when seen in feces.

response followed by possible ulceration. Symptoms include cough with discolored or bloody sputum and difficulty breathing. These cases are rarely fatal, but may last a couple of decades. Occasionally, the wandering juveniles end up in other tissues, such as the brain or spinal cord, which can cause paralysis or death.

Schistosoma mansoni

Schistosoma mansoni (Figure 12-46) is found in Brazil, some Caribbean islands, Africa, and parts of the Middle

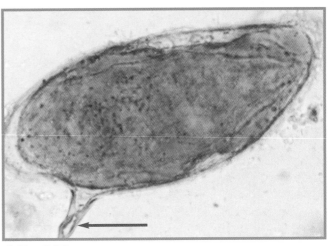

12-47 SCHISTOSOMA MANSONI EGG IN A FECAL SPECIMEN (X1000, D'ANTONI'S IODINE STAIN) ✦ *Schistosoma mansoni* eggs are large (114 to 175 μm long by 45 to 70 μm wide) and contain a larva called a miracidium. They are thin-shelled, lack an operculum, and have a distinctive lateral spine (arrow).

East. Infection occurs via contact with fecally contaminated water containing juveniles of the species. The juveniles penetrate the skin, enter circulation and continue development in the intestinal veins. Eggs (Figure 12-47) from the adults penetrate the intestinal wall and are passed out with the feces. Presence of eggs in the feces indicates infection. Some patients are asymptomatic, whereas others have bloody diarrhea, abdominal pain, and lethargy.

Cestode Parasites Found in Clinical Specimens

Dipylidium caninum

Dipylidium caninum (Figure 12-48) is a common parasite of dogs and cats. Human infection usually occurs in

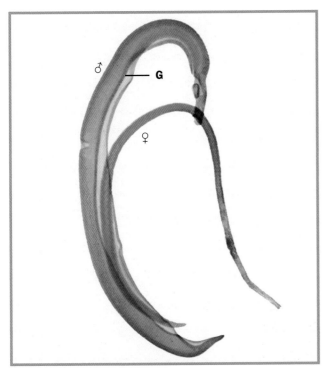

12-46 SCHISTOSOMA MANSONI ADULTS (X20) ✦ The male is larger and has a gynecophoric groove (**G**) in which the slender female resides during mating.

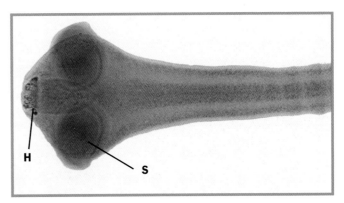

12-48 DIPYLIDIUM CANINUM SCOLEX (X4) ✦ Adult worms may reach a length of 50 cm with a width of 3 mm. The scolex has four suckers (**S**) and a retractable rostellum with rows of hooks (**H**).

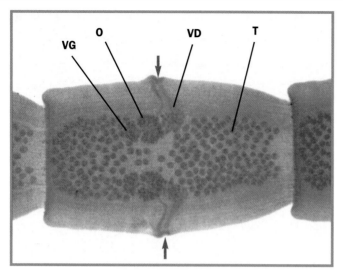

12-49 DIPYLIDIUM CANINUM PROGLOTTIDS (X40) ✦ Visible are the testes (**T**), vasa deferentia (**VD**), ovaries (**O**), and vitelline glands (**VG**). The reproductive openings (arrows) on each side of the proglottid give this worm its common name—the "double-pored tapeworm."

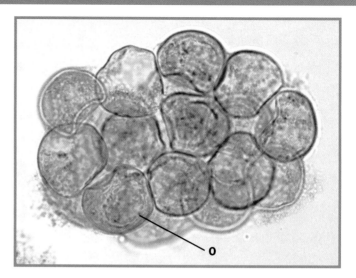

12-50 DIPYLIDIUM CANINUM EGG PACKET (X1000) ✦ Each Dipylidium caninum egg packet is composed of 5 to 15 eggs each with an onchosphere (**O**). The onchosphere contains six hooklets.

children. The adult worms reside in dog or cat intestines and release proglottids (Figure 12-49) containing egg packets that migrate out of the anus. When these dry, they look like rice grains. Larval fleas may eat the eggs and become infected. If the dog, cat, or child ingests one of these fleas, the life cycle is completed in the new host. Infection may be asymptomatic or produce mild abdominal discomfort, loss of appetite and indigestion. Diagnosis is made by identifying the egg packets (Figure 12-50).

Echinococcus granulosus

The definitive host of *Echinococcus granulosus* (Figure 12-51) is a carnivore, but the life cycle requires an intermediate host, usually an herbivorous mammal. Humans involved in raising domesticated herbivores (*e.g.,* sheep with their associated dogs) are most susceptible as intermediate hosts and develop hydatid disease. Ingestion of a juvenile *E. granulosus* leads to development of a hydatid cyst in the lung, liver, or other organ, a process that may take many years. The cyst (Figure 12-52) has a thick wall and develops many protoscolices within (Figure 12-53).

The protoscolices, if ingested, are infective to the definitive host. Symptoms depend on the location and size of the hydatid cyst, which interferes with normal organ function. Due to sensitization by the parasite's antigens, release of fluid from the cyst can result in anaphylactic shock of the host. Diagnosis is made by detection of the cyst by ultrasound or X-ray.

Hymenolepis (Vampirolepis) nana

Hymenolepis nana (Figures 12-54 and 12-55) is the dwarf tapeworm and is the most common cestode parasite of humans in the world. When eggs (Figure 12-56) are ingested, the oncospheres develop into juveniles in the lymphatics of intestinal villi. These juveniles are then released into the lumen within a week and attach to the mucosa to mature into adults. Infection may involve hundreds of worms, yet symptoms are usually mild: diarrhea, nausea, loss of appetite, or abdominal pain. Eggs may reinfect the same host or pass out with the feces to infect a new host. Eggs in the feces are used for identification, but proglottids are not as they are rarely passed.

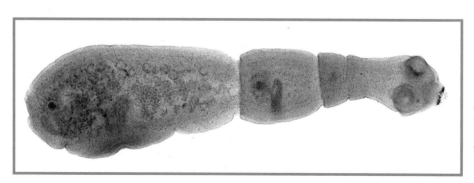

12-51 ECHINOCOCCUS GRANULOSUS ADULT (X20) ✦ Adult worms are about 0.5 cm in length. There is a scolex with a ring of hooks, a neck, and one proglottid that contains up to 1500 eggs.

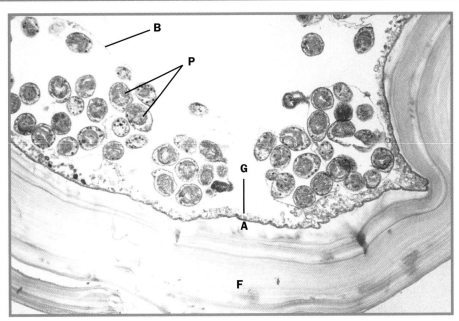

12-52 *ECHINOCOCCUS GRANULOSUS* CYST IN LUNG TISSUE (SEC., X96) ✦ The cyst wall consists of a fibrous layer of host tissue (**F**), an acellular layer (**A**), and a germinal epithelium (**G**) that gives rise to stalked brood capsules (**B**). Brood capsules produce many *E. granulosus* protoscolices (**P**).

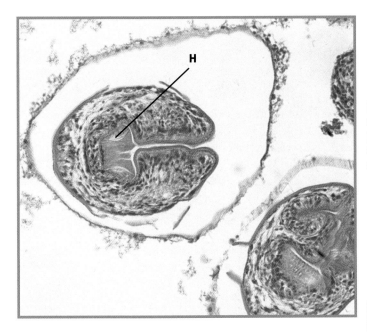

12-53 *ECHINOCOCCUS GRANULOSUS* PROTOSCOLEX (L.S., X320) ✦ The protoscolex contains an invaginated scolex with hooks (**H**). Upon ingestion by the host, the protoscolex evaginates and produces an infectious scolex that attaches to the intestinal wall, matures and produces eggs.

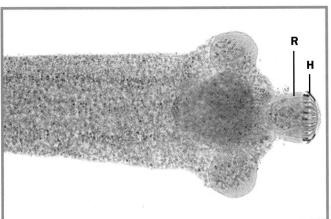

12-54 *HYMENOLEPIS NANA* SCOLEX (X40) ✦ Adults gain a length of up to 10 cm, but are only 1 mm in width. The rostellum (**R**) is armed with up to 30 hooks (**H**).

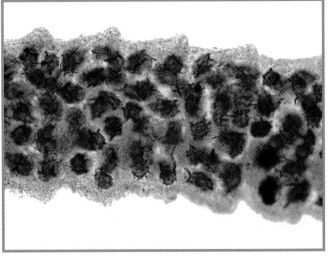

12-55 *HYMENOLEPIS NANA* PROGLOTTIDS (X40) ✦ *H. nana* is hermaphroditic, but in this specimen, the numerous testes obscure most other structures.

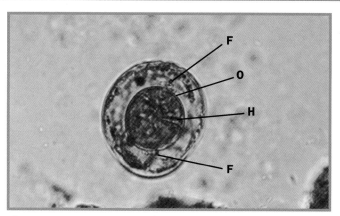

12-56 HYMENOLEPIS NANA EGG IN FECES (X1000, D'ANTONI'S IODINE STAIN) ✦ The *Hymenolepis nana* egg is 30 to 47 μm in diameter and has a thin shell. The oncosphere (**O**) is separated from the shell and contains six hooks (**H**). Another distinguishing feature is the presence of between four and eight filaments (**F**) arising from either end of the oncosphere.

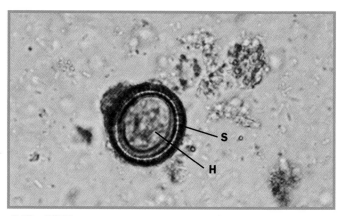

12-57 TAENIA EGG IN FECES (X1000) ✦ Taeniid eggs are distinctive looking enough to identify to genus, but not distinctive enough to speciate. Eggs are spherical and approximately 40 μm in diameter with a striated shell (**S**). The oncosphere contains six hooks (**H**).

Taenia spp.

Two taeniid worms are important human pathogens. These are *Taenia saginata* (*Taeniarhynchus saginatus*)—the beef tapeworm—and *Taenia solium*—the pork tapeworm.

T. saginata infects humans who eat undercooked beef containing juvenile worms. In the presence of bile salts, the juveniles develop into adults and begin producing gravid proglottids within a few weeks. Symptoms of infection are usually mild nausea, diarrhea, abdominal pain, and headache. Diagnosis to species is impossible with only the eggs (Figure 12-57); specific identification requires a scolex or gravid proglottid (Figure 12-58).

The *T. solium* life cycle is similar to *T. saginata*, but the host is pork, not beef, so human infection occurs when undercooked pork is eaten. If eggs are ingested, a juvenile form called a cysticercus develops. Cysticerci may be found in any tissue, especially subcutaneous connective tissues, eyes, brain, heart, liver, lungs, and coelom. Symptoms of cysticercosis depend on the tissue infected, but mostly they are not severe. However, death of a cysticercus can produce a rapidly fatal inflammatory response. As with *T. saginata*, diagnosis to species is impossible with only the eggs; specific identification requires a scolex or gravid proglottid (Figures 12-59 and 12-60).

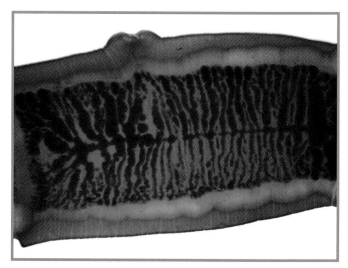

12-58 TAENIA SAGINATA PROGLOTTID (X12) ✦ The uterus of *Taenia saginata* proglottids consists of a central portion with 15 to 20 lateral branches (compare to the *T. solium* uterus in Figure 12-60).

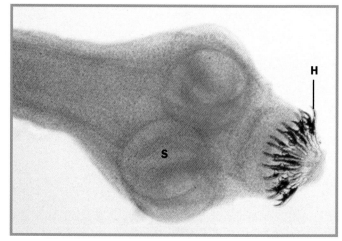

12-59 TAENIA SOLIUM SCOLEX (X64) ✦ The *Taenia solium* scolex has two rings of hooks (**H**) and four suckers (**S**).

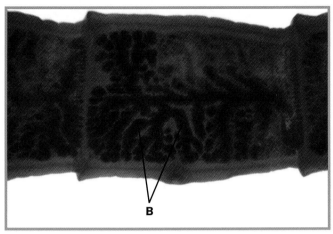

12-60 TAENIA SOLIUM PROGLOTTID (X33) ✦ The uterus of *Taenia solium* proglottids consists of a central portion with 7 to 13 lateral branches (compare to the *T. saginata* uterus in Figure 12-58).

Nematode Parasites Found in Clinical Specimens

Ascaris lumbricoides

Ascaris lumbricoides (Figures 12-61 and 12-62) is a large nematode—females may reach a length of 49 cm! Human infection occurs when eggs in fecally contaminated soil or food are ingested. Juveniles emerge in the intestine, penetrate its wall, and then migrate to the lungs and other tissues. After a period of development in the lungs, the juveniles move up the respiratory tree to the esophagus and are swallowed again. Adults then reside in the small intestine and produce eggs (Figures 12-63 and 12-64). Infection may result in inflammation in organs other than the lungs where juvenile worms settled incorrectly. *Ascaris* pneumonia occurs in heavy infections due to the lung damage caused by the juveniles. If secondary bacterial infections occur, the pneumonia can be fatal. Blockage of the intestines and malnutrition

also are possible in heavy infections. Lastly, under certain conditions, worms can wander to other body locations and cause damage or blockage. Identification of an *Ascaris* infection is made by observing the eggs in feces.

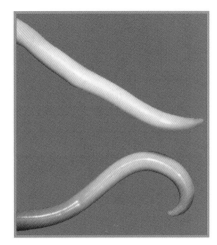

12-62 ASCARIS LUMBRICOIDES ADULT WORMS ✦ *A. lumbricoides* males (bottom) are shorter than females (up to 31 cm vs. 35 cm) and have a curved posterior.

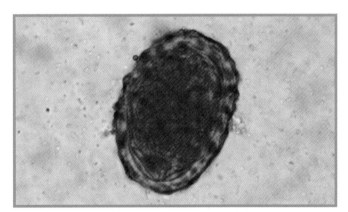

12-63 FERTILE ASCARIS LUMBRICOIDES FERTILE EGG IN FECES (X1000, D'ANTONI'S IODINE STAIN) ✦ Fertile *Ascaris lumbricoides* eggs are 55 to 75 µm long and 35 to 50 µm wide and are embryonated. Their surface is covered by small bumps called mammillations.

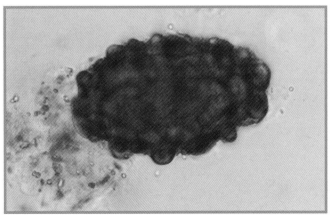

12-64 INFERTILE ASCARIS LUMBRICOIDES EGG IN FECES (X1000, D'ANTONI'S IODINE STAIN) ✦ Infertile eggs are longer (up to 90 µm) than fertile eggs. There is no embryo inside.

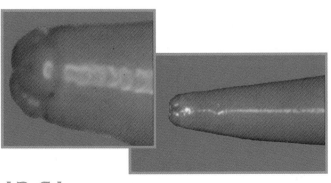

12-61 ASCARIS LUMBRICOIDES ANTERIOR ✦ *A. lumbricoides* has a cylindrical shape with three prominent mouth parts (see inset).

Enterobius vermicularis

Enterobius vermicularis (Figure 12-65) is the human pinworm. It is found worldwide and is especially prevalent among people in institutions (such as orphanages and mental hospitals) because conditions favor fecal–oral transmission of the parasite. Bedding, clothing, and the fingers (from scratching) become contaminated and may be involved in transmission. Poor sanitary habits of children make them especially prone to infecting others. Transmission may also involve eggs (Figure 12-66) being carried on air currents and then inhaled by a susceptible host. After ingestion, eggs hatch in the duodenum and mature in the large intestine where the adults reside. Adult females emerge from the anus at night to lay between 4,600 and 16,000 eggs in the perianal region. About one-third of pinworm infections are asymptomatic. The other two-thirds usually do not produce serious symptoms. Diagnosis is made by identifying the eggs. Since the eggs are laid externally, they are rarely found in feces. Instead, they are collected on cellophane tape from the perianal region and examined microscopically.

Hookworms (*Ancylostoma duodenale* and *Necator americanus*)

The hookworm *Ancylostoma duodenale* (Figure 12-67) and *Necator americanus* (Figure 12-68) have very similar morphologies and life cycles, and the eggs are indistinguishable, so they are considered together here. Infection occurs when juveniles penetrate the skin, enter the blood and travel to the lungs. They penetrate the respiratory membrane and are carried up and out of the lungs by ciliary action to the pharynx, where they are swallowed. When they reach the small intestine, they attach and

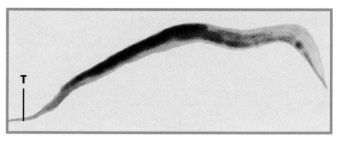

12-65 *ENTEROBIUS VERMICULARIS* ADULT FEMALE ✦ Female pinworms are about 1 cm long and have a pointed tail (T) from which this group derives its common name—pinworm. Males are about half that size and have a hooked tail.

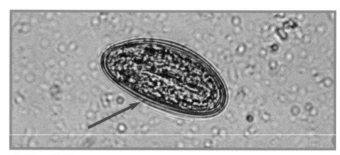

12-66 *ENTEROBIUS VERMICULARIS* EGG (X1000, D'ANTONI'S IODINE STAIN) ✦ The eggs of *Enterobius vermicularis* are 50 to 60 µm long and 20 to 40 µm wide with one side flattened (arrow). They are usually embryonated in typical preparations.

mature into adults that feed on blood and tissues of the host. Adults are rarely seen as they remain attached to the intestinal mucosa. Eggs (Figures 12-69) are passed in the feces and are diagnostic of infection. The severity of hookworm disease symptoms is related to the parasite load, and most infections are asymptomatic. As a rule, severe symptoms of bloody diarrhea and iron deficiency anemia are only seen in acute heavy or chronic infections.

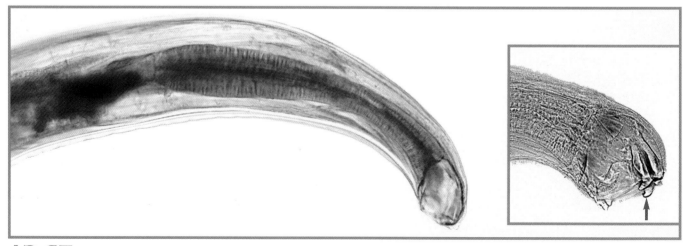

12-67 ANTERIOR OF *ANCYLOSTOMA DUODENALE* (X100) ✦ *Ancylostoma duodenale* head showing the mouth and thick-walled esophagus. The bend in the head gives this group its common name—hookworm. In the inset, the chitinous teeth (arrow) are visible (compare with Figure 12-68).

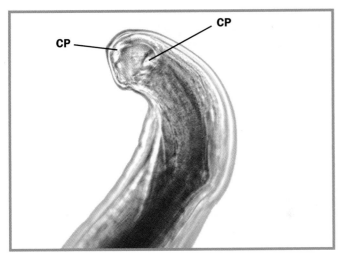

12-68 NECATOR AMERICANUS HEAD (X160) ✦ This detail of *N. americanus* shows the cutting plates (**CP**) that help to differentiate it from *A. duodenale* (compare with Figure 12-67). Also notice the hooked head.

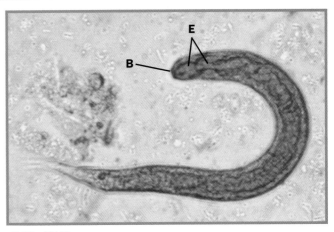

12-70 STRONGYLOIDES STERCORALIS RHABDITIFORM LARVA IN FECES (X1000, D'ANTONI'S IODINE STAIN) ✦ These larvae may be distinguished from hookworm larvae (which are rarely in feces) by their short buccal cavity (**B**). The name "rhabditiform" refers to the esophagus (**E**) shape, which has a constriction within it.

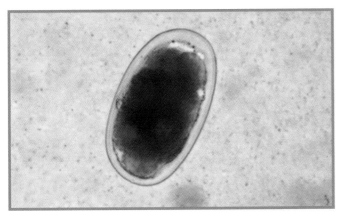

12-69 HOOKWORM EGG IN FECES (X1000, D'ANTONI'S IODINE STAIN) ✦ Hookworm eggs are 55 to 75 μm long and 36 to 40 μm wide. They have a thin shell and contain a developing embryo (seen here at about the 16 cell stage) that is separated from the shell when seen in fecal samples.

Strongyloides stercoralis

Strongyloides stercoralis is the intestinal threadworm. Infection occurs by penetration of the skin by infective juveniles from fecally contaminated soil. The juveniles then migrate to the lungs and develop into parthenogenetic females that migrate to the pharynx, are swallowed, and then burrow into the intestinal mucosa. Each day they release a few dozen eggs that develop into juveniles (Figure 12-70) before they are passed in the feces. These juveniles may become infective or may follow a developmental path that produces free-living

adults. These adults eventually produce more infective juveniles and the cycle is completed. Symptoms of infection may be itching or secondary bacterial infection at the site of entry by the infective juveniles, a cough and burning of the chest during the pulmonary phase, and abdominal pain and perhaps septicemia during the intestinal phase. Diagnosis is by finding rhabditiform larvae in fresh fecal samples.

Wuchereria bancrofti

Wuchereria bancrofti is a filarial worm that causes lymphatic filariasis. Infection occurs from the bite of a mosquito harboring infective juveniles. Upon injection into the host, the worms migrate into the large lymphatic vessels of the lower body and mature. Adults are found in coiled bunches and the females release microfilariae (Figure 12-71) by the thousands. Microfilariae enter the blood and circulate there, often with a daily periodicity—most abundant at night when the mosquito vector is active and hidden away in lung capillaries during the day when it is hot. Some infections are asymptomatic, whereas others result in acute inflammation of lymphatics associated with fever, chills, tenderness, and toxemia. In the most serious cases, obstruction of lymphatic vessels occurs and results in elephantiasis, a disease caused by accumulation of lymph fluid in the tissues, an accumulation of fibrous connective tissue, and a thickening of the skin. Diagnosis of infection is made by identifying microfilariae in blood smears.

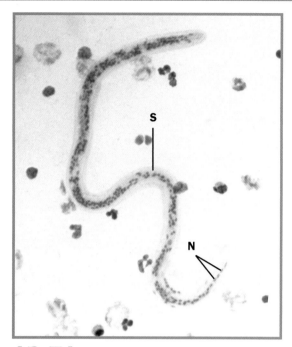

12-71 WUCHERERIA BANCROFTI MICROFILARIA IN A BLOOD SAMPLE (X480) ✦ The microfilariae of *Wuchereria bancrofti* can be distinguished from others in the blood by the sheath (**S**) and the single column of nuclei (**N**) not extending to the tip of the tail.

✦ Materials

✦ Prepared slides of:
- *Ascaris lumbricoides* eggs in a fecal smear
- *Clonorchis sinensis* in a fecal smear
- *Dipylidium caninum* eggs in a fecal smear
- *Echinococcus granulosus* hydatid cyst in section
- *Enterobius vermicularis* eggs in a fecal smear
- Hookworm (*Ancylostoma duodenale* or *Necator americanus*) eggs in a fecal smear
- *Hymenolepis nana* eggs in a fecal smear
- *Paragonimus westermani* eggs in a fecal smear
- *Schistosoma mansoni* eggs in a fecal smear
- *Strongyloides stercoralis* rhabditiform larva in a fecal smear
- *Taenia solium* proglottid (whole mount)
- *Taenia solium* scolex (whole mount)
- *Taenia spp.* eggs in a fecal smear
- *Wuchereria bancrofti* microfilariae in a blood smear

 Procedure

Observe the prepared slides provided of the helminth specimens in fecal smears and other tissues. Scanning on low power (10× objective) is best for most preparations, then move to high-dry or oil immersion to see detail. Most egg specimens are in fecal smears, so there will be a lot of other material on the slide besides the eggs. You must search carefully and with patience. Sketch what you see in the spaces provided on the Data Sheet and measure dimensions. Be able to identify each species if shown an unlabeled specimen.

References

Ash, Lawrence R. and Thomas C. Orihel. 1991. Chapters 15 and 16 in *Parasites: A Guide to Laboratory Procedures and Identification.* American Society for Clinical Pathology (ASCP) Press, Chicago, IL.

Forbes, Betty A., Daniel F. Sahm, and Alice. S. Weissfeld. 2002. Chapter 52 in (Pages 687–698) *Bailey & Scott's Diagnostic Microbiology,* 11th Ed. Mosby-Year Book, Inc. St. Louis.

Garcia, Lynne Shore. 2001. Chapters 10, 12, 13, 14, 16, and 17 in *Diagnostic Medical Parasitology,* 4th Ed. ASM Press, Washington, DC.

Koneman, Elmer W., Stephen D. Allen, William M. Janda, Paul C. Schreckenberger and Washington C. Winn, Jr. 1997. Chapter 20 in *Color Atlas and Textbook of Diagnostic Microbiology,* 5th Ed. J. B. Lippincott Company, Philadelphia.

Orihel, Thomas C., and Lawrence R. Ash. 1999. Chapter 112 in *Manual of Clinical Microbiology,* 7th Ed., edited by Patrick R. Murray, Ellen Jo Baron, Michael A. Pfaller, Fred C. Tenover, and Robert H. Yolken. American Society for Microbiology, Washington, DC.

Roberts, Larry S. and John Janovy, Jr. *Foundations of Parasitology,* 5th Ed. Wm. C. Brown Publishers, Dubuque, IA.

Biochemical Pathways

So much of what is done in microbiology relies on an understanding of basic biochemical pathways. It's not as important to memorize them (although, with exposure they will become second nature) as it is to understand their importance in metabolism and to interpret diagrams of them when available. The following discussion is provided so you can see how the various biochemical tests presented in this manual fit into the overall scheme of cellular chemistry.

Oxidation of Glucose: Glycolysis, Entner-Doudoroff, and Pentose-Phosphate Pathways

Most organisms use **glycolysis** (also known as the "Embden-Meyerhof-Parnas pathway, Figure A-1) in energy metabolism. It performs the stepwise disassembly of glucose into two pyruvates, releasing some of its energy and electrons in the process. The exergonic (energy-releasing) reactions are associated with ATP synthesis by a process called **substrate phosphorylation**. Although a total of four ATPs are produced per glucose in glycolysis, two ATPs are hydrolyzed early in the pathway, leaving a net production of two ATPs per glucose. In one glycolytic reaction, the loss of an electron pair (oxidation) from a three-carbon intermediate occurs simultaneously with the reduction of NAD^+ to $NADH + H^+$. The $NADH + H^+$ then may be oxidized in an electron transport chain or a fermentation pathway, depending on the organism and the environmental conditions. The former yields ATP, and the latter generally does not. In summary, each glucose oxidized in glycolysis yields two pyruvates, $2 NADH + 2 H^+$, and a net of 2 ATPs (Table A-1).

Although the intermediates of glycolysis are carbohydrates, many are entry points for amino acid, lipid, and nucleotide catabolism. Many glycolytic intermediates also are a source of carbon skeletons for the synthesis of these other biochemicals. Some of these are shown in Figure A-1. *Note*: For clarity, many details have been omitted from these other pathways in Figure A-1. Single arrows may represent several reactions, and other carbon compounds not illustrated may be required to complete a particular reaction.

The **Entner-Doudoroff pathway** (Figure A-2) is an alternative means of degrading glucose into two pyruvates. This pathway is found exclusively among prokaryotes (*e.g.*, *Pseudomonas* and *E. coli*, as well as other Gram-negatives and certain Archaeans). It allows utilization of a different category of sugars (aldonic acids) than glycolysis and therefore improves the range of resources available to the organism. It is less efficient than glycolysis because only one ATP is phosphorylated and only one NADH is produced. Table A-2 summarizes this pathway.

The **pentose-phosphate pathway** is a complex set of cyclic reactions that provides a mechanism for producing five-carbon sugars (**pentoses**) from six-carbon sugars (**hexoses**). Pentose sugars are used in ribonucleotides and deoxyribonucleotides, as well as being precursors to aromatic amino acids. Further, this pathway produces NADPH, which is used as an electron donor in anabolic pathways. Unlike NADH, produced in glycolysis and Entner-Doudoroff, NADPH is not used as an electron donor in an electron transport chain for oxidative phosphorylation of ADP.

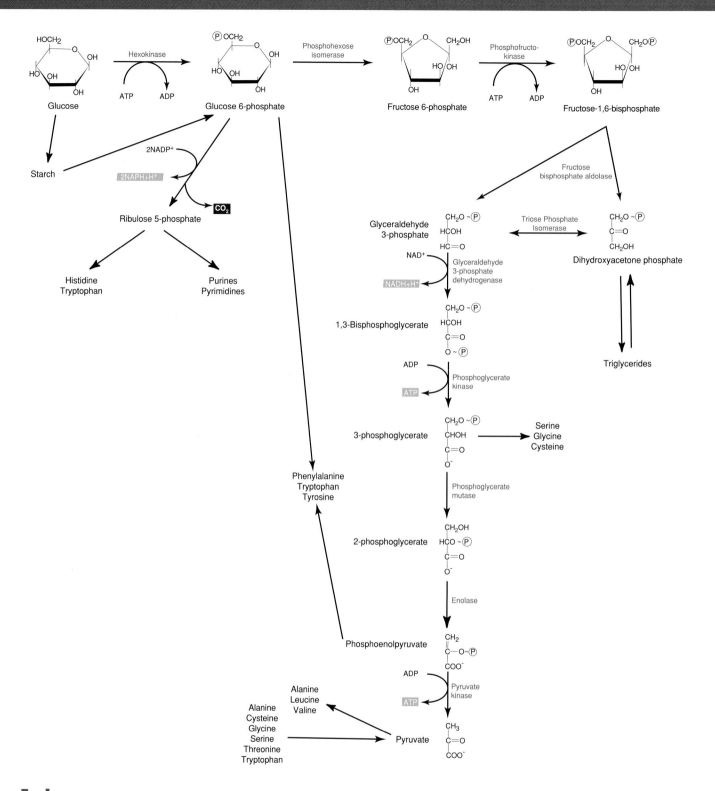

A-1 GLYCOLYSIS AND ASSOCIATED PATHWAYS ✦ The names of glycolytic intermediates are printed in black ink; the enzyme names are in red. Reducing power (in the form of NADH + H⁺) and ATP are highlighted in blue. The major key to getting product yields correct is to recognize that *both* C₃ compounds (Glyceraldehyde 3-phosphate and Dihydroxyacetone phosphate) produced from splitting Fructose 1,6-bisphosphate can pass through the remainder of the pathway because of the triose phosphate isomerase reaction. The conversion of each into pyruvate results in the formation of 2 ATPs and 1 NADH + H⁺ (Table A-1).

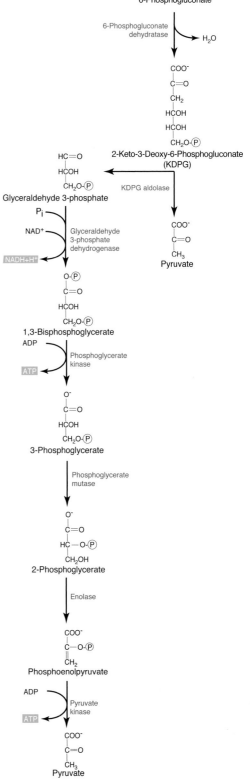

A-2 ENTNER-DOUDOROFF PATHWAY ✦

Notice the similarities between this pathway and glycolysis (Figure A-1). The main difference is in the six-carbon compound that is split into two three-carbon compounds. The result of this split is pyruvate and glyceraldehydes-3-phosphate, which is oxidized as in glycolysis to pyruvate. Because only one three-carbon compound goes through the sequence of reactions leading to pyruvate, the ATP and NADH yield is one-half that of glycolysis. But one NADPH is produced that is not made in glycolysis.

TABLE **A-1** Summary of Glycolytic Reactants and Products per Glucose

TABLE OF RESULTS	
Reactant	Product
Glucose ($C_6H_{12}O_6$)	2 Pyruvates ($C_3H_3O_3$)
2 ATP	2 ADP
4 ADP	4 ATP
NET: 2 ADP	NET: 2 ATP
2 NAD$^+$	2 NADH + 2 H$^+$

TABLE **A-2** Summary of Entner-Doudoroff Reactants and Products per Glucose

TABLE OF RESULTS	
Reactant	Product
Glucose ($C_6H_{12}O_6$)	2 Pyruvates ($C_3H_3O_3$)
1 ATP	1 ADP
2 ADP	2 ATP
NET: 1 ADP	NET: 1 ATP
1 NAD$^+$	1 NADH + 1H$^+$
1 NADP$^+$	1 NADPH + 1H$^+$

TABLE **A-3** Summary of Pentose-Phosphate Reactants and Products per Glucose-6-phosphate

TABLE OF RESULTS

Reactant	Product
Glucose-6-phosphate (C_6)	$6CO_2 + 1\ P_i$
12 NADP$^+$	12 NADPH + 12 H$^+$

The pentose-phosphate reactants and products are listed in Table A-3, and the overall path is shown in Figure A-3. To completely oxidize one hexose to $6CO_2$, a total of six hexoses must enter the cycle as glucose-6-phosphate and follow one of three different routes (notice the symmetry of pathways as drawn). Notice in Figure A-3 that each hexose loses a CO_2 upon entry into the cycle, but at the end, five hexoses are produced. Thus, the net reaction is one hexose being oxidized to $6CO_2$. Notice also the reactions that transfer two-carbon and three-carbon fragments between the five-carbon intermediates. **Transketolase** catalyzes the two-carbon transfer, whereas **transaldolase** catalyzes the three-carbon transfer. Alternatively, the five-carbon intermediates can be redirected into pathways for synthesis of aromatic amino acids and nucleotides (not shown).

Oxidation of Pyruvate: The Krebs Cycle and Fermentation

Pyruvate represents a major crossroads in metabolism. Some organisms are able to further disassemble the pyruvates produced in glycolysis and Entner-Doudoroff and make more ATP and NADH + H$^+$ in the **Krebs cycle**. Other organisms simply reduce the pyruvates with electrons from NADH + H$^+$ without further energy production in **fermentation**.

The Krebs cycle is a major metabolic pathway used in energy production by organisms that respire aerobically or anaerobically (Figure A-4). Pyruvate produced in glycolysis or other pathways is first converted to acetyl-coenzyme A during the **entry step** (also known as the **intermediate** or **gateway step**). Acetyl-CoA enters the Krebs cycle through a condensation reaction with oxaloacetate. Products for each pyruvate that enters the cycle via the entry step are: 3 CO_2, 4 NADH + H$^+$, 1 FADH$_2$, and 1 GTP. (Because two pyruvates are made per glucose, these numbers are doubled in Table A-4). The energy released from oxidation of reduced coenzymes (NADH + H$^+$ and FADH$_2$) in an electron transport chain is then used to make ATP. ATP yields are summarized in Table A-5.

Like glycolysis, many of the Krebs cycle's intermediates are entry points for amino acid, nucleotide and lipid catabolism, as well as a source of carbon skeletons for synthesis of the same compounds. These pathways are shown, but details have been omitted. Single arrows may represent several reactions, and other carbon compounds, not illustrated, may be required to complete a given reaction.

Figure A-5 illustrates some major fermentation pathways exhibited by microbes (though no single organism is capable of all of them). Pyruvate (shown in the blue box) is typically the starting point for each. End products of fermentation are shown in red. Fermentation allows a cell living under anaerobic conditions to oxidize reduced coenzymes (such as NADH + H$^+$ and shown in blue) generated during glycolysis or other pathways. Some bacteria (aerotolerant anaerobes) rely solely on fermentation and do not use oxygen even if it is available. Table A-6 summarizes major fermentations and some representative organisms that perform each.

Notice that fermentation end products typically fall into three categories: acid, gas, or an organic solvent (an alcohol or a ketone). The specific fermentation performed is the result of the enzymes present in a species and often is used as a basis of classification.

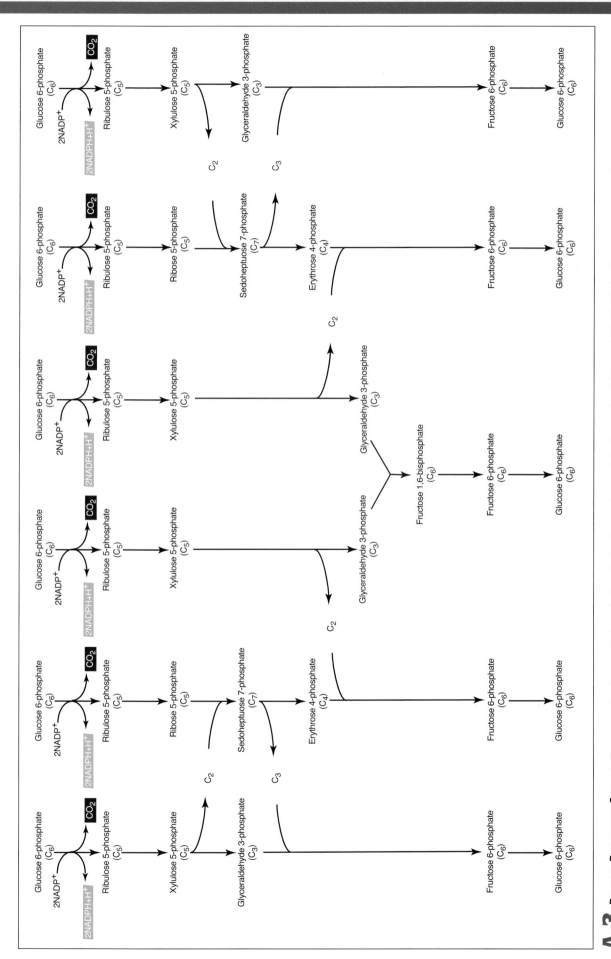

A-3 **PENTOSE-PHOSPHATE CYCLE** ✦ For every six glucose-6-phosphates that enter and complete the cycle, 6CO_2 and 12 NADPH + H$^+$ are produced. Some of the five-carbon intermediates, however, may be redirected into synthesis of aromatic amino acids and nucleotides. If the cycle is performed as shown, 36 carbons enter as six glucose 6-phosphate (6 × C_6 = 36C). Six CO_2 are immediately lost, leaving a total of 30C to get shuffled around by the remaining reactions to form five glucose 6-phosphates (5 × C_6 = 30C).

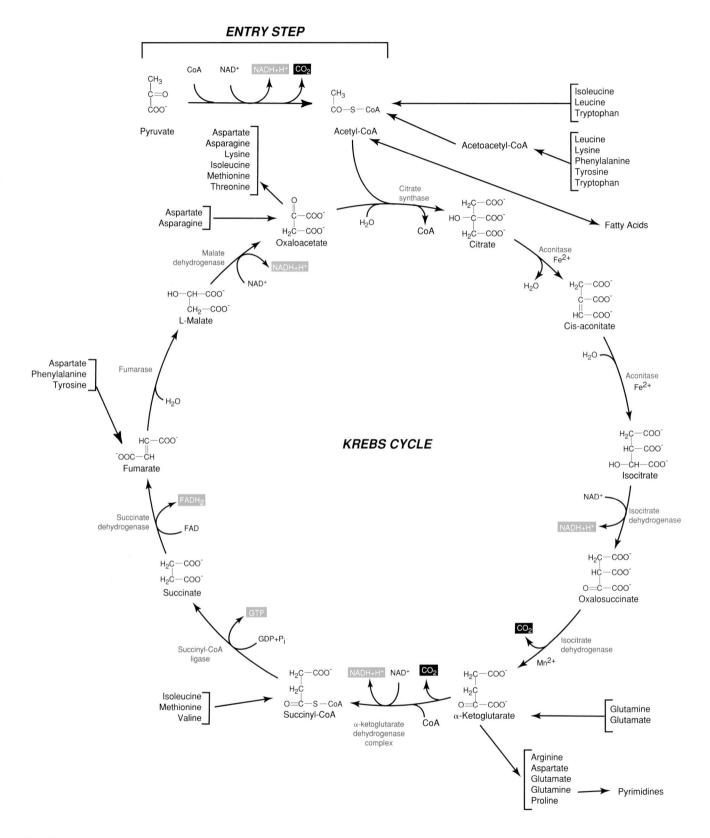

A-4 THE ENTRY STEP AND KREBS CYCLE ✦ The names of intermediates are printed in black ink; enzymes are in red. Reducing power (in the form of NADH+H$^+$ and FADH$_2$) and GTP are highlighted in blue. CO$_2$ produced from the oxidation of carbon is highlighted in black.

TABLE **A-4** Summary of Reactants and Products per Glucose in the Entry Step and the Krebs Cycle

Entry Step		Krebs Cycle	
Reactant	Product	Reactant	Product
2 Pyruvates	2 Acetyl CoA + 2 CO_2	2 Acetyl CoA	4 CO_2
2 Coenzyme A			2 Coenzyme A
2 NAD^+	2 NADH + 2H^+	6 NAD^+	6 NADH + 6H^+
		2 GDP + 2 P_i (= 2 ADP + 2 P_i)	2 GTP (= 2 ATP)

TABLE **A-5** ATP Yields from Complete Oxidation of Glucose to CO_2 by a Prokaryote Using Glycolysis, Entry Step, and the Krebs Cycle with O_2 as the Final Electron Acceptor

Compound	Number Produced	ATP Value	Total ATPs per Glucose
NADH + H^+	10	3	30
$FADH_2$	2	2	4
ATP (by substrate phosphorylation)	4		4

TABLE **A-6** Major Fermentations, Their End-Products, and Some Organisms That Perform Them

Fermentation	Major End Products	Representative Organisms
Alcoholic fermentation	Ethanol and CO_2	Saccharomyces cerevisiae
Homofermentation	Lactate	Streptococcus and some Lactobacillus
Heterofermentation	Lactate, ethanol, and acetate	Streptococcus, Leuconostoc, and Lactobacillus
Mixed acid fermentation	Acetate, formate, succinate, CO_2, H_2, and ethanol	Escherichia, Salmonella, Klebsiella, and Shigella
2,3-Butanediol fermentation	2,3-Butanediol	Enterobacter, Serratia, and Erwinia
Butyrate/butanol fermentation	Butanol, butyrate, acetone, and isopropanol	Clostridium, Butyrivibrio, and some Bacillus
Propionic acid fermentation	Propionate, acetate and CO_2	Propionibacterium, Veillonella, and some Clostridium

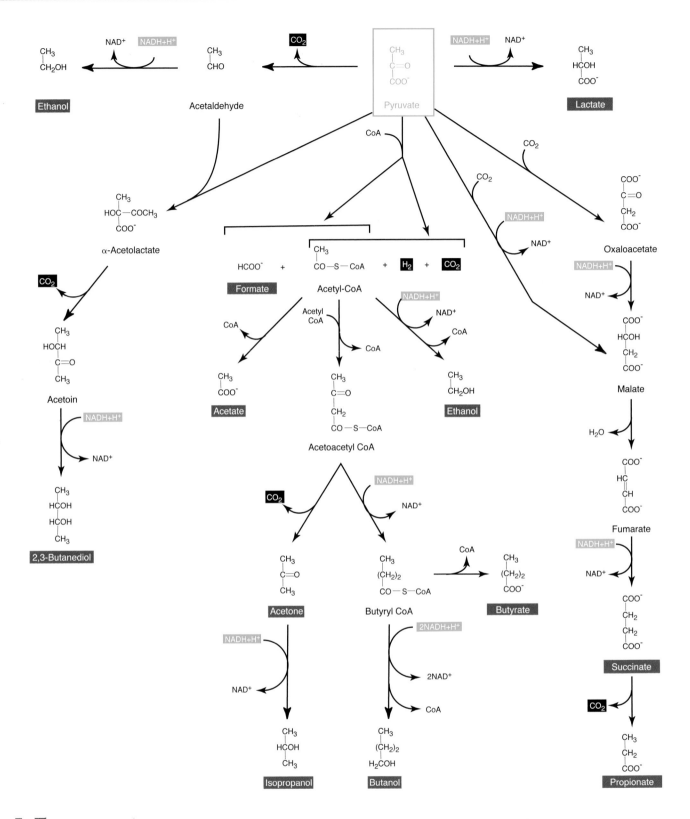

A-5 A SAMPLING OF FERMENTATION PATHWAYS ✦ Note that all pathways start with pyruvate, have a step(s) where NADH + H⁺ (in blue) is oxidized to NAD⁺, and produce end-products falling into one of three categories: acid, gas, or alcohol.

Miscellaneous Transfer Methods

Following are instructions for transfer methods that are performed less routinely than those in Exercise 1-3. As in Exercise 1-3, new skills in each process are printed in blue type.

Transfers Using a Sterile Cotton Swab

A sterile swab generally is often used to obtain a sample from a primary source—either a patient or an environmental site. Occasionally, swabs are used to transfer pure cultures. Sterile swabs may be dry, or they may be in sterile water, depending on the source of the sample. In either case, care must be taken not to contaminate the swab by unintentionally touching other surfaces with it. Your instructor may provide specific instructions on collection of samples from sources other than the ones below.

Obtaining a Sample from a Patient's Throat with a Swab

1 Use a sterile tongue depressor and swab prepared in sterile water to obtain a sample from the throat. Have the patient open his/her mouth, then gently press down on the tongue (Figure B-1).

2 With the swab in your dominant hand, carefully sample the patient's throat with a swirling motion. Touching other parts of the oral cavity is likely to cause contamination. Also, avoid touching the soft palate or it may initiate a gag reflex!

3 Transfer the sample to an appropriate plated medium (see Exercise 1-3) as quickly as possible.

4 If plating is to be done at a later time, place the swab in an appropriate sterile container (such as a sterile, capped test tube).

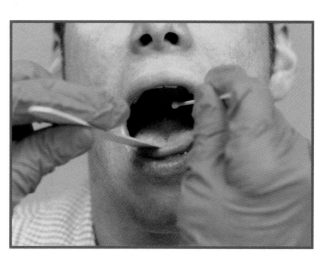

B-1 TAKING A THROAT SAMPLE ✦ Use a sterile tongue depressor and swab to obtain a sample from a patient's throat. Be careful not to touch other parts of the oral cavity or the sample will get contaminated. Transfer the sample to a sterile medium as soon as possible.

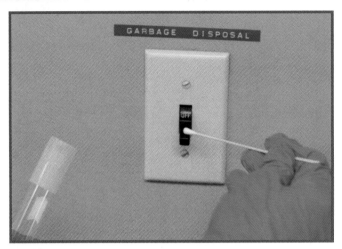

B-2 TAKING AN ENVIRONMENTAL SAMPLE ✦ Use a spinning motion of a sterile swab to sample inanimate objects in the environment. The swab may be placed in a sterile test tube until it is convenient to transfer the sample to a growth medium.

Obtaining an Environmental Sample with a Cotton Swab

1 Use a sterile swab prepared in sterile water to obtain a sample from an environmental source.

2 Rotate the swab to collect from the area to be sampled (Figure B-2).

3 Transfer the sample to an appropriate plated medium (see Exercise 1-3) as quickly as possible.

4 If plating is to be done at a later time, place the swab in an appropriate sterile container (such as a sterile, capped test tube).

Stab Inoculation of Agar Tubes Using an Inoculating Needle

Stab inoculations of agar tubes are used for several types of differential media (usually to examine growth under anaerobic conditions or to observe motility). A stab is *not* used to produce a culture of microbes for transfer to another medium.

1 Obtain the specimen with a sterile inoculating needle.

2 Remove the cap of the sterile medium with the little finger of your inoculating needle hand, and hold it there.

3 Flame the tube by quickly passing it through the Bunsen burner flame a couple of times. Keep your needle hand still.

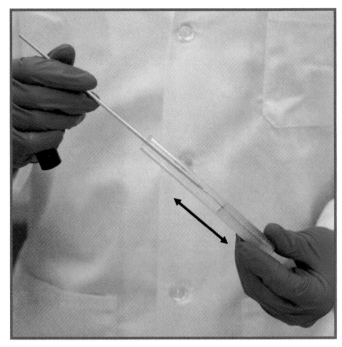

B-3 AGAR DEEP STAB ✦ Use the inoculating needle to stab the agar to a depth about 1 cm from the bottom. It is generally desirable to remove the needle along the original stab line and not create a new one. Upon completion, sterilize the needle.

4 Hold the open tube on an angle to minimize airborne contamination. Keep your needle hand still.

5 Carefully move the agar tube over the needle wire (Figure B-3). Insert the needle into the agar to about 1 cm from the bottom.

6 Withdraw the tube carefully so the needle leaves along the same path it entered. (When removing the tube, be especially careful not to catch the needle tip on the tube lip. This springing action of the needle creates bacterial aerosols.)

7 Flame the tube lip as before. Keep your needle hand still.

8 Keeping the needle hand still (remember, it has growth on it), move the tube to replace its cap.

9 Sterilize the needle as before by incinerating it in the Bunsen burner flame. It is especially important to flame it from base to tip now because the needle has lots of bacteria on it.

10 Label the tube with your name, date, and organism. Incubate at the appropriate temperature for the assigned time.

Spot Inoculation of an Agar Plate

Sometimes an agar plate may be used to grow several different specimens at once. This is a typical practice with plated *differential* media (*i.e.*, media designed to differentiate organisms based on growth characteristics). Prior to beginning the transfer, divide the plate into as many as four sectors, using a marking pen. (Some plates already have marks on the base for this purpose.) Then each may be inoculated with a different organism. Inoculation involves touching the loop to the agar surface once so growth is restricted to a single spot—hence the name "spot inoculation."

1 Lift the lid of the sterile agar plate and use it as a shield to prevent airborne contamination.

2 Touch the agar surface toward the periphery of the sector (Figure B-4).

3 Remove the loop and replace the lid.

4 Sterilize your loop as before. It is especially important to flame it from base to tip now because the loop has lots of bacteria on it.

5 Label the base of the plate with your name, date, and organism(s) inoculated.

6 Incubate the plate in an inverted position for the assigned time at the appropriate temperature.

B-4 SPOT INOCULATION OF A PLATE ✦ Each of four sectors is spot-inoculated with a different organism by touching the loop to the surface and making a mark about 1 cm in length. Generally, spot inoculations are done toward the edge (rather than the crowded middle) to prevent overlapping growth and/or test results.

Transfer from a Broth Culture Using a Glass Pipette

Glass pipettes are used to transfer a known volume of liquid diluent, media, or culture. Originally pipettes were filled by sucking on them like a drinking straw, but mouth pipetting is dangerous and has been replaced by mechanical pipettors. Three examples are shown in Figure C-1, each with its own method of operation. Your instructor will show you how to properly use the type of pipettor available in your lab.

To use a pipette correctly, you must be able to correctly read the calibration. Examine Figure C-2. The numbers at the right indicate the pipette's *total volume* and its *smallest calibrated increments*. This is a 10.0 mL pipette divided into 0.1 mL increments.

When reading volumes, use the base of the meniscus (Figure C-3). The volume in the center pipette is read as exactly 3.0 mL because the meniscus is resting on the line. The left pipette is read as 2.9 mL and the right pipette is read as 3.1 mL (0 is always at the top of the pipette). Although the difference in volume between these three pipettes may seem negligible (1 part in 70,[1] about a 1.5% error), it may be enough to introduce significant error into your work.

Two pipette styles are used in microbiology (Figure C-4): the **serological pipette** and the **Mohr pipette**. A serological pipette is calibrated *to deliver* (TD) its volume by completely draining it and blowing out the last drop. The tip of a Mohr pipette is not graduated, so fluid flow

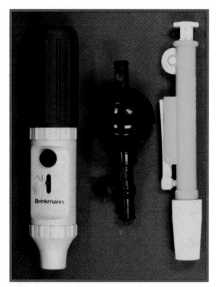

C-1 MECHANICAL PIPETTORS ✦ Three examples of mechanical pipettors are shown here, each with its own method of operation. Your instructor will show you how to properly use the style of pipettor available in your lab. From left to right: A pipette filler/dispenser, a pipette bulb, and a plastic pump.

---

[1] This number presumes the pipette shown has a total volume of 10 mL. If the meniscus is on the 3, the pipette will dispense 7 mL. If emptied completely, an error of 0.1 mL above or below the 3 would represent 0.1 mL in 7 mL or 1 part in 70.

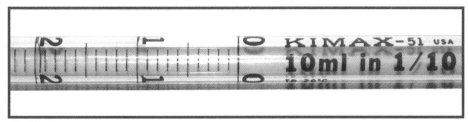

C-2 PIPETTE CALIBRATION ✦ Prior to using a pipette, read its calibration. The numbers indicate the pipette's *total volume* and its *smallest calibrated increments*. This is a 10.0 mL pipette divided into 0.1 mL increments.

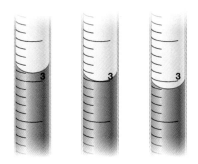

C-3 READ THE BASE OF THE MENISCUS
✦ When reading volumes, use the base of the meniscus. The volume in the center pipette is read at exactly 3.0 mL because the meniscus is resting on the line. The left pipette is read as 2.9 mL and the right pipette is read as 3.1 mL (0 is always at the top of the pipette).

must be stopped at a calibration line. Stopping the fluid beyond the last line on a Mohr pipette results in an unknown volume being dispensed. In either case, volumes are read at the bottom of the fluid meniscus.

Important: If pipetting a bacterial culture, be careful not to allow any to drop from the pipette before disposing of it in the autoclave container. Clean up any spills.

Following are instructions for using a glass pipette. As in Exercise 1-3, new skills in each process are printed in blue type.

Filling a Glass Pipette

1   Bacteria should be suspended in the broth with a vortex mixer (Figure 1-15) or by agitating with your fingers (Figure 1-16). Be careful not to splash the broth into the cap or lose control of the tube.

2   Pipettes are sterilized in metal canisters, individually in sleeves, or as multiples in packages (if disposable). They typically are stored in groups of a single size (Figure C-5). *Be sure you know what volume your pipette will deliver.* Set the canister at the table edge and remove its lid. (Canisters should not be stored in an upright position, as they may fall over and break the pipettes or become contaminated.) If using

C-4 TWO TYPES OF PIPETTES ✦ Two pipette styles—the *serological pipette* (left) and the *Mohr pipette* (right)—are used in microbiology. A serological pipette is calibrated *to deliver* (TD) its volume by completely draining it and blowing out the last drop. The tip of a Mohr pipette is not graduated, so fluid flow must be stopped at a calibration line. Stopping the fluid beyond the last line on a Mohr pipette results in an unknown volume being dispensed.

C-5 GETTING THE STERILE PIPETTE ✦ Pipettes of the same size are autoclaved in canisters, which then are opened and placed flat on the table. Pipettes are removed as needed.

pipettes in a package, open the end *opposite the tips.* Grasp *one pipette only* and remove it.

3   Carefully insert the pipette into the mechanical pipettor (Figure C-6). It's best to grasp the pipette near the end with your fingertips. This gives you more control and reduces the chance that you will break the pipette and cut your hand. *Do not touch any part of the pipette that will contact the specimen or the medium or you will risk introducing a contaminant. Also, do not lay the pipette on the tabletop while you continue.*

4   While keeping your pipette hand still, bring the culture tube toward it. Use your little finger to remove and hold its cap.

5   Flame the open end of the tube by passing it through a Bunsen burner flame two or three times.

6   Hold the tube at an angle to prevent contamination from above.

7   Insert the pipette and withdraw the appropriate volume (Figure C-7). Bring the pipette to a vertical position briefly to read the meniscus accurately. (Remember: The volumes in the pipette are correct only if the meniscus of the fluid inside is resting *on* the line, not below it.) Then carefully remove the pipette from the tube.

8   Flame the tube lip as before. Keep your pipette hand still.

9   Keeping the pipette hand still (remember, it may contain fluid with microbes in it), move the tube to replace its cap.

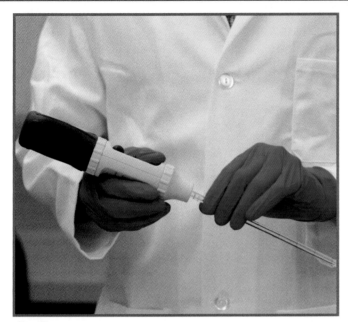

C-6 ASSEMBLING THE PIPETTE ✦ Carefully insert the pipette into a mechanical pipettor. Notice that the pipette is held near the end with the fingertips. For safety, the hand is out of the way in case the pipette breaks.

C-7 FILLING THE PIPETTE ✦ Carefully draw the fluid into the pipette. Briefly bring it to a vertical position and read the volume. Notice how the tube's cap is being held by the pinky of the pipettor hand.

10 What you do next depends on the medium to which you are transferring the liquid. Please continue with the appropriate inoculation section.

Inoculation of Broth Tubes with a Pipette

Pipettes are often used to inoculate a known volume of culture into a known volume of diluent during serial dilutions.

1 While keeping the pipette hand still, bring the broth culture tube toward it. Use your little finger to remove and hold its cap.

2 Flame the tube by quickly passing it through the Bunsen burner flame two or three times. Keep your pipette hand still.

3 Hold the open tube on an angle to minimize airborne contamination. Keep your pipette hand still.

4 Insert the pipette tip and dispense the correct volume of inoculum.

5 Withdraw the tube from over the pipette. Before completely removing it, touch the pipette tip to the glass to remove any excess broth.

6 Completely remove the pipette, but avoid waving it around. This can create aerosols.

7 Flame the tube lip as before. Keep your pipette hand still.

8 Keeping the pipette hand still, move the tube to replace its cap.

9 The pipette is contaminated with microbes and must be disposed of correctly. Each lab has its own specific procedures, and your instructor will advise you what to do. Glass pipettes typically are placed in a pipette disposal container containing a small amount of disinfectant until they are autoclaved and reused (Figure C-8). Disposable pipettes must be placed in an appropriate biohazard container. In either case, be careful when removing the pipette from the mechanical pipettor. There is danger of culture dripping from the pipette or of breaking the glass or plastic.

C-8 DISPOSE OF THE PIPETTE ✦ The pipette may be contaminated with microbes and must be disposed of correctly. Each lab has its own specific procedures, and your instructor will advise you what to do. Shown here is a canister used for glass pipettes. Disposable pipettes must be placed in an appropriate biohazard container. In either case, be careful when removing the pipette from the mechanical pipettor. There is danger of culture dripping from the pipette or of breaking the glass or plastic.

Inoculation of Agar Plates with a Pipette

Pipettes also can be used to dispense a known volume of inoculum to an agar plate.

1 Lift the lid of the plate and use it as a shield to protect it from airborne contamination.

2 Hold the pipette over the agar and dispense the correct volume (often 0.1 mL) onto the center of the agar surface (Figure C-9). From this point, the remainder of steps should be completed within about 15 seconds to prevent the inoculum from soaking into the agar.

3 The pipette is contaminated with microbes and must be disposed of correctly. Each lab has its own specific procedures, and your instructor will advise you what to do. Glass pipettes typically are placed in a pipette disposal container containing a small amount of disinfectant until they are autoclaved and reused (Figure C-8). Disposable pipettes must be placed in an appropriate biohazard container. In either case, be careful when removing the pipette from the mechanical pipettor. There is danger of culture dripping from the pipette or of breaking the glass or plastic.

4 Continue with the spread plate technique (Exercise 1-5).

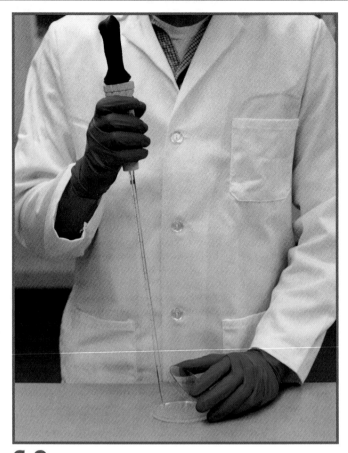

C-9 INOCULATING THE PLATE WITH A PIPETTE ✦ Open the lid and dispense the inoculum onto the surface of the agar near the middle. Notice how the plate's lid is held to prevent aerial contamination.

Transfer from a Broth Culture Using a Digital Pipette

Modern molecular biology procedures often involve transferring extremely small volumes of liquid with great precision and accuracy. This has led to the development of digital pipettors (Figure D-1) that can be set up to dispense microliter volumes through milliliter volumes (recall that 1 mL equals 1000 µL). Common digital pipettors are calibrated to dispense volumes of 1 to 10 µL, 10 to 100 µL, 100 to 1000 µL, or 1 to 5 mL.

Filling a Digital Pipettor

Many manufacturers make digital pipettors, but they all work basically the same way. The following instructions are for Eppendorf Series 2100 models.

1 Growth may be suspended in the broth with a vortex mixer (Figure 1-15) or by agitating with your fingers (Figure 1-16). Be careful not to splash the broth into the cap or lose control of the tube.

2 Determine which digital pipettor should be used to dispense the desired volume.

3 Turn the setting ring to set the desired volume (Figure D-2). If you turn the dial past the volume range of the pipettor, you will damage it.

4 Hold the digital pipettor in your dominant hand.

5 Open the rack of appropriate pipette tips for your pipettor (these are often color-coded and match the pipettor) and push the pipettor into a sterile tip (Figure D-3). Close the rack. Never touch the pipette tip with your hands or leave the rack open. *Never use a digital pipettor without a tip.*

6 Remove the cap from the culture tube with the little finger of your dominant hand, and flame the tube.

7 Press down the control button with your thumb to the first stop. This is the measuring stroke.

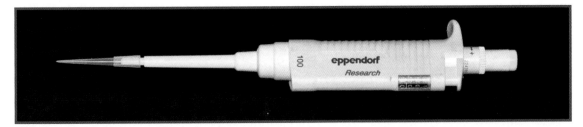

D-1 DIGITAL PIPETTOR ✦ This is a digital pipettor to be used for dispensing volumes betwen 10 µL and 100 µL. Never set the pipettor to volumes above or below the intended volumes as damage may occur. Also, always use a digital pipettor with the appropriate tip. As shown here, tips and pipettors are often color-coded.

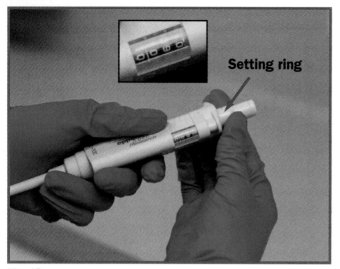

D-2 SETTING THE VOLUME ✦ Rotate the adjustment knob to set the volume on a digital pipettor. This pipettor has been set at 90.0 µL (see inset—the horizontal line between the third and fourth numerals is a decimal point). Never rotate the adjustment knob beyond the volume limits of the pipettor.

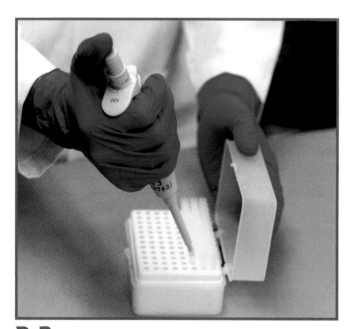

D-3 DIGITAL PIPETTOR TIPS ✦ Digital pipettors must be fitted with a sterile tip of appropriate size (often color-coded to match the pipettor). Open the case and press the pipettor into a tip, then close the case to maintain sterility. Do not touch the pipettor tip.

8 Insert the tip into the broth approximately 3 mm while holding it vertically (Figure D-4).

9 Slowly release pressure with your thumb to draw fluid into the tip. Be careful not to pull any air into the tip.

10 Remove the pipettor and hold it still as you flame the tube as before.

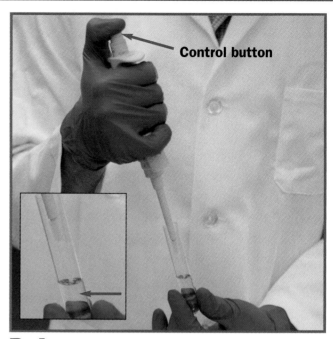

D-4 FILLING THE PIPETTE ✦ Depress the control button with your thumb to the first stop (measuring stroke). Holding the tube and pipettor in a vertical position, insert the tip into the fluid to a depth of 3 mm (see inset). Slowly release pressure with your thumb to fill the pipettor.

11 Keeping the pipettor hand still, move the tube to replace its cap.

12 What you do next depends on the medium to which you are transferring the growth. Please continue with the appropriate inoculation section.

Inoculation of Broth Tubes with a Digital Pipettor

1 Remove the cap of the sterile medium with the little finger of your pipettor hand, and hold it there.

2 Flame the tube by quickly passing it through the Bunsen burner flame a couple of times. Keep your pipettor hand still.

3 Insert the pipette tip into the tube. Hold it at an angle against the inside of the glass (Figure D-5).

4 Depress the control button slowly with your thumb to the first stop and pause until no more liquid is dispensed. Then continue pressing to the second stop to deliver the remaining volume. This is the blow-out stroke.

5 While keeping pressure on the control button, carefully remove the pipettor from the tube by sliding it along the glass. Once it is out of the tube, slowly release pressure on the control button.

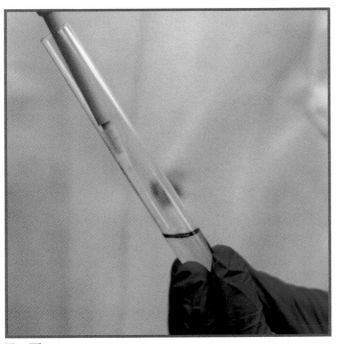

D-6 EJECTING THE TIP ✦ Using the tip ejector button, eject the contaminated tip into an appropriate biohazard container.

D-5 DISPENSING TO A LIQUID ✦ Hold the pipettor at an angle with the tip against the glass. Press the control button to the first stop. When no more fluid comes out, continue pressing to the second stop. Remove the pipettor, then slowly release pressure on the control button, and eject the tip into a biohazard container. Finally, replace the cap on the tube.

6 Flame the tube lip as before. Keep your pipette hand still.

7 Keeping the pipette hand still, move the tube to replace its cap.

8 The pipettor tip is contaminated with microbes and must be disposed of correctly. Use the ejector button to remove the tip into an appropriate biohazard container (Figure D-6). Each lab has its own specific procedures, and your instructor will advise you what to do.

Inoculation of Agar Plates with a Pipettor

1 Lift the lid of the plate and use it as a shield to protect from airborne contamination.

2 Place the pipette tip over the agar surface. Be sure to hold the pipettor in a vertical position (Figure D-7).

3 Depress the control button slowly with your thumb to the first stop, and pause until no more liquid is dispensed. Then continue pressing to the second stop to deliver the appropriate volume. This is the

D-7 DISPENSING TO A PLATE ✦ Hold the pipette in a vertical position and press the control button to the first stop. When no more fluid comes out, continue pressing to the second stop. Slowly release pressure on the control button, then eject the tip into a biohazard container.

blow-out stroke. From this point, the remainder of steps should be completed within about 15 seconds to prevent the inoculum from soaking into the agar.

4 Because the pipettor tip is contaminated with microbes, you must dispose of it correctly. Use the ejector button to remove the tip into an appropriate biohazard container (Figure D-6). Each lab has its own specific procedures, and your instructor will advise you what to do.

5 Continue with the Spread Plate Technique (Exercise 1-5).

Reference

Brinkman Instruments, *Eppendorf Series 2001 Pipette Instruction Manual.* Westbury, New York, NY.

The Spectrophotometer

Theory

The spectrophotometer (Figure E-1) is typically used in quantitative analysis; that is, to determine how much of something is in a liquid sample. In this manual, protocols in Exercises 2-8, 2-9, 2-10, 6-5, 7-2, and 10-1 use spectrophotometry to measure turbidity in liquid for a variety of purposes. Following is a brief introduction to the theory of spectrophotometry and instructions on using the spectrophotometer.

When light strikes a solution, any of several outcomes may occur. The light may be transmitted through it, it may be absorbed, it may be reflected, it may cause fluorescence, or it may be scattered. The spectrophotometer is designed to shine a beam of single wavelength light on a sample and measure **% Transmittance (%T)** or **absorbance** of that light. (Under the conditions in the spectrophotometer, reflection, fluorescence, and scatter are negligible.)

When light is shined through an absorbing medium, each fraction of equal thickness absorbs an equal fraction of the light passing through it (Lambert's law). (For example, each 10% portion

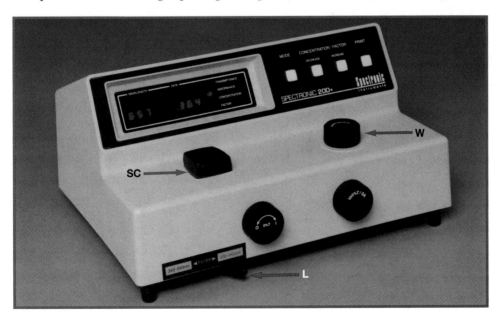

E-1 THE SPECTRONIC 20D+ ✦ This spectrophotometer has a digital display. It is showing an absorbance of 0.364 at a wavelength of 657 nm. The white mode button is used to choose between absorbance and transmittance (note the red light next to absorbance). The black square on the left is the lid to the sample compartment (**SC**). The knob to its right is used to select wavelength (**W**). The lever (**L**) on the lower left positions and removes a filter for use with wavelengths of 600 to 950 nm. The knob on the front left is used to turn the machine on and off, as well as to set 0% Transmittance. The knob on the front right is used to set 100%T with a blank in the sample compartment.

(Photo courtesy of Thermo Spectronic, Rochester, NY)

of the medium absorbs an equal amount.) Further, as long as the light doesn't change the physical or chemical state of the medium,
the fraction of light transmitted by a given thickness is independent of the light's intensity. The fraction transmitted then may be expressed as

$$T = \frac{I}{I_o}$$

where I_o is the intensity of incident light and I is the intensity of the transmitted light. The percent transmittance is calculated by multiplying T by 100, as shown.

$$\%T = \frac{I}{I_o} \times 100$$

Plotting concentration of the absorbing medium against %T produces an exponential curve (Figure E-2A) that is cumbersome to use. Fortunately, concentration of a substance in a solution is directly proportional to the absorbance (A) of the solution (Beer-Lambert law), provided that the incident light is a single wavelength and that the light doesn't change the physical or chemical state of the medium. Figure E-2B illustrates the relationship between concentration and absorbance.

The mathematical relationship between absorbance and percent transmittance is as follows:

$$A = \log\frac{1}{T} = \log\frac{I_o}{I} = \log I_o - \log I$$

This equation may be modified to

$$A = 2 - \log\%T$$

because the spectrophotometer is calibrated such that the value of I_o is 100%. Most spectrophotometers allow reading of either %T or A rather than leaving the calculation up to the user.

A simplified schematic of how a spectrophotometer works is shown in Figure E-3. The beam of white light is broken up into its component colors by either a prism or a diffraction grating. The particular wavelength to be used is selected and allowed to pass through the exit slit. It then is directed through the sample contained in a glass or plastic cuvette with a 1 cm light path. The emerging transmitted light strikes a photodetector, which converts the light energy into an electrical signal proportional to the light intensity. This electrical signal is amplified and displayed by the spectrophotometer as either percent transmittance or absorbance.

In quantitative analysis, absorbance of the compound or particle being examined must be differentiated from absorbance because of the solvent in which it is dissolved and the cuvette itself. This is done by calibrating the spectrophotometer with an identical blank cuvette containing the solvent but lacking the compound or particle being measured.

Instructions for Use of the Spectronic D20+

1 Turn on the spectrophotometer and allow it to warm up for 15 minutes.

2 Choose the wavelength to be used by turning the wavelength control knob on the top of the machine.

3 Make sure the filter is set for the correct range of wavelength to be used.[1]

---

[1] On the older Spectronic 20 machines, a red filter will have to be manually inserted if wavelengths longer than 600 nm are used.

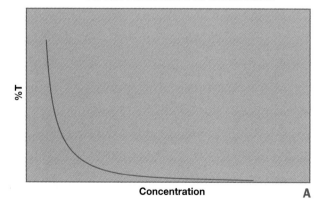

Percent Transmittance vs. Concentration

%T (vertical axis) · Concentration (horizontal axis) · A

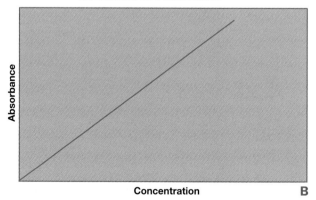

Absorbance vs. Concentration

Absorbance (vertical axis) · Concentration (horizontal axis) · B

E-2 TRANSMISSION AND ABSORBANCE AS A FUNCTION OF CONCENTRATION ✦ **A** Transmittance varies exponentially with concentration. This makes using the graph for interpolation difficult, because many points are required to produce the standard curve. **B** Absorbance and concentration are linearly related, making production and interpretation of a standard curve easier.

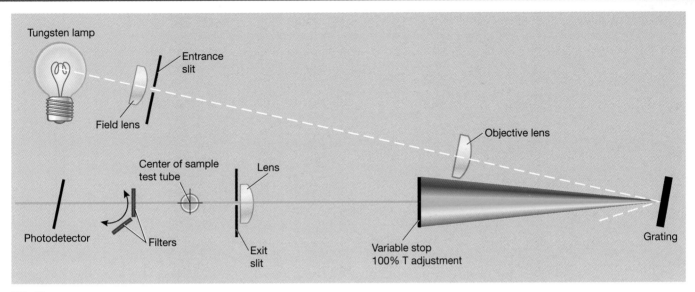

E-3 A SCHEMATIC OF THE SPECTROPHOTOMETER ✦ Light from the source lamp is focused on the diffraction grating by the field and objective lenses. The grating is responsible for breaking up the white light into its component wavelengths (colors). The desired wavelength is chosen by rotating the grating and aiming that wavelength through the exit slit. This incident light (I_o) then passes through the sample (with a light path of 1 cm) and into the phototube as transmitted light (I). The phototube then converts the transmitted light energy into an electrical signal that produces the readout on the display. Other components include the shutter, which closes when no sample is in place so the machine can be "zeroed," and the filter, which is used with wavelengths of 600 to 950 nm.

(Diagram modified from Spectronic 20D+ literature, courtesy of Thermo Spectronic, Rochester, NY.)

4 Set the machine to read % Transmittance by pressing the mode button until the display indicates Transmittance.[2]

5 Prepare a "blank" cuvette identical to the experimental cuvette minus the particles you are measuring. (For instance, sterile medium if you are checking absorbance of bacterial cells.)

6 Use care to handle the cuvettes by the top only, as fingerprints influence results. Use tissue to remove fingerprints.

7 Use the power/zero control knob to set the machine to 0%T.

8 If measuring %T, go to step 9. If measuring A, go to step 10.

9 Be sure the blank is mixed uniformly, but do not introduce bubbles. Insert the blank into the sample compartment and close the cover. Use the Light Control knob to set 100 %T. Continue with Step 11.

10 Select absorbance mode. Be sure the blank is mixed uniformly, but do not introduce bubbles. Then insert the blank into the sample compartment and

close the cover. Use the Light Control knob to set zero A. Continue with Step 11.

11 Remove the blank.

12 Be sure the experimental cuvette is mixed uniformly, but do not introduce bubbles. Insert it into the sample compartment and close the lid. Read the appropriate display, and record the value.

13 It is advisable to calibrate the spectrophotometer prior to each use if there is an interval of more than a few minutes between readings.

14 When finished with all readings, turn off the spectrophotometer.

15 *Note:* Your instructor must be notified immediately if you spill any culture in the spectrophotometer. You will be instructed on how to decontaminate the spill.

References

Abramoff, Peter and Robert G. Thompson. 1982. Appendix A and Appendix B in *Laboratory Outlines in Biology—III*. W. H. Freeman and Co., San Francisco.

Thermo Spectronic. "Theory of UV-Visible Spectrophotometry," downloaded from http://www.thermo.com/eThermo/CDA/KnowledgeBase/Product_Knowledge_Base_Detail/0,1292,12100-119-10023,00.html. Thermo Spectronic, Rochester, NY.

[2] On the Spectronic 20, the display has both %T and A scales.

Alternative Procedures for Section 6

Alternative Procedure for Exercise 6-1, Standard Plate Count (Viable Count)

✦ Materials

Per Student Group

✦ micropipettes (10–100 µL and 100–1000 µL) with sterile tips
✦ six sterile microtubes
✦ flask of sterile water
✦ eight Nutrient Agar plates
✦ beaker containing ethanol and a bent glass rod
✦ hand tally counter
✦ colony counter
✦ 24-hour broth culture of *Escherichia coli* (This culture will have between 10^7 and 10^{10} CFU/mL.)

 Procedure

Refer to the procedural diagram in Figure F-1 as needed.

Lab One

1 Obtain eight plates, organize them into four pairs, and label them A_1, A_2, B_1, B_2, *etc.*

2 Obtain 5 microtubes, and label them 1–5. These are your dilution tubes; make sure they remain covered until needed.

3 Aseptically add 990 µL sterile water to dilution tubes 1 and 2. Cover when finished. Aseptically add 900 µL sterile water to dilution tubes 3, 4, and 5. Cover when finished.

4 Mix the broth culture, and aseptically transfer 10 µL to dilution tube 1; mix well. This is a 10^{-2} dilution.

5 Aseptically transfer 10 µL from dilution tube 1 to dilution tube 2; mix well. This is 10^{-4}.

6 Aseptically transfer 100 µL from dilution tube 2 to dilution tube 3; mix well. This is 10^{-5}.

7 Aseptically transfer 100 µL from dilution tube 3 to dilution tube 4; mix well. This is 10^{-6}.

8 Aseptically transfer 100 µL from dilution tube 4 to dilution tube 5; mix well. This is 10^{-7}.

9 Aseptically transfer 100 µL from dilution tube 2 to plate A_1. Using the spread plate technique (Exercise 1-5), dispense the diluent evenly over the entire surface of the agar. Repeat the procedure with plate A_2 and label both plates "10^{-5} mL original sample."

10 Following the same procedure, transfer 100 µL volumes from dilution tubes 3, 4, and 5 to plates B, C and D, respectively. Label the plates with their appropriate original sample volumes.

11 Invert the plates, and incubate at $35 \pm 2°C$ for 24 to 48 hours.

Lab Two

1 After incubation, examine the plates and determine the countable pair—plates with 30 to 300 colonies. Only one pair of plates should be countable (Figure 6-1).

2 Count the colonies on both plates and calculate the average (Figure 6-3). Record it in the chart provided on the Data Sheet. (*Note:* Because of error, you may have more than one pair that is countable. For the practice, count all plates that have between 30 and 300 colonies, and try to identify which plate(s) you have the most confidence in. If no plates are in the 30–300 range, count the pair that is closest.)

3 Using the formula provided in Data and Calculations on the Data Sheet, calculate the density of the original sample, and record it in the space provided.

Alternative Procedure for Exercise 6-5, Plaque Assay of Virus Titre

✦ Materials

Per Class

✦ 50°C hot water bath containing tubes of liquid soft agar (7 tubes per group)

Per Student Group

✦ micropipettes (10–100 μL and 100–1000 μL) with sterile tips
✦ fourteen sterile capped microtubes (1.5 mL or larger)
✦ seven Nutrient Agar plates
✦ seven tubes containing 2.5 mL soft agar
✦ small flask of sterile normal saline

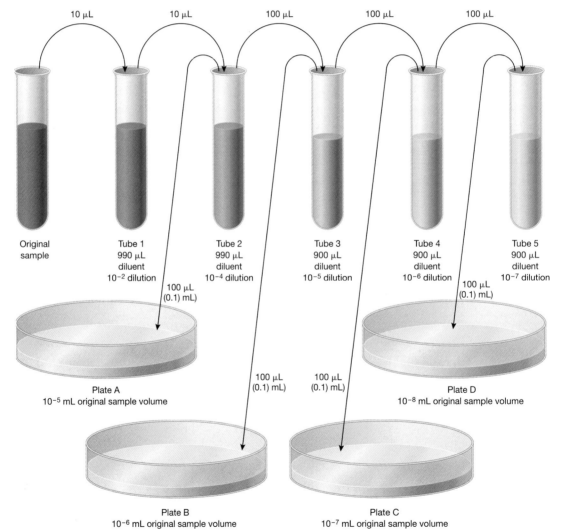

F-1 SERIAL DILUTION PROCEDURAL DIAGRAM (MEASURING VOLUMES WITH A DIGITAL PIPETTE) ✦ This is an illustration of the dilution scheme outlined in the Procedure. The dilution assigned to the dilution tubes and their contents indicate the proportion of original sample present in the tube. (Note that since the final cell density will be expressed in CFU/mL, we have converted the inoculum going to the plates from microliters (100 μL) to milliliters (0.1 mL) for ease of calculation. (A simple conversion now will avoid a more complicated conversion later. It helps to measure in microliters, but try to think in milliliters. Note also that the volume of original sample transferred to a plate is 0.1 mL multiplied by the dilution of the solution from which it came. For example, 0.1 mL of 10^{-4} dilution (which is only 1/10000 original sample) is 10^{-5} mL of original sample.)

✦ seven sterile transfer pipettes
✦ T4 coliphage
✦ 24-hour broth culture of *Escherichia coli B* (T-series phage host)

 Procedure

Refer to the procedural diagram in Figure F-2 as needed.

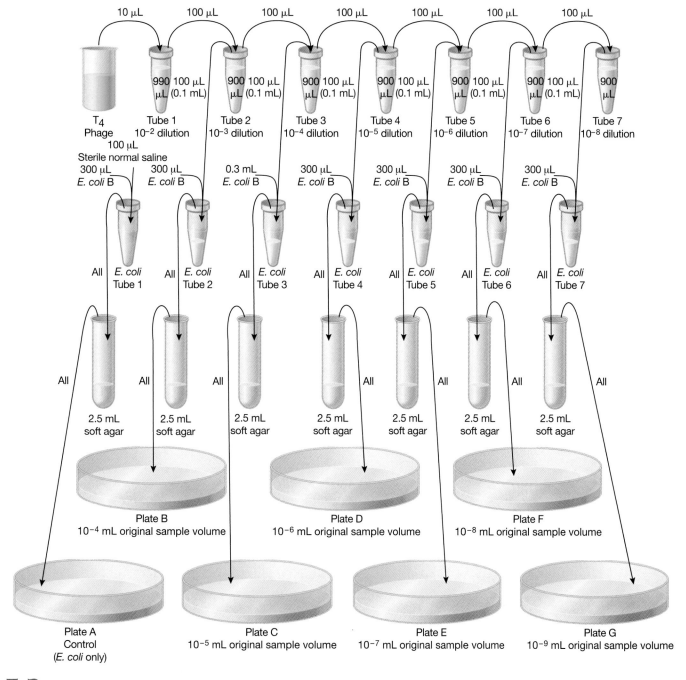

F-2 PLAQUE ASSAY PROCEDURAL DIAGRAM (MEASURING VOLUMES WITH A DIGITAL PIPETTE) ✦ This is an illustration of the dilution scheme outlined in the Procedure. The dilution assigned to the dilution tubes and their contents indicate the proportion of original phage sample present in the tube. (Note that since the final density will be expressed in PFU/mL, we have converted the dilution volume going to the plates (via the *E. coli* tube and soft agar tube) from microliters (100 μL) to milliliters (0.1 mL) for ease of calculation. A simple conversion now will avoid a more complicated conversion later. It helps to measure in microliters, but try to think in milliliters. Note also that the volume of original phage sample transferred to a plate is 0.1 mL multiplied by the dilution of the solution from which it came. For example 0.1 mL of 10^{-4} dilution (which is only 1/10000 original sample) is 10^{-5} mL of original sample.)

Lab One

1. Obtain all materials except for the soft agar tubes. To keep the agar tubes liquefied, leave them in the water bath and take them out one at a time as needed.

2. Label seven microtubes 1 through 7. Label the other seven microtubes *E. coli* 1–7. Place all tubes in a rack, pairing like-numbered tubes.

3. Label the Nutrient Agar plates A through G. Place them in the 35°C incubator to warm them. Take them out one at a time as needed. This will keep the soft agar (added at step 14) from solidifying too quickly and result in a smoother agar surface.

4. Aseptically transfer 990 μL sterile normal saline to dilution tube 1.

5. Aseptically transfer 900 μL sterile normal saline to dilution tubes 2–7.

6. Mix the *E. coli* culture and aseptically transfer 300 μL into each of the *E. coli* microtubes.

7. Mix the T4 suspension and aseptically transfer 10 μL to dilution tube 1. Mix well. This is a 10^{-2} dilution.

8. Aseptically transfer 100 μL from dilution tube 1 to dilution tube 2. Mix well. This is 10^{-3}.

9. Aseptically transfer 100 μL from dilution tube 2 to dilution tube 3. Mix well. This is 10^{-4}.

10. Continue in this manner through dilution tube 7. The dilution in tube 7 should be 10^{-8}.

11. Aseptically transfer 100 μL sterile normal saline to *E. coli* tube 1. This will be mixed with 2.5 ml soft agar and used to inoculate a control plate.

12. Aseptically transfer 100 μL from dilution tube 2 to its companion *E. coli* tube. Repeat this procedure with the remaining five tubes.

13. This is the beginning of the preadsorption period. Let all seven tubes stand undisturbed for 15 minutes.

14. Remove one soft agar tube from the hot water bath and add the entire contents of *E. coli* tube 1. Using a sterile transfer pipette, mix well and immediately pour onto plate A. Gently tilt the plate back and forth until the soft agar mixture is spread evenly across the solid medium. Label the plate "Control."

15. Remove a second soft agar tube from the water bath and add the entire contents of *E. coli* tube 2. Using a sterile transfer pipette, mix well, and immediately pour onto plate B. Tilt back and forth to cover the agar and label it 10^{-4} mL original phage sample.

16. Repeat this procedure with dilutions 10^{-4} thru 10^{-8} and plates C through G. Label the plates with the appropriate original phage sample volumes.

17. Allow the agar to solidify completely.

18. Invert the plates and incubate aerobically at $35 \pm 2°C$ for 24 to 48 hours.

Lab Two

1. After incubation, examine the control plate for growth and the absence of plaques.

2. Examine the remainder of your plates and determine which one is countable (30 to 300 plaques). Count the plaques and record the number in the table provided on the Exercise 6-5 Data Sheet. Record all others as either TNTC (to numerous to count) or TFTC (to few to count).

3. Using the sample volumes on the countable plate and the formula provided on the Exercise 6-5 Data Sheet, calculate the original phage density. (Be sure you have converted the μL volumes to mL before calculating.) Record your results in the space provided on the Data Sheet.

Alternative Procedure for Exercise 6-6, Thermal Death Time Versus Decimal Reduction Value

◆ Materials

Per Class

- ◆ stopwatch
- ◆ thermometers
- ◆ 2 flasks each containing 49.0 mL Nutrient Broth, maintained at 60°C in a water bath
- ◆ 24-hour broth cultures of:
  - ◆ *Escherichia coli*
  - ◆ *Staphylococcus aureus*

Per Group

- ◆ 7 Nutrient Agar plates
- ◆ 16 Nutrient Broth tubes
- ◆ micropipettes
- ◆ sterile 100 μL micropipette tips
- ◆ sterile 1000 μL micropipette tips
- ◆ sterile 1.5 mL or larger microtubes
- ◆ sterile deionized water in a small flask or beaker
- ◆ beaker of alcohol and bent glass rod for spreading organisms

Procedure

Note: This procedure uses the Spread-Plate Technique described in Exercise 1-5. Refer to it as needed. The Plate Subgroups will require at least two or three students for labeling microtubes and plates, performing the serial dilution, and spreading plates. The Broth Subgroups will require one or two students to label and inoculate the 15 Nutrient Broths, keep track of time, and mix the broth prior to all transfers.

Lab One: Group 1 (T_0)

Plate Subgroup

1 Obtain all plating materials. Place seven microtubes in a rack. Label them 1 through 7. Add 900 μL of sterile water to microtubes 2 through 7, and cover them until needed.

2 Label seven Nutrient Agar plates 1 through 7, respectively.

3 When your instructor adds 1.0 mL of the *E. coli* culture to the heated flask, he/she will start the stopwatch. This is T_0. Immediately swirl the flask to disperse the sample, and remove 1000 μL with your pipette. If possible, do this step without removing the flask from the water bath. Begin the following serial dilution:

  ♦ Add the 1000 μL of broth from the flask to tube 1.
  ♦ Without changing the pipette tip, transfer 100 μL from tube 1 to tube 2; mix. Discard the pipette tip. Use a clean pipette tip for each transfer from here forward.
  ♦ Transfer 100 μL from tube 2 to tube 3; mix.
  ♦ Transfer 100 μL from tube 3 to tube 4; mix.
  ♦ Transfer 100 μL from tube 4 to tube 5; mix.
  ♦ Transfer 100 μL from tube 5 to tube 6; mix.
  ♦ Transfer 100 μL from tube 6 to tube 7; mix.
  ♦ Transfer 100 μL from tube 1 to plate 1 and spread according to spread-plate procedure.
  ♦ Transfer 100 μL from tube 2 to plate 2 and spread.
  ♦ Transfer 100 μL from tube 3 to plate 3 and spread.
  ♦ Transfer 100 μL from tube 4 to plate 4 and spread.
  ♦ Transfer 100 μL from tube 5 to plate 5 and spread.
  ♦ Transfer 100 μL from tube 6 to plate 6 and spread.
  ♦ Transfer 100 μL from tube 7 to plate 7 and spread.

4 Calculate the volume of original sample transferred to each plate (volume plated × dilution) and enter it in chart 1 on the Data Sheet.[2]

---

[2] This Procedure requires knowledge of dilutions and calculations (explained fully in Exercise 6-1 Standard Plate Count.) If you haven't completed a standard plate count, or you have any doubt about your ability to complete this exercise, we suggest that you read Exercise 6-1 carefully and do some or all of the practice problems.

5 Tape the plates together, invert and incubate them at $35 \pm 2°C$ for 48 hours.

Broth Subgroup

1 Obtain all broth transfer materials. Label 15 Nutrient Broth tubes T_1 through T_{15} respectively. Label the 16th "Control."

2 When your instructor adds 1.0 mL of the *E. coli* culture to the heated flask, he/she will start the stopwatch. This is T_0. Slightly before T_1, gently swirl the flask to mix the broth. At exactly T_1, transfer 100μL from the flask to the appropriately labeled Nutrient Broth tube. Discard the pipette tip.

3 Repeat this procedure at T_2 through T_{15}. Be sure to mix the broth before each transfer, and use a clean pipette tip.

4 Incubate all broths at $35 \pm 2°C$ for 48 hours.

Lab One: Group 2 (T_{10})

Plate Subgroup

1 Obtain all plating materials. Place six microtubes in a rack. Label them 1 through 6. Add 900 μL of sterile water to microtubes 2 through 6, and cover them until needed.

2 Label seven Nutrient Agar plates 1 through 7, respectively.

3 When your instructor adds 1.0 mL of the *E. coli* culture to the heated flask, he/she will start the stopwatch. This is T_0. At T_{10} remove 1500 μL with your pipette. Unless you have 5000 μL pipettes, this will take two 750 μL withdrawals. In this case, wait for the broth group to withdraw its sample, then make your transfers as quickly as possible.

4 Begin the following serial dilution:

  ♦ Add the 1500 μL of broth from the flask to tube 1.
  ♦ Without changing the pipette tip, transfer 100 μL from tube 1 to tube 2; mix. Discard the pipette tip. Use a clean pipette tip for each transfer from here forward.
  ♦ Transfer 100 μL from tube 2 to tube 3; mix.
  ♦ Transfer 100 μL from tube 3 to tube 4; mix.
  ♦ Transfer 100 μL from tube 4 to tube 5; mix.
  ♦ Transfer 100 μL from tube 5 to tube 6; mix.
  ♦ Transfer 100 μL from tube 6 to tube 7; mix.
  ♦ Transfer 1000 μL from tube 1 to plate 1 and spread according to the spread-plate procedure.
  ♦ Transfer 100 μL from tube 1 to plate 2 and spread.
  ♦ Transfer 100 μL from tube 2 to plate 3 and spread.
  ♦ Transfer 100 μL from tube 3 to plate 4 and spread.

♦ Transfer 100 µL from tube 4 to plate 5 and spread.

♦ Transfer 100 µL from tube 5 to plate 6 and spread.

♦ Transfer 100 µL from tube 6 to plate 7 and spread.

5 Calculate the volume of original sample transferred to each plate (volume plate × dilution) and enter it in the first chart on the Data Sheet.

6 Tape the plates together, invert, and incubate them at $35 \pm 2°C$ for 48 hours.

Broth Subgroup

1 Obtain all broth transfer materials. Label 15 Nutrient Broth tubes T_{16} through T_{30} respectively. Label the 16th "Control." Enter these numbers in the appropriate boxes in the second chart on the Data Sheet.

2 When your instructor adds 1.0 mL of the *E. coli* culture to the heated flask, he/she will start the stopwatch. This is T_0. Slightly before T_{16}, gently swirl the broth flask, and at exactly T_{16}, transfer 100 µL from the flask to the appropriately labeled Nutrient Broth tube. Discard the pipette tip.

3 Repeat this procedure at T_{17} through T_{30}. Be sure to mix the broth before each transfer, and use a clean pipette tip.

4 Incubate all broths at $35 \pm 2°C$ for 48 hours.

Lab One: Group 3 (T_0)

Plate Subgroup

Using the *S. aureus* culture, follow the instructions for the Group 1 Plate Subgroup.

Broth Subgroup

Using the *S. aureus* culture, follow the instructions for the Group 1 Broth Subgroup.

Lab One: Group 4 (T_{10})

Plate Subgroup

Using the *S. aureus* culture, follow the instructions for the Group 2 Plate Subgroup.

Broth Subgroup

Using the *S. aureus* culture, follow the instructions for the Group 2 Broth Subgroup

Lab Two: All Groups

1 Remove all broths and plates from the incubator.

2 Pick the plate containing between 30 and 300 colonies. There should be only one countable plate. All others are either TFTC (too few to count) or TNTC (too numerous to count). Label them as such, and enter this information in Chart 1 on the Data Sheet.

3 If you haven't done so already, count the colonies on the countable plate and enter the number in Chart 1 on the Data Sheet.

4 Examine the broths for growth, comparing each one to the uninoculated control (tube #16). Any turbidity is read as positive for growth. Enter your results in Chart 2 on the Data Sheet. Circle the earliest time that no turbidity appears (i.e., growth did not occur).

5 Calculate the population density of your group's organism for your specified time (T_0 or T_{10}), using the following formula: (**Note:** For best results, convert the volume plated to mL first, then calculate the Final Dilution Factor.)

$$\text{Cell density (CFU/mL)} = \frac{\text{CFU}}{\text{Original sample volume}}$$

6 Enter your results in Chart 1 and again on Chart 3 on the Data Sheet. Also, be sure to convert your results to logarithmic form and enter that in the third chart as well.

7 Collect the data from other groups and enter it in Chart 3 on the Data Sheet. Your instructor will likely provide a transparency or chalkboard space for class data.

8 Follow the directions on the Data Sheet to plot *and* calculate the D values of both organisms.

9 Answer the questions on the Data Sheet.

Agarose Gel Electrophoresis

Gel Preparation and Staining
for Exercises 10-2 and 10-4

This exercise is designed to supplement Exercise 10-2 Restriction Digest and Exercise 10-4 Polymerase Chain Reaction[1]. It is divided into two modules: Module A—Gel Preparation and Module B—Gel Staining. Approximate times for completion are included to facilitate scheduling of lab periods.

✦ Theory

Electrophoresis is a technique in which molecules are separated by size and electrical charge in a gel. Electrophoresis gels typically are prepared from **agarose** or **polyacrylamide,** depending on the molecules to be separated. Agarose is used for large DNA molecules and polyacrylamide is used for small DNA molecules and proteins. In preparation for the procedure the gel is cast as a thin slab containing tiny wells at one end and placed in a buffered solution to maintain proper electrolyte balance. Samples to be examined are loaded into the different wells and electrodes are attached to create an electrical field in the gel.

Under the influence of the electrical field molecules in the samples migrate through the gel. (Proteins and DNA are negatively charged and migrate toward the positive pole.) The molecules travel different distances because of differences in size and electrical charge. At the end of the run, the gel is stained to show the location of the separated molecules as bands in each lane. **Coomassie blue** is commonly used for protein, whereas **ethidium bromide**, a fluorescent dye, or methylene blue can be used for nucleic acids. If the nucleic acid has been radioactively labeled, the gel can be placed on X-ray film to produce an **autoradiograph**.

✦ Application

Gel electrophoresis is a means of isolating proteins or fragments of DNA for a variety of purposes. Unique band patterns can be used for taxonomic or identification purposes. Once isolated, the DNA or proteins can be removed from the gel and used in biochemical studies. Other techniques, such as DNA fingerprinting and Southern, Western, and Northern Blotting, begin with electrophoresis.

Module A–Gel Preparation (~ 40 minutes)[2]

Follow the procedural diagram in Figure G-1.

1 Convert the gel bed (including comb) to a casting tray by sealing the ends with rubber dams or tape. Place it on a level surface.

---

[1] This procedure is adapted, with permission, from Edvotek kit #213 Cleavage of DNA with Restriction Enzymes and kit #330 The Molecular Biology of DNA Amplification by Polymerase Chain Reaction, for which the company holds copyrights. All materials and equipment required for this and the accompanying exercises can be purchased at Edvotek, Inc., PO Box 341232, Bethesda, MD, 20827, USA. email: info@edvotek.com www.edvotek.com

[2] This procedure is for preparation of a 7 × 7 cm 0.8% agarose gel. Adjust the volumes accordingly for different size gels. The time includes preparation, pouring, and cooling to solidify.

2 In a 250 mL flask, add:
   - 25 mL buffer (premixed 50X concentrate diluted 1:49 with distilled or deionized water)
   - 0.4 g agarose

3 On the outside of the flask, mark the level of the solution with a marking pen. (If using a hotplate to dissolve the agarose, add a magnetic stir bar before marking the flask.)

4 Heat the solution until it boils and completely dissolves the agarose. Be sure to watch the flask and remove it from the heat as needed to prevent a boil-over. Also, if using a hotplate, be sure to stir while heating.

5 When the agarose is fully dissolved (completely clear with no crystals or particles), remove the flask from the heat and allow it to cool to 50-60°C. Cover with foil to prevent excessive evaporation.

6 When the gel has cooled, check the volume and add water, if necessary, to bring the level back up to the mark placed on the outside of the flask in #3. Stir gently until the mixture is uniformly clear. Remove the stir bar to avoid pouring it into the gel casting tray.

7 Steadily pour the liquid gel into the casting tray, trying to avoid creating bubbles (Figure G-1). Allow the tray to remain undisturbed until the gel is completely cool and solid.

8 When the gel is solid, carefully remove the tape or dams and the comb. In removing the comb, lift straight up very slowly until it is free from the gel.

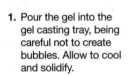

1. Pour the gel into the gel casting tray, being careful not to create bubbles. Allow to cool and solidify.

2. After the gel has solidified, carefully remove the comb by pulling straight up.

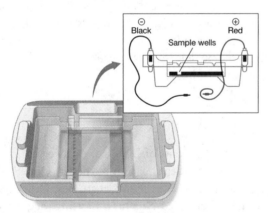

3. Place the entire gel box containing the gel into the electrophoresis chamber and cover with buffer. Orient properly with the wells nearest the black terminal.

4. Fill wells.

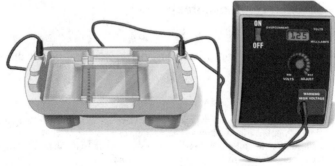

5. Plug in and turn on the power supply to the appropriate voltage.

G-1   GEL PREPARATION AND LOADING PROCEDURAL DIAGRAM ✦

Module B—Gel Staining

The electrophoresis gels in this section can be stained with Instastain® Blue, FlashBlue™ DNA Stain, or Instastain® EtBr stain. (*Note:* Instastain® EtBr stain contains the mutagen, ethidium bromide. Always wear gloves when handling materials containing ethidium bromide and dispose of them according to the safety guidelines of your institution.)

Option 1—Staining with InstaStain® EtBr card (~30 minutes)

Follow the procedural diagram in Figure G-2.

✦ Materials

✦ one pipette
✦ one InstaStain® EtBr card
✦ one gel bed + small beaker for weight
✦ spatula for handling gel
✦ one UV Transilluminator
✦ one pair of UV protective goggles
✦ protective gloves

1 Place the gel on a piece of plastic wrap on a flat surface. Moisten the gel with a few drops of buffer.

2 Wearing gloves, remove the plastic cover from the unprinted side of an InstaStain® EtBr card and place it (unprinted side down) on the gel.

3 Run your fingers over the entire surface of the card to assure good surface contact with the gel.

4 Place the gel casting tray and small beaker on top of the card and allow it to stand undisturbed for 10–15 minutes.

5 Remove the card and, using the spatula, carefully place the gel on the transilluminator.

6 *Put on your goggles.*

7 Turn on the transilluminator and view the gel.

8 Dispose of all materials in appropriate containers as designated by your instructor.

1. Moisten the gel.

2. Place the InstaStain® card on the gel.

3. Press firmly.

4. Place a small weight to ensure good contact.

5. View on U.V. (300 nm) transilluminator.

G-2 INSTASTAIN® ETBR CARD STAINING PROCEDURAL DIAGRAM ✦

Option 2—Staining with InstaStain® Blue card (2 to 3 hours)

Follow the procedural diagram in Figure G-3.

✦ Materials

✦ one InstaStain® Blue card
✦ one gel bed + small beaker for weight
✦ spatula for handling gel
✦ one light box
✦ protective gloves

1 Place the gel on a piece of plastic wrap on a flat surface.

2 Wearing gloves, place the InstaStain® Blue card stain side down on the gel.

3 Run your fingers over the entire surface of the card to assure good surface contact with the gel.

4 Place the gel casting tray and small beaker on top of the card and allow it to stand undisturbed for 10–15 minutes.

5 Remove the card and transfer the gel to a staining tray or a large weigh boat.

6 Add distilled or deionized water to cover the gel.

7 Discard the destaining water and repeat with fresh water up to three times total. Watch for the bands to become visible. (*Note:* It may help to place the destaining tray on the light box to see the bands as they become visible.)

8 Dispose of all materials in appropriate containers as designated by your instructor.

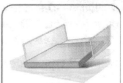

1. Place gel on a flat surface covered with plastic wrap.

2. Place the InstaStain® card on the gel.

3. Press firmly.

4. Place a small weight for approx. 5 minutes.

5. Transfer to a small tray for destaining.

6. Destain with 37°C distilled water.

G-3 INSTASTAIN® BLUE CARD STAINING PROCEDURAL DIAGRAM ✦

Option 3—One-Step Staining and Destaining with InstaStain® Blue Card (for use with 7 × 7 cm gels) (4 hours to overnight)

Follow the procedural diagram in Figure G-4.

✦ Materials

✦ one tray or weigh boat
✦ one InstaStain® Blue card
✦ spatula for handling gel
✦ one light box
✦ protective gloves

1 Place the gel in a small tray or weigh boat.

2 Cover the gel with 75 mL of distilled or deionized water or buffer. If the gel is not completely covered with 75 mL of liquid, place the gel in a smaller tray.

3 Float the InstaStain® Blue card stain side down on the water or buffer.

4 Let stand for at least 3 hours. It can be left overnight, but must be covered with plastic wrap.

5 Remove the water or buffer and view the gel on a light box.

6 Dispose of all materials in appropriate containers as designated by your instructor.

One step stain and destain.

G-4 ONE-STEP STAINING AND DESTAINING WITH INSTASTAIN® BLUE STAINING PROCEDURAL DIAGRAM ✦

Option 4—Staining with FlashBlue™ DNA Stain (2 hours to overnight)

✦ Materials

✦ FlashBlue™ DNA Stain (diluted 1:9)
✦ one tray or weigh boat large enough to hold 600 mL
✦ spatula for handling gel
✦ one light box
✦ protective gloves

1 Place the gel in a small tray or weigh boat.

2 Completely cover the gel with stain for at least 30 minutes. Agitate occasionally.

3 Remove that stain and destain by completely covering it with 600 mL of 37°C (not warmer) distilled or deionized water. Let it destain for 15 minutes with occasional agitation.

4 Repeat with clean water for an additional 15 minutes. It can be left overnight, but must be covered with plastic wrap.

5 Remove the water and view the gel on a light box.

6 Dispose of all materials in appropriate containers as designated by your instructor.

References

Edvotek, Inc. 2008. *EDVO-Kit #330 The Molecular Biology of DNA Amplification by Polymerase Chain Reaction* Instruction Booklet. Edvotek, Inc., PO Box 341232, Bethesda, MD, 20827, USA.

Edvotek, Inc. 2001. *EDVO-Kit #213 Cleavage of DNA with Restriction Enzymes* Instruction Booklet. Edvotek, Inc., PO Box 341232, Bethesda, MD, 20827, USA.

Hendrickson, William and Don Walthers. 2007. Chapter 27 in *Methods for General and Molecular Microbiology, 3rd Edition*, Edited by C. A. Reddy (Editor In Chief), T. J. Beveridge, J. A. Breznak, G. A. Marzluf, T. M. Schmidt, and L. R. Snyder. ASM Press, Washington, DC.

Madigan, Michael T. and John M. Martinko. 2006. Chapter 7 in *Brock Biology of Microorganisms, Eleventh Edition*. Pearson Prentice Hall. Upper Saddle River, NJ.

Nelson, David L. and Michael M. Cox. 2008. Chapters 8 and 9 in *Lehninger: Principles of Biohemistry, Fifth Edition*. W. H. Freeman and Company, New York, NY.

Medium, Reagent, and Stain Recipes

Media[1]

Ames Test: Complete Medium

- Beef extract   3.0 g
- Peptone   5.0 g
- Sodium chloride   5.0 g
- Agar   20.0 g
- Distilled or deionized water   1.0 L

1 Suspend the dry ingredients in the water, mix well, and boil until completely dissolved.

2 Cover loosely and sterilize in the autoclave at 15 lbs. pressure (121°C) for 15 minutes.

3 Remove from the autoclave and cool slightly.

4 Aseptically pour into sterile Petri dishes (20 mL/plate) and cool to room temperature.

Ames Test: Minimal Medium

- Dextrose (glucose)   20.0 g
- 50x Vogel-Bonner salts   20.0 mL
- Histidine   0.00016 g
- Biotin   0.00025 g
- Agar   20.0 g
- Distilled or deionized water   1.0 L

1 Prior to preparation of the medium:

a. Add 1.6 mg histidine to 10.0 mL distilled or deionized water and filter sterilize.

b. Add 2.5 mg biotin to 10.0 mL distilled or deionized water and filter sterilize.

2 Prepare the 50x Vogel-Bonner salts solution by adding the ingredients to just enough water to dissolve them while heating and stirring. After the ingredients are dissolved, add enough water to bring the total volume up to exactly 1 liter.

3 Suspend, mix, and boil the agar in 500.0 mL of the water until completely dissolved.

4 Suspend and mix the dextrose in the remaining 500.0 mL of water until completely dissolved.

5 Cover the agar and dextrose containers loosely and sterilize in the autoclave at 121°C for 15 minutes.

6 Remove from the autoclave and allow to cool to 80°C.

7 Aseptically add 1.0 mL histidine solution, 1.0 mL biotin solution, and 20 mL 50x Vogel-Bonner salts to the glucose solution. Mix well.

8 Add the glucose solution to the agar solution, mix well and aseptically pour into sterile Petri dishes (20 mL/plate). Allow the medium to cool to room temperature.

---

[1] Unless otherwise specified, the volumes used in these recipes are for 16 × 150 mm test tubes or 100 mm Petri dishes.

Ammonium Broth

- Ammonium sulfate — 1.0 g
- $MgSO_4$ — 0.5 g
- NaCl — 2.0 g
- K_2HPO_4 — 1.0 g
- $CaCO_3$ — 10.0 g
- $FeCl_3$ — trace
- Distilled or deionized water — 1 L

    pH 7.3 at 25°C

Bile Esculin Agar

- Beef extract — 3.0 g
- Peptone — 5.0 g
- Oxgall — 40.0 g
- Esculin — 1.0 g
- Ferric citrate — 0.5 g
- Agar — 15.0 g
- Distilled or deionized water — 1.0 L

    pH 6.6–6.8 at 25°C

1. Suspend the dry ingredients in the water, mix well, and boil until completely dissolved.
2. Dispense 7.0 mL volumes into test tubes and cap loosely.
3. Sterilize in the autoclave at 15 lbs. pressure (121°C) for 15 minutes.
4. Remove from the autoclave, slant, and allow the medium to cool to room temperature.

Blood Agar

- Infusion from beef heart (solids) — 2.0 g
- Pancreatic digest of casein — 13.0 g
- Sodium chloride — 5.0 g
- Yeast extract — 5.0 g
- Agar — 15.0 g
- Defibrinated sheep blood — 50.0 mL
- Distilled or deionized water — 1.0 L

    pH 7.1–7.5 at 25°C

1. Suspend the dry ingredients in the water, mix well, and boil until completely dissolved. This is Blood Agar base.
2. Cover loosely and sterilize in the autoclave at 15 lbs. pressure (121°C) for 15 minutes.
3. Remove from the autoclave and cool to 45°C.
4. Aseptically add the sterile, room-temperature sheep blood to the blood agar base and mix well.
5. Pour into sterile Petri dishes and allow the medium to cool to room temperature.

Brain-Heart Infusion Agar

- Calf brains, infusion from 200 g — 7.7 g
- Beef heart, infusion from 250 g — 9.8 g
- Proteose peptone — 10.0 g
- Dextrose — 2.0 g
- Sodium chloride — 5.0 g
- Disodium phosphate — 2.5 g
- Agar — 15.0 g
- Distilled or deionized water — 1.0 L

    pH 7.2–7.6 at 25°C

Plates

1. Suspend the dry ingredients in the water, mix well, and boil until completely dissolved.
2. Cover loosely and sterilize in the autoclave at 15 lbs. pressure (121°C) for 15 minutes.
3. Remove from the autoclave, allow the medium to cool slightly, and aseptically pour into sterile Petri dishes (20 mL/plate).
4. Allow the medium to cool to room temperature. Store upside down.

Slants

1. Suspend the dry ingredients in the water, mix well, and boil until completely dissolved.
2. Dispense 7 mL portions into test tubes, and cap loosely.
3. Autoclave for 15 minutes at 121°C to sterilize the medium.
4. Slant the tubes and cool to room temperature.

Brain-Heart Infusion Broth

- Calf brains, infusion from 200 g — 7.7 g
- Beef heart, infusion from 250 g — 9.8 g
- Proteose peptone — 10.0 g
- Dextrose — 2.0 g
- Sodium chloride — 5.0 g
- Disodium phosphate — 2.5 g
- Distilled or deionized water — 1.0 L

    pH 7.2–7.6 at 25°C

1. Suspend the dry ingredients in the water, mix well, and warm until completely dissolved.
2. Transfer 7.0 mL portions to test tubes and cap loosely.
3. Sterilize in the autoclave at 15 lbs. pressure (121°C) for 15 minutes.
4. Remove from the autoclave and allow the medium to cool to room temperature.

Brilliant Green Lactose Bile Broth

- Peptone — 10.0 g
- Lactose — 10.0 g
- Oxgall — 20.0 g
- Brilliant green dye — 0.0133 g
- Distilled or deionized water — 1.0 L

    pH 7.0–7.4 at 25°C

1. Suspend the dry ingredients in the water, mix well, and boil until completely dissolved.
2. Dispense 10.0 mL portions into test tubes.
3. Place an inverted Durham tube in each broth and cap loosely.
4. Sterilize in the autoclave at 15 lbs. pressure (121°C) for 15 minutes.
5. Remove the medium from the autoclave and allow it to cool before inoculating.

Citrate Agar (Simmons)

◆ Ammonium dihydrogen phosphate	1.0 g
◆ Dipotassium phosphate	1.0 g
◆ Sodium chloride	5.0 g
◆ Sodium citrate	2.0 g
◆ Magnesium sulfate	0.2 g
◆ Agar	15.0 g
◆ Bromthymol blue	0.08 g
◆ Distilled or deionized water	1.0 L

pH 6.7–7.1 at 25°C

1 Suspend the dry ingredients in the water, mix well, and boil until completely dissolved.

2 Dispense 7.0 mL portions into test tubes and cap loosely.

3 Sterilize in the autoclave at 15 lbs. pressure (121°C) for 15 minutes.

4 Remove from the autoclave, slant, and cool to room temperature.

Columbia CNA with 5% Sheep Blood Agar

◆ Pancreatic digest of casein	12.0 g
◆ Peptic digest of animal tissue	5.0 g
◆ Yeast extract	3.0 g
◆ Beef extract	3.0 g
◆ Corn starch	1.0 g
◆ Sodium chloride	5.0 g
◆ Colistin	10.0 mg
◆ Nalidixic acid	10.0 mg
◆ Agar	13.5 g
◆ Sheep blood (defibrinated, sterile)	50.0 mL
◆ Distilled or deionized water	1.0 L

pH 7.1–7.5 at 25°C

1 Suspend the dry ingredients in the water, mix well, and boil until completely dissolved.

2 Cover loosely and sterilize in the autoclave at 15 lbs. pressure (121°C) for 15 minutes.

3 Remove from the autoclave and cool to 45 to 50°C.

4 Add the sheep blood to 950 mL of the mixture and aseptically pour into sterile Petri dishes (20 mL/plate).

5 Allow the medium to cool to room temperature. Store upside down.

Complete Medium (*See* Ames Test)

Decarboxylase Medium (Møller)

◆ Peptone	5.0 g
◆ Beef extract	5.0 g
◆ Glucose (dextrose)	0.5 g
◆ Bromcresol purple	0.01 g
◆ Cresol red	0.005 g
◆ Pyridoxal	0.005 g
◆ L-Lysine, L-Ornithine, or L-Arginine	10.0 g
◆ Distilled or deionized water	1.0 L

pH 5.8–6.2 at 25°C

1 Suspend the dry ingredients in the water, mix well, and boil until completely dissolved. (Use only one of the listed L-amino acids.)

2 Adjust pH by adding NaOH if necessary.

3 Dispense 7.0 mL volumes into test tubes and cap.

4 Sterilize in the autoclave at 15 lbs. pressure (121°C) for 10 minutes.

5 Remove from the autoclave and cool to room temperature.

Desoxycholate Agar (Modified Leifson)

◆ Peptone	10.0 g
◆ Lactose	10.0 g
◆ Sodium desoxycholate	1.0 g
◆ Sodium chloride	5.0 g
◆ Dipotassium phosphate	2.0 g
◆ Ferric citrate	1.0 g
◆ Sodium citrate	1.0 g
◆ Agar	16.0 g
◆ Neutral red	0.033 g
◆ Distilled or deionized water	1.0 L

pH 7.1–7.5 at 25°C

1 Suspend the dry ingredients in the water, mix well, and boil until completely dissolved.

2 When cooled to 50°C, pour into sterile plates.

3 Allow the medium to cool to room temperature.

(Modified) Desulfovibrio Medium

◆ Yeast extract	5.0 g
◆ Wolfe's mineral solution (see Modified Thiobacillus medium)	10.0 mL
◆ Ammonium phosphate, dibasic	0.3 g
◆ Sodium lactate	3.0 g
◆ Sodium sulfite	2.0 g
◆ Dipotassium phosphate	1.7 g
◆ L-Cysteine	0.5 g
◆ Agar	3.0 g
◆ DI H$_2$O	1.0 L

pH 7.5 ± 0.2 at 25°C

1 Mix all ingredients and boil to dissolve the agar.

2 Transfer 8 mL volumes to screw-cap test tubes and cap loosely.

3 Autoclave at 121°C for 15 minutes. (***Note:*** To ensure anaerobic conditions in the medium, it must be autoclaved immediately before class, and placed in a 50°C water bath.)

Ferrous Ammonium Sulfate Solution for Sulfur Reducing Bacteria

◆ Ferrous ammonium sulfate	0.5 g
◆ dH$_2$O	100.0 mL

1 Add the ferrous ammonium sulfate to the water and mix.

2 Filter-sterilize when needed using a syringe and sterile 0.22 or 0.44 m filter.

DNase Test Agar with Methyl Green

- ◆ Tryptose   20.0 g
- ◆ Deoxyribonucleic acid   2.0 g
- ◆ Sodium chloride   5.0 g
- ◆ Agar   15.0 g
- ◆ Methyl green   0.05 g
- ◆ Distilled or deionized water   1.0 L
  pH 7.1–7.5 at 25°C

1   Suspend the dry ingredients in the water, mix well, and boil until completely dissolved.

2   Cover loosely and sterilize in the autoclave at 15 lbs. pressure (121°C) for 15 minutes.

3   Aseptically pour into sterile Petri dishes (20 mL/plate) and cool to room temperature.

EC Broth

- ◆ Tryptose   20.0 g
- ◆ Lactose   5.0 g
- ◆ Dipotassium phosphate   4.0 g
- ◆ Monopotassium phosphate   1.5 g
- ◆ Sodium chloride   5.0 g
- ◆ Distilled or deionized water   1.0 L
  pH 6.7–7.1 at 25°C

1   Suspend the dry ingredients in the water, and mix well until completely dissolved.

2   Dispense 10.0 mL portions into test tubes.

3   Place an inverted Durham tube in each broth and cap loosely.

4   Sterilize in the autoclave at 15 lbs. pressure (121°C) for 15 minutes.

5   Remove the media from the autoclave and allow it to cool before inoculating.

Endo Agar

- ◆ Peptone   10.0 g
- ◆ Lactose   10.0 g
- ◆ Dipotassium phosphate   3.5 g
- ◆ Agar   15.0 g
- ◆ Basic Fuchsin   0.5 g
- ◆ Sodium Sulfite   2.5 g
- ◆ Distilled or deionized water   1.0 L
  pH 7.3–7.7 at 25°C

1   Suspend the dry ingredients in the water, mix well, and boil until completely dissolved.

2   Autoclave for 15 minutes at 15 lbs. pressure (121°C).

3   When cooled to 50°C, pour into sterile plates.

4   Allow the medium to cool to room temperature.

Enriched TSA (*See* Tryptic Soy Agar)

Eosin Methylene Blue Agar (Levine)

- ◆ Peptone   10.0 g
- ◆ Lactose   10.0 g*
- ◆ Dipotassium phosphate   2.0 g
- ◆ Agar   15.0 g
- ◆ Eosin Y   0.4 g
- ◆ Methylene blue   0.065 g
- ◆ Distilled or deionized water   1.0 L
  pH 6.9–7.3 at 25°C

1   Suspend the dry ingredients in the water, mix well, and boil until completely dissolved.

2   Autoclave for 15 minutes at 15 lbs. pressure (121°C).

3   When cooled to 50°C, pour into sterile plates.

4   Allow the medium to cool to room temperature.

Glucose Broth

- ◆ Peptone   10.0 g
- ◆ Glucose   5.0 g
- ◆ NaCl   5.0 g
- ◆ Distilled or deionized water   1.0 L

1   Suspend the dry ingredients the water. Agitate and heat slightly (if necessary) to dissolve completely.

2   Dispense 7.0 mL portions into test tubes and cap loosely.

3   Autoclave for 15 minutes at 121°C to sterilize the medium.

Glucose Salts Medium

- ◆ Glucose   5.0 g
- ◆ NaCl   5.0 g
- ◆ $MgSO_4$ 0.2 g
- ◆ $(NH_4)H_2PO_4$ 1.0 g
- ◆ K_2HPO_4 1.0 g
- ◆ Distilled or deionized water   1.0 L

1   Suspend the dry ingredients in the water. Agitate and heat slightly (if necessary) to dissolve completely.

2   Dispense 7.0 mL portions into test tubes and cap loosely.

3   Autoclave for 15 minutes at 121°C to sterilize the medium.

Glycerol Yeast Extract Agar

- ◆ Glycerol   5.0 mL
- ◆ Yeast extract   2.0 g
- ◆ Dipotassium phosphate   1.0 g
- ◆ Agar   15.0 g
- ◆ Distilled or deionized water   1.0 L

1   Suspend the dry ingredients in the water, mix well, and boil until completely dissolved.

2   Autoclave for 15 minutes at 15 lbs. pressure (121°C).

3   When cooled to 50°C, pour into sterile plates.

4   Allow the medium to cool to room temperature.

---

*An alternative recipe replaces the 10.0 g of lactose with 5.0 g of lactose and 5.0 g of sucrose.

Hektoen Enteric Agar

♦ Yeast extract	3.0 g
♦ Peptic digest of animal tissue	12.0 g
♦ Lactose	12.0 g
♦ Sucrose	12.0 g
♦ Salicin	2.0 g
♦ Bile salts	9.0 g
♦ Sodium chloride	5.0 g
♦ Sodium thiosulfate	5.0 g
♦ Ferric ammonium citrate	1.5 g
♦ Bromthymol blue	0.064 g
♦ Acid fuchsin	0.1 g
♦ Agar	13.5 g
♦ Distilled or deionized water	1.0 L

pH 7.4–7.8 at 25°C

1 Suspend the dry ingredients in the water, mix well, and boil until completely dissolved.

2 Do not autoclave.

3 When cooled to 50°C, pour into sterile plates.

4 Cool to room temperature with lids slightly open.

Kligler's Iron Agar

♦ Beef extract	3.0 g
♦ Yeast extract	3.0 g
♦ Peptone	15.0 g
♦ Proteose peptone	5.0 g
♦ Lactose	10.0 g
♦ Dextrose (glucose)	1.0 g
♦ Ferrous sulfate	0.2 g
♦ Sodium chloride	5.0 g
♦ Sodium thiosulfate	0.3 g
♦ Agar	12.0 g
♦ Phenol red	0.024 g
♦ Distilled or deionized water	1.0 L

pH 7.2–7.6 at 25°C

1 Suspend the dry ingredients in the water, mix well, and boil until completely dissolved.

2 Transfer 7.0 mL portions to test tubes and cap loosely.

3 Sterilize in the autoclave at 15 lbs. pressure (121°C) for 15 minutes.

4 Remove from the autoclave and slant in such a way as to form a deep butt.

5 Allow the medium to cool to room temperature.

Lauryl Tryptose Broth

♦ Tryptose	20.0 g
♦ Lactose	5.0 g
♦ Dipotassium phosphate	2.75 g
♦ Monopotassium phosphate	2.75 g
♦ Sodium chloride	5.0 g
♦ Sodium lauryl sulfate	0.1 g
♦ Distilled or deionized water	1.0 L

pH 6.6–7.0 at 25°C

1 Suspend the dry ingredients in the water until completely dissolved. Heat slightly if necessary.

2 Dispense 10.0 mL portions into test tubes.

3 Place an inverted Durham tube in each broth and cap loosely.

4 Sterilize in the autoclave at 15 lbs. pressure (121°C) for 15 minutes.

5 Remove the medium from the autoclave and allow it to cool before inoculating.

Litmus Milk Medium

♦ Skim milk	100.0 g
♦ Azolitmin	0.5 g
♦ Sodium sulfite	0.5 g
♦ Distilled or deionized water	1.0 L

pH 6.3–6.7 at 25°C

1 Suspend and mix the ingredients in the water and heat to approximately 50°C to dissolve completely.

2 Transfer 7.0 mL portions to test tubes and cap loosely.

3 Sterilize in the autoclave at 113–115°C for 20 minutes.

4 Remove from the autoclave and allow the medium to cool to room temperature.

Luria-Bertani Agar

♦ Tryptone	10.0 g
♦ Yeast extract	5.0 g
♦ NaCl	10.0 g
♦ Agar	15.0 g
♦ Distilled or deionized water	1.0 L

pH 7.4 at 25°C

1 Suspend the dry ingredients in the water, mix well, and boil until completely dissolved.

2 Cover loosely and sterilize in the autoclave at 15 lbs. pressure (121°C) for 15 minutes.

3 Aseptically pour into sterile Petri dishes (20 mL/plate) and cool to room temperature.

Luria-Bertani Broth

♦ Tryptone	10.0 g
♦ Yeast extract	5.0 g
♦ NaCl	10.0 g
♦ Distilled or deionized water	1.0 L

pH 7.4 at 25°C

1 Suspend the ingredients in one liter of distilled or deionized water. Agitate and heat slightly (if necessary) to dissolve completely.

2 Dispense 7.0 mL portions into test tubes and cap loosely.

3 Autoclave for 15 minutes at 121°C to sterilize the medium.

Lysine Iron Agar

- Peptone   5.0 g
- Yeast extract   3.0 g
- Dextrose   1.0 g
- L-Lysine hydrochloride   10.0 g
- Ferric ammonium citrate   0.5 g
- Sodium thiosulfate   0.04 g
- Bromcresol purple   0.02 g
- Agar   15.0 g
  pH 6.5–6.9 at 25°C

1 Suspend the dry ingredients in the water, mix well, and boil until completely dissolved.

2 Transfer 8.0 mL portions to test tubes and cap loosely.

3 Sterilize in the autoclave at 121°C for 15 minutes.

4 Remove from the autoclave and slant in such a way as to produce a deep butt. Allow the medium to cool to room temperature.

MacConkey Agar

- Pancreatic digest of gelatin   17.0 g
- Pancreatic digest of casein   1.5 g
- Peptic digest of animal tissue   1.5 g
- Lactose   10.0 g
- Bile salts   1.5 g
- Sodium chloride   5.0 g
- Neutral red   0.03 g
- Crystal violet   0.001 g
- Agar   13.5 g
- Distilled or deionized water   1.0 L
  pH 6.9–7.3 at 25°C

1 Suspend the dry ingredients in the water, mix well, and boil until completely dissolved.

2 Autoclave for 15 minutes at 15 lbs. pressure (121°C).

3 When cooled to 50°C, pour into sterile plates.

4 Allow the medium to cool to room temperature.

Malonate Broth

- Yeast extract   1.0 g
- Ammonium sulfate   2.0 g
- Dipotassium phosphate   0.6 g
- Monopotassium phosphate   0.4 g
- Sodium chloride   2.0 g
- Sodium malonate   3.0 g
- Dextrose   0.25 g
- Bromthymol blue   0.025 g
- Distilled or deionized water   1.0 L
  pH 6.5–6.9 at 25°C

1 Suspend the dry ingredients in the water. Agitate and heat slightly (if necessary) to dissolve completely.

2 Dispense 7.0 mL portions into test tubes and cap loosely.

3 Autoclave for 15 minutes at 121°C to sterilize the medium.

Mannitol Salt Agar

- Beef extract   1.0 g
- Peptone   10.0 g
- Sodium chloride   75.0 g
- D-Mannitol   10.0 g
- Phenol red   0.025 g
- Agar   15.0 g
- Distilled or deionized water   1.0 L
  pH 7.2–7.6 at 25°C

1 Suspend the dry ingredients in the water, mix well, and boil until completely dissolved.

2 Autoclave for 15 minutes at 15 lbs. pressure (121°C).

3 When cooled to 50°C, pour into sterile plates.

4 Allow the medium to cool to room temperature.

Microgradient Tubes

Top Layer (Mineral Agar)

- Basal medium   1.0 L
- Wolfe's Vitamin mix   1.0 mL
- Agar   2.5 g

Bottom Layer (Plug)[1]

- Basal medium   1.0 L
- $Na_2S_2O_3$ 15.0 g
- Agar   15.0 g

Basal medium

- Wolfe's Mineral solution   10.0 mL
- 1M K_2HPO_4 5.7 mL
- 1M KH_2PO_4 3.3 mL
- 1M $NaHCO_3$ 10.0 mL
- 1M $(NH_4)_2SO_4$ 9.0 mL
- 1M $MgCl_2$ 1.0 mL
- 1M $CaCl_2$ 0.5 mL
- NaCl   1.0 g
- KCl   0.1 g
- dH_2O to 1.0 L
  pH adjusted to 7

1 Add 500 mL of basal medium to two flasks labeled "mineral agar" and "plug."

2 Add 1.25 g of agar to the mineral agar flask. Cover and boil to dissolve the agar.

3 To the plug flask, add 7.5 g $Na_2S_2O_3$ and 7.5 g of agar. Cover and boil to dissolve the agar.

4 Transfer ~1 mL of the plug mix to medium size test tubes and cap loosely.

5 Autoclave all tubes and the flask containing the mineral agar mixture at 121°C for 15 minutes.

6 When finished autoclaving, place the flask in a 50°C water bath and allow the temperatures to equilibrate.

7 Place the tubes in a test tube rack in an ice water bath.

8 When the mineral agar has cooled to 50°C, add 0.5 mL of Wolfe's vitamin solution and mix.

9 When the plugs have solidified and cooled, pour (or pipette) ~ 9 mL volumes of the molten mineral agar into each tube. Keep the tubes cold so as not to loosen or melt the plugs. Try to avoid creating bubbles. The final plug to mineral agar ratio should be 1:9.

---

[1] Due to the volatility and toxicity of sulfide gas, $Na_2S_2O_3$ is substituted for Na_2S.

(Skim) Milk Agar

- ♦ Beef extract — 3.0 g
- ♦ Peptone — 5.0 g
- ♦ Agar — 15.0 g
- ♦ Powdered nonfat milk — 100.0 g
- ♦ Distilled or deionized water — 1.0 L
  pH 7.0–7.4 at 25°C

1 Suspend the powdered milk in 500.0 mL of water in a 1-liter flask, mix well, and cover loosely.

2 Suspend the remainder of the ingredients in 500.0 mL of water in a 1-liter flask, mix well, boil to dissolve completely, and cover loosely.

3 Sterilize in the autoclave at 113–115°C for 20 minutes.

4 Remove from the autoclave and allow the mixtures to cool slightly.

5 Aseptically pour the agar solution into the milk solution. Mix *gently* (to prevent foaming).

6 Aseptically pour into sterile Petri dishes (15 mL/plate).

7 Allow the medium to cool to room temperature.

Minimal Medium (*See* Ames Test)

Modified Desulfovibrio Medium (*See* Desulfovibrio Medium)

Modified Thiobacillus Medium (*See* Thiobacillus Medium)

Motility Test Medium

- ♦ Beef extract — 3.0 g
- ♦ Pancreatic digest of gelatin — 10.0 g
- ♦ Sodium chloride — 5.0 g
- ♦ Agar — 4.0 g
- ♦ Triphenyltetrazolium chloride (TTC) — 0.05 g
- ♦ Distilled or deionized water — 1.0 L
  pH 7.1–7.5 at 25°C

1 Suspend the dry ingredients in the water, mix well, and boil until completely dissolved.

2 Dispense 7.0 mL portions into test tubes and cap loosely.

3 Sterilize in the autoclave at 15 lbs. pressure (121°C) for 15 minutes.

4 Remove from the autoclave and allow it to cool in the upright position.

MR-VP Broth

- ♦ Buffered peptone — 7.0 g
- ♦ Dipotassium phosphate — 5.0 g
- ♦ Dextrose (glucose) — 5.0 g
- ♦ Distilled or deionized water — 1.0 L
  pH 6.7–7.1 at 25°C

1 Suspend the dry ingredients in the water, mix well, and warm until completely dissolved.

2 Transfer 7.0 mL portions to test tubes and cap loosely.

3 Sterilize in the autoclave at 15 lbs. pressure (121°C) for 15 minutes.

4 Remove from the autoclave and allow the medium to cool to room temperature.

Mueller-Hinton II Agar

- ♦ Beef extract — 2.0 g
- ♦ Acid hydrolysate of casein — 17.5 g
- ♦ Starch — 1.5 g
- ♦ Agar — 17.0 g
- ♦ Distilled or deionized water — 1.0 L
  pH 7.1–7.5 at 25°C

1 Suspend the dry ingredients in the water, mix well, and boil until completely dissolved.

2 Cover loosely and sterilize in the autoclave at 121°C (15 lbs.) for 15 minutes.

3 Remove from the autoclave, allow to cool slightly.

4 Aseptically pour into sterile Petri dishes to a depth of 4 mm.

5 Allow the medium to cool to room temperature.

Nitrate Broth

- ♦ Beef extract — 3.0 g
- ♦ Peptone — 5.0 g
- ♦ Potassium nitrate — 1.0 g
- ♦ Distilled or deionized water — 1.0 L
  pH 6.6–7.0 at 25°C

1 Suspend the ingredients in the water; mix well, and warm until completely dissolved.

2 Transfer 7.0 mL portions to test tubes, and cap loosely. (Add inverted Durham tubes before capping, if desired.)

3 Sterilize in the autoclave at 15 lbs. pressure (121°C) for 15 minutes.

4 Remove from the autoclave and allow the medium to cool to room temperature.

Nitrogen Free Broth

- ♦ Glucose — 10.0 g
- ♦ $CaCO_3$ — 5.0 g
- ♦ $CaCl_2$ — 0.1 g
- ♦ KH_2PO_4 — 0.1 g
- ♦ K_2HPO_4 — 0.9 g
- ♦ $MgSO_4$ — 0.1 g
- ♦ $FeSO_4$ — 10.0 mg
- ♦ Na_2MoO_4 — 5.0 mg
- ♦ Distilled or deionized water — 1.0 L
  pH 7.3 at 25°C

Nutrient Agar

- ♦ Beef extract — 3.0 g
- ♦ Peptone — 5.0 g
- ♦ Agar — 15.0 g
- ♦ Distilled or deionized water — 1.0 L
  pH 6.6–7.0 at 25°C

Plates

1 Suspend the dry ingredients in the water, mix well, and boil until completely dissolved.

2 Cover loosely and sterilize in the autoclave at 15 lbs. pressure (121°C) for 15 minutes.

3 Remove from the autoclave, allow to cool slightly, and aseptically pour into sterile Petri dishes (20 mL/plate).

4 Allow the medium to cool to room temperature.

Tubes

1 Suspend the dry ingredients in the water, mix well, and boil until completely dissolved.

2 Dispense 7 mL or 10 mL portions into test tubes, and cap loosely.

3 Autoclave for 15 minutes at 121°C to sterilize the medium.

4 Cool to room temperature with the tubes in an upright position (10 mL) for agar deep tubes. Cool with the tubes on an angle (7 mL) for agar slants.

Nutrient Broth

◆ Beef extract	3.0 g
◆ Peptone	5.0 g
◆ Distilled or deionized water	1.0 L

pH 6.6–7.0 at 25°C

1 Suspend the dry ingredients in the water. Agitate and heat slightly (if necessary) to dissolve completely.

2 Dispense 7.0 mL portions into test tubes, and cap loosely.

3 Autoclave for 15 minutes at 121°C to sterilize the medium.

Nutrient Broth w/thiosulfate
(Flasks for Reduction Potential)

◆ Beef extract	3.0 g
◆ Peptone	5.0 g
◆ Sodium thiosulfate	0.3 g
◆ Distilled or deionized water	1.0 L

1 Suspend the dry ingredients in the water. Agitate and heat slightly (if necessary) to dissolve completely.

2 Dispense 15.0 mL portions into 25 mL flasks, and cap loosely.

3 Autoclave for 15 minutes at 121°C to sterilize the medium.

Nutrient Gelatin

◆ Beef extract	3.0 g
◆ Peptone	5.0 g
◆ Gelatin	120.0 g
◆ Distilled or deionized water	1.0 L

pH 6.6–7.0 at 25°C

1 *Slowly* add the dry ingredients to the water while stirring.

2 Warm to >50°C and maintain temperature until completely dissolved.

3 Dispense 7.0 mL volumes into test tubes and cap loosely.

4 Sterilize in the autoclave at 15 lbs. pressure (121°C) for 15 minutes.

5 Remove from the autoclave immediately and allow the medium to cool to room temperature in the upright position.

O-F Basal Medium
(See O-F Carbohydrate Solution)

◆ Pancreatic digest of casein	2.0 g
◆ Sodium chloride	5.0 g
◆ Dipotassium phosphate	0.3 g
◆ Agar	2.5 g
◆ Bromthymol blue	0.03 g
◆ Distilled or deionized water	1.0 L

pH 6.6–7.0 at 25°C

O-F Carbohydrate Solution
(See O-F Basal Medium)

◆ Carbohydrate (glucose, lactose, sucrose)	1.0 g
◆ Distilled or deionized water to total	10.0 mL

1 Suspend the dry ingredients, *without the carbohydrate*, in the water, mix well, and boil to dissolve completely. This is basal medium.

2 Divide the medium into ten aliquots of 100.0 mL each.

3 Cover loosely and sterilize in the autoclave at 121°C for 15 minutes.

4 Prepare carbohydrate solution, cover loosely, and autoclave at 118°C for 10 minutes.

5 Allow both solutions to cool to 50°C.

6 Aseptically add 10.0 mL sterile carbohydrate solution to a basal medium aliquot, and mix well.

7 Aseptically transfer 7.0 mL volumes to sterile test tubes, and allow the medium to cool to room temperature.

Peptone Water

◆ Peptone	40.0 g
◆ Distilled or deionized water	1.0 L

Phenol Red (Carbohydrate) Broth

◆ Pancreatic digest of casein	10.0 g
◆ Sodium chloride	5.0 g
◆ Carbohydrate (glucose, lactose, sucrose)	5.0 g
◆ Phenol red	0.018 g
◆ Distilled or deionized water	1.0 L

pH 7.1 ± 7.5 at 25°C

1 Suspend the dry ingredients in the water; mix well, and warm slightly to dissolve completely.

2 Dispense 7.0 mL volumes into test tubes.

3 Insert inverted Durham tubes into the test tubes and cap loosely.

4 Sterilize in the autoclave at 116–118°C for 15 minutes.

5 Remove from the autoclave and allow the medium to cool to room temperature.

Phenylalanine Deaminase Agar

◆ DL-Phenylalanine	2.0 g
◆ Yeast extract	3.0 g
◆ Sodium chloride	5.0 g
◆ Sodium phosphate	1.0 g
◆ Agar	12.0 g
◆ Distilled or deionized water	1.0 L

pH 7.1–7.5 at 25°C

1 Suspend the dry ingredients in the water, mix well, and boil until completely dissolved.

2 Dispense 7.0 mL volumes into test tubes and cap loosely.

3 Sterilize in the autoclave at 15 lbs. pressure (121°C) for 10 minutes.

4 Remove from the autoclave, slant, and allow the medium to cool to room temperature.

Phenylethyl Alcohol Agar

♦ Tryptose	10.0 g
♦ Beef extract	3.0 g
♦ Sodium chloride	5.0 g
♦ Phenylethyl alcohol	2.5 g
♦ Agar	15.0 g
♦ Distilled or deionized water	1.0 L

pH 7.1–7.5 at 25°C

1 Suspend the dry ingredients in the water, mix well, and boil until completely dissolved.
2 Autoclave for 15 minutes at 15 lbs. pressure (121°C).
3 When cooled to 50°C, pour into sterile plates.
4 Allow the medium to cool to room temperature.

Photobacterium Broth

♦ Tryptone	5.0 g
♦ Yeast extract	2.5 g
♦ Ammonium chloride	0.3 g
♦ Magnesium sulfate	0.3 g
♦ Ferric chloride	0.01 g
♦ Calcium carbonate	1.0 g
♦ Monopotassium phosphate	3.0 g
♦ Sodium glycerol phosphate	23.5 g
♦ Sodium chloride	30.0 g
♦ Distilled or deionized water	1.0 L

pH 6.8–7.2 at 25°C

1 Suspend the dry ingredients in the water, mix well, and boil until completely dissolved.
2 Dispense into flasks to form a shallow layer.
3 Autoclave for 15 minutes at 15 lbs. pressure (121°C).
4 Allow the medium to cool to room temperature.

Potato Dextrose Agar

♦ Potato flakes	20.0 g
♦ Dextrose	10.0 g
♦ Agar	15.0 g
♦ Distilled or deionized water	1.0 L

1 Suspend the dry ingredients in the water, mix well, and boil until completely dissolved.
2 Autoclave for 15 minutes at 15 lbs. pressure (121°C).
3 Remove the agar mixture from the autoclave and cool to 50°C.
4 Mix and pour into sterile Petri dishes. *Note:* Swirl the flask frequently to keep the flakes uniformly distributed.

Purple Broth

♦ Peptone	10.0 g
♦ Beef extract	1.0 g
♦ Sodium chloride	5.0 g
♦ Bromcresol purple	0.02 g
♦ Carbohydrate (glucose, lactose, or sucrose)	10.0 g
♦ Distilled or deionized water	1.0 L

pH 6.6–7.0 at 25°C

1 Suspend the dry ingredients in the water, mix well, and warm slightly to dissolve completely.
2 Dispense 9.0 mL volumes into test tubes.
3 Insert inverted Durham tubes into the test tubes and cap loosely.
4 Sterilize in the autoclave at 118°C for 15 minutes.
5 Remove from the autoclave and allow the medium to cool to room temperature.

Sabouraud Dextrose Agar

(with antibiotics added to inhibit bacterial growth)

♦ Peptone	10.0 g
♦ Dextrose	40.0 g
♦ Agar	15.0 g
♦ Penicillin*	20000.0 units
♦ Streptomycin*	0.00004 g
♦ Distilled or deionized water	1.0 L

pH 5.2–5.6 at 25°C

1 Suspend the peptone, dextrosen, and agar in the water, mix well, and boil until completely dissolved.
2 Autoclave for 15 minutes at 15 lbs. pressure (121°C).
3 Remove the agar mixture from the autoclave and cool to 50°C.
4 Aseptically add antibiotics. Mix and pour into sterile Petri dishes.
5 Allow the medium to cool to room temperature.

Saline Agar: Double Gel Immunodiffusion

♦ Sodium chloride	10.0 g
♦ Agar	20.0 g
♦ Distilled or deionized water	1.0 L

1 Suspend the dry ingredients in the water, mix well, and boil until completely dissolved.
2 Pour into Petri dishes to a depth of 3 mm. Do not replace the lids until the agar has solidified and cooled to room temperature.

---

*To obtain the desired proportions of antibiotics in the medium, prepare as follows:

1. Dissolve 100,000 units penicillin in 10 mL sterile distilled or deionized water. Add 2 mL to 1.0 liter of agar medium.

2. Dissolve 1.0 g streptomycin in 10 mL sterile distilled or deionized water. Add 1.0 mL of this mixture to 9.0 mL sterile distilled or deionized water. Add 4 mL of this diluted mixture to 1.0 liter of agar medium.

Seawater Agar

- Peptone   5.0 g
- Yeast extract   5.0 g
- Beef extract   3.0 g
- Agar   15.0 g
- Seawater (synthetic)   1.0 L

1 Suspend the ingredients in the seawater, mix well, and boil to dissolve completely.

2 Sterilize in the autoclave at 15 lbs. pressure (121°C) for 15 minutes.

3 When cooled to 50°C, pour into sterile plates.

4 Allow the medium to cool to room temperature.

SIM (Sulfur-Indole-Motility) Medium

- Pancreatic digest of casein   20.0 g
- Peptic digest of animal tissue   6.1 g
- Ferrous ammonium sulfate   0.2 g
- Sodium thiosulfate   0.2 g
- Agar   3.5 g
- Distilled or deionized water   1.0 L
  pH 7.1–7.5 at 25°C

1 Suspend the dry ingredients in the water, mix well, and boil until completely dissolved.

2 Dispense 7.0 mL volumes into test tubes and cap loosely.

3 Sterilize in the autoclave at 15 lbs. pressure (121°C) for 15 minutes.

4 Remove from the autoclave and allow the medium to cool to room temperature.

Skim Milk Agar—*See* (Skim) Milk Agar

Snyder Test Medium

- Pancreatic digest of casein   13.5 g
- Yeast extract   6.5 g
- Dextrose   20.0 g
- Sodium chloride   5.0 g
- Agar   16.0 g
- Bromcresol green   0.02 g
- Distilled or deionized water   1.0 L
  pH 4.6–5.0 at 25°C

1 Suspend the dry ingredients in the water, mix well, and boil until completely dissolved.

2 Transfer 7.0 mL portions to test tubes and cap loosely.

3 Sterilize in the autoclave at 118–121°C for 15 minutes.

4 Remove from the autoclave and place in a hot water bath set at 45–50°C. Allow at least 30 minutes for the agar temperature to equilibrate before beginning the exercise.

Soft Agar

- Beef extract   3.0 g
- Peptone   5.0 g
- Sodium chloride   5.0 g
- Tryptone   2.5 g
- Yeast extract   2.5 g
- Agar   7.0 g
- Distilled or deionized water   1.0 L

1 Suspend the dry ingredients in the water, mix well, and boil until completely dissolved.

2 Transfer 2.5 mL portions to test tubes and cap loosely.

3 Sterilize in the autoclave at 15 lbs. pressure (121°C) for 15 minutes.

4 Remove from the autoclave and place in a hot water bath set at 45°C. Allow 30 minutes for the agar temperature to equilibrate.

Sporulating Agar

- Pancreatic digest of gelatin   6.0 g
- Pancreatic digest of casein   4.0 g
- Yeast extract   3.0 g
- Beef extract   1.5 g
- Dextrose   1.0 g
- Agar   15.0 g
- Manganous sulfate   0.3 g
- Distilled or deionized water   1.0 L
  Final ph 6.6 ± 0.2 at 25°C

Starch Agar

- Beef extract   3.0 g
- Soluble starch   10.0 g
- Agar   12.0 g
- Distilled or deionized water   1.0 L
  pH 7.3–7.7 at 25°C

1 Suspend the dry ingredients in the water, mix well, and boil until completely dissolved.

2 Sterilize in the autoclave at 15 lbs. pressure (121°C) for 15 minutes.

3 Remove from the autoclave and allow to cool slightly.

4 Aseptically pour into sterile Petri dishes (20 mL per plate). Allow the medium to cool to room temperature.

(Modified) Thiobacillus Medium

✦ $Na_2S_2O_3$	10.0 g
✦ Na_2HPO_4	1.2 g
✦ KH_2PO_4	1.8 g
✦ $(NH_4)_2SO_4$	0.1 g
✦ $MgSO_4 \cdot 7H_2O$	0.1 g
✦ $CaCl_2$	0.03 g
✦ $FeCl_3$	0.02 g
✦ $MnSO_4$	0.02 g
✦ $NaHCO_3$	0.5 g
✦ KNO_3	5.0 g
✦ Wolfe's mineral solution	10.0 mL
✦ Wolfe's vitamin solution (filter-sterilized and added after autoclaving)	1.0 mL
✦ Distilled or deionized H_2O	to 1.0 L

pH 7.5 adjusted with KOH

Base Solution

1 Add 990 mL dH_2O to a 2 L flask.

2 Add all compounds except the vitamin solution. Mix well.

3 Adjust the pH to 7.5.

4 Autoclave at 121°C for 20 minutes.

5 Cool to approximately 50°C and add filter-sterilized vitamins.

Wolfe's Mineral Solution

✦ Nitrilotriacetic acid	1.5 g
✦ $MgSO_4 \cdot 7H_2O$	3.0 g
✦ $MnSO_4 \cdot H_2O$	0.5 g
✦ NaCl	1.0 g
✦ $FeSO_4 \cdot 7H_2O$	0.1 g
✦ $CoCl_2 \cdot 6H_2O$	0.1 g
✦ $CaCl_2$	0.1 g
✦ $ZnSO_4 \cdot 7H_2O$	0.1 g
✦ $CuSO_4 \cdot 5H_2O$	0.01 g
✦ $AlK(SO_4)_2 \cdot 12H_2O$	0.01 g
✦ H_3BO_3	0.01 g
✦ $Na_2MoO_4 \cdot 2H_2O$	0.01 g
✦ $dH2O$	1.0 L

1 Add nitrilotriacetic acid to ~500 mL of dH_2O and adjust to pH 6.5 with KOH.

2 Bring volume to 1.0 L.

3 Add the remaining ingredients one at a time and mix well.

4 Autoclave at 121°C for 20 minutes.

5 Cool and store refrigerated in a glass or Nalgene reagent bottle.

Wolfe's Vitamin Solution

(This solution must be filter-sterilized, therefore reduce proportions as necessary for easier handling.)

✦ Biotin	2.0 mg
✦ Folic acid	2.0 mg
✦ Pyridoxine hydrochloride	10.0 mg
✦ Thiamine • HCl	5.0 mg
✦ Riboflavin	5.0 mg
✦ Nicotinic acid	5.0 mg
✦ Calcium D– (+) –pantothenate	5.0 mg
✦ Vitamin B_{12}	0.1 mg
✦ p-Aminobenzoic acid	5.0 mg
✦ Thioctic acid	5.0 mg
✦ dH_2O	1.0 L

1 Add ingredients one at a time to 1.0 L distilled or deionized water and mix well.

2 Filter sterilize and store in the refrigerator.

Thioglycollate Medium (Fluid)

✦ Yeast extract	5.0 g
✦ Casitone	15.0 g
✦ Dextrose (glucose)	5.5 g
✦ Sodium chloride	2.5 g
✦ Sodium thioglycolate	0.5 g
✦ L-Cystine	0.5 g
✦ Agar	0.75 g
✦ Resazurin	0.001 g
✦ Distilled or deionized water	1.0 L

pH 7.1–7.5 at 25°C

1 Suspend the dry ingredients in the water, mix well, and boil until completely dissolved.

2 Dispense 10.0 mL into sterile test tubes.

3 Autoclave for 15 minutes at 15 lbs. pressure (121°C) to sterilize. Allow the medium to cool to room temperature before inoculating.

Tributyrin Agar

✦ Beef extract	3.0 g
✦ Peptone	5.0 g
✦ Agar	15.0 g
✦ Tributyrin oil	10.0 mL
✦ Distilled or deionized water	1.0 L

pH 5.8–6.2 at 25°C

1 Suspend the dry ingredients in the water, mix well, and boil until completely dissolved.

2 Cover loosely, and sterilize together with the tube of tributyrin oil in the autoclave at 15 lbs. pressure (121°C) for 15 minutes.

3 Remove from the autoclave, and aseptically pour agar mixture into a sterile glass blender.

4 Aseptically add the tributyrin oil to the agar mixture and blend on "High" for 1 minute.

5 Aseptically pour into sterile Petri dishes (20 mL/plate). Allow the medium to cool to room temperature.

Triple Sugar Iron Agar (TSIA)

- Pancreatic Digest of Casein   10.0 g
- Peptic Digest of Animal Tissue   10.0 g
- Dextrose (glucose)   1.0 g
- Lactose   10.0 g
- Sucrose   10.0 g
- Ferrous Ammonium Sulfate   0.2 g
- Sodium chloride   5.0 g
- Sodium thiosulfate   0.2 g
- Agar   13.0 g
- Phenol red   0.025 g
- Distilled or deionized water   1.0 L
   pH 7.1–7.5 at 25°C

1 Suspend the dry ingredients in the water, mix well, and boil until completely dissolved.

2 Transfer 7.0 mL portions to test tubes and cap loosely.

3 Sterilize in the autoclave at 15 lbs. pressure (121°C) for 15 minutes.

4 Remove from the autoclave and slant in such a way as to form a deep butt.

5 Allow the medium to cool to room temperature.

Tryptic Nitrate Medium

- Tryptose   20.0 g
- Dextrose   1.0 g
- Disodium phosphate   2.0 g
- Potassium nitrate   1.0 g
- Agar   1.0 g
- Distilled or deionized water   1.0 L
   pH 7.0–7.4 at 25°C

1 Suspend the dry ingredients in the water, mix well, and boil until completely dissolved.

2 Transfer 10.0 mL portions to test tubes, and cap loosely.

3 Sterilize in the autoclave at 15 lbs. pressure (121°C) for 15 minutes.

4 Remove from the autoclave and allow the medium to cool to room temperature.

Tryptic Soy Agar

- Tryptone   15.0 g
- Soytone   5.0 g
- Sodium Chloride   5.0 g
- Agar   15.0 g
- Distilled or deionized water   1.0 L
   pH 7.1–7.5 at 25°C

1 Suspend the dry ingredients in the water, mix well, and boil until completely dissolved.

2 Transfer 7.0 mL portions to test tubes and cap loosely.

3 Sterilize in the autoclave at 121°C for 15 minutes.

4 Slant the tubes and allow the medium to cool to room temperature.

Tryptic Soy Agar
(Enriched with Yeast Extract)

- Tryptone   15.0 g
- Soytone   5.0 g
- Sodium Chloride   5.0 g
- Yeast Extract   5.0 g
- Agar   15.0 g
- Distilled or deionized water   1.0 L
   pH 7.1–7.5 at 25°C

1 Suspend the dry ingredients in the water, mix well, and boil until completely dissolved.

2 Transfer 10.0 mL portions to test tubes and cap loosely.

3 Sterilize in the autoclave at 121°C for 15 minutes.

4 Allow the medium to cool to room temperature with the tubes in an upright position (to be used for agar stabs).

Tryptic Soy Broth

- Tryptone   17.0 g
- Soytone   3.0 g
- Sodium chloride   5.0 g
- Dipotassium phosphate   2.5 g
- Distilled or deionized water   1.0 L
   pH 7.1–7.5 at 25°C

1 Suspend the dry ingredients in the water. Agitate and heat slightly (if necessary) to dissolve completely.

2 Dispense 7.0 mL portions into test tubes and cap loosely.

3 Autoclave for 15 minutes at 121°C to sterilize the medium.

Tryptic Soy Broth Plus 1% Glucose

- Tryptic Soy Broth (see above)
- Glucose   10 g

Urease Agar

- Peptone   1.0 g
- Dextrose (glucose)   1.0 g
- Sodium chloride   5.0 g
- Potassium phosphate, monobasic   2.0 g
- Agar   15.0 g
- Phenol red   0.012 g
- Distilled or deionized water   1.0 L
   pH 6.6–7.0 at 25°C

1 Suspend the agar in 900 mL of the water, mix welln and boil to dissolve completely.

2 Cover loosely and sterilize by autoclaving at 15 lbs. pressure (121°C) for 15 minutes.

3 Remove from the autoclave and allow to cool to 55°C.

4 Suspend the remaining ingredients in 100 mL of the water, mix well, and filter sterilize. *Do not autoclave.* This is urease agar base.

5 Aseptically add the urease agar base to the agar solution, and mix well.

6 Aseptically transfer 7.0 mL portions to sterile test tubes, and cap loosely.

7 Slant in such a way that the agar butt is approximately twice as long as the slant.

8 Allow to cool to room temperature.

Urease Broth

- ♦ Yeast extract — 0.1 g
- ♦ Potassium phosphate, monobasic — 9.1 g
- ♦ Potassium phosphate, dibasic — 9.5 g
- ♦ Urea — 20.0 g
- ♦ Phenol red — 0.01 g
- ♦ Distilled or deionized water — 1.0 L

  pH 6.6–7.0 at 25°C

1 Suspend the dry ingredients in the water and mix well.

2 Filter sterilize the solution. *Do not autoclave.*

3 Aseptically transfer 1.0 mL volumes to small sterile test tubes and cap loosely.

Xylose Lysine Desoxycholate Agar

- ♦ Xylose — 3.5 g
- ♦ L-Lysine — 5.0 g
- ♦ Lactose — 7.5 g
- ♦ Sucrose — 7.5 g
- ♦ Sodium chloride — 5.0 g
- ♦ Yeast extract — 3.0 g
- ♦ Phenol red — 0.08 g
- ♦ Sodium desoxycholate — 2.5 g
- ♦ Sodium thiosulfate — 6.8 g
- ♦ Ferric ammonium citrate — 0.8 g
- ♦ Agar — 13.5 g
- ♦ Distilled or deionized water — 1.0 L

  pH 7.3–7.7 at 25°C

1 Suspend the dry ingredients in water and mix. Heat only until the medium boils.

2 Cool in a water bath at 50°C.

3 When cooled, pour into sterile plates.

4 Allow the medium to cool to room temperature.

Yeast Extract Broth

- ♦ Beef extract — 3.0 g
- ♦ Peptone — 5.0 g
- ♦ NaCl — 5.0 g
- ♦ Yeast extract — 5.0 g
- ♦ Distilled or deionized water — 1.0 L

  pH 6.6–7.0 at 25°C

1 Suspend the dry ingredients in the water. Agitate and heat slightly (if necessary) to dissolve completely.

2 Dispense 7.0 mL portions into test tubes and cap loosely.

3 Autoclave for 15 minutes at 121°C to sterilize the medium.

Reagents

Direct Count—Staining/Diluting Agents

(*See* Stains)

Lysozyme Buffer

- ♦ NaCl — 22.67 g
- ♦ Na_2HPO_4 — 3.56 g
- ♦ KH_2PO_4 — 6.65 g
- ♦ Distilled or deionized water — 1.0 L

1 Dissolve all ingredients in approximately 900 mL water.

2 Add water to bring the total volume up to 1000 mL.

Lysozyme Substrate

- ♦ Lysozyme Buffer — 200.0 mL
- ♦ *Micrococcus lysodeikticus* (freeze-dried) — 0.1 g

1 Add the freeze-dried *Micrococcus lysodeikticus* to the lysozyme buffer and mix well.

2 Using a spectrophotometer with the wavelength set at 540 nm, adjust the solution's light transmittance to 10% by adding water or *M. lysodeikticus* as needed.

MR-VP (Methyl Red–Voges-Proskauer) Test Reagents

Methyl Red

- ♦ Methyl red dye — 0.1 g
- ♦ Ethanol — 300.0 mL
- ♦ Distilled water to bring volume to — 500.0 mL

1 Dissolve the dye in the ethanol.

2 Add water to bring the total volume up to 500 mL.

VP Reagent A (Barritt's)

- ♦ β-naphthol — 5.0 g
- ♦ Absolute Ethanol to bring volume to — 100.0 mL

1 Dissolve the β-naphthol in approximately 95 mL of ethanol.

2 Add ethanol to bring the total volume up to 100 mL.

VP Reagent B (Barritt's)

- ♦ Potassium hydroxide — 40.0 g
- ♦ Distilled water to bring volume to — 100.0 mL

1 Dissolve the potassium hydroxide in approximately 60 mL of water. (*Caution:* This solution is highly concentrated and will become hot as the KOH dissolves. It should be prepared in appropriate glassware on a stirring hot plate.) Allow it to cool to room temperature.

2 Add water to bring the volume up to 100 mL.

McFarland Turbidity Standard (0.5)

- Barium chloride ($BaCl_2 \cdot 2\ H_2O$)   1.175 g
- Sulfuric acid, concentrated (H_2SO_4)   1.0 mL
- Distilled or deionized water $\cong$ 200.0 mL

1 Pour approximately 90 mL of water into a small Erlenmeyer flask.

2 Add the $BaCl_2$ and mix well.

3 Remeasure and add water to bring the total volume up to 100 mL.

4 Add the H_2SO_4 to approximately 90 mL of water.

5 Remeasure and add water to bring the total volume up to 100 mL.

6 Add 0.5 mL of the $BaCl_2$ solution to 99.5 mL of H_2SO_4 and mix well.

7 While keeping the solution well mixed (the barium sulfate will precipitate and settle out) distribute 7 to 10 mL volumes into very clean screw cap test tubes.

Methyl Red (*See* MR-VP)

Methylene Blue Reductase Reagent

- Methylene blue dye   8.8 mg
- Distilled or deionized water   200.0 mL

Nitrate Test Reagents
Reagent A

- Sulfanilic acid   1.0 g
- 5N Acetic acid   125.0 mL

Reagent B

- Dimethyl-β-naphthylamine   1.0 g
- 5N Acetic acid   200.0 mL

Oxidase Test Reagent

- Tetramethyl-*p*-phenylenediamine dihydrochloride   1.0 g
- Distilled or deionized water   100.0 mL

Phenylalanine Deaminase Test Reagent

- Ferric chloride   10.0 g
- Distilled or deionized water $\cong$ 90.0 mL

1 Dissolve the ferric chloride in approximately 90 mL of distilled or deionized water.

2 Add water to bring the total volume up to 100 mL.

10X Tris-Acetate-EDTA buffer

- Trisma base   48.4 g
- Glacial acetic acid   11.42 mL
- 0.5 *M* EDTA, pH 8.0   20.0 mL
- Distilled or deionized water   1.0 L

Voges-Proskauer Reagents A and B
(*See* MR-VP)

Solutions

10x Phosphate Buffered Saline (PBS) Stock Solution

- Na_2HPO_4, anhydrous, reagent grade   12.36 g
- $NaH_2PO_4 \cdot H_2O$, reagent grade   1.80 g
- NaCl, reagent grade   85.00 g

Dissolve ingredients in distilled water to a final volume of 1 L.

1x PBS Working solution (0.01 M phosphate, pH 7.6)

- Stock solution   100 mL
- dH_2O 900 mL

Stains

Acid-Fast, Cold Stain Reagents (Modified Kinyoun)

Carbolfuchsin

- Basic fuchsin   1.5 g
- Phenol   4.5 g
- Ethanol (95%)   5.0 mL
- Isopropanol   20.0 mL
- Distilled or deionized water   75.0 mL

1 Dissolve the basic fuchsin in the ethanol and add the isopropanol.

2 Mix the phenol in the water.

3 Mix the solutions together and let stand for several days.

4 Filter before use.

Decolorizer

- H_2SO_4 1.0 mL
- Ethanol (95%)   70.0 mL
- Distilled or deionized water   29.0 mL

Brilliant Green

- Brilliant green dye   1.0 g
- Sodium azide   0.01 g
- Distilled or deionized water   100.0 mL

Acid-Fast, Hot Stain Reagents (Ziehl-Neelson)

Carbolfuchsin

- Basic fuchsin   0.3 g
- Ethanol   10.0 mL
- Distilled or deionized water   95.0 mL
- Phenol   5.0 mL

1 Dissolve the basic fuchsin in the ethanol.

2 Dissolve the phenol in the water

3 Combine the solutions and let stand for a few days.

4 Filter before use.

Decolorizer

- Ethanol   97.0 mL
- HCl (concentrated)   3.0 mL

Azotobacter Cyst Stain

- Glacial Acetic Acid   8.5 mL
- Sodium Sulfate   3.25 g
- Neutral Red   200.0 mg
- Light Green   200.0 mg
- Ethanol   50.0 ml
- Distilled or deionized water   100.0 mL

Basic Fuchsin Stain (0.5%)

- Basic Fuchsin   0.5 g
- dH$_2$O   99.5 mL

Methylene Blue Counterstain

- Methylene blue chloride   0.3 g
- Distilled or deionized water   100.0 mL

Capsule Stain
Congo Red

- Congo red dye   5.0 g
- Distilled or deionized water   100.0 mL

Maneval's Stain

- Phenol (5% aqueous solution)   30.0 mL
- Acetic acid, glacial   10.0 mL
  (20% aqueous solution)
- Ferric chloride   4.0 mL
  (30% aqueous solution)
- Basic fuchsin (1% aqueous solution)   2.0 mL

Direct Count: Staining/Diluting Agents
Agent A

- 100% saturated crystal
  violet-ethanol solution   1.0 mL
- Ethanol   40.0 mL
- NaCl   0.9 g
- Distilled or deionized water   ≅58.1 mL

1 Dissolve the crystal violet in ethanol and filter.

2 Mix 0.9 g NaCl in approximately 55 mL of distilled or deionized water.

3 Add the Ethanol and crystal violet solutions.

4 Add water to bring the total volume up to 100 mL.

Agent B

- Ethanol   40.0 mL
- NaCl   0.9 g
- Distilled or deionized water   ≅60.0 mL

1 Dissolve 0.9 g NaCl in approximately 55 mL of distilled or deionized water.

2 Add the mixture to 40 mL of ethanol.

3 Add water to bring the total volume up to 100 mL.

Flagella Stain (Ryu Method)
Solution I (Mordant)

- 5% Phenol (aqueous)   10.0 mL
- Tannic acid   2.0 g
- Aluminum potassium sulfate • 12 H$_2$O 10.0 mL
  (saturated aqueous solution)

Solution II (Stain)

- Crystal violet (saturated ethanolic   1.0 mL
  solution—12 g / 100 mL 95% ethanol)

1 Mix 10 parts of solution I with one part solution II.

2 Filter through Whatman No. 4 filter paper.

3 Keep the final stain at room temperature in a syringe fitted with a 0.22 μm pore-size membrane filter. No need to use a needle.

4 Cap the syringe to prevent drying of the stain. This preparation will keep for several weeks.

Rose Bengal Stain (1%)

- Rose Bengal   0.1 g
- Distilled or deionized water   9.9 mL

Gram Stain Reagents

Gram Crystal Violet (Modified Hucker's)
Solution A

- Crystal violet dye (90%)   2.0 g
- Ethanol (95%)   20.0 mL

Solution B

- Ammonium oxalate   0.8 g
- Distilled or deionized water   80.0 mL

1 Combine solutions A and B. Store for 24 hours.

2 Filter before use.

Gram Decolorizer

- Ethanol (95%)
- Can use 75% ethanol and 25% acetone, but decolorization time must be reduced

Gram Iodine

- Potassium iodide   2.0 g
- Iodine crystals   1.0 g
- Distilled or deionized water   300.0 mL

1 Dissolve the potassium iodide in the water *first*.

2 Dissolve the iodine crystals in the solution.

3 Store in an amber bottle.

Gram Safranin

- Safranin O   0.25 g
- Ethanol (95%)   10.0 mL
- Distilled or deionized water   100.0 mL

1 Dissolve the safranin O in the ethanol.

2 Add the water.

Negative Stain Nigrosin

- Nigrosin   10.0 g
- Distilled or deionized water   100.0 mL

Simple Stains

Crystal Violet (*See* Gram Stain)

Methylene Blue (*See* Acid Fast, Hot)

Safranin (*See* Gram Stain)

Carbolfuchsin (*See* Acid Fast, Hot)

Spore Stain
Malachite Green

- Malachite green dye   5.0 g
- Distilled or deionized water   100.0 mL

Safranin (*See* Gram Stain)

Vogel-Bonner Salts (50x)

- Magnesium sulfate   10.0 g
- Citric acid   100.0 g
- Dipotassium phosphate   500.0 g
- Monosodium ammonium phosphate   175.0 g
- Distilled or deionized water
  to bring volume to   1.0 liter*

References

Baron, Ellen Jo, Lance R. Peterson, and Sydney M. Finegold. 1994. *Bailey & Scott's Diagnostic Microbiology,* 9th ed. Mosby–Year Book, St. Louis.

Eisenstadt, Bruce, C. Carlton, and Barbara J. Brown. 1994. *Methods for General and Molecular Bacteriology,* edited by Philipp Gerhardt, R. G. E. Murray, Willis A. Wood, and Noel R. Krieg. American Society for Microbiology, Washington, DC.

Farmer, J. J. III. 2003. Chapter 41 in *Manual of Clinical Microbiology,* 8th ed. Edited by Patrick R. Murray, Ellen Jo Baron, James H. Jorgensen, Michael A. Pfaller, and Robert H. Yolken. ASM Press, American Society for Microbiology, Washington, DC.

Forbes, Betty A., Daniel F. Sahm, and Alice S. Weissfeld. 2002. Chapter 19 in *Bailey & Scott's Diagnostic Microbiology,* 11th ed. Mosby, St. Louis.

Koneman, Elmer W., *et al.* 1997. *Color Atlas and Textbook of Diagnostic Microbiology,* 5th ed. Lippincott-Raven Publishers, Philadelphia.

MacFaddin, Jean F. 1980. *Biochemical Tests for Identification of Medical Bacteria,* 2nd ed. Williams & Wilkins, Baltimore.

Zimbro, Mary Jo, and David A. Power, editors. 2003. *Difco™ and BBL™ Manual —Manual of Microbiological Culture Media.* Becton Dickinson and Co., Sparks, MD.

Data Sheets

Name _____ Date _____

Lab Section _____ I was present and performed this exercise (initials) _____

Glo Germ™ Hand Wash Education System

OBSERVATIONS AND INTERPRETATIONS

Record the degree of hand contamination before and after washing in the table below. Use this qualitative scale for evaluation:

      +++ means "a lot of contamination"
       ++ means "moderate contamination"
        + means "little contamination"
        0 means "no contamination"

There is no absolute cutoff between any of these categories, and what you call "moderate contamination" might be called "little contamination" by another student. Just try to be consistent within your evaluation.

Body Region	Left Hand		Right Hand	
	Before Washing	After Washing	Before Washing	After Washing
Palm				
Back of Hand				
Fingers				
Between Fingers				
Tops of Fingernails				
Under Fingernails				
Front of Wrist				
Back of Wrist				

QUESTIONS

1 *What areas were cleaned most thoroughly with your washing technique?*

_____

_____

_____

_____

_____

2   *What areas were most difficult for you to clean with this washing technique?*

3   *In general, were your two hands cleaned an equal amount, or was one cleaned more than the other? What could account for any differences?*

4   *How do your answers to Questions 1, 2, and 3 compare to your lab partner's answers? Why might they differ?*

5   *Why were you instructed to have your lab partner turn the water on and off and operate the UV lamp rather than you doing these actions yourself?*

6   *Why might it be advisable to modify the procedure and use the UV light to check your hands prior to application of lotion and the paper towels prior to drying?*

7   *Using the same qualitative scale as before, how much lotion was transferred to this Data Sheet and your writing instrument?*

Name _____ Date _____

Lab Section _____ I was present and performed this exercise (initials) _____

Nutrient Broth and Nutrient Agar Preparation

OBSERVATIONS AND INTERPRETATIONS

Record the number of each medium type you prepared, then record the number of apparently sterile ones. Calculate your percentage of successful preparations for each. In the last column, speculate as to probable/possible sources of contamination.

Medium	Total Number Prepared	Number of Sterile Preparations	Percentage of Successful Preparations	Probable Sources of Contamination (if any)
Nutrient Agar Tubes (Slant or Deep)				
Nutrient Agar Plates				
Nutrient Broths				

QUESTIONS

1 *Which medium was most difficult to prepare without contamination? Why do think this might be so?*

_____

_____

_____

_____

_____

_____

_____

2 For each of the following types of contamination, suggest the most likely point in preparation (or later) at which the contaminant was introduced.

a. Growth in all broth tubes.

b. Growth in one broth tube.

c. Growth only on the surface of a plate.

d. Growth throughout the agar's thickness on a plate.

e. Growth only in the upper 1 cm of agar in an agar deep tube.

f. All plates in a batch have the same type and density of contaminants.

g. Only a few plates in a batch are contaminated, and each looks different.

Name _____　Date _____

Lab Section _____ I was present and performed this exercise (initials) _____

Common Aseptic Transfers and Inoculation Methods

OBSERVATIONS AND INTERPRETATIONS

Describe the appearance of growth on/in each medium. Draw representative samples of each growth type.

Organism	Medium Inoculated	
	Nutrient Agar Slant	**Nutrient Broth**
Escherichia coli on NA slant		
Micrococcus luteus on NA slant		
Micrococcus luteus in NB broth		
Micrococcus luteus on NA plate		

QUESTIONS

1 *Considering the cultures used to inoculate each medium in this exercise, how many different microbial types should you expect to see in/on each medium?*

2 *You were asked to describe differences in the appearance of growth in each culture, if present. In which medium was this the most difficult to determine? What made this difficult?*

3 *Which medium was most difficult for you to transfer from? Which medium was most difficult for you to inoculate? Explain your difficulties.*

Name _____   Date _____

Lab Section _____   I was present and performed this exercise (initials) _____

Streak Methods of Isolation

RESULTS AND INTERPRETATIONS

1 *Using your pencil, perform a quadrant streak on the "practice plate" below.*

2 *Examine the streak plate made from the mixture of two organisms. Have your lab partner write a critique of your isolation technique in the space below. The following should be addressed: Did you obtain isolation? Were the first three streaks near the edges of the plate? Did any streaks intersect streaks they shouldn't have? Was the whole surface of the agar used? Was the agar cut by the loop?*

_____

_____

_____

_____

_____

_____

DATA SHEET 1-4

3 *Did you achieve isolation using the quadrant streak? If so, in which streak (1, 2, 3, or 4) did it occur? If you did not achieve isolation, what might you do differently next time to improve your results?*

4 *Most colonies on streak plates grow from isolated colony-forming units (CFUs). On rare occasions, however, a colony can be a mixture of two different organisms. If a culture is started from this colony (thinking it is pure), correct identification will be next to impossible because the extra organism could confound the identifying test results. How could you verify the purity of a colony? (The answers may vary depending on what experience you have had prior to performing this exercise.) If you found the colony to be a mixture of organisms, what could you do to purify it?*

5 *Examine the environmental sample. Were the different streak methods appropriate to the cell densities recovered?*

Name _____ Date _____

Lab Section _____ I was present and performed this exercise (initials) _____

Spread Plate Method of Isolation

OBSERVATIONS AND INTERPRETATIONS

Record your observations in the table below.

Organism	Plate(s) With Isolation	Comments
E. coli		
S. marcescens		

QUESTIONS

1 *On which plate did you obtain isolation with* E. coli? *How about* S. marcescens? *Do you have reason to suspect that they should become isolated on the same dilution plate? Why or why not?*

_____

_____

_____

_____

_____

_____

_____

_____

2 Once you obtained isolation at a particular dilution, did you continue to have isolation on subsequent dilution plates? Is this what you would expect? Why or why not?

3 What is the consequence of not spreading the inoculum adequately over the agar surface?

4 To get isolated colonies on a plate, only about 300 cells can be in the inoculum. What will happen if the cell density of the inoculum significantly exceeds this number?

5 Suppose you have two organisms in a mixture and Organism A is 1000 times more abundant than Organism B. Will you (without counting on good luck!) be able to isolate Organism B using the spread plate technique? Explain your answer.

Name _____ Date _____

Lab Section _____ I was present and performed this exercise (initials) _____

Ubiquity of Microorganisms

OBSERVATIONS AND INTERPRETATIONS

1 Use the circles below as Petri dishes. Then, for each plate, choose two different colonies and draw each as seen from above and from the side. Label the plates according to incubation time, temperature, and source of inoculum. Also include other useful colony information, such as color and relative abundance.

2 Save the plates for Exercise 2-2.

QUESTIONS

1   *What was the purpose of incubating the unopened plates? Be specific. What is an appropriate name for these plates?*

_____

_____

_____

_____

_____

_____

Name _____ Date _____

Lab Section _____ I was present and performed this exercise (initials) _____

2 *If growth appears on both unopened plates, what are some likely explanations? What if growth appears on only one plate? How does growth on the unopened plates affect the reliability (your interpretation) of the other plates?*

3 *Why were the specific types of exposure (air, hair, tabletop, etc.) chosen for this exercise?*

4 *Why were you asked to incubate the plates at two different temperatures? Be specific. What is the likely source (reservoir) of organisms that grew best at 37°C, and how do they survive at room temperature without nutrients?*

5   *Explain why you might have gotten different appearing colonies on plates 2 and 3.*

6   *The plates you are using for this lab will be autoclaved eventually to completely sterilize them. The measures taken to disinfect the tabletops (the source of the organisms on plates 2 and 3) are not as extreme. Why?*

Name _____ Date _____

Lab Section _____ I was present and performed this exercise (initials) _____

Colony Morphology

OBSERVATIONS AND INTERPRETATIONS

Using the terms in Figure 2-3, describe and sketch representative colonies (including side views) on your plates, including those from Exercise 2-1. Use a colony counter if necessary. Measure colony diameters, and include them with your descriptions.

Organism/Plate	Growth Description

QUESTIONS

1 *A description of colony morphology provides important information about an organism. What other information should you include when describing physical growth characteristics?*

2 *Three critical aspects of a description of bacterial growth are colony size, color, and shape. At least three other important factors—not physical descriptions—typically are included when describing bacterial growth. What do you suppose they are and why they are important?*

Name _____   Date _____

Lab Section _____   I was present and performed this exercise (initials) _____

Growth Patterns on Slants

OBSERVATIONS AND INTERPRETATIONS

In the chart below, describe the growth on your slants, including shape, margin, texture, and color. Draw a picture if necessary, and include other information about the conditions of incubation, such as time, temperature, and the medium used.

Organism	Growth Description
Control	

DATA SHEET 2-3

QUESTIONS

1 List some reasons why growth characteristics are more useful on agar plates than on agar slants.

2 Why are agar slants better suited than agar plates to maintain stock cultures?

3 Match the following:

_____ Filiform   1. Produces colored growth

_____ Spreading edge   2. Smooth texture with solid edge

_____ Transparent   3. Solid growth seeming to radiate outward

_____ Friable   4. Almost invisible or easy to see light through

_____ Pigmented   5. Rough texture with a crusty appearance

Name _____ Date _____

Lab Section _____ I was present and performed this exercise (initials) _____

Growth Patterns in Broth

OBSERVATIONS AND INTERPRETATIONS

Examine the tubes and enter the descriptions in the chart below. Draw a picture if necessary, and include other information about the conditions of incubation, such as time, temperature, and the medium used.

Organism	Description of Growth in Broth
Control	

QUESTIONS

1   What factors besides physical growth characteristics are important when recording data about an organism? Why?

2   Match the following:

_____ Flocculent   a. *Evenly cloudy throughout*

_____ Sediment   b. *Growth at top around the edge*

_____ Ring   c. *Growth on the bottom*

_____ Pellicle   d. *Membrane at the top*

_____ Uniform fine turbidity   e. *Suspended chunks or pieces*

Name _____ Date _____

Lab Section _____ I was present and performed this exercise (initials) _____

Evaluation of Media

OBSERVATIONS AND INTERPRETATIONS

1 On the top line of the data table below, write "defined" or "undefined" ("complex") for each medium.

2 Score the relative amount of growth in each tube. Compare the five tubes for each medium with each other, as well as the amount of growth for each organism in the various media. Use "0" for no growth, "3" for abundant growth, and "1" and "2" for degrees of growth in between.

Organism	Nutrient Broth	Brain–Heart Infusion Broth	Glucose Salts Medium	Interpretation (relative fastidiousness)
Defined/Undefined (Complex)				
Uninoculated Broth				
Ability of medium to support a wide range of microorganisms (good, fair, poor)				

QUESTIONS

1   What does comparing growth of the four organisms in a given medium tell you? (That is, are you gaining information primarily about the organism or about the medium?)

2   What does comparing growth of a given organism in the three media tell you? (Again, are you gaining information primarily about the organism or about the medium?)

3   Evaluate the media.

   a.   Which medium supports growth of the widest range of organisms?

   b.   Which medium supports the fewest organisms?

   c.   Is there a correlation between your answers to Questions 3a and 3b, and the terms "defined medium" and "undefined medium?" If so, what is it? If not, attempt to explain the relationship you observe.

4   Evaluate the organisms.

   a.   Which organism appears to be most fastidious? How can you tell?

   b.   Which organism appears to be least fastidious? How can you tell?

5   What is the biochemical basis for the spectrum of fastidiousness seen in the microbial world? (That is, why are some organisms fastidious and others are nonfastidious?)

Name _____   Date _____

Lab Section _____   I was present and performed this exercise (initials) _____

Agar Deep Stabs

OBSERVATIONS AND INTERPRETATIONS

1 Examine all tubes and enter a description of each in the chart below.

2 Categorize each organism with respect to its aerotolerance group based on the location of growth in the medium.

Organism	Location of Growth in Medium	Aerotolerance Category
Sterile stab		

QUESTIONS

1 *Why is it important to use this medium soon after preparation?*

_____

_____

_____

_____

_____

_____

2 If you have a tube that indicates uniform growth throughout the agar, can the aerotolerance category of the organism be determined? Explain.

3 When inoculating agar deep stabs, why do you have to insert and remove the needle along the same stab line?

4 If you inoculate an organism known to chemically reduce sulfur (as done by organisms capable of anaerobic respiration), where would you expect to see its growth in the test medium used today? Could it be seen in more than one zone? Explain.

5 If you have no growth in the stab after 24 hours' incubation, what are some possible explanations?

Name _____ Date _____

Lab Section _____ I was present and performed this exercise (initials) _____

Fluid Thioglycollate Medium

OBSERVATIONS AND INTERPRETATIONS

Draw a diagram of each broth showing the location of growth. Indicate the amount of growth in the broth.

"0" = no growth   "3" = abundant growth   "1" and "2" = degrees of growth in between

Organism	Location of Growth in Medium	Aerotolerance Category
Control		

QUESTIONS

1 *Why is there a colored band at the surface of Fluid Thioglycollate Medium? Which is more desirable, a thick-colored or a thin-colored band?*

_____

_____

_____

_____

_____

_____

_____

_____

2   *Where would you expect to see growth of a strict aerobe? Anaerobe? Microaerophile? Facultative anaerobe?*

3   *Why is it important that this medium be fresh? Which type of organism (aerobe, anaerobe, microaerophile, facultative anaerobe) would most likely be affected negatively by the use of old media? Which would most likely be affected positively?*

Name _____ Date _____

Lab Section _____ I was present and performed this exercise (initials) _____

Anaerobic Jar

OBSERVATIONS AND INTERPRETATIONS

Indicate the amount of growth on each plate.

"0" = no growth   "3" = abundant growth   "1" and "2" = degrees of growth in between

Organism	Growth on Aerobic Plate	Growth on Anaerobic Plate	Aerotolerance Category

QUESTIONS

1 If, after incubation, you observed that the methylene blue indicator strip inside the jar was blue, what would you guess the internal environment to be—aerobic or anaerobic? How would you expect the growth on the plate inside the jar to differ from the plate incubated outside the jar?

_____

_____

_____

_____

_____

_____

_____

_____

_____

_____

2 Which of the three organisms would be most affected by the conditions described in question 1?

3 An alternative to the anaerobic jar is a candle jar, in which a candle is placed in the jar, lit, and the lid closed to enable the flame to use the available oxygen. Typically, in this system, not all of the oxygen is used. Which types of organisms would most likely benefit from this environment?

4 Considering the gaseous changes resulting from the burned candle, which group would benefit the most from a functional candle jar? (**Hint:** Think about a gas not involved in aerotolerance.)

Name _____ Date _____

Lab Section _____ I was present and performed this exercise (initials) _____

The Effect of Temperature on Microbial Growth

OBSERVATIONS AND INTERPRETATIONS

1 Record your numeric values for each organism at each temperature.

Broth Data						
Organism	10°C	20°C	30°C	40°C	50°C	Classification

Enter your observations as 0, 1, 2, or 3 (0 is clear and 3 is very turbid).

2 Record the growth characteristics of the *Serratia marcescens* incubated at two temperatures.

Plate Data	
Incubation Temperature	Description of Growth
20°C	
35°C	

QUESTIONS

1 Using the data from the chart, determine the cardinal temperatures for each of the four organisms. Circle the optimum temperature for each organism. Use brackets to designate the range for each.

_____

_____

_____

_____

2 Plot the numeric values versus temperature on the graph paper provided for this Data Sheet. See pages 7 through 9 for proper graphing technique.

3 Why is it not advisable to connect the data points for each organism in your graph?

_____

_____

_____

_____

_____

_____

4 In what way(s) could you adjust incubation temperature to grow an organism at less than its optimal growth rate?

_____

_____

_____

_____

_____

_____

5 Why do different temperatures produce different growth rates?

_____

_____

_____

_____

Name _____ Date _____

Lab Section _____ I was present and performed this exercise (initials) _____

Name _____ Date _____

Lab Section _____ I was present and performed this exercise (initials) _____

The Effect of pH on Microbial Growth

OBSERVATIONS AND INTERPRETATIONS

Record the numeric values for each organism at each pH.

Organism	pH 2	pH 4	pH 6	pH 8	pH 10	Classification

Enter visual readings as 0, 1, 2, or 3 (0 is clear; 3 is very turbid).

QUESTIONS

1 *Circle the pH optimum for each organism. Place brackets around the range. Is there any overlap between species?*

_____

2 *Account for the inability of organisms to grow outside their pH ranges. Why, for instance, are alkaliphiles able to survive at high pHs when neutrophiles cannot?*

_____

_____

_____

_____

_____

_____

_____

3   *Where is the pH optimum relative to the pH range for each organism? Do you see any parallels between these data and the data produced in Exercise 2-9? Explain.*

4   *Plot the data (numeric values versus pH) for each organism on the graph paper provided. See pages 7 through 9 for proper graphing technique.*

5   *Why is it not advisable to connect the data points for each organism in your graph?*

Name _____ Date _____

Lab Section _____ I was present and performed this exercise (initials) _____

DATA SHEET 2-11

Name _____ Date _____

Lab Section _____ I was present and performed this exercise (initials) _____

Effect of Osmotic Pressure on Microbial Growth

OBSERVATIONS AND INTERPRETATIONS

Record the relative growth in each medium, using 0, 1, 2, and 3 to represent relative growth.

Organism	NaCl Concentration					
	0%	5%	10%	15%	20%	25%
Control						

QUESTIONS

1 Circle *the optimum salinity for each organism.*

2 *Place* brackets *around the range.*

3 *Which organism(s) exhibit(s) the greatest tolerance range?*

_____

_____

_____

_____

_____

_____

4 Use your textbook or other available reference to look up the habitat of each organism. Do the habitats make sense in light of the tolerance ranges you obtained? Explain your answers.

5 Plot your data ("number" values vs. salt concentration) on the graph paper provided. See pages 7 through 9 for proper graphing technique.

6 Why is it not advisable to connect the data points for each organism in your graph?

Name _____ Date _____

Lab Section _____ I was present and performed this exercise (initials) _____

Name _____ Date _____

Lab Section _____ I was present and performed this exercise (initials) _____

Steam Sterilization

OBSERVATIONS AND INTERPRETATIONS

Examine all vials and record your results in the table below.

Indicator Vial	Color Result	Interpretation
Vial #1		
Vial #2		
Vial #3		
Vial #4		

QUESTIONS

1 *Why are the bacterial spores placed on the paper strip and not directly in the fermentation broth?*

_____

_____

_____

_____

_____

_____

_____

_____

_____

2   What is the purpose of the unautoclaved/unbroken vial?

_____

_____

_____

_____

_____

3   How would you interpret the following combinations of results? Vial numbers indicate the treatment as in this experiment.

a. Vial #1—purple, vial #2—purple, vial #3— purple, vial #4—purple?

_____

_____

b. Vial #1—purple, vial #2—purple, vial #3— yellow, vial #4—purple?

_____

_____

c. Vial #1—purple, vial #2—yellow, vial #3—yellow, vial #4—purple?

_____

_____

d. Vial #1—yellow, vial #2—yellow, vial #3—yellow, vial #4—purple?

_____

_____

e. Vial #1—yellow, vial #2—yellow, vial #3—yellow, vial #4—yellow?

_____

_____

4   What changes would you make to avoid repeating the faulty scenarios illustrated in question 3?

_____

_____

_____

_____

_____

DATA SHEET 2-13

Name _____ Date _____

Lab Section _____ I was present and performed this exercise (initials) _____

The Lethal Effect of Ultraviolet Light on Microbial Growth

OBSERVATIONS AND INTERPRETATIONS

Enter your class data in the chart below. Score the relative amount of growth on each plate. Use the numbers 0, 1, 2, or 3 (0 is clear; 3 is very turbid).

Organism	No UV	5 Minutes	10 Minutes	15 Minutes	20 Minutes	25 Minutes	30 Minutes
B. subtilis							
E. coli							

QUESTIONS

1 The purpose of this exercise is to demonstrate the comparative effect of UV on two bacterial populations. This could have been accomplished without the cardboard cover. Why was the cover used?

_____

_____

_____

_____

_____

_____

_____

_____

_____

_____

_____

_____

2 *This is not a quantitative exercise. Keeping this in mind, can you see a general trend between bacterial death and UV exposure time?*

3 *Which organism survived the longest exposure? Why?*

4 *Why were you told to remove the plate covers prior to exposing them to UV?*

5 *Using the graph paper provided, construct a single graph of growth (0, 1, 2, 3) versus UV exposure time for the two organisms. See pages 7 through 9 for proper graphing technique.*

Name _____ Date _____

Lab Section _____ I was present and performed this exercise (initials) _____

Name _____   Date _____

Lab Section _____ I was present and performed this exercise (initials) _____

Chemical Germicides: Disinfectants and Antiseptics

OBSERVATIONS AND INTERPRETATIONS

1 Enter your individual data below.

Controls	Result
#1	
#2	
#3	
#4	

G = Growth, NG = No Growth

Germicide/Concentration	Growth
/	
/	
/	

G = Growth, NG = No Growth

2 Enter the class data in the chart below.

Organism	Household Bleach			Hydrogen Peroxide			Lysol® Brand II Disinfectant			Isopropyl Alcohol		
	0.01%	0.1%	1%	0.03%	0.3%	3%	25%	50%	100%	10%	30%	50%
S. aureus												
E. coli												

G = Growth, NG = No Growth

QUESTIONS

1   *Compare your results with the class data. Which germicide was most effective and at what concentration? Which was least effective? Defend your choices.*

2   *Which organism seemed to be most resistant to the germicides?*

3   *What purposes did the positive control, the negative control, and tubes 2 and 4 serve?*

Name _____ Date _____

Lab Section _____ I was present and performed this exercise (initials) _____

Introduction to the Light Microscope

DATA AND CALCULATIONS

Record the relevant values off your microscope and perform the calculations of total magnification for each lens.

Lens System	Magnification of Objective Lens	Magnification of Ocular Lens	Total Magnification	Numerical Aperture
Scanning				
Low Power				
High-Dry				
Oil Immersion				
Condenser Lens				

Sketch your observations of the Letter "e" slide in the chart below. Be sure the slide is right side up with the label at the left.

Appearance of the "e" with the Naked Eye	Appearance of the "e" Under the Microscope	When the stage moves to the right, the image moves to the . . .	When the stage moves toward you, the image moves . . .

Record your observations of the Colored Thread slide below. Check it under low and high power and see if your answer changes.

Low Power (usually the 10× objective)	Top:
	Middle:
	Bottom:
High Power (usually the 40× objective)	Top:
	Middle:
	Bottom:

QUESTIONS

1 Why aren't the magnifications of both ocular lenses of a binocular microscope used to calculate total magnification?

2 What is the total magnification for each lens setting on a microscope with 15× oculars and 4×, 10×, 45×, and 97× objectives lenses?

3 Assuming that all other variables remain constant, explain why light of shorter wavelength will produce a clearer image than light of longer wavelengths.

4 Why is wavelength the main limiting factor on limit of resolution in light microscopy?

5 On a given microscope, the numerical apertures of the condenser and low power objective lenses are 1.25 and 0.25, respectively. You are supplied with a filter that selects a wavelength of 520 nm.

a. What is the limit of resolution on this microscope?

b. Will you be able to distinguish two points that are 330 nm apart as being separate, or will they blur into one?

Name _____ Date _____

Lab Section _____ I was present and performed this exercise (initials) _____

6 On the same microscope as in Question #5, the high dry objective lens has a numerical aperture of 0.85.

   a. What is the limit of resolution on this microscope?

   _____

   _____

   _____

   _____

   b. Will you be able to distinguish two points that are 250 nm apart as being separate, or will they blur into one?

   _____

7 Calculate the limit of resolution for the oil lens of your microscope. Assume an average wavelength of 500 nm.

   _____

   _____

   _____

   _____

8 Examine Figure 3-3 and explain the results you observed with the Letter "e" slide.

   _____

   _____

   _____

9 With which objective was it easier to determine the sequence of colored threads?

   _____

   _____

10 Why should closing the iris diaphragm improve your ability to determine thread order?

   _____

   _____

   _____

11 *What does the Colored Thread slide demonstrate about specimens you will be observing later in the class?*

Name _____ Date _____

Lab Section _____ I was present and performed this exercise (initials) _____

Calibration of the Ocular Micrometer

DATA AND CALCULATIONS

Record two or three values where the ocular micrometer and the stage micrometer line up for the scanning, low, high dry, and oil immersion objective lenses. Then calculate the calibration for each. Be sure to include proper units in your calibrations.

Scanning Objective Lens

Stage Micrometer (μm)	Ocular Micrometer (OU)	Calibration

Low Power Objective Lens

Stage Micrometer (μm)	Ocular Micrometer (OU)	Calibration

High Dry Objective Lens

Stage Micrometer (µm)	Ocular Micrometer (OU)	Calibration

Oil Immersion Objective Lens

Stage Micrometer (µm)	Ocular Micrometer (OU)	Calibration

Calculate the average value for each calibration and record them in the chart below. If necessary, use one of the values to calculate the calibration of the oil lens. Be sure to include proper units in your answer.

Average Calibrations for My Microscope

Objective Lens	Average Calibration
Scanning	
Lower Power	
High Dry Power	
Oil Immersion	

Name _____ Date _____

Lab Section _____ I was present and performed this exercise (initials) _____

Examination of Eukaryotic Microbes

OBSERVATIONS AND INTERPRETATIONS

Fill in the chart for each eukaryotic microbe you observe.

Organism (Include wet mount or prepared slide)	Sketch (Include magnification and stain)	Dimensions	Identifying Characteristics (List only those that you observed)

QUESTIONS

1 What features did the cells you observed have in common? How were they different?

2 If you observed the same organism on a prepared slide and a wet mount, how did the images compare?

Name _____ Date _____

Lab Section _____ I was present and performed this exercise (initials) _____

Examination of Pond Water

OBSERVATIONS

Carefully draw and label as much structural detail as you can of four representative organisms you *actually observed.* Draw only a single organism in each circle, and take pride in your artwork!

Organism _____

Magnification _____

Dimensions _____

Organism _____

Magnification _____

Dimensions _____

Organism _____

Magnification _____

Dimensions _____

Organism _____

Magnification _____

Dimensions _____

QUESTIONS

1 *For each of the four organisms you identified, list the characteristics you used in making your identification.*

a. Organism: _____

    Characteristics: _____

_____

_____

_____

_____

_____

b. Organism: _____

    Characteristics: _____

_____

_____

_____

_____

_____

c. Organism: _____

    Characteristics: _____

_____

_____

_____

_____

_____

d. Organism: _____

    Characteristics: _____

_____

_____

_____

_____

Name ______________________________ Date ______________

Lab Section ______________________ I was present and performed this exercise (initials) ________

DATA SHEET 3-5

Simple Stains

OBSERVATIONS AND INTERPRETATIONS

Record your observations in the chart below.

Organism	Stain and Duration	Cellular Morphology and Arrangement (Include a sketch)	Cell Dimensions

QUESTIONS

1 What is the consequence of leaving a stain on the bacterial smear too long (overstaining)?

_____

_____

2 What is the consequence of not leaving a stain on the smear long enough (understaining)?

_____

_____

3 Choose a coccus and a bacillus from the organisms you observed, and calculate their surface-to-volume ratios. Consider the coccus to be a perfect sphere and the bacillus to be a rectangular block in which height and width are the same dimension. Use the equations supplied.

Cell Morphology	Surface Area	Volume
Coccus	$SA = 4\pi r^2$	$V = \dfrac{4}{3}\pi r^3$
Bacillus	$SA = 2(H \times W) + 4(L \times H)$	$V = H \times W \times L$

r = radius, H = height, L = length, W = width, π = 3.14

Surface-to-Volume Ratio of Sample Cells

Organism	Cell Morphology	Surface Area (μm^2)	Volume (μm^3)	Surface-to-Volume Ratio

4 Consider a coccus and a rod of equal volume.

a. Which is more likely to survive in a dry environment? Explain your answer.

_____

_____

_____

b. Which would be better adapted to a moist environment? Explain your answer.

_____

_____

Name _____ Date _____

Lab Section _____ I was present and performed this exercise (initials) _____

Negative Stains

OBSERVATIONS AND INTERPRETATIONS

Record your observations in the chart below.

Organism	Stain	Cellular Morphology and Arrangement (Include a detailed sketch of a few representative cells)	Cell Dimensions

QUESTIONS

1 *Why doesn't a negative stain colorize the cells in the smear?*

_____

_____

_____

_____

_____

_____

_____

_____

2 *Eosin is a red stain and methylene blue is blue. What should be the result of staining a bacterial smear with a mixture of eosin and methylene blue?*

3 *Compare the diameter of* M. luteus *cells as measured using a basic stain (Exercise 3-5) and an acidic stain. What might account for any difference?*

Name _____ Date _____

Lab Section _____ I was present and performed this exercise (initials) _____

Gram Stain

OBSERVATIONS AND INTERPRETATIONS

Record your observations in the chart below.

Organism or Source	Cellular Morphology and Arrangement (Include a detailed sketch of a few representative cells)	Cell Dimensions	Color	Gram Reaction (+/−)

DATA SHEET 3-7

QUESTIONS

1   *Predict the effect of the following "mistakes" made when performing a Gram stain. Consider each mistake independently.*

    a. Failure to add the iodine.

    b. Failure to apply the decolorizer.

    c. Failure to apply the safranin.

    d. Reversal of crystal violet and safranin stains.

2   *Both crystal violet and safranin are basic stains and may be used to do simple stains on Gram-positive and Gram-negative cells. This being the case, explain how they stain different cell types in the Gram stain.*

3   *If you saw large, eukaryotic cells in the preparation made from your gumline, they were most likely your own epithelial cells. Are you Gram-positive or Gram-negative? (You can make a good guess about this even if you didn't see your cells.)*

Name _____ Date _____

Lab Section _____ I was present and performed this exercise (initials) _____

Acid-Fast Stains

OBSERVATIONS AND INTERPRETATIONS

Record your observations in the chart below.

Organism	Staining Method (ZN or K)	Cellular Morphology and Arrangement (Include a detailed sketch of a few representative cells)	Cell Dimensions	Color	Acid-Fast Reaction (+/−)

QUESTIONS

1 How does heating the bacterial smear during a ZN stain promote entry of carbolfuchsin into the acid-fast cell wall?

2 Are acid-fast negative cells stained by carbolfuchsin? If so, how can this be a differential stain?

3 Why do you suppose the acid-fast stain is not as widely used as the Gram stain? When is it more useful than the Gram stain?

Name _____ Date _____

Lab Section _____ I was present and performed this exercise (initials) _____

DATA SHEET 3-9

Capsule Stain

OBSERVATIONS AND INTERPRETATIONS

Record your observations in the chart below.

Organism	Cellular Morphology and Arrangement (Include a detailed sketch of a few representative cells)	Cell Dimensions	Capsule (+ / −)	Width of Capsule (If present)

QUESTIONS

1 Capsules are neutrally charged. This being the case, what is the purpose of emulsifying the sample in serum in this staining procedure?

2 Some oral bacteria produce an extracellular "capsule." Of what benefit is a capsule to these cells?

3 Sketch any cells from your mouth sample that display an unusual morphology or arrangement.

Name _____   Date _____

Lab Section _____   I was present and performed this exercise (initials) _____

Endospore Stain

OBSERVATIONS AND INTERPRETATIONS

Record your observations in the chart below.

Organism (Include culture age)	Cellular Morphology and Arrangement (Include a detailed sketch of a few representative cells and spores)	Cell Dimensions	Spores (Present or absent)	Spore Shape and Position (If present)

QUESTIONS

1 *Why does this exercise call for an older (5-day) culture of* Bacillus?

2 *What does a positive result for the spore stain indicate about the organism? What does a negative result for the spore stain indicate about the organism?*

3 *Why is it not necessary to include a negative control for this stain procedure?*

4 *Spores do not stain easily. Perhaps you have seen them as unstained white objects inside* Bacillus *species in other staining procedures. If they are visible as unstained objects in other stains, of what use is the endospore stain?* (**Hint:** *See Figure 3-107.*)

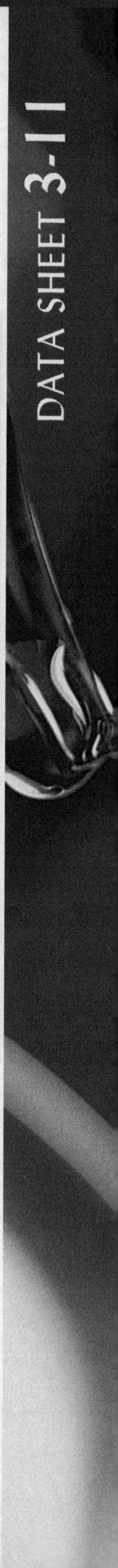

Name _____ Date _____

Lab Section _____ I was present and performed this exercise (initials) _____

Parasporal Crystal Stain

OBSERVATIONS AND INTERPRETATIONS

Record your stain results in the chart below.

Organism	Presence of Crystals (+ / −)	Sketch a Few Representative Cells (if any)

QUESTIONS

1 *It is expected that you saw crystals in* Bacillus thuringiensis *and not in* B. subtilis. *What would be your interpretation if you saw*

a. crystals in both?

b. crystals in neither?

2 *Why would the use of Bt-toxin as an insecticide be more "earth friendly" than that of traditional chemical pesticides?*

DATA SHEET 3-12

Name _____ Date _____

Lab Section _____ I was present and performed this exercise (initials) _____

Wet Mount and Hanging Drop Preparations

OBSERVATIONS AND INTERPRETATIONS

Record your observations in the chart below.

Organism	Procedure (Wet Mount or Hanging Drop)	Cellular Morphology and Arrangement (Include a detailed sketch of a few representative cells)	Cell Dimensions	Motility (+/−)

QUESTIONS

1 *In a wet mount, each of the following complications could lead to a false interpretation of motility. For each, write "false positive" or "false negative," depending on how it could interfere with your reading of a motile organism and a nonmotile organism.*

Complication	Motile Organism	Nonmotile Organsim
Over-inoculation of the slide with organisms		
Cells attaching to the glass slide or cover glass		
Receding water line		
Using an old culture		

2 *You are told that viewing is best done with as little illumination as possible. Why will transparent cells be easier to view with less light?*

3 *Would you expect Brownian motion to increase the longer you observe a hanging drop or wet mount preparation? Why?*

Name _____ Date _____

Lab Section _____ I was present and performed this exercise (initials) _____

Flagella Stain

OBSERVATIONS AND INTERPRETATIONS

Record your observations in the chart below.

Organism	Cellular Morphology and Flagellar Arrangement (Include a detailed sketch of a few representative organisms)	Cell Dimensions	Flagellar Length

QUESTIONS

1   *Why can't flagella be observed in action?*

2   *Flagella have a diameter of about 1 nm. A hypothetical question: To resolve flagella, what is the maximum wavelength of the electromagnetic spectrum that would have to be used to create the image (given numerical apertures of 1.25 for both the condenser and the oil lens)? Refer to Exercise 3-1 for help with this question.*

3   *Why are cells allowed to swim into and out of the loop's water film rather than scraping the agar surface of growth and then emulsifying cells on the slide?*

Name _____ Date _____

Lab Section _____ I was present and performed this exercise (initials) _____

Morphological Unknown

OBSERVATIONS AND INTERPRETATIONS

1 Complete the flowchart using the information in Table 3-4. Each path should end with a single organism.

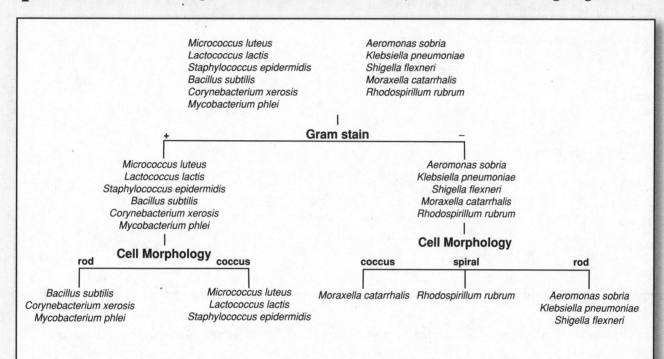

2 Record your stain results and the date each was run in the chart below. Include sketches (as appropriate) along with your written descriptions.

Unknown number: _____

	Gram Reaction Cell Morphology and Arrangement Cell Dimensions	Acid-Fast Stain	Motility (wet mount)	Capsule Stain	Spore Stain (shape and location)
Date(s) Run					
Results	Gram Reaction: Cell Morphology and Arrangement: Cell Dimensions: Sketch:			Sketch:	Sketch:

3 Match your results with those in the flowchart. Highlight the path on the flowchart that leads to identification of your unknown.

4 Highlight your confirmatory test in the chart above.

5 Write the identity of your unknown organism in the space below.

My Unknown is:

Name _____ Date _____

Lab Section _____ I was present and performed this exercise (initials) _____

Phenylethyl Alcohol Agar

OBSERVATIONS AND INTERPRETATIONS

Refer to Table 4-1 when recording your results and interpretations in the chart below.

Organism	Growth (P/G)		Interpretation
	PEA	NA	

QUESTIONS

1 *You were instructed to compare the growth on the NA and PEA plates. Because the two media used in this exercise were completely different and likely to produce differing amounts of growth, why not simply compare the organisms to each other on the PEA plate? In your explanation, include the information provided by the growth on the NA plate.*

_____

_____

_____

_____

_____

_____

_____

_____

2 PEA contains only 0.25% phenylethyl alcohol because high concentrations inhibit both Gram-negative organisms and Gram-positive organisms. List some possible reasons why this is true.

3 If you observed growth of Gram-negative organisms on your PEA plate, does this negate the usefulness of PEA as a selective medium? Why or why not? (**Hint:** Compare the NA and PEA plates.)

4 Is PEA a defined medium or an undefined medium? Why is this formulation desirable?

5 Which ingredient(s) in PEA supply(ies)
   a. Carbon?

   b. Nitrogen?

Name _____ Date _____

Lab Section _____ I was present and performed this exercise (initials) _____

Columbia CNA With 5% Sheep Blood Agar

OBSERVATIONS AND INTERPRETATIONS

Refer to Table 4-2 when recording your results and interpretations in the chart below.

| Organism | Growth (P/G) | | Hemolysis (α, β, γ) | Interpretation |
	Columbia CNA Agar	NA		

QUESTIONS

1 You were instructed to compare the growth on the NA and Columbia CNA Agar plates. Because the two media used in this exercise were completely different and likely to produce differing amounts of growth, would it have been better to compare the organisms to each other on the CNA plate? Include what information is provided by the NA plate in your explanation.

2 Earlier formulations of Columbia CNA Agar were made using a slightly higher concentration of colistin and nalidixic acid, but this was reduced to improve recovery of Gram-positive organisms. List some possible reasons why the higher concentration was less effective at allowing growth of Gram-positive organisms.

Name _____ Date _____

Lab Section _____ I was present and performed this exercise (initials) _____

Bile Esculin Test

OBSERVATIONS AND INTERPRETATIONS

Refer to Table 4-3 when recording your results and interpretations in the chart below.

Organism	Color Result	+ / −	Interpretation

QUESTIONS

1 *Sometimes Bile Esculin Agar is prepared as slants. With that form of the medium, only organisms able to darken more than half of the agar within 48 hours are considered positive (in contrast to the plated medium where any darkening is positive). Why do you think this is so?*

_____

_____

_____

_____

_____

_____

_____

_____

_____

_____

_____

_____

_____

2 *In terms of the selectivity and differential capabilities of the medium, how would its utility be affected by not including oxgall? Ferric citrate? Explain.*

3 *Would it be acceptable to read an early negative result for this test? Why or why not?*

Name _____ Date _____

Lab Section _____ I was present and performed this exercise (initials) _____

Mannitol Salt Agar

OBSERVATIONS AND INTERPRETATIONS

Refer to Table 4-4 when recording your results and interpretations in the chart below. Use abbreviations or symbols as needed.

Organism	Growth (P/G)		MSA Growth Color (Y/R)	Interpretation
	MSA	NA		

QUESTIONS

1 *What purpose does the Nutrient Agar plate serve? In what way does it increase the validity of the test result?*

_____

_____

_____

_____

_____

_____

_____

_____

_____

2 What would be the likely consequences of omitting the NaCl in Mannitol Salt Agar? Why?

3 Would omitting the NaCl alter the medium's specificity or sensitivity? Explain. (See pages 6 and 7 for assistance.)

4 Which ingredient(s) supply(ies)

a. Carbon?

b. Nitrogen?

5 With the diversity of microorganisms in the world, how can a single test such as MSA be used to confidently identify Staphylococcus aureus?

Name _____   Date _____

Lab Section _____   I was present and performed this exercise (initials) _____

MacConkey Agar

OBSERVATIONS AND INTERPRETATIONS

Refer to Table 4-5 when recording your results and interpretations in the chart below. Use abbreviations or symbols as needed.

Organism	Growth (P/G)		MAC Growth Color (R/C)	Interpretation
	MAC	NA		

QUESTIONS

1 *What purpose does the Nutrient Agar plate serve? In what way does it increase the validity of the test results?*

_____

_____

_____

2 *With respect to the MacConkey Agar, what would be the possible consequence of:*

a. replacing the lactose with glucose?

_____

_____

b. replacing the neutral red with phenol red (yellow when acidic; red or pink when alkaline)?

_____

_____

3 How would removing crystal violet from MacConkey Agar alter the sensitivity and specificity of the medium? (See pages 6 and 7 for assistance.)

4 Compare the recipes of Nutrient Agar and MacConkey Agar. If an organism can grow on both media, on which would you expect it to grow better? Why?

5 MacConkey Agar is a selective medium. Is it also a defined or an undefined medium? Why is that formulation desirable?

DATA SHEET 4-6

Name _____ Date _____

Lab Section _____ I was present and performed this exercise (initials) _____

Eosin Methylene Blue Agar

OBSERVATIONS AND INTERPRETATIONS

Refer to Table 4-6 when recording your results and interpretations in the chart below. Use abbreviations or symbols as needed.

Organism	Growth (P/G)		EMB Growth Color (Pi/D/C)	Interpretation
	EMB	NA		

QUESTIONS

1 What purpose does the Nutrient Agar plate serve? In what way does it increase the validity of the test result?

_____

_____

_____

_____

_____

2 Dipotassium phosphate is a buffer added to EMB that adjusts the pH to the proper starting level. What would be a possible consequence of adding buffers to raise the starting pH to 7.8?

_____

_____

_____

3 Would the change in starting pH suggested in question #2 alter the medium's sensitivity or specificity? Explain.

4 You are becoming aware of the diversity and abundance of bacteria, and that we always have some uncertainty about the identity of an isolate. With this in mind, how can we be certain that growth with a green metallic sheen is truly a coliform bacterium?

5 Compare the recipes of Nutrient Agar and EMB Agar. If an organism can grow on both media, on which would you expect it to grow better? Why?

6 EMB is a selective medium. Is it also a defined or an undefined medium? Why is that formulation desirable?

7 Which ingredient(s) in EMB supply(ies)

a. Carbon?

b. Nitrogen?

Name _____ Date _____

Lab Section _____ I was present and performed this exercise (initials) _____

Hektoen Enteric Agar

OBSERVATIONS AND INTERPRETATIONS

Refer to Table 4-7 when recording your results and interpretations in the chart below. Use abbreviations or symbols as needed.

Organism	Growth (P/G)		HE Growth Color (Pi/Bppt/B)	Interpretation
	HE	NA		

QUESTIONS

1 What purpose does the Nutrient Agar plate serve? In what way does it increase the validity of the test result?

_____

_____

_____

_____

2 Which ingredient(s) in this medium supply(ies)

a. Carbon?

_____

b. Nitrogen?

_____

3 Compare the recipes of Nutrient Agar and HE agar. If an organism can grow on both media, on which would you expect it to grow better? Why?

4 HE Agar is a selective medium. Is it also a defined or an undefined medium? Why is that formulation desirable?

5 This medium was designed to differentiate Salmonella and Shigella from other enterics. Salmonella species sometimes produce a black precipitate in their growth; Shigella species do not. If you were designing a medium to differentiate these two genera, which ingredients from this medium would you include? Explain.

6 All enterics ferment glucose. What would be some consequences of replacing the sugars in this medium with glucose? What color combinations would you expect to see?

7 List all the things you know about a viable organism that produces green colonies with black centers on the medium in question #6.

Name _____   Date _____

Lab Section _____ I was present and performed this exercise (initials) _____

Xylose Lysine Desoxycholate Agar

OBSERVATIONS AND INTERPRETATIONS

Refer to Table 4-8 when recording your results and interpretations in the chart below. Use abbreviations or symbols as needed.

Organism	Growth (P/G)		XLD Growth Color (Y/RB/R)	Interpretation
	XLD	NA		

QUESTIONS

1 *What purpose does the Nutrient Agar plate serve? In what way does it increase the validity of the test result?*

_____

_____

_____

_____

_____

_____

2 *Which ingredient(s) in this medium supply(ies)*
   a. Carbon?

_____

   b. Nitrogen?

_____

3   Compare the recipes of Nutrient Agar and XLD Agar. If an organism can grow on both media, on which would you expect it to grow better? Why?

4   XLD is a selective medium. Is it also a defined or an undefined medium? Why is that formulation desirable?

5   What would be the likely consequence of:

a. incubating the medium for 48 hours?

b. not adding desoxycholate?

c. not adding sucrose and lactose?

d. not adding lysine?

e. not adding ferric ammonium citrate?

6   How would each of the changes in question #5 affect the sensitivity and specificity of the medium?

7   Shigella and Providencia species virtually never ferment xylose, sucrose or lactose. Since the medium was designed to directly identify these organisms, why were three unusable carbohydrates included when only one would seem to accomplish the goal?

Name _____ Date _____

Lab Section _____ I was present and performed this exercise (initials) _____

Reduction Potential

OBSERVATIONS AND INTERPRETATIONS

Record your results and interpretations in the chart below. (Nutrient Broth = NB, Nutrient Broth with thiosulfate = NBT)

| Flask Contents | Amount of growth/turbidity | | Interpretation |
	P. mirabilis	Control	
Nutrient Broth			
Nutrient Broth with oil			
Nutrient Broth + thiosulfate			
Nutrient Broth + thiosulfate with oil			

Enter the amount of growth as 0, +, ++, +++, or ++++

QUESTIONS

1 In the table below in descending order of turbidity, list your predicted results and your actual results. Defend your prediction and offer possible reasons for any disagreements between the expected and actual results.

Expected Result	Actual Result

_____

_____

_____

_____

2   *What is the purpose of the controls used in this test? Regarding the controls, what is the desired result? How many and what combinations of results are acceptable?*

3   *Using Figure 5-1 as a reference (without regard to equation balance), circle all of the following reactions that will go forward without the addition of energy.*

$$H_2 + \tfrac{1}{2} O_2 \longrightarrow H_2O$$

$$Fe^{2+} + SO_4^{2-} \longrightarrow FeSO_4$$

$$H_2 + NO_3^- \longrightarrow NO_2^- + H_2O$$

$$\text{cytochrome } a_{red} + \text{cytochrome } c_{ox} \longrightarrow \text{cytochrome } a_{ox} + \text{cytochrome } c_{red}$$

Name _____ Date _____

Lab Section _____ I was present and performed this exercise (initials) _____

Oxidation–Fermentation Medium

OBSERVATIONS AND INTERPRETATIONS

Refer to Table 5-1 when recording your results and interpretations in the chart below.

| Organism | Color Results | | Symbol | Interpretation |
	Sealed	Unsealed		
Uninoculated Control				
.				

QUESTIONS

1 *What is the purpose of the uninoculated control tubes used in this test? Because only one uninoculated tube is sufficient to show the green color unchanged, why is it necessary to use two controls? Be specific.*

_____

_____

_____

_____

_____

_____

_____

_____

_____

_____

2 Some microbiologists recommend inoculating a pair of O–F basal media (without carbohydrate) along with the carbohydrate media. Why do you think this is done?

_____

_____

_____

_____

_____

_____

3 All enterics are facultative anaerobes; that is, they have both respiratory and fermentative enzymes. What color results would you expect for organisms in O–F glucose media inoculated with an enteric? Remember to describe both sealed and unsealed tubes.

_____

_____

_____

_____

_____

_____

4 Suppose that when you examined your tubes (in this exercise) after incubating them, you noticed that the unsealed control contained slight yellowing at the top. Suppose further that pair #1 showed complete yellowing of both tubes and pairs #2 and #3 showed slight yellowing of the unsealed tube. Assuming all other tubes were green, what conclusions could you safely make?

_____

_____

_____

_____

Which results, if any, are reliable? Why?

_____

_____

_____

Which results, if any, are not reliable? Why not?

_____

_____

Name _____ Date _____

Lab Section _____ I was present and performed this exercise (initials) _____

Phenol Red Broth

OBSERVATIONS AND INTERPRETATIONS

Enter your results in the chart below, using the symbols shown in Table 5-2.

Organism	Results				Interpretation
	PR Base	PR Glucose	PR Lactose	PR Sucrose	
Uninoculated Control					GLU
					LAC
					SUC
					GLU
					LAC
					SUC
					GLU
					LAC
					SUC
					GLU
					LAC
					SUC
					GLU
					LAC
					SUC

QUESTIONS

1 What purpose did the uninoculated controls serve in this test? What purpose did the PR Base Broths serve?

2 Early formulations of this medium used a smaller amount of carbohydrate and occasionally produced false (pink) results after 48 hours. This phenomenon is called a reversion. Why do you think this happened? List at least two steps, as a microbiologist, you could take to prevent the problem. (**Hint:** Look at "A Word About Biochemical Tests And Acid-Base Reactions" at the beginning of this section.)

3 Suppose you inoculate a PR broth with an organism known to be a slow-growing fermenter. After 48 hours, you see slight turbidity but score it as (–/–). Is this result a false positive or a false negative? Is this false result caused by poor specificity or poor sensitivity of the test system?

4 Assuming, as in Question 1, you tested an organism using the three carbohydrate broths and a base broth, which of the combinations of results in the following table would be reliable? Interpret each of the combinations and explain why the results are reliable or not.

PR Base	PR Glucose	PR Lactose	PR Sucrose	Interpretation	Reliable? Y/N
–/–	A/G	A/G	A/G		
–/–	A/–	A/–	–/–		
A/G	A/G	A/G	A/G		
A/–	A/G	A/G	K		
K	K	K	K		
–/–	K	A/–	–/–		
K	A/G	A/G	–/–		
A/–	–/–	–/–	–/–		

Name _____ Date _____

Lab Section _____ I was present and performed this exercise (initials) _____

Methyl Red and Voges-Proskauer Tests

OBSERVATIONS AND INTERPRETATIONS

Refer to Tables 5-3 and 5-4 as you record your results and interpretations in the chart below.

Organism	MR Result	VP Result	Interpretation
Uninoculated Control			

QUESTIONS

1 Some protocols call for a shorter incubation time for the MR and VP tests. Other protocols allow for up to 10 days incubation with virtually no risk of producing a false positive.

   a. Which of the two tests would likely produce more false negatives with a shorter incubation time. Why?

_____

_____

_____

_____

   b. Which test would likely benefit most from a longer incubation time? Why?

_____

_____

_____

2 *Would a false negative result for the VP test more likely be attributable to poor sensitivity or to poor specificity of the test system?*

3 *Why were you told to shake the VP tubes after the reagents were added?*

4 *Why is the Methyl Red Test read immediately and the Voges-Proskauer read after 60 minutes?*

5 *Some microbiologists recommend reincubating organisms producing Methyl Red-negative results for an additional 2 to 3 days. Why do you think this is done?*

Name _____ Date _____

Lab Section _____ I was present and performed this exercise (initials) _____

Catalase Test

OBSERVATIONS AND INTERPRETATIONS

Using Table 5-5 as a guide, record your results and interpretations in the chart below.

Organism	Bubbles? Y / N	+ / −	Interpretation
Uninoculated control			

QUESTIONS

1 *When flavoprotein transfers electrons directly to the final electron acceptor, hydrogen peroxide is produced. What other consequences might result from electron carriers in the ETC being bypassed?* (**Hint:** *See Appendix A.*)

_____

_____

_____

_____

_____

_____

_____

_____

_____

_____

_____

2 Reduction often is referred to as the addition of hydrogens to a compound. It is more chemically correct to refer to reduction as the addition of electrons. Provide an example of a reduction reaction performed by cells in which hydrogen is not added to the reduced compound.

3 Would a false positive from the reaction between the inoculating loop and hydrogen peroxide be caused by poor specificity or poor sensitivity of the test system? Explain.

4 Why is it advisable to perform this test on a known catalase-positive organism along with the organism you are testing?

5 What is the purpose of adding hydrogen peroxide to the uninoculated tube?

Name _____ Date _____

Lab Section _____ I was present and performed this exercise (initials) _____

Oxidase Test

OBSERVATIONS AND INTERPRETATIONS

Using Table 5-6 as a guide, enter your results in the chart below.

Organism	Color Result	+ / −	Interpretation

QUESTIONS

1 *Why is it advisable to include a known oxidase-positive control in any test of an unknown organism?*

_____

_____

_____

_____

_____

_____

_____

_____

_____

_____

2 Suppose in a test of an organism that possesses cytochrome c oxidase, the reagent begins to turn blue at 45 seconds. Consider the following possibilities, check all that apply, and defend your choices.

This is an example of (a):

	YES	NO
Valid test	☐	☐
Reliable result	☐	☐
False positive	☐	☐
False negative	☐	☐
Poor sensitivity	☐	☐
Poor specificity	☐	☐

3 Provide a possible explanation as to why this test identifies the presence of cytochrome c oxidase and not other oxidases.

Name ______________________________ Date ______________

Lab Section ______________________ I was present and performed this exercise (initials) ________

DATA SHEET 5-7

Nitrate Reduction Test

OBSERVATIONS AND INTERPRETATIONS

Using Table 5-7 as a guide, enter your results and interpretations in the chart below.

Organism	Results			Interpretation
	Gas	Color		
	Y / N	After Reagents	After Zinc	
Uninoculated Control				

QUESTIONS

1 *Why is gas production not recognized as nitrate reduction when the organism is a known fermenter?*

2 *Suppose you remove your test cultures from the incubator and notice that one of them—a known fermenter—has a gas bubble in the Durham tube. Knowing that fermenters frequently produce gas, you ignore the bubble and proceed to the next step. Adding reagents produces no change, and neither does adding zinc. Is this occurrence consistent with what you have learned about this test? Why?*

3 *Would you change your answer to #2 above if the control broth did not change color after the addition of reagents? What if the control broth had changed color only after the addition of reagents and zinc?*

4 *When testing microaerophiles, some microbiologists prefer to use a semisolid nitrate medium that contains a small amount of agar. Why do you think this is done?*

Name _____ Date _____

Lab Section _____ I was present and performed this exercise (initials) _____

Citrate Test

OBSERVATIONS AND INTERPRETATIONS

Using Table 5-8 as a guide, enter your results and interpretations in the chart below.

Organism	Color Result	+ / −	Interpretation
Control			

QUESTIONS

1 *Many bacteria that are able to metabolize citrate (as seen in the Krebs cycle) produce negative results in this test. Why? Be specific. Refer to Appendix A for help.*

_____

_____

_____

_____

_____

_____

_____

_____

_____

2 If an organism is able to convert the sodium citrate in Simmons Citrate Agar to pyruvate, list some possible reasons why an organism might ferment it rather than metabolize it oxidatively in the Krebs cycle. Refer to Appendix A for help.

_____

_____

_____

_____

_____

_____

_____

_____

_____

3 Explain how an organism that possesses the citrase enzyme might not test positively on Simmons Citrate Agar. Is this a false negative result? Why or why not?

_____

_____

_____

_____

_____

_____

_____

_____

_____

_____

_____

Name _____ Date _____

Lab Section _____ I was present and performed this exercise (initials) _____

Malonate Test

OBSERVATIONS AND INTERPRETATIONS

Using Table 5-9 as a guide, enter your results and interpretations in the chart below.

Organism	Color Result	+ / −	Interpretation
Control			

QUESTIONS

1 *Examine the structural formulas below. Ordinarily, in biochemical reactions there is specificity between an enzyme and its substrate. That is, enzymes typically attach to one substrate exclusively. Explain why it is possible for succinate dehydrogenase to bind to two different substrates. Be specific.*

```
    COOH              COOH
     |                 |
    CH₂               CH₂
     |                 |
    COOH              CH₂
                       |
                      COOH
  Malonic Acid      Succinic Acid
```

_____

_____

_____

_____

_____

2 *The cell has hundreds of succinate dehydrogenase enzymes. How do you suppose malonate concentration affects cell growth?*

3 *How can a yellow color and no color change both be considered negative results for this test?*

4 *What is the purpose of the uninoculated control?*

Name _____ Date _____

Lab Section _____ I was present and performed this exercise (initials) _____

Decarboxylation Test

OBSERVATIONS AND INTERPRETATIONS

Using Table 5-10 as a guide, record your results and interpretations in the chart below.

| Organism | Results | | | | | | | | Interpretation |
| | Lysine | | Ornithine | | Arginine | | Base | | |
	Color	+/−	Color	+/−	Color	+/−	Color	+/−	
Control									

QUESTIONS

1 *What was the purpose of the controls?*

_____

_____

_____

_____

_____

_____

_____

_____

2 What was the purpose of inoculating the base broths? Would a positive result in any of them affect the rest of the exercise? Why?

3 How can no change *and a* conversion to a yellow color *of the broth both be considered negative results?*

4 Incubation time for this medium is 1 week. Under what circumstances would an early reading be allowable? Not allowable? Explain.

5 How does adding the sterile mineral oil affect the specificity or the sensitivity of the test?

Name _____ Date _____

Lab Section _____ I was present and performed this exercise (initials) _____

Phenylalanine Deaminase Test

OBSERVATIONS AND INTERPRETATIONS

Using Table 5-11 as a guide, record your results and interpretations in the chart below.

Organism	Color Result	+ / −	Interpretation
Control			

QUESTIONS

1 *What is the purpose of the control? What other types of controls would be useful to increase reliability of the test?*

_____

_____

_____

_____

_____

_____

_____

_____

_____

_____

_____

2 *If you are performing this test on an unknown organism, why is it a good idea to run simultaneous tests on known phenylalanine-positive and phenylalanine-negative organisms?*

3 *What do deamination and decarboxylation reactions have in common?*

4 *Phenylalanine medium sometimes is prepared in broth form with a pH indicator similar to that used in Decarboxylase Medium so that increases in pH then can be detected by development of purple color. When testing an organism for the ability to deaminate phenylalanine, would you expect to overlay this medium with mineral oil? Why or why not?*

Name _____ Date _____

Lab Section _____ I was present and performed this exercise (initials) _____

Starch Hydrolysis

OBSERVATIONS AND INTERPRETATIONS

Using Table 5-12 as a guide, enter your results and interpretations in the chart below.

Organism	Result	+ / −	Interpretation
(Control) Sector			

QUESTIONS

1 *Suppose you had poured iodine on your plate and noticed clearings in the uninoculated area, as well as around both of your transferred cultures. What are some possible explanations for this occurrence? Was integrity of the exercise compromised? What kinds of things might be done to avoid this problem in future exercises?*

2 *How would you expect the results of this exercise to change if you were to add glucose to the medium?*

3 *In many tests it is acceptable to read a positive result before the incubation time is completed. Why is this not the case with Starch Agar?*

4 *Suppose you could selectively prevent production of α-amylase or oligo-1,6-glucosidase in an organism that normally hydrolyzes starch. Which enzyme would the organism miss the most?*

Name _____ Date _____

Lab Section _____ I was present and performed this exercise (initials) _____

Urea Hydrolysis

OBSERVATIONS AND INTERPRETATIONS

1 Using Table 5-13 as a guide, enter your results for the agar test each day for 6 days or until pink color appears.

Urea Agar								
	Color						+/−	Interpretation
Organism	24 Hrs.	2 Days	3 Days	4 Days	5 Days	6 Days		
Control								

2 Using Table 5-14 as a guide, enter your results for the broth test after 24 hours.

Urea Broth			
Organism	Color Result	+ / −	Interpretation
Control			

QUESTIONS

1 Suppose you ran this test with Providencia stuartii *but it took 48 hours to turn pink. Do you think this is a false result? If so, is it a false positive or a false negative? If not, why not? Give some possible reasons for this occurrence.*

2 *Explain why it is acceptable to record positive tests before the suggested incubation time is completed but it is not acceptable to record a negative result early.*

3 *Typically, liquid medium is filter-sterilized because autoclaving breaks down the urea. However, even unsterilized broth rarely produces false-positive results. Why do you think this is true?*

4 *Did your results from the broth and solid media agree for each organism? Did you expect them to? Why or why not? How could you explain any differences?*

Name _____ Date _____

Lab Section _____ I was present and performed this exercise (initials) _____

Casein Hydrolysis Test

OBSERVATIONS AND INTERPRETATIONS

Using Table 5-15 as a guide, record your results in the chart below.

Organism	Result	+ / −	Interpretation
(Control) Sector			

QUESTIONS

1 *What does the enzyme casease have in common with amylase?*

_____

_____

_____

_____

_____

_____

_____

_____

_____

_____

_____

_____

_____

DATA SHEET 5-14

2   *How do we know that casease is an exoenzyme and not a cytoplasmic enzyme?*

3   *Is it acceptable to read a positive test before the incubation time is completed? How about an early negative result?*

4   *Why is the uninoculated control relatively unnecessary in this test?*

5   *Why is it advisable to use a positive control along with organisms that you are testing?*

Name _____ Date _____

Lab Section _____ I was present and performed this exercise (initials) _____

Gelatin Hydrolysis Test

OBSERVATIONS AND INTERPRETATIONS

Using Table 5-16 as a guide, record your liquefaction results in the chart below.

Organism	Result	+ / −	Interpretation
Control			

QUESTIONS

1 *Some microbiologists recommend incubating this medium at 37°C, along with an uninoculated control, and then transferring all tubes to the refrigerator prior to reading them. Why might this be the preferred technique in some situations? What potential problems can you see with this method?*

_____

_____

_____

_____

_____

_____

_____

_____

_____

_____

2 *If the control is solid and an inoculated tube is liquid, is it acceptable to read the result before the complete incubation time has elapsed? Why?*

3 *If the control is solid and an uninoculated tube is also solid, is it acceptable to read the result before the complete incubation time has elapsed? Why?*

4 *Suggest some ways by which an organism could be a slow gelatin liquefier.*

5 *Suppose that after 7 days a tube inoculated with a slow liquefier shows no evidence of liquefaction. Is this a failure of the test system? If yes, why? If not, why not?*

Name _____ Date _____

Lab Section _____ I was present and performed this exercise (initials) _____

DNA Hydrolysis Test

OBSERVATIONS AND INTERPRETATIONS

Using Table 5-17 as a guide, record your DNAse results in the chart below.

Organism	Result	+ / −	Interpretation
(Control) Sector			

QUESTIONS

1 In Theory on the first page of this exercise, the disassembly of DNA was described as a "depolymerization." What other term applies to the process?

_____

_____

_____

_____

_____

_____

_____

_____

_____

DATA SHEET 5-16

2 A positive result for the DNA hydrolysis test does not distinguish between Staphylococcus DNase and the DNase produced by Serratia. If you had the expertise to correct this weakness in the system, would you improve the test's sensitivity or its specificity?

3 Suggest a reason why this test is read after only 24 hours while other tests (e.g., gelatinase test) may take a week.

4 Why is the uninoculated control relatively unnecessary in this test?

5 Why is it advisable to use a positive control along with organisms that you are testing?

Name _____ Date _____

Lab Section _____ I was present and performed this exercise (initials) _____

Lipid Hydrolysis Test

OBSERVATIONS AND INTERPRETATIONS

Using Table 5-18 as a guide, record your results and interpretations in the table below.

Organism	Result	+ / −	Interpretation
(Control) Sector			

QUESTIONS

1 *Many organisms possessing many different lipases produce positive results on Tributyrin Agar. Is the inability of this medium to distinguish between these different enzymes a weakness in its specificity or its sensitivity?*

_____

_____

_____

_____

_____

_____

_____

_____

_____

_____

2 Tributyrin Agar has a shelf life of only a few days before it loses its opacity.

a. With this in mind, explain the importance of positive and negative controls in this test.

b. How would expired Tributyrin Agar affect the results of lipase (+) and lipase (−) organisms?

3 Imagine this situation: Species #1 and Species #2 are unrelated, but each produces an enzyme capable of hydrolyzing soybean oil. How can this be reconciled with the fact that enzymes are highly specific to their substrates?

4 Why do you think Tributyrin Agar is prepared in the blender as an emulsion rather than simply stirred as a solution?

DATA SHEET 5-18

Name _____ Date _____

Lab Section _____ I was present and performed this exercise (initials) _____

ONPG Test

OBSERVATIONS AND INTERPRETATIONS

Using Table 5-19 as a guide, enter your results and interpretations in the chart below.

Organism	Result	+ / −	Interpretation
Control			

QUESTIONS

1 β-galactosidase is an inducible enzyme.

a. What does "inducible" mean?

_____

_____

_____

_____

b. What is the purpose of inoculating the ONPG tubes with growth from Triple Sugar Iron Agar cultures?

_____

_____

_____

_____

_____

c. What would be a possible consequence of inoculating the ONPG medium with an ONPG-positive organism grown on Nutrient Agar instead of Kligler's Iron Agar or Triple Sugar Iron Agar?

2 Examine the phenotypes of the organisms listed below. For each organism, predict the ONPG result.

Organism	β-galactoside permease	β-galactosidase	Predicted ONPG Result (+ or −)
A	Present	Present	
B	Present	Absent	
C	Absent	Present	
D	Absent	Absent	

3 When running this test on an unknown organism, why should you run a simultaneous test on a known ONPG-positive organism?

Name _____ Date _____

Lab Section _____ I was present and performed this exercise (initials) _____

PYR Test

OBSERVATIONS AND INTERPRETATIONS

Using Table 5-20 as a guide, enter your results and interpretations in the table below.

Organism	Result	+ / −	Interpretation

QUESTIONS

1 *Deep red is the only color reaction that is considered positive for this test. Why do you think yellow or orange results are not positive (or weak positive)?*

_____

_____

_____

_____

_____

_____

_____

_____

_____

_____

_____

_____

_____

_____

_____

_____

2 *Suppose you wanted to perform this test on an unknown organism but you were out of PYR reagent. Suppose further that a fellow lab employee said that you could substitute formaldehyde for PYR reagent and get good results. Would you be tempted to believe him? Why or why not?*

_____

_____

_____

_____

_____

_____

_____

_____

_____

_____

_____

_____

_____

_____

_____

3 *If you were unsure of your PYR reagent's quality and wanted to test it, which of the following chemicals would you try?*

☐   A. Mineral oil

☐   B. Ethanol

☐   C. Lysine

☐   D. Acetone

Name _____ Date _____

Lab Section _____ I was present and performed this exercise (initials) _____

SIM Medium

OBSERVATIONS AND INTERPRETATIONS

1 Using Table 5-21 as a guide, record your results and interpretations in the chart below.

Sulfur Reduction			
Organism	Black PPT? Y / N	+ / −	Interpretation
Uninoculated Control			

2 Using Table 5-22 as a guide, record your results and interpretations in the chart below.

Indole Production			
Organism	Red Color? Y / N	+ / −	Interpretation
Uninoculated Control			

3 Using Table 5-23 as a guide, record your results and interpretations in the chart below.

Organism	Growth Pattern	Motility	
		+ / −	Interpretation
Uninoculated Control			

QUESTIONS

1 *The sulfur reduction test is not able to differentiate H₂S produced by anaerobic respiration and H₂S produced by putrefaction. Is this inability the result of poor sensitivity or poor specificity of the test system?*

_____

_____

_____

_____

2 *What factors dictate the choice of tests included in a combination medium?*

_____

_____

_____

3 *Which ingredient could be eliminated if this medium were used strictly for testing motility in indole producers? Explain.*

_____

_____

_____

_____

_____

Name _____ Date _____

Lab Section _____ I was present and performed this exercise (initials) _____

Triple Sugar Iron Agar / Kligler Iron Agar

OBSERVATIONS AND INTERPRETATIONS

Refer to Table 5-24 when recording and interpreting your results.

Organism	Color Result	Symbol	Interpretation
Control			

QUESTIONS

1 *As mentioned in Theory, the fermentation readings with TSIA and KIA must take place between 18 and 24 hours after inoculation. Why is this true? Is timing as critical with H₂S readings? Why or why not?*

_____

_____

_____

_____

_____

_____

_____

_____

_____

2 You learned in Theory that if the black precipitate obscures the color of the butt that it must be acidic and scored as "A." Why do you think this is true? **Hint:** See Figure 5-63.

3 TSIA and KIA are complex media with many ingredients. What would be the consequences of the following mistakes in preparing this medium? Consider each independently.

a. 1% glucose is added rather than the amount specified in the recipe.

b. Ferrous ammonium sulfate (or ferric ammonium citrate in KIA) is omitted.

c. Casein and animal tissue are omitted.

d. Sodium thiosulfate is omitted.

e. Phenol red is omitted.

f. The initial pH is 8.2.

g. The agar butt is shallow rather than deep.

Name _____ Date _____

Lab Section _____ I was present and performed this exercise (initials) _____

Lysine Iron Agar

OBSERVATIONS AND INTERPRETATIONS

Refer to Table 5-25 when recording and interpreting your results.

Organism	Color Result	Symbol	Interpretation
Control			

QUESTIONS

1 *LIA is a combination medium. Which test(s) does it replace?*

_____

_____

_____

_____

_____

_____

_____

_____

_____

2   Lysine Iron Agar is not designed to identify ability of organisms to ferment. Why, then, is glucose added to the medium (beyond being just another nutrient)?

_____

_____

_____

_____

_____

_____

_____

3   LIA is a complex medium with many ingredients. What would be the consequences of the following mistakes in preparing this medium? Consider each independently.

a. Inclusion of 1% dextrose rather than the amount called for in the recipe.

_____

_____

_____

b. Omission of L-Lysine hydrochloride.

_____

_____

_____

c. Omission of Ferric ammonium citrate.

_____

_____

_____

d. Omission of Sodium thiosulfate.

_____

_____

_____

e. Omission of Bromcresol purple.

_____

_____

Name _____ Date _____

Lab Section _____ I was present and performed this exercise (initials) _____

Litmus Milk Medium

OBSERVATIONS AND INTERPRETATIONS

Refer to Table 5-26 when recording and interpreting your results below.

Organism	Results	Interpretation
Uninoculated Control		

QUESTIONS

1 *An acid clot can appear as pink or white with a pink band at the top. Explain the different conditions that would produce each of these occurrences. If the reaction to produce a white color occurs, why does the surface remain pink?*

_____

_____

_____

_____

_____

_____

_____

2 What do you think is the principal difference between a bacterial species that produces a curd in litmus milk and a species that does not?

3 What reaction would you predict from an organism growing in litmus milk that has the following results in other media?

a. a clear zone around the growth on a milk agar plate

b. A/– in PR lactose broth

c. A/G in purple lactose broth

d. K in PR glucose broth

Name _____ Date _____

Lab Section _____ I was present and performed this exercise (initials) _____

Bacitracin, Novobiocin, and Optochin Susceptibility Tests

OBSERVATIONS AND INTERPRETATIONS

Refer to Tables 5-27, 28, and 29 when recording and interpreting your results in the chart below.

Antibacterial Agent	Organism	Susceptible (S)	Resistant (R)
Bacitracin			
Novobiocin			
Optochin			

QUESTIONS

1 *The susceptibility tests included in this unit are typically used to differentiate between Gram-positive cocci. Would you predict them to be an effective differential test for Gram-negative organisms? Why or why not?*

_____

_____

_____

_____

_____

_____

_____

_____

_____

_____

_____

2   *Why is it important to get a bacterial lawn rather than isolated colonies on the plate?*

3   *Does the zone of inhibition's edge indicate the limit of antibacterial agent diffusion into the agar? Give reasons or evidence to support your answer.*

Name _____ Date _____

Lab Section _____ I was present and performed this exercise (initials) _____

Blood Agar

OBSERVATIONS AND INTERPRETATIONS

Choose four different colonies (preferably with a diversity of hemolysis reactions) and fill in the chart. Refer to Table 5-30 when recording and interpreting your results.

Source of Culture	Colony Morphology	Hemolysis Result	Interpretation

QUESTIONS

1 *The streak-stab technique, used to promote streptolysin activity, is preferred over incubating the plates anaerobically. Why do you think this is so? Compare and contrast what you see as the advantages and disadvantages of each procedure.*

_____

_____

_____

_____

_____

_____

_____

_____

2 Assuming that all of the organisms cultivated in this exercise came from the throats of healthy students, why is it important to cover and tape the plates?

3 Why is the streak plate preferred over the spot inoculations in this procedure?

Name _____ Date _____

Lab Section _____ I was present and performed this exercise (initials) _____

CAMP Test

OBSERVATIONS AND INTERPRETATIONS

Refer to Table 5-31 when recording and interpreting your results in the chart below.

Organism (interacting with *S. aureus*)	Result + / −	Interpretation

QUESTIONS

1 *You were instructed to not allow streak I and streak III to touch. Why do you think this is important?*

_____

_____

_____

_____

_____

_____

_____

_____

_____

_____

_____

_____

_____

_____

2 Why do you think the clearing in a positive CAMP test is an arrowhead shape and not some other shape? Would it likely make an important difference if you began streak II near S. aureus *and streaked across the plate in the other direction? Why?*

3 What result would you expect to get if you accidentally reverse the organisms in the procedure?

Name _____ Date _____

Lab Section _____ I was present and performed this exercise (initials) _____

Coagulase Tests

OBSERVATIONS AND INTERPRETATIONS

Refer to Tables 5-32 and 5-33 when recording and interpreting your results in the charts below.

Slide Test Results			
Organism	**Slide**	**Result**	**Interpretation**
	A		
	B		
	A		
	B		

Tube Test Results		
Organism	**Result**	**Interpretation**
Uninoculated Control		

QUESTIONS

1 Why is it more important to use fresh cultures in the Coagulase Test than in a test medium such as Milk Agar?

2 How would you interpret a negative Slide Test and a positive Tube Test using the same organism?

3 Consider the Slide Test.

a. What is the role of sterile saline plus organism on the Slide Test?

b. Why is it advisable to run a known coagulase-positive organism along with your unknown organism?

c. How will the validity of the test be affected if clumping occurs on both smears of the known coagulase-positive organism?

4 List possible reasons why the Slide Test is not appropriate for detecting free coagulase.

Name _____ Date _____

Lab Section _____ I was present and performed this exercise (initials) _____

Motility Test

OBSERVATIONS AND INTERPRETATIONS

Refer to Table 5-34 when recording and interpreting your results in the chart below.

Organism	Result	+ / −	Interpretation

QUESTIONS

1 *Why is it important to carefully insert and remove the needle along the same stab line?*

_____

_____

_____

_____

_____

_____

_____

_____

_____

_____

_____

_____

_____

2 What are some possible ways that you might obtain false positive and negative results using motility test medium?

3 Why is it essential that the reduced TTC be insoluble? Why is there less concern about the solubility of the oxidized form of TTC?

Name _____   Date _____

Lab Section _____   I was present and performed this exercise (initials) _____

API 20 E Identification System for *Enterobacteriaceae* and Other Gram-negatve Rods

OBSERVATIONS AND INTERPRETATIONS

Tape your API 20 E Result Sheet here.

QUESTIONS

1 Why is it important to perform the reagent tests last?

2 In clinical applications of this test system, reagents are added only if the glucose (oxidation/fermentation) test result is yellow or at least three other tests are positive. If these conditions are not met, a MacConkey Agar plate is streaked and additional tests are performed confirming glucose metabolism, nitrate reduction, and motility. Why do you think this is so? Be specific.

3 Suppose, after 24 hours incubation, you notice no growth in the tubes containing mineral oil. Assuming that it is behaving properly under these conditions, what do you know about the organism and what predictions can you safely make about its performance in the decarboxylase tests, fermentation tests, and nitrate reduction test? Is it a member of Enterobacteriaceae?

Name _____ Date _____

Lab Section _____ I was present and performed this exercise (initials) _____

Enterotube® II

OBSERVATIONS AND INTERPRETATIONS

1 Tape your BBL® Enterotube® II Result Sheet here.

2 Enter the results from the CCIS booklet below.

CCIS Five-Digit Code	Possible Organisms	VP Result + / −	Identified Organism

QUESTIONS

1 *Fecal coliforms such as* Escherichia coli, Enterobacter aerogenes, *and* Klebsiella pneumoniae *are enterics that ferment lactose to acid and gas at 35°C within 48 hours. For most strains of these organisms, the chart below summarizes reactions in the Enterotube® II. Fill in the missing information and, using colored pencils, fill in the appropriate colors for each positive test.*

	Glu	Gas	Lys	Orn	H₂S	Ind	Adon	Lact	Arab	Sorb	VP	Dulc	PA	Urea	Citrate
E. coli			+	−	−	+	−		+	+	−	−	−	−	−
E. aerogenes			+	+	−	−	+		+	+	+	−	−	−	+
K. pneumoniae			+	−	−	−	+		+	+	+	−	−	+	+

2 *Based on the information above, enter the five-digit codes for the three organisms.*

E. coli _____ E. aerogenes _____ K. pneumoniae _____

3 *Examine the following chart. Fill in the color reactions for each test based on the ID value given.*

ID Value	Glucose/ Gas	Lysine	Ornithine	H₂S/ Indole	Adonitol	Lactose	Arabinose	Sorbitol	VP	Dulcitol/ PA	Urea	Citrate
31122												
27345												
04004												
05511												

4 *Which of the four organisms are not members of* Enterobacteriaceae? *Why?*

_____

_____

5 *Which of the four ID values is questionable? Why?*

_____

_____

_____

Name _____   Date _____

Lab Section _____   I was present and performed this exercise (initials) _____

Gram Positive Unknown

OBSERVATIONS AND INTERPRETATIONS

Unknown Number _____

Isolation Procedure (Please record all activities associated with isolation of your organisms—from mixed culture to pure culture. Always include the date, source of inoculum, destination, incubation temperature, and any other relevant information. Also, make note of transfers made to keep your pure culture fresh. This log must be kept current.)

_____

_____

_____

_____

_____

_____

_____

_____

_____

_____

_____

_____

_____

_____

_____

_____

_____

_____

_____

Preliminary Observations

Colony Morphology (include medium) _____

_____

_____

Gram Stain _____   Cell Dimensions _____   Optimum Temperature _____

Cellular Morphology and Arrangement _____

DATA SHEET 5-31

Differential Tests (Include all information through the confirmatory test. This log must be kept current.)

Test #1: _____ Date Begun: _____ Date Read: _____ Result: _____

Comments: _____

Test #2: _____ Date Begun: _____ Date Read: _____ Result: _____

Comments: _____

Test #3: _____ Date Begun: _____ Date Read: _____ Result: _____

Comments: _____

Test #4: _____ Date Begun: _____ Date Read: _____ Result: _____

Comments: _____

Test #5: _____ Date Begun: _____ Date Read: _____ Result: _____

Comments: _____

Test #6: _____ Date Begun: _____ Date Read: _____ Result: _____

Comments: _____

Test #7: _____ Date Begun: _____ Date Read: _____ Result: _____

Comments: _____

Test #8: _____ Date Begun: _____ Date Read: _____ Result: _____

Comments: _____

Test #9: _____ Date Begun: _____ Date Read: _____ Result: _____

Comments: _____

Test #10: _____ Date Begun: _____ Date Read: _____ Result: _____

Comments: _____

Test #11: _____ Date Begun: _____ Date Read: _____ Result: _____

Comments: _____

Test #12: _____ Date Begun: _____ Date Read: _____ Result: _____

Comments: _____

Name _____ Date _____

Lab Section _____ I was present and performed this exercise (initials) _____

Gram-Negative Unknown

OBSERVATIONS AND INTERPRETATIONS

Unknown Number _____

Isolation Procedure (Please record all activities associated with isolation of your organisms—from mixed culture to pure culture. Always include the date, source of inoculum, destination, incubation temperature, and any other relevant information. Also, make note of transfers made to keep your pure culture fresh. This log must be kept current.)

_____

_____

_____

_____

_____

_____

_____

_____

_____

_____

_____

_____

_____

_____

_____

_____

_____

_____

Preliminary Observations

Colony Morphology (include medium) _____

_____

Gram Stain _____ Cell Dimensions _____ Optimum Temperature _____

Cellular Morphology and Arrangement _____

_____

[1] Your instructor will write what tests to rerun in this space if you misidentify your unknown.

Differential Tests (Include all information through the confirmatory test. This log must be kept current.)

Test #1: _____   Date Begun: _____   Date Read: _____   Result: _____

Comments: _____

Test #2: _____   Date Begun: _____   Date Read: _____   Result: _____

Comments: _____

Test #3: _____   Date Begun: _____   Date Read: _____   Result: _____

Comments: _____

Test #4: _____   Date Begun: _____   Date Read: _____   Result: _____

Comments: _____

Test #5: _____   Date Begun: _____   Date Read: _____   Result: _____

Comments: _____

Test #6: _____   Date Begun: _____   Date Read: _____   Result: _____

Comments: _____

Test #7: _____   Date Begun: _____   Date Read: _____   Result: _____

Comments: _____

Test #8: _____   Date Begun: _____   Date Read: _____   Result: _____

Comments: _____

Test #9: _____   Date Begun: _____   Date Read: _____   Result: _____

Comments: _____

Test #10: _____   Date Begun: _____   Date Read: _____   Result: _____

Comments: _____

Test #11: _____   Date Begun: _____   Date Read: _____   Result: _____

Comments: _____

Test #12: _____   Date Begun: _____   Date Read: _____   Result: _____

Comments: _____

My Unknown is: _____   Rerun:[2] _____

_____

[2] Your instructor will write what tests to rerun in this space if you misidentify your unknown.

Name _____ Date _____

Lab Section _____ I was present and performed this exercise (initials) _____

Standard Plate Count

OBSERVATIONS AND INTERPRETATIONS

1 Enter the number of colonies on each countable plate. Only one pair of plates should be countable, but for practice, record all countable plates anyway. For all plates containing more than 300 colonies, enter TNTC ("too numerous to count"). For plates containing fewer than 30, enter TFTC ("too few to count"). If no plates are countable, use the TFTC plate closest to 30 colonies for practice with the calculations. Make a note of this in Part 3 below.

2 Take the average number of colonies from the two (or more) countable plates and record it below.

Plate	A_1	A_2	B_1	B_2	C_1	C_2	D_1	D_2
Colonies Counted								
Average # Colonies								

3 Calculate the original density in CFU/mL using the following formula:

$$OCD = \frac{CFU}{\text{Original sample volume}}$$

Original density of *E. coli* in the broth	

QUESTIONS

1 *Suppose your professor handed you a test tube with 2.0 mL of an* E. coli *broth culture in it and told you to make a 10^{-1} dilution of the entire culture. Explain how you would do this. Show your calculations.*

_____

_____

_____

_____

2 *Suppose your professor handed you a test tube with 2.0 mL of an* E. coli *broth culture in it and told you to make a 10^{-2} dilution of the entire culture. Explain how you would do this. Show your calculations.*

_____

_____

_____

3 How would you produce a 10^{-2} dilution of a 5 mL bacterial sample using the full 5 mL volume?

4 You have 0.05 mL of an undiluted culture at a concentration of 3.6×10^6 CFU/mL. You then add 4.95 mL sterile diluent. What is the dilution, and what is the final concentration of cells?

5 What would be the dilution if 96 mL of diluent is added to 4 mL of a bacterial suspension?

6 You were instructed to add 1.0 mL out of 5.0 mL of an undiluted sample to 99 mL of sterile diluent. Instead, you add all 5.0 mL to the 99 mL. What was the intended dilution and what was the actual dilution?

7 Suppose you were instructed to add 0.2 mL of sample to 9.8 mL of diluent, but instead added 2.0 mL of sample. What was the intended dilution, and what was the actual dilution?

8 Plating 1.0 mL of a sample diluted by a factor of 10^{-3} produced 43 colonies. What was the original concentration in the sample?

9 Plating 0.1 mL of a sample diluted by a factor of 10^{-3} produced 43 colonies. What was the original concentration in the sample?

10 The plate has 72 colonies, with a sample volume of 10^{-7} mL. What was the original concentration in the sample?

11 The plate has 259 colonies, with a sample volume of 10^{-6} mL. What was the original concentration in the sample?

Name _____   Date _____

Lab Section _____   I was present and performed this exercise (initials) _____

12 How many colonies should be on the plate inoculated with a sample volume of 10^{-7} mL using the same sample as in Question #11?

_____

13 How many colonies should be on the plate inoculated with a sample volume of 10^{-5} mL using the same sample as in Question #11?

_____

14 A plate inoculated with a sample volume of 10^{-7} mL produced 170 colonies. What was the original concentration in the sample?

_____

15 After incubation, how many colonies should be on the 10^{-8} mL plate from the dilution series in Question #14?

_____

16 After incubation, how many colonies should be on the 10^{-6} mL plate from the dilution series in Question #14?

_____

17 You have inoculated 100 µL of a sample diluted by a factor of 10^{-3} on a nutrient agar plate. After incubation, you count 58 colonies. What was the original cell density?

_____

18 A plate that received 1000 µL of a bacterial sample diluted by a factor of 10^{-6} had 298 colonies on it after incubation. What was the original cell density?

_____

19 A Nutrient Agar plate labeled 10^{-5} mL had 154 colonies after incubation. What was the cell density in the original sample? What volume was used to inoculate this plate?

_____

_____

20 The original concentration in a sample is 2.79×10^6 CFU/mL. Which sample volume should yield a countable plate? (Express your answer as 10^x mL.)

_____

21 The original concentration in a sample is 5.1×10^9 CFU/mL. Which sample volume should yield a countable plate?

_____

22 A sample has a density of 1.37×10^5 CFU/mL. What sample volume should yield a countable plate? Which two dilution tubes could be used to produce this sample volume? How?

23 A sample has a density of 7.9×10^9 CFU/mL. What sample volume should yield a countable plate? Which two dilution tubes could be used to produce this sample volume? How?

24 You are told that a sample has between 2.5×10^6 and 2.5×10^9 cells/mL. Devise a complete but efficient (that is, no extra plates!) dilution scheme that will ensure getting a countable plate.

25 A sample has between 3.3×10^4 and 3.3×10^8 CFU/mL. Devise a complete but efficient (that is, no extra plates!) dilution scheme that will ensure getting a countable plate.

26 Two plates received 100 µL from the same dilution tube. The first plate had 293 colonies, whereas the second had 158 colonies. Suggest reasonable sources of error.

27 Two parallel dilution series were made from the same original sample. The plates with sample volumes of 10^{-5} mL from each dilution series yielded 144 and 93 colonies. Suggest reasonable sources of error.

28 What are the only *circumstances* that would correctly *produce countable plates from two different dilutions?*

Name _____ Date _____

Lab Section _____ I was present and performed this exercise (initials) _____

Urine Culture

OBSERVATIONS AND INTERPRETATIONS

Enter your colony count and loop volume data in the chart below. Then calculate the original cell density using the following formula.

$$OCD = \frac{CFU}{loop\ volume}$$

Use the extra rows to calculate cell densities of different colony types on your plate or for urine samples of other students. Label these appropriately.

Urine Sample	Colonies Counted	Loop volume (0.01 mL or 0.001 mL)	Original Cell Density (CFU/mL)

QUESTIONS

1 *The plate pictured in Figure 6-5 was inoculated with a 0.01 mL volumetric loop and contains approximately 75 colonies. What was the original cell density?*

_____

_____

_____

2 *The equation shown in the Theory explanation is used for calculating cell density in urine when using a 0.001 mL calibrated loop. The urine transferred in the loop is not literally diluted, yet its volume is equivalent to a dilution factor. What is the dilution factor (based on loop volume) expressed as a fraction? What is the dilution factor in scientific notation?*

_____

_____

_____

3   Calculation of original density in this exercise differs slightly from that offered in Exercise 6-1. Compare and contrast the formula used today with that used in Exercise 6-1. How do they differ? Could you have used the formula in Exercise 6-1 for today's calculations? Explain.

4   Using a volumetric loop is a semiquantitative technique. Why is it not quantitative? Design a procedure that would make it quantitative. (**Hint:** Refer to Exercise 6-1 if necessary.)

Name _____ Date _____

Lab Section _____ I was present and performed this exercise (initials) _____

Direct Count

DATA AND CALCULATIONS

1 Calculate your dilution using the following formula. For an explanation of dilution, refer to Exercise 6-1.

$$D_2 = \frac{V_1 D_1}{V_2}$$

2 Enter your data below.

Total Cells Counted	Squares Counted	Dilution	Original Cell Density (calculated)

3 Calculate the original cell density of the *P. vulgaris* culture using the following equation. Record your answer in mL in the table.

$$\text{Original cell density} = \frac{\text{Total cells counted}}{(\text{Squares counted})(\text{Dilution})(5 \times 10^{-8} \text{ mL})}$$

QUESTIONS

1 *What are the advantages and disadvantages of the direct count over the plate count technique?*

_____

_____

_____

_____

_____

_____

_____

_____

2   *What would be the original cell density of a sample having a 10^{-1} dilution and a count of 75 cells in five small squares? Record your answer in cells/mL.*

3   *Suppose you had to count a culture that was very turbid and found that the recommended dilution of 0.4 mL stain A and 0.5 mL stain B per 0.1 mL culture (10^{-1} dilution) was not enough. How would you adjust the volumes to make a 10^{-2} dilution? Show your work.*

4   *Suppose you were given a diluted broth to count that already had a 10^{-3} dilution. Assuming that you used the staining procedure described in the Procedure, what would the dilution of the solution be when it reached the counting chamber? What would the original cell density be if you counted a total of 116 cells in 16 small squares? Record your answer in cells/mL.*

Name _____ Date _____

Lab Section _____ I was present and performed this exercise (initials) _____

Closed System Growth

DATA AND CALCULATIONS

1 Enter the absorbance values for all groups in Chart A below.

2 On a computer or graph paper, plot your absorbance values versus time. This will produce a growth curve including growth phases as far as the stationary phase. (**Note:** Some cultures may not even reach the stationary phase. Record what you get!). Also plot the growth curves of the other four samples on the same set of axes.

3 From your graph, determine the approximate length of time spent in the various growth stages for all samples, and enter those values in Chart B.

4 From your graph, determine the absorbance for the lag phase and the stationary phase of each culture, and enter those in Chart B.

5 Calculate the mean growth rate constant for each sample. Do this by choosing two points clearly on the linear part of exponential growth. These are A_1 and A_2. Determine the absorbance values of each, and the time in minutes (t) between the two points. Substitute your values in the equation and solve for (k). Enter your results in Chart C.

$$k = \frac{\log A_2 - \log A_1}{0.301t}$$

6 On a computer or the graph paper provided, plot the mean growth rate versus temperature of the different samples.

7 Calculate generation times of the different samples. Enter your results in Chart C.

Chart A — Absorbance Readings																
Temp (°C)	T_0	T_{15}	T_{30}	T_{45}	T_{60}	T_{75}	T_{90}	T_{105}	T_{120}	T_{135}	T_{150}	T_{165}	T_{180}	T_{195}	T_{210}	T_{225}	T_{240}
22																	
25																	
31																	
37																	
40																	

Chart B — Growth Phase Duration					
Temp (°C)	Lag Phase (Minutes)	Exponential Phase (Minutes)	Stationary Phase (Minutes)	Lag Phase (Absorbance)	Stationary Phase (Absorbance)
22					
25					
31					
37					
40					

Chart C — Calculations (Include Appropriate Units)						
Temp (°C)	A_2 / log A_2		A_1 / log A_1		t	k	g
22						
25						
31						
37						
40						

QUESTIONS

1 Examine the microbial growth curve in Figure 6-9. In what ways would you expect it to be different from a growth curve of an organism living in a natural environment?

_____

_____

_____

_____

_____

Name _____ Date _____

Lab Section _____ I was present and performed this exercise (initials) _____

2 *Aside from temperature, what factors in today's experiment likely had an effect on the growth? What factors likely kept the organism from realizing true exponential growth?*

_____

_____

_____

_____

3 *If an organism has a mean generation time of 22 minutes, what is its mean growth rate?*

_____

_____

4 *If an organism has a mean generation time of 47 minutes, what is its mean growth rate?*

_____

_____

5 *If an organism has a mean growth rate constant of 1.7 generations per hour, what is its generation time?*

_____

_____

6 *If an organism has a mean growth rate constant of 0.012 generations per minute, what is its generation time?*

_____

_____

7 *How would you calculate mean growth rate of an organism that quadrupled every generation?*

_____

_____

_____

_____

8 *Compare your growth rate vs. temperature graph with those shown in Figure 2-40. How would* Vibrio nariegens *be classified?*

_____

Name _____ Date _____

Lab Section _____ I was present and performed this exercise (initials) _____

Plaque Assay

OBSERVATIONS AND INTERPRETATIONS

1 Enter the number of plaques counted on the countable plate. Only one plate should be countable. If there are more than 300 plaques, enter TNTC. If fewer than 30 plaques, enter TFTC.

Plate	A	B	C	D	E	F	G
Plaques Counted							
Sample volume							

2 Calculate the original density using the following formula, and enter the result below. Record your answer in PFU/mL.

$$\text{Phage titer} = \frac{\text{PFU}}{\text{Original sample volume}}$$

If you don't have any countable plates, use the TFTC plate closest to 30 in your calculation for practice. Make a note of this in your results.

Phage titer	

QUESTIONS

1 *In this exercise, there must be enough bacteria inoculated to produce a lawn of growth. Why is that important?*

_____

_____

_____

_____

_____

_____

_____

_____

_____

_____

2 How might the results be altered if you had skipped the preadsorption phase?

3 Suppose you followed all the necessary steps outlined in the Procedure and found no plaques on any of your plates after incubation. Suppose further that you knew with certainty that the bacteriophage was viable and had worked in other labs prior to yours. What possible explanations could there be for this occurrence?

4 Why was the water bath set at 50°C? What might be some consequences of changing the temperature?

5 Why was Soft Agar used for the agar overlay? What would you expect to see if standard Nutrient Agar had been used instead?

Name _____ Date _____

Lab Section _____ I was present and performed this exercise (initials) _____

Thermal Death Time Versus Decimal Reduction Value

DATA AND CALCULATIONS

It works best to reproduce the Data Sheet charts on the blackboard or a transparency so that the entire class can enter and share data.

Chart 1 — Group Plate Data							
Organism:			$T_?$:				
Plate Number	1	2	3	4	5	6	7
Plate Number							
Volume of Original Sample Plated (vol. plated × dilution)							
Colonies Counted (CFU)							
Cell Density CFU/mL							

Fill in all boxes. For plates containing fewer than 30 or more than 300 colonies, enter TFTC or TNTC respectively.

Chart 2 — Group Broth Data					
Organism:					
Time	G/NG	Time	G/NG	Time	G/NG
T__		T__		T__	
T__		T__		T__	
T__		T__		T__	
T__		T__		T__	
T__		T__		T__	

Circle the first time at which no growth appears in the broth.

Chart 3 — Class Data					
	T_0		T_{10}		
Organism	Cell Density (Cells/ML)	Log G_{10} of Cell Density	Cell Density (Cells/ML)	Log G_{10} of Cell Density	First Broth Without Growth (T_x)
E. coli					
S. aureus					

Chart 4 — Thermal Death Time Versus Decimal Reduction Value			
Organism	Thermal Death Time (min.)	Plotted D_{60} Value (min.)	Calculated D_{60} Value (min.)
E. coli			
S. aureus			

Perform the following for *each* organism.

Constructing the TDT curve

1 On the graph paper provided, construct a graph of Population size (log) versus Time at 60°C, as shown in Figure 6-14. To do this:

    a. Enter a data point on the *y*-axis representing the log of the original cell density of the broth culture at T_0 (from Chart 3).

    b. Enter a second data point on the *x*-axis at the time when growth in the Nutrient Broth tube stopped—thermal death time (from Chart 4).

    c. Draw a straight line between the two points. This is the TDT curve for this culture at 60°C.

2 Use this curve for plotting the D value.

Plotting D value from TDT curve

3 Locate two successive log values on the *y*-axis below the highest population point for the organism, representing one complete log cycle (*e.g.*, 10^6 and 10^5, 10^5 and 10^4, *etc.*).

4 Draw a horizontal line from each of these points on the *y*-axis to where they intersect the TDT curve.

5 Draw a vertical line from these intersection points downward to where they intersect with the *x*-axis.

6 The horizontal distance between these two points reveals the time required to kill 90% of the population. This is the D value of the organism. Enter it in Chart 4 in the box labeled "Plotted D_{60} value."

Name _____ Date _____

Lab Section _____ I was present and performed this exercise (initials) _____

Calculating D value

1 From Chart 3, obtain the logs of the microbial cell densities before and after heating.

2 Using the formula below, calculate the D_{60} value of each organism.

$$D_T = \frac{t}{\log_{10} x - \log_{10} y}$$

3 Enter your results in Chart 4 in the box labeled "Calculated D_{60} Value."

QUESTIONS

1 *Is the D value of either organism affected by the size of the population? Explain.*

_____

_____

_____

2 *How much did your plotted and calculated values differ? How do you explain any differences?*

_____

_____

_____

3 *Using the information from this exercise, how long would it take to kill a population of* E. coli *with a density of 10^3 cells/mL? How long to kill a population of* S. aureus *with the same density?*

_____

_____

4 *Because this is a quantitative procedure, why is it not necessary to know the volume of broth in the 30 Nutrient Broth tubes?*

_____

_____

_____

Name _____ Date _____

Lab Section _____ I was present and performed this exercise (initials) _____

DATA SHEET 6-6

Name _____ Date _____

Lab Section _____ I was present and performed this exercise (initials) _____

Snyder Test

OBSERVATIONS AND INTERPRETATIONS

Refer to Table 7-1 when recording and interpreting your results below.

	TIME		
	24 Hrs.	**48 Hrs.**	**72 Hrs.**
Color			

Based on the results of this test, I have a _____ susceptibility to dental caries.

QUESTIONS

1 *As demonstrated by this exercise,* Lactobacillus *species ferment glucose (dextrose). Considering the Snyder Test ingredients, suggest other metabolic functions that they might perform.*

_____

_____

_____

_____

_____

_____

_____

_____

_____

_____

_____

_____

_____

_____

2   *Considering, again, the Snyder Test, what kinds of dietary items would likely increase the number of lactobacilli in saliva?*

3   *Why isn't the molten Snyder Test Agar allowed to solidify as a slant?*

Name _____ Date _____

Lab Section _____ I was present and performed this exercise (initials) _____

Lysozyme Assay

OBSERVATIONS AND INTERPRETATIONS

1 Enter the standard curve data from Group 1.

Lysozyme Concentration	Actual Time of Transfer (t_0)	Light Absorbance at t_{20}
0.15625 mg/100 mL		
0.3125 mg/100 mL		
0.625 mg/100 mL		
0.125 mg/100 mL		
0.25 mg/100 mL		
0.5 mg/100 mL		

2 Enter class data for all samples.

Body Fluid	Actual Time of Transfer (t_0)		Light Absorbance at t_{20}	
	10^{-1}	10^{-2}	10^{-1}	10^{-2}

3 Using the information in the tables and the graph paper provided, construct a standard curve of lysozyme concentration (x-axis) versus absorbance (y-axis). Don't forget to label the axes and title the graph.

4 Using the standard curve, estimate the concentration of lysozyme in each diluted sample.

5 Calculate the original concentration (before dilution) of lysozyme in all samples, using the following formula (where OC is original concentration, LC is the lysozyme concentration obtained from the standard curve, and D is the dilution):

$$OC = \frac{LC}{D}$$

QUESTIONS

1 Lysozyme alone is not enough to kill bacterial cells. What other condition must be met in order for cell death to occur?

2 Why do you suppose lysozyme is less effective against Gram-negative bacteria than Gram-positives?

3 Assuming the conditions are present for lysozyme to be effective, would it work on actively-growing cells? Non-growing cells?

Name _____ Date _____

Lab Section _____ I was present and performed this exercise (initials) _____

DATA SHEET 7-2

Name _____ Date _____

Lab Section _____ I was present and performed this exercise (initials) _____

Antimicrobial Susceptibility Test

OBSERVATIONS AND INTERPRETATIONS

Record the zone diameters in mm in Chart A. Using Table 7-4 as a guide, enter "S" if the organism is susceptible and "R" if it is resistant to the antibiotic. (Not all combinations of antibiotics and organisms are available.)

Chart A
Interpretation of Zone Diameters

Organism	Chloramphenicol Zone Diameter	S/R	Ciprofloxacin Zone Diameter	S/R	Trimethoprim Zone Diameter	S/R	Penicillin Zone Diameter	S/R
E. coli								
S. aureus								

If you did the optional exercise, record "G" for growth and "NG" for no growth on the TSA/NA plates in Chart B. Then, decide and record if the antibiotic is bactericidal or bacteriostatic.

Chart B
Determination of Antibiotic Activity

Antibiotic	Chloramphenicol NG/G	–static or –cidal	Ciprofloxacin NG/G	–static or –cidal	Trimethoprim NG/G	–static or –cidal	Penicillin NG/G	–static or –cidal
E. coli								
S. aureus								

QUESTIONS

1 *All aspects of the Kirby-Bauer test are standardized to assure reliability.*

a. What might be the consequence of pouring the plates 2 mm deep instead of 4 mm deep?

_____

_____

b. The Mueller-Hinton II plates are supposed to be used within a specific time after their preparation and should be free of visible moisture. Why do you think this is so?

_____

_____

2 In clinical applications of the Kirby-Bauer test, diluted cultures (for the McFarland standard comparison) must be used within 30 minutes. Why is this important?

3 E. coli *and* S. aureus *were chosen to represent Gram-negative and Gram-positive bacteria, respectively. For a given antibiotic, is there a difference in susceptibility between the Gram-positive and Gram-negative bacteria? If so, what difference(s) do you see?*

4 Suppose you do this test on a hypothetical Staphylococcus *species with the antibiotics penicillin (P 10) and chloramphenicol (C 30). You record zone diameters of 25 mm for the chloramphenicol and penicillin disks. Which antibiotic would be more effective against this organism? What does this tell you about comparing zone diameters to each other and the importance of the zone diameter interpretive chart?*

5 How does the antibiotic get from the disk into the agar?

6 Does the agar have an antibiotic beyond the zone of inhibition? How does your answer tie in with MIC?

7 In the optional exercise, what was the purpose of inoculating TSA (or NA) plates with samples taken from the zones of inhibition?

Name _____ Date _____

Lab Section _____ I was present and performed this exercise (initials) _____

Clinical Biofilms

OBSERVATIONS AND INTERPRETATIONS

Tube	Biofilm (Yes/No)
Control	
Staphylococcus mixture	

QUESTIONS

1 *What is the role of the control tube?*

_____

_____

_____

_____

_____

_____

2 *Why were* Staphylococcus aureus *and* S. epidermidis *chosen as experimental organisms?*

_____

_____

_____

_____

_____

_____

_____

3 *Define "nosocomial infection."*

4 *What two factors make a biofilm in an intravenous line especially dangerous to the patient?*

Name _____ Date _____

Lab Section _____ I was present and performed this exercise (initials) _____

Morbidity and Mortality Weekly Report (MMWR) Assignment

DATA AND CALCULATIONS

1 Record the cumulative totals of your selected disease for the past two complete years in the Data Table on the following pages.

2 Calculate the number of *new cases* during each 4-week period (weeks 1–4, weeks 5–8, *etc.*) by subtracting the total at the end of a 4-week period from the total at the end of the next 4-week period. That is, for the second four weeks, subtract the total at the end of week 4 from the total at the end of week 8. Record these in the chart on pages 666–667 in the column labeled "4-Week Totals." Show a sample calculation (indicating the weeks involved) in the space below.

3 Calculate national incidence values of each 4-week period for both years, using the calculated 4-week totals (weeks 1–4, weeks 5–8, *etc.*) in the numerator. Use the U. S. population size on July 1 of each respective year in the denominator. This value can be obtained from the U. S. Census Bureau Web site at http://www. census.gov/popest/national/national.html and click on the "Monthly Population Estimates" line. That will lead you to a Microsoft Excel file for viewing. Be sure to choose an appropriate value for K, and don't abuse the significant figures! Show a sample calculation in the space below. Indicate the 4-week period used in your calculation.

Estimated population size for the year 20____:

Estimated population size for the year 20____:

$$\text{Incidence Rate} = \frac{\text{number of new cases in a time period}}{\text{size of at-risk population at midpoint of time period}} \times K$$

Morbidity Data for _____ during the years 20____ and 20____

Week	Cumulative Totals by Week Year (20___)	4-Week Totals Year (20___)	Incidence Values Year (20___)	Cumulative Totals by Week Year (20___)	4-Week Totals Year (20___)	Incidence Values Year (20___)
1						
2						
3						
4						
5						
6						
7						
8						
9						
10						
11						
12						
13						
14						
15						
16						
17						
18						
19						
20						
21						
22						
23						
24						

Name _____ Date _____

Lab Section _____ I was present and performed this exercise (initials) _____

Week	Cumulative Totals by Week Year (20___)	4-Week Totals Year (20___)	Incidence Values Year (20___)	Cumulative Totals by Week Year (20___)	4-Week Totals Year (20___)	Incidence Values Year (20___)
25						
26						
27						
28						
29						
30						
31						
32						
33						
34						
35						
36						
37						
38						
39						
40						
41						
42						
43						
44						
45						
46						
47						
48						
49						
50						
51						
52						

QUESTIONS

1 *Characterize the disease you have chosen. Be sure to include causative agent (by scientific name), symptoms, mode of transmission, and treatment. Cite your references.*

Name _____ Date _____

Lab Section _____ I was present and performed this exercise (initials) _____

2 Prepare a graph that illustrates the cumulative data for each week over the two years. Be sure to include a title, axis labels with appropriate units, and legends in your graph.

3 Graph the calculated national incidence values over the two years studied. In your graph, be sure to include a title, axis labels with appropriate units, and legends.

4   *Briefly compare national trends of your chosen disease for the two years. Are the number of cases similar? Does the incidence appear to be seasonal?*

Name _____ Date _____

Lab Section _____ I was present and performed this exercise (initials) _____

Epidemic Simulation

DATA AND CALCULATIONS

1 Circle the case number that matches your candy dish number.

2 Compile data for your class population (each case is the student's number in the exercise). The organism spread during contact between student pairs was *Serratia marcescens*, which produces reddish-orange colonies. You will likely see growth on both sides of the plate, but red colonies are the only sign of disease. If there is reddish-orange growth on side B of the plate, consider the person to have the disease and record a "+" in the appropriate box. If there is no red growth on the plate or there is no difference between the two sides, consider the person to be healthy and record a "−" in the appropriate box.

Case	Week	Disease (+ / −)
1	1	
2	1	
3	1	
4	1	
5	2	
6	2	
7	2	
8	2	
9	2	
10	3	
11	3	
12	3	
13	3	
14	3	
15	3	
16	4	
17	4	
18	4	
19	4	
20	4	

Case	Week	Disease (+ / −)
21	4	
22	4	
23	5	
24	5	
25	5	
26	5	
27	5	
28	6	
29	6	
30	6	
31	7	
32	7	
33	7	
34	7	
35	8	
36	8	
37	8	
38	9	
39	9	
40	10	

3 Identify the case number of the index patient.

DATA SHEET 7-6

Use the data obtained to calculate incidence and prevalence rates in your class population. For this purpose, assume that the cases have been identified over a 10-week period (or a shorter period, depending on how many students you have). Also assume that the duration of the disease is 7 days, so new cases in one week are still diseased in the next week but are healthy by the following week.

Record the population size: _____

Record incidence and point prevalence rates for each week in the chart at the right.

Week	Incidence	Prevalence
1		
2		
3		
4		
5		
6		
7		
8		
9		
10		

QUESTIONS

1 *Explain how you determined the index case.*

_____

_____

_____

2 *What was the purpose of side 1 on each plate?*

_____

_____

3 *Suggest possible reasons (based on the execution of the simulation) for cases showing no S. marcescens growth between cases that exhibited growth.*

_____

_____

_____

4 *Suggest how the gaps could represent "subclinical infections" or "carriers" in this simulation.*

_____

_____

_____

5 *Within the context of this simulation, how could you characterize any "non-Serratia marcescens" growth on Side 1 or 2 of your plate (**Hint:** "Contamination" is not an adequate answer!)*

_____

Name _____ Date _____

Lab Section _____ I was present and performed this exercise (initials) _____

Identification of *Enterobacteriaciae*

ISOLATION PROCEDURE

Unknown Number _____

Record all activities associated with isolation of your organisms—from mixed culture to pure culture. Always include the date, source of inoculum, destination, type of inoculation, incubation temperature, and any other relevant information. Also make note of transfers made to keep your pure culture fresh. (This log must be kept current.)

_____

_____

_____

_____

_____

_____

_____

_____

_____

_____

_____

_____

_____

_____

_____

_____

_____

_____

_____

Preliminary Observations

Colony morphology (include medium and incubation temperature) _____

_____

_____

Gram stain _____ Cell dimensions _____ Optimum temperature _____

Cellular morphology and arrangement _____

Differential Tests

Begin recording with your MacConkey result, followed by the Oxidase test, and then your IMViC results. This log must be kept current.

Record the Identification Chart you are using. If you change charts, indicate the figure number of the new chart and the date of the change. _____

Test #1: _____ Date Begun: _____ Date Read: _____ Result: _____

Comments: _____

Test #2: _____ Date Begun: _____ Date Read: _____ Result: _____

Comments: _____

Test #3: _____ Date Begun: _____ Date Read: _____ Result: _____

Comments: _____

Test #4: _____ Date Begun: _____ Date Read: _____ Result: _____

Comments: _____

Test #5: _____ Date Begun: _____ Date Read: _____ Result: _____

Comments: _____

Test #6: _____ Date Begun: _____ Date Read: _____ Result: _____

Comments: _____

Test #7: _____ Date Begun: _____ Date Read: _____ Result: _____

Comments: _____

Test #8: _____ Date Begun: _____ Date Read: _____ Result: _____

Comments: _____

Test #9: _____ Date Begun: _____ Date Read: _____ Result: _____

Comments: _____

Test #10: _____ Date Begun: _____ Date Read: _____ Result: _____

Comments: _____

Test #11: _____ Date Begun: _____ Date Read: _____ Result: _____

Comments: _____

Test #12: _____ Date Begun: _____ Date Read: _____ Result: _____

Comments: _____

My Unknown is: _____ Rerun:[1] _____

_____ Rerun: _____

_____ Rerun: _____

[1] Your instructor will write what tests to rerun in this space if you misidentify your unknown.

Name _____ Date _____

Lab Section _____ I was present and performed this exercise (initials) _____

Identification of Gram-positive Cocci

ISOLATION PROCEDURE

Unknown Number _____

Record all activities associated with isolation of your organisms—from mixed culture to pure culture. Always include the date, source of inoculum, destination, type of inoculation, incubation temperature, and any other relevant information. Also make note of transfers made to keep your pure culture fresh. (This log must be kept current.)

_____

_____

_____

_____

_____

_____

_____

_____

_____

_____

_____

_____

_____

_____

_____

_____

_____

_____

Preliminary Observations

Colony morphology (include medium and incubation temperature) _____

_____

_____

Gram stain _____ Cell dimensions ✦ _____ CO_2 Requirement _____

Cell morphology and arrangement _____

Differential Tests

Begin recording with your Catalase test (and hemolysis result, if determined). This log must be kept current.

Record the Identification Chart you are using. If you change charts,
indicate the figure number of the new chart and the date of the change. _____

Test #1: _____ Date Begun: _____ Date Read: _____ Result: _____
Comments: _____

Test #2: _____ Date Begun: _____ Date Read: _____ Result: _____
Comments: _____

Test #3: _____ Date Begun: _____ Date Read: _____ Result: _____
Comments: _____

Test #4: _____ Date Begun: _____ Date Read: _____ Result: _____
Comments: _____

Test #5: _____ Date Begun: _____ Date Read: _____ Result: _____
Comments: _____

Test #6: _____ Date Begun: _____ Date Read: _____ Result: _____
Comments: _____

Test #7: _____ Date Begun: _____ Date Read: _____ Result: _____
Comments: _____

Test #8: _____ Date Begun: _____ Date Read: _____ Result: _____
Comments: _____

Test #9: _____ Date Begun: _____ Date Read: _____ Result: _____
Comments: _____

Test #10: _____ Date Begun: _____ Date Read: _____ Result: _____
Comments: _____

Test #11: _____ Date Begun: _____ Date Read: _____ Result: _____
Comments: _____

Test #12: _____ Date Begun: _____ Date Read: _____ Result: _____
Comments: _____

My Unknown is: _____ Rerun:[1] _____

_____ Rerun: _____

_____ Rerun: _____

[1] Your instructor will write what tests to rerun if you misidentify your unknown.

Name _____ Date _____

Lab Section _____ I was present and performed this exercise (initials) _____

Identification of Gram-positive Rods

ISOLATION PROCEDURE

Unknown Number _____

Record all activities associated with isolation of your organisms—from mixed culture to pure culture. Always include the date, source of inoculum, destination, type of inoculation, incubation temperature, and any other relevant information. Also make note of transfers made to keep your pure culture fresh. (This log must be kept current.)

Preliminary Observations

Colony morphology (include medium, incubation temperature, and differences between aerobic and anaerobic growth

Gram stain _____ Cellular morphology, arrangement and dimensions _____

Differential Tests

Begin recording with your spore stain result. This log must be kept current.

Record the Identification Chart you are using. If you change charts,
indicate the figure number of the new chart and the date of the change _____

Test #1: _____ Date Begun: _____ Date Read: _____ Result: _____

Comments: _____

Test #2: _____ Date Begun: _____ Date Read: _____ Result: _____

Comments: _____

Test #3: _____ Date Begun: _____ Date Read: _____ Result: _____

Comments: _____

Test #4: _____ Date Begun: _____ Date Read: _____ Result: _____

Comments: _____

Test #5: _____ Date Begun: _____ Date Read: _____ Result: _____

Comments: _____

Test #6: _____ Date Begun: _____ Date Read: _____ Result: _____

Comments: _____

Test #7: _____ Date Begun: _____ Date Read: _____ Result: _____

Comments: _____

Test #8: _____ Date Begun: _____ Date Read: _____ Result: _____

Comments: _____

Test #9: _____ Date Begun: _____ Date Read: _____ Result: _____

Comments: _____

Test #10: _____ Date Begun: _____ Date Read: _____ Result: _____

Comments: _____

Test #11: _____ Date Begun: _____ Date Read: _____ Result: _____

Comments: _____

Test #12: _____ Date Begun: _____ Date Read: _____ Result: _____

Comments: _____

My Unknown is: _____ Rerun:[1] _____

Rerun: _____

Rerun: _____

---

[1] Your instructor will write what tests to rerun if you misidentify your unknown.

DATA SHEET 8-1

Name _____ Date _____

Lab Section _____ I was present and performed this exercise (initials) _____

Winogradsky Column

OBSERVATIONS AND INTERPRETATIONS

Soil Sample Site: _____

Record your observations in the chart below.

Week	Observations (Include distinctive color development and its level in the column)
1	
2	
3	
4	
5	
6	
7	
8	

QUESTIONS

1 *What is/are the carbon source(s) in the Winogradsky column?*

_____

_____

_____

_____

_____

_____

2   *Based on the metabolic reaction, fill in the mode of metabolism column. Use the definitions under "Theory" as a guide, and the aerobic heterotroph as an example.* **Note:** *hv represents light and CH_2O represents organic carbon. Most reactions are simplified to show the "essence" of the process without excessive detail.*

Organism	Aerotolerance	Mode of Metabolism (*e.g.*, phototroph, autotroph, chemotroph, *etc.*)	Metabolic Reaction
Aerobic Heterotrophs	Obligate Aerobe	Chemoheterotrophic	$C_6H_{12}O_6 + 6O_2 \longrightarrow 6CO_2 + 6H_2O$
Cyanobacteria	Aerobe		$6CO_2 + 6H_2O \xrightarrow{hv} C_6H_{12}O_6 + 6O_2$
H$_2$S Oxidizers (*Beggiatoa*)	Microaerophile		$CO_2 + H_2S \longrightarrow CH_2O + S^0$
Purple nonsulfur bacteria	Microaerophilic Anaerobe		$CH_2O \longrightarrow CO_2 + H_2$, and $CO_2 + H_2 \xrightarrow{hv} CH_2O$
Purple sulfur bacteria	Anaerobe		$CO_2 + H_2S \xrightarrow{hv} CH_2O + S^0$
Green sulfur bacteria	Anaerobe		$CO_2 + H_2S \xrightarrow{hv} CH_2O + S^0$
Sulfate reducers	Anaerobe		$CH_2O + SO_4^{2-} \longrightarrow CO_2 + H_2S$

3   *Why is $CaSO_4$ mixed into the lowermost, enriched slurry when making the column?*

4   *Why does H_2S accumulate in the most anaerobic region of the column?*

5   *In what way does photosynthesis differ between cyanobacteria and the green and purple sulfur bacteria?*

6   *Why can chemoheterotrophs grow throughout the column?*

Name _____ Date _____

Lab Section _____ I was present and performed this exercise (initials) _____

Nitrogen Fixation

OBSERVATIONS AND INTERPRETATIONS

Soil Sample Site: _____

Record your results in the chart below. Also record results from three classmates who worked with different soil.

Tube (Include source of soil samples)	Matching Color on Card	Ammonia Concentration mg/L
Control		
Soil Sample:		
Soil Sample:		
Soil Sample:		
Soil Sample:		

If you did staining, record your observations in the chart below. Include sketches and measurements of cell dimensions.

Organism	Gram Stain	Capsule Stain	Cyst Stain
Broth Mixed Culture			
Pellicle (*Azotobacter*)			

QUESTIONS

1 *Without even looking at the organisms microscopically, how can we be certain that the organisms growing in Nitrogen Free Broth are N-fixers?*

2 *Why are N-fixers so important in food chains?*

3 *Besides plants, what other organisms can directly utilize fixed nitrogen? Using one of each pair of terms chemo/heterotroph and litho/organotroph, to which group do these other organisms belong?*

Name _____ Date _____

Lab Section _____ I was present and performed this exercise (initials) _____

Nitrification: The Production of Nitrate

OBSERVATIONS AND INTERPRETATIONS

Soil Sample Site: _____

Record your results in the chart below. Also record results from three classmates who worked with different soil.

Tube (Include soil sample) sources	Matching Color on Card	Nitrite Concentration mg/L	Nitrate Concentration mg/L	Ammonia Concentration mg/L
AB–Control 1				
AB–Control 2				
Soil Sample:				
Soil Sample:				
Soil Sample:				
Soil Sample:				
AB–Control 3				

QUESTIONS

1 *What is the purpose of AB-Control 1?*

_____

_____

_____

_____

_____

_____

_____

2   *What is the purpose of AB-Control 2?*

3   *What is the purpose of AB-Control 3?*

4   *Why might the results you obtain for your soil show low amounts of nitrite and high amounts of nitrate?*

5   *Why might the results you obtain for your soil show high amounts of nitrite and low amounts of nitrate?*

Name ______________________________ Date ______________

Lab Section ______________________ I was present and performed this exercise (initials) ________

DATA SHEET 8-4

Ammonification

OBSERVATIONS AND INTERPRETATIONS

Soil Sample Site: ______________________________

Record your results in the chart below. Also record results from three classmates who worked with different soil.

Tube (Include source of soil samples)	Matching Color on Card	Ammonia Concentration mg/L
Control		
Soil Sample:		
Soil Sample:		
Soil Sample:		
Soil Sample:		

QUESTIONS

1 *What is the purpose of the control?*

2 If you did the other nitrogen cycle labs, why do you suppose you obtained such a strong reaction for the ammonification test, compared to the others?

3 If you did the other nitrogen cycle labs, what other process besides ammonification could be contributing to the results of this test?

Name _____ Date _____

Lab Section _____ I was present and performed this exercise (initials) _____

Denitrification: Nitrate Reduction

OBSERVATIONS AND INTERPRETATIONS

Soil Sample Site: _____

Record your observations in the chart below. Also record results from three classmates who worked with different soil.

Tube (Include source of soil samples)	Bubble in Durham Tube ("+" or "−")	Nitrate Concentration mg/L
Control		
Soil Sample:		
Soil Sample:		
Soil Sample:		
Soil Sample:		

QUESTIONS

1 *What is the purpose of checking the "NB-Control" tube for nitrate?*

_____

_____

_____

_____

_____

_____

_____

_____

2 How would the presence of a bubble in the Durham tube of the "NB-Control" affect your interpretation of results?

3 This test indicates nitrate reduction to N_2 gas by the accumulation of gas inside the Durham tube. This being the case, why do we perform a nitrate test?

4 Suppose two nitrate broths are inoculated with 2 g of different soil samples. After incubation and testing one tube is yellow-orange and the other is brick red. In which tube would you expect the largest bubble and why?

Name ________________________________ Date ________________

Lab Section ________________________ I was present and performed this exercise (initials) ________

DATA SHEET 8-6

Photosynthetic Sulfur Bacteria

OBSERVATIONS AND INTERPRETATIONS

Sketch the cells you observed. Include the Gram reaction if applicable.

Sample #	Gram Stain Result	Sketch of Organisms Observed

QUESTIONS

1 *Why are the organisms looked for in this exercise found in the lower regions of the column?*

2 In certain anoxic regions there are sulfur oxidizers and sulfur reducers living syntrophically. That is, each species gives the other species something that it needs. Which of the two phototrophs would you expect to see more of in such a habitat?

3 How is it possible for both green sulfur bacteria and purple sulfur bacteria to use ATP in the process of producing ATP without exhausting their energy stores?

4 Suppose that you could genetically engineer a strain of purple sulfur bacteria such that quinone (in the photosystem) was replaced by an electron carrier with a higher reduction potential than the redox pair NADP+/NADPH. Upon cultivating them under identical conditions, what differences would you expect to see between the "wild" strain and the genetically engineered strain?

Name _____ Date _____

Lab Section _____ I was present and performed this exercise (initials) _____

Chemolithotrophic Sulfur-Oxidizing Bacteria

OBSERVATIONS AND INTERPRETATIONS

Sketch the cells you observed. Include the Gram reaction if applicable.

Modified *Thiobacillus* Broth			
Tube #	Gram Stain Result	pH	Sketch of Organisms Observed
1			
2			
3			
4			

Microgradient Tubes—Sketch Your Results		
1	2	3

QUESTIONS

1 *Why is the* Thiobacillus *medium shaken during incubation?*

2 *Many sulfur oxidizers can use organic carbon. In fact enrichment media for several of these organisms include lactate or acetate. Why do you think there was no organic carbon added to the media used in this exercise?*

3 Beggiatoa *was discovered years before it could be cultivated in the laboratory. You likely cultivated it in this exercise. Why do you think it took so long to successfully cultivate it?*

4 *Why are gradient tubes heated prior to inoculation?*

Name _____ Date _____

Lab Section _____ I was present and performed this exercise (initials) _____

Sulfur-Reducing Bacteria

OBSERVATIONS AND INTERPRETATIONS

Enter your observations and data in the table. Include the Gram reaction if applicable. If you are calculating the original concentration of SRB in the sample, use the following formula,

$$OC = \frac{colonies\ counted}{dilution}$$

where OC = original concentration of SRB in cells/mL, colonies counted = the number of distinct colonies, and dilution = the dilution of the tube that was the inoculum source for the countable medium tube. (Refer to Exercise 6-1 for more information on dilutions and calculations.)

Tube #	Colonies Counted	Gram Stain Result	Calculated SRB Concentration
1			
2			
3			
4			
5			
6			

Enter the number of colonies if they are distinguishable from other colonies. If the tube is uncountable, enter "UC."

QUESTIONS

1 *What is the purpose of tube #7?*

_____

_____

_____

_____

_____

_____

_____

2 Assuming during incubation all tubes develop distinct black colonies, how could you verify whether or not your dilution technique was correct?

3 Which organism would you expect to have a higher metabolic rate (faster growth rate), a S^0 reducer or a SO_4^{2-} reducer? Why?

4 How would removing the iron from the formula, affect the number of colonies produced?

Name _____ Date _____

Lab Section _____ I was present and performed this exercise (initials) _____

Bioluminescence

QUESTIONS

1 *Considering the relationship between bioluminescent bacteria and the Flashlight Fish, what evolutionary advantage do you think quorum sensing gives the microorganism? Explain.*

2 *Why does the medium used for this exercise contain seawater?*

3 If after 48 hours incubation, your organism has grown but does not emit light, what is the likely explanation? What should you do to correct the problem?

Name _____ Date _____

Lab Section _____ I was present and performed this exercise (initials) _____

Soil Slide Culture

OBSERVATIONS AND INTERPRETATIONS

Soil Sample Site: _____

Length of incubation: _____

Observe two locations on your slide—one from near the surface and one from a deeper part of the soil—and sketch *relevant* parts of them in the circles below. "Relevant" means something showing a relationship of interest. That is, microbial associations with each other or with soil particles. Label any recognizable microbes.

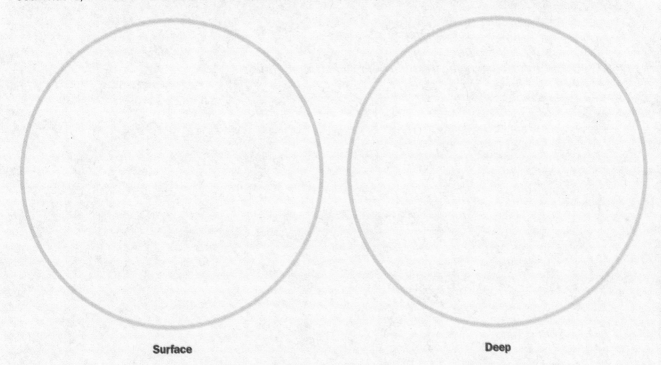

Surface   Deep

QUESTIONS

1 *Why must the soil be kept moist but not too moist?*

_____

_____

_____

_____

_____

2 How can fungal hyphae be distinguished from streptomycetes?

3 Why might spore-forming bacteria be found in your slide communities?

Name _____ Date _____

Lab Section _____ I was present and performed this exercise (initials) _____

Soil Microbial Count

DATA AND CALCULATIONS

1 Count the colonies on the countable plates. Enter that number and the original sample volume in the chart below. (**Note:** Remember that the "original sample volume" is the product of the volume and dilution of the liquid being plated (*e.g.* 0.1 mL $\times$ 10^{-2}).

Group	Colonies Counted	Original Sample Volume (mL)	Cell Density per mL of H_2O	Cell Density per Gram of Soil
1				
2				
3				
4				
5				
6				

2 Calculate the original cell density of the water using the following formula, where OCD is original cell density in cells per milliliter, CFU is colony forming units (colonies counted on the plate), and V is the original sample volume:

$$OCD = \frac{CFU}{V}$$

3 Because your task is to determine the original density of soil, you must convert cell density in milliliters to grams as follows (remembering that the original 100 milliliters of solution contained 10 grams of soil):

$$\frac{CFU}{mL} \times \frac{100 \text{ mL}}{10 \text{ g}} = \frac{CFU}{g}$$

QUESTIONS

1   *Which type of organism was most abundant? Least abundant?*

2   *Calculations for this exercise could have been done in grams and milligrams because 1 milliliter of water weighs 1 gram, but the volume of water in the dilution bottle would have to have been adjusted to do this. To what should the volume be changed for calculations in grams to be correct?*

3   *Considering only the dilution schemes for each type of organism, which would you predict to be the most abundant? Least abundant?*

Name _____ Date _____

Lab Section _____ I was present and performed this exercise (initials) _____

Membrane Filter Technique

OBSERVATIONS AND INTERPRETATIONS

1 Enter the number of colonies counted in each group's sample and calculate the total coliforms per 100 milliliters of water using the following formula:

$$\frac{\text{Total coliforms}}{100 \text{ mL}} = \frac{\text{coliform colonies counted} \times 100 \text{ mL}}{\text{volume of original sample in mL}}$$

2 Enter the results below.

Sample	Number of Colonies	Colonies/100 mL	Potable? Y/N

QUESTIONS

1 *For this test, why is Endo Agar used instead of Nutrient Agar?*

2 How would adding glucose to this medium affect the results? Would it affect its sensitivity or specificity? Explain.

3 Suppose you were to count one coliform colony produced from a 10 mL water sample. What is the coliform density of the water in cells per 100 mL? Is the water potable? Explain.

Name _____ Date _____

Lab Section _____ I was present and performed this exercise (initials) _____

Multiple Tube Fermentation Method for Total Coliform Determination

DATA AND CALCULATIONS

1 Enter your data here.

BGLB Data				
Group	Group A	Group B	Group C	Totals (A + B + C)
Dilution (D)	10^0	10^{-1}	10^{-2}	NA
Portion of dilution added to each LTB tube that is *original sample* (1.0 mL × D)	1.0 mL	0.1 mL	0.01 mL	NA
# LTB tubes in group	5	5	5	NA
# Positive results (Gas)				
# Negative results (No gas)				NA
Total volume of *original sample* in all LTB tubes that produced negative results (D × 1.0 mL × # negative tubes)				
Volume of *original sample* in all LTB tubes inoculated (D × 1.0 mL × # tubes)	5.0 mL	0.5 mL	0.05 mL	5.55 mL

Group	Group A	Group B	Group C	Totals (A + B + C)
EC Data				
Dilution (D)	10^0	10^{-1}	10^{-2}	NA
Portion of dilution added to each LTB tube that is *original sample* (1.0 mL × D)	1.0 mL	0.1 mL	0.01 mL	NA
# LTB tubes in group	5	5	5	NA
# Positive results (Gas)				
# Negative results (No gas)				NA
Volume of *original sample* in all LTB tubes that produced negative results (D × 1.0 mL × # negative tubes)				
Volume of *original sample* in all LTB tubes inoculated (D × 1.0 mL × # tubes)	5.0 mL	0.5 mL	0.05 mL	5.55 mL

2 Enter your final results here.

Total coliform MPN/100 mL (from BGLB data)	
E. coli MPN/100 mL (from EC data)	

QUESTIONS

1 *In the following formula, V_n and V_a symbolize critical volumes in calculating the most probable number of coliforms in 100 milliliters of a sample. Do the symbols represent the amount of dilution added to tubes of LTB or the amount of original sample added to the tubes? Defend your answer in full and explain what would happen if you were to choose the wrong volumes in your calculation.*

$$\text{MPN/100 mL} = \frac{100P}{\sqrt{V_n V_a}}$$

Name _____ Date _____

Lab Section _____ I was present and performed this exercise (initials) _____

2 *Suppose you were to run this test on a water sample and, after 48 hours incubation of the LTB tubes, found turbidity but no gas bubbles. What should you do next?*

3 *What if you found gas in the LTB tubes but none in the BGLB? How does this affect the EC test?*

4 *All coliforms ferment glucose, but none of the media used for this test includes glucose. Why do you think glucose is not used?*

5   *Would adding glucose increase or decrease the sensitivity? Specificity?*

Name _____ Date _____

Lab Section _____ I was present and performed this exercise (initials) _____

Methylene Blue Reductase Test

OBSERVATIONS AND INTERPRETATIONS

Record your results and interpretations in the chart below.

Milk Sample (Brand and raw or processed)	Starting Time T_s (Milk is Blue)	Ending Time T_e (Milk is white)	Elapsed Time ($T_e - T_s$)	Milk Quality

QUESTIONS

1 *Why were you told to cap the tubes tightly? Explain.*

_____

_____

_____

_____

_____

_____

_____

_____

_____

_____

2 *What results would you expect if the tubes were inoculated with a strict aerobe? A strict anaerobe?*

Name _____   Date _____

Lab Section _____ I was present and performed this exercise (initials) _____

Making Yogurt

OBSERVATIONS AND INTERPRETATIONS

Record yogurt made by students groups below.

Culture Organisms	Cell Morphology and Arrangement	Flavor	Consistency	pH

Name _____ Date _____

Lab Section _____ I was present and performed this exercise (initials) _____

Extraction of DNA From Bacterial Cells

OBSERVATIONS AND INTERPRETATIONS

1 In the chart below, record the absorbance values for the wavelengths used.

Wavelength (nm)	Absorbance
220 (optional)	
240 (optional)	
260	
280	
300 (optional)	
320 (optional)	

2 Calculate the DNA concentration in your sample. Be sure to take any dilutions into account.

3 Determine the probable purity of your sample using the equation:

$$\frac{\text{Absorbance}_{260nm}}{\text{Absorbance}_{280nm}}$$

4 Plot the absorption spectrum (Absorption versus Wavelength) of the DNA sample if absorbance at addtional wavelengths was determined.

QUESTIONS

1 *What is the importance of heating the cell lysate?*

_____

_____

_____

2 *Why does the extraction work better with cold 95% isopropanol than with room-temperature 95% isopropanol?*

_____

_____

_____

3 *Based on your results, what wavelength gives maximum absorption of DNA? Does this match the accepted wavelength for maximum absorption? If not, suggest reasons for the discrepancy.*

_____

_____

_____

Name _____ Date _____

Lab Section _____ I was present and performed this exercise (initials) _____

Restriction Digest

OBSERVATIONS AND INTERPRETATIONS

1 Use Table 1 to plan your digestion reactions.

TABLE **1** Digestion Reaction Planner

Reaction Tube/Lane Number	Buffer 5 μL	DNA 1 5 μL	DNA 2 5 μL	DNA 3 5 μL	EcoRI 5 μL	BamHI 5 μL	H₂0 to 50 μL	Reaction Volume	Gel Load 5 μL	Total Volume 55 μL
1										
2										
3										
4										
5										
Markers Lane 6	Markers are ready for electrophoresis. Do not add additional reagents.									

Note: To reach a total digestion volume of 50 μL, calculate water volume (for each tube) by subtracting the sum of all other reagents from 50 μL. Add restriction enzymes after all other reagents have been added.

2 Following gel electrophoresis, measure the migration distances of each of the DNA standards and enter your data in the right-hand column in Table 2. (Refer to the example in Figure 1 as needed.) Measure from the bottom of the well to the bottom of each band. Do not enter the measurement for the first band (closest to the well); this point will not be used for the standard curve.

TABLE **2** Migration Distances (cm)

Reaction 1	Reaction 2	Reaction 3	Reaction 4	Reaction 5	DNA Standards

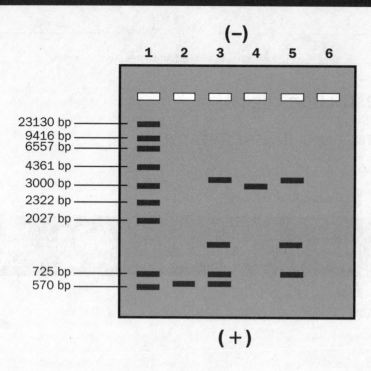

FIGURE 1 GEL SCHEMATIC ✦ This is an idealized drawing of a 7×7 cm restriction digest gel. The base pair sizes are those of the DNA standards used in this exercise. (Drawing adapted, with permission, from Edvotek kit #213 Cleavage of DNA with Restriction Enzymes.)

3 Measure the migration distance of each band in each lane from your reaction mixtures in the same manner as described in #2. Enter the data in Table 2.

4 Using the migration distance data in Table 2 and the base pair data of the DNA standards in Table 3, construct a standard curve on the semi-logarithmic graph paper provided. Refer to the example in Figure 2 as needed. Do not connect the data points; draw a straight trend line through them as centrally located as possible. [**Note:** Do not include the 23130 base pair fragment (shaded) in the curve. It will not properly line up with the other fragments.]

5 Use the standard curve and the data in the table to plot the migration distances of your reaction mixtures. To do this, locate the migration distance on the x-axis and draw a straight vertical line from that point up to the standard curve. Then draw a horizontal line from that point of intersection with the standard curve to the y-axis. The place where the line intersects the y-axis is the number of base pairs in the DNA fragment. Enter this data in Table 3.

TABLE **3** Base Pair Number

Reaction 1	Reaction 2	Reaction 3	Reaction 4	Reaction 5	DNA Standards
					23130
					9416
					6557
					4361
					3000
					2322
					2027
					725
					570

Name _____ Date _____

Lab Section _____ I was present and performed this exercise (initials) _____

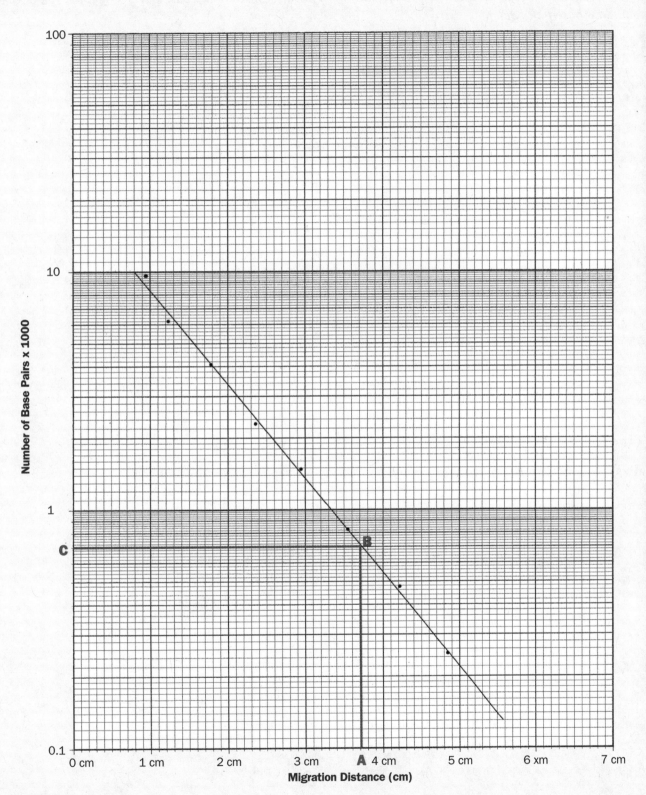

FIGURE **2** STANDARD CURVE ✦ This example of a standard curve demonstrates how to determine base pair sizes represented by the bands in your gel. Assume that a band, measured from the bottom of the well to the bottom of the band, had migrated 3.7 cm. A vertical line is drawn from the 3.7 cm point on the x-axis **A** to the diagonal line (standard curve) and then from that intersection **B**, a horizontal line is drawn to the y-axis. The point where the horizontal line intersects the y-axis **C** reveals the fragment size in number of base pairs. The DNA fragments in this example contained 700 base pairs.

QUESTIONS

1 *Which restriction enzyme is likely to produce more fragments, one with a 4 base-pair recognition site or one with an 8 base-pair site?*

2 *Why is it important to heat your reaction tubes before loading the electrophoresis gel?*

Name _____ Date _____

Lab Section _____ I was present and performed this exercise (initials) _____

3 *Suppose you run a digest and after electrophoresis and staining, you see stained DNA in the gel, but it is more of a long trail of DNA than distinct bands. Assuming that the electrophoresis (including gel preparation and staining) was done correctly, what do you think went wrong with the procedure and what could you have done (or not done) to improve the results?*

4 *Why do you think the largest standard DNA fragment does not fit into the standard curve?*

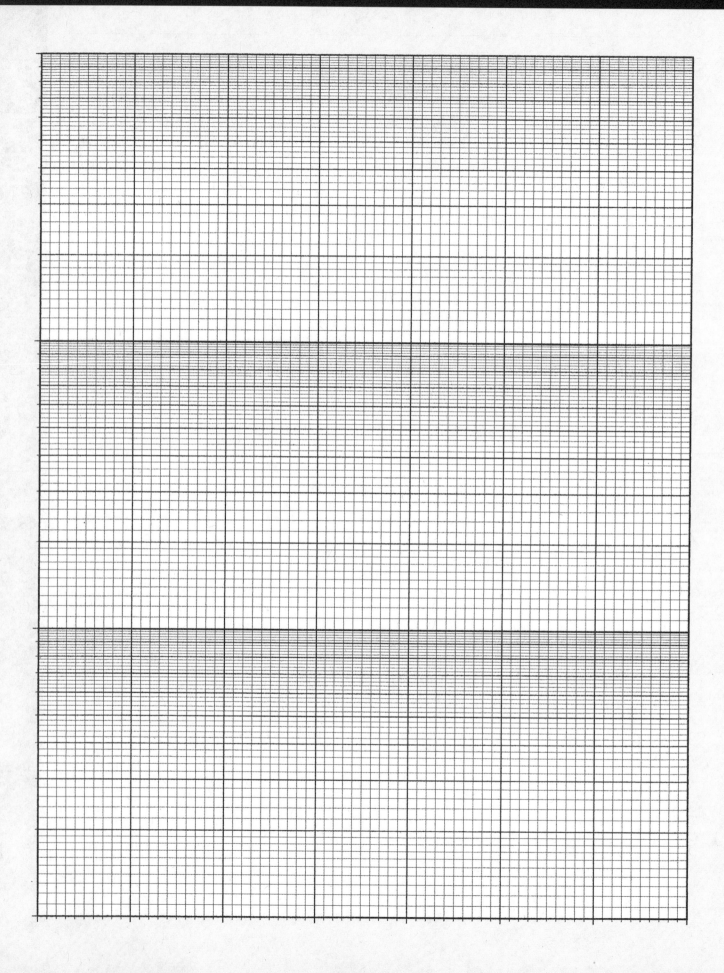

Name _____ Date _____

Lab Section _____ I was present and performed this exercise (initials) _____

Bacterial Transformation: the pGLO™ System

OBSERVATIONS AND INTERPRETATIONS

1 Record the appearance of the pGLO™ plasmid solution with and without UV illumination.

_____

_____

_____

_____

2 In the chart below, record your observations of the plates after incubation.

Plate	Inoculum	Number of Colonies	Appearance in Ambient Light	Appearance with UV Light
LB	*E. coli* − DNA			
LB/amp	*E. coli* − DNA			
LB/amp	*E. coli* + DNA			
LB/amp/ara	*E. coli* + DNA			

QUESTIONS

1 *In the chart below, fill in the genotype of* E. coli *prior to transformation and after transformation.*

E. coli	GFP Gene (GFP⁺ or GFP⁻)	Bla (*Bla⁺* or *Bla⁻*)
Before transformation		
After transformation		

2 *What was the purpose of examining the original pGLO™ solution with and without UV illumination?*

_____

_____

3 What was the purpose of transferring the +DNA and −DNA tubes from ice, to hot water, to ice again?

4 Why were the vials incubated for 10 minutes in LB broth rather than transferring their contents directly to the plates?

5 What information is provided by the LB/−DNA plate?

6 Obviously, transformation could occur only if the pGLO™ plasmid was introduced into the solution. Which plate(s) exhibit transformation?

7 What information is provided by the LB/amp/−DNA plate?

8 Why does the LB/amp/ara/+DNA plate fluoresce when the LB/amp/+DNA plate does not?

9 Use the following information to calculate transformation efficiency.

a. You put 10 µL (one loopful) of a 0.03 µg/µL pGLO™ solution into the +DNA tube. Calculate the µg of DNA you used.

b. The +DNA tube contained 510 µL of solution prior to plating, but you did not use all of it. Calculate the fraction of the +DNA solution you used in each tube.

c. Use your answers to questions 9a and 9b to calculate the micrograms of DNA you plated.

d. Using the number of colonies on the LB/amp/ara/+DNA plate, calculate the transformation efficiency. (**Hint:** The units are transformants/µg of pGLO™ DNA.)

e. According to the manufacturer's (Bio-Rad Laboratories) manual, this protocol should yield a transformation efficiency between 8.0×10^2 and 7.0×10^3 transformed cells per microgram of pGLO™ DNA. How does your transformation efficiency compare? Account for any discrepancy.

Name _____ Date _____

Lab Section _____ I was present and performed this exercise (initials) _____

Polymerase Chain Reaction

OBSERVATIONS AND INTERPRETATIONS

1 Following gel electrophoresis, measure the migration distances of each of the DNA standards and enter your data in Table 1 below. (See the example in Figure 1 as needed.) Measure from the bottom of the well to the bottom of each band. Do not enter the measurement for the first band (closest to the well); this point will not be used in the standard curve.

2 Assuming that all bands from your sample migrated the same distance, measure the distance from the bottom of the well to the bottom of the most prominent band. Enter the data in Table 1.

3 Using the migration distance data in Table 1 and the base pair data of the DNA standards in Table 2, construct a standard curve on the semi-logarithmic graph paper provided. Refer to the example in Figure 2 on page 715 as needed. Do not connect the data points; draw a straight trend line through them as centrally located as possible. Do not include the 23130 base pair fragment (shaded) in the curve. It will not properly line up with the other fragments.

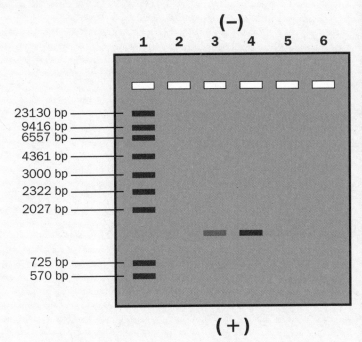

FIGURE 1 **GEL SCHEMATIC** ✦ This is an idealized drawing of a 7 × 7 cm PCR gel. The base pair sizes are those of the DNA standards used in this exercise. (Drawing adapted, with permission, from Edvotek Kit #330–The Molecular Biology of DNA Amplification by Polymerase Chain Reaction.)

4 Use the standard curve and the data in the table to plot the migration distance of your sample. To do this, locate the migration distance on the x-axis and draw a straight vertical line from that point up to the

TABLE 1 DNA Migration Distance (cm)

DNA Standards								
Sample								

TABLE 2 Base Pairs

DNA Standards	23130	9416	6557	4361	3000	2322	2027	725	570
Sample									

standard curve. Then draw a horizontal line from that point of intersection to the y-axis. The place where the line intersects the y-axis is the size of your DNA sample in base pairs. Enter these data in Table 2.

QUESTIONS

1 *Which band was the heaviest? Why?*

2 *Why is TaqDNA, rather than another polymerase, used for the extension phase?*

3 *Genomes are frequently described according to their G + C content. G (guanine) and C (cytosine) make triple hydrogen bonds in the double helix, whereas thymine and adenine make double hydrogen bonds. Would you expect to program a thermal cycler the same way for an organism with 50% G + C as an organism with 25% G + C? If yes, why? If not, what would you change?*

Name _____ Date _____

Lab Section _____ I was present and performed this exercise (initials) _____

4 *Polymerase chain reactions sometimes require an initial 5-minute denaturing period immediately before the first cycle, depending on how the DNA has been processed prior to the PCR. Under what circumstances would you expect to see this preliminary procedure?*

_____

_____

_____

_____

_____

_____

_____

_____

5 *Why do you think it is necessary to wear gloves when running PCR?*

_____

_____

_____

_____

_____

_____

_____

_____

_____

_____

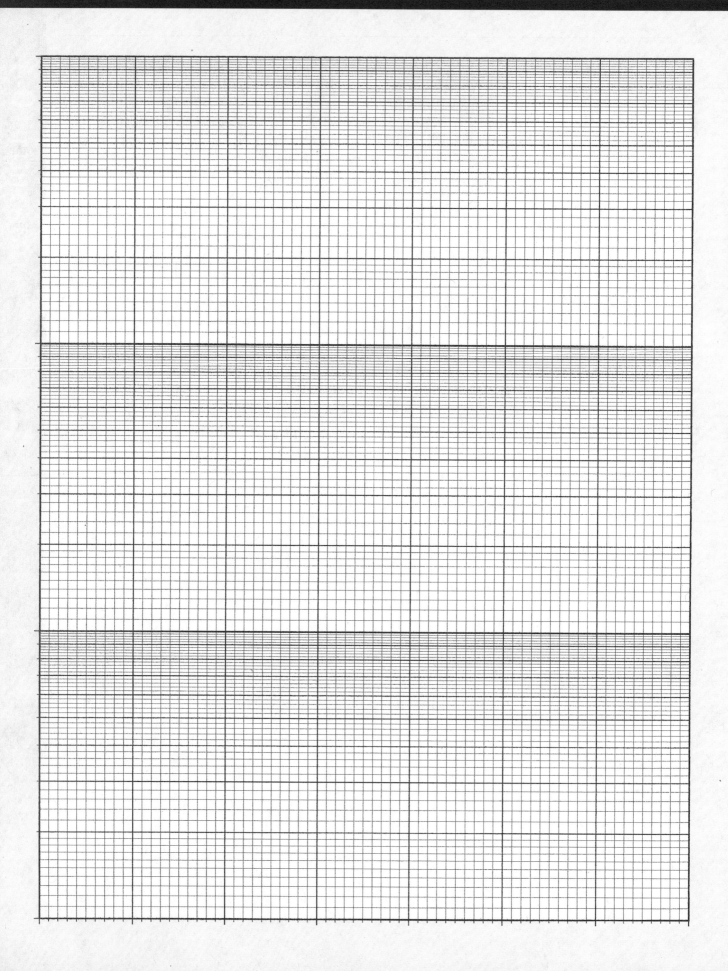

Name _____   Date _____

Lab Section _____ I was present and performed this exercise (initials) _____

Ultraviolet Damage and Repair

OBSERVATIONS AND INTERPRETATIONS

Using the terms "confluent–dense growth," "confluent–sparse growth," "individual colonies," and "none," describe the growth of *Serratia marcescens* in both parts of each plate (*i.e.,* where the mask shielded the growth from UV and where it did not). Record your observations in the chart below. Make a note as to the significance of each type of growth.

Plate	Mask/Lid	UV Exposure Time (min.)	Incubation	Growth on Agar Surface Covered by Mask	Growth on Agar Surface Beneath Opening of Mask
1	Mask, no lid	0.5	Sunlight		
2	Mask, no lid	0.5	Dark		
3	Mask, no lid	3.0	Sunlight		
4	Mask and lid	3.0	Sunlight		
5	Mask and lid	3.0	Dark		
6	Lid only	0	Sunlight		
7	Lid only	0	Dark		

QUESTIONS

1 Consider the exposed part of the plate.

  a. What does the relative sparseness of growth tell you about the effect of UV radiation?

  _____

  _____

  _____

  b. What do the colonies tell you about the cells from which they grew?

  _____

  _____

2 What does the relatively heavy growth on the plates covered by the mask and lid during exposure tell you about UV radiation?

3 What is the effect of longer UV exposure on *Serratia marcescens*? To answer this question, which plates should be compared?

4 What is the effect of the posterboard mask on UV radiation? To answer this question, which plates should be examined?

5 What is the effect of the plastic lid on UV radiation? To answer this question, which plates should be compared?

Name _____ Date _____

Lab Section _____ I was present and performed this exercise (initials) _____

Ames Test

OBSERVATIONS AND INTERPRETATIONS

Record your observations for the Ames Test plates below. On the CM plates, measure the zone of inhibition diameter in millimeters. On the MM plates, count only the large colonies, not the "hazy" background growth. (These are the colonies produced by auxotrophs that did not back-mutate to prototrophs and grew only until the histidine in the minimal medium was exhausted.)

	Zone Diameter		Colonies Counted	
	CM with TA 1535	CM with TA 1538	MM with TA 1535	MM with TA 1538
DMSO				
Test Substance _____				

QUESTIONS

1 *Which plates do you have to examine to determine if your test substance is toxic? What is your conclusion about your test substance?*

_____

_____

_____

_____

_____

_____

_____

_____

_____

_____

2 Which plates do you need to examine to determine if your test substance is mutagenic? What is your conclusion about your test substance?

3 Why can't your test substance contain protein?

4 What is the purpose of including a small amount of histidine in minimal medium?

5 Why are the small colonies on the minimal medium plates not considered back-mutants? Why don't they grow to full size?

6 What are some advantages of the Ames Test over carcinogen tests using rats or mice? What are some disadvantages?

Name _____ Date _____

Lab Section _____ I was present and performed this exercise (initials) _____

Phage Typing of *E. coli* Strains

OBSERVATIONS AND INTERPRETATIONS

Record your results in the Table. Use "+" for clearing and "−" for no clearing.

E. coli Strain	Sterile Water	Phage T$_2$	$\phi\chi$*174*
Wild Type			
B			
C			

QUESTIONS

1 *What is the purpose of phage typing?*

_____

_____

_____

_____

_____

_____

_____

_____

_____

_____

_____

2 *What would a clear zone produced by the sterile water mean? How would that affect the interpretation of your results?*

3 *Suppose you were told these results for a third phage, T₇:*

E. coli Strain	Phage T₇
Wild Type	−
B	+
C	−

a. Could you tell wild type E. coli *and* E. coli C *apart based exclusively on these results? Explain.*

b. *If your answer was "no," suggest what could be done to help you differentiate between them. Be thorough.*

Name _____ Date _____

Lab Section _____ I was present and performed this exercise (initials) _____

Differential Blood Cell Count

OBSERVATIONS AND INTERPRETATIONS

Record your data from the differential blood cell count in the chart below. As you count the 100 white blood cells, make tally marks in the appropriate boxes. Then calculate the percentages of each type and compare them to the expected values.

Normal Blood						
	Monocytes	**Lymphocytes**	**Segmented Neutrophils**	**Band Neutrophils**	**Eosinophils**	**Basophils**
Number						
Percentage						
Expected Percentage	3–7%	25–33%	55–65% (all neutrophils)	—	1–3%	0.5–1%

Abnormal Blood (Condition: _____)						
	Monocytes	**Lymphocytes**	**Segmented Neutrophils**	**Band Neutrophils**	**Eosinophils**	**Basophils**
Number						
Percentage						
Expected Percentage	3–7%	25–33%	55–65% (all neutrophils)	—	1–3%	0.5–1%

QUESTIONS

1 How do the percentages of each WBC compare to the published values? What might account for any differences you noted?

_____

_____

_____

_____

_____

_____

_____

2 If you did a differential count on abnormal blood, how did the percentages compare to normal blood? How can any differences you noted be explained in the context of the disease and/or defense process?

_____

_____

_____

_____

_____

_____

_____

_____

3 What is the purpose of scanning the slide in a systematic pattern (as described in the text)?

_____

_____

_____

_____

_____

_____

_____

_____

Name _____ Date _____

Lab Section _____ I was present and performed this exercise (initials) _____

Precipitin Ring Test

OBSERVATIONS AND INTERPRETATIONS

Draw the tubes after incubation. Label the solutions in each and any precipitation lines.

QUESTIONS

1 *What was the purpose of tube B?*

_____

_____

_____

_____

_____

_____

_____

2 In general terms, describe how equine albumin antiserum could be obtained. What animal is least likely to be a source for it?

3 Suppose you were instructed to make two antiserum solutions: The first is identical to what you used in the lab exercise. The other is a 10^{-6} dilution of the antiserum. After incubation with the antigen, the full-strength antiserum produces a precipitin ring, but the diluted antiserum does not. Explain these results. Is this due to poor specificity or sensitivity of the test?

4 Suppose you were instructed to repeat this experiment, again using equine albumin antiserum in two tubes. You layer equine serum (containing equine albumin) over the antiserum in one tube and pig serum (containing pig albumin) over the antiserum in the other tube. After incubation, you see precipitin rings in both tubes. Explain these results. Is this due to poor specificity or sensitivity of the test?

Name _____ Date _____

Lab Section _____ I was present and performed this exercise (initials) _____

Radial Immunodiffusion

OBSERVATIONS AND INTERPRETATIONS

Draw and interpret the results of your immunodiffusion plate in the diagram below. (**Note:** Do not use this as the template for the wells in your plate; use Figure 11-10.)

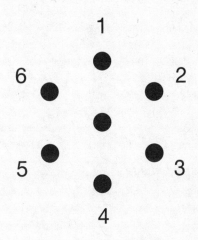

QUESTIONS

1 *What was the purpose of well #5?*

_____

_____

_____

_____

_____

_____

_____

_____

_____

2 *How many lines of identity did your plate demonstrate?*

3 *Between which two wells did you see nonidentity?*

4 *Between which two wells did you see partial identity?*

5 *Examine Figure 11-7. Suppose you performed gel immunodiffusion and all you got was the Zone of Antigen Excess. Is this a false positive or a false negative result? Is it due to poor specificity or sensitivity of the test?*

6 *Suppose you performed the gel immunodiffusion test again using the procedure in this exercise. How could you explain the presence of two precipitation lines between well #1 and the center well?*

Name _____ Date _____

Lab Section _____ I was present and performed this exercise (initials) _____

Slide Agglutination

OBSERVATIONS AND INTERPRETATIONS

Sketch and label your results in the diagram below.

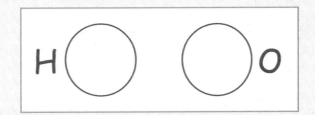

QUESTIONS

1 *What is the purpose of performing this experiment with both H and O antigens?*

_____

_____

_____

_____

_____

_____

_____

_____

_____

_____

_____

_____

_____

_____

2 *Suppose you performed this test and got agglutination in both samples. Eliminating contamination of the antigen and antiserum samples as a possibility, provide an explanation of this hypothetical result. Is this attributable to poor sensitivity or specificity of the test system?*

3 *Suppose you performed this test and neither sample produced agglutination even though you know the antigens and antibodies should react. Would this be attributable to poor sensitivity or specificity of the test system?*

4 *Higher vertebrates produce antibodies; Salmonella is a prokaryote. Given these facts, how can Salmonella antiserum be produced?*

Name _____ Date _____

Lab Section _____ I was present and performed this exercise (initials) _____

Blood Typing

OBSERVATIONS AND INTERPRETATIONS

Record your results below.

Antiserum	Agglutination (+ / −)
Anti-A	
Anti-B	
Anti-Rh	

My blood type is: _____

Record class data in the chart below.

Blood Type	Percentage in U.S. Population	Number in Class	Percentage in Class	Deviation from National Values
O +	37.4			
O −	6.6			
A +	35.7			
A −	6.3			
B +	8.5			
B −	1.5			
AB +	3.4			
AB −	0.6			

Source: Stanford School of Medicine. Copyright © 2009.

QUESTIONS

1 Examine the blood type data obtained from your class. Attempt to explain any deviations from the national values.

2 People with Type "O" are said to be "Universal Donors" in blood transfusions. Explain the reasoning behind this. Which blood type is designated as the "Universal Recipient"?

3 What biological fact makes the concepts of "Universal Donor" and "Universal Recipient" misnomers?

4 Maternal-fetal Rh incompatibility (where the Rh − mother's Anti-Rh antibodies destroy the fetus's Rh + red blood cells) is a well-known phenomenon. Less well known are situations of maternal-fetal ABO incompatibility. Suggest combinations of maternal and fetal blood types that could lead to this situation.

5 Why are red blood cells used in many indirect agglutination tests?

Name _____ Date _____

Lab Section _____ I was present and performed this exercise (initials) _____

Mononucleosis Hemagglutination Test

OBSERVATIONS AND INTERPRETATIONS

We tested Patient _____

Record your group's results in the chart below. Share your data with another group that tested the other patient.

Patient	Reaction	Evidence
1		
2		

QUESTIONS

1 *What is the role of guinea pig kidney antigen (Reagent A) in this test?*

_____

_____

_____

_____

2 *What is the role of horse erythrocytes (Reagent B) in this test?*

_____

_____

_____

_____

3 *You used the positive and negative controls supplied in the kit as "patient" samples. In a medical lab, how would these positive and negative controls be used?*

_____

_____

_____

_____

_____

4 *Why is it important to have the reagent bottles mixed thoroughly?*

5 *Why is it important not to mix the solutions on the card too vigorously?*

6 *Why must Reagent A be mixed with the patient's sample before mixing with Reagent B?*

7 *What would be the outcome of mixing the patient's sample with Reagent B first if the serum contained heterophile antibodies (that is, it is a true positive)?*

8 *What would be the outcome of mixing the patient's sample with Reagent B first if the serum did not heterophile antibodies (that is, it is a true negative)?*

Name _____   Date _____

Lab Section _____   I was present and performed this exercise (initials) _____

Quantitative Indirect ELISA

OBSERVATIONS AND INTERPRETATIONS

Fill in the chart indicating the degree of reaction using symbols "++++" for maximum through "0" for no reaction.

	1 1:100	2 1:400	3 1:1,600	4 1:6,400	5 1:25,600
A Ab1					
B Ab1					
C PBS					
D Ab2					
E PBS					
F Ab1					
G Ab2					
H Ab2					

QUESTIONS

1 *Why is it important not to mix antigen, antibody, or antigen/antibody solutions during the experiment?*

_____

_____

_____

2 *Why is thorough mixing so important in performing the dilutions of Ab1 and Ab2?*

_____

_____

_____

3   *Why are the contents of Rows A, B, and F, and D, G, and H the same?*

4   *Why is Row D placed near A and B, and Row F placed near G and H?*

5   *What is the role of the two PBS rows (C and E)?*

6   *Why is it important to remove Ab1, Ab2, and the 2°Ab from the most diluted sample to the least diluted sample?*

7   *Why must the "Stop Solution" be removed from the least diluted sample to the most diluted sample?*

8   *The mixtures in Column 5 rarely show a reaction. In fact, the kit is designed that way. Yet, the wells were treated with the same antigen, primary antibody, secondary antibody, substrate, and stop solution as the other wells in the same row. How can you account for the results in Column 5 wells?*

9   *Refer to the Procedural Diagram for the following questions.*

a. *In Step 4, what would be the consequence of washing too vigorously with PBS? Of not washing enough?*

b. *What might be the consequence of not adding Blocking Agent in Step 5?*

c. *In Step 6, what would be the consequence of washing too vigorously with PBS? Of not washing enough?*

d. *In Step 13, what would be the consequence of washing too vigorously with PBS? Of not washing enough?*

e. *In Step 16, what would be the consequence of washing too vigorously with PBS? Of not washing enough?*

Name _____ Date _____

Lab Section _____ I was present and performed this exercise (initials) _____

The Fungi—Common Yeasts and Molds

OBSERVATIONS AND INTERPRETATIONS

Sketch your observations of the microscopic structure of the assigned fungi in the spaces below.

Yeasts

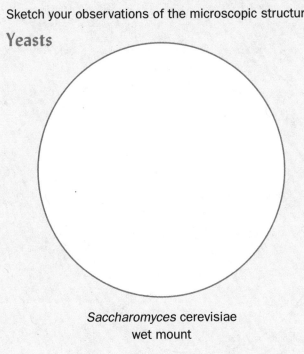

Saccharomyces cerevisiae
wet mount

(X_____)

Cell dimensions _____

Candida albicans
vegetative cells

(X_____)

Cell dimensions _____

Molds

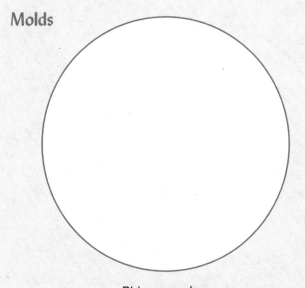

Rhizopus colony

Obverse color: _____   Colony texture: _____

Reverse color: _____   Colony topography: _____

Rhizopus sporangia

(X_____)

Sporangium dimensions _____

Molds (continued)

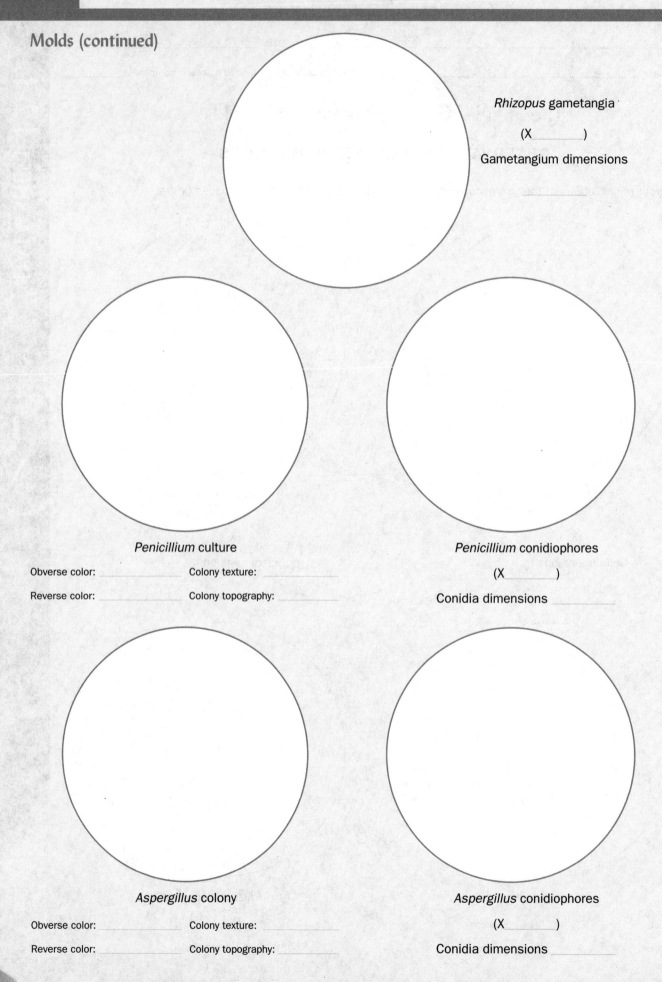

Rhizopus gametangia

(X_____)

Gametangium dimensions

Penicillium culture

Obverse color: _____   Colony texture: _____

Reverse color: _____   Colony topography: _____

Penicillium conidiophores

(X_____)

Conidia dimensions _____

Aspergillus colony

Obverse color: _____   Colony texture: _____

Reverse color: _____   Colony topography: _____

Aspergillus conidiophores

(X_____)

Conidia dimensions _____

Name _____ Date _____

Lab Section _____ I was present and performed this exercise (initials) _____

Fungal Slide Culture

OBSERVATIONS AND INTERPRETATIONS

Record your observations of each fungal slide culture in the circles provided. Take pride in your sketches and label relevant structures. It is more important to draw one or two components in detail than to scribble the entire tangled mass of hyphae.

Organism: _____   Organism: _____

Magnification: _____   Magnification: _____

QUESTIONS

1   *Why is slide culture used rather than a traditional wet mount for examining molds?*

2   *What is the purpose of the moist filter paper in the slide culture? Why isn't it necessary in bacterial cultures?*

Name _____ Date _____

Lab Section _____ I was present and performed this exercise (initials) _____

Examination of Common Protozoans of Clinical Importance

OBSERVATIONS AND INTERPRETATIONS

Sketch your observations of the assigned protozoans in the spaces below.

Entamoeba histolytica
trophozoite

(X_____)

Dimensions _____

Entamoeba histolytica
cyst

(X_____)

Dimensions _____

Entamoeba coli
trophozoite

(X_____)

Dimensions _____

Entamoeba coli
cyst

(X_____)

Dimensions _____

Balantidium coli
trophozoite

(X_____)

Dimensions _____

Balantidium coli
cyst

(X_____)

Dimensions _____

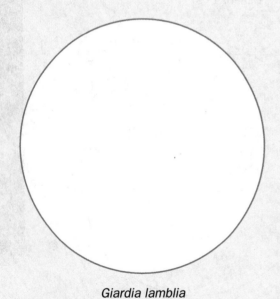

Giardia lamblia
trophozoite

(X_____)

Dimensions _____

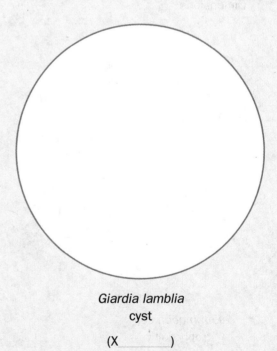

Giardia lamblia
cyst

(X_____)

Dimensions _____

Name _____ Date _____

Lab Section _____ I was present and performed this exercise (initials) _____

Trichomonas vaginalis
trophozoite

(X_____)

Dimensions _____

Trypanosoma spp.

(X_____)

Dimensions _____

Plasmodium spp.
ring and mature trophozoites

(X_____)

Dimensions _____

Plasmodium spp.
schizont

(X_____)

Dimensions _____

Plasmodium spp.
gametocytes

(X_____)

Dimensions _____

Toxoplasma gondii
trophozoite

(X_____)

Dimensions _____

Name _____ Date _____

Lab Section _____ I was present and performed this exercise (initials) _____

Parasitic Helminths

OBSERVATIONS AND INTERPRETATIONS

Sketch your observations of the assigned helminths in the spaces below.

Trematodes

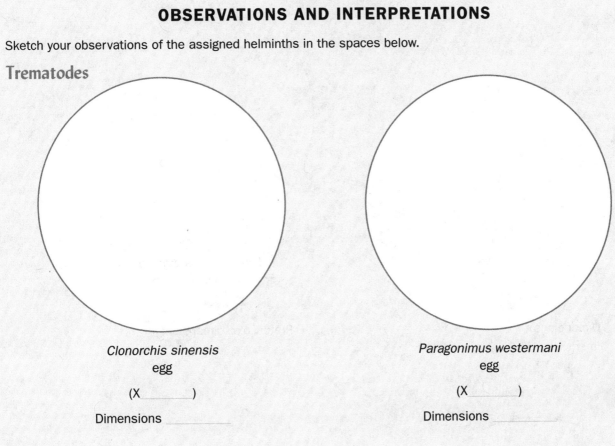

Clonorchis sinensis
egg

(X_____)

Dimensions _____

Paragonimus westermani
egg

(X_____)

Dimensions _____

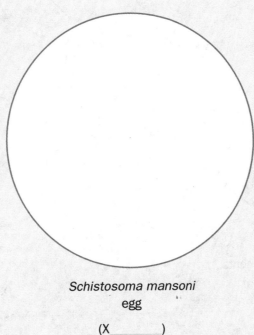

Schistosoma mansoni
egg

(X_____)

Dimensions _____

Cestodes

Dipylidium caninum
egg

(X_____)

Dimensions _____

Echinococcus granulosus
protoscolices in a hydatid cyst

(X_____)

Protoscolex dimensions _____

Hymenolepis nana
egg with oncosphere and filaments

(X_____)

Dimensions _____

Taenia spp.
egg

(X_____)

Dimensions _____

Name _____ Date _____

Lab Section _____ I was present and performed this exercise (initials) _____

Taenia solium
scolex

(X_____)

Taenia solium
proglottid

(X_____)

Nematodes

Ascaris lumbricoides
eggs with mammillations

(X_____)

Dimensions _____

Enterobius vermicularis
egg

(X_____)

Dimensions _____

Nematodes (continued)

Hookworm (*Ancylostoma* or *Necator*)
egg

(X_____)

Dimensions _____

Strongyloides stercoralis
rhabditiform larva

(X_____)

Dimensions _____

Wuchereria bancrofti
microfilariae in a blood smear

(X_____)

Dimensions _____

_____

_____

(X_____)

Dimensions _____

Glossary

A

absorbance A measurement, using a spectrophotometer, of how much light entering a substance is *not* transmitted.

acetoin A four-carbon intermediate in the conversion of pyruvic acid to 2,3-Butanediol; the compound detected in the Voges-Proskauer test.

acetyl-CoA Compound that enters the Krebs cycle by combining with oxaloacetate to form citrate; may be produced from pyruvate in carbohydrate metabolism or from oxidation of fatty acids.

acid clot Pink clot formed in litmus milk from the precipitation of casein under acidic conditions; an indication of lactose fermentation.

acidic stain Staining solution with a negatively charged chromophore.

acidophil *See* eosinophil.

acidophile Microorganism adapted to a habitat below pH 5.5.

acid reaction Pink color reaction in litmus milk from acidic conditions produced by lactose fermentation.

adenosine triphosphate (ATP) In hydrolysis of ATP to adenosine diphosphate (ADP), supplies the energy necessary to perform most work in the cell.

aerobe Microorganism that requires oxygen for growth.

aerobic respiration A type of energy releasing metabolism in which oxygen is the final electron acceptor in the pathway. It also has the highest potential ATP yield of all catabolic and exergonic reactions.

aerosol Droplet nuclei that remain suspended in the air for a long time.

aerotolerance Designation of an organism based on its ability to grow in the presence of oxygen.

aerotolerant anaerobe Anaerobe that grows in the presence of oxygen but does not use it metabolically.

agar overlay Used in plaque assays, the soft agar containing a mixture of bacteriophage and host poured over nutrient agar as an inoculum.

agglutination Visible clumping produced when antibodies and particulate antigens react.

agranulocyte Category of white blood cells characterized by the absence of prominent cytoplasmic granules; includes monocytes and lymphocytes.

akinete A thick-walled resting spore produced by some cyanobacteria. It is derived from a vegetative cell.

alkaliphile Microorganism adapted to a habitat above pH 8.5.

alpha (α) hemolysis The greening around a colony on blood agar as a result of partial destruction of red blood cells; typical of certain members of *Streptococcus*.

ammonification The metabolic production of ammonia. *See* deamination.

amphitrichous Describes a flagellar arrangement with flagella at both ends of an elongated cell.

amylase Family of enzymes that hydrolyze starch.

anaerobe Microorganism that cannot tolerate oxygen.

anaerobic respiration A type of energy releasing metabolism in which an inorganic compound other than oxygen is the final electron acceptor. These include NO_3, NO_2, SO_4, and CO_2. The ATP yield is higher than fermentation, but not as high as aerobic respiration.

annealing The attachment of a primer to a complimentary sequence of DNA. Annealing is done on each strand of the denatured DNA in positions flanking the desired sequence (template) to be replicated. Annealing is the second of three phases in a PCR cycle, following denaturation and prior to extension.

antibiotic Refers to an antimicrobial substance produced by a microorganism such as a bacterium or fungus.

antibody A glycoprotein produced by plasma cells, in response to an antigen, that reacts with the antigen specifically.

antigen A molecule of high molecular weight and complex three-dimensional shape that stimulates the production of antibodies and reacts specifically with them.

antimicrobic(al) Describes any substance that kills microorganisms, natural or synthetic.

antiporter A system that simultaneously transports substances in opposite directions across the cytoplasmic membrane.

antiserum General term applied to a serum that contains antibodies of a specific type.

Ascomycetes A division of fungi that produces sexual spores in a saclike structure called an ascus.

aseptic Describes the condition of being without contamination.

assimilation (nitrogen) The incorporation of NH_4 or NO_3 into organic form (amino acids) by plants.

assimiliatory sulfate reduction The conversion of sulfate (SO_4^{2-}) to a sulfhydryl group (–SH) in the biosynthesis of amino acids cysteine and methionine.

autoinducer Substance secreted by bioluminescent bacteria that, upon reaching sufficient concentration, triggers the bioluminescent reaction; believed to help conserve energy by synchronizing individual cellular reactions.

autotroph An organism that is able to make all of its organic molecules using CO_2 as the source of carbon. *See also* heterotroph.

auxotroph A nutritional mutant; a cell incapable of synthesizing the nutrient that is acting as the marker in a specific genetic experiment; will grow only on a complete medium. *See also* prototroph.

B

bacillus Rod-shaped cell.

back mutation Mutation in a gene that reverses the effect of an original mutation.

bacteriophage Virus that attacks bacteria; usually host-specific.

basic stain Staining solution with a positively charged chromophore.

Basidiomycetes Division of fungi characterized by club-like appendages (basidia) that produce haploid spores; include common mushrooms and rusts.

basophil Category of white blood cells; one of three types of granulocyte (along with eosinophils and neutrophils) characterized by a cytoplasm with dark-staining granules and a lobed nucleus (often difficult to see because of the dark cytoplasmic granules).

β-hemolysis The clearing around a colony on blood agar resulting from complete destruction of red blood cells; typical of certain members of *Streptococcus*.

β-lactamase Enzyme found in some bacteria that breaks a bond in the β-lactam ring of β-lactam antibiotics, rendering the antibiotic ineffective; the source of resistance against these antibiotics.

β-lactam antibiotic A group of structurally related antimicrobial chemicals that interfere with cross-linking of peptidoglycan subunits, which results in a defective wall structure and cell lysis; examples are penicillins and cephalosporins.

biofilm A layer of bacterial cells adhering to and reproducing on a surface; used commercially for acetic acid production and water purification.

bioluminescence Process by which a living organism emits light.

biosafety level (BSL) One of four sets of minimum standards for laboratory practices, facilities, and equipment to be used when handling organisms at each level; BSL-1 requires the least care, BSL-4 the most.

bound coagulase An enzyme (also called clumping factor) bound to the bacterial cell wall, responsible for causing the precipitation of fibrinogen and coagulation of bacterial cells.

C

capnophile Microaerophile that requires elevated CO_2 levels.

capsule Insoluble, mucoid, extracellular material surrounding some bacteria.

carbohydrate One of four families of biochemicals; characterized by containing carbon, hydrogen, and oxygen in the ratio 1:2:1.

carcinogen Any substance that causes cancer.

cardinal temperatures Minimum, optimum, and maximum temperatures for an organism.

casease A family of hydrolytic enzymes that break down casein, often extracellularly.

casein Milk protein that gives milk its white color.

catalase Enzyme produced by some bacteria that catalyzes the breakdown of metabolic H_2O_2.

cestode The class of parasitic worms commonly called tapeworms; characterized by having a head (scolex) with suckers and hooks and numerous segments (proglottids) that are little more than sex organs.

chemoheterotroph Any organism that uses chemicals for energy and organic compounds as a carbon source. Usually the same organic compounds serve both purposes.

chemolithotroph Any organism that gets its energy from chemicals and its electrons from an inorganic source. No eukaryotes are chemolithotrophic, only some prokaryotes such as sulfur and ammonia oxidizing bacteria.

chromogenic reducing agent A substance that produces color when it gives up electrons (becomes oxidized).

chromophore The charged region of a dye molecule that gives it its color.

coagulase *See* bound coagulase; free coagulase.

coagulase-reacting factor Plasma component that reacts with free coagulase to trigger the clotting mechanism.

coccus Spherical cell.

coenzyme Nonprotein portion of an enzyme that aids the catalytic reaction (usually by accepting or donating electrons).

coliform Member of *Enterobacteriaceae* that ferments lactose with production of gas within 48 hours at 37°C.

colony Visible mass of cells produced on culture media from a single cell or single CFU (cell forming unit); a pure culture.

colony forming unit Term used to define the cell or group of cells (*e.g., staphylococci, streptococci*) that produces a colony when transferred to plated media.

commensal Describes a synergistic relationship between two organisms in which one benefits from the relationship and the other is affected neither negatively nor positively.

common source epidemic An epidemic in which the disease is transmitted from a single source (such as water supply) and is not transmitted person-to-person.

compatible solutes Compounds such as amino acids that function both as metabolic constituents and solutes necessary to maintain osmotic balance between the internal and external environments.

competent cell Cell capable of picking up DNA from the environment; some cells are naturally competent, in other cases the cells are made to be competent artificially.

competitive inhibition Process whereby a substance attaches to an enzyme's active site, thereby preventing attachment of the normal substrate.

complete medium A medium used in genetic experiments that supplies all nutrients for growth required by prototrophs and auxotrophs.

complex medium A medium in which at least one nutritional ingredient is of unknown composition or amount. Media with any plant or animal extract are always complex.

confirmatory test One last test after identification of an unknown, to further support the identification.

conjugate acid A compound (one of a conjugate pair) that donates a proton in solution.

conjugate base A compound (one of a conjugate pair) that accepts a proton in solution.

conjugate pair A compound that alternates between acid and base forms by losing or gaining one proton. Example: HCl/Cl⁻.

constitutive Term used to describe an enzyme that is produced continuously, as opposed to an enzyme produced only when its substrate is present. *See also* induction.

cos site (cohesive end sites) In Lambda phage (linear) DNA, the single stranded tails at the 5' ends of both strands are complimentary to each other and can form a circular molecule. These "sticky" ends can be a source of error in DNA fragment size determinations if not properly heated before running gel electrophoresis.

counter-stain Stain applied after decolorization to provide contrast between cells that were decolorized and those that weren't.

culture A liquid or solid medium with microorganisms growing in or on it.

cyst "Resting stage" in life cycle of certain protozoans. *See also* trophozoite.

cytochrome A class of iron-containing enzyme in electron transport chains. Characterized by a porphyrin ring structure.

cytochrome oxidase An electron carrier in the electron transport chain of aerobes, facultative anaerobes, and microaerophiles that makes the final transfer of electrons to oxygen.

cytoplasm The semifluid component of cells inside the cellular membrane in which many chemical reactions take place.

D

deaminase Enzyme that catalyzes the removal of the amine group (NH_2) from an amino acid.

death phase Closed system microbial growth phase immediately following stationary phase; characterized by population decline usually resulting from nutrient deficiencies or accumulated toxins.

decarboxylase Enzyme that catalyzes the removal of the carboxyl group (COOH) from an amino acid.

decimal reduction value Amount of time at a specific temperature to reduce a microbial population by 90% (one log cycle).

defined medium A growth medium in which the exact amount and chemical formula (and thus identity) of each ingredient are known.

denaturation The first of three phases in a PCR cycle, followed by annealing and extension. The separation of the strands in double stranded DNA.

denitrification The bacterial process in which nitrate is reduced to various forms of nitrogen in its role as final electron acceptor of anaerobic respiration.

deoxyribonuclease A family of hydrolytic enzymes that depolymerize DNA into polynucleotides. The precise location of hydrolysis within the sugar-phosphate backbone depends on which DNase is acting. These are usually secreted enzyme.

Deuteromycetes An unnatural grouping of fungi in which the sexual stages are either unknown or are not used in classification.

differential medium Growth medium that contains an indicator (usually color) to detect the presence or absence of a specific metabolic activity.

differential stain Staining procedure that allows distinction between cell types or parts of cells; often involves more than one stain, but not necessarily.

dilution blank A test tube containing a measured volume of sterile diluent (water, saline or buffer) used to serially dilute a concentrated solution or broth.

dilution factor Proportion of original sample present in a new mixture after it has been diluted; calculated by dividing the volume of the original sample by the total volume of the new mixture. (Subsequent dilutions in a serial dilution must be multiplied by the dilution factor of the previous dilution.)

diploid Defines a cell that has two complete sets of genetic information; one stage in the life cycle of all eukaryotic microorganisms (which also have a haploid [one set] stage).

direct agglutination Serological test in which the combination of antibodies and *naturally* particulate antigens, if positive, form a visible aggregate. *See also* indirect agglutination.

disaccharide A sugar made of two monosaccharide subunits; *e.g.*, the monosaccharides glucose and galactose can combine to make the disaccharide lactose.

disk diffusion test (Kirby-Bauer Test) This is one way to check susceptibility and resistance of a microbe to an antibiotic. A paper disk containing an antibiotic is placed on a plate inoculated to produce confluent growth of a bacterium. Susceptibility is determined by the size of the zone of inhibition of growth around the disk.

disproportionation reaction A redox reaction of some sulfur reducers whereby a substrate is split into two molecules, one that is more reduced and one that is more oxidized than the original compound. Example: $S_2O_3^{2-} \rightarrow SO_4^{2-} + H_2S$.

dissimilatory sulfate reduction A respiratory reaction in which sulfate is reduced exclusively for the purpose of gaining energy.

DNA polymerase A group of enzymes that catalyze the addition of deoxyribonucleotides to the 3' end of an existing polynucleotide chain.

DNase *See* deoxyribonuclease.

Durham tube A small, inverted test tube used in some liquid media to trap gas bubbles and indicate gas production.

E

electromagnetic energy The energy that exists in the form of waves (including x-rays, ultraviolet, visible light, and radio waves).

electron donor A compound that can be oxidized to transfer electrons to another compound, which in turn becomes reduced.

electron transport chain A series of membrane-bound electron carriers that participate in oxidation-reduction reactions whereby electrons are transferred from one carrier to another until given to the final electron acceptor (oxygen in aerobes, other inorganic substances in anaerobes) in respiration.

endergonic Any metabolic reaction in which the products have more potential energy than the reactants. These are usually associated with anabolic (synthesis) reactions.

endospore Dormant, highly resistant form of bacterium; produced only by species of *Bacillus*, *Clostridium*, and a few others.

enteric bacteria Informal name given to bacteria that occupy the intestinal tract.

Enterobacteriaceae A group of Gram-negative rods that also are oxidase-negative, ferment glucose to acid, have polar flagella if motile, and usually are catalase-positive and reduce nitrate to nitrite.

enzyme A protein that catalyzes a metabolic reaction by interacting with the reactant(s) specifically; each metabolic reaction has its own enzyme.

eosinophil A category of white blood cells; one of three types of granulocyte (along with basophils and neutrophils) characterized by a cytoplasm with red (in typical stains) granules and lobed nucleus.

epitope That portion of an antigen that stimulates the immune system and reacts with antibodies.

eukaryote Type of cell with membranous organelles, including a nucleus, and having 80S ribosomes and many linear molecules of DNA.

exergonic Any metabolic reaction in which the reactants have more potential energy than the products. These are usually associated with catabolic (degradative) reactions.

exoenzyme Enzyme that operates in the external environment after being secreted from the cells.

exponential phase Closed-system microbial growth phase immediately following the lag phase; characterized by constant maximal growth during which population size increases logarithmically.

extension The third of three phases in PCR, following denaturation and annealing, whereby a polymerase (usually *Taq*I) catalyzes the replication of a DNA template.

extracellular Pertains to the region outside a cell.

extreme halophile Organism that grows best at 15% or higher salinity.

extreme thermophile Organism that grows best at temperatures above 80°C.

F

facultative anaerobe Microorganism capable of both fermentation and respiration; grows in the presence or absence of oxygen.

facultative thermophile Microorganism that prefers temperatures above 40°C but will grow in lower temperatures.

false negative Test result that is negative when the sample is actually positive; usually a result of lack of sensitivity in the test system.

false positive Test result that is positive when the sample is actually negative; usually a result of lack of specificity of the test system.

fastidious (microorganism) Describes a microorganism with strict physiological requirements; difficult or impossible to grow unless specific conditions are provided.

fatty acid A long chain, organic molecule with a carboxylic acid at one end with the rest of the carbons being bonded to hydrogens.

fermentation Metabolic process in which an organic molecule acts as an electron donor and one or more of its organic products act as the final electron acceptor, marking the end of the metabolic sequence (differs from respiration in that an inorganic substance is not needed to act as final electron acceptor).

final electron acceptor (FEA) The molecule receiving electrons (it becomes reduced) at the end of a metabolic sequence of oxidation/ reduction reactions.

flavin adenine dinucleotide (FAD) Coenzyme used in oxidation/ reduction reactions that acts as an electron acceptor or donor, respectively.

flavoprotein A flavin-containing protein in the electron transport chain, capable of receiving and transferring electrons as well as entire hydrogen atoms; sometimes bypasses normal route of transfer and reduces oxygen directly, forming hydrogen peroxide and superoxide radicals.

free coagulase An enzyme produced and secreted by some microorganisms that initiates the clotting mechanism in plasma; seen as a solid mass in the coagulase tube test or clumps in the slide test.

free energy (of formation) The energy released by a molecule in a chemical reaction that is able to do work (*e.g.*, produce ATP). It is equivalent to the energy required to form the molecule from its individual elements.

free living A term used to describe nonparasitic organisms.

G

gametangium A structure that produces gametes; used in describing fungi.

gamete Reproductive cell that must undergo fusion with another gamete to continue the life cycle; usually are haploid. *See also* spore.

gelatinase Enzyme secreted by microorganisms, which catalyzes the hydrolysis of gelatin.

generation time The time required for a population to produce offspring (*i.e.*, in bacteria that reproduce by binary fission, the time needed for the population to double); inverse of mean growth rate.

germ theory Theory holding that diseases and infections are caused by microorganisms; first hypothesized in the 16th century by Girolamo Fracastoro of Verona.

glycolysis Metabolic process by which a glucose molecule is split into two 3-carbon pyruvic acid molecules, producing two ATP (net) and two $NADH_2$ molecules.

gradient bacteria Microaerophiles with very narrow nutritional requirements living in habitats characterized by nutrients diffusing upward from sediments below and oxygen diffusing downward from air above.

granulocyte A category of white blood cells characterized by prominent cytoplasmic granules; includes neutrophils, eosinophils, and basophils.

group A streptococci α-hemolytic members of the genus *Streptococcus*, characterized by possessing the Lancefield group A antigen; belong to the species *S. pyogenes*, an important human pathogen.

group B streptococci β-hemolytic members of the genus *Streptococcus*, characterized by possessing the Lancefield group B antigen; belong to the species *S. agalactiae*.

group D streptococci Streptococci possessing the Lancefield group D antigen; include *S. bovis* and species of the genus *Enterococcus*.

H

halophile Microorganism that grows best at 3% or higher salinity.

haploid Describes a cell that has one complete set of genetic information; all eukaryotic microorganisms have a haploid stage and a diploid (two sets) stage in their life cycle; all prokaryotes are haploid.

hemagglutination General term applied to any agglutination test in which clumping of red blood cells indicates a positive reaction.

hemoglobin Iron-containing protein of red blood cells that is responsible for binding oxygen.

hemolysin A class of chemicals produced by some bacteria that break down hemoglobin and produce hemolysis.

heterocyst This is a hollow-appearing cell found in some cyanobacteria specialized to perform nitrogen fixation.

heterotroph An organism that requires carbon in the form of organic molecules. *See also* autotroph.

host An organism that serves as a habitat for another organism such as a parasite or a commensal.

hydrolysis The metabolic process of splitting a molecule into two parts, adding a hydrogen ion to one part and a hydroxyl ion to the other. (One water molecule is used in the process.)

hyperosmotic (solution) A term used to describe extracellular solute concentration relative to the cell; a solution that contains a higher concentration of solutes than the cell, such that water tends to move down its concentration gradient and diffuse out of the cell.

hypha A filament of fungal cells.

hyposmotic A term used to describe extracellular solute concentration relative to the cell; a solution that contains a lower concentration of solutes than the cell, such that water tends to move down its concentration gradient and diffuse into the cell.

I

IMViC Acronym representing the four tests used in identification of *Enterobacteriaceae*: Indole, Methyl Red, Voges-Proskauer, and Citrate.

incomplete medium *See* minimal medium.

incubation The process of growing a culture by supplying it with the necessary environmental conditions.

index case First occurrence of an infection or disease that results in an epidemic.

indirect agglutination Serological test in which artificially produced particulate antigens or antibodies are used to form a visible aggregate if positive. *See also* direct agglutination.

induction Process by which a substrate (inducer) causes the transcription of the genes used in its digestion.

infectious disease A transmissible illness or infection.

inoculum The organisms used to start a new culture or transferred to a new place.

inorganic molecule A molecule that does *not* contain carbon and hydrogen.

intracellular Within the cell; as an intracellular enzyme catalyzing reactions inside the cell.

iron-sulfur protein Iron-containing electron carrier in electron transport chains. Differentiated from cytochromes by the lack of porphyrin ring structure.

isolate (*v.*) The process of separating individual cell types from a mixed culture; (*n.*) the group of cells resulting from isolation.

isosmotic Describe extracellular solute concentration relative to the cell; a solution that contains the same concentration of solutes as the cell, such that water tends to move equally into and out of the cell.

K

karyogamy The process of nuclear fusion that occurs after fertilization (plasmogamy) in sexual life cycles.

Krebs cycle A cyclic metabolic pathway found in organisms that respire aerobically or anaerobically.

L

lag phase Closed-system microbial growth phase immediately preceding the exponential phase; characterized by a period of adjustment in which no growth takes place.

limit of resolution The closest two points can be together for the microscope lens to make them appear separate; two points closer than the limit of resolution will blur together.

lipase A family of hydrolytic enzymes that break down fats into their component parts: glycerol and up to three fatty acids.

lipid A fat.

lophotrichous Describes a flagellar arrangement with a group of flagella at one end of an elongated cell.

lymphocyte A category of white blood cells; one of two types of agranulocyte (along with monocytes); characterized by a large nucleus and little visible cytoplasm; involved in specific acquired immunity as T-cells and B-cells.

lysozyme A naturally occurring bactericidal enzyme in saliva, tears, urine, and other body fluids; functions by breaking peptidoglycan bonds.

lytic cycle The viral life cycle from attachment to lysis of the host cell.

M

mean growth rate constant The number of generations produced per unit time; the inverse of generation time.

medium A substance used for growing microbes; may be liquid (usually a broth) or solid (usually agar).

meiosis The process in which the nucleus of a diploid eukaryotic cell nucleus divides to make four haploid nuclei.

mesophile A microorganism that grows best at temperatures between 15°C and 45°C.

methylation A protective enzymatic mechanism in many organisms whereby methyl groups ($-CH_3$) attach to DNA, thus blocking attachment and destruction by restriction endonucleases.

microaerophile A microorganism that requires oxygen, but can't tolerate atmospheric levels of it.

microbial growth curve Graphic representation of microbial growth in a closed system, consisting of lag phase, exponential (log) phase, stationary phase, and death phase.

minimal medium A medium used in genetic experiments, supplying all nutrients for growth *except* the one required by auxotrophs, and thus supporting growth of only prototrophs.

minimum inhibitory concentration The lowest concentration of an antimicrobial substance required to inhibit growth of all microbial cells it contacts; on an agar plate, typically the outer edge of the zone of inhibition where the substance has diffused to the degree that it no longer inhibits growth.

mitosis The process in which a nucleus divides to produce two identical nuclei.

mixed acid fermentation Vigorous fermentation producing many acid products including lactic acid, acetic acid, succinic acid, and formic acid, and subsequently lowering the pH of the medium to pH 4.4 or below.

mold Informal grouping of filamentous fungi. *See also* yeast.

monocyte A category of white blood cells; one of two types of agranulocyte (along with lymphocytes); characterized by large size and lack of cytoplasmic granules; the blood form of macrophages.

monotrichous Describes a flagellar arrangement consisting of a single flagellum.

morbidity Epidemiological measurement of incidence of a disease; typically accompanied by "incidence rate," referring to the incidence of a disease over time.

morphology The shape of an organism.

mortality Epidemiological measurement of death caused by a disease; typically accompanied by "incidence rate," referring to the incidence of death from a disease over time.

mutagen A substance that causes mutation in DNA; most mutagens are carcinogens.

mutation Alteration in a cell's DNA.

mutualistic Describes a synergistic relationship between two organisms in which both benefit from the interaction.

mycelium A mass of fungal filaments (hyphae).

N

nematode A class of roundworms; environmentally abundant in some parasitic species.

neutrophil Category of white blood cells; one of three types of granulocyte (along with eosinophils and basophils) characterized by a granular cytoplasm and lobed nucleus; also known as "polymorphonuclear granulocytes" or "PMNs."

neutrophile Microorganism adapted to a habitat between pH 5.5 and 8.5.

nicotinamide adenine dinucleotide (NAD) A coenzyme used in oxidation/reduction reactions that acts as an electron acceptor or donor, respectively.

nisin Antibiotic produced by *Lactococcus lactis*.

nitrate A highly oxidized form of nitrogen; NO_3.

nitrate reductase An enzyme produced by all members of *Entero-bacteriaceae* (and others) that catalyzes the reduction of nitrate (NO_3) to nitrite (NO_2).

nitrification The chemical process in which nitrogen compounds, such as ammonia and nitrite are oxidized to form nitrate. Typically, different bacteria perform these two steps in which ammonia is first oxidized to nitrite, which is subsequently oxidized to nitrate.

nitrite NO_2, an oxidized form of nitrogen.

nitrogen fixation The bacterial process in which gaseous nitrogen (N_2) becomes reduced to NH_4. This is ecologically important because N_2, while abundant in the atmosphere, is not in a form usable by most organisms, whereas NH_4 is usable by plants and can thus enter food chains.

nitrogenase The enzyme involved in nitrogen fixation, that is, converting gaseous N_2 to NH_4. This is performed by free-living bacteria (*e.g., Azotobacter*), symbiotic bacteria (*e.g., Rhizobium*), and many cyanobacteria.

noninfectious diseases Conditions not caused by microorganisms; examples are stroke, heart disease, and emphysema.

O

objective lens The microscope lens that first produces magnification of the specimen in a compound microscope.

obligate (strict) aerobe Microorganism that requires oxygen to survive and grow.

obligate (strict) anaerobe Microorganism for which oxygen is lethal; requires the complete absence of oxygen.

obligate thermophile Microorganism that grows only at temperatures above 40°C.

ocular lens The lens the microscopist looks through; produces the virtual image by magnifying the real image.

ocular micrometer A uniformly graduated linear scale placed in the microscope ocular used for measuring microscopic specimens.

oligonucleotide A short nucleic acid molecule.

operon A prokaryotic structural and functional genetic unit consisting of two or more structural genes that code for enzymes in the same pathway and that are regulated together.

opportunistic pathogen A microorganism not ordinarily thought of as pathogenic (*i.e.,* most enterics) that will cause infection when out of its normal habitat.

organic molecule A molecule made of reduced carbon—that is, containing at least carbon and hydrogen.

osmosis Diffusion of water across a semipermeable membrane.

osmotolerant Microorganism that will grow outside of its preferred salinity range.

oxidation/reduction Chemical reaction in which electrons are transferred. (The molecule losing the electrons becomes oxidized; the molecule gaining the electrons becomes reduced.)

oxidative deamination The metabolic process in which an amino acid has its amine removed, producing ammonia and an organic acid.

oxidative phosphorylation The process by which an electron transport chain is used to add a phosphate to ADP to make ATP.

oxidizing agent A substance that removes electrons from (oxidizes) another. *See also* reducing agent.

P

parasite An organism that lives symbiotically with another, but to the detriment of the other organism (called a host).

parthenogenesis A process in which females produce offspring from an unfertilized egg.

peptidoglycan The insoluble, porous, cross-linked polymer comprising bacterial cell walls; generally thick in Gram-positive organisms and thin in Gram-negative organisms.

peptone A digest of protein used in formulating some bacteriological media.

peritrichous Describes a flagellar arrangement in which flagella arise from the entire surface of the cell.

pH The measure of a solution's alkalinity or acidity; the negative logarithm of the hydrogen ion concentration.

phage *See* bacteriophage.

phage host Bacteria attacked by a virus.

phototaxis Phototaxis is a response to light stimuli. Positive phototaxis is movement toward a light stimulus, whereas negative phototaxis is movement away from light.

plaque The clearing produced in a bacterial lawn as a result of cell lysis by a bacteriophage; used to calculate phage titer (PFU/mL) when accompanied by a serial dilution.

plaque forming unit (PFU) Term that replaces "viral particle" or "single virus" when referring to phage titer; accounts for multiple particle arrangements in which more than one virus is responsible for initiating a plaque.

plasma The noncellular (fluid) portion of blood; consists of serum (including serum proteins) and clotting proteins.

plasmid Small, circular, extrachromosomal piece of DNA found in prokaryotic cells; often carries genes for antibiotic resistance.

plasmogamy The cytoplasmic fusion of gametes at the time of fertilization. *See also* karyogamy.

plasmolysis Shrinking of cell membrane (pulling away from the rigid cell wall) because of loss of water to the environment and reduced turgor pressure.

polar flagellum A single flagellum at one end of an elongated cell.

pour plate technique Method of plating bacteria in which the inoculum is added to the molten agar prior to pouring the plate.

preadsorption period When performing a plaque assay, the time given to allow a bacteriophage to attach to the host before adding the mixture to the molten agar being plated.

precipitate An insoluble material that comes out of a solution during a precipitation reaction.

presumptive identification Tentative identification of an isolate based on one or more key test results.

primary stain The first stain applied in many differential staining techniques; usually subjected to a decolorization step that forms the basis for the differential stain.

primer A short nucleotide sequence (typically 10–20 nucleotides) used in PCR to attach to DNA for the purpose of replicating a desired sequence of nucleotides. Typically, two primers attach to sites on opposite strands, flanking the area (template) to be replicated.

proglottid Tapeworm segments posterior to the scolex, used for absorption of nutrients and containing reproductive organs.

prokaryote Type of cell lacking internal compartmentalization (membranous organelles, including a nucleus) and having 70S ribosomes and a circular molecule of DNA; more primitive than eukaryotes.

promoter site The patch of DNA upstream from the structural gene(s), which binds RNA polymerase to begin transcription.

propagated transmission Conveying a disease person-to-person.

proteolytic Refers to catabolism of protein; *e.g.*, a *proteolytic* enzyme.

prototroph A strain that is capable of synthesizing the nutrient that is acting as the marker in a particular genetic experiment; prototrophs will grow on complete and minimal medium. *See also* auxotroph.

pseudohypha A chain of fungal cells produced by budding (rather than by cytokinesis that produces two equally sized cells) and which is characterized by constrictions at cell junctions.

psychrophile A microorganism that grows only at temperatures below 20°C, with an optimum around 15°C.

psychrotroph A microorganism that grows optimally at temperatures between 20 and 30°C but will grow at temperatures as low as 0°C and as high as 35°C.

pure culture Microbial culture containing only a single species.

purine nucleotide Purines are made of a sugar (ribose or deoxyribose), phosphate, and a nitrogen-containing base composed of two joined rings. Examples are adenine (A) and guanine (G). These are found in DNA and RNA.

putrefaction The process of digesting dead organic material; decay.

pyrimidine nucleotide Pyrimidines are made of a sugar (ribose or deoxyribose), phosphate, and a nitrogen-containing base composed of a single ring. Examples are cytosine (C) in DNA and RNA, thymine (T) in DNA, and uracil (U) in RNA.

pyruvic acid (pyruvate) A three-carbon compound produced at the end of glycolysis that may enter a respiration or a fermentation pathway; also serves as a starting point for synthesis of certain amino acids and an entry point for their digestion.

Q

quinoidal (compound) A color-producing compound containing quinone as its central structure.

quorum sensing The phenomenon in bioluminescing bacteria whereby the light-emitting reaction of all cells takes place simultaneously when a threshold concentration of secreted autoinducer is reached.

R

real image Magnified image of a specimen produced by the objective lens of a microscope; the real image is magnified again by the ocular lens to produce the virtual image.

recognition site A short DNA sequence where a restriction enzyme attaches. Each enzyme recognizes and attaches to its own specific sequence.

reducing agent A substance that donates electrons to (reduces) another. *See also* oxidizing agent.

reductase An enzyme that catalyzes the transfer of electrons from donor molecule to acceptor molecule, thereby reducing the acceptor.

reduction *See* oxidation/reduction.

refraction The bending of light as it passes from a medium with one refractive index into another medium with a different refractive index.

reservoir A nonhuman host or other site in nature serving as a perpetual source of pathogenic organisms.

resolution The clarity of an image produced by a lens; the ability of a lens to distinguish between two points in a specimen; high resolution in a microscope is desirable.

resolving power *See* limit of resolution.

respiration Metabolic process by which an organic molecule acts as an electron donor and an inorganic substance—such as oxygen, sulfur, or nitrate—acts as the final electron acceptor in an electron transport chain, marking the end of the metabolic sequence (differs from fermentation, which uses one of its own organic products as the final electron acceptor).

restriction enzyme Enzymes involved in cutting DNA at specific recognition sites, unique to the enzyme. Involved in removing damaged DNA and destroying foreign (*e.g.*, phage) DNA.

reticuloendothelial system Combination of macrophages and associated cells located in the liver, spleen, bone marrow, and lymph nodes.

reverse citric acid cycle An energy consuming process used by green sulfur bacteria whereby molecules ordinarily associated with the oxidative respiratory reactions of the Krebs cycle perform reactions in the opposite direction to reduce carbon dioxide and form pyruvate.

reverse electron flow An energy consuming process used by purple sulfur bacteria whereby electrons transferred down the ETC of the photosystem are boosted upward against their thermodynamic gradient to reduce $NADP^+$ needed by the Calvin cycle.

reversion In carbohydrate fermentation tests, the phenomenon of a microorganism fermentively depleting the carbohydrate and reverting to amino acid metabolism, thereby neutralizing acid products with alkaline products; produces a false negative.

rhizoid A root-like structure used for attachment of some fungi to the substrate.

RNA polymerase A group of enzymes that catalyze the addition of ribonucleotides to the 3' end of an existing polynucleotide chain.

S

saltern Salterns are low pools of saltwater. Evaporation of the water leaves salt, which can then be harvested.

saprophyte A heterotroph that digests dead organic matter; a decomposer.

scolex The "head" of a tapeworm, often with suckers and hooks for attachment.

selective medium Growth medium that favors growth of one group of microorganisms and inhibits or prevents growth of others.

sensitivity This is a measure of a test's ability to detect small amounts of the item being tested for. The better the sensitivity, the fewer the false negatives (due to smaller amounts triggering a positive reaction), and the more useful the test is.

serial dilution Series of dilutions used to reduce the concentration of a culture and thereby produce between 30 and 300 colonies when plated, providing a means of calculating the original concentration.

serology A discipline that utilizes a serum containing antibodies (antiserum) to detect the presence of antigens in a sample; also refers to identification of antibodies in a patient's serum.

serum Fluid portion of blood minus the clotting factors.

soft agar A semisolid growth medium containing a reduced concentration of agar; used in plaque assay to allow diffusion of bacteriophage while arresting movement of the bacteriophage host.

solute The dissolved substance in a solution.

solution The mixture of dissolved substance (solute) and solvent (liquid).

solvent The liquid portion of a solution in which solute is dissolved.

specificity This is a measure of a test's ability to produce a positive response only when reacting with the particular item being tested for. The better the test's specificity, the fewer the false positives (due to better discrimination), and the more useful it is.

spirillum Spiral-shaped cell.

sporangium Structure that produces spores.

spore In bacteria, a dormant form of a microbe protected by specialized coatings produced under conditions of, and resistant to, adverse conditions; also known as an endospore; in fungi and plants, spores are specialized reproductive cells; frequently a means of dissemination.

spread plate technique Method of plating bacteria in which the inoculum is transferred to an agar plate and spread with a sterile bent-glass rod or other spreading device.

stage micrometer A microscope ruler used to calibrate an ocular micrometer.

standard curve A graph constructed from data obtained using samples of known value for the independent variable; once made, can be used to experimentally determine the value of the independent variable when the dependent variable is measured on an unknown.

stationary phase Closed system microbial growth phase immediately following exponential phase; characterized by steady, level growth during which death rate equals reproductive rate.

stolons Surface hyphae of some molds (*e.g., Rhizopus*) that attach to the substrate with rhizoids.

stormy fermentation Vigorous fermentation produced in litmus milk by some species that produce an acid clot but subsequently break it up because of heavy gas production (members of *Clostridium*).

streptolysin Hemolysin (blood hemolyzing exotoxin) produced and secreted by members of *Streptococcus*.

superoxide dismutase Enzyme produced by some bacteria that catalyzes the conversion of superoxide radicals to hydrogen peroxide.

syntrophy A situation where two or more organisms, having differing metabolic capabilities, derive mutual benefit by providing and/or receiving essential nutrients otherwise not available to them.

T

thermal death time The amount of time required to kill a population of a specific size at a specific temperature.

thermophile A microorganism that grows best at temperatures above 40°C.

titer A measurement of concentration of a substance or particle in a solution; used in measurements of phage concentration.

transformant cells Cells that have undergone transformation by picking up foreign DNA in a genetic engineering experiment.

transformation A form of genetic recombination performed by some bacteria in which DNA is picked up from the environment and incorporated into its genome.

trematode A class of parasitic flatworms; also known as "flukes."

trend line A line drawn on a graph to show the general relationship between X and Y variables; also known as a regression line.

triacylglycerol *See* triglyceride.

tricarboxylic acid cycle *See* Krebs cycle.

trichome A filament of cyanobacterial cells. It does not include the sheath, if present.

triglyceride A molecule composed of glycerol and three long chain fatty acids.

trophozoite "Feeding" stage in the life cycle of certain protozoans. *See also* cyst.

tryptophan An amino acid.

tryptophanase Enzyme that catalyzes the hydrolysis of tryptophan into indole and pyruvic acid; detected in the indole test.

turgor pressure The pressure inside a cell that is required to maintain its shape, tonicity, and necessary biochemical functions.

2,3-butanediol fermentation The end-product of a metabolic pathway leading from pyruvate through acetoin with the associated oxidation of NADH; detected in the Voges-Proskauer test.

U

undefined medium *See* complex medium.

urease Enzyme that catalyzes the hydrolysis of urea into two ammonias and one carbon dioxide.

utilization medium A differential medium that detects the ability or inability of an organism to metabolize a specific ingredient.

V

variable A factor in a scientific experiment that is changed; the control and experimental groups in a good experiment differ in only one variable, the one being tested.

vector In genetic engineering, a means, often a plasmid or a virus, of introducing DNA into a new host.

vegetative cell An actively metabolizing cell.

virtual image The image produced when the ocular lens of a microscope magnifies the real image; appears within or below the microscope.

Voges-Proskauer test A differential test used to identify organisms that are capable of performing a 2,3-butanediol fermentation.

W

wavelength Measurement of a wave from crest to crest, usually in nanometers—as in electromagnetic energy.

whey Watery portion of milk as seen upon coagulation of casein in the production of a curd.

X

X-Y scatter plot Graph presenting the relationship between two variables.

Y

yeast An informal grouping of unicellular fungi. *See also* mold.

Z

zone of inhibition On an agar plate, the area of nongrowth surrounding a paper disc containing an antimicrobial substance. (The zone typically ends at the point where the diffusing antimicrobial substance has reached its minimum inhibitory concentration, beyond which it is ineffective.)

zygospore The product of fertilization and the site of meiosis in some molds.

zygote The product of gamete fusion (plasmogamy) and nuclear fusion (karyogamy); a fertilized egg.

Index